MODERN MATERIALS AND MANUFACTURING PROCESSES

Third Edition

R. Gregg Bruce
Purdue University School of Technology

William Dalton
Purdue University School of Technology

John E. Neely
Instructor, Machine Technology, Ret.
Lane Community College, Eugene, Oregon

Richard R. Kibbe
Managing Director, Machining and Tooling Development
R.R.K. Machine and Manufacturing

Upper Saddle River, New Jersey
Columbus, Ohio

Library of Congress Cataloging in Publication Data

Modern materials and manufacturing processes / R. Gregg Bruce . . . [et al.].—3rd ed.
p. cm.
ISBN 0–13–094698–X
1. Manufacturing processes. 2. Materials. 3. Metal work. I. Bruce, R. Gregg.

TS183.M617 2004
670—dc22 2003053695

Editor in Chief: Stephen Helba
Executive Editor: Debbie Yarnell
Editorial Assistant: Jonathan Tenthoff
Developmental Editor: Katie E. Bradford
Production Editor: Louise N. Sette
Production Supervision: Carlisle Publishers Services
Design Coordinator: Diane Ernsberger
Cover Designer: Ali Mohrman
Production Manager: Brian Fox
Marketing Manager: Jimmy Stephens

This book was set in Times Roman by Carlisle Communications, Ltd. It was printed and bound by Courier Kendallville, Inc. The cover was printed by Phoenix Color Corp.

Pearson Education Ltd.
Pearson Education Singapore Pte. Ltd.
Pearson Education Canada, Ltd.
Pearson Education–Japan
Pearson Education Australia Pty. Limited
Pearson Education North Asia Ltd.
Pearson Educación de Mexico, S.A. de C.V.
Pearson Education Malaysia Pte. Ltd.

10 9 8 7 6 5 4 3 2 1
ISBN 0-13-094698-2

Preface

Modern Materials and Manufacturing Processes, Third Edition, presents state-of-the art technology in materials and manufacturing processes. Written for engineering technology and vocational students, the text is descriptive in nature and relies more on descriptive material than analysis. The book covers a broad range of materials and processes and includes plentiful illustrations.

This text discusses both traditional and recently developed materials and processes. In addition to presenting the specific materials and processes of manufacturing normally included in a survey course, the text, in a unique approach, shows how those materials and processes are integrated into today's functioning manufacturing industry. This makes the text a good fit for a survey course for engineering technology and vocational students covering both materials and manufacturing processes.

Modern Materials and Manufacturing Processes is organized to present clear, sequential course material for any teaching system, whether competency-based or traditional. Objectives and key words at the beginning of each chapter clarify the goals of each chapter and familiarize the student with important terminology encountered on each topic. Over 1000 illustrations and photographs complement the text. Case studies and case problems are included where appropriate to promote student thought and discussion. Questions and case problems may be used for review or examination purposes. An extensive and newly enhanced glossary, including approximately 160 new terms, concludes the book.

New to This Edition

- New and updated illustrations throughout.
- Complete glossary: approximately 160 new terms/definitions.
- Glossary terms appear in boldface in text.
- New Key Words feature begins each chapter.
- Each chapter has introductory summary of the chapter's content to improve the pedagogy of the material.
- New Introduction chapter.
- Expanded coverage of ceramics, plastics, and composites.
- New Chapter 7, Design Specifications and the Capability of the Manufacturing Process, explains the importance of precision before the subsequent chapters cover manufacturing processes.
- Parts II and III have been reversed so that manufacturing processes are covered before measurement and quality assurance.

- New Chapter 22 covers automation and mass production.
- Material has been reordered and updated throughout to be more logical and complete.

Content and Organization

Part I, Materials of Manufacture, discusses the basic materials used in manufacturing: metals, plastics, and ceramics. Chapters 1–3 identify and classify these materials by their atomic makeup, similarities, and differences. Chapters 4 and 5 explain how the materials are extracted from raw materials, and how the materials are processed. Chapter 6 reviews the many classification systems used for common metallic materials, and then discusses the properties and characteristics that lead to their selection and application.

The detailed coverage in Part I allows the student to proceed to a study of product manufacturing that uses these materials. Where appropriate, newly developed materials, processes, and methods are presented throughout the text.

Part II, Survey of Manufacturing Processes, discusses specific manufacturing processes and their capabilities. Chapter 7 is a new chapter on the specifications used in design and their relationship to the capabilities of the manufacturing process to be used. Subsequent chapters cover casting, cold and hot forming, forging, powdered metal processes, machining, plastics manufacturing, the joining of materials, and non-traditional manufacturing processes. Chapter 18 on the corrosion and protection of metals concludes Part II.

Part III, Measurement and Quality Assurance, Chapters 19 and 20 introduce measurement and quality assurance, with particular emphasis on tolerance, measurement, and calibration.

Part IV, Manufacturing Systems, Chapters 21 and 22 discuss the design, tooling, and production aspects of manufacturing. These final chapters also introduce modern automation methods, including computer-aided design (CAD) and rapid prototyping, as designers' tools; plus computer aided manufacturing (CAM) topics, including computer numerical control (CNC) and industrial robotics. The integration of these technologies is discussed as computer-integrated manufacturing (CIM) and flexible manufacturing systems (FMS), culminating in total factory automation.

As mentioned above, a Glossary updated to feature 160 additional terms completes this edition of the text.

Acknowledgments

The authors express their appreciation to Patricia J. Olesak, Assistant Professor, Purdue University School of Technology, for lending her expert knowledge of materials to the second edition.

Appreciation is also expressed to the following reviewers who provided valuable feedback and helpful suggestions for preparing this revision: Thomas J. Beck, Texas A&M University; Ken Ekegren, North Central Technical College, Mansfield, Ohio; Ramamurthy Prabhakaran, Old Dominion University; and Bryan Reaka, Department of Technology, Southern Illinois University–Carbondale.

The authors are most grateful to the companies and individuals who have contributed materials and help in preparing this test. Special thanks go to Eugene Aluminum & Brass Foundry, Hergert's Industry, Lane Community College, and Lockheed Martin Company, who granted access to their facilities for photography and for supplying illustrations and technical information.

3M Company, St. Paul, MN

AGCO Corporation, Duluth, GA
Ajax Magnethermic Corp., Warren, OH
Alcoa Inc., Alcoa Center, PA
Allied Uniking, Memphis, TN
American Die Casting Institute, Des Plains, IL
American Iron & Steel Institute, Washington, DC
Apex Broach and Machine Co., Detroit, MI
ASM International, Materials Park, OH
Automated Finishing, Menomonee Falls, WI

Baird Corp., Stratford, CT
Baltec Corporation, Canonsburg, PA
Barrington Automation, Ltd., Crystal Lake, IL
Belden Inc., St. Louis, MO
Bird-Johnson Company, Rolls-Royce Marine, Ulsteinvik, Norway
Boeing, Inc., Seattle, WA
Robert Bosch Power Tool Corp., Charlotte, NC
Bourn and Koch, Fellows Corp., Rockland, IL
Brown and Sharpe, North Kingstown, RI
Buck Forkhardt, Erkrath, Germany
Buehler Ltd., Lake Bluff, IL
Burlington Industries, Inc., Greensboro, NC

Clausing Industrial, Inc., Kalamazoo, MI
Coherent, Inc., Santa Clara, CA
Consolidated Engineering Company, Kennesaw, GA
Consolidated Metco, Portland, OR
Crucible Materials Corporation, Camillus, NY

DaimlerChrysler Corporation, Auburn Hills, MI
Disogrin Industries, Manchester, NH
Do All Company, Des Plaines, IL
Dover Publications, Mineola, NY
Dupont Corporation, Wilmington, DE

Ethan Allen, Inc., Danbury, CT
Eugene Aluminum and Brass Foundry, Eugene, OR
Exxon Corporation, Houston, TX

Federal Products Corporation, Providence, RI
Foote-Jones/Illinois Gear, Chicago, IL
Ford Motor Co., Dearborn, MI

Giddings & Lewis, Fond du Lac, WI
The Glass Group, Inc., Millville, NJ
Gleason Corporation, Rochester, NY
The Goodyear Tire and Rubber Company, Akron, OH

Haas Automation, Inc., Oxnard, CA
Hamilton Caster & Mfg. Co., Hamilton, OH
Heck Industries, Hartland, MI
Hergert's Industries, Inc., Drain, OR
Howden Buffalo, Camden, SC

Ibarmia, Azkoitia, Spain

JG Engine Dynamics, Alhambra, CA

Kennametal Inc., Latrobe, PA
Kevex Corporation, San Carlos, CA

Landis Tool, Columbia City, IN
Laser Fare, Smithfield, RI
Laserdyne Prima, Champlin, MN
Likdoping Mekaniska Verkstads AB, Lidkoping, Sweden
Lockheed Martin Co., Bethesda, MD
Louis Levin & Son, Inc., Santa Fe Springs, CA
LTV Steel, Cleveland, OH

M. L. Sheldon Plastics Corp., New York, NY
Magic North America, Markham, Ont.
Magnaflux, Glenview, IL
Maine Industrial Tires Limited, Gorham, ME
McGraw-Edison Company, Columbia, MO
Milacron Inc., Cincinnati, OH
Miller Electric Mfg. Co., Appleton, WI
The Minster Machine Company, Minster, OH
Mitutoyo America Corporation, Aurora, IL
Mueller-Phipps International, Inc., Poughkeepsie, NY

National Machinery LLC, Tiffin, OH
Newage Testing Instruments, Southhampton, PA
Niagara Machine & Tool Works, Buffalo, NY
Nikon, Inc., Instrument Division, Melville, NY
Northrop Grumman Corp., El Segundo, CA

Okamoto Corporation, Buffalo Grove, IL

Pacific Machinery and Tool Steel Company, Portland, OR
Parker Hannifin Corp., Cleveland, OH
Pennsylvania Pressed Metals, Inc., Emporium, PA
Pierce Corporation, Eugene, OR
Pilkington, Toledo, OH
Portland Cement Association, Skokie, IL
Positech, Laurens, IA

Raytheon Aircraft Company, Wichita, KS

Sandusky Foundry & Machine Co., Sandusky, OH
SECO/WARWICK Corporation, Meadville, PA
Sheffield Measurement Division, Fond du Lac, WI
Siempelkamp Corporation, Marietta, GA
Sodick Inc., Mt. Prospect, IL
L. S. Starrett Company, Athol, MA
Subaru-Isuzu Automotive, Lafayette, IN

Teledyne Wah Chang, Albany, OR
Testing Machine Inc., Ronkonkoma, NY
Tinius Olsen Testing Machine Co., Philadelphia, PA

United Calibration Corporation, Stanton, CA
United States Steel Corp., Pittsburgh, PA
Unitron Inc., Bohemia, NY

Wellcraft Marine Corp., Sarasota, FL
John Wiley & Sons, New York, NY

Brief Contents

Contents

Introduction: The Importance of Training and Education in the High-Technology World

Working knowledge of the newest skills in the world of high technology will, to a great extent, determine who will be employable at adequate pay levels in the manufacturing organizations of the future. Progressive manufacturing industries, schools, colleges, and universities will be responsible for a continuing supply of adequately prepared technicians, technologists, and engineers to meet the needs of the technology age. In-service training programs to teach new methods will also be a high priority for manufacturers. Students of materials and manufacturing processes in the high-technology world will need hands-on experience in the application, operation, and support of new manufacturing systems.

For those who are properly prepared, the world of high-technology manufacturing presents many opportunities. The technology of tomorrow will require trained personnel to repair, service, and implement the manufacturing systems of an ongoing high-technology revolution. New product designs and the manufacturing systems required to produce them will be an ever-expanding field for those who acquire the proper training. The production technician, technologist, engineer, and designer will contribute greatly to maintaining a nation's international technological position. Will you be ready to meet these challenges?

The following chapters will aid you in answering this important question. Before we begin, however, we think students should have an understanding of how we have arrived at where we are in the manufacturing process, and what the process consists of.

MATERIALS AND MANUFACTURING

Few human enterprises represent a more aggressive and dedicated use of time, energy, money, and material resources than does modern manufacturing—the processing of raw materials into the endless variety of products that make the modern lifestyle possible. Since the first appearance of humans on this planet the processes and methods of converting materials into different forms and products have progressed hand in hand with human development. Many examples of toolmaking by ancient humans have been discovered, and with tools ancient peoples were able to shape materials into clothing, weapons, shelters, and other necessities of life. These items could be traded or sold, generating commerce and demand for more, different, and improved materials and products.

The desire and demand for exotic materials and products stimulated explorers and traders to travel to the far corners of the earth. These exploits created a need for higher technology in manufactured products, including, for example, navigational instruments and accurate maps, items that were essential to ensure that the trader would return safely with the materials that were in such demand. Such incentives stimulated the development, through improvements in design and better application of materials, of advanced products that were superior to other goods available at the time.

Manufacturing societies began to develop as ancient peoples gathered together to avail themselves of products, services, and materials. The structure of villages, towns, and, ultimately, cities was established, and the increasing concentrations of people created needs for improved manufacturing methods to satisfy growing demands. Specialty trades and crafts developed rapidly, in turn creating a need for faster methods of production. Animals, water, steam, and, finally, nuclear energy were harnessed to power manufacturing machinery.

The lure of employment in manufacturing industries and the availability of manufactured products that would make life easier and more pleasant contributed greatly to the industrialization and urbanization of much of the world. Great manufacturing centers developed, usually located close to fundamental sources of raw materials such as coal, oil, and iron ore, prime ingredients on which industrialization was built. Today, a vast international manufacturing economy exists that aggressively makes use of newly developed technologies in materials development and application, as well as in *manufacturing processes.*

In today's world of exotic materials, high technology, and agile manufacturing enterprises, technology is constantly changing and growing. New production methods and new materials applications become available almost daily and rapidly become obsolete as new technology comes online. Although most of the well-established processes for manufacturing products will continue to exist, new applications and materials require new processing methods. The manufacturing technologist of today and tomorrow must be well versed in conventional materials applications and processes, but he or she will be engaged in the search for new applications of materials and processes and the development of entirely new processes. Owing to this trend, manufacturing is continually becoming more specialized, making it difficult for the technologist to keep informed of rapidly changing and expanding technology.

The central purpose of this text is to prepare future manufacturing engineers, engineering technology graduates, and industrial technologists by presenting an overview of materials science, a survey of traditional as well as high-technology manufacturing processes, and a look at manufacturing systems in the computerized age.

THE MANUFACTURING ECONOMY: PAST, PRESENT, AND FUTURE

By the middle of the nineteenth century, goods formerly made by hand were being made by machine. A truly remarkable revolution, the Industrial Revolution, which had begun two centuries earlier, was well established, and the era of large-scale mass production had begun. The effects of the Industrial Revolution have been far-reaching. Cheap production of goods in large quantities, now the order of the day, raised the standard of living for people in industrialized societies. The Industrial Revolution formed the basis of the modern lifestyle that most of us take for granted, and general industrialization has provided expanded employment.

In the latter half of the twentieth century a new industrial revolution took place, one of computers and space-age high technology. The effects of this computer-driven revolution will be much more far-reaching than anything even dreamed of in past centuries. Modern industrial technology has given us the power and the tools to be far more productive with less direct involvement of people in the manufacturing work. It is not to be denied that modern technology has made the lifestyle of the modern industrial society more comfortable and more convenient and that it is filled with incredible products, most within the purchasing power of the average person; however, modern technology may displace those who are not prepared to work in a high-technology, computerized, and automated manufacturing world. As lower-skill-level jobs are automated the employee of today must look to the future and acquire the job skills necessary to qualify for the higher-skilled positions that are created by advancing technology.

The gradual industrialization of less technologically oriented societies continues today throughout the world. For much of the past century the established industrial world could afford, due to the distinct advantage provided by superior technology, to take little notice of the advances of less developed countries; however, as the wage demands of the established industrial societies rose over time, manufacturers sought to take advantage of the lower wage structures generally prevalent in emerging economies. The eager approach of workers in less developed countries toward their jobs and the resulting level of productivity and quality have created industrial revolutions in their economies. The continuing migration of manufacturing to less developed countries presents intense competition to the employment levels and economies of established industrial societies.

What are the results of this migration? It is true that manufacturing industries that have moved abroad have been able to operate under more favorable wage structures; however, the very industries that these manufacturers sought to staff with cheaper labor have, by a natural evolution, raised the standard of living for workers in the countries where the industries are now located. More jobs have been created, and with jobs have come higher pay and the ability of industrial workers to buy not just necessities but goods that make life easier and more pleasant. The effects have been and will continue to be far-reaching.

Global competition has placed the technical leadership of the established industrial world (those countries where industrialization first occurred) in question. For example, the influx of high-quality goods manufactured abroad and imported into the United States for sale at a low price has continued unabated for many years. To remain competitive, American manufacturing industries are required to counter by improving productivity through modern practices in automation, innovative manufacturing processes, superior quality control, and rapid response to changes in markets. Industries that do not respond are only too evident as jobs in the domestic economy are lost. Foreign manufacturing, often operating in an unrestricted business environment with the favorable support of their government, has taken advantage of modern technology and is engaged in an effective program of operating modern, efficient industries.

Industrial economies will remain intact and the standard of living will remain at a high level only if the industrial and manufacturing community maintains favorable attitudes toward improving manufacturing efficiency and competitiveness. Quality and productivity must be maintained and improved, and the worker, who often is directly responsible for industrial output, must be treated fairly. Manufacturers must respond to market demand for defect-free products through the application of modern quality assurance methods and must continually strive for simplification (aided by computerization) of products, services, systems, and manufacturing processes. As the pace of change accelerates to a dizzying level manufacturing enterprises must remain agile, poised to react instantly to changes in markets, competitors, technology, society, and governments.

It is a sure bet that future products will be manufactured at a high level of automation. As new products are designed to meet the needs of society increasing numbers of high-technology manufacturing systems and methods will be needed to produce them. Manufacturing industries must put to use all the modern tools necessary to accomplish efficient, quality production in the high-technology age.

THE COMPUTER'S INFLUENCE IN MANUFACTURING

The history of technological development encompasses many contributions. Since the first recorded manufacturing efforts manufacturing industries have continually applied the latest materials, tools, and methods to satisfy manufacturing requirements. First, metallurgy made available a wide selection of metals for toolmaking and for manufactured products. The Industrial Revolution of the eighteenth and nineteenth centuries brought the age of machine-made mass-produced goods. Electronics, especially solid-state and integrated circuit (IC) chip technology, has made instant communication available to practically everyone on earth; thus, the knowledge of available manufactured goods has reached ever-widening markets. Although these technological advances have transformed industry and society as a whole, the computer has had a more extensive effect than any other technological device yet developed.

The concept of computer control of processes is revolutionary, and it will continue to profoundly change the way manufacturing is accomplished. As computer technology and software have become more sophisticated the decision-making powers of computers have increased proportionally, enhanced by advances such as *artificial intelligence (AI)* technology. More sophisticated computers and software provide versatile, robust, and effective control of both routine processes and rapidly changing production operations.

Until recently, however, most manufacturing industries have needed people to operate machines to turn out product units. Today the computer, with its considerable ability to make decisions in process control applications, is displacing people in the manufacturing process. Flexible manufacturing systems (FMS) (Figure 1), the forerunners of completely automated factories, have become

FIGURE 1
The flexible manufacturing system (FMS) with automated workpiece handling and transfer (Giddings and Lewis).

commonplace as computer-controlled machines continue to take over the mundane work previously done by manual labor. The need for the manual handling of materials is minimized in this system, as automated transporters move workpieces between manufacturing stations, and industrial robots (Figure 2) perform the routine tasks of loading and unloading parts for computer numerically controlled (CNC) machining centers. Automation, already highly utilized in manufacturing, will increasingly become the preferred method of production.

Does this mean that computers will control everything, including the actions of people? The answer to this question is a definite no; however, the computer helps organize the actions of people not only in manufacturing but in many other phases of life as well. The total impact of computers and computerized automation trends in manufacturing on the future worker is unknown, but it is certain that the worker of the future will have to become more of an electromechanical technician, replacing the machine tender of the past.

The technician of today and tomorrow must have a more complete knowledge of electrical/electronic and hydraulic systems, automated mechanisms, computer systems, and computer programming. For those not willing to learn new skills or accept retraining, future employment prospects at a suitable pay level may be limited. People with an open mind and a natural desire to understand how things work will continue to be in demand for the design, operation, maintenance, and repair of complex electronic/mechanical machines.

MANUFACTURING TECHNOLOGY TODAY AND TOMORROW

Manufacturing technology will always be in a continuing state of development. As new materials, methods, and computer aids come into play, the manufacturing engineer

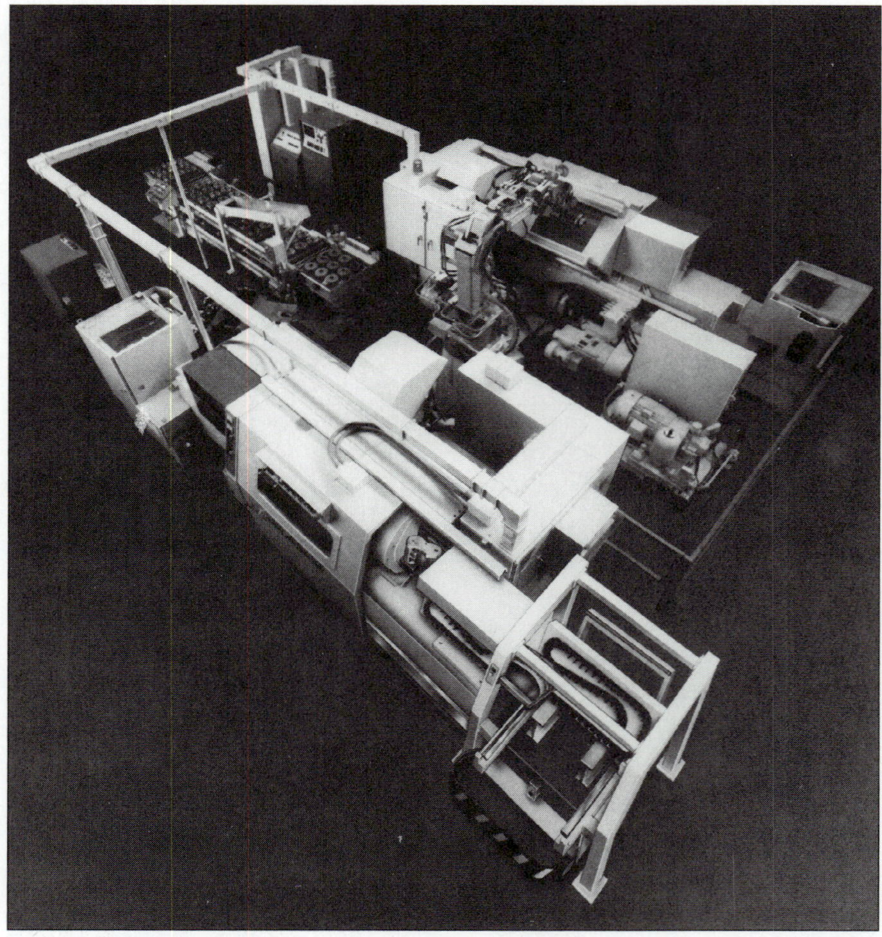

FIGURE 2
FMS work cell for CNC turning. The robot (center) loads and off-loads parts for both machine tools (Cincinnati Milacron).

and technologist will apply them to solve production problems.

Materials

The product designer will continually seek new materials that are lighter and easier and cheaper to process while meeting design requirements for strength and durability. For example, plastic and composite materials permit energy savings both in manufacturing processes and in powering road, air, and space vehicles (Figure 3).

FIGURE 3
High-technology designs and materials for tomorrow's Mach 5 high-altitude aircraft. The leading edges of the airframe will glow red hot at a temperature of 1000°F (Lockheed Martin Co.).

Design

The computer continues to play an ever-increasing role in initial product design and design evaluation (Figure 4), and *computer-aided design (CAD)* is an absolutely essential tool. Control programs for automated manufacturing equipment are generated from the CAD database, thus forming integrated CAD and *computer-aided manufacturing (CAM)* systems. The day is fast approaching when the product design created at the CAD station will quickly appear as finished products at the shipping dock.

Computer-aided engineering (CAE) tools are commonly utilized for evaluating product designs, often eliminating the expensive building and testing of physical prototypes. Complex contoured shapes are studied and evaluated using CAE methods, as illustrated in Figure 5.

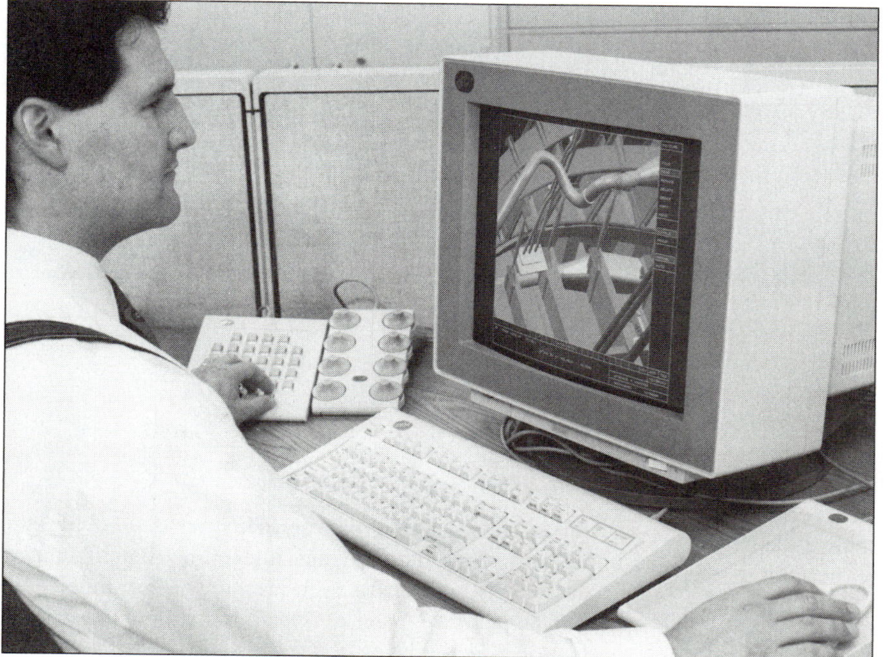

FIGURE 4
The CAD engineering workstation is today's preferred method of designing (Boeing Commercial Airplane Group).

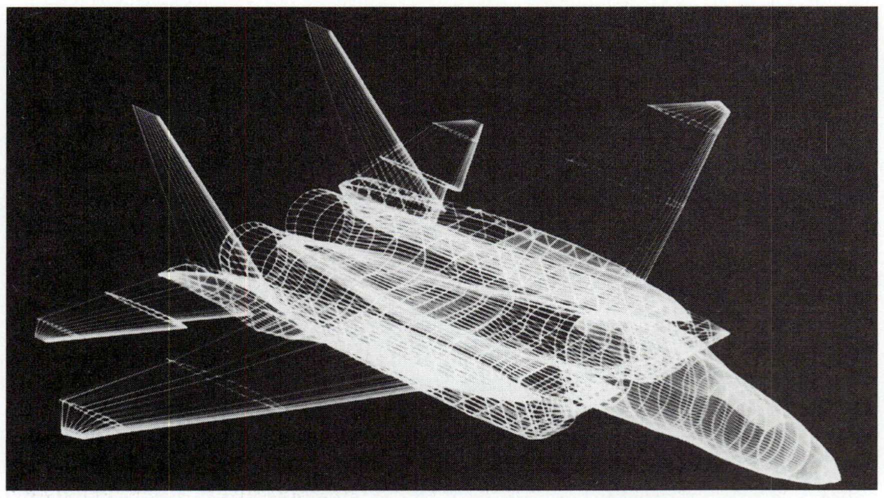

FIGURE 5
CAE for design evaluation of aerodynamics utilizing a CAD model (Evans and Southerland Computer Corporation).

FIGURE 6
A new look at old and proven technology—high-technology turbojet propellers (Lockheed Martin Co.).

Proven designs of the past are reevaluated with an eye toward energy and fuel consumption efficiency. For example, Figure 6 shows a high-technology application of the airscrew. Driven by the efficient turboprop jet engine, the innovative propeller design can propel aircraft with the same speed and altitude capability of jet engines but with a large reduction in fuel consumption.

Manufacturing Processes

Manufacturers in all types of industries search for and implement new, unique, and interesting manufacturing processes. In Figure 7 a high-pressure jet of water containing an abrasive material is used to cut titanium aircraft parts. Because almost no heat is developed in this process, distortion of the part is practically eliminated.

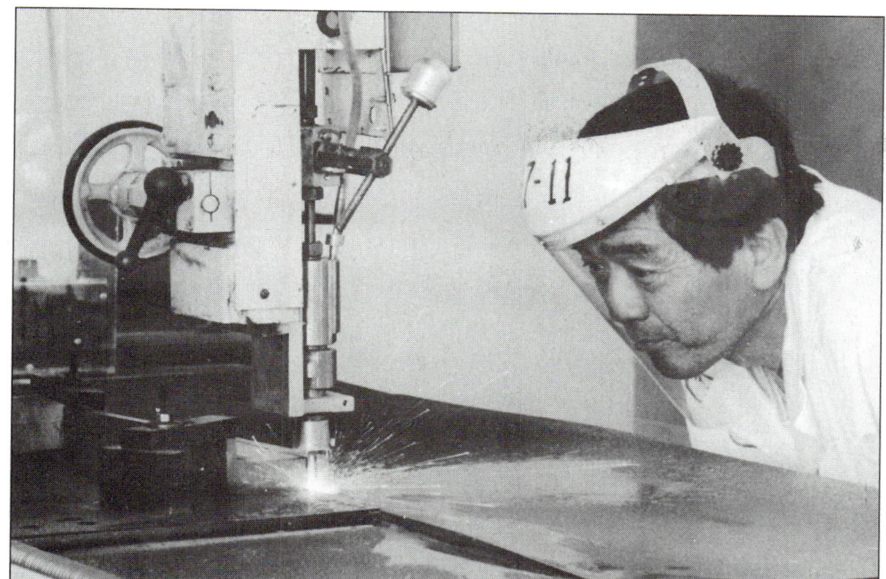

FIGURE 7
High-pressure water containing an abrasive is used to cut titanium aircraft parts (Lockheed Martin Co.).

FIGURE 8
Electronic components are automatically inserted onto circuit boards.

FIGURE 9
Computer-driven spot welding robots spot weld front body assemblies with great speed and accuracy (Ford Motor Company).

Computer control of automated processes is found in nearly every aspect of manufacturing including electronic component insertion (Figure 8), robotic spot welding (Figures 9 and 10), and precision measurement and inspection (Figure 11). Other high-technology applications of available technology include the use of lasers for many applications (Figure 12).

THE MANUFACTURING INDUSTRY

Let us now look at how people and jobs are organized in the typical manufacturing enterprise. As manufacturing organizations formed and grew during the Industrial Revolution the various job functions within the organization became increasingly specialized, leading to the multilevel structure of today's manufacturing companies. Although the makeup of manufacturing enterprises varies greatly, most fit the basic structure outlined next.

Manufacturing industries make up a large portion of total business and are the places of employment for a

FIGURE 10
An underbody and two side panels are secured in a large fixture for production spot welding.

large segment of the labor force. There are many types of manufacturing industries, ranging in size from one- or two-person businesses to major corporations employing several hundred thousand employees. Major manufacturing corporations operate on a worldwide scale, and the dollar value of sales can range from a few thousand dollars in a small business to several billion dollars for a large multinational corporation.

Many large companies are parts of larger groups or conglomerates, and companies engaged in the same or similar product manufacturing may form technology groups. In this manner, duplication of product lines is eliminated and technical expertise may be pooled for more efficient business operations, product design, and production. In other cases large manufacturers of complex products such as automobiles may own subsidiaries that supply them with component parts. This ensures that a supply of required parts is always available and permits the manufacturer to tap diverse expertise for all phases of business. Regardless of the size of the business, all have a basic management structure in which job responsibilities are delegated in much the same way. A typical manufacturing corporation of average size and having most of the typical departments and job titles will be organized along the lines shown in the organizational chart of Figure 13.

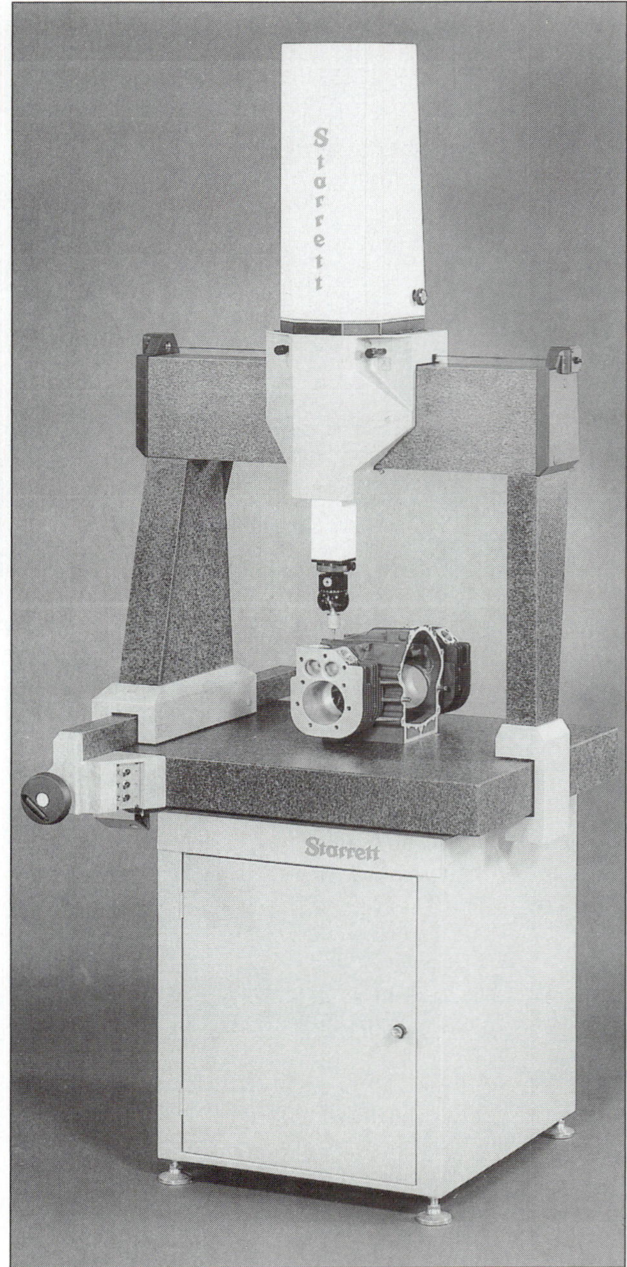

FIGURE 11
Computerized coordinate measuring machine (Fellers and Hunt, *Manufacturing Processes for Technology,* 2d ed. Reprinted by permission of Pearson Education, Inc., Upper Saddle River, NJ.).

FIGURE 12
Laser used in precision measurement (Giddings and Lewis).

Upper Management

Upper management includes the company president, chief executive officer (CEO), chairman of the board of directors, and one or more vice presidents. The responsibilities of the CEO, president, and/or chairman of the board of directors are to oversee the entire operation of the business. The head of the company is responsible to the stockholders, and in many cases, upper management is elected by the stockholders to run and manage the business.

Vice presidents (VPs) are likely to be specialists responsible for specific areas. Examples include manufacturing, finance, advertising, research and development, engineering, quality assurance, and marketing. The VP of manufacturing is responsible for all aspects of the manufacturing operation. The VP for finance is responsible for company loans, stockholder dividends, stock and bond issues, and accounts payable and receivable. The advertising VP is responsible for advertising the company's products if this is done in-house. In many cases, advertising is contracted out to advertising agencies that specialize in product promotion.

If the company does research and development of new products or is in the business of product development for outside customers, the VP for research and development oversees this aspect of the business. Engineering involving product design is the responsibility of the VP for engineering.

Middle Management

Employees in middle management positions translate the directives of upper management into manageable tasks, assign those tasks to individuals or functional groups, and assure that the work is completed on schedule. A wide range of responsibilities are found in middle management encompassing many job titles, including line positions (directly responsible for manufactured output) such as plant managers and heads of departments who provide support to manufacturing operations.

An important issue for middle managers is the span and scope of their responsibilities. Middle management is a frequent target of cost-cutting manpower reductions, but reducing the number of middle managers increases the number of employees that each manager must direct. Too wide a reporting span (the number of employees reporting to each manager) reduces the effectiveness of middle management. Conversely, too many layers of middle managers reporting to other middle managers impedes communication within the company and slows its reactions to changes in the business and technical environment.

One justification for reductions in middle management positions is the increased use of computers for communication. Computers have improved managers' access to the information needed to make rapid, accurate decisions, and they have provided more and faster means of communication. Through reductions in management positions, the middle manager who is able to communicate better and

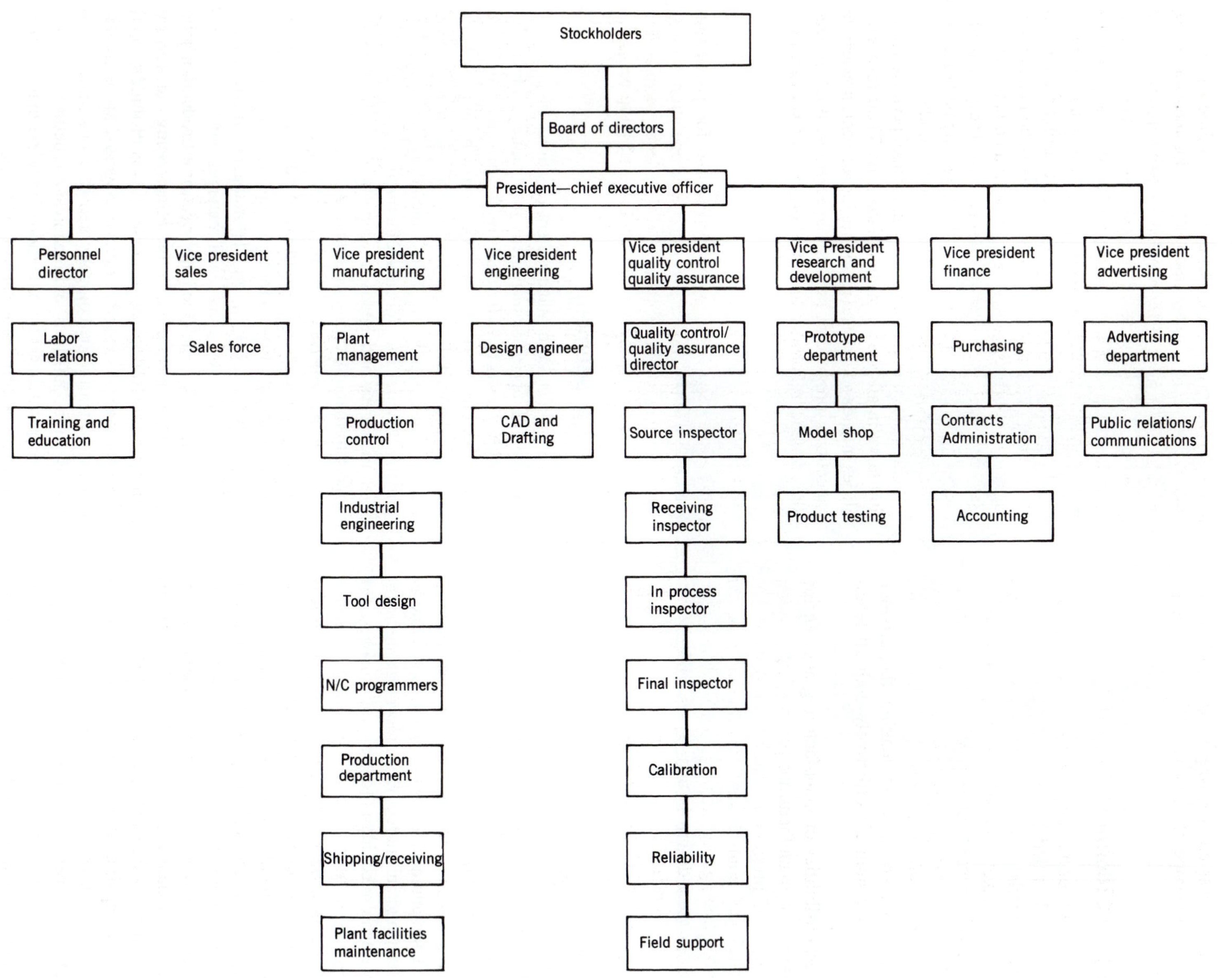

FIGURE 13
Basic organizational framework of a typical manufacturing corporation.

faster in today's business environment is frequently asked to undertake a wider scope of responsibilities and greater reporting span.

Engineering and Technical Support

Manufacturing engineering and industrial engineering positions are typically filled by engineering, engineering technology, and industrial technology graduates. Personnel that staff these jobs make an invaluable contribution to the smooth running of a manufacturing industry. Trained in broad areas of manufacturing, these individuals know both the business end of manufacturing and the technical details of the processes that are used in making the product, enabling them to communicate effectively with all levels of management.

Typical responsibilities of manufacturing engineering include designing the manufacturing process; jig, fixture, and tool design; CNC programming; and many aspects of quality assurance and control. Industrial engineers often deal with cost estimating, ergonomics, and facilities planning. Although product design is the responsibility of the design engineer, manufacturing engineers and industrial engineers are often the individuals who ultimately make the manufacture of the product possible.

Production Operations Support

Many other individuals also contribute to the production process. Although their contributions may be indirect, they are nonetheless vital to the successful function of the business. Job titles in this area include purchasing agents, who are responsible for purchasing tools, equipment, and raw material required for the production process, and production schedulers, who make sure that the proper material is ordered and arrives at the production facilities in a timely manner. This permits production to continue in an orderly fashion with no delays due to a lack of material. The company personnel department is responsible for hiring necessary individuals to meet the staffing needs of the company.

Another vital function in any manufacturing business is filled by individuals involved in plant facilities and production equipment maintenance. Their job is to keep the production line moving. Manufacturing is a costly business—and if the production line stops, pay for the employees does not. The line must be restored to service as soon as possible. Maintenance personnel are therefore always on call in many high-production manufacturing industries. In modern computer-automated manufacturing, the electromechanical and the computer numerical control electronics technicians are often directly responsible for uptime on expensive production equipment. *Uptime* refers to the time that a piece of manufacturing equipment is actually producing product units. *Downtime* refers to times when a machine is not producing. Machine downtime can occur after breakdown, during required maintenance, or when the machine is waiting for material to arrive and be loaded for processing. It is naturally desirable to have as little downtime as possible, so in fast-paced production plants the mechanical and electronic maintenance technicians are often very busy. They often are called to work during night shifts or weekend periods in order to keep production equipment operating at peak efficiency. As computer automation trends continue to predominate in modern manufacturing the maintenance technician will play an ever-increasing part in keeping manufacturing equipment in an uptime mode.

Quality assurance and control personnel perform many vital tasks in all phases of manufacturing. Their responsibilities include source and receiving inspections, first article inspections, production process inspections, final inspections, product reliability, and after-production warranties.

Line Supervision

Line supervisors are directly responsible for production worker supervision and the hour-by-hour operation of manufacturing processes. Job titles at this level include line supervisors and foremen, team leaders, and setup persons.

The Production Worker

No discussion of business structure would be complete without mentioning those who do the actual work of production and production support. Job titles are many and varied in production work and include machine operators, machinists, toolmakers, toolroom attendants, plumbers, sheet metal workers, mechanics, inspectors, electronics assemblers, helpers, electricians, and electronic and mechanical system maintenance technicians.

MANUFACTURING AND YOU

Trade-Level Opportunities

At the trade level many of the repetitive production jobs held by minimum-wage employees are being changed by automation such as computer numerical control (CNC) and industrial robots. As robots become more and more sophisticated and capable they move further into skilled trades such as welding (Figure 14). On many repetitive welding tasks the robot welder can outperform its human counterpart in both quantity and quality.

The general machinist who can operate highly computerized systems will continue to be needed. Higher-level skilled trades, such as tool and die making, will continue to be in demand, in part because preparation for these jobs requires several years of diligent training.

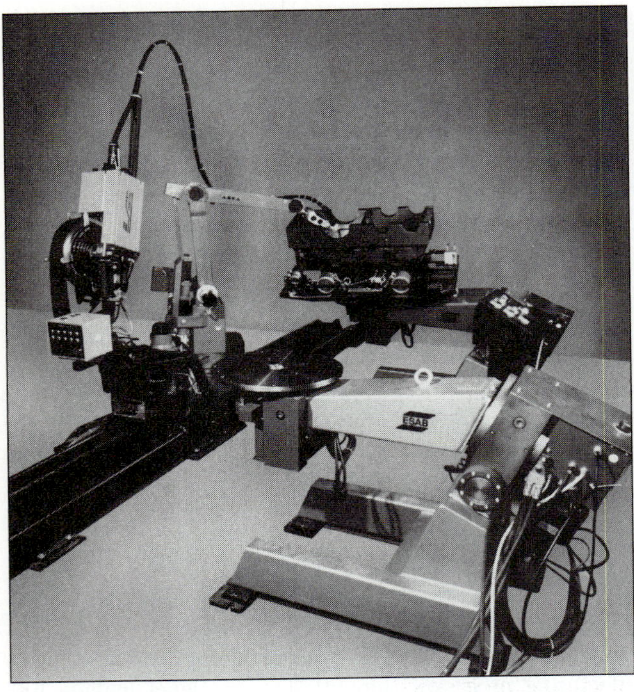

FIGURE 14
Robot welder welds a gear case mounted on one of two workpiece positioners.

FIGURE 15
Production technician assembles laser computer disks in an ultra-clean-room environment.

The production technician, an individual who has broad training in electric/electronics, mechanical/hydraulic systems, and computers, probably has the brightest future in trade-level manufacturing jobs. The capability of modern automated manufacturing systems is truly amazing, and the equipment and control systems are complex, requiring skilled maintenance. Production technicians must be able to diagnose problems quickly and effect repairs rapidly and efficiently so that costly equipment can be placed back into production as soon as possible. They assemble and test delicate high-precision equipment, often in controlled environments, as illustrated in Figure 15, where technicians are assembling computer disks that must be handled in an ultra-clean-room environment.

Technical and Management Levels

The future in manufacturing for the college-trained engineer, engineering technologist, and industrial technologist is bright. As automated manufacturing systems are put in place the engineer and technologist will be needed for plant facilities; tool, jig, and fixture design (Figure 16); robot and numerical control programming; and implementation of all aspects of the manufacturing system (Figure 17). Like the production technician, the engineer and technologist will need broad training in materials applications, computers, and manufacturing processes. Such individuals will also be heavily involved with quality assurance as the extremely precise manufacture of miniaturized high-technology products becomes more commonplace.

FIGURE 16
Automatic wing spar assembly tooling. New methods of tooling will challenge the manufacturing engineer (Boeing).

Design Engineering

Creative product designers will always be needed as the capacity of high-technology manufacturing makes possible the production of a vast array of incredible new products. The designer must be thoroughly knowledgeable in modern manufacturing methods and their capabilities, using CAD to smoothly integrate new design ideas with new materials and new processes as they become available. Preparation for jobs in this area requires college-level education plus on-the-job industrial training.

Preparation

For those who avail themselves of proper education and training, opportunities in manufacturing will abound; however, the future employee at any level in manufacturing will have to undergo broader and more complex training than at any time in the past. We conclude the introduction where we began—will you be ready to meet these challenges? The chapters that follow will help you prepare.

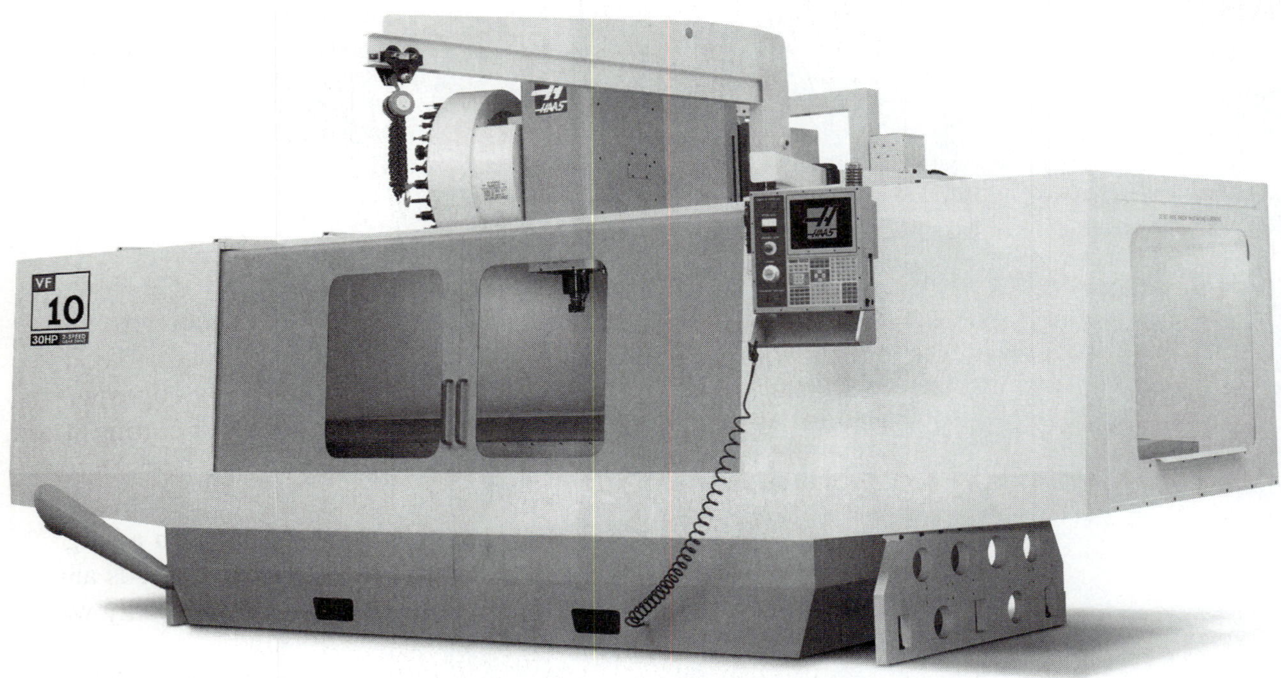

FIGURE 17
CNC machining center (Haas Automation Inc.).

Review Questions

1. Outline the structure of a typical manufacturing industry.
2. Discuss some of the responsibilities of the manufacturing engineer or technologist.
3. Who is most likely to be responsible for installing a new electrical service to a piece of production equipment?
4. The company needs 10 million rivets. Who places the order for them?
5. Whom would you approach if you were seeking employment with the company?
6. Who is finally responsible for company financial affairs?
7. Who designs production tooling?
8. Who designs products?
9. Who do you think is the most important person in the company? Why?

PART I

MATERIALS OF MANUFACTURE

At the dawn of history, the only materials used by humans were taken directly from natural sources. Stones were used to erect crude shelters or structures; harder stones with cutting edges were used for arrowheads and spear points and for tools for cutting wood, bone, and leather. Clay was shaped into containers and bricks for building. The discovery that firing these clay items made them stronger and water resistant was probably the first step in a long procession of the adaptation of natural materials to the needs and uses of humanity. However, humans' creativity and ingenuity were limited by the properties of the available materials.

Today, humans make use of literally thousands of materials that are derived from nature, but they also use many more synthetic materials that do not come directly from natural sources. All these amazing developments have come about because of increasing knowledge and understanding of the properties of materials. In the late 1800s we began to understand the arrangement of atoms and crystal structures and their behavior in various conditions. Today, our knowledge and understanding of materials make it possible to develop and design materials to meet property requirements, challenging our creativity and ingenuity even further. Plastics, elastomers, metals, ceramics, and composites are some of the important materials of today and tomorrow that will be studied in Part I, Chapters 1–6, and later in Chapters 11 and 16.

CHAPTER 1

The Atomic Structures of Materials

Objectives

This chapter will enable you to:

1. Explain a simplified model of the atom.
2. Understand several bonding arrangements and explain the role of valence in bonding.
3. Describe the crystalline structure of metals and how this affects their behavior.
4. Give an account of the structure of some polymers for the manufacture of certain plastic materials and elastomers.
5. Describe the crystal structure of clays and other ceramics and their bonding arrangements.

Key Words

metal
ceramic
valence
atomic bonding
space lattice
thermoplastic
traditional ceramic
polymer
allotropy
grain
polycrystalline
thermoset
engineered ceramic

Materials can be divided into a number of different families; in this chapter we will consider three: metals, polymers, and ceramics. Each of these families has properties that are different from those of the others, related primarily to how their atoms are bonded. The unique metallic bond of **metals** makes them hard and strong but allows them to be stamped, rolled, drawn, and worked in other ways. In some situations metals can form compounds. **Polymers** are compounds that can be linked together into very long chains. **Ceramics** are also compounds, typically a combination of a metal and a nonmetal, that are so brittle that they cannot be formed and so hard that they cannot be machined by traditional means.

Today, we know much about the structure of the physical universe, and through the endeavors of the scientific community, many more facts are constantly being discovered. Engineers in turn design new products based on these new discoveries, often using radically new concepts. The phenomenal growth in solid-state electronics and high technology is an example. Therefore, technologists must be able to understand some of the new terminology and technology of the scientist and engineer in order to communicate in the manufacturing field. It is useful to understand the basic structure of the atom, the smallest unit of an element, prior to studying the properties of materials and their behaviors under various conditions. Some of these principles and concepts are included in this chapter.

Although *atoms* are too small to be seen with the aid of ordinary light microscopes, their outline can now be detected with such devices as the electron microscope and the field ion emission microscope. A *molecule* is defined as the smallest particle of any substance that can exist free and still exhibit all the chemical properties of that substance. A molecule can consist of one or more atoms of the same kind. For example, oxygen gas is found in nature as a molecule consisting of two oxygen atoms (O_2) or of three oxygen atoms, ozone (O_3). A *compound* is composed of two or more elements combined chemically; for example, hydrogen and oxygen combine to produce a molecule of water (H_2O). A *mixture* consists of two or more elements or compounds that are physically, not chemically, combined.

All matter in our world is composed of one or more of the 92 naturally occurring chemical elements. Each element is a pure material, such as a metal or nonmetal, unlike any other; yet every element is composed of protons, neutrons, and electrons; only the number and arrangement differ. So, before we study the families of materials, we must examine the atom, how it is classified, and how it bonds with other atoms.

STRUCTURE AND CLASSIFICATION OF ATOMS

The Atom

Atoms, and in fact all materials, are composed of the same basic components: *protons, neutrons,* and *electrons* (Figure 1.1). The **nucleus** of the atom is composed of positively charged protons, and neutrons. The latter have almost the same mass as protons but are without electrical charge. Surrounding the nucleus, orbiting it in what are called *shells,* somewhat like the orbits of planets in our solar system, are the electrons, equal in number to the protons, which have a negative charge equal to and opposite of the charge of the proton in a neutral atom. It is this electrical charge that holds the atom together and is responsible for much of the chemical and electrical behavior of the elements. Almost all the atomic mass is contained in the nucleus; an electron has only about 1/1839 the mass of a proton or neutron.

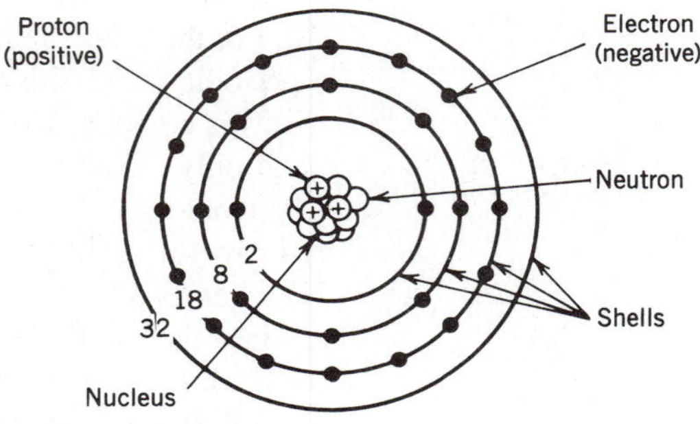

FIGURE 1.1
The basic parts of the atom, also showing the maximum number of electrons possible in each of the first four shells.

Periodic Table

The chemical elements in the periodic table are classified according to the number of electrons and protons in their atoms. For example, the atoms of the element hydrogen have in their normal state a single electron moving about a nucleus that has a positive charge equal to the negative charge of the electron. Likewise, the atom of the element helium has two electrons moving about its nucleus, which has a positive charge equal to twice one electronic charge. An atom of the element lithium has three electrons and a positive nuclear charge equal to three times that of a hydrogen atom. This is the basis for the numbers on the periodic table (Table 1.1), called *atomic numbers.* These numbers range from 1 for hydrogen to 92 for uranium, and still higher for the transuranic elements. These numbers are important because they determine the chemical behavior of the atoms.

The table of elements is called *periodic* because elements with similar properties appear at regular intervals in the arrangement of the elements. The horizontal rows of elements are known as *periods,* and the vertical columns are known as *groups.* The greatest similarities exist in the vertical columns. For example, copper (Cu), silver (Ag), and gold (Au) have similar crystal structures and readily combine as alloys. The vertical group of chromium (Cr), molybdenum (Mo), and tungsten (W) has similar characteristics as alloying elements in steel for promoting strength and hardenability. Also, sulfur (S), selenium (Se), and tellurium (Te) have all been used to promote machinability in steels. Besides the relationship of the elements in the vertical groups, there are resemblances in several horizontal triads: iron (Fe), cobalt (Co), and nickel (Ni) are all ferromagnetic, for example.

The six metals of the platinum group consist of two horizontal triads: ruthenium (Ru), rhodium (Rh), palladium (Pd); and osmium (Os), iridium (Ir), platinum (Pt). The first subgroup (Ru, Rh, Pd) has atomic weights and specific gravities that are about half those of the second subgroup (Os, Ir, Pt). Furthermore, each metal of the first group has a member in the second group that seems more closely related to it; for example, Ru and Os, Rh and Ir, and Pd and Pt may be considered as pairs. All these metals are chemically resistant and have high melting points. These six metals along with gold and silver are called *precious metals.* They are also called the *noble metals* because of their resistance to atmospheric conditions and to most acids. In general, crystal structures, melting points, densities, and other properties of elements are functions of the atomic number.

In the foregoing discussion, the electrons were spoken of as moving around the nucleus like planets in orbit around the sun. We know now by the **uncertainty principle** that this

TABLE 1.1
The periodic table of the elements

Light Metals												*Nonmetals*					*Inert gases*
IA *H* *1*	*IIA*	*Heavy Metals*										*IIIA*	*IVA*	*VA*	*VIA*	*VIIA*	*O* *He* *2*
Li 3	Be 4	←Brittle metals→					←Ductile metals→				Low melting	B 5	C 6	N 7	O 8	F 9	Ne 10
Na 11	Mg 12	IIIB	IVB	VB	VIB	VIIB	VIII			IB	IIB	Al 13	Si 14	P 15	S 16	Cl 17	Ar 18
K 19	Ca 20	Sc 21	Ti 22	V 23	Cr 24	Mn 25	Fe 26	Co 27	Ni 28	Cu 29	Zn 30	Ga 31	Ge 32	As 33	Se 34	Br 35	Kr 36
Rb 37	Sr 38	Y 39	Zr 40	Nb 41	Mo 42	Tc 43	Ru 44	Rh 45	Pd 46	Ag 47	Cd 48	In 49	Sn 50	Sb 51	Te 52	I 53	Xe 54
Cs 55	Ba 56	* 57–71	Hf 72	Ta 73	W 74	Re 75	Os 76	Ir 77	Pt 78	Au 79	Hg 80	Tl 81	Pb 82	Bi 83	Po 84	At 85	Rn 86
Fr 87	Ra 88	† 89–103															

*	La 57	Ce 58	Pr 59	Nd 60	Pm 61	Sm 62	Eu 63	Gd 64	Tb 65	Dy 66	Ho 67	Er 68	Tm 69	Yb 70	Lu 71
†	Ac 89	Th 90	Pa 91	U 92	Np 93	Pu 94	Am 95	Cm 96	Bk 97	Cf 98	Es 99	Fm 100	Md 101	No 102	Lr 103

In general, it can be stated that the properties of the elements are periodic functions of their atomic numbers. The greatest similarities between elements exist in the vertical columns. There are also some similarities between the A and B groups on either side of the VIII groups. For example, the scandium group, IIIB, is in some respects similar to the boron group, IIIA. The most important properties in the study of metallurgy that vary periodically with the atomic numbers are crystal structure, atomic size, electrical and thermal conductivity, and possible oxidation states.

(John E. Neely, *Practical Metallurgy and Materials of Industry,* 2d ed., Wiley, New York, ©1984.)

cannot be true of the electrons. The exact orbit of an electron about the nucleus has been replaced by a mathematical function called an *orbital.* Orbitals are electronic energy levels within the shells to which quantum numbers have been assigned. An electron can become excited, and will jump to a higher energy orbit, by absorbing energy in the form of electromagnetic radiation. This radiation is absorbed in energy packets called *photons* (that is, a photon is a quantum of radiant energy). If an electron returns to its original energy orbit or, shell, it will release a photon of energy (such as heat or light) that is characteristic of that jump.

In Figure 1.1 the maximum number of electrons possible in a row or shell, *N,* is determined from the relationship $N = 2n^2$, where *n* is the shell number. For example, for the second shell (which is also the second row in Table 1.1), 2×2^2 equals 8, and from Li (lithium) to Ne (neon) there are 8 elements. Since the first row has two elements, neon has an atomic number of 10. For the third shell, 2×3^2 equals 18. Inspection of the periodic table, Table 1.1, shows that the third row of elements is completed with an atomic number of 18, or a total of only 18 electrons. To completely fill the first three shells would take 2 + 8 + 18, or 28 electrons. Thus, in the third row (or shell) there are 10 spaces for electrons that did not get filled.

As we begin to put electrons into the fourth row and form potassium (K) and calcium (Ca), the next electrons do fill the spaces left in the *third* shell. From calcium (Ca) through zinc (Zn), there are exactly 10 elements. This means that the two electrons that were added to form calcium are by themselves in the fourth shell. These electrons in the outermost shell are called **valence** electrons, and they can easily be stripped off to form compounds or to engage in other bonding arrangements. The valence electrons are responsible for the electrical and chemical behavior of the elements. Also, because the added electrons occupy an inner shell in the center part of the periodic table, the valence of metals will normally be 2; there are some exceptions, but 2 is a good general rule.

Looking at the fourth row of the periodic table you can see that the addition of 10 electrons to the third row carries you to zinc (Zn). After zinc the electrons resume filling the fourth shell (or row), so that the valence of gallium (Ga) is 3 (note the Roman numerals at the top of the columns or groups). Krypton (Kr), in the fourth row (period), contains 8 electrons in its outermost shell which is then considered filled. This sequence illustrates that the *outermost,* or valence, shell is considered filled when it contains 8 electrons; this is referred to as "the rule of eight."

Metals and Nonmetals

As can be seen in Table 1.1, the chemical elements can roughly be divided into three categories: metals, nonmetals, and inert gases. Approximately three quarters of all elements are considered to be metals. A metal must have some of the following properties:

1. Ability to donate electrons to form a positive ion
2. Crystalline structure
3. High thermal and electrical conductivity
4. Ability to be deformed plastically
5. Metallic luster or reflectivity

Although metallic elements are more numerous and fulfill many requirements, plastics, rubber, and other synthetic products are produced for the most part, from nonmetals. Some nonmetals, such as carbon, germanium, and silicon, resemble metals in one way or another and are sometimes called *metalloids.* They can have a metallic luster and crystalline microstructure, and they can be semiconductors of electricity, but they cannot be deformed without fracturing because they are brittle, as are many nonmetals. The growth in solid-state electronic components has largely been based on the metalloids germanium and silicon.

ATOMIC BONDING

In our study of the periodic table, we saw that the number of electrons in the outer or valence shell was a minimum of one and a maximum of 8. When an element has its proper complement of valence electrons, it is electrically neutral; when an atom has more or fewer than that number, it is an *ion* and has an electric charge. The charge is negative when extra electrons are present and positive when some are missing. Metal atoms are easily stripped of their valence electrons and thus form positive ions. When the outer shell has a full complement of 8 valence electrons, it is said to have zero valence. The inert gases have zero valence and do not readily combine with other elements to form compounds.

Three basic types of primary **atomic bonding** arrangements hold atoms together, namely, *ionic* (also called electrovalent), *covalent* (polar or nonpolar shared electrons), and *metallic.*

The Ionic Bond

The simplest type of linking or bonding of atoms is the ionic bond. Ionic bonding links dissimilar atoms by the attraction of negative and positive ions. Sodium chloride (NaCl) is an example of ionic bonding (Figure 1.2). The sodium loses an electron and thus is a positive ion; the chlorine atom gains one electron and is a negative ion. The two charges offset, so the compound NaCl (salt) is electrically neutral (Figure 1.3).

Ionic bonds tend to have the characteristics of brittleness, moderate strength, and high hardness and are good electrical insulators when they are in pure solid form because there is little opportunity for ion or electron movement. It should be noted, however, that many ionic compounds, such as sodium chloride, are water soluble and that solutions of these compounds will conduct electricity very well.

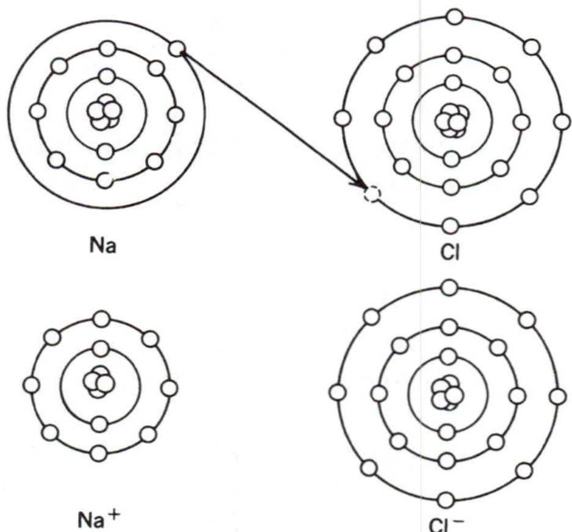

FIGURE 1.2
Ionic bonding in sodium chloride (Neely and Bertone, *Practical Metallurgy and Materials of Industry,* 5th ed. © 2000 Prentice Hall Inc.).

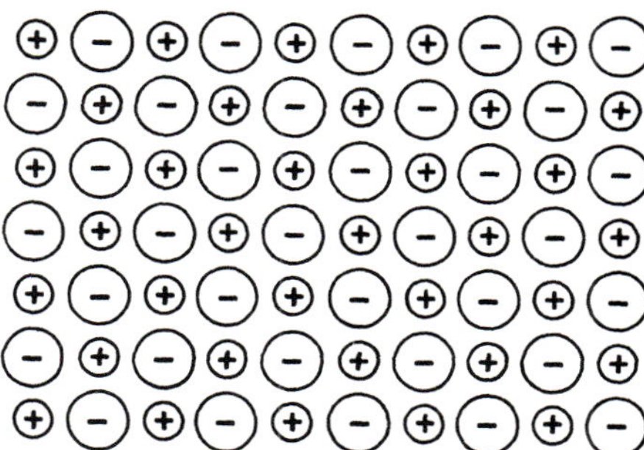

FIGURE 1.3
Traditional view of the lattice structure of sodium chloride (Neely and Bertone, *Practical Metallurgy and Materials of Industry,* 5th ed. © 2000 Prentice Hall, Inc.).

The Covalent Bond

The covalent bond, in which valence electrons are *shared,* is a very strong atomic bond whose strength depends on the number of shared electrons. It is interesting to note, however, that a variety of covalent bonding patterns exist. In *network* covalent substances such as diamond and silicon carbide (carborundum), the sharing of electrons forms a large, strongly bonded 3-dimensional lattice (see Figure 1.25 on p. 29). On the other hand, *molecular* covalent substances such as sugars, polymers, or waxes satisfy their sharing of electrons *within* the molecule and have very little attraction to other molecules surrounding them (see Figure 1.17 on p. 26). Network covalent substances have high melting points and hardnesses, whereas molecular covalent substances have low melting points and hardnesses.

Atoms attempt to achieve the inert gas electronic structure of eight electrons in the outer shell. (In the case of hydrogen it tries to add one.) Chlorine, having a complement of only seven electrons in its outermost shell, has a valence of -1 and therefore tends to join with any element that will fill its outermost shell to eight electrons. As we have seen, chlorine will join with sodium to obtain an extra electron and form an ionic substance; it will also join with another chlorine atom to gain an extra electron by sharing it to form a covalent molecule (Figure 1.4). The element oxygen has the same tendency, but it must share two valence electrons (Figure 1.5). Negative valences greater than 4 do not occur.

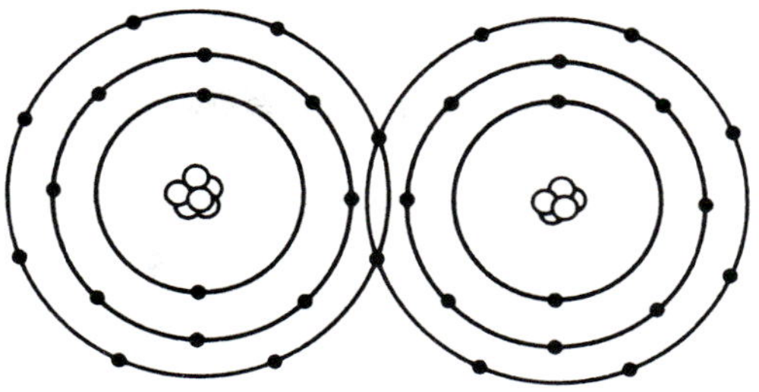

FIGURE 1.4
Chlorine atoms form a covalent bond by sharing atoms and become a molecule of Cl_2.

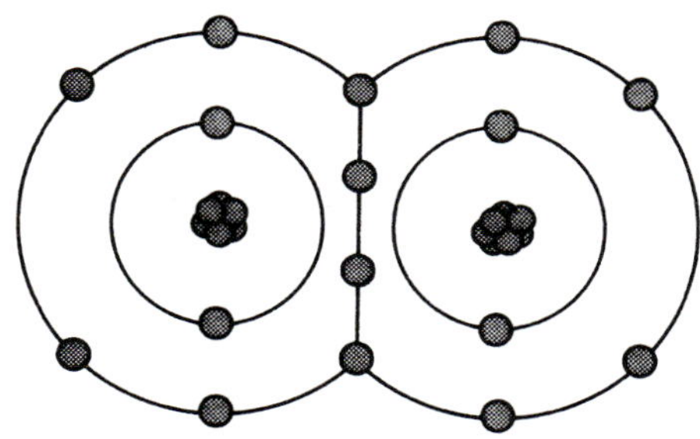

FIGURE 1.5
The covalent bond of the oxygen molecule (Adapted from Neely and Bertone, *Practical Metallurgy and Materials of Industry,* 5th ed. © Prentice Hall, Inc.).

The Metallic Bond

The metallic bond is formed among metallic atoms when some electrons in the valence shell separate from their orbital path and exist in a free state like a cloud or gas. Surrounding the positive nuclei in this highly mobile state, the negative cloud bonds the atoms together (Figure 1.6). Because the electrons in the cloud do not "belong" to any particular atoms, the bond strengths can vary, deformation can take place without rupture, and good heat and electrical conductivity can occur owing to the easy movement of electrons from one atomic site to the next. This type of bonding makes possible some of the useful characteristics of metals such as plasticity and ductility, which will be discussed in Chapter 2.

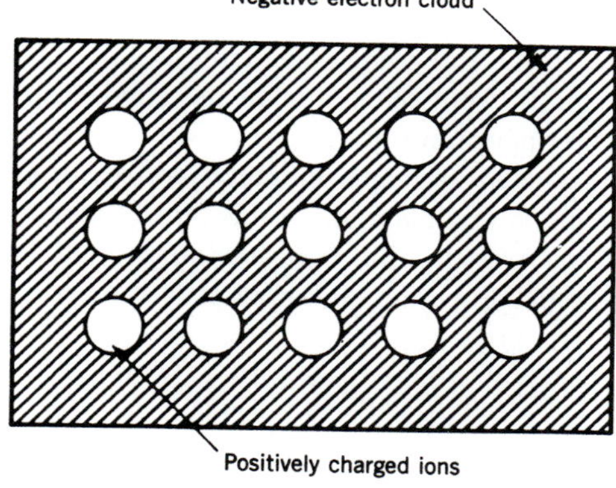

FIGURE 1.6
The metallic bond (Neely and Bertone, *Practical Metallurgy and Materials of Industry,* 5th ed. © Prentice Hall, Inc.).

Van der Waals bonding is found between neutral atoms of elements such as the inert gases and between neutral molecules of covalent molecular compounds, such as long-chain polymers. In polymers the van der Waals bond is a secondary bond *between* the chains of large molecules that provides a type of cross-linking that improves the strength of the thermoplastic polymers. In metals the van der Waals bond is only a weak attractive force and is important in the study of metals only at very low temperature.

METALS

Solid metals exist in distinctive latticelike crystals or space lattices. This structure is first encountered as the metal solidifies or freezes from its molten state. The heat removal process requires that the atoms freeze into unique forms called *dendrites.* As the crystalline dendrites grow they run into other dendrites that limit their growth and form a boundary. The completely grown dendrite is called a **grain,** and its junction with its neighbor is called a *grain boundary.*

Crystalline Structure

When metals cool from the molten state and solidify, they generally form a crystalline structure called a **space lattice.** If imaginary lines are drawn connecting the centers of the atoms together, the figure is called a *lattice structure* (Figure 1.7). A recent development has created an exception to the foregoing statement. Metals can now be cooled rapidly from the liquid to the solid state so that there is no opportunity for a crystal structure to develop. These products are appropriately called **amorphous metals**. Only a small amount of this type of metal is being produced at present. Since it has very good electrical and magnetic properties, amorphous metal is useful for making transformer cores. However, when metals cool at normal rates, they always form a space lattice that consists of orderly rows of atoms.

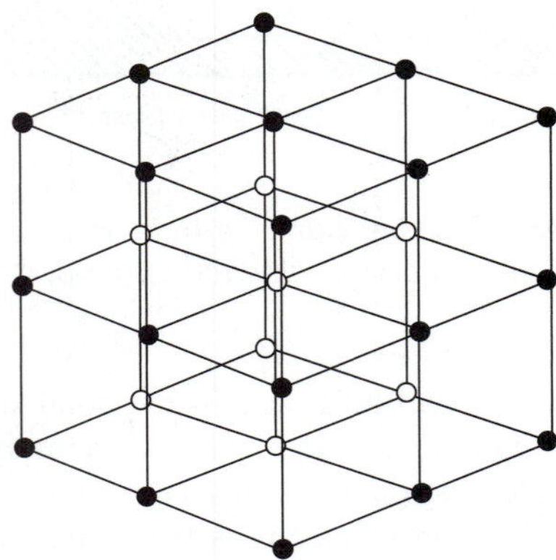

FIGURE 1.7
Lattice structure of atoms in solids.

Interatomic Distances

Because similar electrical charges tend to repel and unlike charges attract, the electron-filled outer shells of the metallic ions tend to have a repulsive force. If it were not for the free valence electrons vibrating among the atoms to produce forces of attraction, metal atoms would fly apart. There is a distinct balance of forces in metallic bonding causing the atoms to maintain a distance. If we assume an atom to be a solid ball, then the distance between the centers of two adjacent atoms represents the effective diameter of the atom. This distance can be determined experimentally and is shown on some versions of the periodic table.

A comparison of the diameters of atoms shows that the largest atoms are in the lower left corner of the periodic table, and the smallest are in the upper right corner of the table. On the other hand, the metals in the center of the table have very similar sizes, which is one of the reasons why many metallic elements can easily combine to form alloys. One comparison of interest is that the diameter of the iron atom is approximately 1.75 times that of the carbon atom; this fact will be important in our later discussion about steel.

Interatomic distances can be altered by external forces, such as thermal energy and electrical or mechanical forces. This is why materials expand when heated and contract when cooled.

Unit Cells

A space lattice can be described as having repeating unit building blocks throughout its structure. These are called *unit cells* (Figure 1.8). Metals solidify into 14 known types of space lattices, or crystal structures, eight of which are shown in Table 1.2. Most of the commercially important

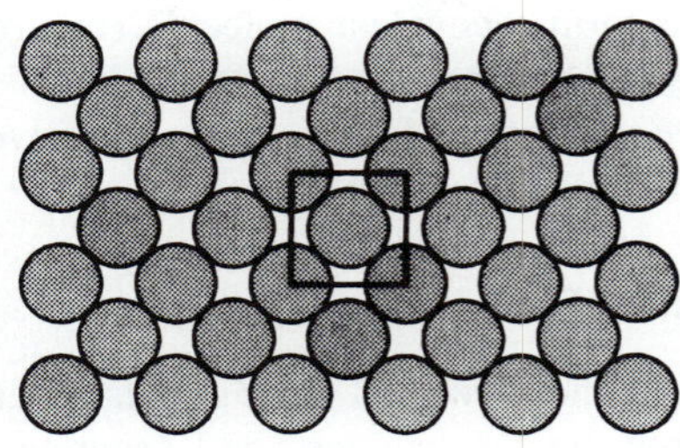

FIGURE 1.8
Body-centered cubic unit cell shown in the lattice (Neely and Bertone, *Practical Metallurgy and Materials of Industry,* 5th ed. © 2000 Prentice Hall, Inc.).

TABLE 1.2
Crystal structures of some common metals

Symbol	*Element*	*Crystal Structure*
Al	Aluminum	FCC
Sb	Antimony	Rhombohedral
Be	Beryllium	CPH
Bi	Bismuth	Rhombohedral
Cd	Cadmium	CPH
C	Carbon (graphite)	Hexagonal
Cr	Chromium	BCC (above 26° C, 78.8° F)
Co	Cobalt	CPH
Cu	Copper	FCC
Au	Gold	FCC
Fe	Iron (alpha)	BCC
Pb	Lead	FCC
Mg	Magnesium	CPH
Mn	Manganese	Cubic
Mo	Molybdenum	BCC
Ni	Nickel	FCC
Nb	Niobium (columbium)	BCC
Pt	Platinum	FCC
Si	Silicon	Cubic, diamond
Ag	Silver	FCC
Ta	Tantalum	BCC
Sn	Tin	Tetragonal
Ti	Titanium	CPH
W	Tungsten	BCC
V	Vanadium	BCC
Zn	Zinc	CPH
Zr	Zirconium	CPH

(Neely and Bertone *Practical Metallurgy and Materials of Industry,* 5th ed., © 2000 Prentice Hall, Inc.).

metals solidify into one of three types of lattice structures (Figure 1.9)—body-centered cubic (BCC), face-centered cubic (FCC), and close-packed hexagonal (CPH).

If we again assume that atoms are solid balls that touch their nearest neighbors, two common structures will look like those in Figure 1.10. The metallic bond, to have maximum plasticity and other needed properties, is best when the atoms are close-packed as in body-centered cubic structures that form a cube with one atom at its center. Even better for plastic deformation is the more closely packed, face-centered cubic structure. The close-packed hexagonal unit structure does have some planes that are close-packed, but the location of other planes causes them to interfere with deformation, so these metals have poor formability.

Some materials have two or more of these structures under different conditions such as temperature change. Such a material is called **allotropic.** For example, the three allotropic forms of carbon are amorphous (charcoal, soot, and coal), graphite, and diamond. Iron also exists in three allotropic forms that are temperature dependent. (The term *iron* refers to elemental or pure iron unless specified as wrought or cast.) At room temperature iron is BCC and is called alpha. At higher temperatures it is first FCC, called gamma, and then BCC, called delta. We will encounter these forms again in Chapter 3.

SOLIDIFICATION OF METALS

Dendrite Formation

During the **solidification** of a molten metal its atoms must form into the space lattice for that metal. The heat removal process is such that it requires the atoms to solidify in a three-dimensional treelike form called a **dendrite** (after the

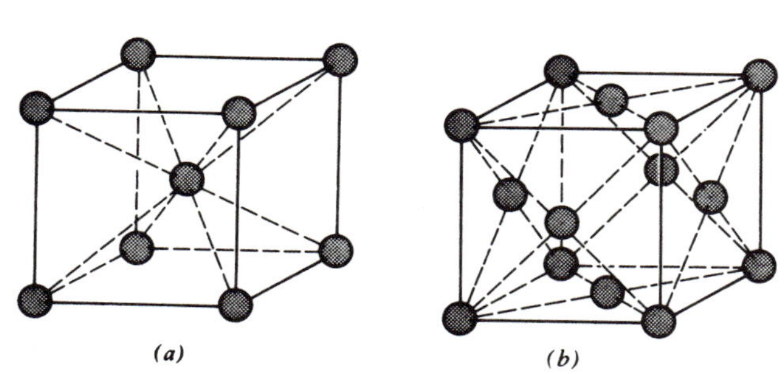

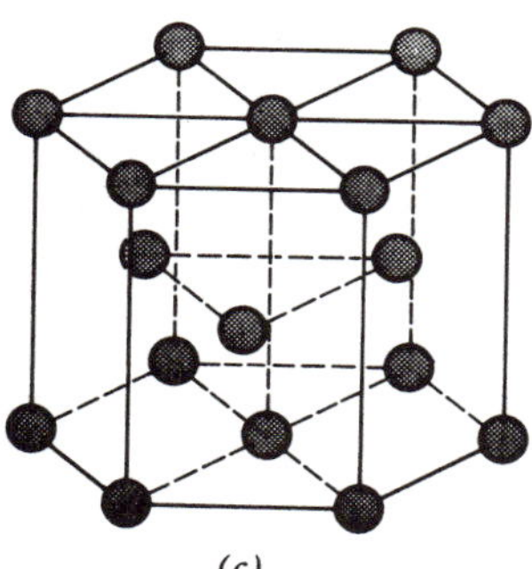

FIGURE 1.9
Three common unit cells of the lattice structure are *(a)* body-centered cubic, *(b)* face-centered cubic, and *(c)* close-packed hexagonal (Neely and Bertone, *Practical Metallurgy and Materials of Industry,* 5th ed., © 2000 Prentice Hall, Inc.).

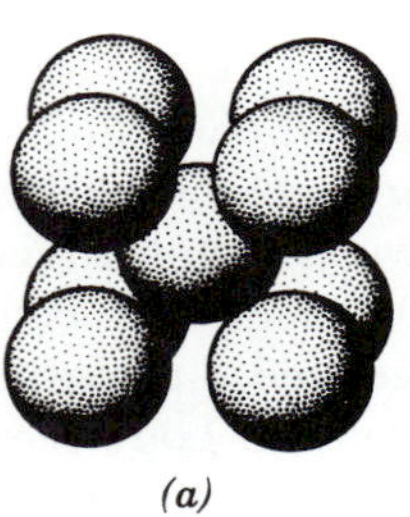

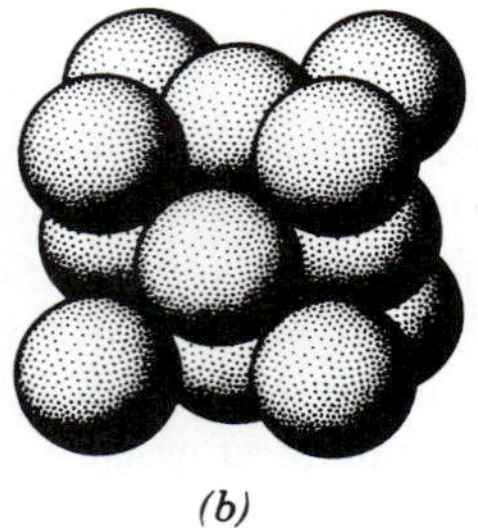

FIGURE 1.10
Model of atoms in unit cells. *(a)* Body-centered cubic (BCC). *(b)* Face-centered cubic (FCC) (Neely and Bertone, *Practical Metallurgy and Materials of Industry,* 5th ed. © 2000 Prentice Hall, Inc.).

FIGURE 1.11
Dendrite growth in metal as shown in a photomicrograph.

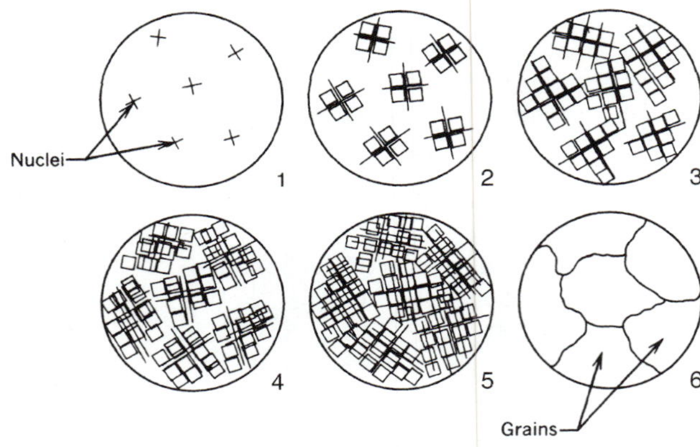

FIGURE 1.12
The formation of grains during solidification of molten metal (Neely and Bertone, *Practical Metallurgy and Materials of Industry,* 5th ed. © 2000 Prentice Hall, Inc.).

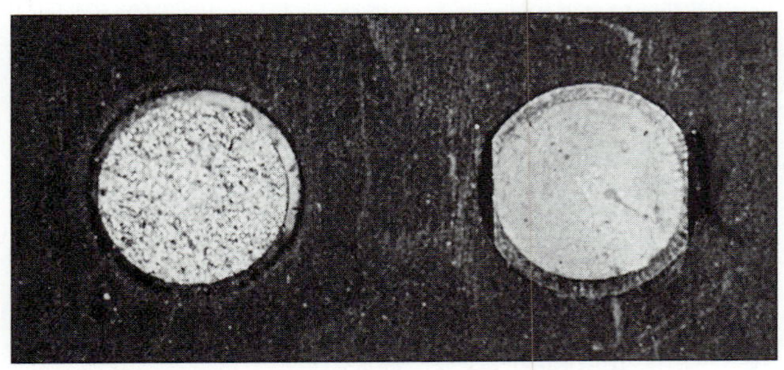

FIGURE 1.13
Broken sections of steel showing its crystalline structure. The one on the left is coarse and can be seen clearly. The one on the right is fine grained and requires magnification to see the grains.

Greek word for tree). In the upper left portion of Figure 1.11 can be seen a "tree" lying horizontally on its limbs, as if it had been split through its trunk. Dendrites begin to grow from seed crystals or nucleating particles such as impurities and purposely added grain refiners. For example, aluminum is often added to molten steel; in reducing the amount of oxygen dissolved in the molten metal the aluminum provides aluminum oxide nucleating sites. This nucleating process is illustrated in view 1 of Figure 1.12.

Grains and Grain Boundaries

As dendrites grow they form three-dimensional solid crystals called **grains.** Most metals encountered in everyday manufacturing have more than one grain and are referred to as **polycrystalline.** Because the axes of these grains are randomly oriented the lattices of the grains formed do not line up (views 2 through 5 of Figure 1.12), so when they meet they form **grain boundaries** (view 6). All metals when normally cooled are composed of these tiny grain structures, which can be seen with the aid of a microscope when a specimen is cross-sectioned, polished, and etched. Often the grain structure in a broken section of a piece of metal can be seen with the naked eye as small crystals (Figure 1.13).

Since the crystal lattice or dendrite of each grain grows in a different direction, the atoms are jammed together in a misfit pattern at the grain boundary (Figure 1.14). Although the nature of the atomic forces in the grain boundaries is not entirely understood, the grain boundary is generally assumed to be an interlocking border in a highly strained condition. Grain boundaries are only about one or two atoms

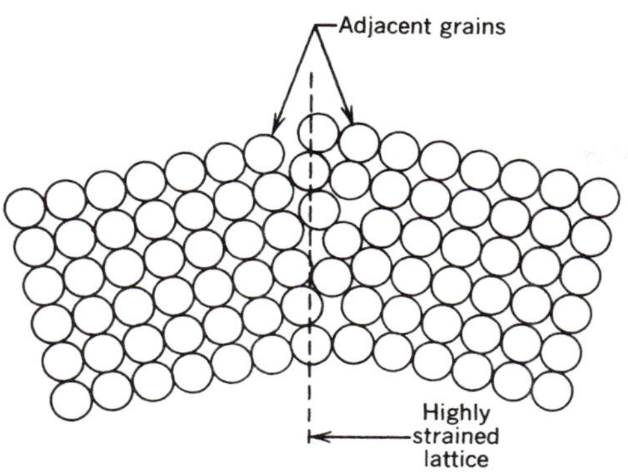

FIGURE 1.14
The grain boundary is in a highly strained condition (Neely and Bertone, *Practical Metallurgy and Materials of Industry,* 5th ed. © 2000 Prentice Hall, Inc.).

wide, but their strained condition causes them to etch darker than the less strained portion of the grain, allowing the boundaries to be seen with a microscope.

Grain Size

The size of a grain is determined when it runs into adjacent grains in the freezing process; that is, the more grains there are to interfere with a grain's formation the smaller it will be. Because the number of grains depends on the number of nucleating sites, the more nucleating sites there are, the smaller the grains will be. There are three means of encouraging grains to form, thereby achieving smaller grain size: (1) Grain refiners can be used as discussed earlier. (2) Increasing the cooling rate increases the ability of the metal to nucleate. Metals cooled rapidly tend to have smaller grains; conversely, those cooled slowly will have larger grains. Some metals when slowly cooled have extremely large grains. Galvanized articles will often show large grains of zinc, and doorknobs and push plates can display large copper grains. (3) Plastic deformation (also termed *cold working*) creates nucleating sites. The heating of a metal that has been previously cold worked can cause grains to form in sizes that are *inversely* related to the amount of cold work. That is, a large amount of cold work will result in smaller grains; a small amount of cold work can cause larger grains to form. This topic will be discussed in Chapter 2.

The role of grain size in influencing mechanical properties, for example tensile strength and elongation, will also be discussed in Chapter 3. In general, a smaller grain size is preferred for higher strength, but a larger grain size improves ductility and workability.

POLYMERS

Polymers are very large long-chain molecules from which plastics are made. **Plastics** are polymers that have been shaped by some process while in a softened state. The distinction between plastics and polymers is often lost, and the terms are often used interchangeably, but because our topic is materials we will more often talk about polymers in this chapter.

Many plastics are isomers. In organic chemistry and in the study of structural polymers, the phenomenon of **isomerism** is of some importance. Some compounds having the same number and kind of atoms can join to form different structures having completely different properties, roughly analogous to various crystalline phases in metals (allotropism). A simple example of structural isomerism is the arrangement of the atoms C, H, O to form both ethyl alcohol and dimethyl ether (Figure 1.15).

Natural organic materials such as leather, wood, cotton, and natural rubber are typically **polymers,** made up of long-chain molecules. Early discoveries in organic chemistry produced substitutes for organic materials that were not widely accepted and proved to be poor replacements. For example, in 1866 John Wesley Hyatt first produced a synthetic material (cellulose nitrate) he called *celluloid* by combining camphor with nitrocellulose (gun cotton) that was made with slightly less nitric acid than usual. This new product was used as a substitute for the ivory that was obtained from the tusks of animals and was quite expensive. Soon, products such as combs, handles, and billiard balls were made of this new synthetic material. Celluloid had two bad qualities: it had a tendency to burn rapidly when ignited, and it deteriorated (discolored and cracked) with age. Movie film was first made of celluloid, but later a nonflammable cellulose acetate was developed for movie film.

Since this early discovery a wide range of plastics, synthetic rubbers, textile fibers, adhesives, and paints have been developed by chemists. These plastics are considered by some to be among the greatest scientific achievements of modern times. These substances, often called *high polymers,* have molecular weights that are often hundreds of times greater than those of ordinary materials. These giant

```
   H  H               H     H
   |  |               |     |
H—C—C—O—H         H—C—O—C—H
   |  |               |     |
   H  H               H     H
```

Ethyl alcohol **Dimethyl ether**

FIGURE 1.15
An example of structural isomerism.

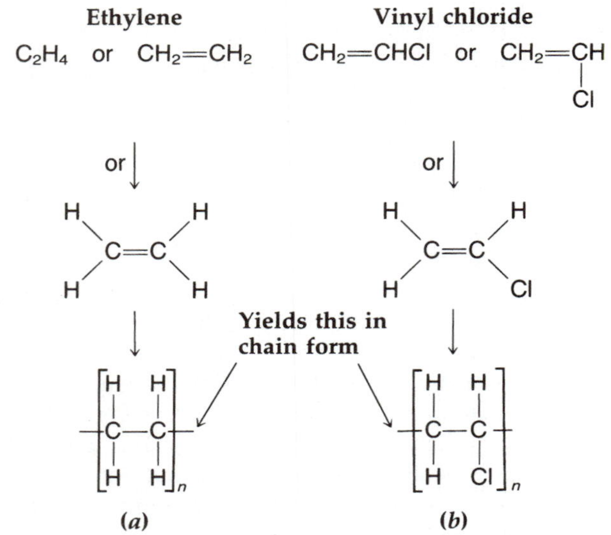

Ethylene

C_2H_4 or $CH_2{=}CH_2$

Vinyl chloride

$CH_2{=}CHCl$ or $CH_2{=}CH{-}Cl$

Yields this in chain form

(*a*) (*b*)

FIGURE 1.16
The repeating unit is formed from the compound. The subscript *n* indicates the indefinite length of the polymer.

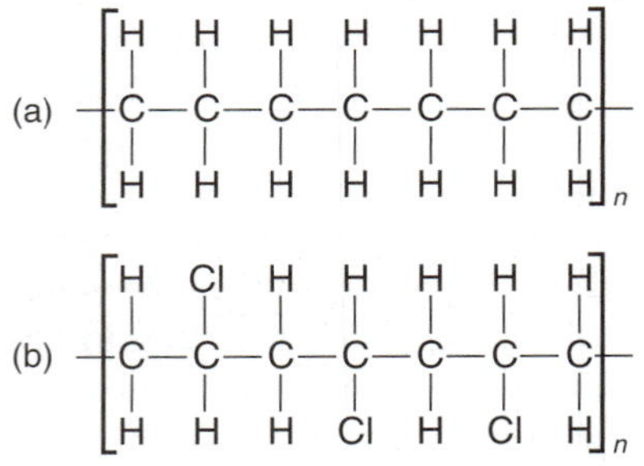

FIGURE 1.17
The molecular forms (structural formulas) that serve as monomers to produce (*a*) polyethylene and (*b*) polyvinyl chloride (PVC). Thus, giant chain molecules may be formed from the monomers of Figure 1.16.

molecules are made up of small molecules called *monomers* (Figure 1.16). These are the building blocks from which the polymer chain is built (Figure 1.17).

Polymerization is a chemical process in which many small molecules are linked together to form one large molecule. The first high polymer to be manufactured commercially from small molecules of raw materials was the strong, hard plastic called *Bakelite,* which was developed by the Belgian chemist Leo Baekeland in 1908. He was able to join the small molecules of phenol and formaldehyde to produce large molecules. The basic structure of Bakelite is shown in Figure 1.18.

There are two basic types of polymerization: *addition* and *condensation.* The bonds created by these polymerization processes are strong covalent bonds, called primary bonds rather than the weak van der Waals secondary bonds. Addition polymerization, which is also called linear polymerization, takes place when similar monomers join to form a chain (Figures 1.17 and 1.19). *Copolymerization* takes place when two or more *different* kinds of monomers are combined. For example, the polyethylene and polyvinyl chloride of Figure 1.16 *alternate* linearly in one chain; there are no links between or across chains. Polymers of these two types are termed **thermoplastic** because they become soft and malleable as temperature increases.

Condensation polymerization (which has a by-product, usually water; see Figure 1.18) is a process capable of forming long chains of molecules as well as forming primary bonds *between* or *across* chains; that is, the chains are **cross-linked** (Figure 1.20). In Figure 1.18 the cross-links would form where the present hydrogen atoms are shown and would extend to one or more nearby chains. Thus, polymers formed by condensation polymerization, such as Bakelite, can be very rigid. These constitute the family of polymers known as **thermosets.** Heat is used to activate the cross-links, and the starting material becomes "set," hard and strong. On reheating, thermosets do not become soft but will deteriorate (char) if the temperature is raised high enough. Charring is what happens to a cooking pot handle that is left over a hot burner on the stove.

Although thermoplastics normally do not have primary-bond cross-links, there are van der Waal bonds of attraction between adjacent chains of these large molecules. On heating, this bond largely disappears and is one of the reasons these polymers are *thermo*plastic. Thermoplastics also can have chains that are in random coils (like a bowl of spaghetti), and are said to be amorphous; however, their chains also can become aligned, and the polymer is said to exhibit **crystallinity**. Some polymers such as polyethylene can be highly directional or crystalline.

Natural rubber is an example of a very pliable long-chain material (see Figure 1.21*a*). When this natural rubber is treated with sulfur and heat it is **vulcanized** and becomes cross-linked (see Figures 1.21*b* and 1.22) and much more rigid. A general formula for vulcanized rubber is shown in Figure 1.22.

A great variety of synthetic rubbers have been developed. Many of these have properties different from those of natural rubber and are superior for some specific applications. Most of these synthetic rubbery materials are derived from petrochemicals (petroleum distillates) such as acetylene, butadiene, isoprene, and chloroprene (Figure 1.23). These chemicals and their derivation will be further discussed in Chapter 5.

CH_2O

Formaldehyde molecule

C_6H_5OH

Phenol molecule

Bakelite polymer

Plus H_2O

FIGURE 1.18
The basic chemical structure of Bakelite as it is produced from formaldehyde and phenol monomers by condensation polymerization.

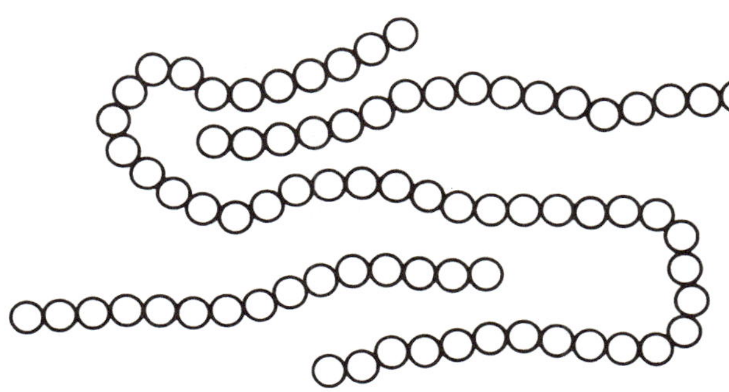

FIGURE 1.19
The spheres in these long-chain molecules represent monomers. Some chains are more rigid, as in PVC, whereas others are elastic, as in polyethylene, nylon, and cured rubber.

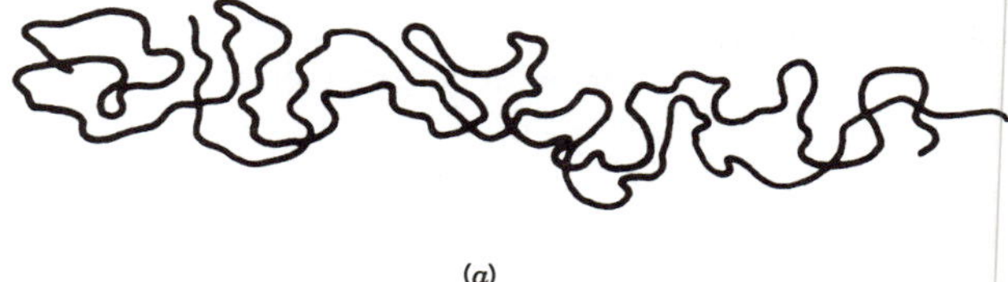

(a)

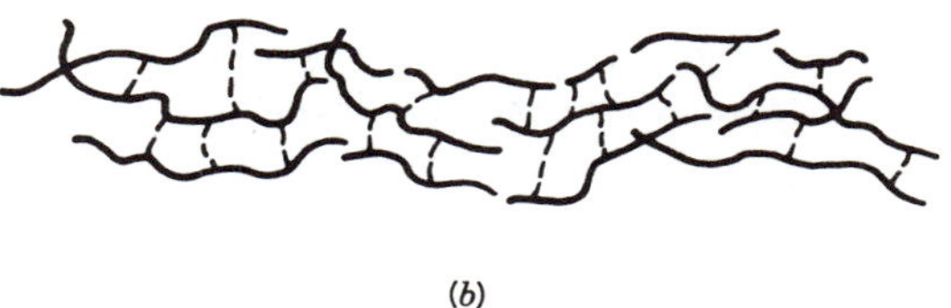

(b)

FIGURE 1.21
(a) Natural rubber long-chain molecules; wire model, unstretched. *(b)* Vulcanized rubber, cross-linked.

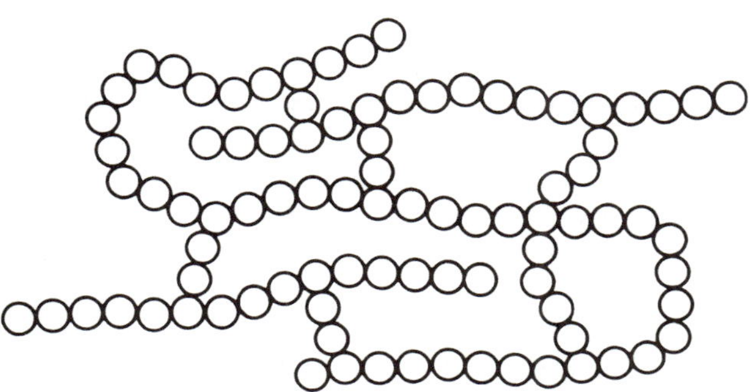

FIGURE 1.20
Cross-linking is only one of several ways in which polymers can be strengthened.

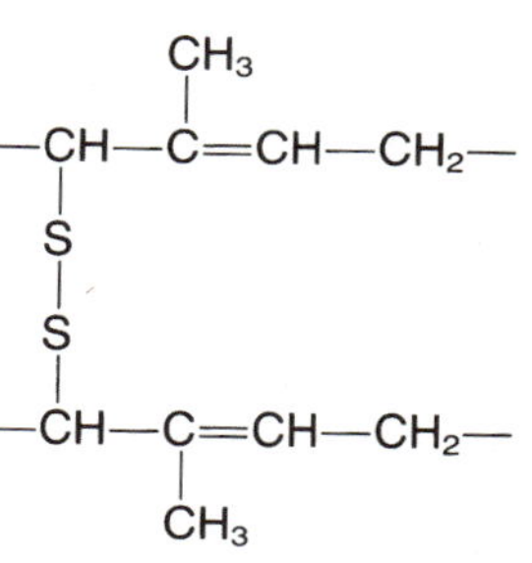

FIGURE 1.22
General formula for vulcanized rubber, showing sulfur atoms providing the cross-links.

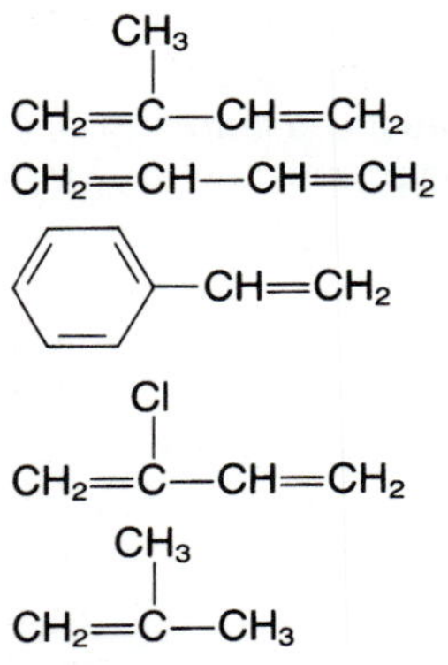

$CH_2{=}C(CH_3){-}CH{=}CH_2$	**Isoprene**
$CH_2{=}CH{-}CH{=}CH_2$	**Butadiene**
$C_6H_5{-}CH{=}CH_2$	**Styrene**
$CH_2{=}C(Cl){-}CH{=}CH_2$	**Chloroprene**
$CH_2{=}C(CH_3){-}CH_3$	**Isobutylene**

FIGURE 1.23
Some of the petrochemicals from which rubbers and elastomers are derived.

CERAMICS

The term *ceramics* is used to define a wide range of products that fall into two major divisions. Pottery, dishes, flowerpots, bricks, concrete blocks, clay drain tiles, glass, and similar items fall into a division we'll call **traditional ceramics**; these are the products that have been used in and around our homes for years. More sophisticated products, such as spark plugs, cutting tools, turbocharger impellers, internal combustion engine valves, valve seats, and cam followers are newer, highly engineered products that are relatively new; we'll refer to these items as **engineered ceramics**. Because this division of ceramics is fairly new, there is no general agreement on its name; prefixes such as *fine, advanced,* and *new* are also used to describe a ceramic of this type.

As a general rule ceramics are compounds containing a metal and a nonmetal. Some examples follow:

Al_2O_3, alumina	Cr_2O_3, chromium oxide
ZrO_2, zirconia	VC, vanadium carbide
TiC, titanium carbide	SiC, silicon carbide
WC, tungsten carbide	MoS_2, molybdenum disulfide
BN, boron nitride	Si_3N_4, silicon nitride

Note that these examples encompass oxides, carbides, sulfides, and nitrides and include uses for electrical insulators, wear coatings, thermal and electric insulation, pigments, abrasives, and cutting tools. Other ceramics are intermetallics. Examples are Ni_3Al, nickel aluminide, which is used for wear coating and high-temperature applications, and metalloids such as germanium and silicon, which are used in electronic devices.

Because ceramics are compounds, their chemical elements are in ratios of small whole numbers, as shown in the preceding list, and they have ionic or covalent bonds. The solid ceramics exist in a lattice that is often more complicated than the types we discussed earlier in regard to metals. Figure 1.24 depicts a four-sided (tetrahedron) lattice structure of silicon dioxide wherein one silicon atom is nestled between four oxygen atoms. For covalent bonding each oxygen has space in its outer shell to share two valence electrons. The silicon, however, has only four valence electrons to share with these four oxygens. During the firing of the ceramic material the four extra available valence spaces will be shared with other silicon atoms, and this bonding is what forms the chain of molecules shown in Figure 1.25. Note also that in Figure 1.25 there are two oxygen atoms in each tetrahedron that have one bonding site available to share with other silicon atoms from other chains not shown. The silicon–oxygen chains can cross-link to form a three-dimensional crystalline structure.

All the bonds just described are strong, primary bonds, and they occur in rather rigid space lattices. Ceramics are not ductile like metals but are very hard and brittle, and their melting temperatures are much higher than those of metals. Since there is no cloud of free electrons as in metals, ceramics are poor conductors of both heat and electricity. A glance at the list of uses given previously illustrates the characteristics just described.

With such mechanical characteristics it is evident why ceramics are very difficult to fabricate. They usually cannot be machined, cannot be softened by heating, and cannot be melted and cast. Thus, they have to be formed in their softened state and then fired. A low-tech example is the production of bricks. When the structures of Figure 1.25 are combined with alumina they become clay, a very plastic substance when wet and rigid when dried. Clay is composed of very small crystalline, flat, or platelike grains that tend to be stacked like packs of playing cards, each pack oriented in a different direction. Each crystal plane is separated and surrounded by water that acts as a lubricant and a bond, allowing the grains to slide easily over one another (Figure 1.26*a*). When the water is dried out, as in sun-dried bricks, the lubrication is gone. The material becomes rigid but has low tensile and compressive strength. Clay in this state can be easily wetted and brought back to its plastic state; however if it is fired it loses that ability, and it becomes somewhat stronger and quite brittle. The plates of hydrated silicate of alumina ($Al_2O_3 \cdot 2SiO_2 \cdot 2H_2O$) when fired combine to form new crystalline phases called glasses that permeate the entire structure, bonding the new crystals together into one fused mass (Figure 1.26*b*).

Glass The basic ingredient of glass is quartz (silica) SiO_2, and it contains other substances such as lime and sodium. When quartz is cooled, it forms a crystalline structure that is brittle and semitransparent. Glass is slow cooled

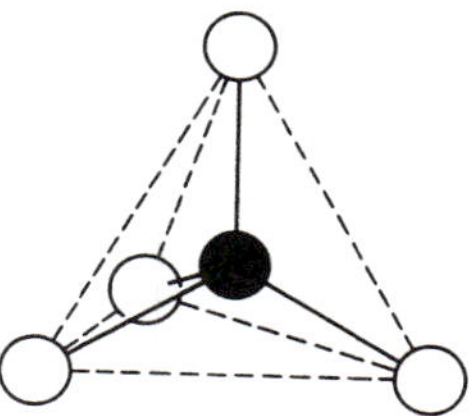

FIGURE 1.24
Unit structure of silicon dioxide (SiO_4). The solid circle is silicon, and the open circles represent oxygen atoms. The silicon atom has a valence of 4; that is, it is capable of making four single bonds. In this molecule it uses all four to bond with four oxygen atoms and form a silicate tetrahedron; however, each oxygen atom has two bonds available, only one of which is used to bond to the silicon. Therefore each oxygen atom has one bond available to bond to another atom. (Neely and Bertone, *Practical Metallurgy and Materials of Industry,* 5th ed., © 2000 Prentice Hall, Inc.).

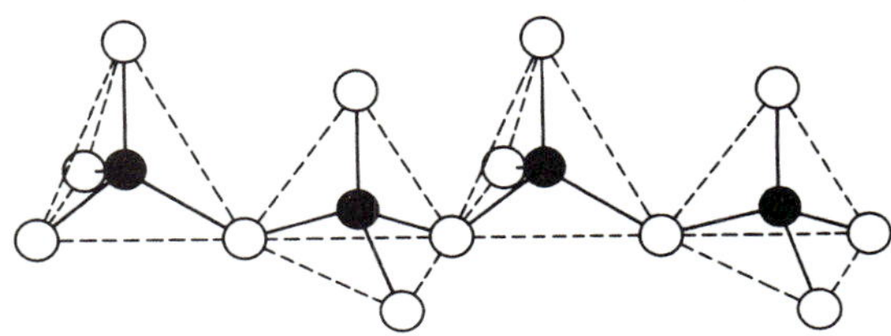

FIGURE 1.25
The unused oxygen bonds, mentioned in Figure 1.24, can link with other silicon atoms, forming a chain of silicate tetrahedra. The bond of the oxygen atoms not involved in this chain can bond with adjacent silicon atoms to form cross-links between chains. A crystalline structure is thus formed. The solid circles represent silicon atoms, and the open circles are oxygen atoms (Neely and Bertone, *Practical Metallurgy and Materials of Industry,* 5th ed., © 2000 Prentice Hall, Inc.).

to keep it from crystallizing. The molecular structure of glass is amorphous, like a liquid; in fact, glass is considered to be a solid liquid. The raw material is silica sand, which is quite plentiful, making glass a relatively inexpensive material.

Portland Cement Concrete that is used to make roads, sidewalks, dams, buildings, and other permanent structures is composed of portland cement, sand, and graded gravel. The cement when combined with water hardens to a rocklike consistency, binding the sand and gravel into one homogenous mass. This hardening takes place without the presence of air. In fact, portland cement must be kept from drying out during the hardening process. This process is called *hydration.* Portland cement is basically a calcium silicate mixture predominantly containing tricalcium and dicalcium silicates. These materials are basically derived from shales, clays, and limestone with small amounts of iron oxide and dolomite. The alumina (Al_2O_3) content of clay has an effect on the long-term strength of concrete. Other than impurities, the important ingredients of portland cement can be written as $CaO \cdot SiO_2 \cdot Al_2O_3$. During hydration and hardening, fibers grow out of the silica gel–cement particles and join with adjacent particles to form a solid interlocked mass. The production and uses of portland cement and concrete will be covered in later chapters.

In Figure 1.27*a* the schematic shows the tiny cement particles to which mixing water has been added as they would appear magnified many thousands of times. A silica gel begins to form on their outer surfaces. Soon, whiskerlike fibers begin to grow from the particles, and after 7 days they are interconnected, as can be seen in Figure 1.27*b.* The electron micrograph in Figure 1.27*c* reveals this condition at 7 days in hydrated portland cement. After 28 days this cement and the concrete made with it is about as hard as it will ever get. In moist conditions, as found in dams, the concrete may get slightly harder and stronger over the years. Figure 1.27*d* shows how these fibers develop into coarser and stronger bars or plates that interlock to form a solid rocklike mass. The electron micrograph, Figure 1.27*e,* shows hydrated portland cement at 28 days.

Review Questions

1. Not considering the many subatomic particles, name the three basic components of which atoms are made. What differences are there among the three?
2. Briefly define *valence.*
3. Atoms with positive or negative valence are called ions. What causes this condition?
4. Name four kinds of bonding arrangements of atoms.

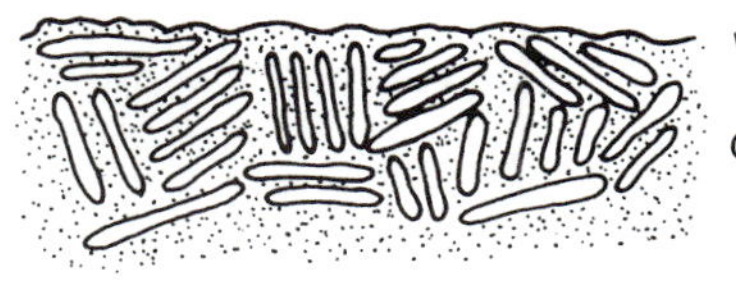

(a)

(b)

FIGURE 1.26
Platelike grains are like decks of cards. *(a)* These are wet and plastic. *(b)* These are dried, hard, and brittle.

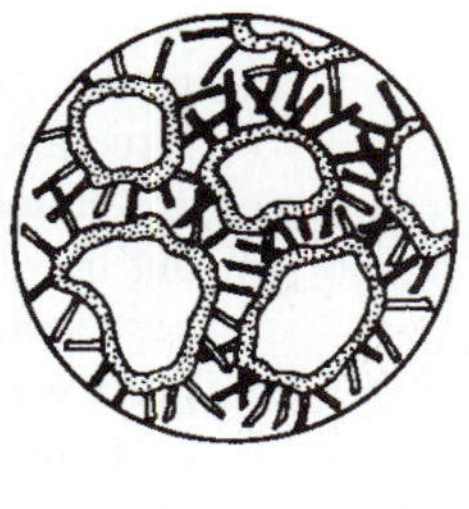

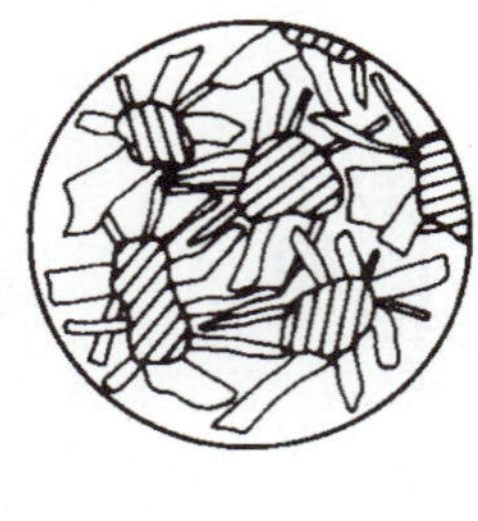

FIGURE 1.27
The hydration of portland cement. *(a)* Schematic of the formation of silica gel in the presence of water, 1 day. *(b)* Schematic showing fibers beginning to grow, 7 days. *(c)* Electron micrograph showing the growth of fibers at 7 days (3000×). *(d)* Schematic showing the final development of the strong fibers, 28 days. *(e)* Electron micrograph of completely hydrated portland cement at 28 days (5000×). (*c* and *e* are SEM photographs by Hugh Love, Portland Cement Association, Skokie, IL).

5. What does the type of bond in metals have to do with their elastic, plastic, thermal, and electrical behavior?
6. What is a space lattice in solid materials?
7. What are unit cells in metals?
8. When a metal changes from one unit cell structure to another at certain temperatures, what is the metal said to be?
9. When a metal begins to solidify, many grains begin to grow. These are called *dendrites.* At what location do they stop growing?
10. All other conditions being the same, which metal would be more plastic and ductile, a fine-grained steel or a coarse-grained steel?
11. Name the process that links small molecules together to form a long chain. What are the three methods of accomplishing this process? What forms of bonding occur as a result of this process?
12. What are the similarities and differences between thermoplastics and thermosets?
13. A gummy substance such as natural rubber latex can be made into useful products by the process of cross-linking. In rubber, what is this process called?
14. The first high polymer was produced by joining the small molecules of phenol and formaldehyde. What is this well-known plastic called?
15. Ceramics are typically compounds of what type elements? What type(s) of bonding are used in ceramics?
16. How can clay be plastic when wet, brittle when dry, and hard and strong when fired?
17. What is the basic ingredient of glass?
18. Portland cement hardens and bonds together in the presence of water. What is this process called and how does it progress to form a solid mass?

CHAPTER 2

Properties of Metals

Objectives

This chapter will enable you to:

1. Understand the behavior of metals in terms of their mechanical properties.
2. Describe the method of testing metals to determine their various properties.
3. Make some material selections on the basis of mechanical and physical properties.
4. Explain how the microstructure of metals can be seen and photographed.
5. Show how mechanical parts, steel structures, and pipelines can be tested for flaws without damaging them for use.

Key Words

stress, strain
tension, compression
torsion
shear
elasticity, plasticity
ductility
creep
ductile, brittle
ductile–brittle transition temperature
impact or notch toughness
malleability
fatigue
hardness

In the field of manufacturing in which metals are cast, shaped, or formed, the *mechanical* properties of the metal selected for processing must be known. Scientific equipment such as light and electron microscopes is used to view and analyze the crystal structure of metals and testing machines are used to pull, push, bend, twist, fatigue, and impact metal specimens to determine their properties. Some of these properties are strength, ductility, malleability, hardness, brittleness, creep, and toughness. Various tests and testing machines are used to determine the mechanical properties. In this chapter each test will be identified and described, and then the properties determined by the test will be discussed. These tests are usually destructive; that is, the specimen or part is destroyed by testing.

Engineers and scientists also need to have an understanding and a means of measurement of the *physical* properties of metals. Some of these properties are the modulus of elasticity, melting point, coefficient of thermal expansion, electrical and thermal conductivity, specific gravity, magnetic susceptibility, and reflectivity. Tests for determining the physical properties of metals are often nondestructive. Nondestructive testing is discussed later in this chapter.

MECHANICAL PROPERTIES

Tensile Test

The tensile test involves the relationship between tensile stress and strain. Although these terms are sometimes confused, they have separate and distinct meanings. **Stress** is defined as the resistance of a material to external elements such as force, load, or weight measured in pounds per square inch (PSI) (Figure 2.1). It is calculated by dividing the applied load to a specimen by the cross-sectional area of the specimen. When the stress is applied it will cause the specimen to deform or stretch. **Strain** is the amount of deformation or stretch that occurs over a standard gage length (usually 2 in.) expressed in inches/inch or as a percentage. This gage length may also be referred to as "the original length."

In this study we are primarily interested in tensile stresses and strains, but it is important to recognize that other types of loadings exist. There are four types of stresses involved in the strength of materials: **tension, compression, torsion,** and **shear** (Figure 2.2). Torsional shear is found in rotating parts such as axles and machinery shafting. A torsional shear measurement, torque, is shown in Figure 2.3. Some metals such as steel have nearly equal compressive and tensile strengths, but cast iron has a relatively low tensile strength compared to its higher compressive strength. Shear strength is lower than tensile strength in virtually all metals (Table 2.1).

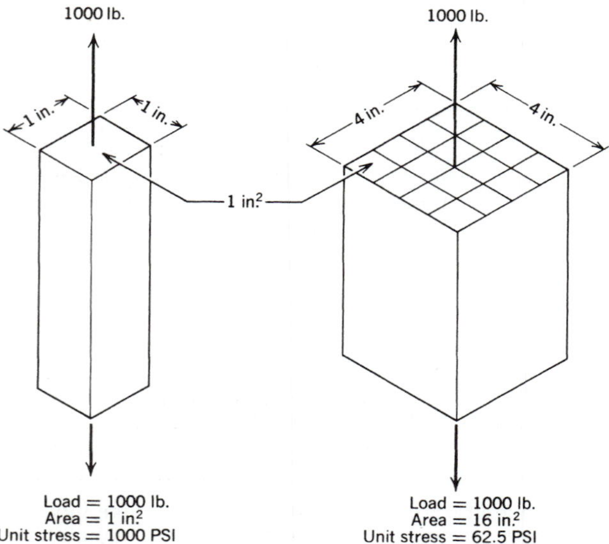

FIGURE 2.1
Unit stress (Neely and Bertone, *Practical Metallurgy and Material of Industry,* 5th ed., © 2000 Prentice Hall, Inc.).

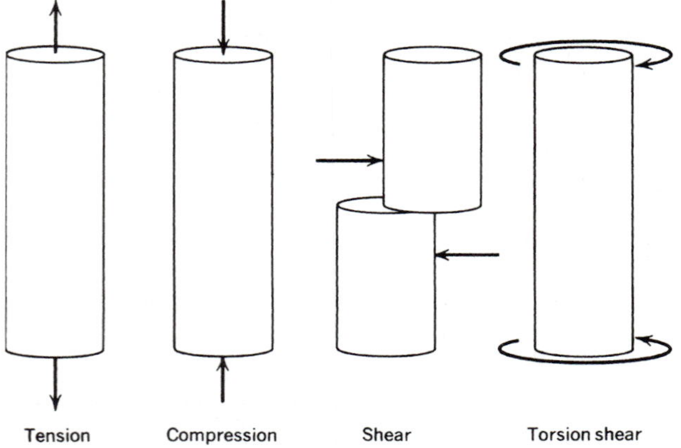

FIGURE 2.2
The four types of stresses (Neely and Bertone, *Practical Metallurgy and Materials of Industry,* 5th ed., © 2000 Prentice Hall, Inc.).

FIGURE 2.3
Digital readout of torque values on a torque testing machine (Photograph courtesy of Tinius Olsen Testing Machine Co., Inc., Willow Grove, PA.).

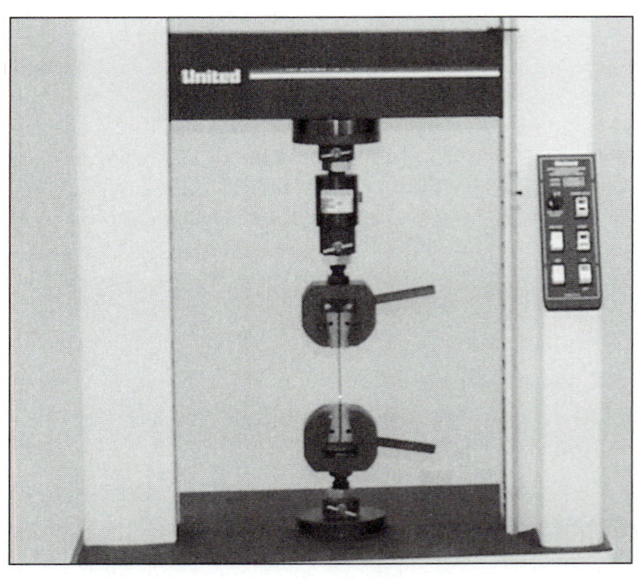

FIGURE 2.4
A testing machine equipped to do tensile testing of a flat (as opposed to a round) specimen. To grip the specimen the two jaws are operated by the horizontal handles projecting to the right (Courtesy of United Testing Systems, Inc.).

To perform the tensile test metal specimens are "pulled" on a machine called a *tensile tester* (Figures 2.4 and 2.5) until they break. Figure 2.4 shows such a tester with a specimen in its jaws; the jaws grip the flat specimen and are operated manually by the handles seen projecting to the side. Above the upper jaw is a load cell that converts the pull of the tester into a voltage proportional to that pull. The electrical signal from the load cell is fed to a plotter that records the force on graph paper. Dividing these values by the cross-sectional area of the specimen gives the stresses carried by the specimen, usually to the breaking point of the specimen.

One method of determining the strain is to use an extensometer, as shown in Figure 2.5. As the specimen is stretched it causes the strain gage mounted in the right side of the extensometer to send a signal (note the electrical wire in the photo) to a plotting device. Because the knife edges are 2 in. apart, the stretch is measured relative to 2 in., a standard gage length, so the output of the extensometer and plotter is the strain value.

The load cell (the stress value) drives the plotter on the vertical or *Y*-axis, and the extensometer (the strain value) drives it along the horizontal or *X*-axis. The result is a

TABLE 2.1
Material strength

Material	Modulus of Elasticity (PSI)	Allowable Working Unit Stress: Tension	Compression	Shear	Extreme Fiber in Bending
Cast iron	15,000,000	3,000	15,000	3,000	
Wrought iron	25,000,000	12,000	12,000	9,000	12,000
Steel, structural	29,000,000	20,000	20,000	13,000	20,000
Tungsten carbide	50,000,000				

Material	Elastic Limit (PSI): Tension	Compression	Ultimate Strength (PSI): Tension	Compression	Shear
Cast iron	6,000	20,000	20,000	80,000	20,000
Wrought iron	25,000	25,000	50,000	50,000	40,000
Steel, structural	36,000	36,000	65,000	65,000	50,000
Tungsten carbide	80,000	120,000	100,000	400,000	70,000

(Neely and Bertone, *Practical Metallurgy and Materials of Industry,* 5th ed., © 2000 Prentice Hall, Inc.)

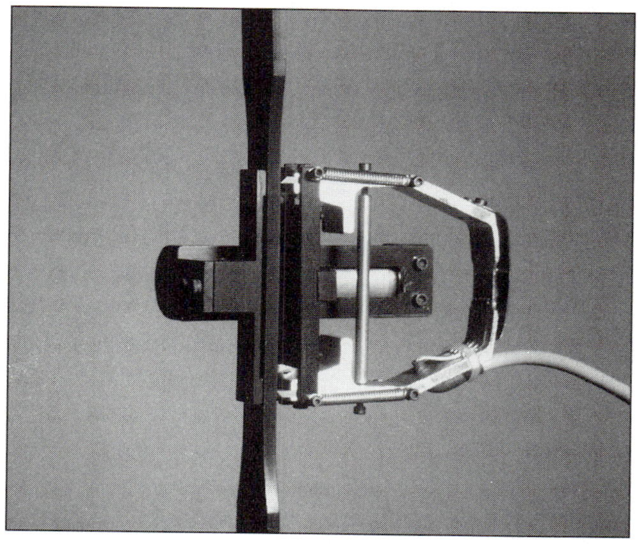

FIGURE 2.5
A test specimen (oriented vertically) has an extensometer clamped to it by means of a spring (which is hidden from view). Knife edges can be seen in contact with the specimen. At no load they are 2 in. apart (Courtesy of United Testing Systems, Inc.).

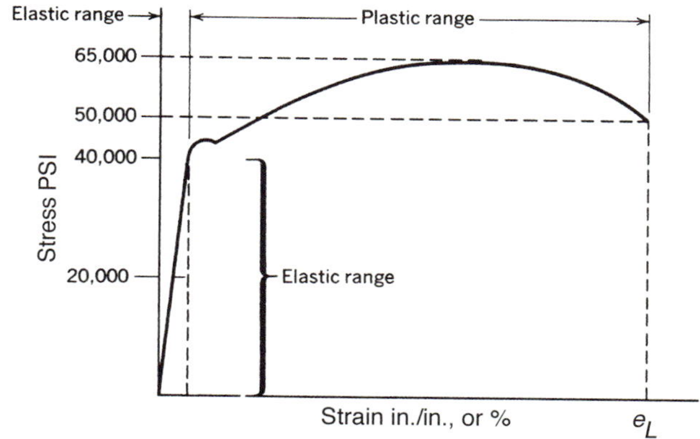

FIGURE 2.6
Stress–strain diagram for a ductile steel (White, Neely, Kibbe, Meyer, *Machine Tools and Machining Practices,* Vol. II, © 1977 John Wiley & Sons, Inc.).

stress–strain diagram, which for many metals will look something like Figure 2.6. Computerization of the whole process is shown in Figure 2.7. Here the test is controlled, and data are collected and displayed, by use of a computer system.

Standard-sized specimens are usually used. Figure 2.5 showed a flat specimen. Such specimens are obviously used with materials that are flat with a uniform thickness, most often extrusion and sheet products. The width of the test section is approximately 0.500 in. The overall length is 8 in., and the specimen is 1 in. wide where it fits in the grips. Figure 2.8 describes the standard round specimen; it typically has a diameter of 0.505 in., making the area 0.200 in.^{2}

When a tensile test is conducted it will be noticed that in the initial loading of the specimen the stress and strain are proportional, or the metal is said to behave *elastically.* This

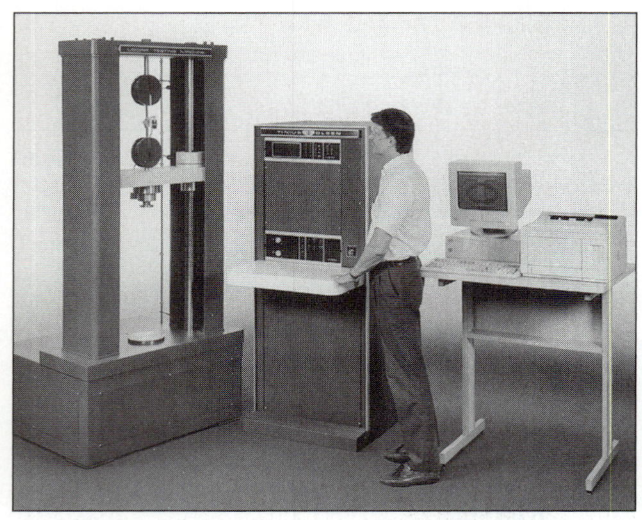

FIGURE 2.7
A tensile test being done utilizing a computer to control the test, to collect the data, and to display the results (Courtesy of Tinius Olsen Testing Machine Co., Inc., Willow Grove, PA.).

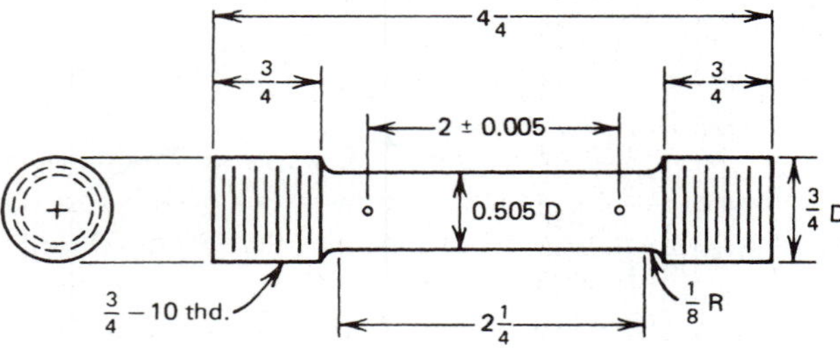

FIGURE 2.8
This standard test specimen has a cross-sectional area of 0.200 in^2. and a gage length of 2 in. Round numbers make calculations for percent reduction of area and percent elongation easier (Neely and Bertone, *Metallurgy,* 5th ed., © 2000 Prentice Hall, Inc.).

phenomenon is described by Hooke's law and is illustrated in Figure 2.9; equal increments of added load cause equal amounts of additional stretch. This is the portion of Figure 2.6 labeled "Elastic range." Designers are very interested in the stress value at the end of the elastic region; if they keep loads less than that number (with appropriate allowances) they can be reasonably sure their structure will not deform permanently under load. That is, when the load is removed it will elastically return to its original dimensions.

The slope of the stress–strain curve in the elastic region is a measure of the *modulus of elasticity.* Because this value is a constant for any metal it is considered a physical property and will be discussed later in this chapter.

Proportional Limit, Elastic Limit The *proportional limit* is technically the stress value at which the elastic portion of the curve loses its proportionality between stress and strain. Since it is so difficult to determine (when is the line no longer straight?), it is rarely used in a practical manner. As the name implies, the *elastic limit* is the highest stress (at the limit) at which the specimen will return to its original length. Because this value can be found only by trial and error it is valuable as a concept only. Neither of these values is included among handbook data.

Yield Point, Yield Strength In Figure 2.6, at the top of the elastic region, the amount of strain increases with very little increase in stress. That is, the metal yields, and this yield point is a very obvious end of the elastic region and is a good value on which to base a design. If the metal was cold-worked or heat-treated, or was a nonferrous metal, it would not have a yield point. In Figure 3.24 the curve for the high-carbon steel does not have a yield point; in that case a line is constructed parallel to the elastic region at a

FIGURE 2.9
Diagram of the elastic properties of metals as described by Hooke's law. Four weights, one of which represents zero load, are suspended by identical metal wires. The three loads toward the right of the diagram are heavier in an equal progression. The resulting stretch *a, b,* and *c,* is also equal for the three loads.

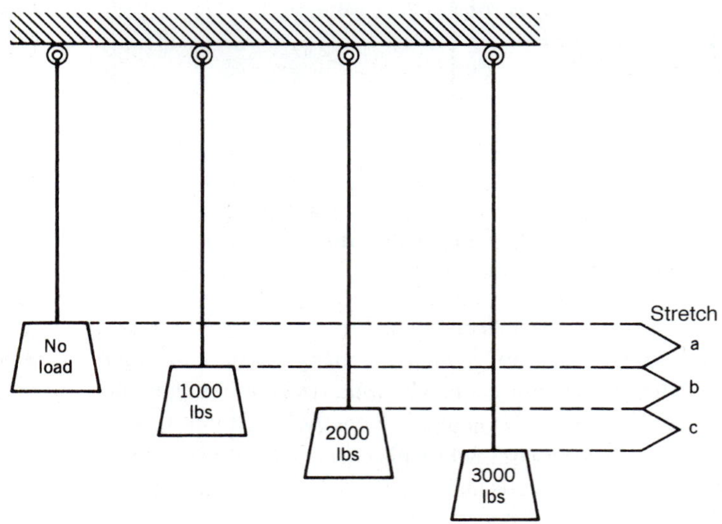

FIGURE 2.10
Brittle fracture of a shaft.

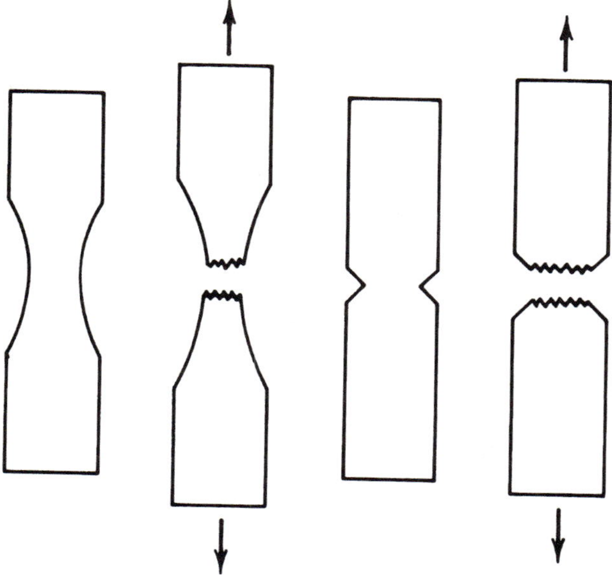

FIGURE 2.11
Notching and its effect on plasticity. When pulled, the specimen (left) becomes deformed before breaking, showing ductility. The notched specimen of the same material (right) exhibits complete brittleness.

strain value of 0.002 in./in. Where that line intersects the stress–strain curve is known as the *yield strength.* The yield point or yield strength, whichever is appropriate, is a handbook value for metals and alloys and is commonly used in design; units are PSI.

Tensile Strength, Ultimate Tensile Strength The tensile strength in pounds per square inch is determined by dividing the maximum load (in pounds) by the original cross-section (in square inches) before testing, or

$$\text{tensile strength} = \frac{\text{maximum load}}{\text{original area}}$$

For the tensile test of Figure 2.6 the tensile strength is 65,000 PSI.

Percent Elongation, Percent Reduction of Area Elongation, or percent elongation, is defined as the value of strain at failure. On Figure 2.7 this value is indicated by e_L. It is found by determining the amount the specimen stretched during the test, dividing by the original length, and expressing the result as a percentage. The percent reduction of area is determined in a similar fashion; the reduction in the cross-sectional area of the specimen (at the point where it failed) is divided by the original area and expressed as a percentage. Both these measures are used as indicators of **ductility** (which is the property that allows a metal to deform permanently when loaded in tension). Any metal that can be drawn into wire is ductile. Soft steel, aluminum, gold, silver, and nickel are ductile metals.

Malleability is the ability of metals to be deformed permanently when loaded in compression. Metals that can be rolled, hammered, or pressed into flat pieces or sheets are malleable. Ductile metals are usually also malleable, but there are some exceptions. Lead is very malleable but not very ductile and cannot be drawn into wire very easily, but it can be extruded in wire or bar form. Gold, tin, silver, iron, and copper are malleable metals. **Plasticity,** a more general property, is the ability of a metal to be permanently deformed without failing. On the stress–strain curve of Figure 2.6 it would mean the capability to be worked in the region labeled there as "Plastic range."

A material that will not deform plastically under a load is said to be **brittle** (Figure 2.10). A brittle break or failure in metals can be identified by a crystalline surface, often with a tiny rock candy appearance. Sometimes the individual crystals are so small that the surface has a somewhat smooth appearance to the eyes, but under magnification the crystals can be seen. Of course, metals and some other materials may exhibit other properties such as ductility or malleability along with brittleness. When a metal specimen is pulled apart in a tensile tester, a considerable amount of deformation can be seen in a ductile metal before a brittle failure takes place. However, very hard steel and cast iron show little or no ductility when pulled apart, so we say those metals are brittle, meaning they are completely brittle. A normally ductile metal can exhibit complete brittleness under some circumstances. A sharp notch that concentrates the load in a small area or at one point can reduce plasticity

(Figure 2.11). Also, low temperatures can have an embrittling effect on some steels. The behavior of metals at low temperatures is discussed later in this chapter.

Hardness Test

Hardness is generally defined as the resistance to penetration or indentation. The greater the hardness, the smaller the indentation with the same penetrator tip and pressure. However, the property of hardness is related to the elastic and plastic properties of metals, and certain hardness tests are based on rebound or elastic hardness. Also, certain scratch tests are used to determine abrasive hardness or resistance to abrasion. Hardness is also related to tensile strength, so it can be assumed that a hard steel is also strong and resistant to wear.

Hardness tests that use a load applied to a penetrater or indenter are the most widely used for industrial applications. The Rockwell testing machines, shown in Figures 2.12 and 2.13, can be used with a variety of loads (designated in kilograms) and types of penetraters (diamond point, $\frac{1}{16}''$ and $\frac{1}{8}''$ balls), so materials with a wide range of hardnesses can be tested. The combination of load and indenter determines a "scale" identified by a letter, usually A through K. The test is conducted by forcing the indenter into the material and then measuring the depth of penetration; this value is read directly by the machine. The Rockwell hardness number is a comparative value that varies between zero and 100 and is indicated by the abbreviation HR_, where the blank is the letter of the scale used. Generally, loads and penetraters should be chosen so that the values fall between 20 and 80.

The Brinell test (Figures 2.14 and 2.15) is performed by forcing a steel ball of known diameter into the material by a known force. This creates a cavity or crater, the diameter of which is read by a calibrated microscope, shown in Figure 2.13. The Brinell hardness number (HB) is found by dividing the load by the impressed area of the crater, so the hardness value is a load per area, or a stress value.

Microhardness testers (Figure 2.16) are used in laboratories to measure hardness using very small indenters that may make impressions less than 0.100 mm long. There are two scales—the Knoop, which leaves an elongated diamond-shaped impression, and the Vickers, which makes an impression like an inverted pyramid. The values are expressed as HK or HV, but like the Brinell number, the result is really a load divided by the

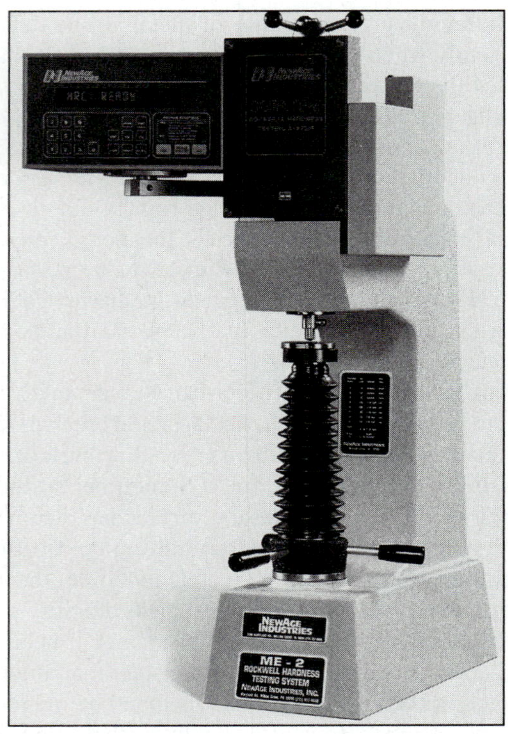

FIGURE 2.12
Rockwell-type hardness testing machine equipped with an electronic digital read out (Photo courtesy of Newage Testing Instruments, Inc.).

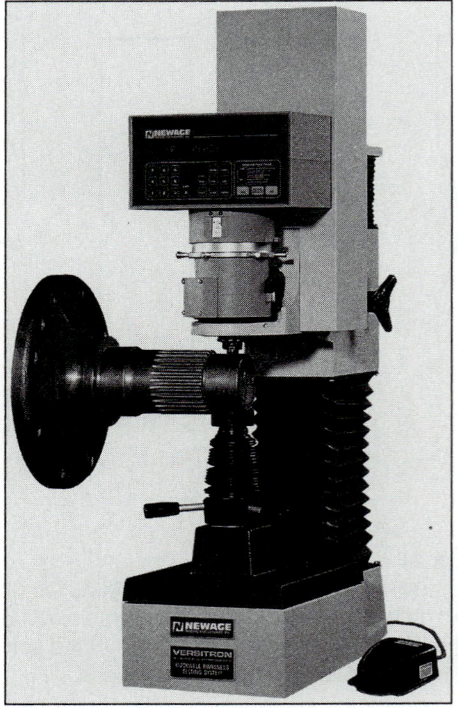

FIGURE 2.13
A workpiece clamped, ready for a Rockwell hardness test by a Versitron® Tester (Photo courtesy of Newage Testing Instruments, Inc.).

FIGURE 2.14
A large Brinell hardness tester with a separate readout panel (Photograph courtesy of Tinius Olsen Testing Machine Co., Inc., Willow Grove, PA.).

FIGURE 2.15
The Brinell microscope is used for measuring indentation diameters for simple Brinell hardness testers. Some Brinell testers, such as the one in Figure 2.14, are fully automatic and do not require a microscope (Courtesy of Newage Testing Instruments, Inc.).

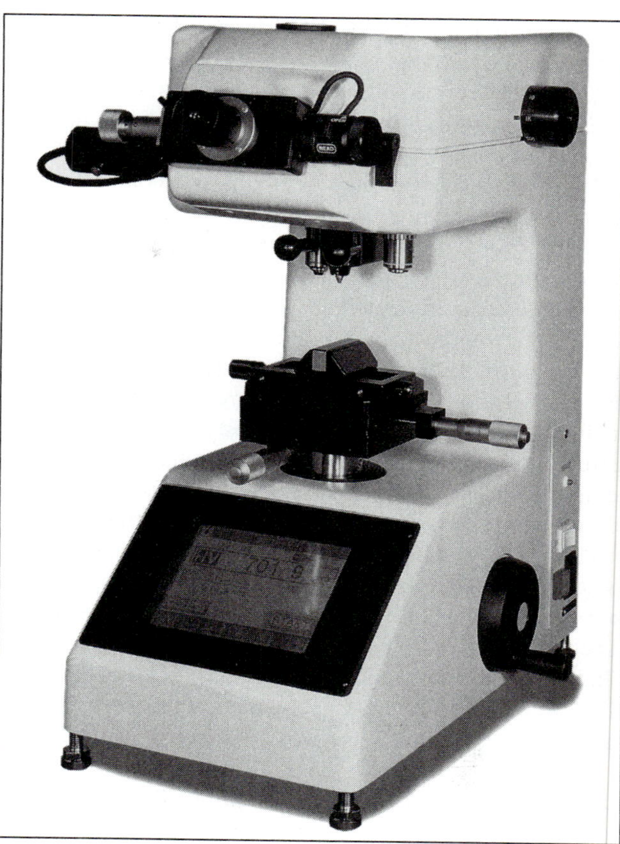

FIGURE 2.16
A microhardness tester, used to test surface hardness using the Knoop and Vickers scales (Buehler, Ltd.).

area of the impression, or a stress. The impressions can be made small enough that the hardness of portions of a microstructure, or the hardness measured across very thin materials such as a razor blade, can be measured. The method requires that the metal specimen be metallographically prepared, a process to be described shortly.

One measurement of hardness is based not on penetration but rather on the rebound of a dropped hammer with a diamond point. The higher the rebound, the harder the material. This rebound is based on elastic resistance to penetration. Because of its simplicity of operation, this test can be performed very quickly.

Impact Test

This test is performed to measure the **impact strength** of a metal; since a specimen with a notch (stress raiser) is used, this test is also said to measure **notch toughness.** Thus, this test measures the ability of a metal to resist rupture from impact loading (also called mechanical shock) when there is a notch or stress raiser present. The property of toughness is not necessarily related to ductility, hardness, or tensile strength; however, as a rule, a brittle metal with low tensile strength such as cast iron will fail under low shock loads; hardened, tempered tool steels show a high impact strength; and coarse-grained metals have lower shock resistance than fine-grained metals. A notch or groove will lower the shock resistance of a metal. Test specimens are made with given dimensions and a specific notch shape (Figure 2.17).

Notch toughness is measured on a device called an Izod–Charpy testing machine in which either of two styles of testing can be used (Figure 2.18). A weighted, swinging arm or pendulum strikes a specimen that is clamped in a vise, breaks the specimen, and continues swinging. It will not swing as high as the starting position because of the energy absorbed in breaking the specimen (Figure 2.19). The difference in starting and ending height is measured in foot-pounds. The pendulum will not swing as high for tough metals as for brittle metals. The difference between the two styles of testing, Izod and Charpy, is in the mounting of test specimens (Figures 2.20 and 2.21). The Izod specimen is mounted horizontally, and the Charpy specimen is mounted vertically.

Ductile–Brittle Transition Temperature Another function of the impact test is to determine the effect of temperature change on various metals. Metals that have the body-centered cubic atom lattice, for example, iron and steel, show a temperature range in which ductility and, more important, toughness drop rapidly. This *transition zone* can be seen in Figure 2.22. When the notch-bar specimens show half brittle and half ductile failures, the transition temperature has been reached.

Izod test specimens

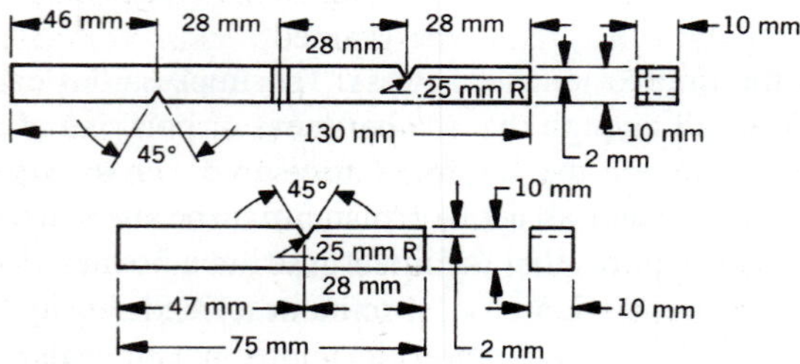

Charpy test specimen

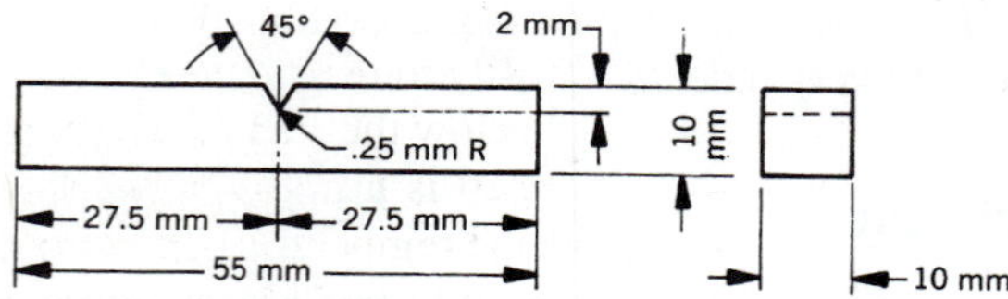

FIGURE 2.17
Test specimen specifications for Charpy and Izod tests (Photograph courtesy of Tinius Olsen Testing Machine Co., Inc., Willow Grove, PA.).

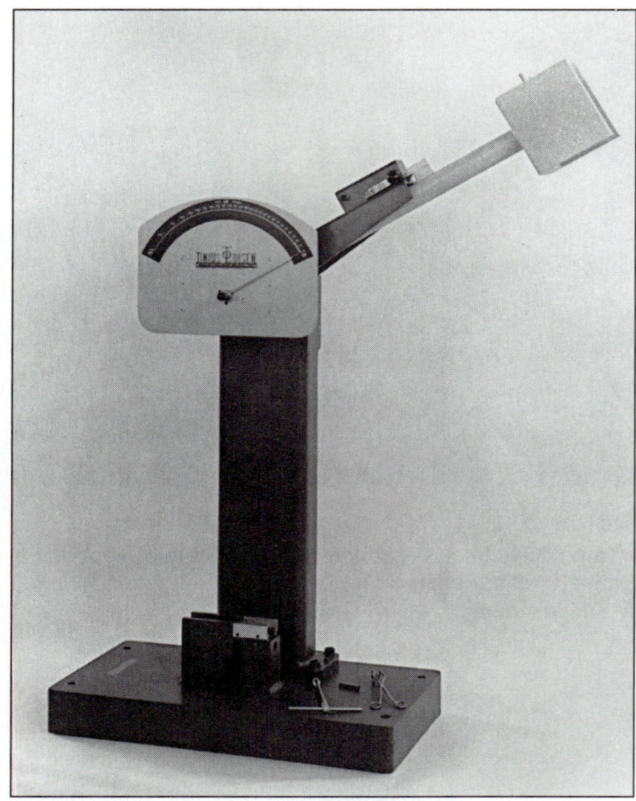

FIGURE 2.18
Izod–Charpy testing machine (Photograph courtesy of Tinius Olsen Testing Machine Co., Inc., Willow Grove, PA.).

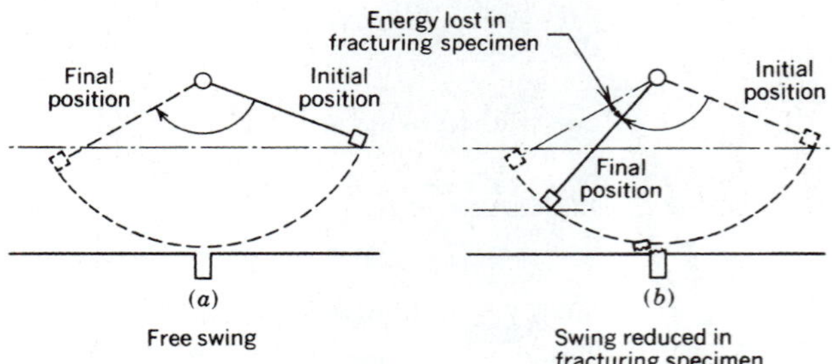

FIGURE 2.19
The method by which (Izod) impact values are determined (White, Neely, Kibbe, Meyer, *Machine Tools and Machining Practices,* Vol. II, © 1977 John Wiley & Sons, Inc.).

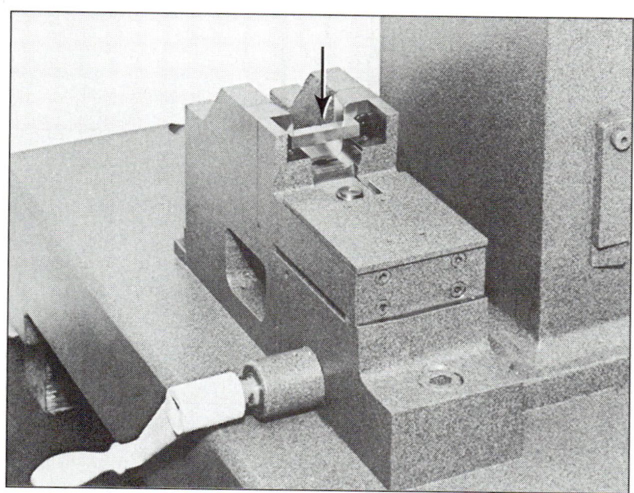

FIGURE 2.20
Mounting of the Izod specimen (Photograph courtesy of Tinius Olsen Testing Machine Co., Inc., Willow Grove, PA.).

FIGURE 2.21
Mounting of the Charpy specimen (Photograph courtesy of Tinius Olsen Testing Machine Co., Inc., Willow Grove, PA.).

Metals for parts that are designed for low-temperature service should be selected from among metals with a low ductile-brittle transition temperature.

This transition brittle zone of metals has caused some spectacular failures in the past, before it was known that some low-carbon steels would fail at low temperatures. Some of these failures occurred in pipelines, storage tanks, bridges, and ships that broke in half at sea. These failures are almost instantaneous, with the split moving along pipelines at 600 feet per minute.

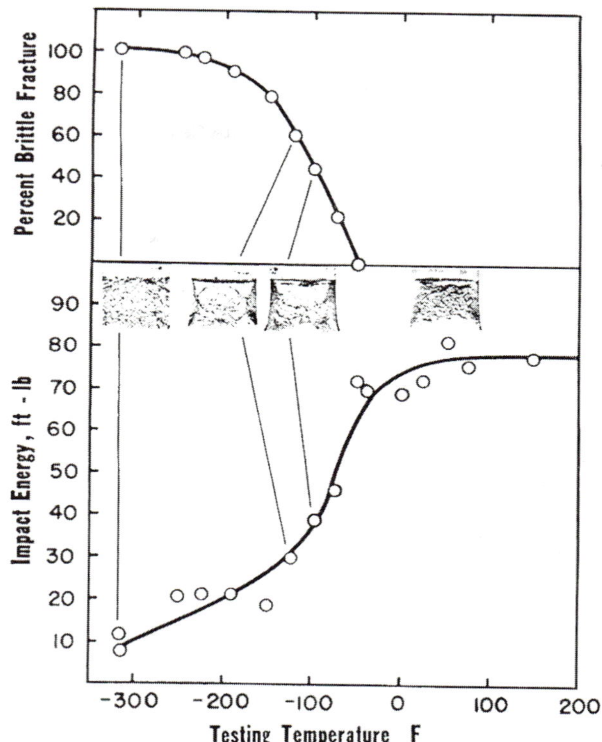

FIGURE 2.22
Appearance of Charpy V-notch fractures that were obtained in a series of tempered martensite of hardness 30Rc (Illustration courtesy of LTV Steel Company).

Plain carbon steels having coarse grain have higher transition temperatures (that is, will fail in brittle fashion at higher temperatures) than fine-grained and alloy steels. Certain alloying elements in steel, particularly nickel, tend to lower the transition temperature. Metals with the face-centered cubic (FCC) atom lattice do not go through the ductile–brittle transition. Austenitic stainless steel (FCC) is used to contain liquefied gases at temperatures below $-150°$ F ($-101°$ C). Aluminum and monel, both FCC, are also used at cryogenic temperatures without embrittlement.

Creep Test

Creep strength is a metal's ability to resist **creep,** a continuing plastic flow at a stress below the yield strength of a metal. For the most part, creep is a high-temperature phenomenon that increases as the temperature increases. Table 2.2 gives some stress loads at various temperatures required to cause elongation by creep. Some metals have been designed for high-temperature service such as for aircraft jet turbine engine blades that show no signs of creep at temperatures that would melt many metals.

TABLE 2.2
Creep strengths for several alloys

Alloy	70° F Tensile Strength (PSI)	Creep Strength (PSI) 800° F—Stress for 1 percent Elongation per 10,000 hr	1200° F—Stress for 1 percent Elongation per 100,000 hr	1500° F—Stress to Failure
0.20 percent carbon steel	62,000	35,100	200	1,500
0.50 percent molybdenum	64,000	39,000	500	2,600
0.08 percent to 0.20 percent carbon steel				
1.00 percent chromium 0.60 percent molybdenum 0.20 percent C steel	75,000	40,000	1,500	3,500
304 stainless steel 19 percent chromium 9 percent nickel	85,000	28,000	7,000	15,000

(Neely and Bertone, *Practical Metallurgy and Materials of Industry,* 5th ed., (p. 108) © 2000 Prentice Hall Inc.)

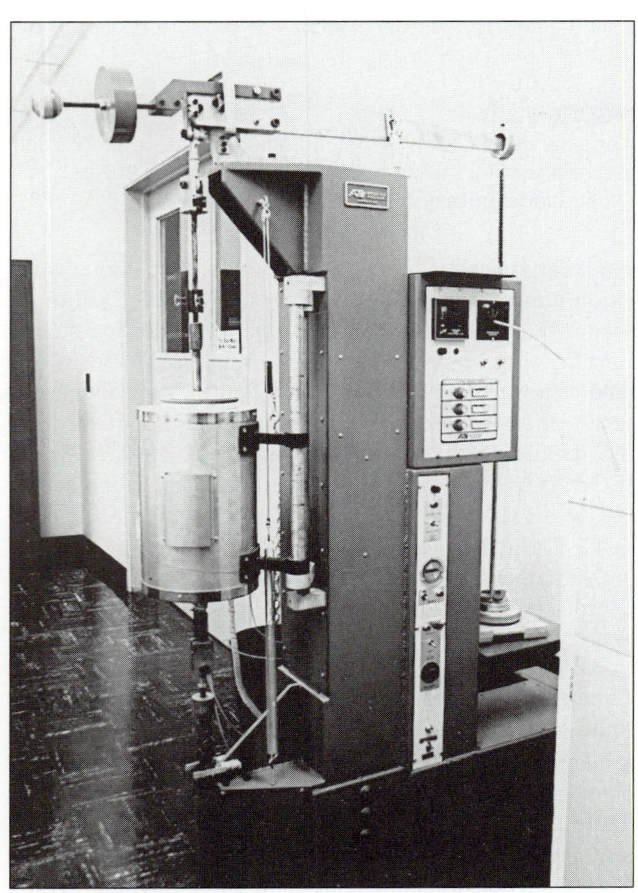

FIGURE 2.23
Creep-testing machine in Oregon Bureau of Mines testing laboratory (Oregon Bureau of Mines).

Creep tests are made at elevated temperatures under a certain stress in creep-testing machines (Figure 2.23) over a period of many hours.

Fatigue Test

Endurance Limit, Fatigue Strength **Fatigue** in metals can occur when metal parts are subjected to repeated loading and unloading, particularly in cyclic reversals of stress, such as is seen in a rotating shaft having transverse or one-side loading. Fatigue failures can occur at stresses far below the yield strength of a material with no sign of plastic deformation. When machine parts that are subject to cyclic loading are designed, their resistance to fatigue may be more important than yield or ultimate strength. Fatigue is not time dependent and can be initiated by any number of factors such as machining tool marks or welding undercuts and localized stress caused by a welding bead. Fatigue can be accelerated by a corrosive atmosphere and by higher frequency stress reversals. A typical fatigue break pattern of a rotating shaft can be seen in Figure 2.24. Fatigue break patterns on a shaft can usually be seen in three stages; first, a slow progression of striations that has a smooth surface; second, coarse striations (like on a clam shell) (Figure 2.25) that progress more rapidly; and third, the brittle, crystalline section that suddenly fails when the shaft strength finally becomes less than the applied stress. Metals are tested for their resistance to fatigue (endurance test) on a fatigue testing machine that rotates a specimen under various loads and conditions. Fatigue life can be consid-

FIGURE 2.24
Fatigue on a shaft.

FIGURE 2.25
Fatigue of Ti-6Al-4V, 28,000 cycles at 8000 lb (10 Hz). Transmission electron microscope replica (5400×) (Oregon Bureau of Mines).

erably increased by surface hardening or by carburizing or shot peening.

Figure 2.26 is a representation of the results of fatigue tests for steel and nonferrous metals displayed in typical S-N curves. For the steel if the stress is kept below the value of *e* the part will not fail; this is termed the **endurance limit** of that steel. For the nonferrous metal, however, a loading that creates a stress shown as *a* can be

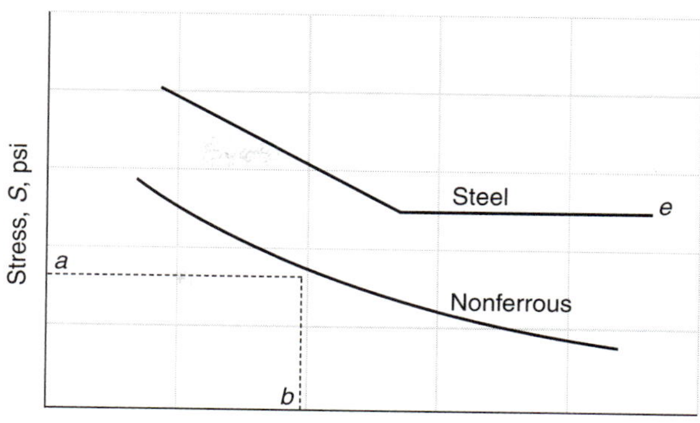

FIGURE 2.26
Typical S-N curves, the result of fatigue tests of various metals.

sustained only for the *b* number of cycles; that is, *a* is the metal's **fatigue strength** for that number of cycles.

PHYSICAL PROPERTIES

Our discussion of mechanical properties was related to the test performed, since the test played a role in developing the property. In the discussion that follows the physical property will be our focus rather than the test used. Nonferrous metals do not have an endurance limit.

Modulus of Elasticity

The **modulus of elasticity** is actually an expression of the stiffness of a material. It is also called *Young's modulus.* If, *within the elastic range,* the stress is divided by the corresponding strain at any given point, the result will be the modulus of elasticity for that material; thus,

$$\text{modulus of elasticity} = \frac{\text{stress}}{\text{strain}}$$

Values of the modulus of elasticity for some common metals were given in Table 2.1. Carbon, alloy, and hardened or soft steels all have about the same modulus, which varies between 29 million and 30 million PSI. This means a steel piece of a given size and shape will twist or bend about the same amount with a given load within the elastic range. Therefore, if a mild steel part is replaced with an alloy steel part, its tensile strength may be increased but not its stiffness, and it will still deflect the same amount if it is the same size; however, if it is exchanged for a metal having a higher modulus, such as tungsten carbide, which is 50 million PSI, then it will not bend or deflect so easily. Surprisingly, cast iron will deflect to some degree within its elastic range, showing a stiffness that is about half that of steel. Values of the modulus of elasticity of the common nonferrous metals are much less than that of steel: brass, 15 to 17 million PSI; aluminum, 10 million PSI: magnesium alloys, 6.5 million PSI.

Conductivity

The **thermal conductivity** and **electrical conductivity** of metals are both related to the fact that metals have free electrons in their atom lattices. (See Atomic Bonding, Chapter 1.) Thus, if a metal such as silver or copper is a good conductor of electricity, it is also a good conductor of heat. Pure metals are better conductors than their alloys. For this reason copper or aluminum used for electrical wiring must be uncontaminated. Electrical copper grades are 99.97 percent pure. Resistance to the flow of electricity can be altered by several factors:

1. Resistance increases as the temperature increases.
2. It increases as a result of cold-working or heat treating the metal.
3. It increases with the amount of impurities and alloying elements present.

In the case of electrical heating elements, resistance is an advantage because the electric current produces and liberates heat.

Metals may look alike but have very different physical properties. For example, one can scarcely tell the difference between zinc, nickel, stainless steel, and silver by appearance, but if you held a bar of equal length of each in your hand and heated the other ends, you would quickly drop the silver bar but could hold the others for several minutes before feeling any heat. Figure 2.27 compares the thermal conductivity of some common metals.

Thermal Expansion and Contraction

Most metals expand when heated and contract when cooled. When heated or cooled each metal expands and contracts at a different rate that is related to its **coefficient of thermal expansion** (Figure 2.28). This physical property of metals can cause difficulties in construction and manufacturing. Engineers plan use of expansion joints in paved highways and bridges to avoid buckling due to heating. When a metal device is to be subjected to repeated heating and cooling, dissimilar metals that expand at different rates can cause abnormal stresses and subsequent damage when rigidly fastened together. If a machinist measures a finished dimension while the part is still hot from machining, the part may be under the tolerance limit when it is cool and will have to be scrapped. For example, brass expands at the rate of 0.0001 in. per degree Fahrenheit per inch of length (or diameter). If a 3 in. diameter brass sleeve is measured immediately after it is machined when its temperature is 100° F (38° C) above room temperature, then when it cools back to room temperature, it has shrunk (0.00001 in. × 100 × 3 = 0.003 in.) and may be undersize. The coefficient of thermal expansion for some common metals can be seen in Figure 2.28.

Reflectivity

Some metals have a high degree of **reflectivity** and are used as light reflectors and in insulation as heat deflectors. Aluminum, nickel, chromium, tin, and bronze are among the metals having high reflectivity, whereas lead has virtually none. In spite of that exception reflectivity is one of the characteristics that distinguishes metals from other substances.

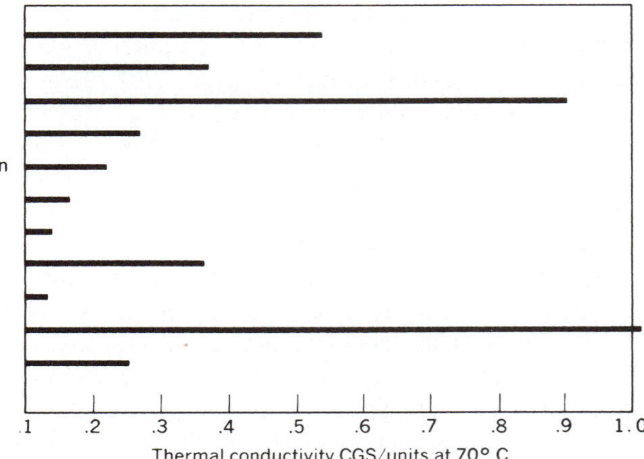

FIGURE 2.27
Comparison of thermal conductivity (Neely and Bertone, *Practical Metallurgy and Materials of Industry,* 5th ed., © 2000 Prentice Hall, Inc.).

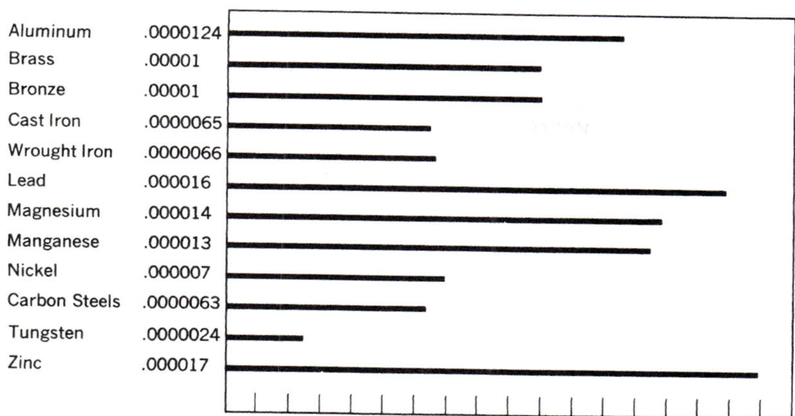

FIGURE 2.28
Coefficient of thermal expansion per degree Fahrenheit per unit length (Neely and Bertone, *Practical Metallurgy and Materials of Industry,* 5th ed., © 2000 Prentice Hall, Inc.).

Ferromagnetism

Ferromagnetism is characterized by an attraction between certain metals that have the ability to retain a residual magnetic force. The ferromagnetic metals are iron, cobalt, and nickel. Of these, iron is the most commercially important metal used extensively in electrical machinery. Soft iron does not retain much residual magnetism when removed from a magnetic field. This quality makes it useful for electromagnets and electric motors in which the field is turned on and off or reversed repeatedly. Powerful permanent magnets are made by alloying metals such as cobalt, nickel, iron, and aluminum.

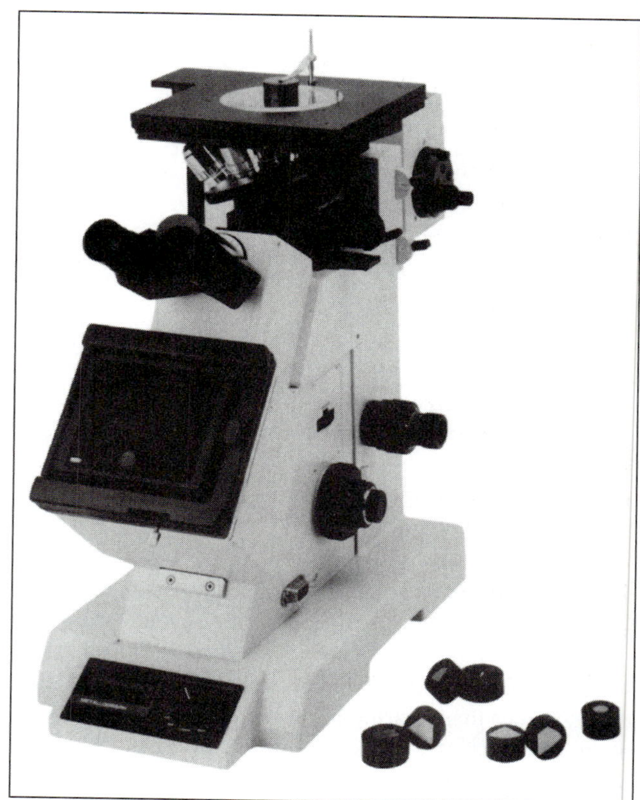

FIGURE 2.29
The Versamet metallographic microscope has the advantage of having a display screen, so more than one person may observe the magnified specimen. Also, photographs may be taken of the enlarged specimen (*Versamet* is a trademark and a copyright of Unitron Inc.).

METALLURGICAL MICROSCOPY

Metal surfaces are examined by means of metallurgical microscopes and associated techniques of surface preparation. Structures in metals not readily visible to the naked eye may be seen at high-power magnification from 10 times the diameter (10×) to 3000× for light microscopes and many more thousands of diameters with the electron microscope. Techniques of photomicrography in which a camera is attached to the microscope make possible photographic illustrations of specimens (Figure 2.29). Metallurgical microscopes, unlike those used to see *through* a specimen, use reflected light from an internal source (Figure 2.30). Inverted stage microscopes are widely used to study metals (Figure 2.31).

Specimens to be analyzed are cut from a metal part with a metallurgical saw (Figure 2.32). The specimen is mounted in plastic in a press (Figures 2.33 and 2.34) and rough ground (Figure 2.35). Next, the mounted specimen is ground on successively finer grit abrasive surfaces (Figure 2.36), after which it is polished on a rotating cloth-covered disk (Figure 2.37) or by other techniques such as electropolishing or vibratory polishing. When the specimen has a mirror finish, it is etched in a reagent that is most commonly a dilute acid, varying with the type of metal being prepared. Etching causes surface characteristics of the metal to become visible; some of these in steels are grain boundaries, grains, impurities, and

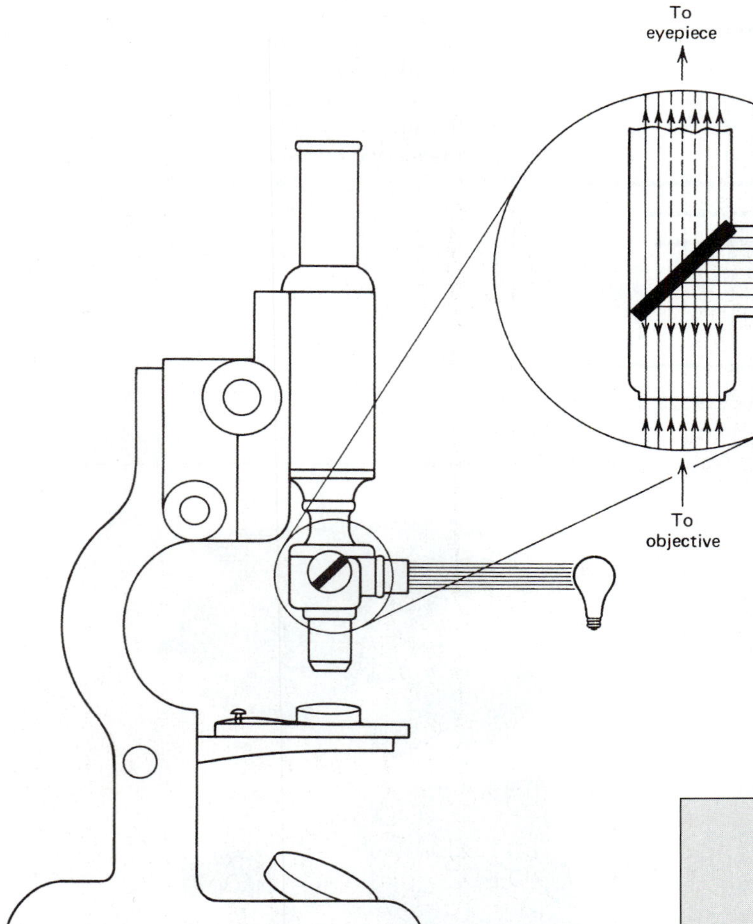

FIGURE 2.30
Illumination in a metallurgical microscope (Neely, *Metallurgy,* 3d ed., © 1989 Prentice Hall, Inc.).

microstructures in the grains such as pearlite, bainite, ferrite, iron carbide (cementite), and martensite (Figure 2.38).

Ferrite is more or less pure iron. Pearlite is a relatively soft form of alternating layers of cementite and ferrite. Bainite is harder than pearlite and is actually a tough form of martensite. Acicular martensite is an extremely hard form of steel. The microscopic study of metals can help us to readily determine the condition of the specimen. Of course, much more information about metals is gained by metallurgists by studying microstructures in metals with powerful microscopes.

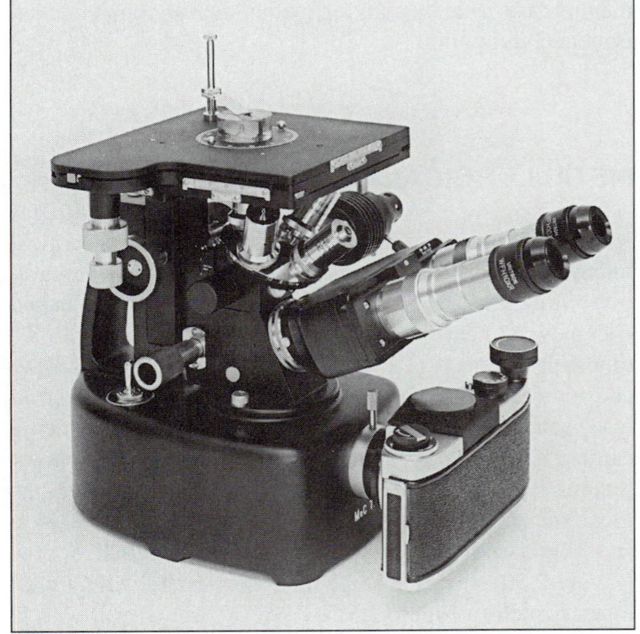

FIGURE 2.31
BMEC inverted stage microscope. The specimen is placed upside down on the stage on the top of the microscope. A 35mm camera can also be attached to this microscope (BMEC is a trademark and a copyright of Unitron Inc.).

NONDESTRUCTIVE TESTING

As the name implies, nondestructive testing in no way impairs the part for further use. It does not measure mechanical or physical properties but instead identifies defects such as voids, inclusions, or cracks that might later lead to failure of the part. Also, improper handling of

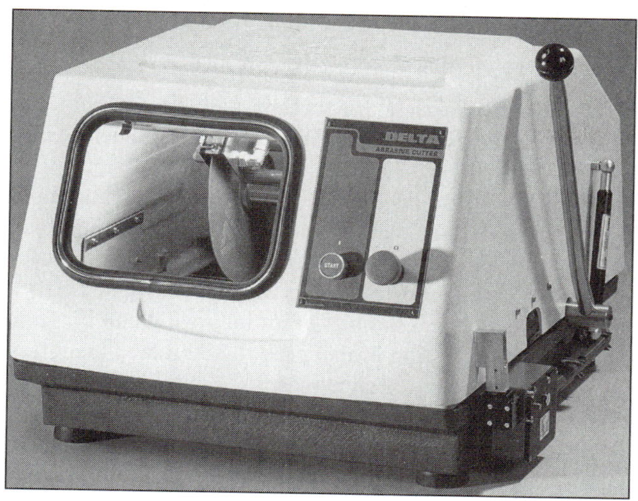

FIGURE 2.32
Metallurgical saw. The specimen is cut off with a thin abrasive blade to which coolant is applied. This keeps the specimen from becoming overheated and prevents damage to the metallurgical structure (Buehler, Ltd.).

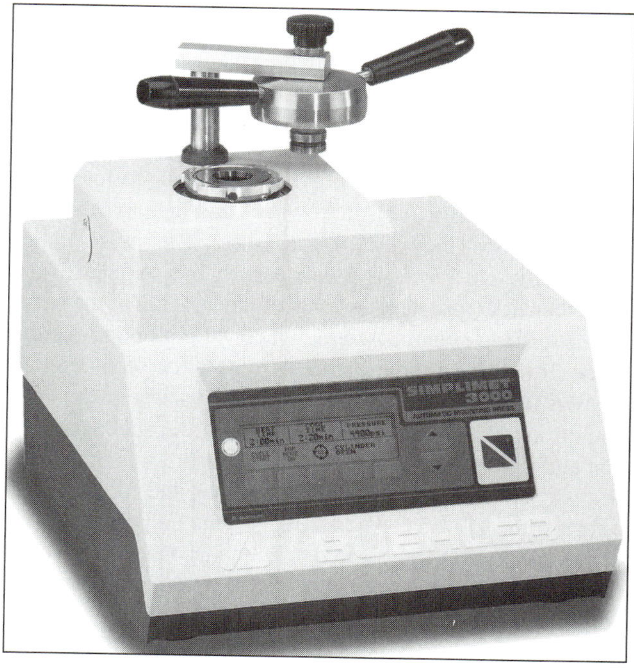

FIGURE 2.33
Metallurgical press. After the specimen is placed in the press, granulated thermoplastic is poured in. It is then heated under pressure, forming a solid cylindrical shape with the specimen embedded on one side (Buehler, Ltd.).

FIGURE 2.34
Mounted specimen.

FIGURE 2.35
Rough grinding the specimen using a belt surfacer (Buehler, Ltd.).

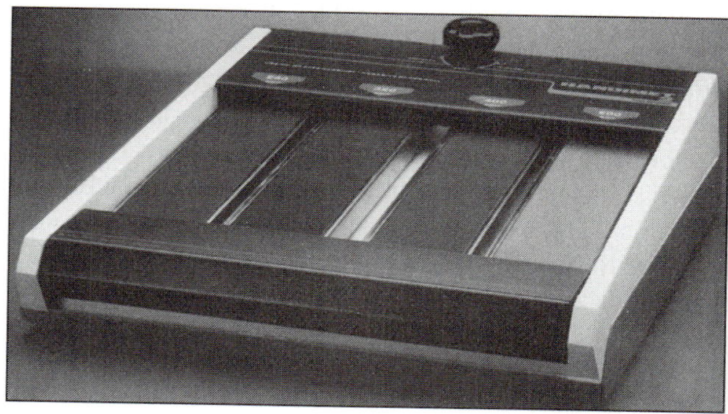

FIGURE 2.36
Finish grinding the specimen on HandiMet® (Buehler, Ltd.).

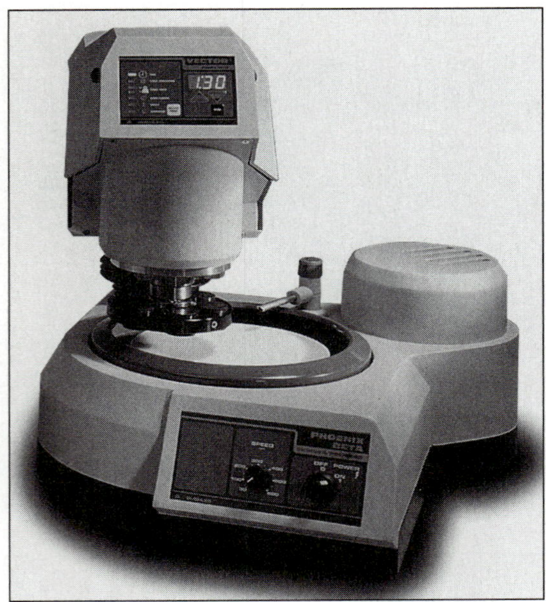

FIGURE 2.37
A specimen polishing wheel; this model is able to polish one specimen manually or multiple specimens semiautomatically (Buehler, Ltd.).

materials that causes changes in the microstructure can lead to failure. For example, a nick or dent in a highly stressed part such as a spring can cause early failure. At present, even these aberrations can be detected on a production basis. Nondestructive methods can also verify hardness or heat treatment in many cases. Since testing for defects in the early stages of the manufacturing process can substantially reduce rejects, nondestructive testing is closely related to quality control. (See Chapter 19.) In the study of fracture mechanics, the damage-tolerant philosophy has assumed that all metals have microscopic flaws such as discontinuities and microcracks on the surface that will lead to ultimate failure. With inspection and testing methods, a safe service life or stress-life curve can be determined. Along with a safety factor, the design is known as a safe-life design. With the increasing use of materials that lack the ability to strengthen as they are deformed, such as composites, nondestructive testing will be needed even more in manufacturing and maintenance operations.

The following are the most common types of nondestructive testing:

1. Magnetic-particle inspection
2. Fluorescent penetrant inspection
3. Dye penetrants
4. Ultrasonic inspection
5. Radiography (X-ray and gamma ray)
6. Eddy-current inspection

Magnetic-particle inspection methods are used to detect surface flaws in ferromagnetic materials such as iron and steel (Figures 2.39 to 2.42). One method starts with sprinkling dry iron powder on a magnetized part. A more widely used method starts with pouring a liquid in which the iron powder is suspended over the magnetized workpiece; this is then viewed with a fluorescent light. Surface cracks only a few millionths of an inch wide can be seen. This method of testing is also used in automated manufacturing systems with advanced laser scanning techniques to sort out questionable parts from good ones.

Fluorescent penetrants are available in sealed pressure spray cans to apply for visual examination of any material such as metals, ceramics, and plastics. A fluorescent light is also used with this process to detect cracks. Testing systems range from huge automated systems to portable test kits for field inspection. Zyglo is the Magnaflux copyrighted name for this method (Figure 2.43). Dye penetrants also depend on visual inspection but do not use fluorescent light. Essentially, a dye is caused to seep into cracks and then is wiped off, and a developer is applied. A bright color then indicates the location of cracks (Figures 2.44 and 2.45). This method, called Spotcheck®, is also useful for field testing.

Ultrasonic inspection makes use of high-frequency sound waves that are sent through a test piece by means of a transducer, which is usually a piezoelectric crystal that converts electrical energy to mechanical energy (Figures 2.46 and 2.47). A transceiver that has the opposite function of converting mechanical energy (vibrations) to electrical pulses picks up the ultrasound waves and displays them on an oscilloscope (Figure 2.48) (p. 50) as signal pulses called *pips.* In the pulse-echo system, part of the sound wave is reflected back from the other side of the material to the transducer/transceiver and is seen as a second pip on the oscilloscope. If there is a flaw, a smaller pip will be seen between the two, indicating the distance from the surface to the flaw. The through-transmission method uses a transducer on one side and a transceiver on the other. Ultrasonics is one of the best ways to check for cracks in pipelines and for structural weld testing (Figure 2.49) (p. 50). This method of inspection is very adaptable to automated manufacturing systems and is used widely to inspect for internal defects, cracks, laminations, variation in thickness, and weld bonds.

Radiography (X-ray and gamma ray) **inspection** has been used widely on pipelines and manufactured parts to detect flaws, especially internal defects such as porosity. There are several disadvantages to this system, however. Because X-ray and gamma radiation are both hazardous,

(text continues on p. 50)

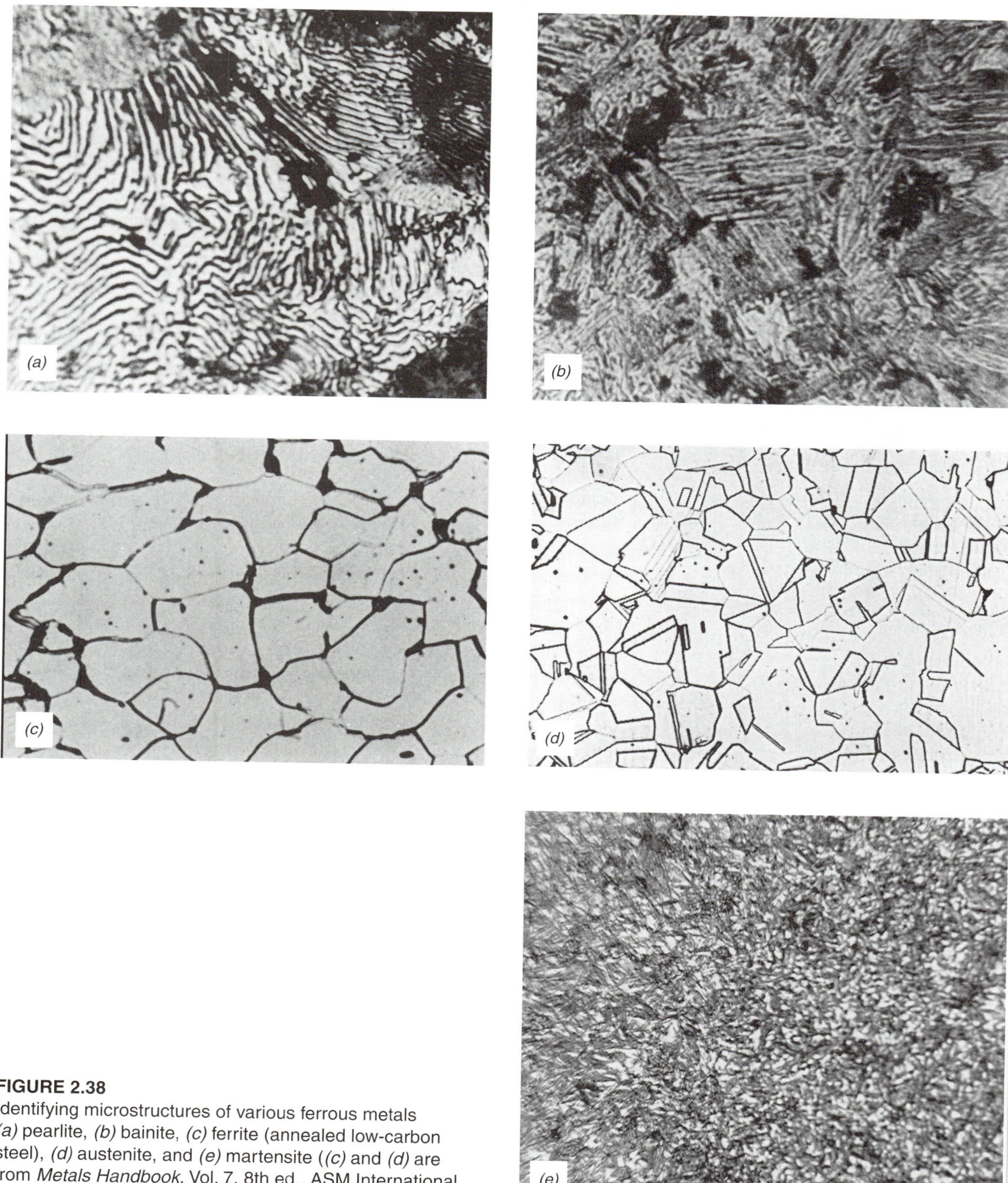

FIGURE 2.38
Identifying microstructures of various ferrous metals *(a)* pearlite, *(b)* bainite, *(c)* ferrite (annealed low-carbon steel), *(d)* austenite, and *(e)* martensite (*(c)* and *(d)* are from *Metals Handbook,* Vol. 7, 8th ed., ASM International, 1972, pp. 9 and 86. With permission.).

FIGURE 2.39
Magnetic-particle testing machine (Courtesy of Magnaflux Corporation, Chicago).

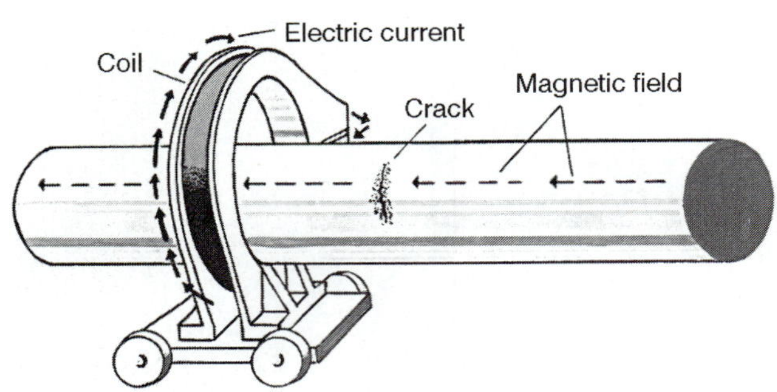

FIGURE 2.41
Longitudinal method of magnetization (Courtesy of Magnaflux Corporation, Chicago).

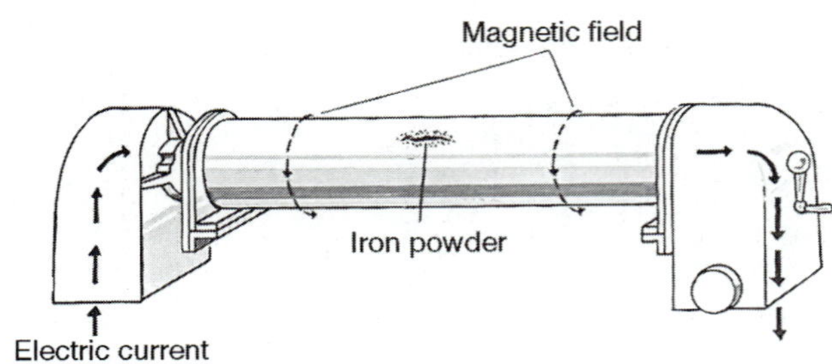

FIGURE 2.42
Circular method of magnetization (Courtesy of Magnaflux Corporation, Chicago).

FIGURE 2.40
By inducing a magnetic field within the part to be tested and applying a coating of magnetic particles, surface cracks are made visible; in effect, the cracks form new magnetic poles. Particles cling to the defect as tacks would to a simple magnet (Courtesy of Magnaflux Corporation, Chicago).

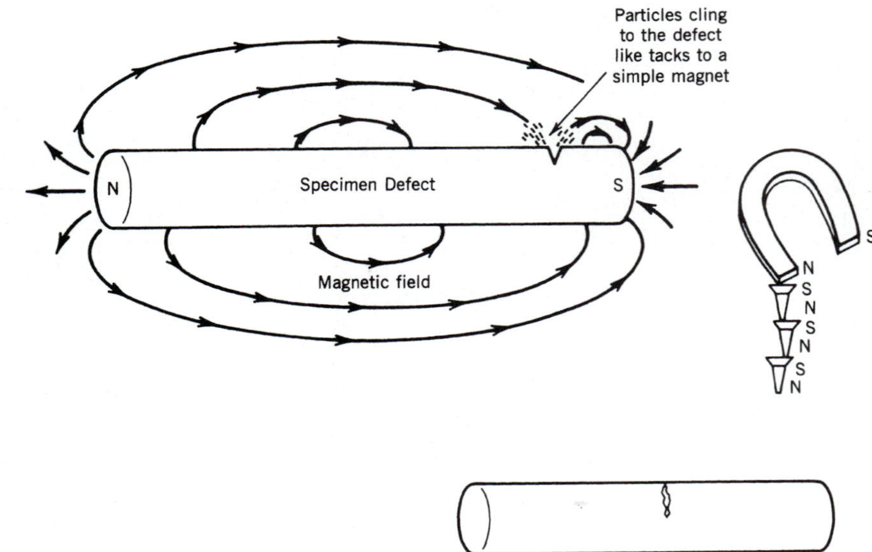

FIGURE 2.43
With the Zyglo®, this vanadium alloy stainless forged hip-ball joint used in a medical prosthesis is revealed to have a crack, detected in manufacture, that could result in severe injury to a patient (Courtesy of Magnaflux Corporation, Chicago).

FIGURE 2.44
Spraying Spotcheck developer on a gear (Courtesy of Magnaflux Corporation, Chicago).

FIGURE 2.45
Defect is now visible on the gear (Courtesy of Magnaflux Corporation, Chicago).

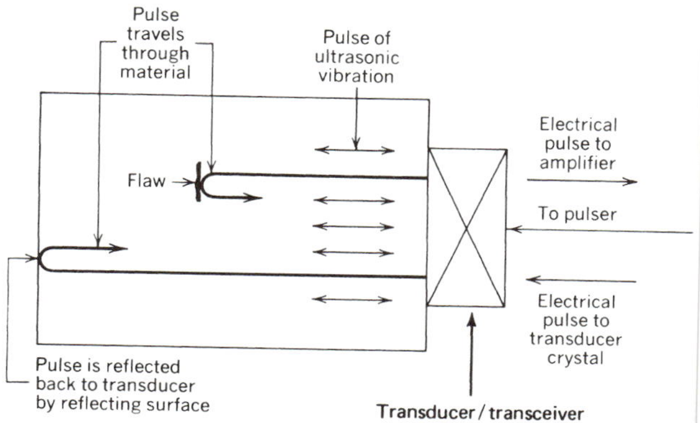

FIGURE 2.46
How ultrasonic sound waves are used to locate flaws in material (White, Neely, Kibbe, Meyer, *Machine Tools and Machining Practices,* Vol. I, © 1977 John Wiley & Sons, Inc.).

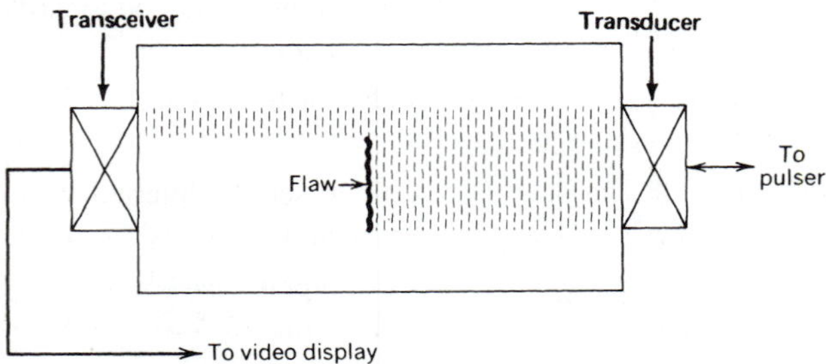

FIGURE 2.47
Ultrasonic through-transmission (White, Neely, Kibbe, Meyer, *Machine Tools and Machining Practices,* Vol. I, © 1977 John Wiley & Sons, Inc.).

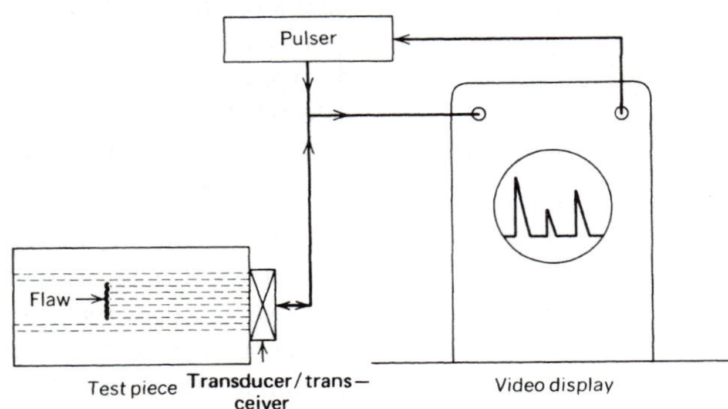

FIGURE 2.48
Ultrasonic pulse-echo system (White, Neely, Kibbe, Meyer, *Machine Tools and Machining Practices,* Vol. I, © 1977 John Wiley & Sons, Inc.).

technicians using this process are required to be trained. Radiography is also expensive and time consuming, and there is a considerable time delay until the film is developed. The test is also two-dimensional, possibly missing flaws viewed at another angle. However, a new process has been developed in which a video image and tape is made in real time (instantly). In the past, attempts at real-time imaging of X rays had poor resolution and were not practical.

Eddy-current inspection systems can test only electrically conducting materials. When a test piece is placed in or near a coil in which alternating current produces a magnetic field, an eddy current is induced in the test piece. This current causes a change in impedance in the magnetic field that is read on a meter or oscilloscope (Figure 2.50). Variations such as seams, thicknesses, and other physical differences can be detected quickly with this method. Eddy-current testing is quite adaptable to automated systems.

FIGURE 2.49
Portable ultrasonic machine instruments can be used for routine manual inspection under field conditions. The instruments are very useful for structural weld testing and corrosion surveys (Courtesy of Magnaflux Corporation, Chicago).

FIGURE 2.50
Eddy-current testing instrument for nondestructive comparison of magnetic materials. It is used to separate materials according to their metallurgical properties, such as alloy variations, heat-treat condition, hardness differences, internal stress, and tensile strength (Courtesy of Magnaflux Corporation, Chicago).

Review Questions

1. What is the difference between elasticity and plasticity? Define each.
2. How can the tensile strength of a specific metal be determined?
3. What property of a metal describes its ability to be easily drawn into a wire but not rolled into a sheet without splitting?
4. Define hardness as it is measured on a Rockwell or Brinell testing instrument.
5. When a material breaks suddenly with no sign of deformation, what is it said to be?
6. The properties of some very ductile metals, such as certain low-carbon steels, change at low temperatures at their particular transition zone. How can this information help an engineer design machinery that will be operating in subzero temperatures?
7. Steel has a modulus of elasticity of 29 million PSI and tungsten carbide of 50 million PSI. If a steel and a tungsten carbide bar of the same dimensions were both loaded on one end and supported on the other as cantilever beams, which would bend more?
8. Name the phenomenon that occurs when a metal is loaded well within the elastic range at a temperature of 800° F (427° C) and it slowly deforms permanently.
9. Welds, machining tool marks, and stress reversals, such as those in rotating parts, can cause a progressive failure even on a lightly loaded part such as a shaft. There is no sign of ductility, only a smooth surface having striations like a clamshell. How would you identify this kind of failure? What is it called?
10. Which is the better conductor of heat or electricity, a pure metal or an alloy?
11. By what means is it possible to see the internal structure of metals at high magnification? How is this done?
12. How can machinery parts and pipeline welds be tested for flaws without destroying them? Name several processes.

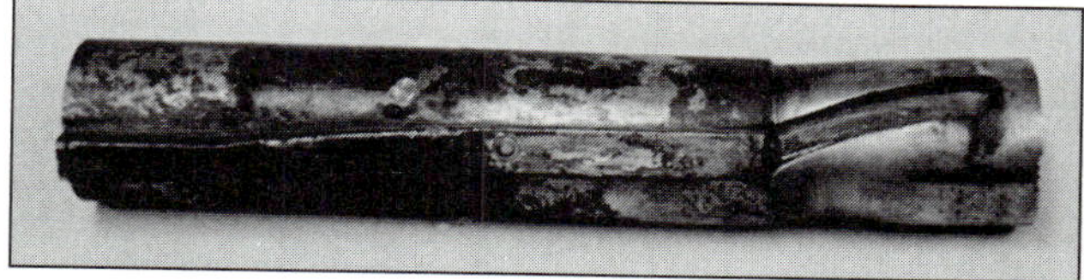

FIGURE 2.51
See Case 1.

Case Problems

Case 1: Case of the Twisted Shaft

The shafting shown in Figure 2.51 was taken from a mill transfer chain that was frequently overloaded, causing the shaft (made of low-carbon steel) to fail by torsion shear. It has obviously been loaded in torsion beyond its elastic limit into the plastic range and has taken a permanent set. Since the sprocket and bearing bores are not easily changed to substitute a larger diameter shaft, what type of replacement shaft will solve the problem? What different mechanical properties must it possess so it will not again fail by torsion shear?

Case 2: Broken Shaft

A hammer mill that was located outside a building was driven by an 800-hp electric motor whose main shaft was 8 in. in diameter. It had operated for years; the only maintenance needed was to lubricate bearings and replace hammers and grates. One winter morning the temperature dropped to −10° F. When the machine operator turned the switch to start the machine, which was unloaded and free-running, the main shaft broke in two instantly in a brittle-type fracture. What could have been the problem and how could it be avoided in the future?

Case 3: Piston Clearance

An aluminum piston 4 in. in diameter that moves in the cylinder of a cast iron engine block must have a clearance diameter of 0.003 in. when hot. The temperature rises from 70 to 160° F. What should the clearance between the bore diameter and the piston be if the linear coefficient of expansion per unit length (inch) per degree Fahrenheit is 0.00001244 for aluminum and 0.00000655 for cast iron when the temperature is 70° F?

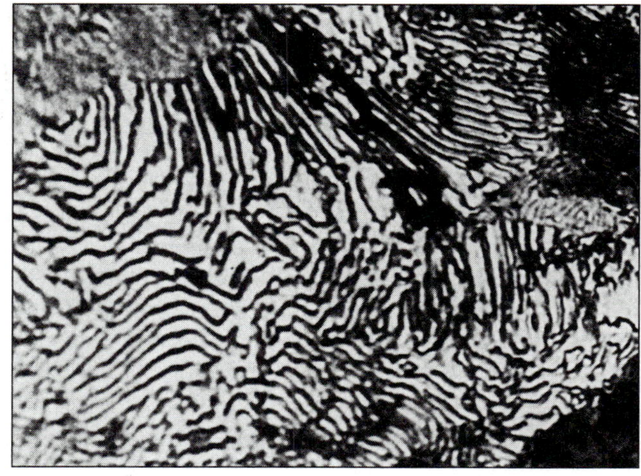

CHAPTER 3

Changing the Properties of Metals

Objectives

This chapter will enable you to:

1. Use an iron–carbon phase diagram to explain the basic principles involved in heat-treating steel.
2. Use an I-T diagram and cooling curves to explain the various microstructures found in carbon steel as a result of cooling rates.
3. Describe how steel is hardened and tempered.
4. Explain several methods or techniques to surface harden steel and the significance of selective surface hardening in manufacturing.
5. State the principles involved in precipitation hardening.
6. List the various types of annealing and stress relief and explain their purposes and uses for industrial applications.
7. Explain the factors involved in recrystallizing low-carbon work-hardened steel.

Key Words

phases and phase changes
solution heat treating
plastic deformation
solid solution strengthening
time–temperature transformation
solutionizing
quenching
annealing
solubility and insolubility
precipitation hardening
recrystallization

From ancient times, the immense value of metals to the progress of civilization could be attributed to their ability to be shaped easily at ordinary temperatures and even more easily at high temperatures. Second only to their elastic and plastic behavior is the ability of carbon-bearing iron, called *steel,* to become very hard when heated and suddenly cooled as in a quenching medium like water. Other heat treatments such as temper drawing, annealing, and stress relieving also have a profound effect upon metals. The high quality of swords made in Damascus has been legendary since Alexander the Great was said to have cut the Gordian knot with one and the Saracens wielded them against the Crusaders. The process of making those swords was a well-kept secret that was never discovered. It is now believed that the forging process had two important steps that were responsible for this unusually hard and tough grade of carbon steel: the blade was folded and refolded between hammerings, and the heating was done at a low enough temperature that the hard **cementite** formed in the high-carbon steel was not dissolved.

Today, there are many kinds of alloy and tool steels, each of which often requires special heat treatments in the manufacturing process to give it the qualities necessary to fit the required specifications. In addition, many nonferrous metals such as aluminum and titanium are heat-treated to increase their strength and durability. The manufacture of machinery, automobiles, modern aircraft and space vehicles, and consumer products would not be possible without the technology of heat treatment of metals.

Still other products are dependent for their properties on the extent to which they are plastically deformed or cold-worked. The aluminum beverage container or pop can is produced by such a strengthening process.

This chapter begins by considering the effects of heating and cooling metals. This leads to a discussion of phase diagrams for the iron–carbon alloys (steels and cast irons) and for **nonferrous** alloys. The phase diagram is basic to a discussion of heat treating, which is the next topic. Another method of strengthening, by cold working, is considered, and the softening process of **annealing** is described. The chapter concludes with a brief look at some of the equipment used in heat treating.

HEATING AND COOLING OF METALS

When materials are heated or cooled, heat, measured in calories or British thermal units (BTUs), is added or removed. Each material requires a certain amount of heat to change its temperature one degree, however, if the material, say water, also changes its state—goes from solid ice to liquid water—an additional amount of heat must be *added;* or if the liquid water freezes and forms ice—thus changing its state—the same amount of heat must be *removed.*

Temperature versus Time Curves

This same phenomenon occurs when metals are melted or frozen and is illustrated in Figure 3.1*a;* from *a* to *b* the metal is cooling as a liquid. At *b* it begins to freeze, from *b* to *c* the heat required to solidify the metal is removed, and from *c* to *d* the metal cools as a solid. These curves are typically shown for cooling, but if the metal were heated, the same figure would be generated but in the opposite direction, as shown in Figure 3.1*b.* Because in this example the metal is a pure metal (not an alloy) the heat needed to solidify (*b* to *c*) or melt (*c* to *b*) occurs at one temperature, or there is an "arrest" in the curve. The length of the *b* to *c* line is the time required for the heat to cause the solidification or melting to occur.

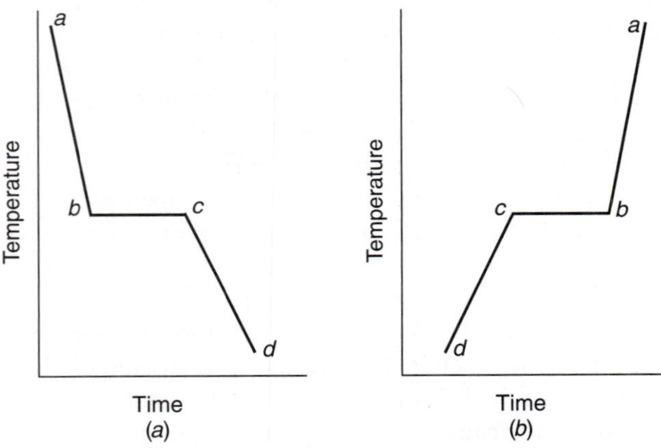

FIGURE 3.1
Temperature versus time curves for a pure metal being *(a)* cooled and *(b)* heated.

Allotropic Forms of Iron

As mentioned in Chapter 1, iron is one of the relatively few metals that have the property of **allotropy,** or is allotropic. In addition to having an arrest at their freezing/melting temperature, these metals also have arrests at the temperatures at which they change their atom lattice. Figure 3.2 is a cooling curve for iron, showing a number of arrests related to changes in atom lattice as well as the pause in temperature at 2800° F where the metal changes from a liquid to a solid. From 2800° F (1538° C) down to 2554 °F (1401 °C), iron is BBC, non-magnetic, called delta iron, and is of little importance in manufacturing. Between 2554° F (1401° C) and 1666° F(908° C), iron is FCC, non-magnetic, and called **austenite** or gamma iron. At 1666° F (908° C), iron changes to the BBC lattice, continues to be non-magnetic, and is called **ferrite** or alpha iron. The ferrite becomes magnetic when its temperature falls below 1414° F (768° C), known as the **Curie temperature.**

The states (liquid and solid) and the different atom lattices are referred to as phases (a **phase** being something separate, distinct, and homogeneous), so when these changes occur they are referred to as **phase changes,** and a diagram of where these changes take place is called a **phase**

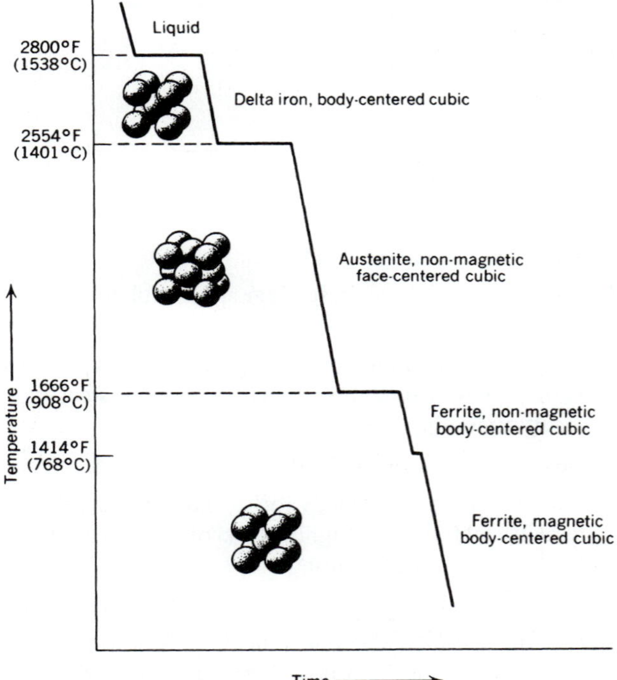

FIGURE 3.2
Cooling curve for pure iron. As iron is cooled slowly from the liquid phase (or heated from solid to liquid) it undergoes these allotropic transformations (Neely and Bertone, *Practical Metallurgy and Materials of Industry,* 5th ed., © 2000 Prentice Hall, Inc.).

diagram. On the iron–carbon phase diagram of Figure 3.3 are shown lines marked **liquidus** and **solidus.** Liquidus indicates the temperatures at which the various compositions of the alloy *begin* to become solid as the temperature is reduced. Solidus indicates the temperatures at which the various compositions of the alloy are completely solid, and thus all liquid is absent. The ferrite and austenite phases are very important in heat treating and manufacturing.

THE IRON–CARBON PHASE DIAGRAM

It was stated previously that the phase diagram is important to an understanding of the principles of heat treating. Figure 3.4 is the iron–carbon phase diagram, a plot of temperature versus the percent carbon that has been added to iron.

Since Figure 3.2 represents pure iron, the same changes in solid iron can be seen at the left-hand vertical line on the

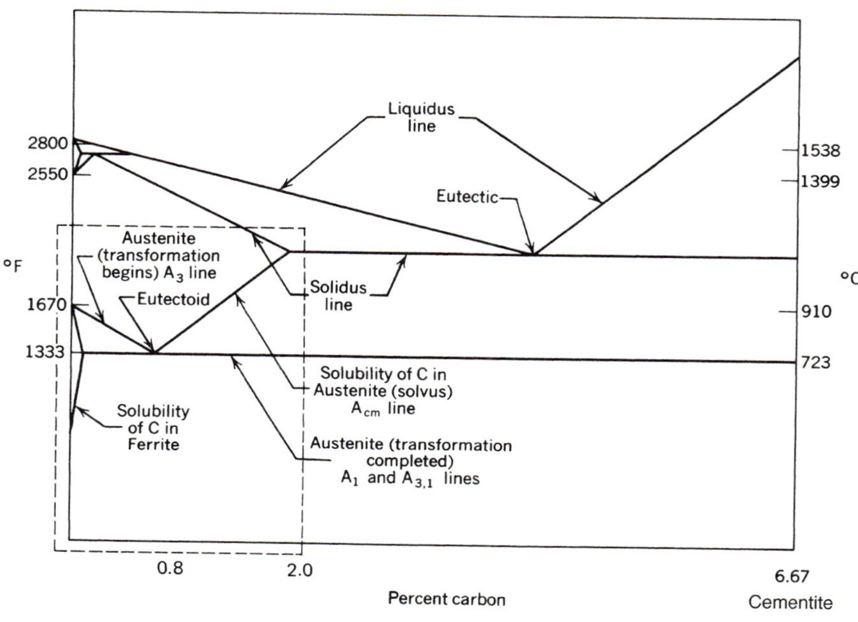

FIGURE 3.3
The iron–carbon phase diagram showing the steel portion (Neely and Bertone, *Practical Metallurgy and Materials of Industry,* 5th ed., © 2000 Prentice Hall, Inc.).

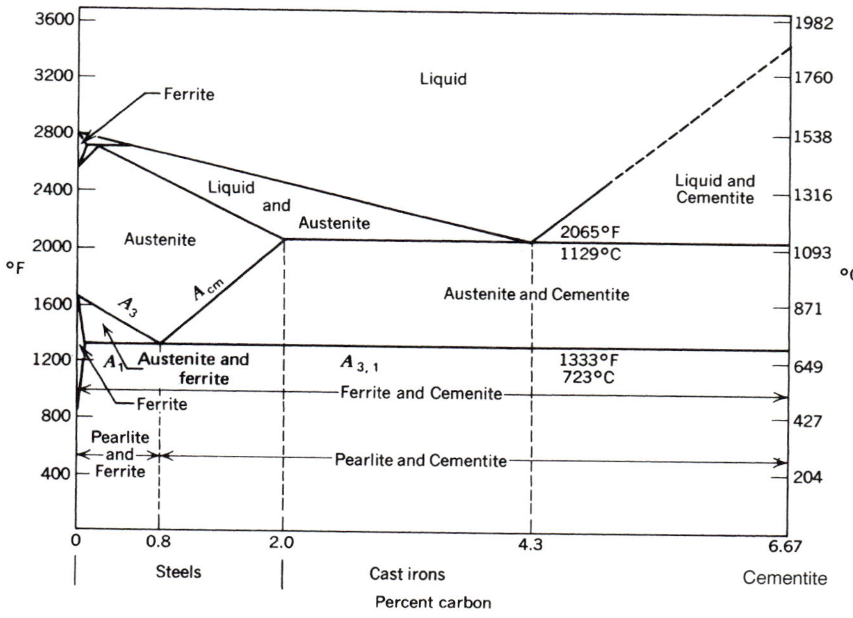

FIGURE 3.4
Simplified iron–carbon equilibrium phase diagram, that is, based on very slow heating and cooling (Neely and Bertone, *Practical Metallurgy and Materials of Industry,* 5th ed., © 2000 Prentice Hall, Inc.).

iron–carbon diagram. The percentage of carbon in the iron increases toward the right side of the diagram, ending at 6.67 percent carbon. The diagram ends there because 6.67 percent C by weight is the amount of carbon in the compound iron carbide (Fe_3C), or cementite, and no other useful information would be given by extending the diagram to the right. Much of the diagram is devoted to cast irons because 2 to 4 percent carbon produces cast iron; 2 percent carbon and under produces steel. The area within the dotted lines (Figure 3.3) is the steel portion of the iron–carbon diagram.

Eutectic The root of this word means the lowest melting point. In Figure 3.4 the lowest melting point can be seen at 4.3 percent carbon, and this point is identified as the eutectic in Figure 3.3. A **eutectic** can occur because the two metals involved in the alloy system lack complete **solubility;** that is, they have **partial solubility,** which also means they are **partially insoluble.** (Metals can also be completely insoluble, analogous to oil and water.) Insolubility of this type usually occurs because the metallic atoms do not have the proper size relationship to fit into the same lattice. (For metals to be soluble as solids their atoms must be able to reside in the same atom lattice.) In the case of the iron–carbon system, carbon has a small-sized atom that can fit in the spaces between the iron atoms in the FCC lattice of austenite. It seems reasonable that there would be a limit to how much carbon can fit into the spaces of the austenite lattice, and that limit is 2 percent; in Figure 3.4 it can be seen that the austenite region ends at 2 percent carbon and 2065° F.

Because of this insolubility we can see in Figure 3.4 that at the eutectic point the single-phase liquid solidifies as two phases, austenite and cementite. On either side of the eutectic point the metal solidifies over a temperature range in which the metal is somewhat "mushy," but at the eutectic point, it solidifies at one temperature like a pure metal.

Eutectoid This word has the same root as *eutectic* but now refers to metals that are solids, so instead of being the lowest melting point it is the lowest temperature at which one solid phase transforms into other solid phases. The eutectoid also occurs because of insolubility, and in Figure 3.4 it can be seen that the single-phase austenite transforms into two phases, ferrite and cementite. The eutectoid indicated in Figure 3.3 occurs at 0.8 percent carbon, as seen in Figure 3.4. Again, on either side of the **eutectoid** the transformation is gradual over a temperature range, but at the eutectoid, it occurs at one temperature. As noted above, up to 2 percent carbon can be contained in austenite as a solution. We can see this in Figure 3.3 in the area above the A_3 and A_{cm} lines.

When austenite containing 0.8 percent carbon (eutectoid) is slowly cooled there is just enough carbon present to produce a 100 percent pearlitic microstructure (Figure 3.5). **Pearlite** consists of alternating plates of an iron–carbon compound called cementite (Fe_3C) and relatively pure iron, called ferrite (Fe). **Hypoeutectoid** steels, which contain less carbon than the eutectoid amount, when slowly cooled from above the A_3 line, form grains of ferrite that contain virtually no carbon. Instead, the carbon content of the remaining austenite increases until the eutectoid temperature (A_1) is reached, at which temperature the carbon content is 0.8 percent, the eutectoid amount. As heat is extracted the temperature plateaus here while the remaining austenite transforms to pearlite (Figure 3.6). Full regular grains of ferrite will

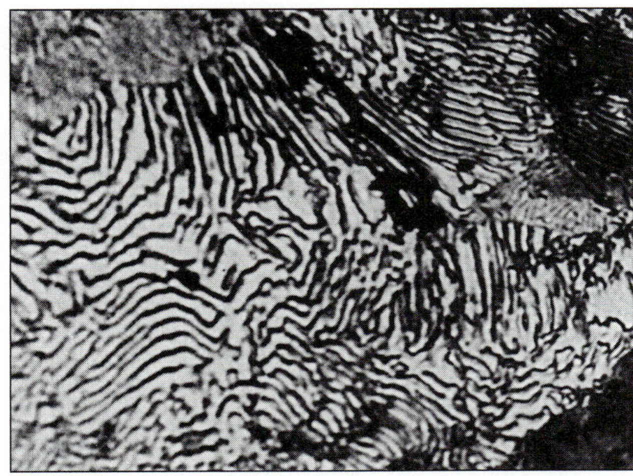

FIGURE 3.5
SAE 1090 steel slowly cooled, showing very coarse pearlite (500×).

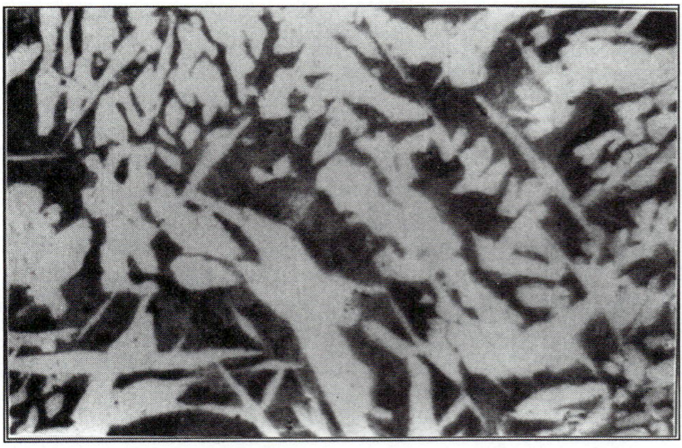

FIGURE 3.6
Ferrite and pearlite in cast 0.4% carbon steel (200×). Blades of ferrite have transformed from the austenite and the remaining austenite has transformed to pearlite (Clyde W. Mason, *Introductory Physical Metallurgy.* Materials Park, OH: ASM International, 1947, Fig. 53.90).

form depending on the cooling rate and carbon content. The carbon content of the steel of Figure 3.7 is so low that very little pearlite forms. When **hypereutectoid** steel (steel containing more carbon than eutectoid) is slowly cooled from a point on the A_{cm} line, the excess carbon over 0.80 percent begins to form cementite at the prior austenite grain boundaries, since the A_{cm} line is the limit of solubility of cementite. Within the grains the eutectoid composition is formed and at the $A_{3,1}$ line, the grains become 100 percent pearlite with a network of cementite at the grain boundaries (Figure 3.8).

Microstructures other than those shown here, for which the phase diagram was used to study the slow cooling of steels, will occur if the cooling rates are more rapid. That, however, involves the topic of heat treating, which will be dealt with shortly.

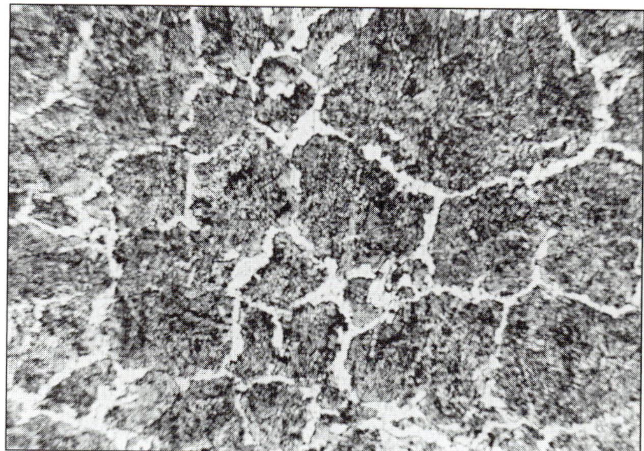

FIGURE 3.8
SAE 1095 hypereutectoid steel showing cementite at the grain boundaries. Massive cementite and ferrite both etch white (500×).

NONFERROUS PHASE DIAGRAMS

The number of possible nonferrous binary (two-metal) alloy systems is quite large, but here we consider only those combinations that can be heat treated. As in the iron–carbon system the nonferrous metals in which we are interested are those that are only partially soluble. (Completely soluble and completely insoluble metals cannot be heat treated, as will be discussed when we cover that topic.)

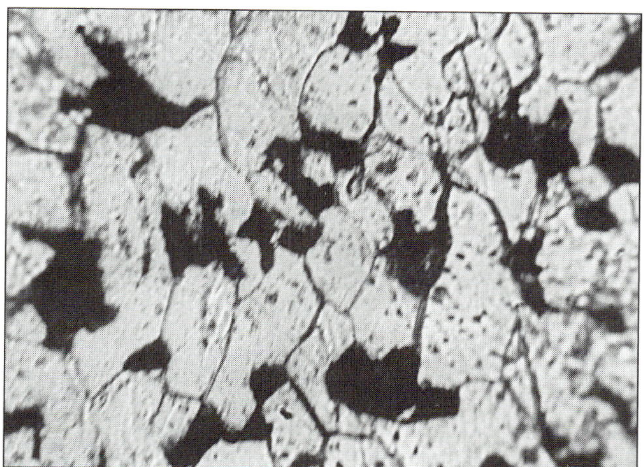

FIGURE 3.7
Low-carbon steel with large ferrite grains and very little pearlite (500×).

In the iron–carbon system the carbon went into solution in the solid iron by fitting between the atoms of the austenite's FCC lattice. In forming solid solutions in nonferrous alloys the element added must become a part of the lattice of the parent metal. Thus, the size of the atoms of the two metals must be close enough that they can *substitute* for each other but not so close that they become completely soluble (and thus not able to be heat treated). The nonferrous alloys capable of heat treatment, and that are in common usage in fields such as transportation, manufacturing, and consumer products, are aluminum, copper, magnesium, and titanium. These all form alloys with other metals that exhibit phase diagrams such as shown in Figure 3.9. A comparison of Figure 3.9 with the iron–carbon diagrams of Figures 3.3 and 3.4 shows their similarity, especially if you ignore the lower

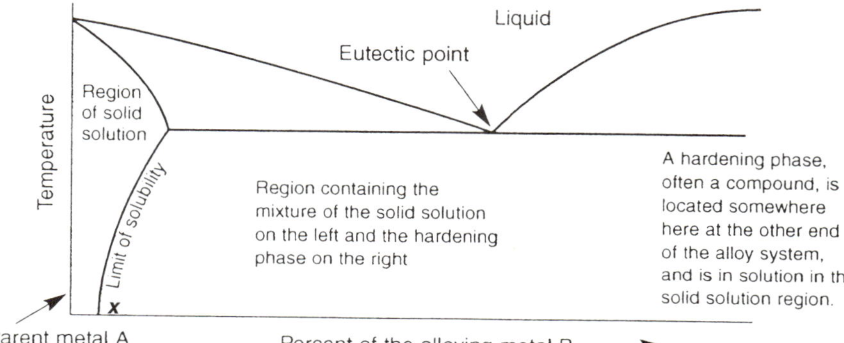

FIGURE 3.9
A generic phase diagram typical of nonferrous metals that are capable of being heat treated.

portion of Figures 3.3 and 3.4 where iron goes through its allotropic change; you can identify that each has a eutectic point and a solid solution region along its left (temperature) axis. Thus, the allotropy of iron is the major difference.

PRINCIPLES OF HEAT TREATING

Heat treating is generally identified with processes in which a metal is heated to an elevated temperature from which it is cooled very rapidly, and in the process the metal somehow gets harder and stronger. But what is really going on?

The phase diagrams of Figures 3.3 and 3.9 tell us what happens when the various compositions of alloys are heated or cooled under "equilibrium" conditions; for our purposes we can interpret equilibrium to mean that the alloys are heated or cooled very slowly. If, instead of cooling slowly, say we cool very rapidly by **quenching** in water, what can the phase diagram tell us? If our attempts at cooling are completely successful, we will retain at room temperature whatever phase existed at the higher temperature. That is, we will have made the metal do something that by its original nature it was not supposed to do. Thus, for Figure 3.3 if we heat a steel (iron with 2 percent or less of carbon) into the austenite range and then quench it we will have austenite at room temperature containing much more carbon than iron should at that temperature. Note that the ferrite phase that normally exists at room temperature contains almost zero carbon.

A similar situation can occur with nonferrous alloys. If an alloy that has a composition "x" as marked on the phase diagram of Figure 3.9 is heated above the line marked "limit of solubility," it will move from a region of two phases (a mixture) to a region of only one phase; if cooled rapidly from that temperature the one phase will be retained at room temperature, and it will have dissolved in it more of the hardening phase than it should have. Thus, in both these cases the intent of the heating and rapid cooling is to make the alloy contain at room temperature phases that would normally not be there.

HEAT TREATING FERROUS METALS

The microstructures that result from the cooling processes used in heat treating of ferrous metals are best understood using an analysis that takes into account time, which is not a factor in the phase diagrams of Figures 3.3 and 3.4.

TTT or I-T Diagrams

Such an analysis is shown in a **time–temperature–transformation** (TTT) diagram, or an **isothermal transformation** (I-T) diagram, Figure 3.10. Both names suggest that the steel is held at one temperature (isothermal) until it transforms into another structure. It can be seen on Figure 3.10 that when austenite is cooled to a point below its stable temperature (A_1 and A_{31} on Figures 3.3 and 3.4) it will begin to transform into any of a number of transformation products, depending on the cooling rate. If it is held at a constant temperature (isothermal) for a given length of time, it will transform into one particular kind of microstructure. If, on the other hand, it is cooled so suddenly that the cooling curve No. 1, as shown in Figure 3.11, does not cut into the "nose" of the I-T diagram, no transformation can take place that will form microstructures that are relatively soft, such as **pearlite** or **bainite;** instead, the austenite will be cooled to an area below the M_s line (about 420° F in Figure 3.10) where **martensite** is formed. Therefore, the formation of martensite, if cooled rapidly enough, is temperature and not time dependent, as are the other microstructures.

A tool steel having about 1 percent carbon can contain all the carbon at 1500° F (816° C) in the interstices of the FCC lattice of the austenite because there is more open space than in the BCC lattice. There is room for only a very small percentage of carbon in the BCC structure. Therefore, when the steel is suddenly quenched to room temperature, there is no time for the carbon atoms to leave the interstices and form a compound of iron carbide (Fe_3C), which is what happens when steel is slowly cooled. The result of sudden cooling is that the carbon atoms are trapped in the BCC structure and cause the crystal to elongate to form the *body-centered tetragonal* (BCT) unit crystal formation (Figure 3.12) called **martensite**.

Martensite is very hard and brittle and must be tempered to be useful (see the section on tempering carbon steel later in this chapter). The 0.90 percent C steel I-T diagram (Figure 3.11) shows cooling curve No. 2 with a very slow cooling rate, as when steel is annealed to soften it. This produces a soft, coarse pearlitic microstructure in this carbon steel that is easily machined. Cooling curve No. 3 represents a steel that is air-cooled or **normalized** in which mixed microstructures are formed.

Hardening Steels

Plain carbon steel contains no alloying elements other than carbon and small percentages of elements such as manganese that are necessary in steel manufacture. It is used for knives, files, and fine cutting tools such as wood chisels because it will hold a keen edge. The hardness and strength of alloy steels is determined by the carbon they contain, and other elements contribute properties to the steel such as corrosion resistance, and high- or-low-temperature strength. One of the major reasons for using alloying elements is to gain **hardenability,** or as the word suggests, the ability to become hard. When alloying elements are added they usually slow down the rate at which austenite can change into the softer products that result

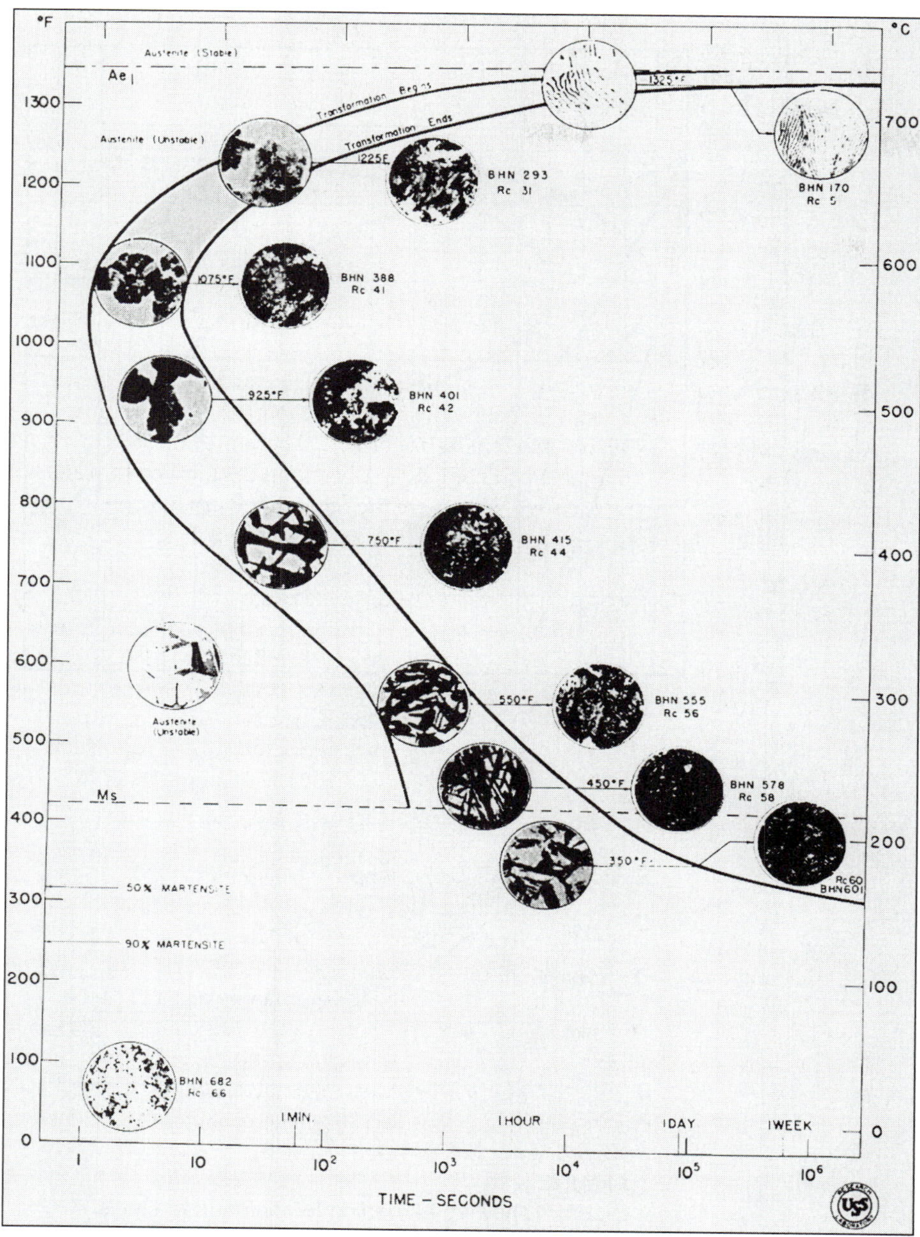

FIGURE 3.10
TTT or I-T diagram of 0.89 percent carbon steel (©United States Steel Corp.).

from cooling, for example, pearlite. The effect of the alloying elements is to move the TTT diagram to the right, giving the steel the time needed to cool to the Ms temperature and transform to martensite.

The process of hardening steel is carried out in two operations. The first step is to heat the steel to a temperature that is slightly above the A_3 and $A_{3,1}$ lines on the iron–carbon phase diagram (Figure 3.13). This operation is called *austenitizing* by metallurgists. The austenitized steel (FCC crystal structure) contains all the carbon in the interstices (spaces or voids in the lattice structure). The second step is to cool the red-hot metal so quickly that it has no opportunity to transform into softer microstructures but still holds the carbon in solution in the austenite. This operation is called *quenching*. **Quenching media,** such as brine, tap water, fused salts, oil, and air, all have different cooling rates. Slower cooling is necessary for tool steels, and rapid rates are needed for plain carbon steel. Rapid quenching can produce cracking in thicker sections and therefore is normally used on small or thin sections with low mass and for plain carbon steels.

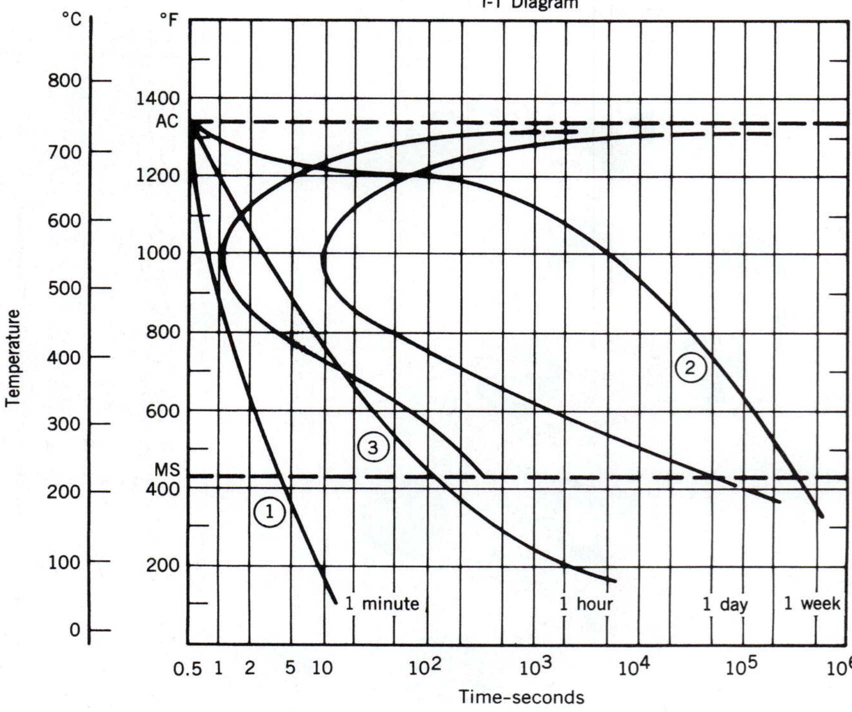

FIGURE 3.11
Three cooling curves on an I-T diagram of 0.90 percent carbon steel.

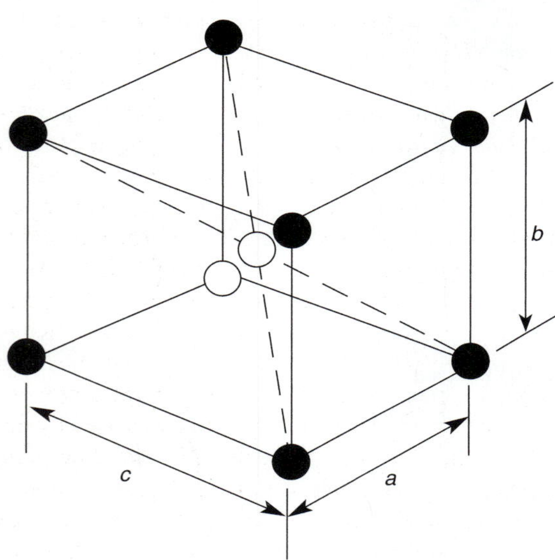

FIGURE 3.12
The body-centered tetragonal lattice of martensite. The atoms shown are iron atoms; the carbon atoms would be between them. The "extra" carbon contained in the FCC lattice of austenite, which cannot be dissolved in the BCC lattice, causes the BCC lattice to elongate, forming a BCT lattice, where $c > a = b$.

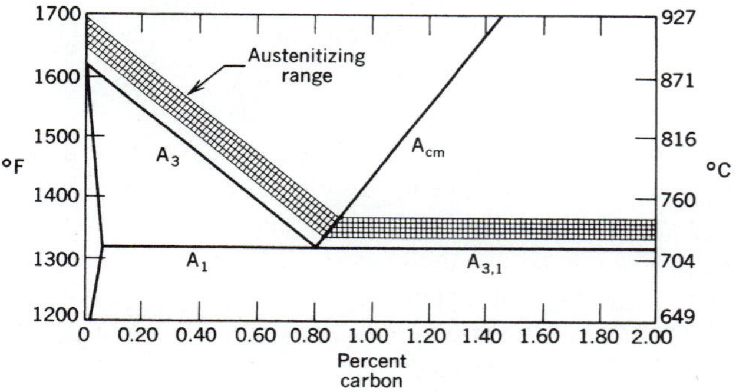

FIGURE 3.13
Diagram showing hardening temperature of steels (austenitizing range) (White, Neely, Kibbe, Meyer, *Machine Tools and Machining Practices,* Vol. II © 1977 John Wiley & Sons, Inc.).

Figure 3.14 is an example of the variables that exist in the quenching process. With its temperature above the A_3 line this gear was lowered into the liquid quenchant, that is, down, from the top of the page. The darkest of the temperature contours represents the depth from the surface that cooled to 600° F in a certain amount of time. Note that on the underside and root of the center tooth this contour is very thin, meaning the cooling was not as effective there, because the vapor, created by the boiling of the quenchant,

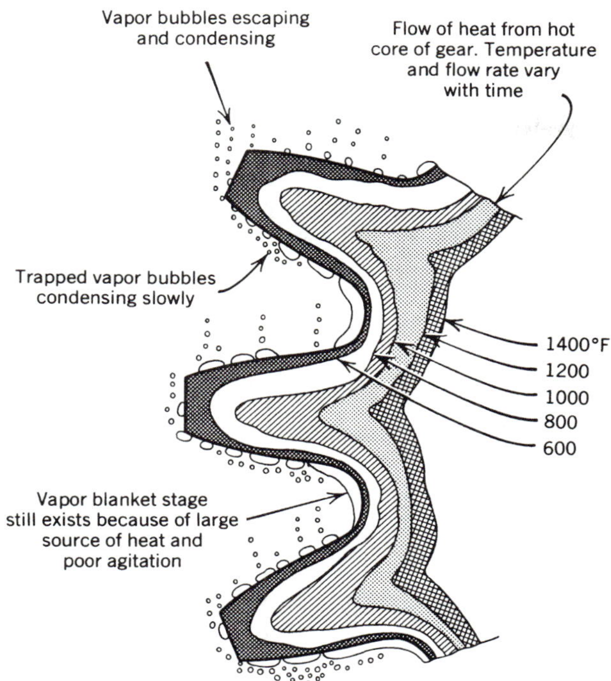

FIGURE 3.14
Temperature gradients and other major factors affecting the quenching of a gear. The gear was quenched edgewise in a quiescent (not agitated) volatile liquid (*Metals Handbook,* Vol. 2, 8th ed., ASM International, 1964, p. 16. With permission.).

was not able to escape and formed a barrier, slowing the removal of the heat. Quenching would have been more uniform if the gear had been lowered into the quenchant with the gear teeth parallel to the vertical. Agitation of the workpiece or quenchant, apparently purposely not done in this case, would have increased the rate of quenching.

The austenitized steel, when quickly cooled to a point below the M_s temperature, attempts to transform to its natural BCC crystal structure, which can contain almost no carbon in its interstices. The result is the distorted, elongated cubic structure of martensite that is very hard (Figure 3.12). Its microstructure is shown in Figure 3.15.

Hardening Cast Irons

In Figure 3.4 it can be seen that in the cast iron region, above the A_{31} transformation line, is an area containing austenite. This austenite can be cooled slowly to form pearlite, or rapidly to form martensite. Thus, all the forms of cast iron (white, gray, ductile, malleable; see Chapter 6) can be produced with the same options as steels. Austenite can also be cooled on a path that will hold it above the Ms temperature and allow it to transform to **bainite**. This

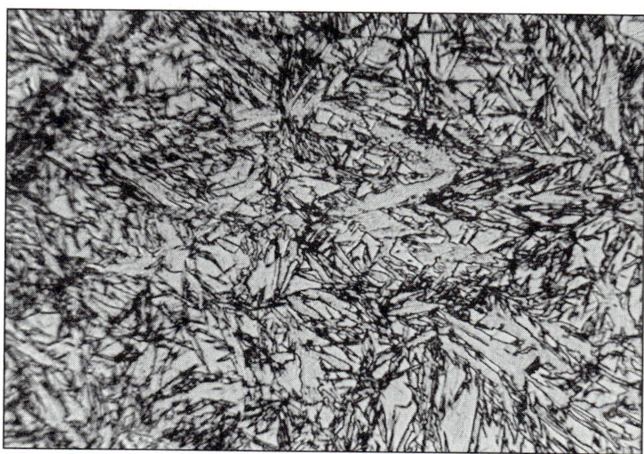

FIGURE 3.15
The microstructure of martensite (500×).

process is known as **austempering** and a currently popular form of cast iron is **ADI**, or **austempered ductile iron**.

Tempering

Martensite, whether in a carbon or alloy steel, is very brittle until it is tempered. A tool that is hardened and not tempered will break in pieces when first used. Tempering involves reheating the hardened steel to a much lower temperature than that used for hardening, but it is more than just a low-temperature anneal. During tempering some of the carbon leaves the BCT martensite lattice and forms a complex carbide. In this process the steel loses some of its hardness, depending on the temperature, and gains toughness, reducing brittleness. Hardening and tempering temperatures are given in Table 3.1. In Table 3.2 it can be seen that the higher the tempering temperature used, the softer the metal becomes. Oxide colors that form on the clean surfaces of steel in a given temperature range show heat treaters the approximate temperature of the metal. This color method of tempering was used by blacksmiths to determine the temperature before they plunged the part into a water tank to stop the heating action. It is still used to some extent in small shops, but more exact methods are used in the manufacturing of heat-treated steel parts.

Surface Hardening

Often, it is desirable to harden only the outer surface of a steel part, or to **surface harden** it to create a hard case around a softer core. A gear, for example, needs to be hard on its surface to resist wear but tough and impact-resistant

TABLE 3.1
Temperatures and colors for heat treating and tempering steel

Colors		*Degrees Fahrenheit*	*Process*	
	White	2500		
		2400		High-Speed Steel Hardening (2250–2400° F)
	Yellow to white	2300		
		2200		
		2100		
	Yellow	2000		
		1900		
	Orange to red	1800		Alloy Steel Hardening (1450–1950° F)
		1700		
Heat Colors	Light cherry red	1600		
		1500		Carbon Steel Hardening (1350–1550° F)
	Cherry red	1400		
	Dark red	1300		
		1200		
		1100		
	Very dark red	1000		
		900		
	Black red in dull light or darkness	800		High-Speed Steel Tempering (350–1100° F)
		700	Carbon Steel Tempering (300–1050° F)	
	Pale blue (590° F) Violet (545° F)	600		
Temper Colors	Purple (525° F) Yellowish brown (490° F)	500		
	Straw (465° F) Light straw (425° F)	400		
		300		
		200		
		100		
		0		

(Pacific Quality Steels, "Stock List and Reference Book," No. 85, Pacific Machinery and Tool Steel Company, 1981.)

TABLE 3.2
Temper color chart

C°	F°	Oxide Color	Suggested Uses for Carbon Tool Steels	
220	425	Light straw	Steel-cutting tools, files, and paper cutters	Harder
240	462	Dark straw	Punches, dies	
258	490	Gold	Shear blades, hammer faces, center punches, and cold chisels	
260	500	Purple	Axes, wood-cutting tools, and striking faces of tools	
282	540	Violet	Springs, screwdrivers	
304	580	Pale blue	Springs	
327	620	Steel gray	Cannot be used for cutting tools	Softer

(Neely and Bertone, *Practical Metallurgy and Materials of Industry,* 5th ed., © 2000 Prentice Hall, Inc.).

in its interior so it can resist sudden and repetitive loads. There are two basic approaches to meeting this requirement: (1) adding carbon at the surface of an otherwise low carbon steel to change the chemistry of the surface and then heat treating the whole gear or (2) starting with sufficient carbon in the steel to achieve the hardness required and heat treating only the outer surface. In the first method the *surface chemistry* of the steel *is changed,* and in the second it is *selectively heat treated.*

Changing the Surface Chemistry **Carburization** has been used for many years as a means of raising the carbon content of the surface. This is done by diffusing carbonaceous or nitrogenous substances into the surface followed by heating and quenching in most cases. Some of these processes are carburizing, **nitriding,** carbonitriding, and cyaniding. Low-carbon steel can be carburized and surface hardened to a depth of about 0.003 in. by heating it with a torch to about 1700° F (927° C) and rolling it in a carbon compound such as Kasenit® followed by reheating and water quenching. In order to harden to 1/16 in. deep, the part must be packed in the carburizing compound and held at that temperature for about 8 hours. Nitriding produces a harder case with a lower temperature and less distortion. Other methods produce a more uniform, harder case than carburizing in a shorter time. Some of the disadvantages to these methods of surface hardening are that the entire part must often be heated and quenched, altering its entire chemical structure. Rising energy costs and the need for increased production efficiency have brought about the development of better surface hardening methods.

Selective Heat Treating of the Surface Although induction hardening has been used for many years to harden small parts or the ways on machine tools (Figure 3.16 and 3.17), newer processes make use of this hardening process in a selective manner so that wear surfaces are hardened only in stripes moving progressively along the surface (Figures 3.18 and 3.19). This is done on flat surfaces, inside cylinders, and for bearing races on shafts. In these quick-heating processes the heated area is self-quenched by the adjacent cold metal, resulting in a shallow hardened area in

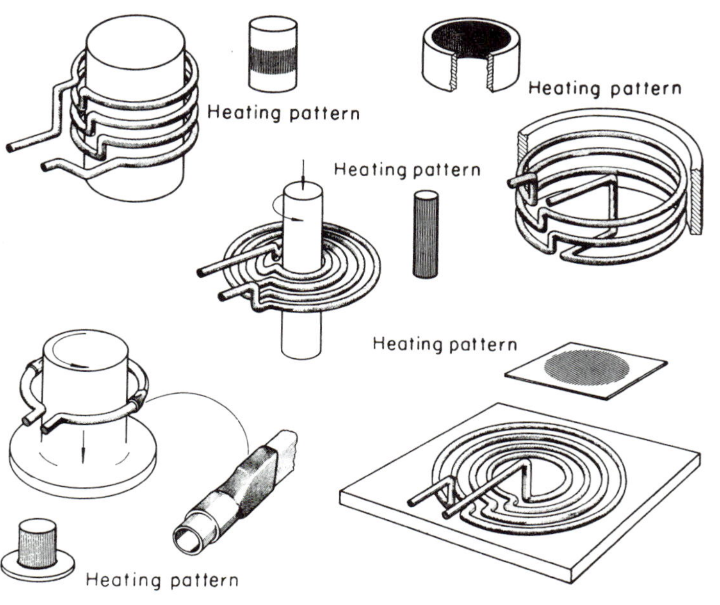

FIGURE 3.16
Induction hardening. Typical work coils for high-frequency units (*Metals Handbook,* Vol. 2, 8th ed., ASM International, 1964, p. 172. With permission.).

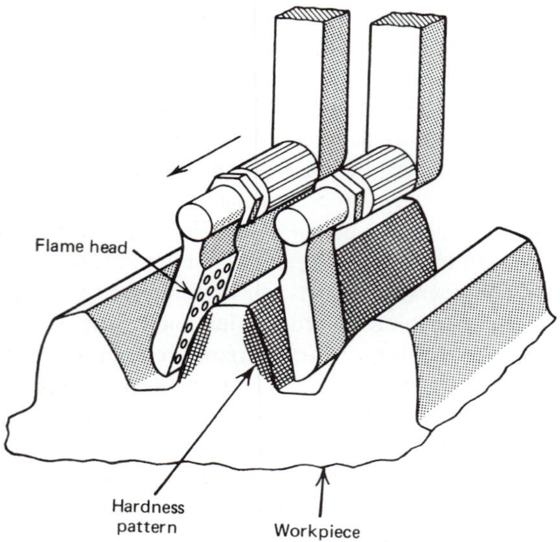

FIGURE 3.17
Hardness pattern developed in sprocket teeth when standard flame tips were used for heating. When space permits this method to be used, hardening one tooth at a time results in low distortion (*Metals Handbook,* Vol. 2, 8th ed., ASM International, 1964, p. 200. With permission.).

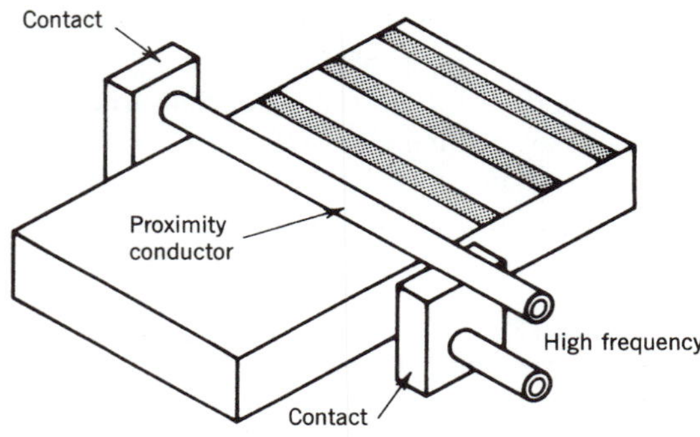

FIGURE 3.18
Selective high-frequency resistance heating on a flat surface. A water-cooled "proximity conductor" is placed close to the surface to be heated, through which a high frequency current is applied.

the form of a line (stripe) or spiral. Electron beam equipment is also capable of producing selectively hardened areas, but it usually is done in a vacuum. Laser systems can operate in ambient conditions for selective heat treating (Figure 3.20). Their primary advantage is accessibility to selective areas on a metal part where other heating methods cannot be easily applied. A longer part life is claimed for laser-hardened parts. One disadvantage to laser hardening is that a nonreflective coating is sometimes needed on the part to be hardened. High-frequency resistance heating is

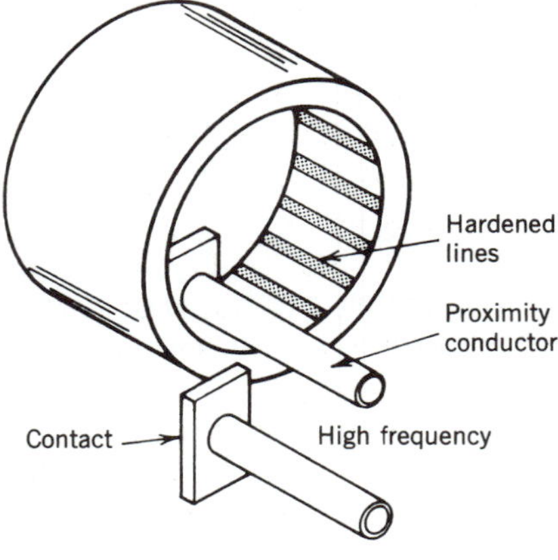

FIGURE 3.19
Internal surfaces are selectively hardened by high-frequency heating.

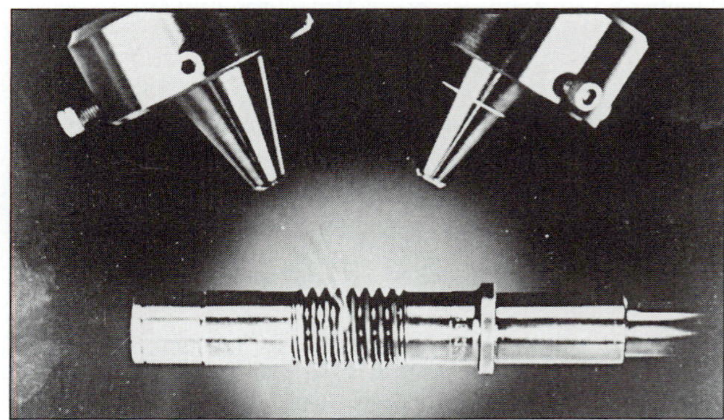

FIGURE 3.20
Laser hardening a worm gear. This is an example of selective heat treating. Instead of the entire part being hardened, which could cause distortion, just the contact faces of the gear are hardened (Coherent, Inc., Palo Alto, CA.).

an alternative method that can produce a hardened line or stripe in one "shot" without the traversing operation. The cycle time is also shorter and is quite adaptable to conventional automation. Tempering is not normally required in surface-hardening processes.

SOLUTION HEAT TREATING AND PRECIPITATION HARDENING (HARDENING NONFERROUS METALS)

Many nonferrous metals are part of alloy systems that are partially soluble and have phase diagrams similar to Figure 3.9. A few ferrous alloys, such as the precipitation hardening stainless steels (see the section on stainless steels in Chapter 6), also use the precipitation mechanism for hardening.

If the alloy is heated from room temperature, where it consists of two phases, to above its solubility limit line (see Figure 3.9), the two phases coexist as a solid solution (this is termed **solutionizing**). If the alloy is cooled rapidly, that is, quenched from that elevated temperature, the major portions (if not all) of the two phases stay in solution at room temperature (the alloy has now been **solution heat treated**). In this condition the alloy violates the equilibrium conditions that should exist at room temperature. As a result, the element or compound that is in solution will begin to come out of solution or *precipitate.* If the precipitation process is carefully controlled, the precipitate will be evenly distributed within the grains of the alloy, interfering with the slip planes and raising the mechanical strength; **precipitation hardening** occurs.

Figure 3.21 illustrates this process; the length of time parts are held at temperature varies with their thickness. Table 3.3 contains data for some frequently used aluminum alloys. Often, fabrication processes are performed between solution heat treating and precipitation hardening. Precipitation hardening is commonly referred to as **aging** because some alloys gain strength while at room temperature (RT), but the process takes days to occur. Thus, the metal appears to gain strength by getting older, thus aging. Note that alloy 2014 requires "artificial" aging at 320° F, whereas 2024 is aged "naturally" at room temperature but requires at least 4 days for enough precipitation to occur to reach a reasonable level of properties. For each alloy there is an optimum temperature and time for the aging treatment. Generally, lower temperatures and longer times produce the highest mechanical strength. Often, other considerations such as improving corrosion resistance may dictate a practice that produces lower mechanical strength.

Some copper, nickel, magnesium, and titanium alloys, as well as alloys of some other metals, can be hardened by precipitation heat treatments.

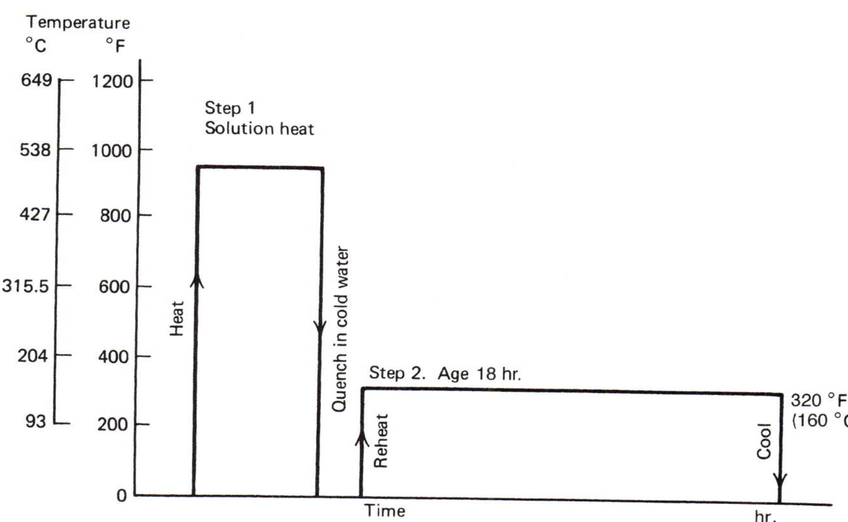

FIGURE 3.21
Process of solution heat treating and artificial aging of 2014-T6 aluminum alloy. (Adapted from Neely and Bertone, *Practical Metallurgy and Materials of Industry,* 5th ed., © 2000 Prentice Hall, Inc.).

TABLE 3.3
Solution heat treatment and precipitation hardening practices for some hardenable commercial aluminum alloys

Alloy Designation	*Solution Heat Treating Temperature (° F)*	*Precipitation Hardening (Aging)*		*Alloy-Temper Designation*
		Temperature (° F)	*Time (hrs)*	
2014	935	320	18	2014 T6
2017	935	RT	96	2017 T4
2117	935	RT	96	2117 T4
2024	920	RT	96	2024 T3
6061	985	320	18	6061 T6
		340	8	6061 T6
7075	900	250	24	7075 T6

(*Metals Handbook,* Vol. 4, 1991, 9th ed. ASM International, pp. 845–7. Materials Park, OH.)

STRENGTHENING BY PLASTIC DEFORMATION AND ALLOYING

Plastic Deformation

Two of the most important properties of metals for manufacturing purposes are elastic deformation and plastic deformation. Stress (load/area) is a material's resistance to the applied load or force. When metals are initially placed under a tension, torsion, or compression stress, a slight stretch or compression takes place in the crystal lattice. This movement is called **strain.** This elastic distortion of the crystal lattice is illustrated in Figure 3.22. Elastic deformation is not permanent, so as soon as the stress is removed, the structure returns to its former shape. Also, its mechanical properties remain the same as before it was deformed.

Figure 3.23*a* shows atoms with shear forces applied. In the case of elastic deformation the upper row of atoms can be pictured as *starting* to slide or slip over the bottom row, but not getting over the next atom, and then returning to its original location when the stress is removed. If, however, a larger stress is applied, the upper row can slide for some distance from its original position; in the process the mechanical strength of the material will increase. This phenomenon is termed **plastic deformation,** and it tends to be favored along planes of highest atomic density (close-packed planes) and the greatest parallel separation (Figure 3.23). It can be seen in this figure that the more dense or closely packed the atoms are, the better opportunity there is for slip along planes or rows of atoms. Forging, drawing, forming, extruding, rolling, stamping, and pressing all involve plastic deformation. Plastic deformation in metals can take place only at a stress or load higher than the elastic limit (Figure 3.24). As the applied load is increased the atoms will slide over each other and produce a permanent change in shape, eventually breaking the atomic bonds, resulting in total rupture. These deformation processes and their nomenclature as related to the tension test are described in more detail in Chapter 2.

The behavior of metal crystals under load depends on a number of factors:

- the interatomic bonding strength;
- irregularities in the lattice—vacancies and discontinuities; and
- the lattice type.

The third factor, lattice type, determines two other very important factors:

- the density of the atoms in the atom planes of the lattice and
- the space or distance between the planes of atoms in the lattice.

As depicted in Figure 3.25, slip occurs within the grain between the adjacent parallel planes of the lattice; it does not occur along grain boundaries between grains. It can be shown that shearing forces are maximized on a plane that is 45° to applied tensile forces. Therefore, since slip occurs

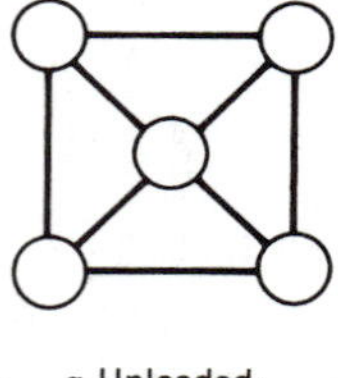

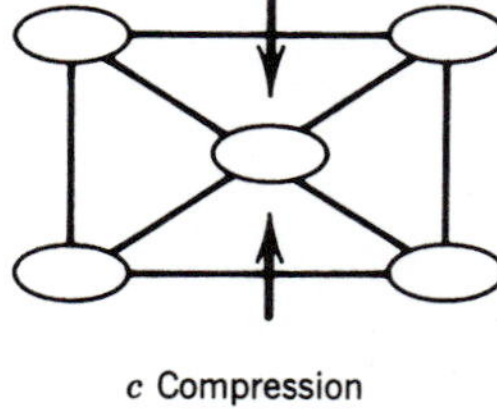

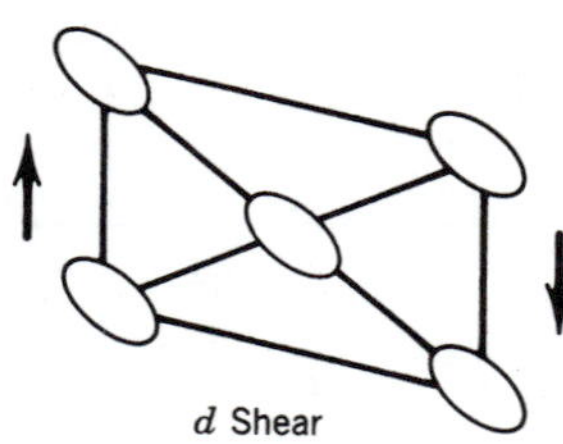

FIGURE 3.22
Distortion of crystal lattice under various kinds of elastic stresses. *(a)* Unloaded. *(b)* Tension. *(c)* Compression. *(d)* Shear.

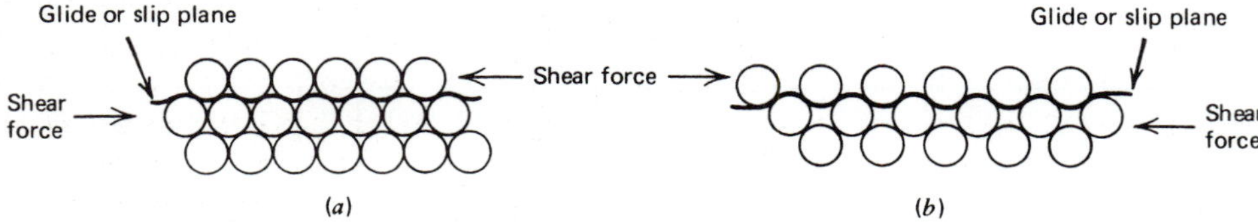

FIGURE 3.23
Rows of atoms can glide along slip planes as shown. BCC and FCC lattices have rows of atoms closely spaced as in *(a),* making movement easier than in *(b).* CPH lattices are less dense and thus less ductile, since many slip planes, as in *(b),* are spaced closer (Neely and Bertone, *Practical Metallurgy and Materials of Industry,* 5th ed., © 2000 (Prentice Hall, Inc.).

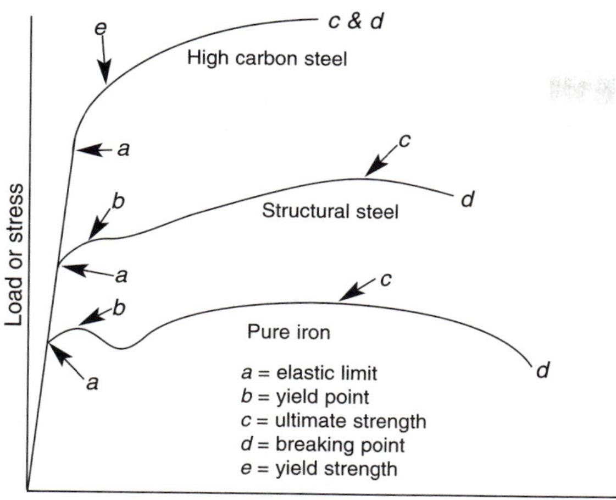

FIGURE 3.24
Load-deformation (stress–strain) diagrams for ductile steels.

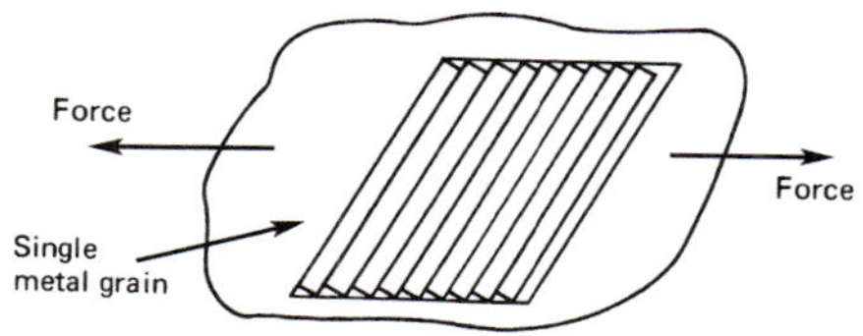

FIGURE 3.25
When stress is applied to a metal grain above its elastic range, shearing stresses cause slip to occur, and the grain flattens and becomes elongated. As stress increases, planes that did not slip initially may rotate so that they are favorably oriented (Neely and Bertone, *Practical Metallurgy and Materials of Industry,* 5th ed., © 2000 Prentice Hall, Inc.).

due to shearing forces, slip will first occur on those planes that are at or near 45° to the applied tensile force as illustrated in Figure 3.25. Figure 3.26 is a photomicrograph of a specimen that was plastically deformed after it was polished. The parallel lines seen within the grains are the edges of slip planes and are evidence that there was physical movement of the atom planes.

If planes of atoms are going to slide, they must have somewhere to go. The rows of atoms fill in defects in the atom lattice that remain from the original solidification of the metal; these defects are called **vacancies** (where one atom is missing) and **dislocations** (where a large number of atoms are absent or "dislocated"). Thus, the atoms do not have to push whole rows or planes of atoms, but just fill in empty spaces. As the spaces are filled in, however, more and more atoms will have to be pushed to the next open space, so the force required increases. This process is what is known as **work hardening** or **cold working.** The more a metal is plastically

FIGURE 3.26
Slip planes as they appear in a micrograph resulting from the cold working of low-carbon steel (*Metals Handbook,* Vol. 7, 8th ed., ASM International, 1972, p. 135. With permission.).

deformed the more force is required to continue the deformation. This relationship can be seen in the shape of the curves of Figure 3.24; the force or stress required increases up to a maximum, from point *b* (or *e*) to point *c*. At point *c* there are no more open spaces to push atoms into and the metal begins to deform locally, the force (and stress) decreases from point *c* to *d,* and the metal fails, or ruptures at *d.*

Work hardening or strain hardening is a problem to be dealt with in manufacturing where cold working processes are involved. After a certain amount of cold working, a metal must be subjected to a process called annealing to restore its grain structure to a softer, more plastic condition. This topic will be discussed later in this chapter.

Grain Size

Because slip occurs within the grains, it can be understood that large-grain metals have a greater capacity to undergo slip, to be deformed, and have more plasticity than fine-grain metals. A fine-grain steel, for example, is stronger than an equivalent coarse-grain steel; that is, it can withstand greater stresses without being permanently deformed. Since each grain has a random orientation (direction of dendrite growth during solidification), slip in each grain ends at the grain boundary; small grains have a lesser capacity for slip than large grains do. For this reason, grain size is an important principle to consider when describing the nature of metals. Steel manufacturers provide mechanical properties charts and tables for each carbon, alloy, or tool steel that give, among other specifications, the grain size in terms of standard grain sizes (Figure 3.27). Ordinary machine steel may have a grain size of about 6 to 8. When a specimen of steel is fractured, the broken cross section often reveals the grains

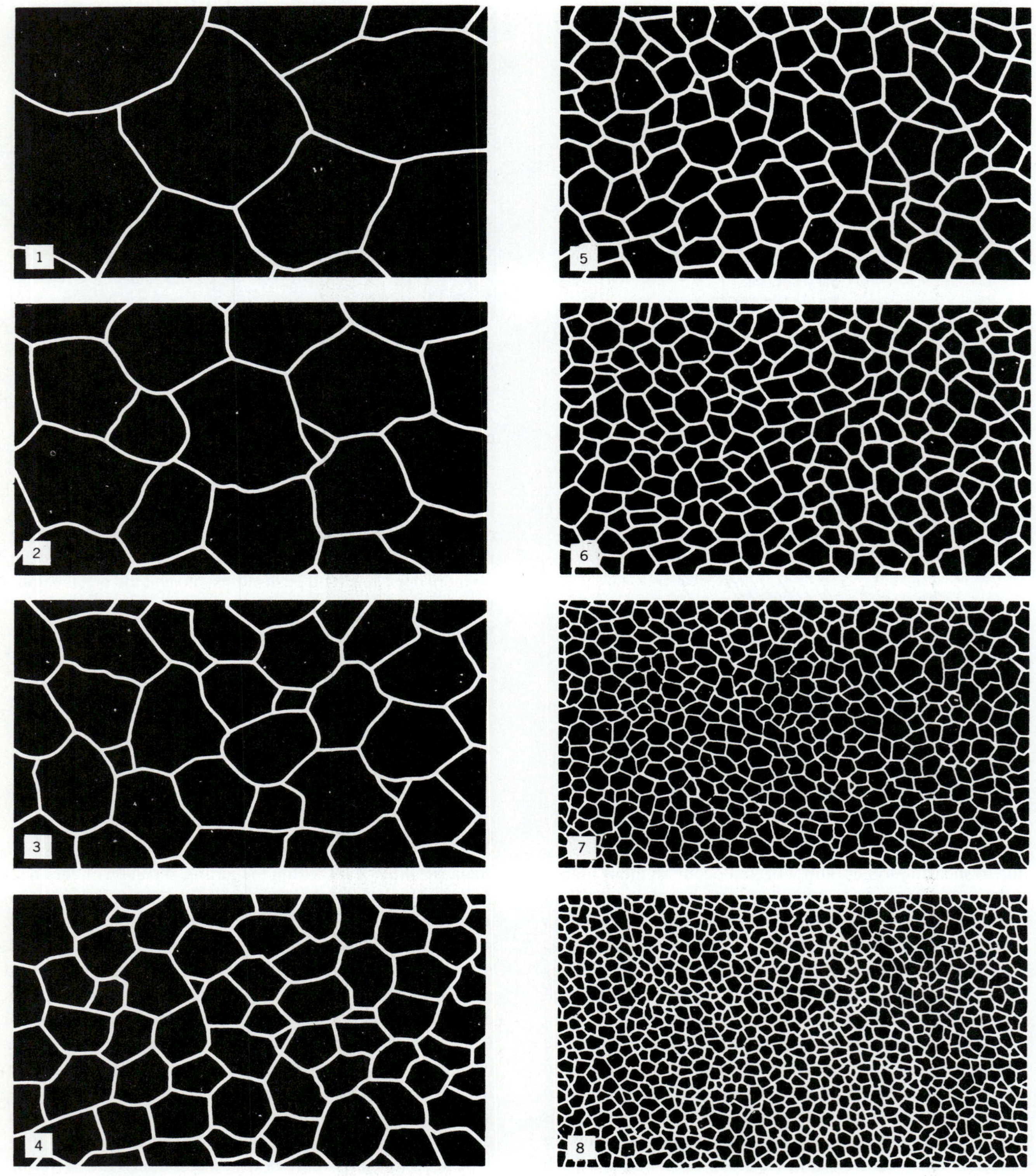

FIGURE 3.27
Standard grain size numbers. The grain size per square inch at 100× (Courtesy of Bethlehem Steel Corporation).

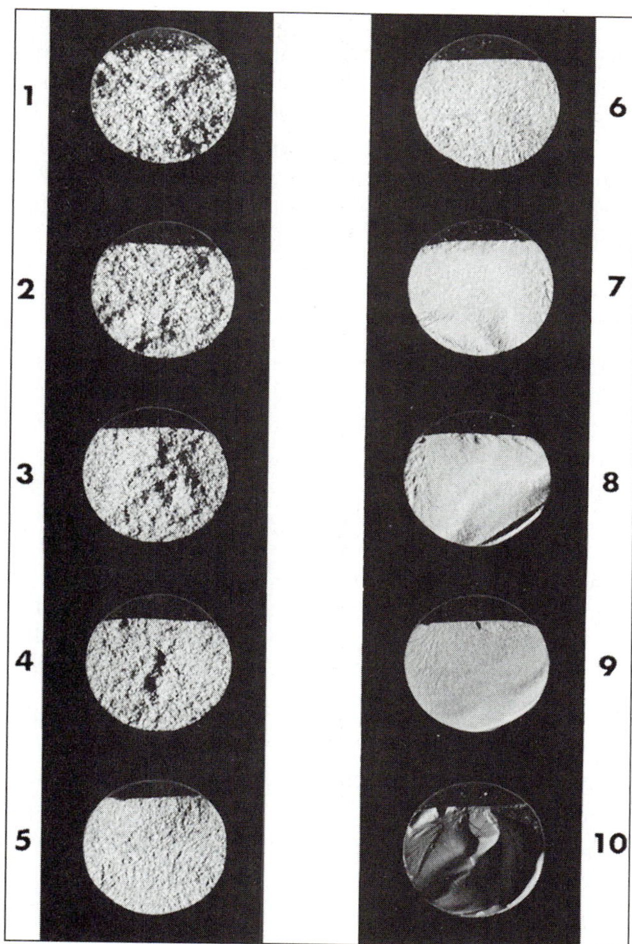

FIGURE 3.28
Shepherd grain size fracture standards. The fracture grain size test specimen can be compared visually with this series of 10 standards. Number 1 is the coarsest and number 10 is the finest fracture grain size (Photo courtesy of Republic Steel Corporation).

to the naked eye. These can be compared with the Shepherd grain size fracture standards; number 1 is coarsest and number 10 is finest (Figure 3.28).

Alloying

Metals may also be hardened by blocking the slip planes with atoms of other elements or compounds, by alloying. The diameters of the atoms of two metals can vary by as much as ± 14 percent and the two metals will still have some solubility. Although these differences may seem small, the combining of two such atoms in the same lattice can double the strength of the alloy. This phenomenon is referred to as **solid solution strengthening.** Such an increase is not as dramatic as what can be accomplished with heat treating, but alloying can still improve properties enough to make some alloys useful as engineering materials. For example, the alloy monel (2/3 nickel, 1/3 copper) finds many applications in which corrosion resistance is important. Because nickel and copper are completely soluble in each other the alloy cannot be heat treated, but combining the two metals increases the tensile strength well above the strength of either metal alone. Figure 3.29 illustrates how atoms of differing sizes interfere with slip within the lattice.

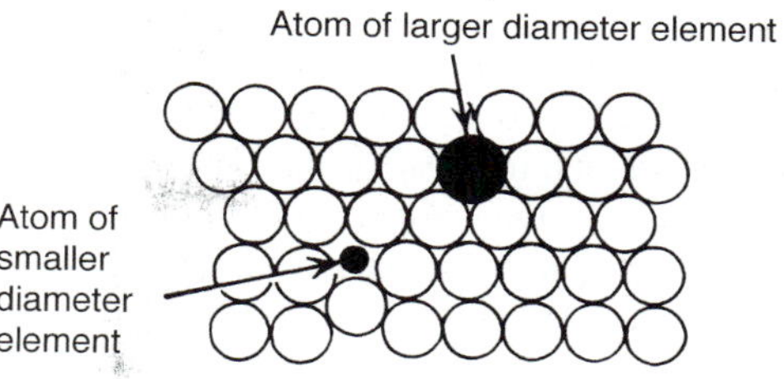

FIGURE 3.29
Atoms of different diameters can interfere with slip (Neely and Bertone, *Practical Metallurgy and Materials of Industry,* 5th ed., © 2000 Prentice Hall, Inc.).

ANNEALING

The term **annealing** refers to any one of several thermal processes: **stress relieving, process annealing, normalizing, full annealing,** or **spheroidizing.** In general, the purpose of these processes is to return a metal to a softer, more workable condition than before the treatment. Compared with heat treating annealing involves much slower cooling rates; in effect it is the opposite of heat treating. It should be appreciated that metals do not have to undergo one of these controlled thermal treatments for the effects to occur. That is, if a part is heated to cure an epoxy adhesive, and the temperature and time are sufficient, then the part can experience stress relief. Also, a heat-treated part may be fully annealed if it is welded.

Metals go through three stages in turn as they are heated at increasing temperatures: stress relief, **recrystallization,** and **grain growth.** These stages occur in all metals, ferrous and nonferrous. We shall therefore consider these stages first and then apply that knowledge to special applications with *ferrous* metals including normalizing, **full annealing,** and **spheroidizing.**

Stress Relief

As its name suggests the stress relief process requires that the metal has experienced a forming or heating process (for example, welding) that has left behind stresses that are called **residual stresses.** Such stresses are not the stresses that produced the plastic deformation of a rolling or forging process, for example, but rather these are *elastic* stresses left residing in the metal by these operations. Please refer to the stress–strain diagrams of Figure 3.24 during the following explanation. (Although these diagrams are for tensile forces, and the rolling and forging operations mentioned are compressive, the discussion is still valid.) At a low level of stress the loading is within the elastic region; as the stress is increased the metal moves through the elastic region and begins to become plastic, which on these diagrams means the stress is just above *b* (or *e*). Assume that the process being used causes deformation equivalent to moving to the right 1 in. on the diagram. This would represent rolling or forging of the metal. When the rolling or forging force is removed the stress is also removed, corresponding to zero on the diagram. This means the overall stress level of the metal is zero, but the metal is made up of many grains, whose atom lattices were randomly oriented relative to the forces applied during rolling or forging, and thus grains received varying amounts of cold work. Therefore, although the overall stress is zero, every grain is not at a zero stress level; these are the residual stresses. They are elastic stresses; they are still alive and can cause distortion and warping during later operations such as machining, and they can contribute to corrosion.

Stress relief is often needed for castings and weldments. Large welded structures such as tanks are sometimes stress relieved by covering them on the outside with thermal insulation blankets and heating them on the inside with propane burners.

Thermal stress relief is preferred for most manufacturing processes; however, vibratory stress relief (VSR) is often used for cast or welded structures that are too large to fit into a heat-treat furnace. To use VSR effectively, there must be

1. loading of a structure by means of resonance by close control of a vibrator's frequency,
2. proper instrumentation to display the pertinent VSR data.

In order to relieve stresses using VSR, the entire part must be made to resonate, and the induced vibration must be very near to its natural resonant frequency that is generally in the 50-Hz to 100-Hz range. The dynamic loading causes plastic flow where any concentration of high-level residual stress is present, such as in welds. Flexure at these places combined with the residual stress is sufficient to momentarily exceed the yield point. Electrically powered portable vibrators are attached with clamps to the part that is to be stress relieved and, in addition, elastic isolation cushions support the part at strategic locations. It seems that there is no upper limit on size for this stress-relieving technique.

Welds are sometimes shot-peened to relieve residual weld stresses. The difference compared with other methods is that shot-peening induces surface compressive stresses to counteract the tensile stresses in the weld.

Stress-relief anneals are performed at relatively low temperatures, with each metal having its own range. If done correctly stress relief anneals *will not* reduce properties such as hardness and tensile strength.

Recrystallization

Recrystallization takes place when cold-worked metals are heated to their specific recrystallization temperatures (Table 3.4). The stored energy from cold working combines with the heat energy of the annealing furnace, enabling small nucleating sites to form that contain unstrained atom lattices. With time additional atoms form up on these lattices, and gradually the whole cold-worked structure is replaced with a new "recrystallized" structure. Because the number of nucleating sites is determined by the amount of cold work, highly cold worked metals will have the smaller grains after annealing. The following factors are important in recrystallization:

1. A minimum amount of deformation is necessary for recrystallization to occur, regardless of the temperature.
2. Similarly, a minimum temperature is required for recrystallization to occur regardless of the amount of cold work present.
3. The larger the grain size before cold working the greater the amount of cold work, or temperature, is required to cause a given amount of recrystallization.

TABLE 3.4
Recrystallization temperatures of some metals

Metal	*Recrystallization Temperature (° F)*
99.999% aluminum	175
Aluminum bronze	660
Beryllium copper	900
Cartridge brass	660
99.999% copper	250
Lead	25
99.999% magnesium	150
Magnesium alloys	350
Monel	100
99.999% nickel	700
Low-carbon steel	1000
Tin	25
Zinc	50

(Neely and Bertone, *Practical Metallurgy and Materials of Industry,* 5th ed., © 2000 Prentice Hall, Inc.).

4. Increasing the time of anneal decreases the temperature necessary for recrystallization.
5. The recrystallized grain size depends mostly on the degree of deformation and, to some extent, on the annealing temperature.
6. Continued heating after recrystallization (re-forming of grains) is complete increases the grain size.
7. The higher the temperature at which the cold work is done, the larger the amount of deformation required to cause an equivalent percentage of cold work.

Recrystallization anneals are often used to return a highly cold worked metal to a softened state so it can be additionally cold worked. Products such as thin-wall, cold-drawn tubing and thin-gage sheet (both ferrous and nonferrous) often go through multiple sequences of working and annealing to get to very small dimensions. These anneals are commonly referred to as **process anneals** since they are done within the process as opposed to being a final operation. Products produced by cold-working processes may achieve their final properties by

- being cold worked to the desired level of properties; or
- being cold worked and then given a **partial anneal** that reduces properties to a desired lower level; or
- being cold worked and then given a **full anneal** that reduces strength properties to their lowest level.

Grain Growth

In performing such annealing treatments it is important to avoid heating for too long a time and/or at too high a temperature that causes the metal to go into the third stage of heating—grain growth. The large grains that result improve a metal's ductility but may cause the surface to be roughened in a condition called **orange peel.** Full annealing, with some amount of grain growth, is sometimes done to facilitate a difficult forming operation, in which case the orange peel will probably be removed. Full annealing is usually accomplished by cooling the metal from the annealing temperature in the furnace with the doors closed to achieve a very slow cooling rate. The resulting metallurgical structure is very similar to what is predicted by equilibrium cooling on the phase diagram.

Application to Ferrous Metals

Ferrous metals require slightly different slow cooling thermal treatments because these metals are allotropic. That is, for recrystallization to occur the temperature has to be raised only above the transformation lines; however, the recrystallization-type anneals discussed earlier are used with steels. In steel cold working it is typical to use process anneals at temperatures *below* the transformation temperatures, using the stored energy from the cold work to initiate recrystallization (Figures 3.30 and 3.31).

The importance of the transformation temperatures can be seen in Figure 3.32, which shows three treatments that more or less fall in the category of full annealing, but each does something a little different. First, it is important to remember that at room temperature the carbon in slow-cooled steels is in the form of hard iron-carbide called cementite.

Full Annealing The microstructure of hypoeutectoid steels consists of soft ferrite and pearlite. Heating to above the A_1 line causes only the pearlite (which is ferrite plus cementite) to transform to austenite. In order for the remaining ferrite to be changed to austenite the steel must be heated above the second transformation line, A_{31}. The requirement for hypereutectoid steels is somewhat different since their microstructure consists of pearlite plus more cementite; that is, as soon as the temperature reaches A_{31} the ferrite transforms to austenite. Thus, on the hypereutectoid side it is necessary only to heat slightly above A_{31}. If a hypereutectoid steel was completely transformed to austenite (heated above A_{cm}) the cementite would be dissolved in the austenite, so if it was slow cooled it would form very hard cementite structures around the austenite grains and would often not be machinable. Spheroidizing, to be discussed shortly, is a solution to the machinability problem for high-carbon steels.

FIGURE 3.30
Microstructure of 0.10 percent carbon steel, cold rolled. Both ferrite and pearlite grains have been flattened by plastic deformation (*Metals Handbook,* Vol. 7, 8th ed., ASM International, 1972, p. 10. With permission.).

FIGURE 3.31
The same 0.10 percent carbon steel as in Figure 3.30 but annealed at 1025° F (552° C), the recrystallization temperature of ferrite. Ferrite grains are mostly re-formed to their original state, but the pearlite grains are still distorted because the A_1 transformation temperature was not reached (1000×) (*Metals Handbook,* Vol. 7, 8th ed., ASM International, 1972, p. 10, With permission.).

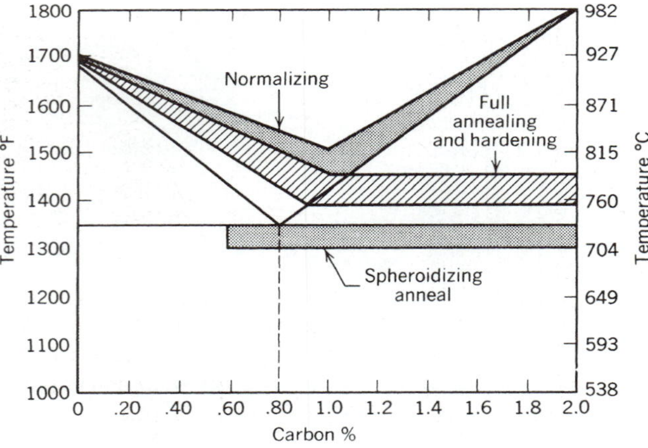

FIGURE 3.32
Temperature ranges used for heat treating carbon steel (*Source:* White, Neely, Kibbe, Meyer, *Machine Tool Practices,* 7th ed. 2002 Prentice Hall, Inc.).

Normalizing Normalizing is a treatment that requires heating both hypo- and hypereutectoid steels above the upper transformation lines and then air cooling. This is the condition that would "normally" occur with steels that were hot worked, say rolled, and left to cool in the air. On the hypoeutectoid side this treatment is just like full annealing except for the cooling rate. The normalized structure will contain ferrite and pearlite, but the pearlite laminations will be finer than with full annealing. On the hypereutectoid side the air-cooling reduces the amount of cementite that forms at the austenite grain boundaries and also causes the pearlite to have finer laminations as compared with full annealing.

Lower-carbon steels, between 0.30 and 0.60 percent carbon, that have been hardened can be normalized to soften them sufficiently for machining. The normalizing process does not completely soften these steels but leaves them just hard enough for efficient machining. A metal that is too soft will not machine well because it is too "gummy," leaving a rough torn surface finish, whereas a somewhat harder metal will have a much smoother surface. Normalizing is also used to refine castings and forgings, since it recrystallizes the coarse, irregular grains to produce a stronger, fine-grained metal.

Spheroidizing To make higher-carbon steels more machinable, something must be done about the rings of hard cementite at the grain boundaries and the sheets of hard cementite present in the pearlite. Spheroidizing is a treatment that can be of help. As the name suggests, the cementite is formed into spheroids or balls that are much more easily machined than the rings and sheets on the normal structure. In effect, the cementite becomes hard particles in a matrix of soft ferrite; the tool can now chip them out rather than try to cut through them.

To accomplish spheroidizing the steel is held at temperatures near the A_1/A_{31} temperature for a period of about 24 hours (Figure 3.31). Some treatments cycle the temperature above and below the transformation line. The result of the procedure can be seen in Figure 3.33.

HEATING EQUIPMENT

Almost any method can be used that will heat a metal to the temperature required for the various heat treatments. Formerly, blacksmiths used a charcoal fire to heat their metal before hand hammering it on an anvil. In an earlier section we saw that certain surface hardening methods use a torch, but more often today heating is done where the temperature can be very accurately controlled.

Annealing equipment such as that shown in Figures 3.34 and 3.35 can be used to anneal, normalize, harden, or temper relatively small steel tools and dies. When furnaces are to be used in the production of large volumes of product, their size and complexity increases. Figure 3.36 shows a furnace being used to homogenize cast aluminum "logs" that will be used to produce extrusions. This process requires heating the metal to a temperature of about 1000° F (538° C) for 10 to 20 hours.

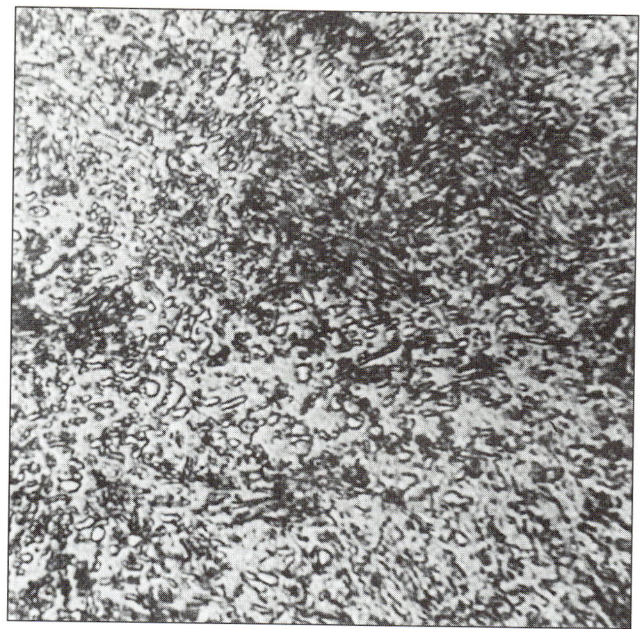

FIGURE 3.33
Micrograph of spheroidized carbon steel (500×).

FIGURE 3.34
A small furnace used for annealing, hardening, or tempering steel tools and dies. (Bureau of Mines).

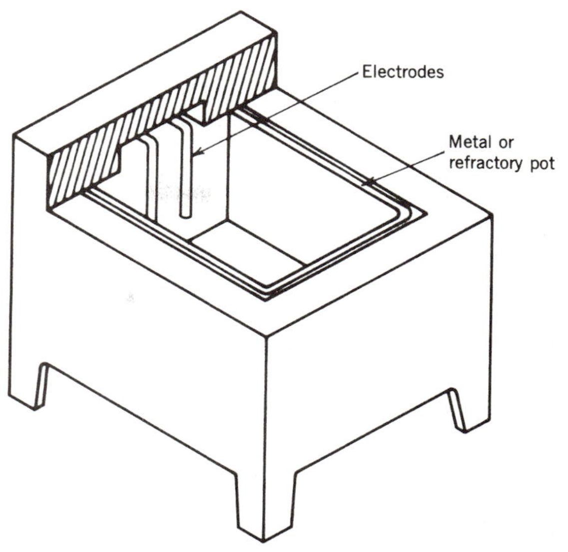

FIGURE 3.35
Drawing of one type of salt bath furnace. Electrodes immersed in the salt produce the heating effect. Parts are heated by submerging them in the molten salt.

FIGURE 3.36
As-cast aluminum logs waiting to go into a homogenizing furnace. The logs will later be cut into lengths for use as extrusion ingots. Furnaces of this type can be used for many purposes, depending on their heat input and interior structure (Courtesy of SECO/WARWICK Corp.).

A furnace of similar configuration but with the proper temperature capability could be used, for example, to anneal fabricated aluminum parts, to age or precipitation harden aluminum products, or to anneal coiled steel sheet. The capability of the furnace is determined by the materials used to build it and the heating equipment provided.

Figure 3.37 is a schematic representation of a continuous furnace. It is continuous in that parts to be heated are put on a conveyor at one end of the furnace and are carried through the heating zone and out the other end. The

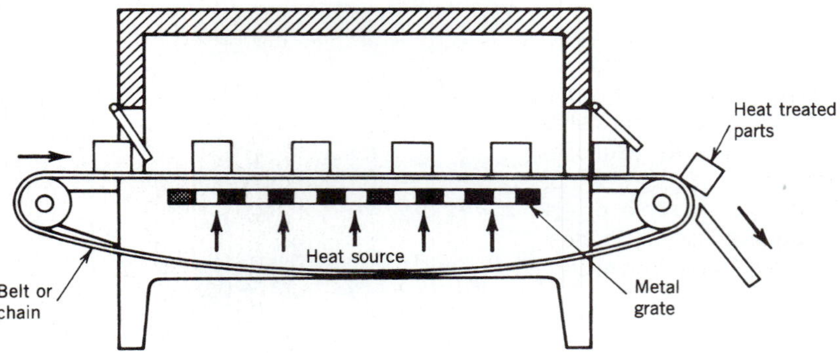

FIGURE 3.37
Continuous heat treating furnace.

FIGURE 3.38
Vertical solution heat treating furnace used for maximum 34-ft-long aluminum extrusions. The quench tank extends below floor level (Courtesy of Consolidated Engineering Company).

FIGURE 3.39
A furnace capable of heat treating long extrusions horizontally. The parts will enter the furnace through the door at this end of the furnace and exit through a door at the other end before passing through a water spray for quenching (Courtesy of SECO/WARWICK Corp.).

temperature setting of the furnace and the speed of the conveyor must be coordinated so the metal is at temperature for the required length of time. Furnaces of this type are often used for annealing of many metals and austempering of steel.

Sometimes the unique aspect of a furnace arrangement is the way it facilitates quenching, as in solution heat treating aluminum products. Figure 3.38 shows a vertical furnace; aluminum extrusions are fastened to a lifting device, pulled up inside the silo-like furnace, heated to their solutionizing temperature, and lowered (rapidly) into a water tank in the floor. Such a process works well if the length of the product is reasonable, but aircraft wing spans over 100 ft in length require a horizontal furnace. In Figure 3.39 long extrusions wait to enter the heating chamber at the rear of the picture. The metal will be held for the proper length of time at temperature in the heating chamber and then exit the furnace out the other end before passing through a spray quench chamber. This furnace operates somewhat as shown in Figure 3.37, but in a semicontinuous way rather than continuously.

FIGURE 3.40
By assembling heating, cooling, and quenching equipment into an integrated unit this company normalizes, cools, heat treats, quenches, and tempers 25,000 lb/hr of forged crankshafts and suspension parts (Courtesy of Consolidated Engineering Company).

When large volumes of product are to be heated for whatever reason the thermal treatments may be integrated with other operations. Figure 3.40 is a picture of a unit capable of processing 25,000 lb per hour of crankshaft and suspension forgings. The unit is capable of normalizing, slow cooling, heating and quenching to harden the steel, and then tempering it. Note the presence of many conveyors to move the product almost continuously through the facility.

All heating furnaces are built of materials that resist the temperatures at which they operate. Furnaces that are designed for 1000° F and less are generally made of steels, with the higher end of the range requiring stainless steels and other heat-resisting steels. Above 1000° F refractory materials such as ceramics are required. Electronic systems are used to control the furnace at the set temperature, with the temperature usually being sensed by a bimetallic thermocouple (Figure 3.41).

FIGURE 3.41
Thermocouple sensor. The twisted wires of a thermocouple produce a feeble electric current when heated. Furnace temperatures are controlled from this sensor.

Review Questions

1. To what temperature should hypoeutectoid (less than eutectoid) plain carbon steel be heated for hardening purposes according to the iron–carbon diagram?
2. If a 0.80 percent plain carbon steel is austenitized (heated to a red heat) and quenched in water to room temperature so that the cooling curve does not cut into the "nose" of the I-T diagram, what will the resultant microstructure be?
3. On the right side of the iron–carbon diagram, what kind of metal is found when the carbon content is over 2 percent?
4. State the difference between the eutectic and eutectoid points.
5. What happens to a quench-hardened tool or machine part that was not tempered when it is subjected to shock loads or bending stress?
6. What does the process of temper drawing do for hardened carbon steel?
7. Name two kinds of equipment that are used for heating steel to a precise hardening temperature.

8. How can a steel that contains insufficient carbon to be hardened be surface hardened?
9. By what means can a small area or stripe be selectively hardened on a large steel part?
10. How can some nonferrous metal alloys such as aluminum or monel be hardened?
11. What annealing method should be used for softening a hardened high-carbon tool steel?
12. A low-carbon steel part has been partly formed toward its final shape and now is in a work-hardened condition. Any further cold working will cause it to tear. At this point, heating to a red heat will distort it too much to recover its shape with further work. How can this part be restored to a soft condition so it can be cold worked to its final shape?

Case Problems

Case 1: The Case of the Badly Worn Pin

The soft steel pin in Figure 3.42 was removed by a millwright in a manufacturing plant from a hydraulic cylinder clevis at the end of a cylinder rod. It was originally a straight cylindrical pin, but it wore almost completely through before a mill shutdown allowed it to be removed. The holes in the clevis, which is made of soft cast steel, were also elongated accordingly. The clevis was periodically lubricated but the extreme forces still caused excessive wear. Considering what you have learned in this chapter, what could be a possible solution to this problem of excessive wear?

FIGURE 3.42
See Case 1.

Case 2: Strength versus Weight Problem

A structure for an aircraft part (made of 1000 series aluminum) has a certain weight limitation. After all weight reduction by engineering design had been done, the part was still too heavy. No metal other than aluminum is permissible in this case. How can the part weight be further reduced and still have sufficient strength?

CHAPTER 4

Mining and Extraction of Metals

Objectives

This chapter will enable you to:

1. Describe methods of mining and processing metallic ores.
2. Give an account of the development of and methods used in the ironmaking and steelmaking processes.
3. Explain some of the principles and methods used in smelting several nonferrous metals.
4. Show the construction and principles of operation for several types of furnaces and refining vessels.

Key Words

ferrous	ore
conferrous	extraction
reduction	pig iron
killed steel	basic oxygen furnace (BOF)
rimmed steel	

Metallic ores and other important minerals are found on virtually every continent; however, there is an unequal distribution of many of these minerals. For example, tin (Sn) is a metal that is abundant only in Africa, Bolivia, and Southeast Asia and is found in a few other locations that are not major producers. Chromium ore is extensively mined in South Africa and in the former Soviet Union; there are only minor sources of this strategic metal in the United States, so most of our chromium is supplied from these two areas.

Our modern civilization is probably more dependent on strategic metals than on oil. These metals include cobalt, chromium, manganese, platinum, niobium, strontium, titanium, and tantalum. Iron and aluminum are relatively plentiful, but even these materials are affected by the increasing cost of energy.

Mining originated in prehistoric times. Some metals were found in small quantities in their "native" or metallic state. Native copper, silver, and gold were found in streambeds and in outcroppings. These were shaped by hammering to make jewelry and crude tools. The need for larger quantities of metals brought about underground and surface mining methods that have been considerably improved in modern times.

In Chapters 1 and 3 we learned about the atomic structure of metals and how metals can be modified by heating and working. In Chapter 2 we studied how metals can be tested to verify the properties generated by the modifications performed in Chapter 3. Now that we know something about what metals are, in this chapter we will study how the metal itself is produced. The starting point is an ore that has to be obtained by some process, then usually treated. Operations are then performed to separate the metal from an undesired chemical or chemicals. Finally, a metal is obtained, often in a molten state, that can be used to produce a product.

The discussion will center on iron and steel, the metals we refer to as **ferrous** metals. We will also cover aluminum, copper, magnesium, lead, and zinc, as well as a number of the space-age metals, and all of these are **nonferrous** metals.

ORES AND MINING

An ore is any geological deposit of nonrenewable material having a sufficiently high percentage of the needed element or metal that can be recovered at a profit. Most elements are thinly distributed over the earth's crust. For example, aluminum is found in clays (aluminosiliates) in substantial quantities, but it would not be economical to process that material for the metal. Ores that occur in narrow veins are usually mined by underground methods (Figure 4.1). This is done by excavating vertical or sloping shafts and horizontal openings or rooms where the ore is removed, broken up, and hauled to the surface in mine cars or bucket and cable devices. When ores are found in large deposits, a surface mining technique, called *open-pit mining* (Figure 4.2), is used. Ore bodies are almost always covered by soil or other

FIGURE 4.1
Mining machine uses whirling cutting teeth to rip coal from underground seams. Coal is conveyed backward to shuttle cars. Water sprays entrap the coal dust (American Iron and Steel Institute).

FIGURE 4.2
In an open-pit mine on Minnesota's Mesabi Range, a power shovel dumps a 13-ton dipperful of iron ore into a truck. Although this is a natural ore, it must be beneficiated or improved in some way before it is shipped to blast furnaces (American Iron and Steel Institute).

waste material, called *overburden,* which must be removed prior to mining the ore. In a copper mine near Bingham, Utah (Figure 4.3), an entire mountain has been excavated and removed for its copper ore. In some surface mining operations the topsoil is removed, the ore is removed, and the soil is returned to its former place. This method helps prevent excessive damage to the ecology of the region.

FIGURE 4.3
An entire mountain has been removed from this gigantic open-pit copper mine near Salt Lake City, Utah.

Mining Iron Ore

Before the large deposits of rich *hematite* and *magnetite* ores in the Lake Superior Mesabi Range were depleted, as they are now, the 60 to 70 percent iron in those ores was quite profitable to extract. There are still vast deposits of iron ore in these regions, but the iron is bound up in a very hard rock formation that is difficult to mine and process and that has a relatively low percentage of iron. This material, called *taconite,* is now the major source of iron ore in that region. The most common natural ores are given in Table 4.1.

TABLE 4.1
Some common ores of industrial metals

Metal	*Ore*	*Color*	*Chemical Formula*
Aluminum	Bauxite	Gray white	Al_2O_3 (40 to 60%)
Cadmium	Sphalerite (zinc blende)	Colorless to brown to black	ZnS + 0.1 to 0.2% Cd
Chromium	Smithsonite	White	$Zn_4Si_2O_7OH \cdot H_2O$
Cobalt	Chromite	Green	$FeO \cdot Cr_2O_3$
	Cobaltite	Gray to silver white	$CoAsS$
	Linnaeite (Cobalt pyrite)		Co_3S_4
	Smaltite	Bluish white or gray	$CoAs_2$
	Erythrite	Transparent, crimson, or peach red	$Co_3(AsO_4)_2 \cdot 8H_2O$
Copper	Chalcopyrite	Brass yellow	$CuFeS_2$
	Chalcocite	Lead gray	Cu_2S
	Cuprite	Red	Cu_2O
	Azurite	Blue	$2CuCo_3 \cdot Cu(OH)_2$
	Bornite	Red-brown	$FeS \cdot 2Cu_2S \cdot As_2S_5$
	Malachite	Green	$CuCo_3 \cdot Cu(OH)_2$
	Native copper	Reddish	Cu
Gold	Native (alluvial or in quartz)	Yellow	Au
Indium	Sphalerite	White to yellow	ZnS + low % of In
Iron	Magnetite	Black	Fe_3O_4
	Hematite	Reddish brown	Fe_2O_3
	Limonite	Brown to yellow	$2Fe_2O_3 \cdot 3H_2O$
	Goethite	Brown	$Fe_2O_3 \cdot H_2O$
	Siderite	Blue	$FeCO_3$
	Taconite	Flintlike	Fe_3O_3
Lead	Galena	Black	PbS
Magnesium	Seawater		$Mg(OH)_2$
Manganese	Pyrolusite	Black	MnO_2
	Manganite	Steel gray to black	$Mn_2O_3 \cdot H_2O$
	Rhodochrosite	Rose red	$MnCO_3$
Molybdenum	Molybdenite	Blue	MoS_2
Nickel	Niccolite	Pale copper-red	$NiAs$
	Pentlandite	Bronze yellow	$(FeNi)_9S_8$
	Garnierite	Bright apple-green	$(MgNi)_3Si_2O_5(OH)_4$
Niobium	Niobite	Black	$(FeMn)(NbTa)_2O_6$
Silver	Argentite	Black	Ag_2S
	Native	White	Ag
Tin	Cassiterite	Brown or black	SnO_2
Titanium	Anatase, Brookite, and Rutile	White Reddish brown	TiO_2
Tungsten	Wolframite	Brownish to grayish black	$(FeMn)WO_4$
	Scheelite	Various	$CaWO_4$
	Ferberite		$FeWO_4$
	Huebnerite		$MnWO_4$
Uranium	Uraninite	Black	UO_2
	Pitchblende	Brown to black	U_3O_8
Vanadium	Vanadinite	Brown, yellow, red	$Pb_5(VO_4)_3Cl$
	Carnotite	Yellow	$K_2(UO_2)_2(VO_4)_23H_2O$
Zinc	Sphalerite	Black, brown, yellow, red	ZnS
	Franklinite	Black	$ZnFe_2O_4$
	Willemite	Varying	Zn_2SiO_4
	Zincite	Red to orange yellow	ZnO
Zirconium	Zircon (zirconium silicate sand)	Brown or gray square prisms or transparent	$ZrSiO_4$

Mining Aluminum Ore

Aluminum is the third most abundant element on the earth, after oxygen and silicon; however, only a few deposits of two aluminum-bearing minerals are considered to be profitable to mine. These are gibbsite and boehmite, which are composed of aluminum oxide (alumina, Al_2O_3) and water mixed with sand or clay and other minerals that must be removed. This ore is called *bauxite,* named for a French village, Les Baux, where the first deposit was found in 1821. Bauxite is usually mined by the open-pit method and is found on every continent except Antarctica. Major deposits, however, tend to be concentrated near the equator.

EXTRACTION OF IRON

Ore Treatments

The cost of shipping iron ores from the mine to the smelter can be greatly reduced if the unwanted rock and other impurities are removed prior to shipment. This is accomplished with several processes called *beneficiation.* These processes include crushing, screening, tumbling, flotation, and magnetic separation. The refined ore is enriched to over 60 percent iron by these methods and is often formed into pellets before shipment. In the case of taconite, about 2 tons of unwanted minerals are removed for each ton of pellets shipped (Figure 4.4), and the tailings are returned to the mining site. Growing vetches, grasses, and trees on some of these barren landscapes has been a project of biologists and foresters.

FIGURE 4.4
Taconite ore powder, now free of much impurity and rich in iron, is mixed with coal dust and a binder, rolled into small balls in this drum pelletizer, and then baked to hardness (American Iron and Steel Institute).

Direct Reduction of Iron

In ancient times, the only known method of separating iron from its ore was by the process of direct reduction in a charcoal forge or furnace. Unlike copper ores, which yield molten copper in these furnaces, iron will not melt at temperatures below 2799° F (1537° C). The highest temperature that could be reached in those primitive smelters appears to have been about 2192° F (1200° C). Iron ore subjected to that temperature yields not a molten puddle of metal but a spongy mass mixed with iron oxide and iron silicate with perhaps other impurities that together make up the slag. The iron worker removed this mass of sponge from the furnace, reheated it in a forge, and then literally squeezed the slag out of it by hammering it on an anvil. The hammering produced a wrought iron with very little carbon interspersed with a few stringers of slag that had not been eliminated. This wrought (worked) iron was not much stronger than copper and was a poor substitute for bronze because it could be hardened somewhat only by hammering, as could copper and bronze; however, the discovery of making steel by combining iron with a small amount of carbon and then heat treating it made steel a far superior metal. The old method of smelting iron and steel has long since been abandoned and replaced by processes using the blast furnace and steelmaking furnaces.

Sponge Iron and Powder

Since the 1920s a new method of direct reduction of ores has been used to produce sponge iron and iron powder, bypassing the blast furnace stage. Many of these "direct-reduced" ores have a lower percentage of impurities than does molten iron from a blast furnace. Pellets formed in this process contain as much as 95 percent iron and can be economically shipped from remote locations to be remelted in steelmaking furnaces. Iron powder is also produced by direct-reduction processes; much of it is used in the powder metallurgy industry. (See Chapter 11.)

Wrought Iron

The wrought (worked) iron of ancient times had some advantages over steel. It had less tendency to corrode (rust) and it had a fibrous quality (from the stringers of slag) that gave it a certain toughness and workability. Although the ancient process of making wrought iron is obsolete, a modern method, called the *Aston process,* is used to produce wrought iron today. Molten pig iron and steel are poured into an open hearth furnace with a prepared slag that removes most of the carbon. The slag cools the molten metal to a pasty mass that is later squeezed in a hydraulic press to remove most of the slag. Wrought iron is used for iron work such as railings for stairways and for the production of pipe and other products subject to deterioration by rusting.

Pig Iron

Until about A.D. 1350 all the iron made in the known Western world was made by the direct-reduction process. The development of the shaft or blast furnace at that time (although it might have been developed earlier in China) made possible the high production rate of iron that was needed for the Industrial Revolution. The blast furnace (Figure 4.5) made possible a molten, castable iron product that contains about 4 percent carbon and some impurities. The high carbon content causes it to be somewhat brittle, and it has little usefulness for tools and cutting instruments or other products. When refined in cupola furnaces, it is an excellent metal for some machinery parts, cooking utensils, and other cast iron products, however, the greatest usefulness of the blast furnace is as an intermediate step in the steelmaking process (Figure 4.6).

The three raw materials needed to make pig iron are iron ore, coke, and limestone. Coke is a residue left after soft coal is heated in the absence of air. The gases are driven off and the resultant hard, brittle, and porous material is coke, which contains 85 to 90 percent carbon with some impurities. Coke is produced in coke ovens (Figures 4.7 and 4.8), where the volatile elements that are driven off the coal are collected. Many useful products are made from these gases: fuel gas, ammonia, sulfur, oils, and coal tars. From the coal tars come many important products such as plastics, dyes, synthetic rubbers, perfumes, sulfa drugs, and aspirin.

The limestone ($CaCO_3$) is present to serve as a **flux,** called **slag,** that will absorb unwanted chemicals and gases.

FIGURE 4.5
One of four blast furnaces at Bethlehem Steel Corporation's Sparrows Point, Maryland, plant can process more than 3000 tons of molten iron per day (Courtesy of Bethlehem Steel Corporation).

During the operation of the blast furnace the slag is drawn off at intervals and hauled away in slag cars to a dumping area. This slag is ground up and used as an aggregate for concrete products and asphaltic concrete roads.

The blast furnace is a **reduction** process; that is, it reduces the third material, iron ore (Fe_2O_3) by removing the oxygen. The carbon provided by the coke reacts with the oxygen to form CO and CO_2, leaving behind molten iron with about 4 percent carbon. To produce 1 ton of iron requires about 2 tons of ore, 1 ton of coke, and 1/2 ton of limestone.

In Figure 4.5 the blast furnace is the very large structure shown in the center of the picture. The coke, limestone, and iron ore are carried to the top of the furnace in skip cars that ride the tracks that go from the lower left of the picture to the top of the furnace. At the top the contents of the skip cars are dumped into a chamber that has upper and lower doors. The upper door opens to accept the materials, then closes; the lower door then opens to let the charge fall into the furnace. Although the furnace is loaded at intervals it operates continuously.

During operation of the furnace the solid materials enter the furnace from the top while heated air is blasted in at the bottom. The air is heated in stoves containing refractory brick that is heated by exhaust gases from the furnace. The stoves alternate between blasting hot air to the furnace to being in a heating cycle. The fuel burns near the bottom of the blast furnace and the rising heat meets the descending charge of coke, limestone, and ore, producing very high temperatures of about 3000° F (1649° C). As the molten iron is released it settles to the bottom and a slag forms on top of it. The molten iron is tapped (drawn off) at intervals and collected in an insulated transfer car (Figure 4.9). It is then taken to a steelmaking furnace and charged into it. Some of the iron is used elsewhere. For shipping, it is cast in pigs (Figure 4.10, p. 84) and later remelted and refined in foundries for casting purposes. In earlier times the pigs were made in open sand molds consisting of a groove or trough with many small molds on each side, reminding one of a sow and pigs, hence the name *pig iron.*

STEELMAKING

From Iron to Steel

Pure iron is a soft, easily bent metal having limited usefulness. Steel is iron combined or alloyed with other metals or nonmetals such as carbon. Plain carbon steel actually contains a small amount of the metal manganese as a scavenger to remove unwanted oxygen and to control sulfur. Sulfur is difficult to remove from steel, and iron sulfide (the form it naturally takes in iron) is deleterious to steel, but if manganese is present, it combines more readily than iron with the sulfur, forming manganese sulfide, which is not harmful

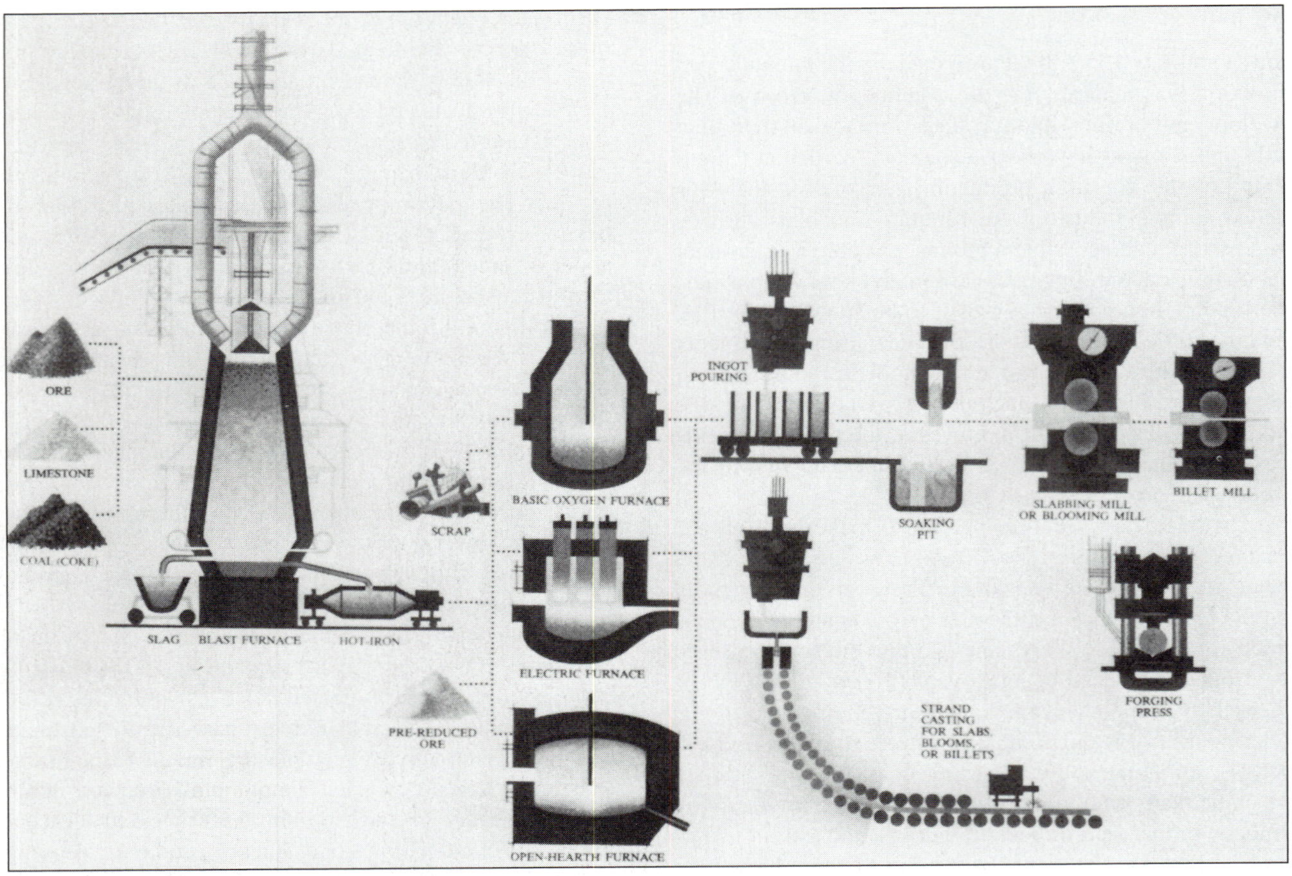

ORE
LIMESTONE
COAL (COKE)
SLAG
BLAST FURNACE
HOT-IRON
SCRAP
BASIC OXYGEN FURNACE
ELECTRIC FURNACE
PRE-REDUCED ORE
OPEN-HEARTH FURNACE
INGOT POURING
SOAKING PIT
SLABBING MILL OR BLOOMING MILL
BILLET MILL
FORGING PRESS
STRAND CASTING FOR SLABS, BLOOMS, OR BILLETS

FIGURE 4.6
Flowchart of the iron and steel production process, showing the route from ore to iron to finished steel (Courtesy of Bethlehem Steel Corporation).

FIGURE 4.7
An 850,000-ton-per-year battery of 80 slot-type, side-by-side coke ovens (Courtesy of Bethlehem Steel Corporation).

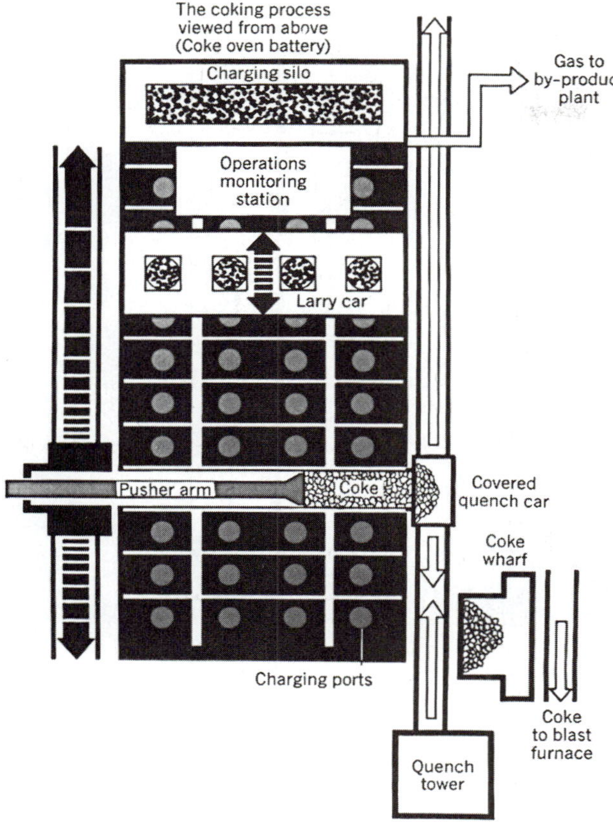

FIGURE 4.8
The operation of a coke oven battery at Bethlehem Steel Corporation's Sparrows Point, Maryland, plant is similar to the drawing. The larry car (or charging machine) is loaded with coal from a charging silo, then charges the coal into the individual ovens through ports in the oven roofs. After baking for approximately 18 hours and reaching a final temperature of at least 1800° F (982° C), coke is pushed from the ovens by the pushing machine at left. Hot coke is then moved by way of a quench car to a quench tower where it is flooded with approximately 16,000 gallons of water to cool it before being unloaded at a coke wharf to await further processing (Courtesy of Bethlehem Steel Corporation).

FIGURE 4.9
The blast furnace is tapped after slag removal, and the molten iron flows into ladles or hot-metal cars. Most iron is used in steelmaking (American Iron and Steel Institute).

to the steel. For this reason no steel is made without some manganese. The element carbon causes steel to be such a useful material because it can be hardened by heat treatments. Only a small amount of carbon is needed to make steel: Up to 0.30 percent for low-carbon steel, 0.40 percent to 0.60 percent for medium-carbon steel, and 0.60 percent to over 1 percent for high-carbon steel. It is uncommon to see steels containing more than 1.2 percent carbon. By definition, steel can contain up to 2 percent carbon, but over that amount it is considered to be cast iron, in which excess carbon forms graphite if sufficient silicon is present. Steel can be given many different and useful properties by alloying the iron with other metals such as chromium, nickel, tungsten, molybdenum, vanadium, and titanium and with nonmetals such as boron and silicon.

Processes and Furnaces Used

The principal purpose of the steelmaking furnace is to remove most of the carbon from pig iron. A measured amount of carbon is then added to the molten steel to give it the desired properties. Steel scrap is also utilized and other elements are added to improve the properties of the steel. A now obsolete furnace, the Bessemer converter, made possible the high-tonnage production of steel in the mid–nineteenth century. Great steel ships, railroads, bridges, and large buildings were built as a result of the availability of this new source of steel, however, many of the impurities remained in the steel, and since a blast of air was used to burn out the carbon, the nitrogen in the atmosphere also became an impurity that weakened the steel.

The three modern processes and furnaces used to produce steel are *basic oxygen, open hearth,* and *electric.* These furnaces are not used interchangeably. Each system requires different energy sources and different sources of

FIGURE 4.10
Automatic pig casting machine. Some blast furnace iron is poured to solidify in molds. Iron foundries are major users of cast pigs (American Iron and Steel Institute).

raw materials. The kind of facility is therefore chosen for economic reasons and the availability of raw materials and energy sources.

The **basic oxygen furnace (BOF)** (Figures 4.11 and 4.12) is designed to produce a high-quality steel in a relatively short time, about one 200- to 300-ton heat (a single batch) per hour as compared with the 100 to 375 tons per heat in 5 to 8 hours of the open-hearth process. The principal raw material fed into the basic oxygen furnace is molten pig iron from the blast furnace. Steel scrap is charged into the furnace, and lime is used as a fluxing agent that reacts with the impurities and forms a slag on top of the molten metal. A water-cooled lance is lowered to within a few feet of the charge to blow a stream of oxygen at a pressure of

FIGURE 4.11
Basic oxygen furnace being charged with molten pig iron (American Iron and Steel Institute).

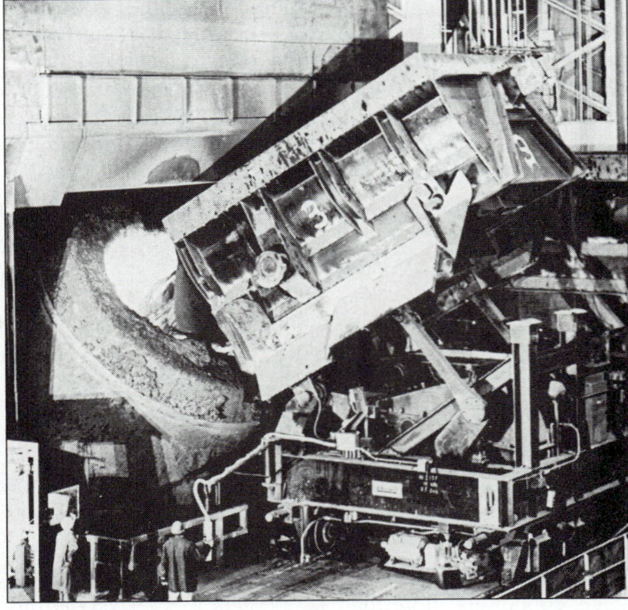

FIGURE 4.12
Charging basic oxygen furnace with scrap (American Iron and Steel Institute).

more than 100 pounds per square inch (PSI) on the surface of the molten bath. The oxidation of the carbon and impurities causes a violent rolling agitation that brings all the metal into contact with the oxygen stream. The ladle is first tipped to remove the slag and then rotated to pour the molten steel into a ladle. Carbon, usually in the form of anthracite coal pellets, is added in carefully measured amounts along with other elements to produce the desired composition and quality in the steel. A schematic cross section of a basic oxygen furnace is shown in Figure 4.13.

Open-hearth furnaces get their name from a shallow steelmaking area called a *hearth* that is exposed or open to a blast of flames that alternately sweeps across the hearth from one end for a period of time and then from the other end of the furnace (Figure 4.14). The furnace is charged (loaded) from a door in the side facing the charging floor. Theoretically, an open-hearth furnace can operate with blast furnace iron alone or with steel scrap alone, but most operations use both in about equal proportions, varying with the price of scrap and the availability of molten iron.

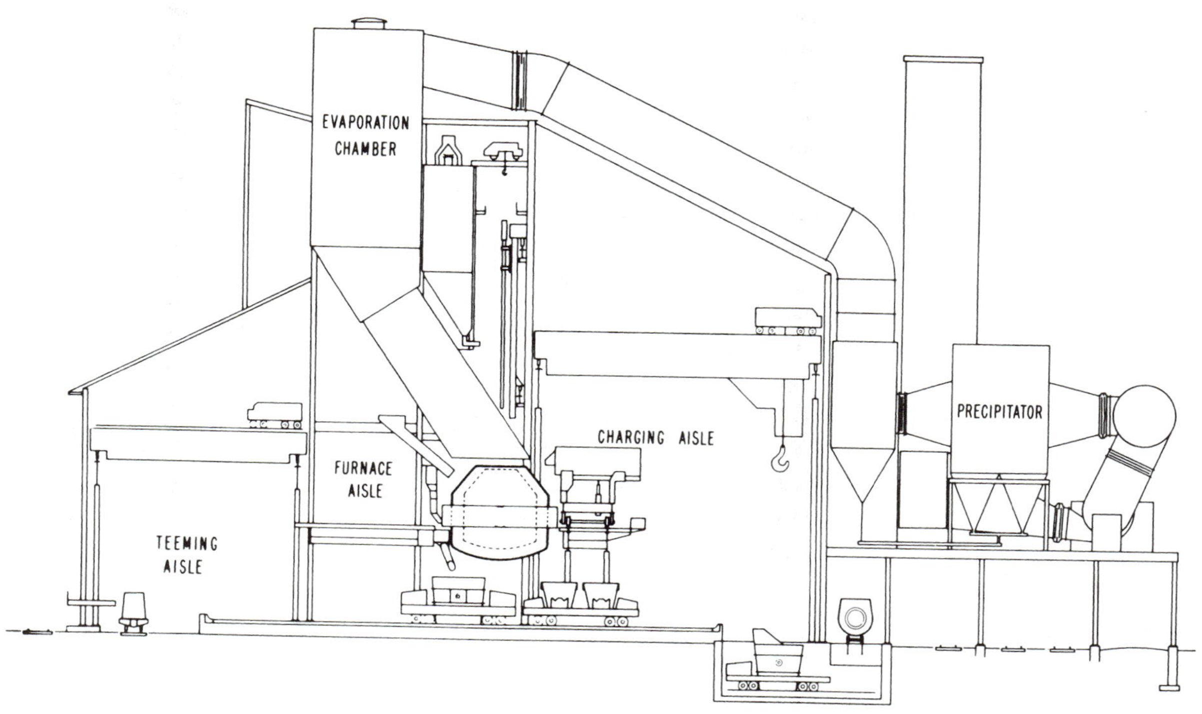

FIGURE 4.13
Schematic of a basic oxygen furnace showing the facilities needed to charge scrap and molten iron into the vessel and receive the steel after the oxygen-blowing process is complete (American Iron and Steel Institute).

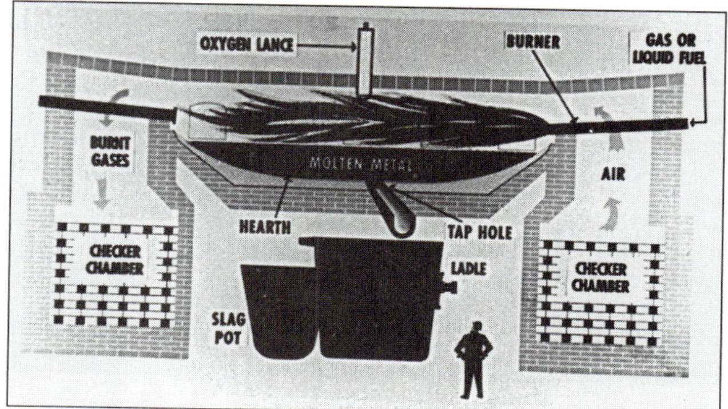

FIGURE 4.14
Simplified cutaway diagram of a typical open-hearth furnace. Oxygen may be injected through one or more lances. In some cases it is introduced through the burners to improve combustion (Courtesy of Bethlehem Steel Corporation).

Scrap is placed in the furnace by a charging machine, which usually serves a line or series of open-hearth furnaces in a single building (Figure 4.15). Other elements are added to the steel as required. In steel mills iron ore is used in steel-making furnaces to reduce carbon content. Aluminum ferrosilicon is added if the steel is to be killed. A **killed steel** is one that has been sufficiently deoxidized to prevent gas evolution in the ingot mold, making a uniform steel. If the steel is not killed, it is called **rimmed steel** because the gas pockets and holes from free oxygen form in the center of the ingot, whereas the rim near the surface of the ingot is free of defects. Most of these flaws, however, are removed by later rolling processes (discussed in Chapter 9). When the contents of the heat or melt are acceptable and the temperature is right, the furnace is tapped and the molten metal is poured into a ladle (Figure 4.16). As with the basic oxy-

FIGURE 4.15
Close-up of charging machine arm thrusting a box of baled scrap through an open-hearth furnace door (American Iron and Steel Institute).

FIGURE 4.16
Tapping an open-hearth furnace. Molten steel fills the ladle; the slag spills over into the slag pot (Courtesy of Bethlehem Steel Corporation).

gen furnace (BOF), carbon is added to the contents of the ladle to adjust the carbon content. Aluminum pellets are added to the ladle to deoxidize the molten metal and produce a finer grain size.

The ladle with its load of molten steel is picked up by a traveling crane and brought over a series of heavy cast iron ingot molds. The steel is teemed (poured) into the molds by means of an opening in the bottom of the ladle (Figures 4.17 and 4.18).

Electric furnaces lined with refractory (heat-resistant ceramic) materials (Figures 4.19 to 4.22) produce a high-quality steel from selected scrap, lime, and mill scale. Most alloy and tool steels are made in electric furnaces, but for these high-quality steels selected scrap must be used. Tin cans and automobile scrap contain contaminants such as tin and copper that spoil the steel for these purposes. Electric furnaces are most competitive where low-cost electricity is available. The charge is melted by the arcing between carbon electrodes and metal scrap. In some cases, the electrodes are 2 ft in diameter and 24 ft long.

Advanced Melting and Refining

Standard steelmaking furnaces are not capable of removing all impurities and unwanted gases. When higher-grade specialty steels are needed, further processing is necessary. There are two types of these refining vessels, those that receive molten metal from conventional steel furnaces and those that remelt to refine it (Figures 4.23*a* to 4.23*h*).

Vacuum Stream Degassing Figure 4.23*a*. Ladles of molten steel from conventional furnaces are taken to vacuum

FIGURE 4.18
Once steel is poured into an ingot mold and allowed to solidify, the mold is stripped off, exposing a glowing hot ingot (Courtesy of Bethlehem Steel Corporation).

FIGURE 4.17
Molten steel from a basic oxygen furnace is teemed into ingot molds (Courtesy of Bethlehem Steel Corporation).

FIGURE 4.19
Scrap is being charged into an electric furnace that produces a 185-ton heat of steel every 2 ½ hours (Courtesy of Bethlehem Steel Corporation).

FIGURE 4.20
In a rare view, the interior of an electric steelmaking furnace resembles an amphitheater. The furnace interior is lined with refractory brick and special water-cooled panels to withstand an interior temperature of 3000° F (1649° C). Steel scrap is dumped into the furnace through the roof and is melted electrically and then refined into high-quality steel (Courtesy of Bethlehem Steel Corporation).

FIGURE 4.21
The entire electric furnace is tilted during a tapping operation in which molten steel flows into a waiting ladle (Courtesy of Bethlehem Steel Corporation).

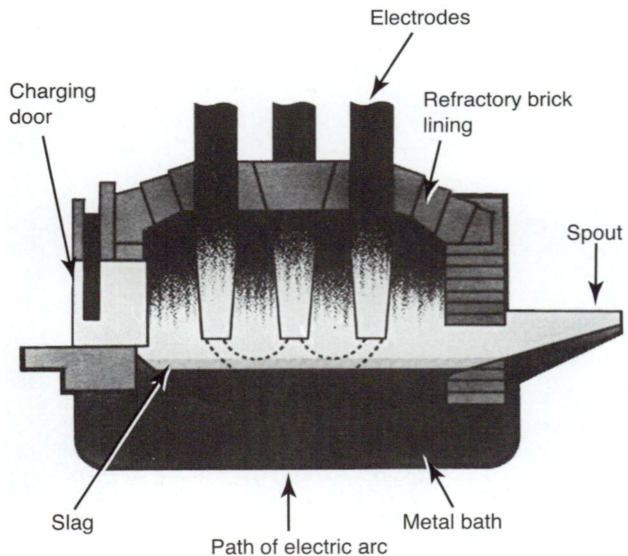

FIGURE 4.22
A cross section of an electric furnace (Courtesy of Bethlehem Steel Corporation).

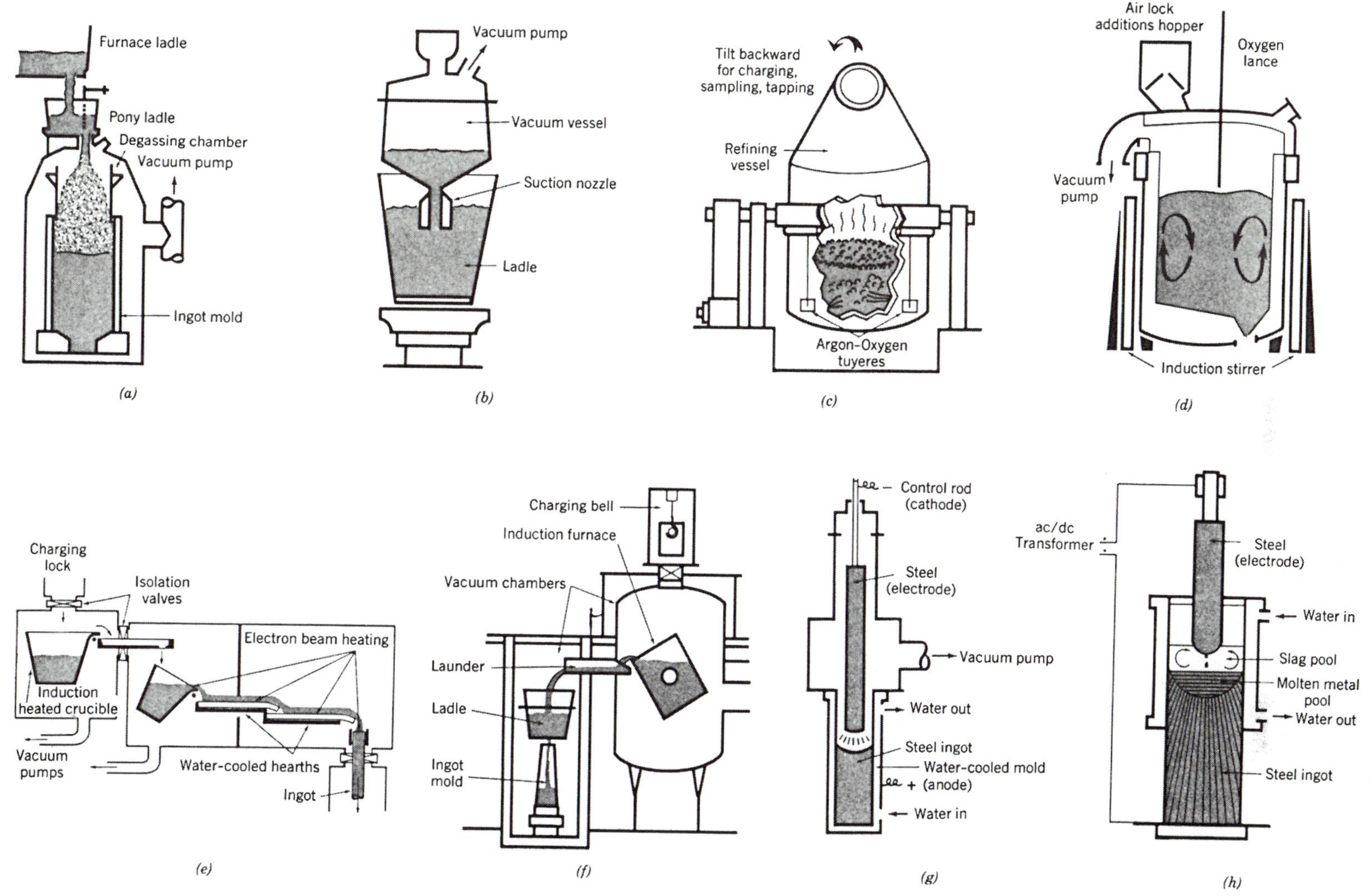

FIGURE 4.23
Remelting processes that are used mostly to produce highly sophisticated alloy and specialty steels for the world's new technologies. *(a)* Vacuum stream degassing. *(b)* Vacuum ladle degassing. *(c)* Argon–oxygen decarburization. *(d)* Vacuum oxygen decarburization. *(e)* Electron beam melting. *(f)* Vacuum induction melting. *(g)* Consumable electrode melting. *(h)* Electroslag melting (American Iron and Steel Institute).

chambers where the molten steel is poured into a pony ladle. There the molten stream of steel is controlled and broken up into droplets when exposed to the vacuum within the chamber. At this point large quantities of undesirable gases escape from the steel and are drawn off by the vacuum pump.

Vacuum Ladle Degassing Figure 4.23*b*. Several kinds of ladle degassing facilities are in current use. One of the most important is shown here. The furnace ladle is placed beneath a heated vacuum vessel that has a descending suction nozzle. The entire vessel is lowered into the ladle of molten steel. The molten steel is forced through the suction nozzle into the heated vacuum chamber. Gases are thus removed in the vacuum chamber, which is then raised so that the molten steel is returned by gravity into the ladle. The process is repeated several times.

Argon–Oxygen Decarburization Figure 4.23*c*. Argon, an inert gas, is used in this refining vessel to dilute the carbon–oxygen atmosphere in the melt. The argon and oxygen gases have separate tuyeres, or nozzles, so they can enter the bath of molten steel simultaneously. In the case of some stainless steels that must have a very low carbon content, this process increases the affinity of carbon for oxygen, producing CO_2. This also minimizes the oxidation of chromium, and a less expensive form of chromium can be used. The argon, being inert, also stirs the molten steel, promoting equilibrium between the molten metal and the slag.

Vacuum Oxygen Decarburization Figure 4.23*d*. The vessel, full of molten steel, is placed in an induction stirrer, which may be mounted or permanently installed on a transfer car. A separate roof is then placed over the ladle. The roof contains an oxygen lance and is equipped for vacuum degassing. The metal is degassed (decarburized) while being stirred. This first roof is then removed, and a second roof (through which additives and alloys are dropped from a hopper into the melt) is placed over the ladle.

Electron Beam Melting Figure 4.23*e*. This purification process involves a cascading action of molten steel carried on in a vacuum. The drawing shows steel melted in an induction-heated crucible inside a vacuum chamber. This crucible tilts, pouring partially degassed steel into an induction-heated ladle in an adjoining vacuum chamber. The electron beam hits the steel imparting thermal energy to keep it molten as it flows over a water-cooled copper hearth. The steel is continuously cast into ingots.

Vacuum Induction Melting Figure 4.23*f*. In this process, molten steel, or more commonly steel scrap, is charged into an induction furnace situated in a vacuum chamber. Undesirable gases are removed by a vacuum pump after which the molten refined steel is poured into a trough by tilting the furnace. The steel is then conveyed to a holding ladle in an adjoining vacuum chamber and ingot molds are filled from this ladle.

Consumable Electrode Melting Figure 4.23*g*. Sometimes called the *vacuum arc* process or furnace, this vessel remelts steels produced by other methods and allows it to solidify into ingots, usually in a cylindrical shape. A solid steel bar is slowly lowered on a control rod into a vertical water-cooled mold. The electric current forms an arc that melts the bar much like the electrode melts at the arc in electric arc welding. The gaseous impurities are drawn off by the vacuum in the chamber as the molten steel progressively fills the mold. The consumable electrode melting process is also used in refining some nonferrous metals such as titanium and zirconium. (See the flowsheet of zirconium production in this chapter, Figure 4.26.)

Electroslag Melting Figure 4.23*h*. There is a strong similarity between the electroslag process and the consumable electrode process, but there are some further technological refinements. As in the consumable electrode process, the electrode is suspended in a mold and connected to an electrical power source. Instead of a vacuum, a powdered slag, which quickly melts, is used to refine the steel. The molten droplets from the steel electrode drop through the liquid slag and collect in a molten pool beneath the slag and quickly solidify to form the ingot. Impurities are removed by flotation in the slag, and a thin layer of slag also solidifies between the ingot and mold walls, resulting in a smooth surface free of defects.

EXTRACTION OF ALUMINUM

Ore Refining

The bauxite ore from which aluminum is obtained contains many unwanted impurities such as iron oxide, silicon, titania, and water. The ore is first crushed to a powder and mixed in a solution of strong caustic soda (sodium hydroxide, NaOH) and dumped into large tanks or digesters, where more caustic soda or soda ash and lime are added to make up for losses.

In the Bayer process the alumina (aluminum oxide, Al_2O_3) is dissolved in the caustic soda, forming a sodium aluminate solution and leaving the impurities, which are insoluble. A series of separation processes involving filtration and precipitation separates the alumina from unwanted solids called *red mud*. This aluminum *hydrate* is then heated to 1800° F (982° C) in kilns to drive off the chemically attached water. The resulting alumina is a fine white powder that is about half aluminum and half oxygen,

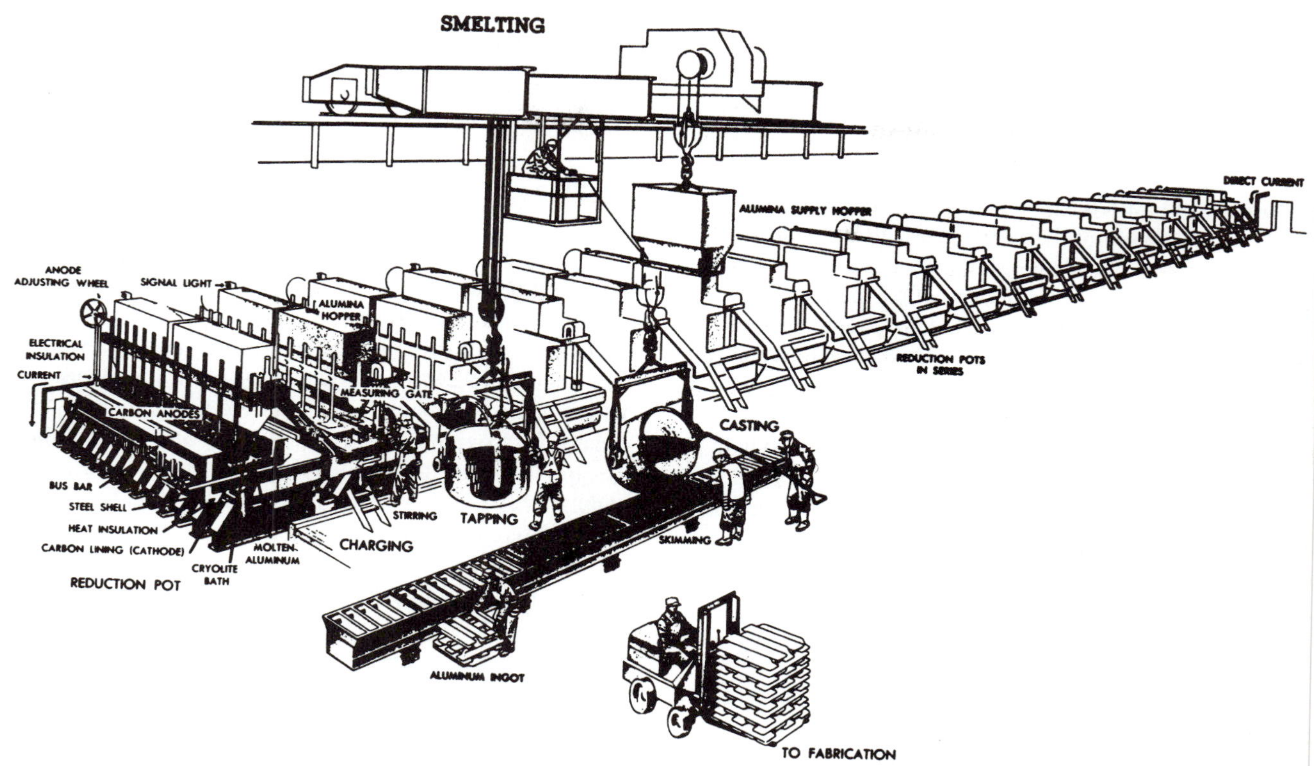

FIGURE 4.24
The process of making aluminum (Courtesy of Alcoa Inc).

bonded so tightly that neither heat nor chemical action alone can separate them.

Smelting of Aluminum

An abundant supply of low-cost electricity is required to produce aluminum competitively. For this reason, aluminum smelters are usually located in areas where hydroelectric or large coal-fired generating plants are established.

Smelting is the process by which alumina is reduced—that is, separated into its two constituents, aluminum and oxygen (Figure 4.24). In a modern smelter, the alumina is dissolved in a bath of molten cryolite in electrolytic furnaces called pots. Cryolite (sodium aluminum fluoride, Na_3AlF_6) is a material, mined in Greenland, that allows the passage of an electric current, whereas alumina does not conduct electricity. The smelting process is continuous; low voltage–high amperage current is passed through the cryolite bath from carbon anodes suspended in the melt to a carbon lining in the bottom of the pot, which serves as a cathode. This powerful electric current separates the dissolved alumina into pure aluminum and oxygen.

The molten aluminum collects at the bottom of the pot and is periodically siphoned off while the oxygen combines with the carbon anode and is released as carbon dioxide. A small portion of aluminum fluoride is added from time to time to compensate for small fluoride losses. About 2 pounds of alumina will yield 1 pound of aluminum, and about half a pound of carbon anode is consumed for every pound of aluminum produced.

In Figure 4.24 the aluminum metal is shown being produced in the form of "pigs," which will be later remelted into some usable form. It is more common to convert the molten metal into the form of an ingot that is used to produce sheet or extrusions directly without remelting.

EXTRACTION OF COPPER AND NICKEL

The two main classes of copper ores are oxides and sulfides. The oxidized ores can simply be heated with carbon to release the copper metal; however, most available ores are sulfides (known as blendes), and perhaps half the world's copper deposits are in the form of chalcopyrite ore

($CuFeS_2$). These sulfide ores require more complex treatment in the melting process.

The ore is first crushed and then ground to a powder in ball mills. The powder is enriched by a flotation process that removes much of the unwanted material. Silica is added and the mix is melted in a furnace (usually a reverberatory type), producing two liquid layers: a lower layer of copper matte (cuprous sulfide plus sulfides and oxides) and an upper layer of silicate slag, which is drawn off, leaving cuprous and ferrous sulfides called matte. Air under pressure is blown through the molten copper matte in a converter. This operation is one of oxidation–reduction similar to that of the iron-producing blast furnace. The iron is first removed, and in a second blow the sulfur is removed as sulfur dioxide. The molten copper is poured into ladles. At this stage it is about 98 percent to 99 percent pure and is cast into ingots. When further refinement is necessary, the copper anodes are placed in electrolytic cells that produce 99.95 percent pure copper. The impurities collect at the bottom of the tanks in a sludge from which other metals such as gold, silver, and other rare and precious metals are later removed.

Nickel is often associated with copper and iron ores as a nickel sulfide concentrate. The smelting process for nickel ores is similar to that of copper, in which the ore is roasted, or smelted, in an electric furnace to produce a matte. It is then placed in a converter and further refined before being cast into nickel–copper anodes. The nickel is then separated from other metals by the electrolytic refining process.

EXTRACTION OF LEAD AND ZINC

Somewhat like copper and nickel, lead and zinc are also extracted, primarily, from sulfide ores. These ore deposits often contain the ores for both metals as well as copper. The ores are ground into small pieces, and flotation processes are used to separate the ore from contaminants. Then, using specific reagents in a specified order, the various ores are separated. At this point the methods for processing the two metals diverge. The lead ore is smelted in reverberatory furnaces or in water-jacketed blast furnaces to produce molten lead, which is drawn off at the bottom of the furnace.

The zinc blende (ZnS) is also heated (or roasted), but too much energy is required to completely remove the sulfur and leave molten zinc. Instead, the roasting process produces zinc oxide (ZnO) plus SO_2 as a gas (which is used to produce sulfuric acid as a by-product). Roasting the zinc oxide (ZnO) will remove the oxygen and yield molten zinc, but this is no longer the process of choice. Rather, the modern method is an electrolytic process; the ZnO is dissolved in sulfuric acid, forming zinc sulfate. Electrolysis of this solution causes metallic zinc to plate out on a cathode (see Figure 18.6). The metal is melted off the cathodes (usually made of aluminum) and cast into shapes and sizes dependent on the end use.

EXTRACTION OF MAGNESIUM

There are several important ores of magnesium, but the most important source is seawater, which contains about 0.13 percent dissolved magnesium. The electrolytic method is used to extract magnesium from seawater to which calcium hydroxide ($Ca(OH)_2$) has been added. This causes the magnesium in the water to precipitate as magnesium hydroxide ($Mg(OH)_2$), which is converted to magnesium chloride ($MgCl_2$) by adding hydrochloric acid (HCl). This product is dried and electrolyzed to produce chlorine gas and magnesium metal.

Thermal reduction is used to extract magnesium from ores such as roasted dolomite (MgO-CaO).

SPACE-AGE METALS

Applications

The metals niobium, titanium, hafnium, zirconium, tantalum, and their alloys are used in the aerospace industry, in jet aircraft, rockets, and missiles, and for nuclear reactors.

Niobium is a relatively lightweight metal that can withstand high temperatures and can be used for the skin and structural members of aerospace equipment and missiles. Niobium, a superconductor of electricity, presents a possible way to store and transfer large quantities of energy in the future.

Titanium is extensively used for jet engine components (rotors, fins, and compressor parts) and other aerospace parts because it is as strong as steel and 45 percent lighter. Titanium–aluminum–tin alloys possess high strength at high temperatures, and since titanium forms an alloy with many metals, it can be made to have properties that make it useful for various applications.

Hafnium and *zirconium* are always found together, and their properties are so similar that there has been no commercial reason to separate them except one. Zirconium and hafnium have opposite properties in relation to neutron flow in a nuclear reactor, so they have to be separated for such use. Hafnium has the property of stopping the flow of neutrons by absorbing them, thus stopping the reaction taking place in the reactor core. In contrast, zirconium freely allows neutrons to flow through itself, allowing the fission process to occur. These metals make possible the precise control of nuclear reactors. Zirconium (and its alloys) was originally developed to encapsulate uranium in nuclear reactors, but because of its superior corrosion resistance, it is being used in the chemical industry and for surgical implants. Pure zirconium is a reactive metal that will burn in air with a brilliant white light. It has long been used as the light source in photo flashbulbs.

Tantalum often occurs as the mineral niobite–tantalite. The metal tantalum is very malleable and ductile and has a high melting point 5425° F (2996° C). It is used as a replacement for platinum in chemical, surgical, and dental instruments and equipment. Tantalum is also used in vacuum furnaces and to make electrolytic capacitors.

Tantalum and niobium are probably more closely related in more ways than any other two metals in the periodic table. They occur together in the same ores. Both are highly immune to attack by strong acids, and both are more ductile (easily formed) than any other refractory (high-temperature) metal.

Extraction

Niobite or tantalite ore is usually separated from other minerals by washing and magnetic separation at the mine. Ores processed for use as alloys of steel are normally converted to ferro–niobium–tantalum by reduction in electric furnaces.

The production of pure niobium or tantalum requires a more elaborate treatment in which the ore is ground and fused with caustic soda. The impurities are washed away, and the fused product is dissolved in acids to form salts, which are separated and purified by crystallization. Both niobium and tantalum are made by powder metallurgy in which the metal is prepared as a powder by crushing and washing with acids. The powders are compacted by a hydraulic press to produce bars of about one square inch in cross section and about 30 inches long. These are sintered in a vacuum chamber, clamped on each end to a water-cooled terminal, and subjected to the flow of a heavy electric current, raising it to a fusing temperature. After sintering, the bar is cold worked by rolling or forging.

Titanium ores are converted to titanium tetrachloride and reduced with magnesium or sodium in an inert gas atmosphere. The product of this chemical reaction is a spongy mass of metal that must be melted to form an ingot before it can be further processed. Because titanium is a reactive metal, it will react with oxygen in the air and burn while in the molten state. Therefore, melting must be carried out in a vacuum or in an inert gas atmosphere. The titanium sponge is crushed and then compacted into briquettes in a hydraulic press. These briquettes are welded together in a weld tank to form an electrode, which is melted in a water-cooled copper crucible in a vacuum arc furnace. The resulting ingot is processed in rolling mills in much the same way as steel products are formed.

Zirconium is processed in a manner similar to titanium; however, hafnium and zirconium are separated by precipitation and calcining, which produces zirconium oxide (ZrO_2) and hafnium oxide (HfO_2). The product is chlorinated and then reduced with magnesium, and finally the magnesium is distilled from the zirconium metal, leaving a porous spongelike material. The sponge is broken up, crushed, blended, and then pressed into compact bars. These are welded together to form an electrode (Figure 4.25). An ingot is formed in a vacuum arc furnace by the consumable arc melting process (Figure 4.26, 18). The ingot must be machined on the surface to remove scale and impurities. The ingots are then fabricated into products by forging, hot or cold rolling, extrusion, or cold drawing. An example of the use of zirconium tubing is shown in Figure 4.27.

Zirconium is also a reactive metal that can burn in the presence of oxygen or in the air. This property and others can be greatly altered in reactive metals by alloying them with other metals. For example, niobium–titanium alloys are produced for superconducting wire and other products. Titanium is alloyed with aluminum and molybdenum to give it a high tensile strength of about 170,000 PSI.

FIGURE 4.25
Pressed bars of zirconium sponge have been welded together to form an electrode to be used in the consumable arc melting process; see Figure 4.26, #18. (Courtesy of Teledyne Wah Chang Albany, Albany, OR).

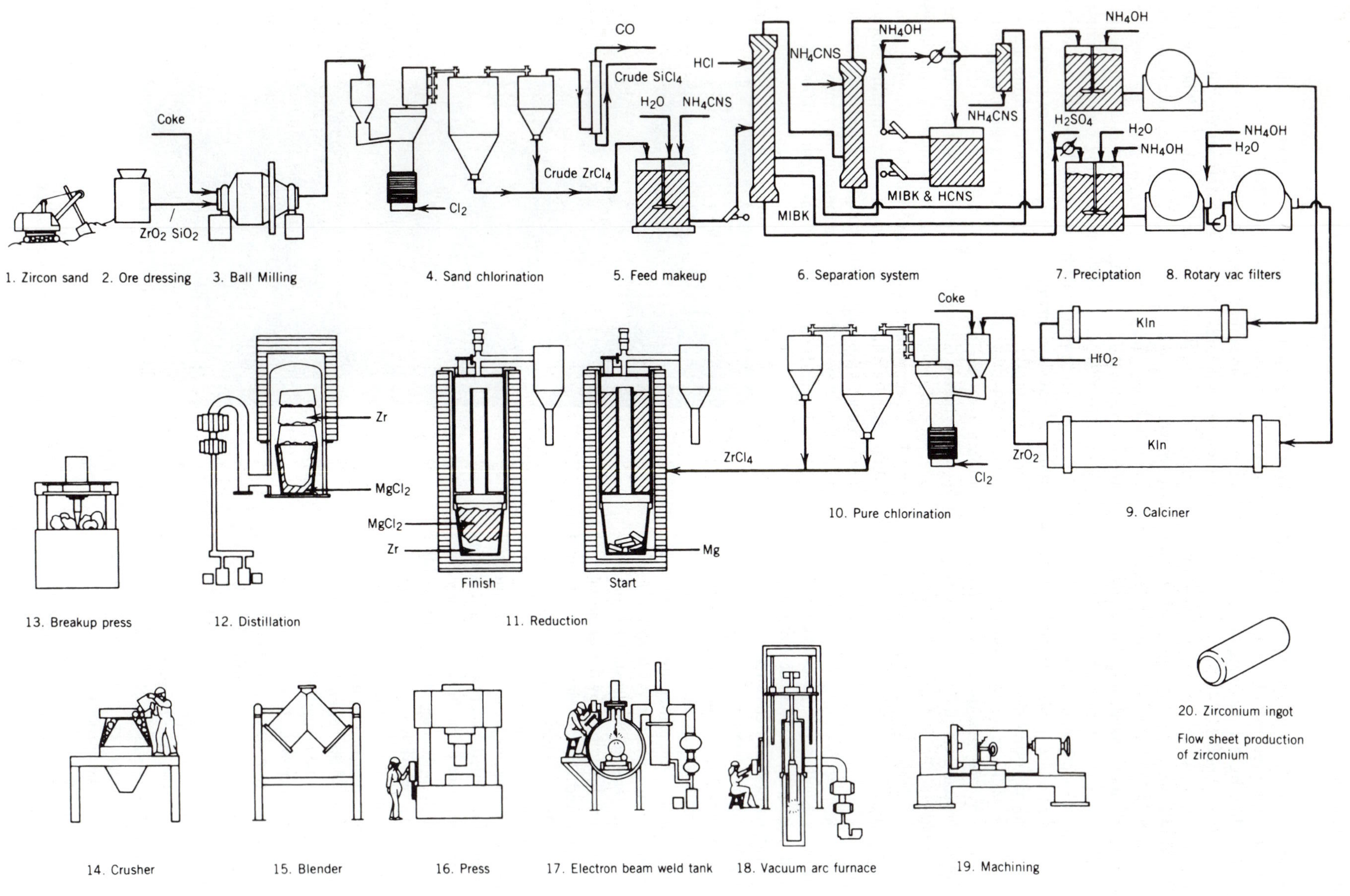

FIGURE 4.26
Flowchart of zirconium production (Courtesy of Teledyne Wah Chang Albany, Albany, OR).

FIGURE 4.27
An employee in a chemical processing plant inspects the more than 10 miles (17 kilometers) of zirconium tubing in this large heat exchanger. Zirconium is used because of its excellent resistance to corrosion (Teledyne Wah Chang).

Review Questions

1. Are titanium and zirconium ferrous or nonferrous metals? Explain.
2. A large proportion of common clay, which is available almost everywhere on earth, is aluminum. Why is bauxite used for the extraction of aluminum instead?
3. Iron ore is usually processed at the mine site to upgrade it and remove unwanted impurities. Why is this done at the site?
4. What is the name of the furnace that first made possible the smelting of iron in a molten state? What product is made in this furnace?
5. What three raw materials are used to make pig iron?
6. Why does all steel contain the element manganese in small quantities?
7. Why is cast iron a poor material to use for making cutting tools such as knives?
8. By definition, how much carbon can a steel contain? What is a more practical or nominal limit?
9. Name the three modern types of steelmaking furnaces.
10. What energy source is a necessity for smelting aluminum?
11. What is the final stage of refinement of 99.95 percent copper called?
12. Space-age metals such as titanium are extracted by complex processes. What is the major energy source for these processes: coal, coke, petroleum, oil, or electricity?
13. As you consider what you've just learned about extracting metals from their ores, read the first two paragraphs of the section titled METAL CORROSION in Chapter 17. What is the relationship between that information and what you've just learned?

Case Problem

Case 1: Cast Iron versus Steel

You are required to choose a metal for manufacturing carpenter's claw hammers. Nonferrous metals have been ruled out because they are all too costly or have insufficient strength; this leaves cast (pig) iron and steel. Making the hammer heads of cast iron is the least expensive method because the metal is cheaper and the hammer head can readily be cast in sand or reusable permanent molds. Thus, their sale price would be much lower than that of hammer heads made of steel forgings or castings. What other factors should be considered in this choice? Would you choose cast iron or steel to make these hammer heads?

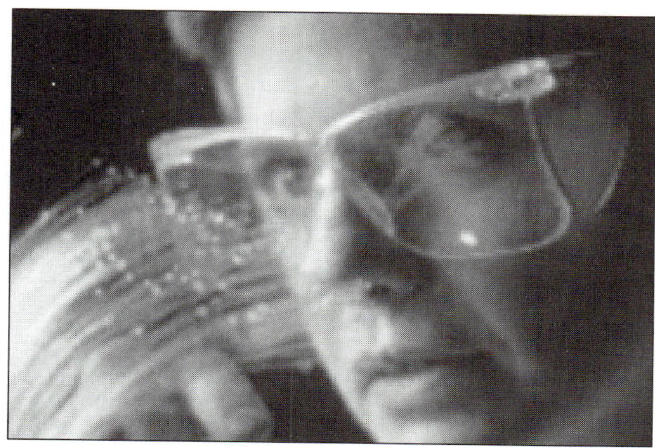

CHAPTER 5

Extraction and Refinement of Common Nonmetallic Materials

Objectives

This chapter will enable you to:

1. Identify the raw materials derived from petrochemicals that are used to produce many synthetic materials.
2. Explain the properties of some thermoplastics and thermosets that are used to manufacture products.
3. State the uses for certain additives and finishes for plastic materials.
4. Relate the means by which elastomers are derived.
5. Name the uses and explain the significance of asphalts and lubricants in the manufacturing industry.
6. Demonstrate how ceramic materials are produced.
7. Assess the significance of wood materials in manufacturing.

Key Words

petrochemicals
phenolics
polyamides
polyamide-imides
kaolin
cellulose
acrylonitrile-butadiene-stylene (ABS)
polymerization
monomer
polyimides
ceramics
aggregate
lignin

Stone, clay, and wood are among the oldest engineering materials used by humans to construct buildings and aqueducts, to make containers to store foodstuffs and contain liquids, and to make tools and weapons. The Romans discovered a way to make fluid stone that would solidify into a hard mass. It was called pozzolan cement, because it was first made from the volcanic ash at Pozzuoli, Italy. After this ash was combined with slaked lime it would begin to harden. Modern portland cement, from which concrete foundations, sidewalks, and dams are made, is similar in some ways to that ancient Roman cement.

The making of glass by fusing silica sand with another substance dates to at least 3000 B.C. Glass was used for making jewelry, containers, small windowpanes, and other artifacts; its use for plate glass windows, however, is a relatively recent occurrence because of the difficulties of producing large, flat sheets with optical quality.

Although wood has been displaced by other materials to some extent in recent times in the manufacture of furniture and in building construction, it continues to be important as a structural material. Wood has the advantage of being a renewable resource, and if forestry programs of replanting harvested trees are maintained, there will always be a nondepletable source.

Chemicals from which the commercially important plastic resins are derived are themselves derived from natural products that are now quite plentiful. Those raw materials include coal, water, limestone, sulfur, salt, and petroleum.

Products produced from petroleum, termed **petrochemicals,** such as plastics and elastomers, have only recently appeared on the scene and have rapidly displaced the use of wood, metal, and glass in some areas. The great advantage of plastics is that a special polymer can be created for a specific purpose that will fit the requirements better than any natural material. Thus, surgical implants, quiet long-lasting gears, and nondeteriorating rubber are some items that are produced by chemical synthesis of raw materials. All polymers are classified as either **thermosetting** or **thermoplastic,** each group having distinctly different qualities under conditions of heating and cooling.

The atomic structure of these materials was examined in Chapter 1, and they will be discussed in detail in this chapter; however parts manufactured from these materials will be discussed in Chapter 16, Processing of Plastics and Composites. Although the distinction is far from universally understood, the term *polymers* is used to describe the material used to make products, which are then referred to as *plastics.*

PETROCHEMICALS

Chemicals made from crude oil, natural gas, and liquefied natural gas are the basic raw materials for many of the synthetic fibers, sealants, paints, synthetic rubbers, plastics, and other products. These products are typically polymers that are made from small molecules called **monomers.** These are the building blocks of these synthetic materials, and they derive their names from the Greek *mono* (one), *poly* (many), and *meros* (parts). Polymer chains are composed of monomer cells (raw materials) that are combined to produce these useful materials (Table 5.1).

By the process called *catalytic cracking,* heat is used to induce a chemical reaction in the crude oil with the help of **catalysts,** usually composed of refractory oxides of aluminum, silicon, and magnesium. High-temperature-boiling components of petroleum are broken down into gasoline and oils. The remaining gaseous material is made up of compounds having from one to four atoms. Among these are ethylene, propylene, and the butylenes. Natural gas, from which the liquefied gases butane and propane are derived, is also a source of propylene, ethylene, hydrogen, and methane obtained by means of the process of thermal cracking. Acetylene gas, made from the hydration of calcium carbide, is also a source of plastic and rubbery materials. Most acetylene gas is now produced from methane rather than from calcium carbide. Chloroprene and polychloroprene (neoprene) are derived from acetylene.

TABLE 5.1
Products from petrochemicals

Products	*Petrochemical Monomer (Raw Materials)*
Antifreeze fluid, Mylar	Ethylene glycol
Butyl rubber	Isobutene
Epoxy resins, polycarbonates (Lexan®) acetone, phenol, bisphenol-A	Isopropylbenzine
Ethylene glycol antifreeze, Mylar (Dacron®) ethyl alcohol	Ethylene
Mylar, poly-para-xylene (Parylene®)	Xylenes
Neoprene rubber	Chloroprene
Nylon synthesis, neoprene, and polybutadiene rubber (source)	Butadiene
Orlon: chloroprene and neoprene rubber	Acetylene
Phenol, phenolic resins, styrene, and polystyrene	Benzene
Phenolic plastics	Toluene
Phenolic resins, epoxy resins	Phenol
Polypropylene, isopropylbenzine, isopropyl alcohol	Propylene
Polystyrene	Styrene
Polystyrene (nonsoluble)	Divinylbenzene
Production of acetylene	Methane
Synthetic rubber	Isoprene

(Neely and Bertone, *Practical Metallurgy and Materials of Industry,* 5th ed., © 2000 Prentice Hall, Inc.)

Nitrogen fertilizer made from natural gas, and pesticides made from other petrochemicals are both vital to food production. About 60 percent of the fibers used for clothing are synthetic and are mostly based on petrochemicals; exceptions are rayon, acetate, and acetic anhydride. Many of today's medicines are derivatives of petrochemicals; examples are antihistamines, antibiotics, sulfa drugs, penicillin, and aspirin. Automobiles use plastic gears and body parts; our homes have synthetic materials in such things as carpets, siding, heating and cooling systems, furnishings, and housewares. The petrochemical materials and derivatives covered in this chapter will be the plastics, rubbers, adhesives, lubricants, and asphalts.

POLYMERS

There are two broad categories of polymers: thermosets and thermoplastics. Both have *thermo* in their name, but the heat referred to has opposite effects on each. When a thermoset polymer is heated it "sets"; it becomes rigid and hard. It achieves its highest strength and hardness. If heated again it will not soften; in fact, if heated long enough and at a high enough temperature, it will char rather than melt. In contrast, a thermoplastic becomes softer and weaker as its temperature is increased. Thermoplastic parts are formed at an elevated temperature and gain their strength when cooled. If they are again subjected to heat they become softer and lose their strength.

The vast majority, about 80 percent, of the volume of polymeric material used in the United States is thermoplastic, but this does not mean that one is better than the other; each fills different requirements and specifications. Also, in each category roughly 80 percent of the usage is in commodities as opposed to applications of an engineering nature.

Thermosets

As explained in Chapter 1, the thermosets are characterized by the formation of cross-links between their long chains of monomers, formed during the heating, so the cross-links create a three-dimensional network. The cross-link bonds are covalent, primary bonds; they are very strong and rigid. Some types of thermosets do not require heat to harden but are cross-linked by combining liquids in definite proportions. The mixture then hardens in a predetermined time. Some examples are epoxy, silicone, polyurethane, and polyester.

At this point in the efforts to recycle materials, thermoset polymers cannot be recycled. There is a possibility that the spent thermoset material may find other uses if ground to a powder.

As a general rule, the thermosets must be combined with fillers and reinforcing agents to give them the mechanical properties needed for molded parts. For example, most of the epoxies are quite brittle, but when combined with glass fibers they become strong and tough materials.

Alkyds The alkyds are polyester derivatives, produced by the reaction of an alcohol and an acid. The name alkyd comes from the words *al*cohol and ac*id*. Their most important use is in paints and lacquers. Also, molded automobile ignition parts and light switches are made from these thermosets because they do not break down at higher temperatures and are not affected by moisture, conditions found around automobile engines.

Allyl Plastics The allyl plastics, produced from an alcohol, are very hard, scratch-resistant, clear plastics used for optical glass—prisms and lenses in eyeglasses (Figure 5.1), radar domes, and window glass. Although very costly thermosets, they are far superior to the clear acrylic plastic that is also used for transparent shapes. Allyl plastics resist most solvents such as gasoline, benzene, and acetone, and have a high tensile and compressive strength.

FIGURE 5.1
A researcher examines a bundle of optic fibers while wearing plastic safety glasses (Photo Researchers).

Aminos Urea and melamine belong to the family of amino plastics that have an unlimited range of colorability. The urea–formaldehyde plastics are produced by reacting urea with formaldehyde in the presence of a catalyst and mixing in a filler. These are used for knobs, handles, and household fixtures. Urea–formaldehyde plastics are used as adhesives for bonding plywood, foundry sand cores, and other materials. Melamines are similar but more complex plastics of the amino family and are harder, more shock-, heat-, and water-resistant than the urea types and have good resistance to acid and alkali attack. Because of their water resistance, they are used to bond exterior plywood. They are used for electrical parts such as circuit breakers and terminal blocks, and because of their ability to resist high temperatures and their chemical stability, the melamines are used for dinnerware and abuse-resistant furniture surfaces.

Epoxies These plastic materials are quite brittle (although a few types are tough and elastic) and have good electrical properties and chemical resistance. They cure at normal temperatures and pressures. Though they have good bonding properties as adhesives, their shock load strength is poor. Epoxies can be cast in molds and are used in paints and for surface coatings for metals.

Phenolics The **phenolic**-formaldehyde plastics are low-cost molding compounds with high strength, high temperature resistance, good dimensional stability, and good electrical properties. The most common type, Bakelite, is made from phenol and formaldehyde with a filler material such as wood flour. Bakelite is used for electrical switches and automobile distributor caps. Canvas, linen, or paper is sometimes used as a lamination to reinforce the plastic. Examples of these phenolic laminates are Micarta® and Formica®. The major drawback of the phenolics is the lack of colorability. They are commonly brown or black.

Polyesters Polyesters belong to a class of plastics related chemically to the alkyd type and contain two different types: "saturated" and "unsaturated." The saturated types are best known as film and fibers such as Dacron® and Mylar®. The unsaturated types are used in molding and casting. There are also many subtypes in each group. Polyester can be cast in molds without heat and pressure, but a disadvantage is that it shrinks considerably when hardening. This plastic is widely used with glass fiber and other reinforcements to provide high strength for such applications as boats, tanks, and automobile bodies. Polyesters can be either thermosetting or thermoplastic depending on formulation.

Polyurethanes Polyurethane plastic materials may be either thermosetting or thermoplastic. Since these are rubbery plastics, they are often made in a foamed structure for mattresses and insulation and are a substitute for foamed rubber. The thermosetting types can be cast as tough, wear-resistant elastomeric materials. They can be rigid or flexible depending on formulation.

Both urethane elastomers and rigid products are produced by the casting, injection molding, and coating process. Because they have good chemical and weather resistance along with high resilience and elasticity, they are used in a number of products such as oil seals (Figure 5.2) that can last up to eight times longer than rubber. Some of these seals are graphite impregnated to make them self-lubricating. Urethanes also have load-damping ability and very good abrasion resistance, making them ideal materials for industrial solid truck tires (Figure 5.3). They are also used for caster wheels (Figure 5.4), gears, drive belts, floor coverings, and roofing materials. Metal objects such as furniture are sometimes covered with a tough, resilient skin of polyurethane that imparts both aesthetic form and a colorful, useful surface to the part while the steel core provides the needed high strength.

A synthetic leather produced by DuPont, called Corfam®, is a polyester-reinforced urethane. It can be made porous or sealed with silicone. It was developed for shoe-upper parts but is used in many other applications, such as to make industrial gaskets and shaft seals.

FIGURE 5.2
New urethane compounds for hydraulic seals will last eight times longer than rubber and provide leak-free sealing up to 10,000 PSI (Disogrin Industries).

FIGURE 5.3
Urethane industrial solid truck tires (Maine Rubber International).

FIGURE 5.4
These solid polyurethane caster wheels are available in many different sizes, shapes, and colors (The Hamilton Caster & Mfg. Co.).

Thermoplastics

Thermoplastic materials have the cost-saving advantage in that they can be injection molded (Figure 5.5) and therefore have faster cycling times than compression- or transfer-molded thermosets (Figures 5.6 to 5.8). Thermoplastics are generally very ductile and can also be bonded by the process of ultrasonic welding. Also, since they retain some flexibility, thermoplastics can be designed with snap-fit parts, eliminating screws and other fasteners. Many thermoplastics can be used as a "living hinge" because they can flex thousands

FIGURE 5.5
Modern injection molding machine with microprocessor that automatically controls cycling times (Hergert's Industries, Inc.).

of times without breaking. Although thermosets have been the workhorse of the electrical industry applications because of their desirable electrical properties and dimensional stability, thermoplastics are rapidly taking over in this application. Electrical and electronic standards are changing, highlighted by the trend toward miniaturized parts that must endure ever higher operating temperatures. Tiny, more complex parts are more easily produced by injection molding techniques, so new thermoplastics have been and are being developed to fulfill the requirements for new products.

Fillers are often added to thermoplastics to gain properties such as dimensional stability, high temperature resistance, heat or electrical conductivity, or magnetic properties. Plasticizers are added to improve their flexibility, and stabilizers are added to prevent deterioration by light, heat, and oxidation. Coloring and fluorescent additives give these thermoplastics more customer appeal. Surface finishes, such as paint or electroplated metals, are also applied to plastics to enhance their appearance.

Acrylonitrile-Butadiene-Styrene (ABS), a Copolymer **ABS** copolymers are very tough, very strong thermoplastics with very high tensile strengths and have an unusual combination of high rigidity and high impact strength even at very low temperatures. These copolymers can be obtained in different combinations of heat resistance, flexibility, and toughness. Their major disadvantages are flammability and solubility in many solvents.

The first of the styrene plastics was introduced in 1938 and was called *polystyrene.* It has a very low strength but is one of the least expensive plastics, and for this reason is used for disposable products such as knives, forks, spoons, and water glasses. Styreneacrylonitrile (SAN) is a styrene copolymer that is stronger and has a better chemical resistance than polystyrene; however, ABS is, by far, the most outstanding plastic, making a wide range of specific materials available. Since it has high impact resistance, it is used in luggage, safety helmets, and such automobile parts as grilles. It can be plated in large volumes and can be blended with polyvinyl chloride (PVC) to become flame resistant.

Acetals Acetal thermoplastics, also called *engineering resins* because of their high strength, have excellent engineering properties, very high tensile strength, stiffness, good dimensional stability, low moisture absorption, and resistance to creep under load. There are two types of acetals, homopolymer and copolymer. They are very similar in most respects, but the homopolymer is stiffer, stronger, and more impact resistant. It also has a self-lubricating quality that

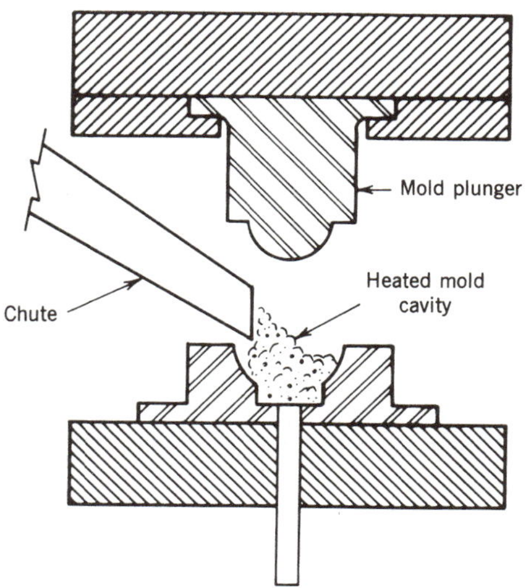

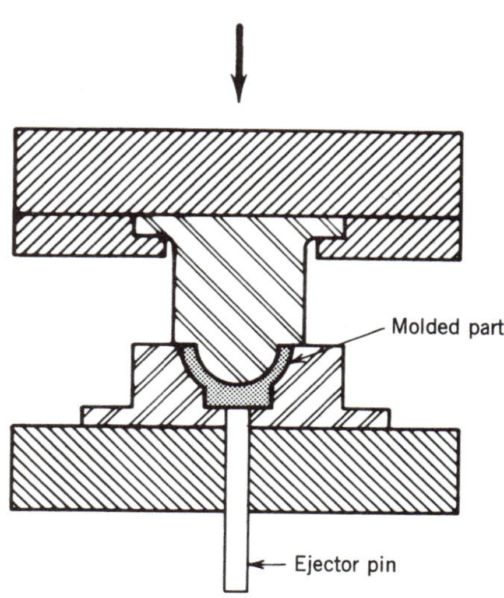

FIGURE 5.6
Compression molding. Granulated plastic is placed in a heated mold (left) and compressed (right) until object is formed. It is then removed with an ejector pin.

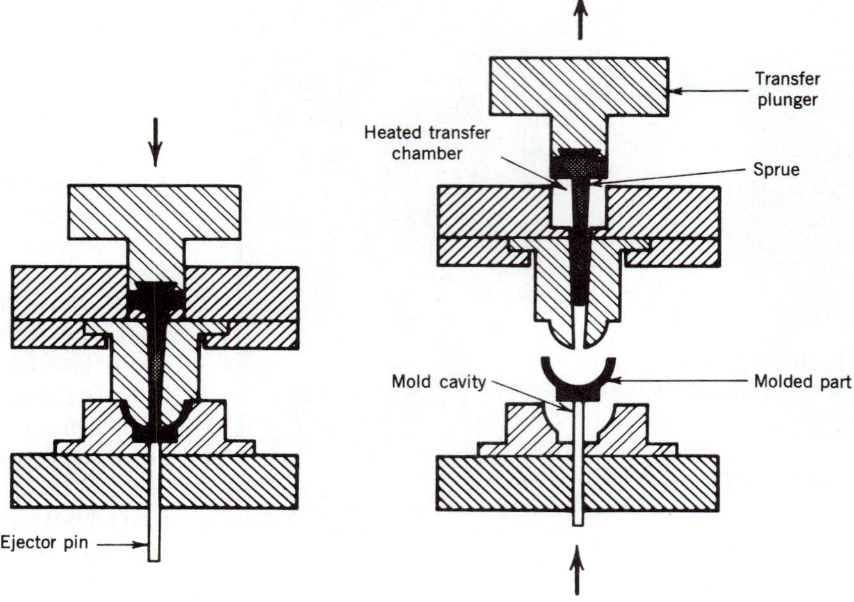

FIGURE 5.7
Transfer molding. Plastic is heated in an upper chamber and forced into the mold (left) by a second plunger. The object is formed and ejected (right).

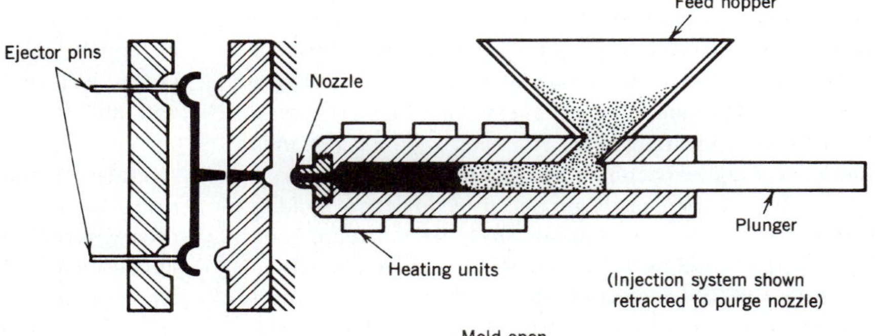

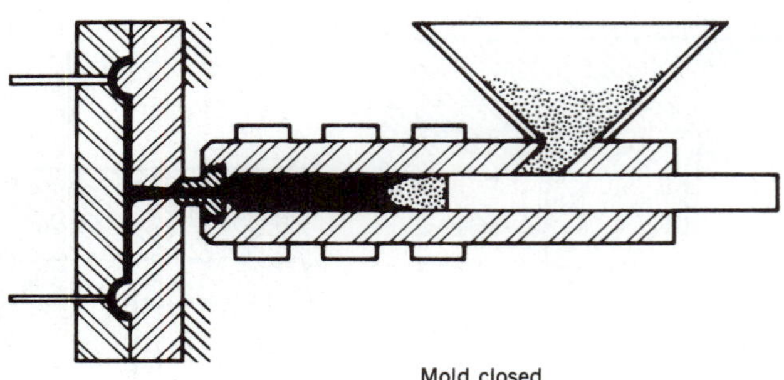

FIGURE 5.8
Principle of injection molding. Granulated thermoplastic is moved forward by the ram and passes through a series of heaters that melt the plastic. It is then injected with pressure into a closed mold where the molten plastic quickly solidifies. The mold is then opened and the part with the sprue is ejected. This cycle is then automatically repeated.

makes it useful for bearings and gears. The DuPont acetal Delrin® is a homopolymer that is used for such products as gears, butane cigarette lighters (because butane lighter fluid will not affect it), and car door handles (Figure 5.9).

Mineral-filled acetal copolymer resins have potential applications in such products as stereocassette cases, spools for videocassettes, toys, and zippers. In the plumbing industry, faucet cartridges, shower mixing valves, and shower heads molded of acetal homopolymer are replacing those made of brass and zinc.

Acrylics Widely used for their light transmission properties and optical clarity, acrylic plastics have excellent resistance to outdoor weathering and dimensional stability; that

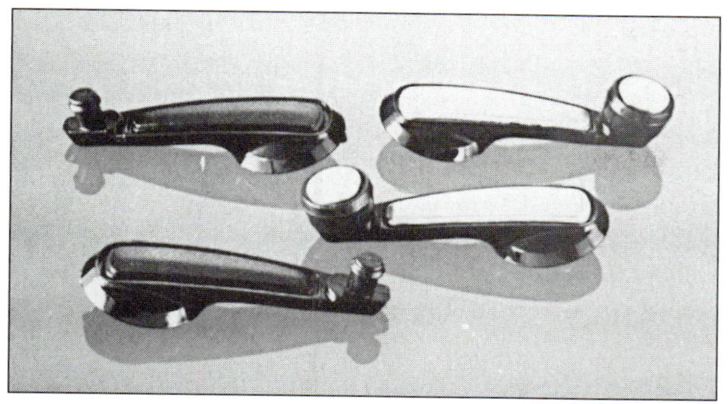

FIGURE 5.9
Delrin® in interior automotive door handles is not only lighter in weight than die-cast metals, it also eliminates the need for any finishing operations (Engineering Polymers Division, DuPont Company).

is, they tend to keep their shape over a wide range of temperatures, however, they will soften at about 200°F (93°C). The acrylics are not very scratch resistant and have a low shock resistance; therefore, they are not as satisfactory for making eyeglasses as are the allyl thermosets. Acrylics can be made in brilliant transparent colors and are used for the manufacture of automobile taillights, jewelry, and novelty items. Lucite® and Plexiglas® are acrylics that are made in the form of tubes, sheets, and rods. Acrylic parts are easily machined from these standard shapes.

Cellulosics In their natural form, cellulosics do not melt, but with proper chemical treatment they can be molded in the same standard equipment as other thermoplastics. Cellulosic plastics are among the toughest of all plastics. There are five basic groups: nitrate, acetate, propionate, butyrate, and ethyl cellulose. Cellulose nitrate is the toughest of these but has the drawback of being very flammable. Celluloid is one product made from this material. Cellulose acetate is nonflammable, but like the nitrate, embrittles with age. The propionate and butyrate types do not have that problem and have better chemical resistance. The ethyl type possesses a very high shock resistance, so it is used for plastic products that may be handled and accidentally dropped, such as flashlight cases and toys. Cellulose plastics are also made into thin sheets and are sold under the name *cellophane.* Cellulosics tend to creep or cold flow when under load.

Fluoroplastics The fluoroplastics family includes polytetrafluoroethylene (TFE), chlorotrifluoroethylene (CTFE), fluorinated ethylene propylene (FEP), and polyvinylidene fluoride (PVF_2). TFE has the greatest chemical resistance of all plastics; however, it is difficult to mold. FEP is more easily molded but still possesses the properties of TFE, yet it is more expensive than the other types to produce. CTFE is similar to the other fluoroplastics but has a much lower permeability to water and gases and is the least permeable of all plastics. PVF_2 has a greater stiffness than the other types and is used for rigid pipe and in load-bearing applications. Also, it is the easiest type to extrude.

Fluoroplastics have the lowest coefficient of friction of any plastic and they have a very high chemical resistance; none of the common chemicals will react with them at all. In fact, adhesives for the most part will not adhere to them, although some adhesives were developed to bond fluoroplastics to metals. Fluoroplastics are useful for nonstick cooking surfaces, such as baking pans and frying pans that need not be greased. Paints made of these materials have exceptional resistance to weather and sunlight. Fluoroplastics are used for labware, valve seats, O rings, and bearings. TFE fibers are sometimes used as fillers in acetal resins to improve thermal and frictional qualities.

Ionomers Ionomers are plastics that have both ionic and covalent bonds. These clear plastics have high optical clarity, excellent toughness, and high chemical resistance, and are easily formed. Films are hard to tear, but once a tear begins, propagation is easy.

Nylon Nylon is the name given by DuPont to a family of **polyamides.** It was first used for the production of fibers for such products as webbing, parachutes, and stockings. The most widely used nylons are Type 6, Type 6/6, Type 6/10, Type 11, and Type 12. All nylons have the properties of toughness, fatigue resistance, low friction, and inertness to aromatic hydrocarbons. Nylons 6 and 6/6 possess some incredible qualities when combined with other materials such as glass fibers. They can replace metals in many applications, such as automobile speedometer gears (Figure 5.10), bicycle wheels, and carburetors (Figure 5.11). Nylon 6/10 is more flexible than nylons 6 or 6/6 and has less tendency to absorb moisture (which can alter the physical dimensions of a plastic part) and thus can be used in water-mixing valves. Nylons 11 and 12 are commonly used in film and tubing.

Polycarbonates Polycarbonate resins can be used for many of the same products that can be made of nylon, since they have similar properties. They are extremely strong and are very difficult to break. For this reason they are used for unbreakable bottles, as a window glass replacement where crime and vandalism are problems, and for eyeglasses. The General Electric polycarbonate Lexan® is an excellent glazing product that is made with a special surface coating that resists ultraviolet light, which causes yellowing and impact degradation in polycarbonates. This nonyellowing coated product is used for architectural glazing and for solar energy systems. When polycarbonate sheets are laminated and fused with proprietary layers to a thickness of 0.75 in., they outperform bullet-resistant glass against medium-power weapons. Polycarbonate foams are other important products that are used for thermal insulation and sound dampening suitable for automobile and business machine use.

Polyester These thermoplastic polyester molding compounds are listed chemically as polybutylene-terephthalate (PBT) and polytetra-methylene terephthalate (PTM). Compared with the thermosets, thermoplastic polyesters possess better molding characteristics and more rapid cycle times because of their rapid crystallization rate. Glass fiber is added to improve heat resistance and strength. Applications include automotive electrical parts, gears, housings, and pump impellers.

Polyimide This engineering resin can exist both as a thermoset and as a thermoplastic, but unlike other thermoplastics it does not melt. Polyimides are difficult to injection mold, requiring high molding temperatures 660°F (349°C), a long cycle time, and high injection pressures. These resins can withstand 500°F (260°C) continuously and 900°F (482°C) for a short time without breaking down. **Polyimides** resist most chemicals, except for strong alkaline solutions and strong inorganic acids, and they have poor weathering resistance; they have almost no creep even at high temperatures, and when certain fillers such as glass or graphite fibers are used, the thermal expansion is close to that of metals. They are used for bearing sleeves, seals, and thrust washers.

Another engineering resin, **polyamide-imide**, can also resist high temperatures of 500°F (260°C) for long periods. Because of its toughness, high strength (30,000 PSI at room temperature), and heat resistance, it has many applications in the automotive industry and in making cookware and electrical equipment. The properties of polyamide-imide are shown in Table 5.2.

Polyolefins The polyolefins include polyallomer, polypropylene, and polyethylene. Polypropylene is a very strong, flexible resin that has an unusually high flex life, which makes it an excellent material for "living hinges." Polyethylene is used for squeeze bottles, ice cube trays, film for boil-in packaging, and vending machine food containers. High-density polyethylene is used to make pressure pipe (Figure 5.12). Since it can be made either flexible or stiff, it is used for a great variety of housewares and toys. Polyethylene terephthalate (PET) is a more recently developed plastic resin that is being increasingly used for beverage bottles. Because beverage bottles usually contain no reinforcing fibers, the thermoplastic resin is highly recyclable, and millions of pounds per year of PET bottles and scrap are recovered and

FIGURE 5.10
Heat-stabilized Zytel® (nylon) has been used for speedometer take-off, driven, and governor gears since 1954. In this difficult environment they have performed as reliably as the metals they have replaced, and they weigh much less and are much less expensive (Engineering Polymers Division, DuPont Company).

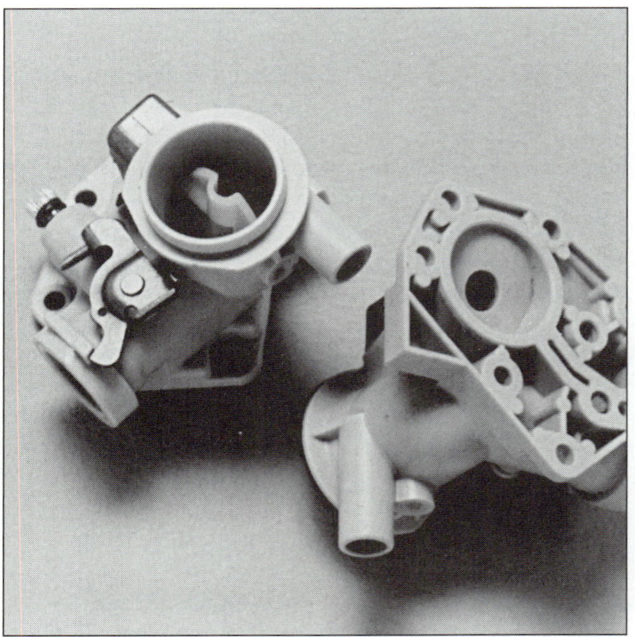

FIGURE 5.11
This lawnmower carburetor molded of Minlon® 10B (a reinforced nylon) is a first among four-cycle engines (Engineering Polymers Division, DuPont Company).

reprocessed. A polycarbonate polyethylene terephthalate blend (PC/PET) was developed and is valued for its high impact strength at ambient and low temperatures as well as its resistance to chemical attack.

Polypropylene is used for electrical insulation because of its outstanding electrical properties. Polyallomer possesses many of the properties of polyethylene and polypropylene, but it can be processed more easily.

Polysulfone The polysulfone engineering resin has a very high constant-use temperature of 400 to 500°F (204 to 260°C) and good impact strength. It is resistant to salt solutions, acids, alkalies, and most solvents and can be processed by extrusion, injection molding, and compression molding. Such resins can also be ultrasonically welded and machined. Because of their good electrical insulating properties, they are appropriate for use in printed circuit boards, switches, lamp housings, and electrical insulation.

Polyvinyl Chloride There are two basic types of polyvinyl chloride (PVC), rigid and flexible. Rigid PVC is used in pipes and pipe fittings and other construction materials. Flexible PVC has replaced rubber in many applications, such as in making gaskets, vacuum tubing, garden hoses, shower curtains, floor mats, and toys. PVC can also be foamed and coated with other solutions, can be fabricated by most processes, and may be joined by heat sealing and solvent welding. Plastic products manufacture will be covered in Chapter 16.

Silicones Silicone plastics are a type of synthetic rubber that is one of the highest-temperature-resistant plastics; they may be used at temperatures up to 500°F (260°C). Because of their excellent electrical properties, they are widely used in the electrical industry for encapsulation and components. Silicones are also used as sealants and caulking because they do not lose their

TABLE 5.2
Properties typical of polyamide-imide polymers

	Torlon 4347[a]	*Torlon 4203L*[b]	*Torlon 5030*[c]
I. Mechanical Properties			
Tensile strength, break (PSI)	17,800	27,800	32,100
Elongation, %	9	15	2
Flexural strength, yield (PSI)	27,000	34,900	48,300
Compressive strength (PSI)	18,300	32,100	38,300
Impact strength, Izod notch (ft-lb/in.)	1.30	2.70	1.50
Hardness, HRE	66	86	94
II. Thermal Properties			
Processing temperature (°F)	675	650	650
Heat distortion (°F @ 264 PSI)	1.50×10^{-5}	1.70×10^{-5}	9.00×10^{-6}
Coefficient of thermal linear expansion (in./in./°F)	532	532	539
III. Electrical Properties			
Dielectric constant @ 1 MHz	6.0	3.90	6.50
Dissipation factor, @ 1 MHz	0.071	0.031	0.023
Dielectric strength, V/10^{-3} in.	–	580	840
IV. Physical Properties			
Tensile modulus (PSI)	8.7×10^5	7.0×10^5	2.11×10^6
Flexural modulus (PSI)	8.9×10^5	7.3×10^5	1.70×10^6
Specific gravity	1.46	1.40	1.57
Density (lb/ft^3)	91.1	89.8	98.0
Water absorption, 24 hr (%)	0.17	0.33	0.24

[a] Torlon 4347 is an unfilled material for applications such as bearings and washers.

[b] Torlon 4203L is a titanium dioxide–filled material that also contains 0.5% PTFE for better mold release.

[c] Torlon 5030 contains 30 percent glass fiber and 1 percent PTFE; it has high stiffness and low creep and is intended as a replacement for metal parts.

(The source of the data is the *International Plastics Selector,* Edition 11, 1990, Volume 2, p. 224. [These polymers are produced by Amoco Performance.])

FIGURE 5.12
Construction workers on street in New York compare section of 48-in.-diameter high-density polyethylene pressure pipe, the largest size previously available, with a section of newly available record 63-in.-diameter "MLSO-ppd/305" high-density polyethylene pressure pipe (M.L. Sheldon Plastics Corp., New York, NY).

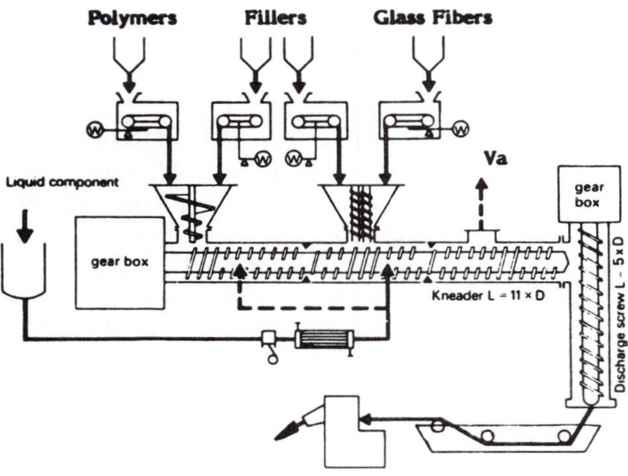

FIGURE 5.13
Flow chart showing continuous compounding of engineering plastics (Buss-Condux, Inc.).

rubbery qualities or crack with age, and they have good bonding characteristics.

Molds can be made with room-temperature vulcanizing silicone rubbers that require a separate catalyst as a hardener that is mixed in at the time of use. Silicone products are also used as lubricants. Because they do not tend to break down in hostile environments, the silicones have been very successful for use as prosthetic devices, such as artificial heart valves, and for reconstructive surgery near the surface of the body.

ADDITIVES

Various additives, pigments, and fillers are combined with polymer resins by mixing and kneading machinery. It is now possible to compound these materials continuously as a manufacturing process (Figure 5.13).

Since the first plastic materials were synthesized the list of uses and applications for products has grown significantly. In many cases plastic products are replacing metal products and competing with other products made from natural materials such as wood and leather. This widespread and diversified use of plastics is partly possible because of additives that alter and strengthen the plastic resins to make them more easily processed and more useful. Some of the additives used are as follows.

Antioxidants These materials impart ultraviolet stability to such resins as ABS, polyethylene, polypropylene, and polystyrene, and they impart melt-flow retention, making them easier to mold.

Blowing Agents When plastics must be processed at low temperatures, as required in sponge plastic and rubber applications, these additives allow production speeds to be increased and create voids to give porosity or low density.

Colorants Pigments and dyes for plastics give them their brilliant colors and are important for giving them sales appeal. Some pigments are designed to provide a solid, opaque color, whereas others tint the clear resin to impart a translucent coloration. Luminous colors are provided for some products, such as bicycle reflectors, to make them easily seen. A new luminous-green pigment was developed that is said to be nontoxic and safe for toys and glows in the dark for about 8 hours after being exposed to a light source.

Fillers Filler materials can reduce the cost of plastic products by reducing resin usage. They also lower the part weight and still retain, or even increase, physical properties such as tensile and compressive strength. Common fillers are wood flour, quartz, limestone, clay, and metal powders.

Plasticizers Some plastics such as vinyls are normally hard, brittle materials, but the addition of a plasticizer can make them soft and flexible.

Reinforcements Resins tend to have rather low physical strengths, but the addition of reinforcements can markedly increase their strength. Some reinforcing materials for plastics are glass fiber, mica, jute, sisal, graphite, carbon, and ceramic. Containers such as large tanks are formed by winding filaments of various materials such as glass, boron, and graphite with a plastic binder such as epoxy over a form that rotates while the filament moves in circumferential, longitudinal, and helical patterns, or a combination of these to produce a very high strength product. This topic will be dealt with further in Chapter 16.

Stabilizers Some plastic resins such as styrene and vinyl are subject to degradation or breakdown in the presence of oxygen or heat. PVC compounds are stabilized by barium–cadmium–zinc compounds, and phenols can be added to styrene to stop degradation.

FIGURE 5.14
The bright metallic surface on this plastic toy appears to have been applied by the process of sputtering or by condensation.

PLASTIC FINISHES

Almost every type of molded plastic part can be decorated to make it wear, scratch, and chemical resistant, as well as to give it aesthetic appeal. The molding process can impart a decorative surface with raised or depressed designs in geometric patterns such as basket weave, pebble, or leather texture. Texturing allows the manufacturer to mask or hide flaws, flow lines, sink marks, pinholes, and swirls, thus lowering the rate of part rejection. Other finishes are made with lacquers, enamels, and decorative overlays. Plastics can be metallized on the surface by the process of vacuum metalizing, in which metal filaments are vaporized onto the plastic surface and then condensed in a vacuum chamber. Gold, silver, nickel, chromium, and pure aluminum may be deposited by this method. Another widely used method of depositing metals on the surface of plastic pans is called *sputtering* (Figure 5.14). In this method the metal is ionized in a vacuum chamber by an electrical discharge between two electrodes. A metallic cloud forms inside the chamber and fogs the plastic parts with a thin layer of metal. Sputtering techniques are widely used to plate copper and other metals on printed circuit boards (PCBs) for the electronics industry. Glass–epoxy laminates are usually used for PCBs, and the copper-plated circuits printed on them are easily soldered to the electronic components.

ENGINEERING PROPERTIES OF PLASTICS

When a plastic material must be selected from the many available types for the manufacture of a new product, the choice must be based on the various engineering properties of the plastic. For example, if a gear was made of a brittle material such as acrylic with no reinforcing material, it would shatter when a load was applied, but a high-performance plastic such as one of the acetals or nylons would make excellent gears for many purposes. However, if a less expensive, disposable plastic product was needed, it would be noncompetitive and wasteful to select an expensive engineering plastic, and a low-cost material such as polystyrene should be selected. Some products must be made highly resistant to impact; some of these are automobile windshields, doors, bumpers, telephones, and toys. The engineer should probably choose the plastic for these items from among the high-impact polymers such as the polycarbonates, polyesters, and the newer polyphenylene oxide–based polymers and polyetherimides. Plastic resin manufacturers make the properties of their plastics available to engineers and designers. See Table 5.2 for an example of a properties chart.

Standard tests are used to assure the engineer that the selected plastic material will meet the requirements of the product. These tests were set up by the American Society of Testing and Materials (ASTM). The many tests for plastics can be divided into several general areas. The most common are mechanical properties, thermal properties, electrical properties, physical properties, and special characteristics. Mechanical properties include tensile strength, hardness, compressive strength, impact strength, and flexural strength (Figures 5.15 and 5.16). Thermal properties include Underwriters Laboratories (UL) ratings, heat distortion, coefficient of thermal expansion, thermal conductivity, and flammability rating. Electrical properties are dielectric constant, dielectric strength, and dissipation factor. Physical properties are specific gravity, density, hardness, water absorption, and radiation resistance. Special

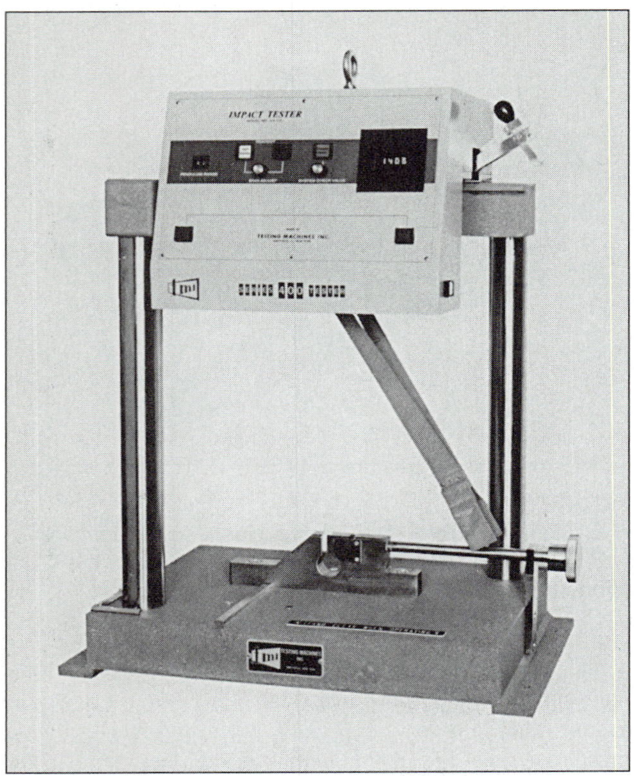

FIGURE 5.15
Izod–Charpy impact tester used for determining impact strength of plastics, ceramics, and light metals. This device uses a sensing system with a rotary transducer and a digital display to replace the analog dial. Its range is between 0 and 1 and 0 and 45 ft-lb (Testing Machines Inc.).

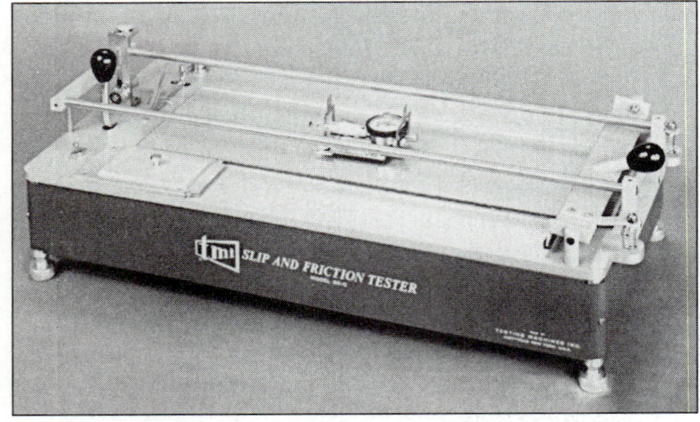

FIGURE 5.16
Slip and friction tester. This device can make several tests, such as tensile tests, measurements of the coefficient of plastic surfaces, and peel tests for plastic laminates (Testing Machines Inc.).

characteristics include weatherability, optical clarity, chemical resistance, and colorability.

Plastics designers selecting plastics for food packaging must consider the government regulations set forth by the National Institute for Occupational Safety and Health (NIOSH). The institute also makes recommendations concerning toxicity in plastics manufacture to the Occupational Safety and Health Administration (OSHA) regulatory agency.

ELASTOMERS AND ADHESIVES

Elastomers are materials, synthetic or natural, possessing rubbery qualities such as high resilience, extensibility, and elastic properties. The term *rubber* was originally used only for the natural product that is obtained from a thick, milky fluid (latex) that oozes from certain plants when they are cut. Most latex comes from the Para rubber tree *(Hevea brasiliensis)* that grows in South America, Southeast Asia, and Sri Lanka. Natural rubber in the form of cured latex is a sticky, gummy substance that has a limited usefulness for rubber products.

In 1839 Charles Goodyear discovered that this latex could be greatly strengthened and its elastic properties improved when he added a small amount of sulfur to it and heated the mixture. The process is called *vulcanizing*. By varying the sulfur content and adding certain pigments such as carbon black, fillers, and softeners, rubber can be made to any hardness from very soft to very hard, such as ebonite. Latex is compounded with additives and vulcanizers in a mixing machine to form a homogeneous mass. The mix is then placed in a mill with chilled rolls (to prevent premature vulcanization). The rubber compounds are made into sheet form on calender rolls (Figure 5.17). When rubber products such as in automobile tires need to be strengthened, textile cords or fabrics are used as reinforcing materials.

Artificial Elastomers

Synthetic rubbers were first created in the early 1930s and were subsequently developed because of the uncertainty of the supply of natural rubber. This first rubbery synthetic was derived from acetylene gas and was a long-chain molecule called polychloroprene, better known as neoprene. Natural rubber is somewhat more resilient than neoprene, but it tends to deteriorate in sunlight (ultraviolet) and swells when in contact with petroleum oil; neoprene has better resistance to these factors. Chemical additives such as antioxidants, plasticizers, and stabilizers are now added to natural rubber and other elastomers to overcome these dif-

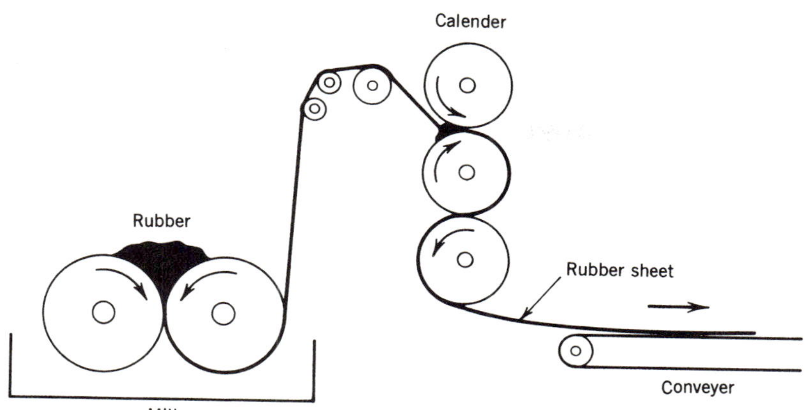

FIGURE 5.17
Rubber calender rolls.

ficulties. One synthetic rubber, polyisoprene, seems to have almost the same properties as natural rubber, and some others such as the rubbery plastics have certain specific advantages over natural rubber.

Adhesives and Sealants

Adhesives bond two surfaces together as a kind of fastener. Sealants are materials that adhere to a crack or joint for the purpose of making a fluid or airtight seal. Sealants are very similar to adhesives in that they must adhere to the sealing surface even though they usually do not have the high bonding strength of adhesives. Some types of sealants can remain soft and flexible for years, maintaining a good seal.

Prior to the development of synthetic adhesives, bonding agents were glues made from the hides and hooves of animals or parts of fish. Pastes were made of tree gums, starch, or casein. All these products were easily loosened by the action of moisture. Since there is little molecular attraction between these materials and the parts to be bonded, they are not very strong adhesives. Now, almost any material can be bonded to another if the right adhesive is used. Most adhesives used today are plastics, either thermoplastic or thermoset.

Because of their flexibility, structures bonded with adhesives resist damage from vibration better than riveted or bolted assemblies. A riveted joint must resist high stress at each rivet hole, and cracking can result, but adhesive bonding is distributed over the entire surface, reducing stress concentration. Also, connectors such as rivets or bolts can be a cause of corrosion, allowing galvanic currents to flow, which allows corrosion to take place. In contrast, adhesives prevent galvanic current flow because they normally have a high electrical resistance.

Nitrile rubber epoxy resins are used for bonding honeycomb structures for aircraft (Figure 5.18). Two-part epoxy kits are available for general use in which a resin and a catalyst are measured out and mixed together. When mixed, this cement hardens in minutes or hours, depending on the formation. Epoxy resins with fiberglass cloth are used on the hulls of ships and boats to give them a strong leakproof surface. Automobile body repairs are often made with fiberglass and resin.

The cellulosic plastics are used in model cement and some household cement. Polyvinyl acetate is a white glue used in woodworking. Polyethylene is often used as a hot-melt adhesive. Polysulfides, polyurethanes, and silicones are used as sealants for caulking of buildings and boat hulls. Phenolic resins have been used for years as a bonding material for abrasive grinding wheels. Most adhesives will harden in the presence of air (aerobic conditions), but some types such as the cyanoacrylates will harden only when they are not exposed to oxygen in the atmosphere (anaerobic conditions) and only when spread out as a thin film between two surfaces. Cyanoacrylates, popularly known as super glue, will instantly bond smooth flat surfaces with a high bonding strength but will not bond uneven or porous surfaces. Table 5.3 lists some of the kinds of adhesives and their uses.

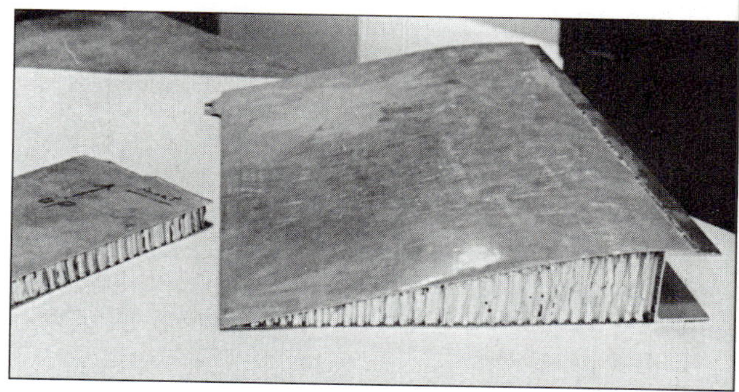

FIGURE 5.18
Bonded honeycomb aluminum structures for aircraft wing parts.

TABLE 5.3
Adhesives and their uses

Name	*Advantages*	*Disadvantages*	*Uses*
Anaerobic cyanoacrylates (popularly called superglue)	High strengths (2000 PSI) in a few minutes	High cost, thicknesses greater than 0.002 in. cure slowly or not at all, do not cure in presence of air	Production speed in bonding small parts, especially metal to metal and metal to plastics
Acrylic	Clear with good optical properties, ultraviolet stability	Moisture sensitive, high cost	Bonding plastics
Cellulose esters and ethers	Very soluble	Moisture sensitive	Bonding of organic substances such as paper, wood, and some fabrics
Epoxies: modified, nitrile, nylon, phenolic, polyamide, vinyl	Low-temperature curing, adheres by contact pressure, good optical qualities for transmitting light, high strength, long shelf life	High cost, low peel strength in some types	Bonding of microelectron parts, aircraft structural parts, cutting tools, and abrasives
Phenolics (Bakelite)	Heat resistance, low cost	Hard, brittle	Abrasive wheels, electrical parts
Polyimides	Resistance to high temperature (600°F, or 315.5°C)	Cost, high cure temperature (500 to 700°F, or 260 to 371°C)	Aircraft honeycomb sandwich assemblies
Polyester (anaerobic)	Hardens without the presence of air	Limited adhesion, cost	Locking threaded joints
Polysulfide	Bonds well to various surfaces found in construction	Low strength	Caulking sealant
Polyurethane (rubbery adhesive)	Strong at very low temperatures, good abrasion resistance	Low resistance to high temperatures, sensitive to moisture	Flexible and rigid foam protective coatings, vibration damping mounts, coatings for metal rollers and wheels
Rubber adhesives: butadienes, butyl, natural (reclaim), neoprene, nitrile	High-strength bonds with neoprene (4000 PSI), oil resistance (nitrile)	Low tensile and shear strength, poor solvent resistance	Adhesives for brake linings, structural metal applications, pressure-sensitive tapes, household adhesives
Silicones	High resistance to moisture, can resist some temperature extremes	Low strength	Castable rubbers, vulcanized rubbers, release agents, water repellents, adhesion promoters, varnishes and solvent-type adhesives, contact cements
Urea–formaldehyde resins	Low cost, cures at room temperatures	Sensitive to moisture	Wood glues for furniture and plywood
Vinyl: polyvinyl-butyralphenolic vinyl plastisols, polyvinyl acetate and acetals, polyvinyl chloride	High room temperature peel, ability to bond to oily steel surfaces (vinyl plastisols)	Sensitive to moisture	Bonds to steel (brake linings), safety glass laminate, household glue, water and drain pipe (PVC), household wrap

(Neely and Bertone, *Practical Metallurgy and Materials of Industry,* 5th ed., © 2000 Prentice Hall, Inc.)

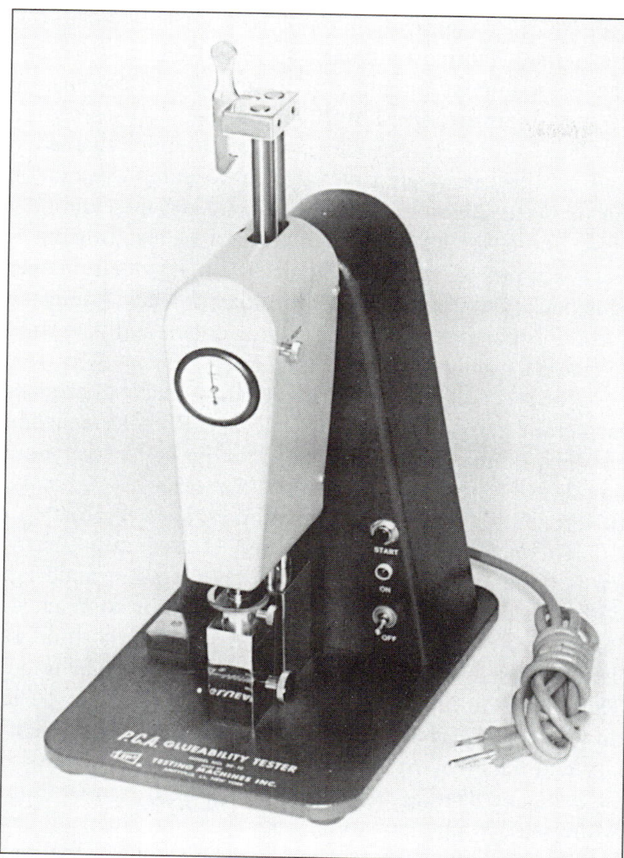

FIGURE 5.19
The glueability tester tests the strength of a bond formed between adhesive and material being tested under specific, carefully controlled circumstances (Testing Machines Inc.).

As with other plastics and rubbers, additives are used in adhesives to impart needed properties such as resilience and toughness. When bonding agents are required to conduct electricity, conducting additives such as carbon black, graphite, or metal flakes or powders are used. Certain destructive tests are used to determine joint strength and efficiency for various adhesives and bonding materials. Among these are shear, tensile, compression, impact, and peel tests. Cleavage (a variation of the peel test), tack (adhesion), creep, and fatigue tests are also very common (Figure 5.19). There are several specifications applicable to adhesive bonding: two of them are the Military, Federal Adhesive Specifications and those of the ASTM.

PETROLEUM PRODUCTS

Lubricants

In 1859 Colonel E. L. Drake struck oil in the little Pennsylvania town of Titusville. Its early impact on the economy was to replace whale oil as fuel for lamps. This discovery, no doubt, helped save the whales for posterity, but today, petroleum has many more uses; it fuels our industries and our transportation systems and provides the mineral oils used in the lubricants that keep our equipment and machinery running. In addition to the mineral oils, present-day lubricants contain synthetic oils (polymers) made from chemical feed stock.

As it comes from the oil wells petroleum crude is composed of a large variety of hydrocarbon molecules in many chemical combinations. By the process of distillation and condensation, these constituents are separated into several classes, namely, gases, gasoline, distillate fuel, and lubricating oil, with the residual heavy end products being used in the production of heavy lubricating-oil stocks, bunker fuels, and asphalt. Chemically, the carbon and hydrogen molecules in petroleum exist in many combinations, giving various properties to oils from different sources of crude. In addition, all petroleum oils contain small amounts of sulfur and nitrogen, which affect the stability and corrosiveness of the oil. The hydrocarbon molecules found in petroleum can be grouped into three classes: paraffins, naphthalenes, and aromatics. Of these, the paraffins and naphthalenes have the greatest stability and are the principal constituents of lubricating oils that are used in automobile crankcases and in other machinery. The aromatic oils have good solvency characteristics and are used for certain chemical formulations and for ink oils.

Greases Greases are actually mineral or synthetic lubricating oils held by absorption and capillary action in the fibrous or granular structure of a thickening agent, for example, clay. The oil used for greases that are intended to withstand high pressures at relatively slow speeds is usually one of higher viscosity than one intended for normal or higher speeds. The most important quality of greases is their adhesiveness to metallic surfaces and their "stay-put" property. Greases are extensively used to pack wheel bearings and steering linkages in automobiles and for lubricating manufacturing machinery.

Asphalts

Asphalt is a black, sticky substance that is also separated from crude petroleum. However, some asphalts occur naturally in pits or lakes that are residues of crude oil in which the lighter fractions (more volatile parts) have evaporated over a period of many thousands of years. Asphalts can generally be divided into two categories, paving asphalts (asphalt cements) and liquid asphalt materials. Thermoplastic asphalt cements are solid but can be heated to a liquid condition for use as roofing tar and for paving. Liquid asphalts are not heated but are diluted with solvents to provide the proper consistency. The liquid asphalts harden over a period of time as the lighter fractions evaporate, but the molten (heated) asphalt hardens when it cools.

Asphalts are used in the manufacture of waterproof papers and textiles and for waterproof linings for irrigation canals and reservoirs; however, plastics are beginning to replace asphalt products in many of these applications. Asphaltic concrete is used extensively for road and highway construction and for parking lots and driveways.

CERAMIC MATERIALS

Traditionally, **ceramic** materials include clay, silica, glass, natural stone, and portland cement (with which concrete is made). To this list must be added the materials used in the newer "engineered" ceramics given in Chapter 1—metal oxides, carbides, nitrides, and sulfides. Ceramics are held together by ionic and covalent bonds, which are very strong but rigid and much less ductile than the metallic bond, making them subject to brittle failure. Because they are good thermal and electrical insulators and can withstand high temperatures, ceramic materials are extensively used in the electrical industry and for furnace linings.

Engineered ceramics, along with plastics, appear to be among the rapidly growing areas of materials technology.

Clays

Clays vary considerably according to where they are mined. Ordinary clays containing impurities such as iron oxide (which gives them their reddish color) are used to manufacture brick for building construction and firebrick for furnaces. **Kaolin,** a purer white clay composed mostly of alumina and silica, is used in the manufacture of earthenware, fine china, porcelain, paper products, and firebrick. Fire clay has less than 10 percent impurities and is used for the manufacture of clay products and for joining firebrick in furnace construction.

Porcelain enamels are composed of quartz, feldspar, borax, soot ash, and other substances. These are used to coat iron, steel, or aluminum for wear and decorative purposes for products such as stoves, refrigerators, pots, and pans.

Engineered Ceramics

Some recently developed kinds of ceramic materials are known as fine ceramics or engineered ceramics. These include insulation materials for electronic substrates, metal oxides (semiconductors), heat-resistant materials for engines, cutting tools, and corrosion-resistant materials. The composition and atomic structure of engineered ceramics was discussed in Chapter 1.

The carbides form one group of ceramics that often find application as abrasives. In the formation of silicon carbides the silicon is provided by sand and the carbon by coke. These two are baked together, and the resulting material is crushed to create small particles. In a similar manner tungsten is carburized, and the resulting powdered tungsten carbide is compacted with powdered cobalt and then fired to create a usable tool chip, for example.

The relatively new sol gel process is capable of producing ceramics with very fine particle size. Rather than using crushing to break up the material, in this process two or more liquids are combined to form a gel; this gel is poured into a suitable mold that when fired creates a shaped ceramic.

The most frequently used process for creating ceramics is compacting plus sintering. The ceramic material, with the desired particle size, is densified or compacted. In this *green state* it has very little strength, so binders may be added so it can be handled, shaped, or machined. The binders are removed later by thermal processes. The disadvantage of using binders is that they lower the density and thus the strength. The last step in the process is sintering or heating the ceramic at a high temperature. Although a simple-sounding operation, the sintering can cause extensive shrinkage, making the sizing of the part difficult, since ceramics at this point are too hard to shape or machine.

The intermetallic compound nickel aluminide (Ni_3Al) is an interesting and promising addition to the ceramic family that is beginning to be used in high-temperature applications. It has the unusual property that its strength *increases* as temperature increases to approximately 1500°F (830°C).

Glass

Glass is a supercooled rigid liquid. It is in a metastable state; that is, it can go to a stable lower energy state of crystallization only by passing through an intermediate higher-energy state. In effect, glass has become too cold to freeze from the liquid. Common types of glasses are based on silicon dioxide (SiO_2) which occurs abundantly in nature as quartz and cristobalite and as a part of many of the silicate minerals. Natural volcanic glass, called *obsidian,* was used by primitive peoples for arrowheads and spear points. Quartz and silica sand are common raw materials for the production of commercial glasses and silica. Additives can alter the glass, giving it special properties. For example, the common soda-lime glass has a lowered softening temperature and is used for windows, containers, and lamp bulbs. Borosilicate glass, with a low thermal expansion, does not tend to crack when unevenly heated; consequently, it is used for laboratory glassware, large telescope mirrors, and household cookware such as Pyrex®. Lead-alkali-silicate glass has an even lower softening point and is known as flint glass for optical purposes and crystal glasses for tableware. For a high refractive index, the lead oxide (PbO) content may be as much as 65 percent.

Glass has excellent colorability and is a versatile material that has excellent chemical resistance to everything except hydrofluoric acid, which is used to etch glass objects. A considerable amount of glass plate is used in modern building construction, and glass fiber is used extensively as a reinforcement material in plastic resins and in the communications industry.

Portland Cement

The invention of portland cement is generally credited to Joseph Aspdin, an English mason who obtained a patent for it in 1824 and named his product portland cement because the concrete made of it resembled a natural limestone on the Isle of Portland. Materials used in the manufacture of portland cement are lime, silica, alumina, and iron. Each of these can come from any of a number of source materials. For example, alumina comes from clay, shale, slag, aluminum ore refuse, and a number of other source materials.

Selected raw materials are pulverized and proportioned to produce the desired chemical composition. The prepared mix is fed into the upper end of a sloping rotating kiln. A powdered coal, fuel oil, or gas is burned at the lower end of the kiln, producing temperatures of 2600 to 3000°F (1427 to 1649°C). The resulting portland cement clinker is cooled and then pulverized, and a small amount of gypsum is added to regulate the setting time of the cement. This finely ground product, which will pass through a sieve with 40,000 openings in a square inch, is the finished portland cement, ready to be shipped. Typical steps in the manufacture of portland cement are illustrated in a flowchart (Figure 5.20). Each manufacturer of portland cement uses a trade or brand name under which the product is sold.

Concrete

Concrete is commonly seen in sidewalks, streets, dams, and building foundations. It is a mixture of portland cement, sand, and gravel plus water and, sometimes, additives. These ingredients (sand and gravel are called **aggregates**) are mixed together in specific proportions to produce a fluid mass called concrete. It is then poured or pumped into molds where it is left to harden. The forms or molds may be removed in a few days. Ordinary concrete takes about 28 days to completely cure, providing it is kept wet, at which time it is nearly as hard and strong as it will ever be.

Specialty Concretes

Among the many admixtures used to improve portland cement concrete is fly ash, a finely divided residue resulting from the combustion of coal; in the presence of lime the fly ash acts like a cement. Its main value lies in its ability to reduce the percentage of portland cement needed in concrete, thus reducing the cost. It also imparts a higher chemical resistance than ordinary concrete. Like fly ash, blast furnace slag can also reduce the percentage of cement needed in concrete; when granulated and activated by sodium hydroxide, it can completely replace portland cement in some cases. Silica fume, a powdery by-product of electric arc furnace production of ferrosilicon alloys and metallic silicon, also behaves like natural pozzolan, and it gains strength earlier in the curing cycle than fly ash and slag concretes do.

Polymer Concretes A growing technology involves the introduction of polymers into concrete. Latex-modified portland cement concretes have high strengths, reduced porosity, and water resistance. Polymer-cement concrete has better abrasion, higher strength, and impact resistance than conventional concrete, making it useful for industrial floor applications. Polymer–portland-cement concrete is mixed with either a monomer or a polymer in a liquid, powdery, or dispersed phase that is allowed to cure along with the cement or is polymerized in place.

Polymer concrete is a composite material in which an aggregate mixture is combined with a monomer that is polymerized in place; no portland cement is used. Monomers used in polymer concrete include polyester-styrene resins, furan, vinyl esters, and epoxy. Also, sulfur concrete and sulfur-impregnated concrete are included in a broad definition of polymer concretes. Polymer concrete materials are excellent for patching because of their rapid curing time (15 to 60 min). A major disadvantage is that they cannot withstand higher temperatures.

Fiber-Reinforced Concrete Steel reinforcing bars have long been in use in portland cement concrete. Concrete mixtures using fibers act as secondary reinforcement in concrete and can greatly increase its tensile strength. Small fibers can prevent crack growth in concrete structures. Typical fibers include cotton, rayon, glass, steel, acrylic, nylon, polyester, polyethylene, and polypropylene.

WOOD PRODUCTS

Forest industries, in which wood products such as lumber, plywood, masonite, and particleboard are manufactured, supply a large portion of the building materials for homes and offices and for furniture manufacture. Paper products produced from wood pulp are extensively used in packaging

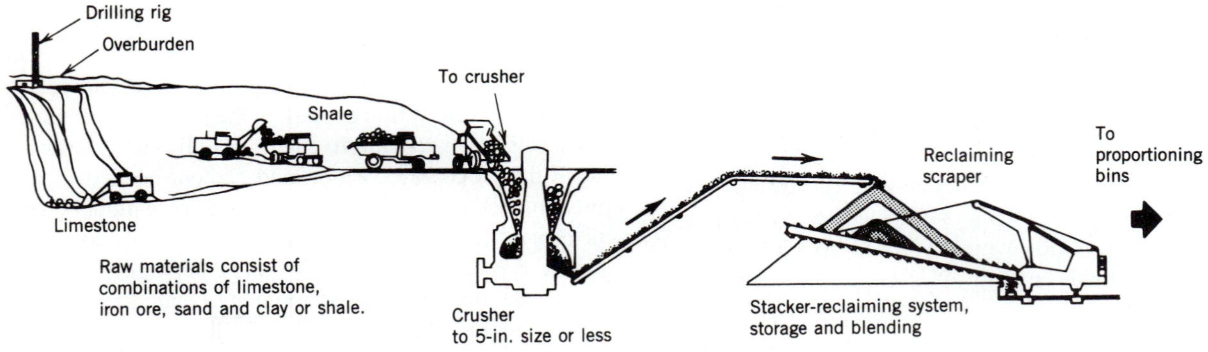

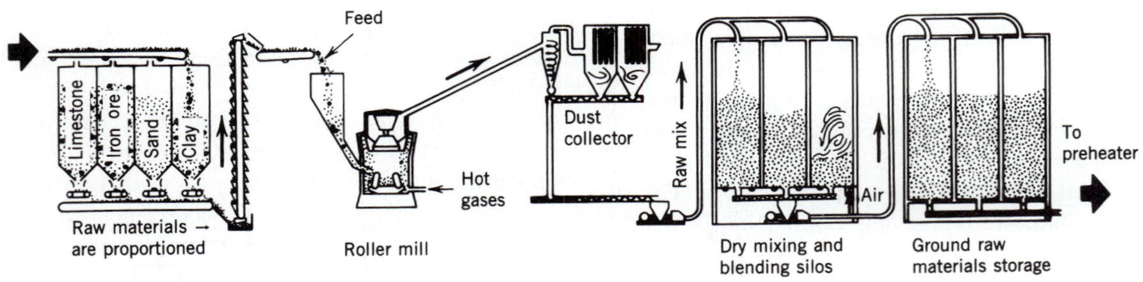

FIGURE 5.20
Flowchart showing the steps in the manufacture of portland cement by the dry process using a preheater. (1) Quarry and blending of raw materials. (2) Proportioning and fine grinding of raw materials. (3) Kiln system. Preheating, burning, cooling, and clinker storage. (4) Finish grinding and shipping (Oregon Portland Cement Company).

and in printing. Wood is composed of **cellulose, lignin,** and small amounts of inorganic materials that make up the ash when wood is burned.

The engineering properties of wood vary a great deal in different species of trees and even within one species because of imperfections such as knots and checks. Moisture markedly decreases the mechanical properties of wood. Because of this and the shrinkage of "green" wood when it dries, construction-grade lumber and other wood products are either air or kiln dried before they are finished (planed or sanded) to size. Wood may be classified into two broad categories, hardwood and softwood.

Hardwoods

Deciduous trees shed their leaves each year and are classed as hardwoods. The definition generally refers to the difficulty of sawing rather than its hardness as the term is used in engineering. In fact, balsa, although it is very lightweight, soft, and porous, is classed as a hardwood because it comes from a deciduous tree.

Oak trees are found in many varieties in the United States, but they generally can be divided into two classes, white oak and red oak. White oak is more resistant to decay and finer grained than red oak, and it is more difficult to work. Red oak is used extensively in furniture manufacture, flooring, and plywood. Oaks, sugar and black maples, and birch grow in the northern forests from Minnesota east and south to Virginia. Sweet gum, ash, and tupelo are found in Texas and some other southern states. Hickory, oak, black walnut, and basswood grow in the prairie states. Oregon maples are not as hard or strong as sugar and black maples. Walnut is found in most eastern and northern states and in some states in the far West.

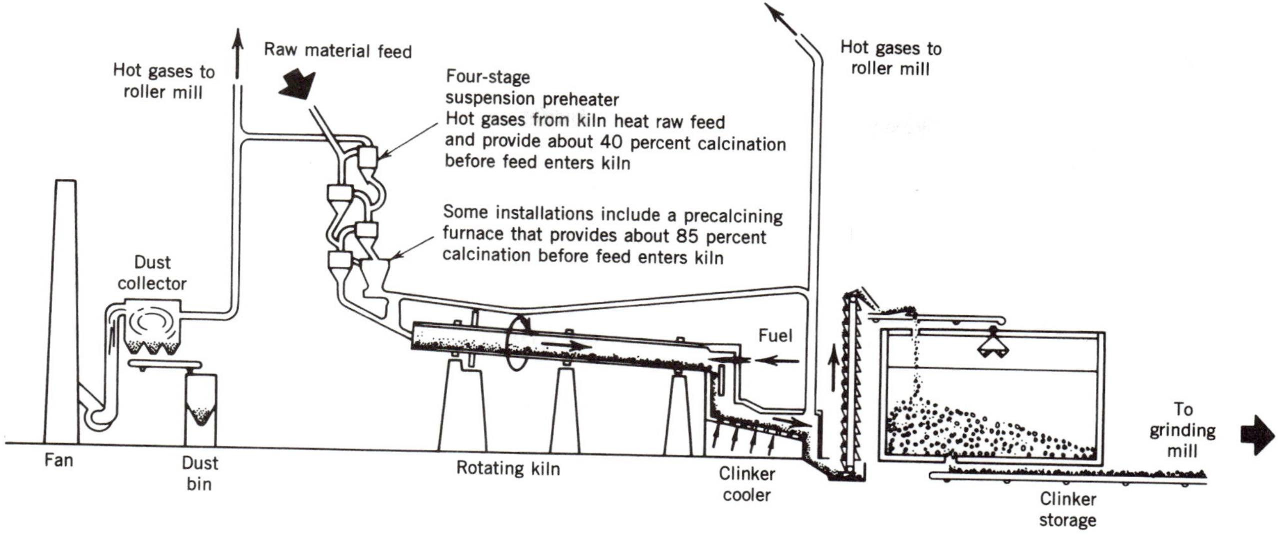

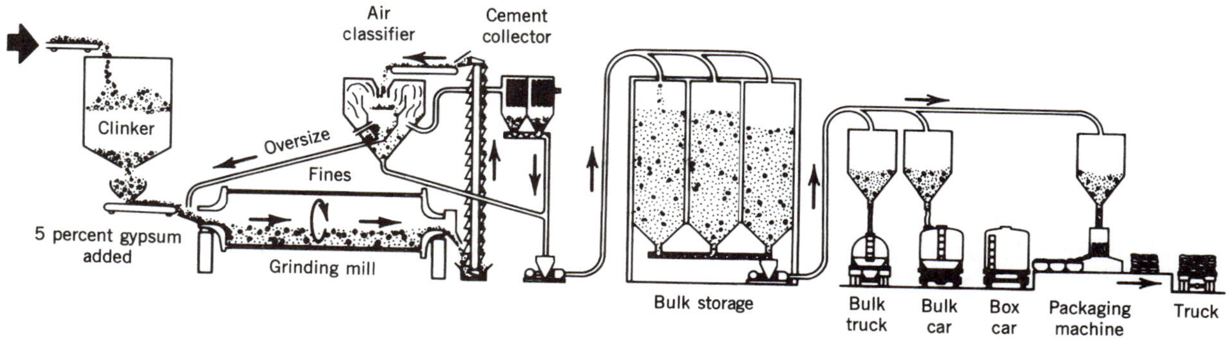

Softwoods

Coniferous trees have cones and needles instead of flat leaves and are sometimes called evergreen trees because they do not lose their needles in the fall but keep them throughout the year. The coniferous trees are classed as softwoods even though some varieties have a very tough, hard wood. For example, the Sitka spruce was once used for structural frames in aircraft because it was considered to have the highest strength-to-weight ratio of any wood. Because of its resonant qualities, it is also used for musical instruments.

The pines are the most widely used softwoods. Western white pine, sugar, ponderosa, and lodgepole pine are found in the western and northwestern parts of the United States. Southern yellow pine is grown in the southeastern part of the United States.

Douglas firs and white fir are grown in the far West and are widely used in building construction. Firs are somewhat harder and stronger than pines. Spruce, hemlock, redwood, and cedar are generally found in the western United States and are used in building construction, boats, mine timbers, and various wood products.

Review Questions

1. From what raw materials are most plastics derived?
2. What is the source of natural rubber?
3. What polymer derived from acetylene gas is called polychloroprene (neoprene)?

4. From what basic source of raw material do we get most of our pesticides, fertilizers, clothing fibers, and many medicines such as antibiotics?
5. What is the major difference between thermosetting and thermoplastic materials?
6. Which type of material, thermosetting or thermoplastic, can be easily injected by pressure into a mold?
7. If you need a plastic material for a product that must resist a variety of chemicals and the effects of sunlight, have a nonstick surface, and withstand high temperatures such as would be found in a baking oven, what would you choose: (a) silicone plastic, (b) polyurethane plastic, (c) Bakelite, or (d) fluorocarbon plastic?
8. Name two items in which glass-reinforced nylon 6/6 can replace metal where considerable loading is involved.
9. Some plastics are colored or tinted with pigments, whereas others have external surface finishes made with lacquers, enamels, or overlays. How can a metal film be applied to a plastic product?
10. In what process is natural latex combined with sulfur and heated to produce the flexible substance we call rubber?
11. Modern aircraft are made with fewer rivets and more adhesive-bonded structures than in the past. Name two advantages of using bonded structures over riveted ones.
12. Of the three classes of hydrocarbon oils, the aromatics are used for solvents and ink oils. What is the principal use of the other two types, paraffins and naphthalenes?
13. The tarry substance used for roofing and highway paving is a petroleum derivative. What is it called?
14. The use of clay for bricks, porcelains, and electrical insulation is well known. New uses for special ceramics, called engineered ceramics or high-technology ceramics, are being developed. Name two applications for these special ceramic materials.
15. Traditional construction materials include brick, stone, metal, and glass. Also, plastic materials are being used for many products in the construction industry. Besides these, what are the two major material types used in construction?

Case Problems

Case 1: Plastic Color Problem

A small plastic product made of Bakelite is black and has had a good sales record for years; however, sales have begun to drop off significantly. The competition is using bright colors in their product and is taking over the market. In what way can the first company improve its product, add color, and increase productivity to lower the price?

Case 2: Air Cylinder Stick-Slip Problem

A small air cylinder was manufactured to move a machine part. The piston seal was made of neoprene rubber, which had a tendency to take a set or stick to the cylinder wall when it was not operated for a given time. This caused the mechanism to stick-slip, that is, to hesitate and then jump forward too fast. Rubber O-rings, cups, and other types of rubber seals did not solve the stick-slip problem, which was obviously a seal problem. From what synthetic material would you choose to make this piston seal in order to eliminate this sticking problem?

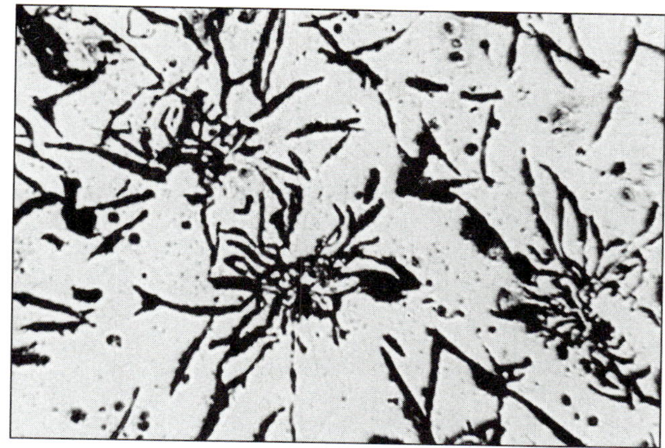

CHAPTER 6

Selection and Application of Materials

Objectives

This chapter will enable you to:

1. Identify many metals and alloys by their numerical classification systems.
2. Describe some of the characteristics of many commonly used metals and alloys.
3. Name some methods used to identify and analyze various metal alloys and materials.
4. State some of the methods and factors involved in selecting materials for manufacturing purposes.

Key Words

AISI	SAE
ASTM	ASM
HSLA	microalloying
stainless steels	cast iron

When only carbon steel, cast iron, wrought iron, and a few nonferrous metals were commonly used, identifying them was a relatively simple process. Spark testing on a grinder was enough to separate the three ferrous metals, and for the most part, the nonferrous metals could be identified by their color. Of course, nonferrous metals such as gold, silver, and copper began to be alloyed (combined) as early as the time of Archimedes. As the story goes, after he made the discovery of the natural law of buoyancy, he immediately applied it to solving the riddle of whether a golden crown contained pure gold or was cheapened with an alloying element having lesser value.

In our modern world there are literally thousands of alloys, both ferrous and nonferrous. Selecting and identifying metals is therefore a far more complex process today, requiring classification systems and sophisticated testing equipment. Even more important than metals identification is the need for methods and equipment to test and analyze other materials of industry, especially those that present radiation hazards and contain toxic materials.

With so many new engineering materials being developed in the areas of plastics, composites, metals, and ceramics, it can be difficult to choose manufacturing materials. Although a new product might have more sales appeal when made of plastic, it might not have enough strength or heat resistance. Other products might be too costly to fabricate from metal but could be made from a selected plastic at a high production rate by the injection molding process and still have satisfactory engineering properties. Chapter 6 covers some of these factors in material selection.

Although this chapter discusses materials in a general way, its primary aim is to help unravel the numerous classification systems used with metals. Metals have been used as engineering materials for so long that their classification systems are very complete and sophisticated. The chapter begins with the various categories of ferrous metals, then covers the many types of nonferrous metals and concludes with some information about other types of materials.

CLASSIFICATION SYSTEMS FOR METALS

The most common numerical system used to classify carbon and low-alloy steels in the United States was developed by the Society of Automotive Engineers (**SAE**) and the American Iron and Steel Institute (**AISI**) for steels used in manufacturing. The American Society of Testing and Materials (**ASTM**) developed specifications for carbon and alloy structural steels. Tool steels are classified under their own system, which covers seven major types. Stainless steels are classified under the chromium and chromium–nickel types, and cast irons are identified by ASTM numbers. Nonferrous metals are also classified under specific systems for each metal or alloy. Color coding of steels has long been used as a means of identification, but there is no universal code; each manufacturer, distributor, or local manufacturing plant has its own code. It is a very useful system on a local basis as long as the color coding is not cut off, leaving the remainder of the stock unbranded.

CARBON AND ALLOY STEELS

The SAE classification system for alloy steels uses a four- or five-digit number (Table 6.1). The first number on the left designates the type of steel. For example, plain carbon steel is denoted by the number 1, 2 is a nickel steel, 3 is a nickel–chromium steel, and so forth. The second digit indicates the approximate percentage of the predominant alloying element. The third and fourth digits (and fifth digit if required), represented by X, always represent the amount of carbon in hundredths of a percent or "points" of carbon. For plain carbon steel it is normally between 0.08 and 1.70 percent. SAE 1040 is plain carbon steel containing 0.40 percent carbon. SAE 4140 is a chromium–molybdenum alloy steel containing 0.40 percent carbon and about 1 percent of the major alloy, molybdenum.

TABLE 6.1
SAE numerical designation of alloy steels*

SAE Number	*Identifying Elements*
Carbon Steels	
10XX	Nonresulfurized, Mn 1.00% max
11XX	Resulfurized
12XX	Rephosphorized and resulfurized
15XX	Nonresulfurized, Mn max > 1.00%
Alloy Steels	
13XX	Manganese steels
23XX	Nickel steels
25XX	Nickel steels
31XX	Nickel–chromium steels
32XX	Nickel–chromium steels
33XX	Nickel–chromium steels
34XX	Nickel–chromium steels
40XX	Molybdenum steels
41XX	Chromium–molybdenum steels
43XX	Nickel–chromium–molybdenum steels
44XX	Molybdenum steels
46XX	Nickel–molybdenum steels
47XX	Nickel–chromium–molybdenum steels
48XX	Nickel–molybdenum steels
50XX	Chromium steels
51XX	Chromium steels
50XXX	Chromium steels
51XXX	Chromium steels
52XXX	Chromium steels
61XX	Chromium–vanadium steels
71XXX	Tungsten–chromium steels
72XX	Tungsten–chromium steels
81XX	Nickel–chromium–molybdenum steels
86XX	Nickel–chromium–molybdenum steels
87XX	Nickel–chromium–molybdenum steels
88XX	Nickel–chromium–molybdenum steels
92XX	Silicon–manganese steels
93XX	Nickel–chromium–molybdenum steels
94XX	Nickel–chromium–molybdenum steels
97XX	Nickel–chromium–molybdenum steels
98XX	Nickel–chromium–molybdenum steels
Carbon and Alloy Steels	
XXBXX	B denotes boron steels
XXLXX	L denotes leaded steels
XXVXX	V denotes vanadium steels

(*2001 SAE Handbook,* Vol. 1, Society of Automotive Engineers, Warrendale, PA.)
*XX represents percent carbon in hundredths or points.

TOOL STEELS

Tool steels are mostly high carbon or high alloy in carbon steel; however, a few types do have low carbon or alloy to give them special properties. As the name implies, these special steels were designed for tools, dies, molds, machinery parts, and for certain specific applications. They all require heat treatments such as hardening and tempering to prepare them for use. A letter has been assigned to each of the nine major groups of tool steels, with one or more numerals used to identify the individual steels. Each group is designated by its special purpose or by the quenching medium used (Table 6.2).

Compared with steels used for structural or machine-part applications, tool steels are very hard with high yield strengths. They usually lack toughness, as would be measured by an ability to absorb shock or undergo plastic deformation; that is, they would *not* have a large percent elongation. Because of the requirements of tools, a tool that deforms plastically (a bent thread tap?) is of little value as a tool.

Some tools must resist wear, therefore, many of the tool steels contain large amounts of carbon that combine with elements such as tungsten, titanium, and vanadium to form very hard carbides. These elements plus others such

TABLE 6.2
SAE designations for tool and die steels

Water-hardening tool steels	W
Shock-resisting tool steels	S
Cold-work tool steels	
Oil-hardening types	O
Medium-alloy air-hardening types	A
High-carbon–high-chromium types	D
Hot-work tool steels	
Chromium-base types	H
Tungsten-base types	H21
High-speed tool steels	
Tungsten-base types	T
Molybdenum-base types	M
Special-purpose tool steels	
Low-alloy types	L

(*2001 SAE Handbook,* Volume 1, Society of Automotive Engineers, Warrendale, PA.)

as chromium and manganese contribute to hardenability, as discussed in Chapter 3.

The volume of tool steels produced is very small compared with the tonnage used for other purposes, and tool steels tend to be manufactured by small companies to their own specifications. Although a tool or die steel may be produced to meet one of the SAE categories just discussed, the steel provided by company A will probably behave differently in an application than the one provided by company B. Caution should be taken in replacing a successfully performing tool/die steel with "the same steel" of another company without extensive trials.

STRUCTURAL STEELS

Standard ASTM specifications are used to designate carbon and alloy structural steels. These steels are produced in steel mills as standard shapes (Figure 6.1) and are used in the construction of buildings, bridges, and pressure vessels, and for other structural purposes. Engineers who design steel structures, and welders, are more likely to use the ASTM specifications for steels than designers of machinery, heat treaters, and machinists who are more likely to need the SAE–AISI system to refer to the steels they will be using. These products are usually in the form of bar stock, strip, plate, and round bars and are used extensively in manufacturing steel products. Some bar stock is alloy steel such as AISI 4140, which is used to produce tough heat-treated parts such as gears and machine parts. AISI 8620 is a carburizing grade used where an extremely tough material with

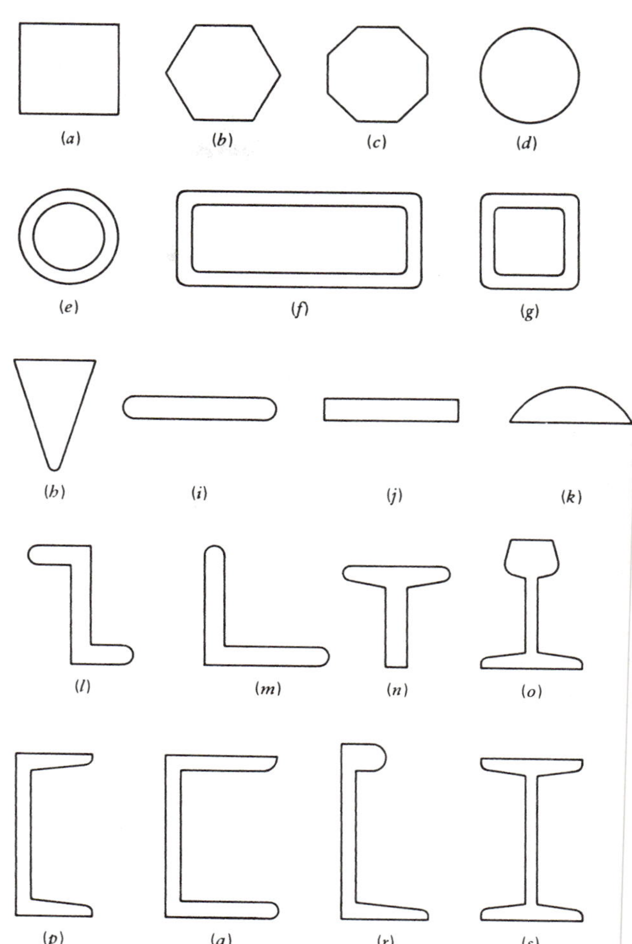

FIGURE 6.1
Steel shapes used in manufacturing. *(a)* Square HR or CR. *(b)* Hexagonal. *(c)* Octagon. *(d)* Round. *(e)* Tubing and pipe (round). *(F)* HREW (hot-rolled electric-welded) rectangular steel tubing. *(g)* HREW square steel tubing. *(h)* Wedge. *(i)* HR flat bar (round edge spring steel flats). *(j)* Flat bar (CR and HR). *(k)* Half round. *(l)* Zee. *(m)* Angle. *(n)* Tee. *(o)* Rail. *(p)* Channel. *(q)* Car and ship channel. *(r)* Bulb angle. *(s)* Beams—I, H, and wide flange. (Neely and Bertone, *Practical Metallurgy and Materials of Industry,* 6th ed., ©2003 Prentice Hall, Inc.).

a high surface hardness is required; however, much of the steel used in manufacturing is mild steel (low-carbon) AISI 1020. Low-carbon steel comes from the mill with either a cold finish or hot-rolled surface. Cold finish (CF) is simply a broad term for cold-rolled (CR) or cold-drawn (CD). These cold-worked steels are tougher and stronger than are hot-rolled (HR) steels that have the same carbon content.

Cold-drawn steels have a relatively smooth finish without any scale. Hot-rolled steels are covered with a black mill scale that is formed when the steel is at rolling or forging heat by the action of oxygen. This scale, which is chemically Fe_3O_4, must be removed before the steel can be subjected to cold-finishing operations. Cold-finished steels are more expensive than hot-rolled steel, so they are not normally used when less accurate dimensions are acceptable and where heating or welding is to be done. Steel shafting is available as low-carbon or resulfurized cold-drawn steel and as alloy steel ground and polished (G and P) which has a brighter surface finish than cold-drawn steel and is held to closer dimensional tolerances. It is also a much higher strength steel and is much more expensive than cold-drawn, low-carbon steel. Cold-finished steel strip is produced in rolling mills for use in punch press work for blanking and drawing operations. It is also provided in a variety of surface finishes and colors that are not damaged by the manufacturing process. Tool steels and alloy steels are usually referred to by a trade name, but they usually are also identified by number.

HSLA STEELS

As a category, the relatively new high-strength, low-alloy (**HSLA**) steels are not very well defined. Most steels in this family do not attain their strength through heat treatment but rather through the control of their grain size on cooling from the austenite region. This is in contrast to the older practice of performing hot work and then letting the steel slow cool from austenite, during which time the steel would undergo grain growth. The steels thus produced had good ductility but relatively low yield strengths, the value upon which most designs would be based.

Improved properties are developed in the HSLA steels by the use of very small amounts, that is, **microalloying** amounts, of such elements as nitrogen, niobium, vanadium, and aluminum. These elements can form carbides and nitrides that act as sites upon which new grains form and thus prevent the growth of the austenite grains. Thus, improved properties are achieved with small amounts of alloying elements and also with lower carbon contents. The lower carbon content has the added advantages of reducing brittleness and improving weldability.

The HSLA were originally developed primarily for automotive weight reduction because a thinner sheet, plate, or structural shape or a smaller-diameter bar will almost always do the work of a thicker, heavier, plain carbon steel. These steels are appropriate for many other applications where high strength, formability, and weldability are needed. Some of these HSLA steels in plate, bar, and structural shapes include A709 and A737 for bridges; A441, A572, and A633 for low-temperature use; and A242 and A588 for weather-resistant construction steels. Full ASTM specifications may be found in materials and welding data handbooks.

STAINLESS STEELS

Early in Chapter 18 (Corrosion and Protection of Materials) chromium is identified as a metal capable of forming an adherent oxide that protects it from further corrosion. It is this property of chromium that is the basis of the **"stainless" steels.** If a minimum of 10 percent chromium is dissolved in the atom lattice of the steel, it will be protected by this oxide in the presence of many chemicals and in most atmospheric environments. Various of the stainless steel types have limited resistance to certain chemicals, for example, chlorine and/or sulfur ions, but others have been specifically designed to combat these and other troublesome environments. The manufacturers and suppliers of these steels, as well as the many handbooks available, usually will provide the information necessary to select the proper steel.

In Table 6.3 it can be seen that the stainless steels can be categorized by their microstructures and by their major alloying elements. Those of the 2XX and 3XX types are austenitic; the large amounts of nickel or nickel plus manganese slow the transformation of austenite to its normal room-temperature ferritic microstructure, and it stays austenitic. Some of the type 4XX are ferritic, but some can be heat treated to form martensite. Between these two are a group of duplex steels that contain enough nickel that they are partially austenitic, but not completely; thus, they are a combination of austenite and ferrite.

The type 3XX steels are in general the most corrosion and heat resistant of the stainless steels. The FCC lattice of austenite makes them more corrosion resistant than the BCC ferrite, and with proper processing also makes them nonmagnetic. Their FCC lattice also makes them very deformable. These steels are often used in food processing machinery, kitchen sinks, and jet engine parts.

The 4XX types, especially if they have sufficient carbon, can be heat treated and are often used where a sharp edge is required. The non-heat-treatable 4XX alloys usually have less carbon and are used where their strength can be developed through cold working. They are frequently used in automotive exhaust systems.

Included in the chromium–nickel types are the precipitation hardening stainless steels that harden over a period of time after solution heat treatment. Stainless steels are more difficult to machine than mild steels, and they tend to work harden quickly when cut with dull tools. With proper

TABLE 6.3
The classifications and basic compositions (percent) of some stainless steels

Type	*Carbon*	*Manganese*	*Silicon*	*Chromium*	*Nickel*	*Molybdenum*
Austenitic (Chromium–Nickel) (Chromium–Manganese–Nickel)						
201	0.15	5.50–7.50	1.00	16.00–18.00	3.50–5.50	
202	0.15	7.50–10.0	1.00	17.00–19.00	4.00–6.00	
301	0.15	2.00	1.00	16.00–18.00	6.00–8.00	
302	0.15	2.00	0.75	17.00–19.00	8.00–10.00	
304	0.08	2.00	0.75	18.00–20.00	8.00–10.50	
310S	0.04–0.10	2.00	0.75	24.00–26.00	19.00–22.00	
316	0.08	2.00	0.75	16.00–18.00	10.00–14.00	2.00–3.00
316L	0.030	2.00	0.75	16.00–18.00	10.00–14.00	2.00–3.00
Duplex (Austenitic–Ferritic)						
329	0.080	1.00	0.75	23.00–28.00	2.50–5.00	1.0–2.0
Ferritic or Martensitic (Chromium)						
410	0.15	1.00	1.00	11.50–13.50	0.75	
440C	0.95–1.20	1.00	1.00	16.00–18.00		0.75
444	0.25	1.00	1.00	17.5–19.5	1.00	1.75–2.50

(*2001 SAE Handbook,* Vol. 1, Society of Automotive Engineers, Warrendale, PA, except for 440C, which is from *Metals Handbook,* 8th ed., Vol. 1, ASM International, Materials Park, OH, 1961.)

tooling and cutting fluids, however, they lend themselves to good production rates, especially the free-machining types.

CAST IRONS

Cast iron is essentially an alloy of iron, carbon, and silicon. Theoretically, it can contain from 2 to 6.67 percent carbon, but commercial cast iron rarely has more than 4 percent carbon. Since austenite can contain only 1.7 percent carbon in solution, any more than that amount will precipitate out to form flakes of graphite (Figure 6.2) in ferrite when cast iron is cooled slowly, or graphite flakes *plus* cementite and ferrite as pearlite if it is cooled rapidly. Also, graphitization is increased as the silicon content is increased (Figure 6.3). The iron–graphite equilibrium diagram (Figure 6.4) identifies the various microstructures found in cast iron.

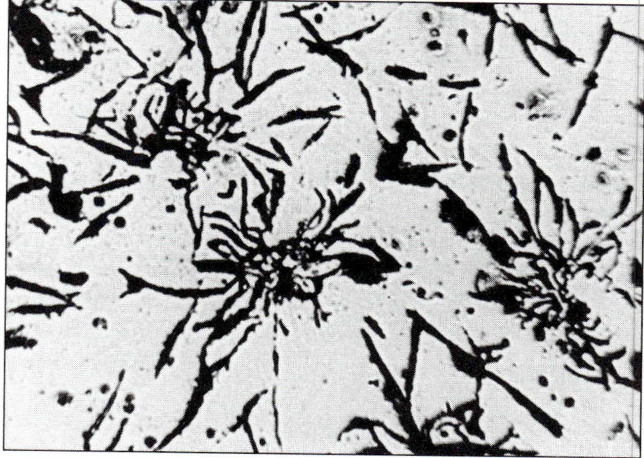

FIGURE 6.2
Gray cast iron showing graphite flakes, unetched (500 ×).

Gray Cast Iron

Gray cast iron is widely used to make machinery parts, since it is easily machined. Machine bases, tables, and slideways are made of gray cast iron because it tends to remain dimensionally stable after an aging period, and it makes an excellent bearing and sliding surface because of the graphite content. Cast iron has a lower melting point than steel and possesses good fluidity, so it is easily cast into sand molds, even those having intricate designs. When low-cost but serviceable machine tool housings and frames are needed, the choice is almost always gray cast iron or one of its alloys because of its ability to deaden vibration and minimize tool chatter. Gray cast irons are often alloyed with elements such as nickel, chromium, vanadium, or

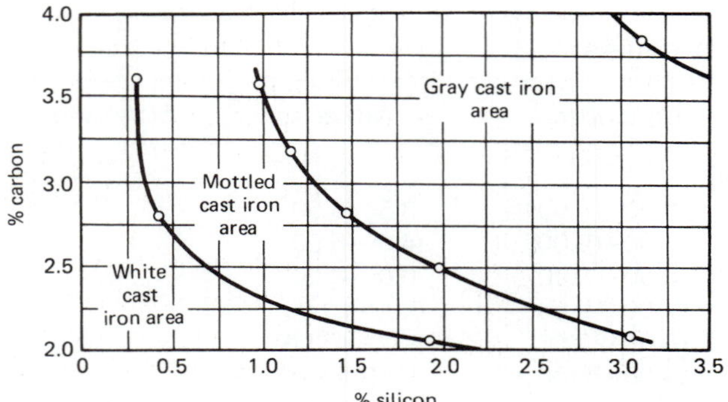

FIGURE 6.3
Composition limits for white, gray, and mottled cast irons (Neely, and Bertone *Metallurgy,* 5th ed., © 2000 Prentice Hall, Inc.).

TABLE 6.4
Classes of gray iron

ASTM Number	*Minimum Tensile Strength (PSI)*
20	20,000
25	25,000
30	30,000
35	35,000
40	40,000
45	45,000
50	50,000
60	60,000

(*1999 Annual Book of ASTM Standards,* ASTM, 100 Bar Harbor Dr., West Conshohocken, PA.)

copper to toughen and strengthen them. Unalloyed gray cast iron has two major drawbacks as an engineering material. It is quite brittle and it has a relatively low tensile strength, so it is not used where high stresses or impact loads are found. Cast irons may be ferritic or pearlitic in microstructure, which affects their strength. Gray cast irons are classified according to their tensile strength by ASTM numbers (Table 6.4). Although they can be welded, their high carbon content and their tendency to form extremely hard, brittle microstructures next to the weld make good welds difficult without preheating the base metal.

White Cast Iron

Unlike gray cast iron, white cast iron is virtually nonmachinable because it is so hard. It contains massive cementite, which is the hardest substance in iron and steel. For this reason, it is mostly used for its abrasion resistance in such applications as wear plates in machinery. The principal reason, however, for producing white cast iron is to convert is to malleable cast iron. There are two ways white cast iron can be formed: by lowering the iron's silicon content and by rapid cooling, in which case it is called chilled cast iron. White cast iron cannot be welded.

FIGURE 6.4
Iron–graphite equilibrium diagram (Adapted from Neely and Bertone, *Practical Metallurgy and Materials of Industry,* 6th ed., © 2003 Prentice Hall, Inc.).

°C °F
1538 2800
Delta
1399 2550
910 1670
738 1360
Liquid
Liquid + austenite
Liquid + graphite
Austenite
Iron + graphite
Iron graphite
Austenite + graphite
or
Austenite + cementite
Ferrite + graphite
or
Ferrite + cementite
0.69
2.0
4.3
6.67
% carbon
Cementite

Malleable Cast Iron

A cast iron that possesses all of the positive qualities of gray cast iron but has increased ductility, tensile strength, and toughness is called malleable cast iron. It is produced from white cast iron by a prolonged heating at or near the lower transformation temperature ($A_{3,1}$) for about 30 hours. This process, with the help of a small amount of silicon, essentially graphitizes the white cast iron to form tiny clumps of graphite instead of flakes as in gray cast iron (Figure 6.5). The iron matrix surrounds the graphite in this form, making it a more homogeneous metal, whereas graphite flakes (which are very soft and weak) cut across the iron matrix in every direction, causing it to be weak and brittle. In order for malleable cast iron to be produced, the thickness of the castings must be such that they can be cooled rapidly enough to form white iron throughout. Thus, there is a limit to this thickness.

Nodular Cast Iron

Essentially the same purpose is achieved in nodular cast iron as in malleable cast iron; graphite spheres are formed, but without the extensive heat treatment. Nodular cast iron is known by several names: nodular iron, ductile iron, and spheroidal graphite iron. Like malleable cast iron, it has toughness, good castability, machinability, good wear resistance, weldability, low melting point, and hardenability. The graphite is formed into spheres in the ladle by adding certain elements such as magnesium and cerium to the melt just prior to casting. Because the nodules are formed while the metal is in the molten state there is no thickness limitation, as with malleable irons. A vigorous, often spectacular, mixing reaction takes place that forms the graphite into tiny balls dispersed throughout the mix. Malleable and nodular cast iron castings are often used for machine parts that experience a higher tensile stress and moderate impact loading. These castings can be a less expensive alternative to steel castings or steel weldments.

NONFERROUS METALS

Nonferrous metals vary considerably in density (weight per unit volume), color, melting points, and mechanical properties. Some, often called noble metals, such as gold and platinum, have a high resistance to corrosion. Many of the so-called space-age metals such as titanium and niobium were only relatively recently extracted from their ores in

FIGURE 6.5
Micrograph of malleable cast iron. The clumps of graphite are shown in a matrix of ferrite. The matrix can also be pearlite. (100 ×) (*Metals Handbook,* Vol. 7, 8th ed., ASM International 1972, p. 95. With permission).

commercial quantities, and even aluminum, which we see everywhere today, was not extracted from ores in quantities until the late 1800s. Most nonferrous metals and alloys are classified in a numerical system peculiar to that metal.

Aluminum

Aluminum has a white or white-gray appearance because its surface is oxidized. Unlike the oxidation or rust of iron, which continues to form, the thin oxidized surface of aluminum prevents further corrosion. Aluminum weighs only 168.5 lb/ft^3 (or about 0.1 lb./in.^{3}) compared with 487 lb/ft^3 (or about 0.3 lb./in.^{3}) for steel, and pure aluminum melts at 1220° F (660° C).

Aluminum products are classed as either wrought or cast. Wrought aluminum products usually begin from alloyed ingots cast by a direct-chill (DC) process; they may then be shaped by hot or cold working processes. Their properties may be determined by cold-working operations or heat treating. For cast products, molten aluminum is cast by gravity into sand and permanent molds or injected at high pressure into dies. Cast aluminum is generally less hard and strong than wrought products, yet some types can be hardened by heat treatments.

Pure aluminum resists corrosion better than aluminum alloys, so where corrosion is a problem, the purer the metal the better, but where strength is a consideration, as it usually is, other solutions must be sought. One solution, often used in the aircraft industry, is to clad a stronger, more corrodible core alloy with sheets of a purer and more anodic aluminum alloy (see Figure 6.6 and Chapter 18 for further information). The cladding and core are bonded or welded together by the heat and pressure of a rolling operation.

There are several classification systems of specifications used to identify aluminum alloys: Federal, Military, the Society of Automotive Engineers (SAE), and the American Society for Testing and Materials (ASTM). The system most used by manufacturers is the one adopted by the Aluminum Association in 1954, which is shown in Table 6.5.

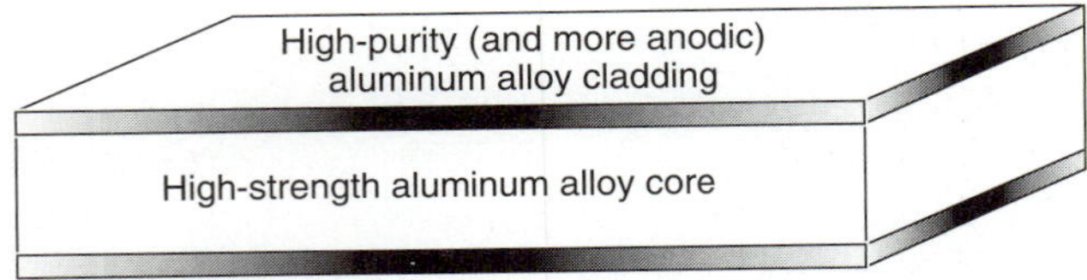

FIGURE 6.6
Metal cladding. Covering the stronger alloy core with a more anodic metal protects it from corrosion.

TABLE 6.5
Designation system for wrought aluminum and aluminum alloys

Code Number	*Major Alloying Element*
1xxx	None
2xxx	Copper
3xxx	Manganese
4xxx	Silicon
5xxx	Magnesium
6xxx	Magnesium and silicon
7xxx	Zinc
8xxx	Other element
9xxx	Unused series

(*2001 SAE Handbook,* J933, Society of Automotive Engineers, Warrendale, PA, © 2001.)

The first digit in Table 6.5 represents the alloy type. The second digit, here represented by the letter X, indicates any modifications made to the original alloy, and the last two digits (also represented by X) identify either the specific alloy or impurity. The first entry in the series represents nearly pure aluminum. An 1120 aluminum contains no alloying element and is 99.2 percent pure aluminum. This type cannot be hardened by heat treatment. In the 2000, 6000, and 7000 series, alloying elements make it possible to harden these alloys by solution heat treatment and aging. The alloys that cannot be heat treated are usually designated in the 1000, 3000, 4000, or 5000 series; however, all these alloys are work hardenable and therefore have an additional designation for temper. In this case, temper denotes hardness from cold working.

Temper Designation Temper designations for those alloys that are not heat treatable follow the four-digit alloy series number when appropriate.

—F As fabricated; no special control over strain hardening or temper designation

—O Annealed, recrystallized wrought products only; softest temper

—H Strain-hardened, wrought products only; strength is increased by work hardening.

The symbol —H is always followed by two or more digits. The first digit, which can be 1, 2, or 3, denotes the final degree of strain (work) hardening.

—H1 Strain hardened only

—H2 Strain hardened and partially annealed

—H3 Strain hardened and annealed

The second digit denotes the final degree of strain hardening.

2 = ¼ hard

4 = ½ hard

6 = ¾ hard

8 = Full hard

9 = Extra hard

For example, 3003—H16 is an aluminum–manganese alloy, strain hardened to ¾ hard temper. A third digit is sometimes used to indicate a variation of a two-digit, —H temper.

Heat Treatment Classifications For aluminum alloys that can be hardened by solution heat treatment and precipitation hardening or aging, the symbol —T follows the four-digit number. Numerals 1 to 10 follow this letter to indicate the sequence of treatment.

—T1 Cooled from an elevated temperature shaping process and naturally aged

—T2 Annealed (cast products only)

—T3 Solution heat treated and cold worked (and natural aging occurs)

—T4 Solution heat treated and naturally aged

—T5 Cooled from an elevated temperature shaping process and artificially aged

—T6 Solution heat treated and artificially aged

—T7 Solution heat treated and stabilized

—T8 Solution heat treated, cold worked, and artificially aged

—T9 Solution heat treated, artificially aged, and cold worked

—T10 Cooled from an elevated shaping process, artificially aged and cold worked.

For example, aluminum rivets may be made of 2024—T3 (an aluminum–copper alloy), which is solution heat treated, held at a low temperature to prevent hardening, riveted into place (which is cold working), and then naturally aged to full strength.

Cast Aluminum Classification

Aluminum casting alloys are generally divided into three categories: sand casting alloys, permanent mold alloys, and die casting alloys. A classification system similar to that of wrought aluminum alloys is used (Table 6.6). The general-purpose sand casting aluminum–copper alloy 212 has an ultimate strength of about 23,000 PSI. The die casting alloy 356—T6 is used for transmission cases and truck parts and has a tensile strength of about 38,000 PSI.

TABLE 6.6
Designation system for cast aluminum and aluminum alloys

Code Number	*Major Alloy Element*
1xx.x	None, 99 percent aluminum
2xx.x	Copper
3xx.x	Silicon with Cu and/or Mg
4xx.x	Silicon
5xx.x	Magnesium
7xx.x	Zinc
8xx.x	Tin
9xx.x	Other element
6xx.x	Unused series

(*2001 SAE Handbook,* J933, Society of Automotive Engineers, Warrendale, PA, © 2001.)

Cadmium

Cadmium is used as an alloying element for low-temperature-melting metals such as solder, metals used to cast typesetting characters, for printing processes, bearing metals, and storage batteries. It has a blue-white color and is commonly used as a protective plating on steel parts such as screws, bolts, and washers. Cadmium compounds are extremely toxic and can cause illness when inhaled. Welding, heating, or machining these plated parts can produce toxic fumes. These fumes can be avoided with proper personal protection equipment and ventilation.

Chromium

Chromium is a hard, slightly grayish metal that can take a brilliant polish and is often plated on metal parts. It is very corrosion resistant and when used as an alloying element in steel in quantities over 10 percent it produces stainless steel; in lesser quantities, it produces a high-strength alloy steel. Chromium is too brittle to have any direct applications in its pure massive state.

Cobalt

This hard silver-white metal is widely used as a matrix or binder in which tungsten carbide particles are combined and formed into cutting tools by the powder metallurgy process. It is also used as an alloying element for high-speed and stellite cutting tools, for resistance wire, and for heat-resistant alloys used for jet engine blades. Dental and surgical alloys contain large percentages of cobalt and chromium; for example, 62 percent Co and 29 percent Cr in alloy A. They are not attacked by body fluids and do not irritate the tissues, as many other metals do. Cobalt is resistant to ordinary corrosion in air. Some iron–nickel–cobalt–chromium alloys are called *superalloys* because they are used where high strength and resistance to high temperature are required. Some of these alloys have the same expansion rate as glass and so are used for glass-to-metal joints and seals. Strong permanent magnets are made with cobalt and other metals. The familiar Alnico magnet material is an alloy of aluminum, nickel, cobalt, and iron. Cobalt and nickel are two of the few nonferrous metals that can be attracted to a magnet.

Copper and Copper-Based Alloys

Copper is a soft, reddish metal that has a number of valuable properties as a pure metal and in various combinations with other metals. Its high thermal and electrical conductivity, which is second only to that of silver, is of major significance. Copper has high formability, corrosion resistance, and medium strength. To U.S. industry, the term *copper* refers to that element with less than 0.5 percent impurities or alloying elements. Copper-based alloys are those having more than 40 percent copper.

Deoxidized Copper This type of copper differs chiefly in its lower electrical and thermal conductivity. It also has a somewhat higher ductility and is more readily formed. For this reason it is the most commonly used copper for the manufacture of tubular products such as those used in domestic and industrial plumbing.

Low-Alloy Copper Copper is often alloyed with very small percentages of other metals (from a fraction of 1 percent to approximately 2 percent) for the purpose of imparting such qualities as corrosion resistance, higher operating temperature, and increasing tensile strength and machinability. These additives include arsenic, silver, chromium, cadmium, tellurium, selenium, and beryllium. Copper cannot be hardened by heat treatments, but with the addition of about 2 percent beryllium it can be sufficiently hardened by precipitation and aging so that it can be used for making springs, flexible bellows, and tools. Because of their nonsparking quality, beryllium–copper tools are used in explosion-hazardous environments such as mines, powder factories, and some chemical plants. Because beryllium is quite expensive, an alloy containing only 0.4 percent Be with 2.6 percent Co was developed that is used as a substitute for some purposes, but the straight beryllium–copper alloy develops a higher strength and hardness by heat treatment than the cobalt-bearing alloy.

Oxygen-Free Copper This type is the purest of commercial coppers (99.92 percent minimum) and therefore is called a high-conductivity copper. It is used in electrical and electronic equipment, radiators, refrigeration coils, and distillers. Industrial copper is classified by a series of SAE numbers; for example, SAE No. 75 is a 99.90 percent deoxidized copper used for tubes.

Tough-Pitch Copper This copper contains a carefully controlled amount of oxygen, between 0.02 and 0.05 percent. Electrolytic refined tough-pitch copper is the most widely used type for electrical conductors and for building trim, roofing, and gutters. It lends itself to high-tonnage production after being cast into wire, bars, and billets for further fabrication. Tough-pitch copper can withstand the ravages of time and weather because it does not harden with age and develop season cracks.

Brass

Brass is essentially an alloy of copper and zinc. As the relative percentages of copper and zinc vary there is a corresponding variation in properties such as color, strength, ductility, and machinability. Brass colors usually range from white to yellow, and alloy brasses range from red to yellow. Alloy brasses rarely contain more than 4 percent of the alloying elements, either singly or collectively. These elements are manganese, nickel, lead, aluminum, tin, and silicon, however, in the plain brasses containing only copper and zinc, the zinc content can be as much as 45 percent. Gilding metal used for jewelry contains 95 percent copper and 5 percent zinc, whereas Muntz metal, which is used for sheet stock and brazing rod, contains 60 percent copper and 40 percent zinc. Lead is sometimes added to brass to increase its machinability. In general, because of their ductility and malleability, brasses are easily cold worked by any of the commercial methods such as drawing, stamping, spinning, and cold rolling.

Bronze

Bronze originated as an alloy of copper and tin and has been used for thousands of years. Tin-bronze today is known commercially as phosphor bronze because a small percentage of phosphorus is generally added as a deoxidizing agent in the casting of these tin–copper alloys. Phosphor-bronze can range from 1.25 to 10 percent in tin content. Bronze colors generally range from red to yellow. Some bronzes have good resistance to corrosion near seawater. There are many types of bronzes that derive their names from an alloying element other than tin. Two of these are the silicon-bronzes and the aluminum-bronzes, which cover a wide range of metals. Phosphor-bronze can be cold worked, and silicon-bronze can be hot and cold worked, but aluminum-bronzes are not severely cold worked. Most bronzes are easily cast into molds and they make good antifriction bearings. As with brasses, lead is often added to improve machinability. Both brasses and bronzes are designated by an SAE standard number system. For example, red brass casting metal is SAE standard No. 40, and SAE standard No. 64 is phosphor-bronze casting metal. These standard alloys used in the automotive industry can be found in reference books.

Die-Cast Metals

Die casting, which will be covered in Chapter 8, is a means of producing identical castings at a rapid rate by injecting molten, low-melting-point alloys into a mold with intense pressure. The mold then opens and the solidified part is ejected. Such small parts as carburetors, toys, and car door handles are mass produced by the process of die casting. Die-cast metals, also called *pot metals,* are classified into six major groups.

1. Tin-based alloys
2. Lead-based alloys
3. Zinc-based alloys
4. Aluminum-based alloys
5. Copper, bronze, or brass alloys
6. Magnesium-based alloys

The specific content of the alloying elements for any of the die-cast alloys may be found in materials reference books.

Indium

Indium is a very soft metal that can be scratched with a fingernail. It is silver-white, similar in appearance to platinum, and has a brilliant metallic luster. Indium adheres to other metals on contact, and the molten metal clings to the surface of glass and wets it. Indium is used in bearing materials for wear reduction by diffusing it into metallic surfaces or as a coating of indium and graphite. Lead–indium alloys are used for solders and brazing materials. Indium in its pure form is not used as a manufacturing material; it is used only as it is alloyed with other metals.

Lead-Based Materials

Lead is a soft, heavy metal that is somewhat silvery when newly cut but gray when oxidized. It has been a very useful metal since ancient times when it was used for plumbing, utensils, and jewelry. Pure lead is so soft that it has limited usefulness, but when alloying constituents are added, its hardness is increased. Antimony in quantities up to 12 percent is added to harden and improve the mechanical properties of lead. Although lead has a high density, it has low tensile strength and low ductility; therefore, it cannot easily be drawn into wire. It does have the property of exceptionally high malleability, which allows it to be rolled or compressed into thin sheets or foil. Lead shapes and wire cable sheathing can be easily extruded (pushed through a die). Lead has good corrosion resistance and is a good shielding material for gamma and X-ray radiation, but lead compounds are very toxic, and adequate protection is necessary in handling and manufacturing lead products.

Babbitt A widely used antifriction lead alloy that is used for antifriction bearings is called babbitt. Babbitts can be lead, tin, or cadmium based. Lead-based babbitts contain up to 75 percent lead with antimony, tin, and some arsenic. Tin babbitts contain from 65 to 90 percent tin with antimony, lead, and some copper added. These are the higher grades and are more expensive than lead-based types. Cadmium-based babbitts resist higher temperatures than the other types. They contain from 1 to 15 percent nickel, a small amount of copper, and up to 2 percent silver.

Solders Tin and lead are combined in various proportions to form low-melting-point solders that are used to join metals. The eutectic of this alloy system melts at 361°F(183°C).The most common general-purpose solders contain about 50 percent of each metal. They are sometimes called 50-50 solders. The canning industry uses solders with only 2 to 3 percent tin. When there is a need to lower the melting point so the metal to be joined will not warp, cadmium or bismuth is added to the lead–tin alloy.

Terne Plate Terne metal is an alloy of tin and lead, the tin content being between 10 and 25 percent. Sheet steel coated with this alloy is called terne plate. The coating has a high resistance to corrosion and terne plate is used for roofing and other architectural needs. Terne plate forms a good base for paints, which is needed to protect the steel core because the terne metal is cathodic to the steel. In the deep drawing of metal articles the plating provides a good die lubricant. These coatings are applied to steel by the hot-dip process, or they are sprayed on or electrodeposited.

Magnesium

When pure, magnesium is a soft silver-white metal that resembles aluminum but is much lighter in weight (108.6 lb/ft^3 compared with 168.5 lb/ft^3 for aluminum and 487 lb/ft^3 for steel). When alloyed with other metals such as aluminum, zinc, or zirconium, magnesium has quite high strength-to-weight ratios, making it a useful metal for some aircraft components. Magnesium is also used in cast iron foundries as an additive in the ladle to produce nodular iron. Household goods, typewriters, and portable tools are some of the many items made of this very light metal. Magnesium when finely divided will burn in air with a brilliant white light. Magnesium alloys are designated by several numerical systems: Military, AMS, SAE, ASTM, and Federal, all of which may be found in reference handbooks. Magnesium can be shaped and processed by practically all the methods used to manufacture other metal products. It can be cast by sand, permanent mold, or die-cast methods. It can be rolled, extruded, forged, and formed by bending, drawing, and other methods.

Manganese

This silver-white metal is seldom seen in its pure state, since it is normally used as an alloying element for other metals. In steel production it is used as a deoxidizer and to control sulfur. All steel contains small percentages of manganese. Large percentages (12 percent or more) cause steel to become austenitic and no longer ferromagnetic. High-manganese steel quickly work hardens and has excellent abrasion resistance as a result, so it is used extensively in making earth-moving machinery, rock crushers, and conveyors. Manganese is also used in some stainless steels.

Molybdenum

As a pure metal, molybdenum is used in high-temperature applications such as for filament supports in lamps and electron tubes. It is used as an alloying element in steel to promote deep hardening and to increase tensile strength and toughness.

Nickel and Nickel-Based Alloys

Nickel is a silvery white metal that is noted for its corrosion resistance. It is widely used for electroplating on other metals, especially steel, as a protection coating. It is also used as an alloying element in other metals and, like manganese, a large percentage (50 percent) of nickel in steel, or 8 percent Ni and 18 percent Cr as in austenitic stainless steel, causes it to become austenitic and lose its ferromagnetism.

Monel An alloy of 67 percent nickel and 28 percent copper plus some other impurity metals such as cobalt, manganese, and iron is called Monel. It is a tough alloy that is machinable, ductile, and corrosion resistant. Monel is widely used to make marine equipment such as pumps, valves, and fittings that are subjected to salt water, which is very corrosive to many metals.

Nichrome and Chromel These nickel–chromium–iron alloys are used as electric resistance elements in electric toasters and heaters. *Inconel* is similar to nichrome and chromel but is used to manufacture parts to be used in high-temperature applications.

Nickel–Silver This alloy could be considered one of the copper alloys, since it contains a high percentage of copper. A common nickel–silver contains about 55 percent copper, 18 percent nickel, and 27 percent zinc. The color is silver-white, making it a substitute for silver tableware and other such products. Because of the nickel content, it has good corrosion resistance.

Niobium and Tantalum

Niobium, tantalum, and vanadium are all related chemically and they appear as a vertical group in the periodic table of the elements; however, niobium and tantalum are truly sister elements in more ways than one. Both occur in the same ores and have the same melting points. Unlike the other refractory (high-temperature) metals, they are both ductile and malleable at room temperature, and therefore they can be easily formed. They are very immune to attack by strong acids, with the exceptions of fuming sulfuric and hydrofluoric acids. The unusual oxide films they form are stable semiconductors and it is because of this property that they are used in capacitors and rectifiers. Niobium is used as a carbon stabilizer in stainless steels and as an alloying element for low-temperature service steels such as those used in Arctic pipelines. Niobium is used to help control the grain size in HSLA structural steels. Tantalum is used in electronics and for surgical implants. Tantalum carbide, along with tungsten carbide and titanium carbide, is widely used for cutting tools in machining operations.

Precious Metals

Gold and silver have long been used for coinage and jewelry because they tend to resist deterioration by corrosion and are relatively rare. Both are fairly evenly distributed over the earth's surface, although not in sufficient amounts for economical recovery except in a few locations. The six metals of the platinum group are palladium (Pd), rhodium (Rh), iridium (Ir), osmium (Os), ruthenium (Ru), and platinum (Pt). Along with gold and silver these metals are generally called the *precious metals group.* Gold is used in dentistry and in the electrical and chemical industries. Silver has many uses in manufacturing and is well known in its use in photographic film. The specialty metals are all used for catalysts, as alloying elements, and in many industrial applications.

Tin

Tin is whiter than either silver or zinc. It has good corrosion resistance, so it is used to plate steel (see Chapter 18). Vast quantities of tin plate are used in the food-processing industry. Tin is one of the constituents of bronze and is used in babbitts and solders. Because of the rapidly increasing price of tin in recent times, other less expensive metals or materials have been developed to replace it in some products.

Titanium

This silver-gray metal weighs about half as much as steel and when alloyed with other metals is as strong as steel. Like stainless steel, it is a relatively difficult metal to machine, but machining is accomplished with rigid setups, sharp tools, and proper coolants. The greater application for titanium and its alloys is in making jet engines and jet aircraft frames; however, many promising civilian, military, naval, and aerospace uses for titanium are being considered because of its high corrosion resistance strength, and because it weighs less than steel. At the present time, its high cost precludes many of these proposed applications.

Tungsten

Tungsten is one of the heaviest of the metals and has the highest known melting point of any metal (6098°F or 3370°C). It is not resistant to oxidation at high temperatures, and when used as lamp filaments it must be in a vacuum or inert gas atmosphere. Tungsten is used for electrical contacts and for welding electrodes. It is used as an alloying element in tool steels and in cutting-tool alloys. The carbides of tungsten are the most widely used and most valuable cutting tools. Tungsten carbide powder combined with cobalt powder is compressed into tool shapes and sintered in a furnace in a process called powder metallurgy.

Uranium and Thorium

Both uranium and thorium are radioactive metals and alpha emitters. Competent authorities should be consulted regarding the handling of these metals and their ores. Uranium is a silver-white metal in its pure state and it can be machined like other metals. If finely divided, uranium will burn in air. Uranium is an important fuel for nuclear reactors. Thorium, a dark gray metal, is also used in nuclear reactors. Thorium oxide with 1 percent cerium oxide is used for gaslight mantles.

Vanadium

Vanadium is a silver-white, very hard metal that oxidizes when exposed to air. It is used almost exclusively as an alloying element in steel to impart improved impact resistance (toughness) and better elastic properties. Vanadium contributes to abrasion resistance and high-temperature hardness in cutting tools, and it promotes finer grain in steels, giving them toughness. Usually not more than 1 percent vanadium is alloyed with steels.

Zinc

Zinc is a white metal that is valued for its corrosion protection of steel. The familiar galvanized steel is coated with zinc by any one of several processes: hot dip, electrogalvanizing, sherardizing (heating steel in zinc dust), metal spraying, or zinc-dust paint. When zinc and iron are galvanically (electrically) in contact under corrosive conditions, the zinc becomes anodic and serves as a sacrificial metal, corroding in preference to the steel. The greater consumption of zinc is for galvanizing steel. Zinc-based die-cast metals also consume a high percentage of zinc production, as do brass and rolled zinc to a lesser extent.

Zirconium

Zirconium and titanium are similar in appearance and physical properties, and like the other reactive metals, zirconium will burn in air or explode when finely divided. In fact, it was once used as an explosive primer and as a flashlight powder for use in photography because of the brilliant white light it gives off when it explodes. Zirconium alloys are still used in flashbulbs. Perhaps one of its most notable uses is in nuclear reactors. It has good corrosion resistance and stability and therefore has been used for surgical implants. Like titanium, it is somewhat difficult to machine.

MATERIALS IDENTIFICATION

We soon learn to identify natural materials such as wood, leather, stone, glass, and earthy materials, such as clay, by sight and touch. Similarly, we can learn to identify many manufacturing materials by sight; however so many varieties of these materials look exactly the same that other means must be used to identify them. For example, in the past when a distinction had to be made between low-carbon and high-carbon steel that look alike on the surface, identification was made by means of spark testing (Figure 6.7). The blacksmith simply touched the two samples on a grinding wheel and observed the sparks: the high-carbon steel produced a brilliant display of lines and sparks, and the low-carbon steel showed few carrier lines and a few sparks. This method is still sometimes used today to compare unknown steels with known samples in steel yards and shops, although its accuracy is directly related to the operator's experience. Although spark testing is still a useful

FIGURE 6.7
Spark testing high-carbon steel.

tool, it is not a reliable method of identifying the alloy content of steel. Also, very few nonferrous metals produce a spark; exceptions are nickel, titanium, and zirconium. Nickel produces short carriers with no spark, similar to those of stainless steel, and titanium produces a brilliant white display of carrier lines, each having a spark at the end. The display of zirconium carrier and spark is identical with that of titanium (Figure 6.8).

Metals can often be identified by their reaction to certain acids and alkaline substances. A clean surface of a particular metal will become darkened or colored when a certain reagent is placed on it. For example, an aluminum sample can be quickly distinguished from a magnesium sample by placing a drop of a copper sulfate solution on each. The aluminum will not be affected but the magnesium will be blackened. Copper sulfate will turn ordinary steel a copper color, but it will not affect any type of stainless steel, nickel, Monel, or Inconel, all of which have a similar appearance. There are also many acid tests. Commercial spot testers are available in kits for the purpose of identifying metals. Magnetic testing is another means of identifying look-alike metals. Most ferrous metals, and those in which nickel or cobalt is the primary constituent, are attracted to a magnet, but some stainless steels, including the 300 series and the precipitation hardening types, 17-4 Ph and 15-5 Ph, may or may not be magnetic. Austenitic stainless steel (300 series) is not attracted to a magnet unless it is cold worked; then it is attracted to the extent to which it is cold worked. Of course, none of these methods will identify all the elements or how much of each is in a sample of metal, but this can be done by spectrographic analysis in the laboratory. Speedier results are obtainable with a portable unit (Figure 6.9). This device identifies and quantifies key elements (alloying and residual) plus carbon in metals. Fuess spectroscopes work on the same principle as the spark test except that the reflected light is refracted through a prism. Because each element produces characteristic emission lines, it is possible to identify which

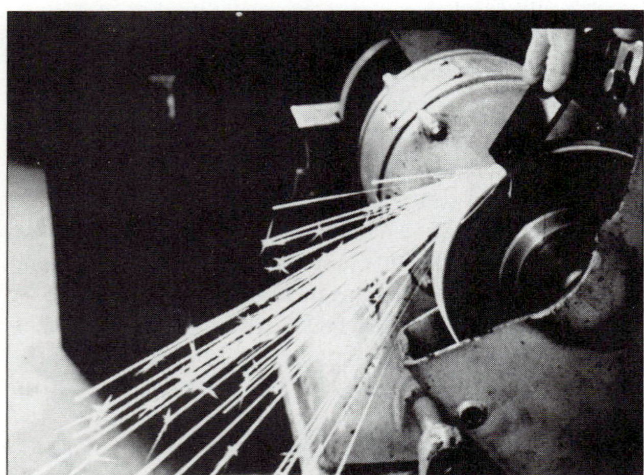

FIGURE 6.8
Spark testing titanium.

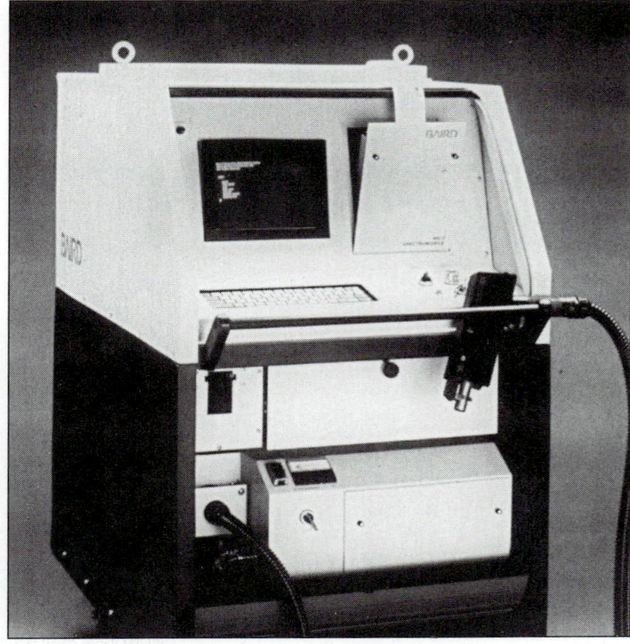

FIGURE 6.9
Spectromobile. A wheel-mounted spectrometer facilitates rapid, on-site identification and sorting of steel. It identifies and quantifies all key elements (alloying and residual) in metals and determines carbon content to within plus or minus 0.05 percent (Photograph courtesy of Spectrochemical Products Division, Baird Corporation).

FIGURE 6.10
In as little as 5 seconds the X-site portable X-ray analyzer can identify metals such as this bar stock. The abbreviated alloy name appears on the LED display within a few seconds (Photo courtesy of Kevex Corporation © 1982).

FIGURE 6.11
Metal parts in a foundry are analyzed for content of various elements with an X-ray source that produces characteristic emissions from every element in the sample. A console with a microprocessor is connected to the handheld unit that processes the data (Photo courtesy of Kevex Corporation © 1982).

elements are present. It is also possible to compare the results with standard spectra as an aid in identifying alloys. Mobile spectrometers work on the Fuess principle and are not able to sort alloys but are preset for a particular matching test. They are calibrated to a known standard and can quickly identify that material with a matching signal.

Using an X-ray analyzer is probably one of the best methods of sorting and verifying metals (Figures 6.10 and 6.11). Its greatest drawback is that it costs more than other systems. Scrap metals are among the most difficult metals to identify and sort. High-integrity steel products must have carefully sorted scrap. All the previously mentioned methods may be applied to this task of sorting scrap metals; however, generally, the more accurate a method is the more the system costs.

Plastics Identification

Because there is such a great variety of plastics, and more are being developed every day, identification of plastics is very difficult by ordinary means. Certain solvents will dissolve certain types of plastics, but this would not be a reliable method of testing because many types of plastics react to the same solvent; however, as discussed earlier, when heat is applied to a thermoplastic material, it will melt and solidify again when cool, and it can be reused, but heat applied to a thermosetting material will cause it to break down instead of melting and it cannot be reused. Of course, in either case, this would be a destructive test for the plastic object involved.

Some plastics with chemical odors can be identified by holding a sample in a flame. An odor of phenol indicates a phenolic material, and an acrid odor with a yellow flame indicates cellulose acetate or vinyl acetate. An odor of burnt wood indicates ethyl cellulose. Melamine has a fishy odor when burned, and casein smells like heated milk. Polystyrenes give off an odor of domestic heating and cooking gas, whereas polyethylene and polyolefins produce little or no odor, burn with a smoky flame, and melt like a candle while burning. **Note:** Cellulose nitrate is very flammable, especially in thin films or sheets. Precautions must be taken to protect hands, face, and eyes as well as to prevent fire when conducting any of these tests.

To identify consumer products made of plastics, the Society of Plastic Industry, Inc. (SPI) has developed a plastic container coding system. Other organizations (ASTM, SAE, ISO) have also proposed similar systems of identification using single digit numbers in a triangle, usually embossed on the bottom of the container or on the object, whatever it may be. The numbers vary somewhat but in general the codes are as follows:

1—PET or PETE, polyethylene terephthalate

2—HDPE, high density polyethylene

3—V or PVC, vinyl or polyvinyl chloride

4—LDPE, low density polyethylene

TABLE 6.7
Telura® numbering systems

The three-digit Telura grade number provides a key to the properties of the oil. The first digit designates the type of product, as follows:

1—Aromatic
2—Intermediate aromatic[a]
3—Naphthenic
4—Extracted naphthenic
5—Paraffinic
6—Extracted paraffinic
7—Special grades

The second digit is indicative of the approximate viscosity in centistokes (cSt) at 40° C, as follows:

0—< 10
1—10 to 19
2—20 to 29
3—30 to 39
4—40 to 49
5—50 to 59
6—60 to 99
7—100 to 999
8—1000 to 1999
9—2000

The third digit is an arbitrary number used to discriminate between grades of similar viscosity.

(Reprinted with the permission of Exxon Corporation, "Telura® Industrial Process Oils," Exxon Corporation, 1984.)
[a]Products in this category are not currently available.

5—PP, polypropylene

6—PS, polystyrene

7—Other, all other types of resins

In many communities, efforts to segregate and sell recyclable materials (newsprint, cardboard, glass, metal containers as well as plastics) have greatly reduced the amount of waste being put into landfills, but our society still has much to accomplish.

FLUID ANALYSIS

Industrial process oils are used in a broad range of applications in manufacturing processes. An overview of these petroleum products is seen in the Telura® numbering systems of the Exxon Company, USA (Table 6.7) The applications for these oils are in Table 6.8.

The Telura line of process oils consists of 19 grades suitable for a broad range of applications. The line consists of six general types of products.

1. **Aromatic** Aromatic oils have good solvency characteristics. They are used to make proprietary chemical formulations and ink oils.
2. **Naphthenic** Naphthenic oils are characterized by low pour points and good solvency properties. They are used to make printing inks, shoe polish, rustproofing compounds, and dust suppressants, and they are used in textile conditioning, leather tanning, and steam turbine flushing. The colors of the low-viscosity Telura oils in this category are relatively light.
3. **Extracted Naphthenic** These oils have outstanding color and color stability and good low-temperature characteristics. They are used in proprietary chemical formulations, dust suppressants, and rustproofing compounds. Of the extracted naphthenic products. Telura 415 and 417 meet the requirements of Federal Food and Drug Administration (FDA) Regulation 21 CFR 178.3620(c) for mineral oils used in products that may have incidental food contact.
4. **Paraffinic** Paraffinic oils are characterized by low aromatic content and light color. They are used in furniture polishes, ink oils, and proprietary chemical formulations.
5. **Extracted Paraffinic** These oils offer light colors, low volatilities, relatively low pour points, and improved color stability. They are used in waste disposal system fluidizing oils, textile oils, and autoclave oils. Telura 607, 612, 613, and 619 are the oils in this category that conform to the requirements of FDA Regulation 21 CFR 178.3620(c).
6. **Special Grade** Telura 797 is a heavy, black residual oil used in newsprint ink formulations.

Oil Contamination

One of the main causes of wear and breakdown of machinery is contamination of lubricating or hydraulic oils by grit, water, and other unwanted materials. Regular inspection of industrial oils can save a great deal of breakdown time. This can be done by means of a fluid analysis kit (Figure 6.12). Examples of SAE oil contamination classification are also provided in the kit. Oil can also be contaminated with hazardous chemicals of which polychlorinated biphenyls (PCBs) are the most common, especially around electrical apparatus such as older transformers. PCBs are toxic and are thought to be a cause of cancer. Field tests for PCB spills or contaminated oil can easily be made with portable testers like the one shown in Figure 6.13.

Water

Aside from municipal water systems, industrial water supply is of major importance. Water that has been used for industrial purposes is often contaminated with heavy metals and

TABLE 6.8
Applications for Telura industrial process oils

	Telura																		
	126	*171*	*309*	*323*	*343*	*401*	*407*	*415*	*417*	*515*	*521*	*607*	*612*	*613*	*619*	*662*	*668*	*671*	*797*
Absorption oil in gas plants			•									•							
Autoclave oil																•			
Canning machinery lubricant								•											
Carbon paper manufacturing				•															
Chain oil		•																	
Coke oven absorption oil			•																
Compressor cylinder oil			•	•															
Concrete form oil				•							•								
Cordage oil				•					•		•								
Cutting oil base		•		•							•								
Drilling mud						•													
Dust suppressant		•	•	•							•								
Fiber finishing												•							
Foam depressant									•					•					
Furniture polish		•								•									
Glass mold lubricant							•	•											
Heat transfer oil for low temperatures						•	•												
Honing oil								•											
Impregnating oil								•											
Ink oil		•		•					•	•									•
Leather tanning				•	•						•								
Oil field chemicals	•					•													
Plywood waterproof coating				•															
Proprietary chemical formulations		•		•	•		•			•					•				
Quench oil											•				•				
Rustproofing			•	•					•		•								
Sealant component																•	•		
Shoe polish				•						•									
Slushing oil			•																
Spray oil base													•						
Steel roll oil base				•					•										
Textile conditioning oils				•			•								•				
Turbine flushing oil			•																
Waste disposal system fluidizing oil																		•	
Water treating chemical				•								•							

(Reprinted with the permission of Exxon Corporation, "Telura® Industrial Process Oils," Exxon Corporation 1984.)

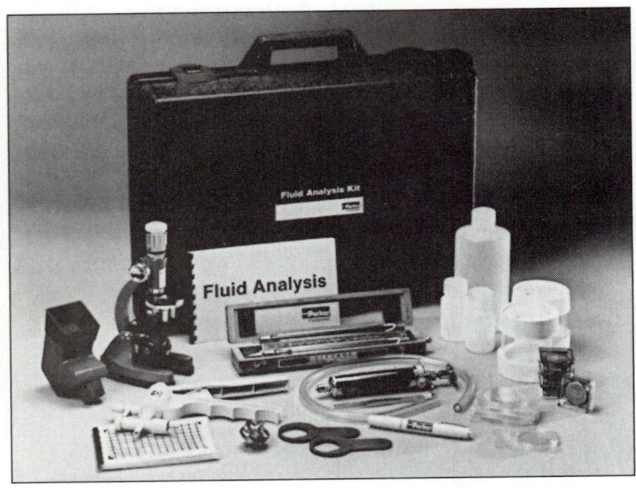

FIGURE 6.12
Fluid analysis kit. Oil samples in field, plant, and laboratory hydraulic systems can be checked easily for contaminants with this portable kit (Courtesy of Parker-Hannifin Corporation).

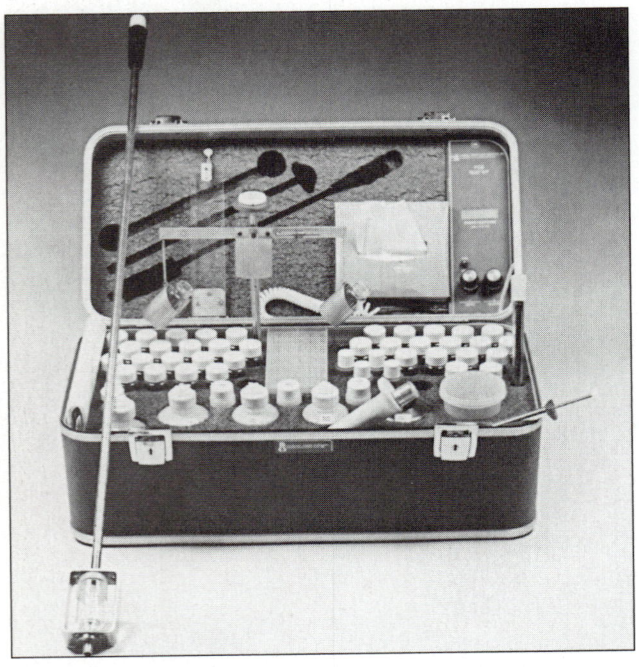

FIGURE 6.13
PCB field test kit for screening oil and soil. The soil collector is shown removed from the case and assembled (Courtesy of McGraw-Edison Company, Power Systems Division, Pittsburgh, PA).

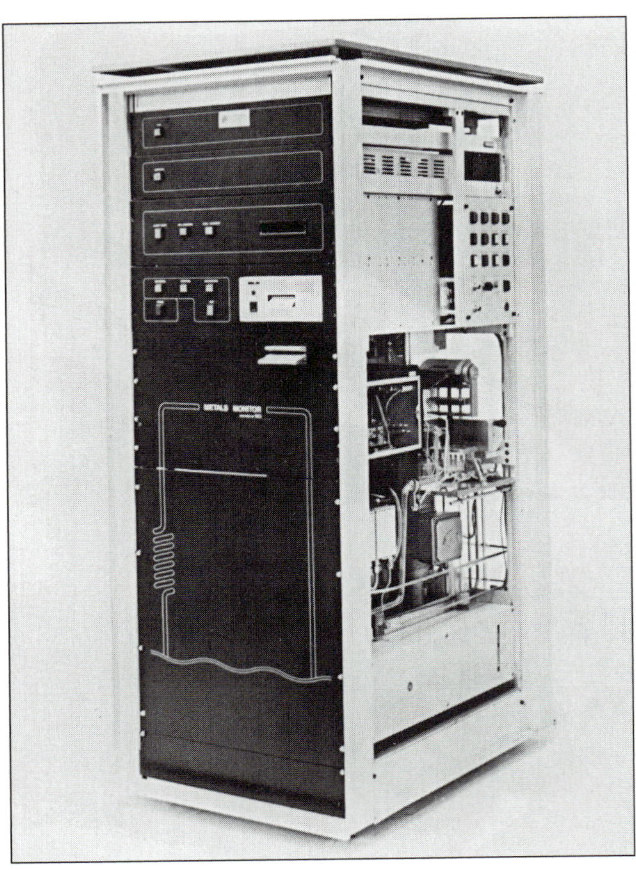

FIGURE 6.14
Water analyzer for testing water for heavy metals (Alewijnse Electrical Engineering, Environment Systems Technology Division, Nijmagen, Holland).

other materials that must be removed before this wastewater is returned to a river or other outflow. Various kinds of tests are made to assure the purity of wastewater; among them is an automatic metals monitor (Figure 6.14) that determines the level of cadmium, copper, lead, and zinc in wastewater.

Heavy metals may pose a serious threat to public health if they occur in sources of potable water such as rivers, lakes, and groundwater. The metals monitor, a completely automatic electrochemical installation, can detect and analyze all types of water without any human supervision.

Besides potable water, the instrument can monitor seawater, rainwater, and industrial wastewater. It can also be used to detect corrosion in power stations and waterworks, to find leaks in pipes, and to analyze the effluents of the metal finishing, plating, and chemical industries. Besides cadmium, copper, lead, and zinc, other metals can be detected by the monitor if the computer program and treatment of the sample are modified.

All the control, monitoring, and calculating functions are performed by a microprocessor. The data are recorded four times an hour on a printer and an LED display. In addition, daily averages of the metal concentrations are printed once a day. If it is connected to a telemetric recording circuit, the metals monitor will analyze concentrations ranging from 0.0005 to 1 mg/L.

MATERIALS APPLICATIONS

Material selection for a particular manufactured product is an involved and sometimes complex process requiring a broad understanding of the nature and behavior of materials. Many factors other than material strength and durability need to be considered. In view of domestic and foreign competition, the cost of materials is of prime consideration, in addition to the manufacturing cost. Product warranties and customer service requirements are to be considered as well as product liability and patent infringement. Many new materials for products are chosen because of the necessity for weight reduction, especially for energy savings in the field of transportation; however there is an inherent risk in making a sudden change in material for a product without extensive evaluation of its long-term performance. Many hasty substitutions of materials have resulted in product liability cases and many more have resulted in higher costs because the new material failed to perform up to expectations. Some of the factors that may be involved in making a material choice are tensile, torsional, and yield strengths, stiffness, fatigue strength, hardness, impact resistance, wear resistance, ductility, machinability, formability, hardenability, weldability, corrosion resistance, availability, and cost. Of course, many other factors relate to the specific selection, such as the prevailing conditions—temperature, atmosphere, and vibration—to which the products will be subjected. Reference handbooks on materials are helpful when making selections. Some of these are the *Metals Handbook* and the *Source Book on Materials Selection,* both published by **ASM International,** and the *Cast Metals Handbook,* published by the American Foundryman's Association.

There are also innumerable handbooks and data sources on polymers, ceramics, lubricants, and the like. Also available are many web sites that provide data at no charge for a wide range of materials; one of the currently more valuable is http://www.matweb.com.

Example: Galvanized versus Plastic-Coated Pipe

A West Coast manufacturer of irrigation equipment, OEM (Original Equipment Manufacturer), had been using galvanized steel pipe to make pivot irrigators. These devices

FIGURE 6.15
Pivot irrigator covers 160 acres in a circular pattern (Pierce Corporation, Eugene, OR).

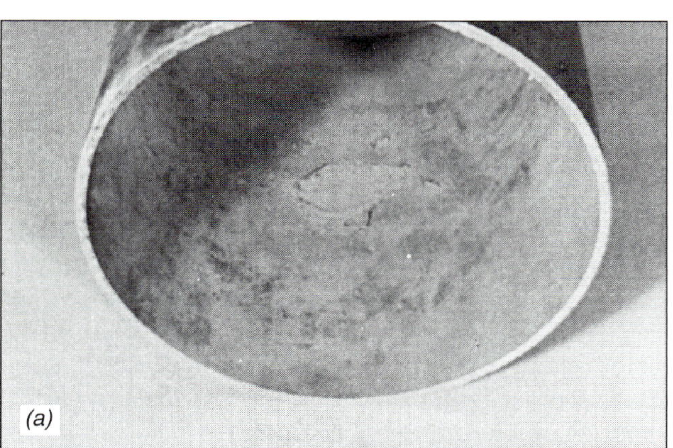

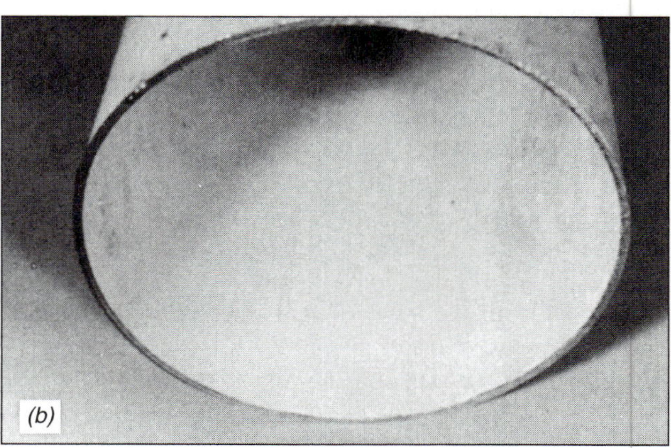

FIGURE 6.16
After just a few years in the field, the section of galvanized pipe *(a)* shows the roughness and discoloration typical of a "worn-out" galvanized coating. The pipe with the electrostatically applied epoxy coating *(b)* still looks like new and is smooth to the touch even after several seasons of use (Pierce Corporation, Eugene, OR).

slowly move a pipe around a center tower in a circular pattern, spraying water that often contains liquid fertilizer (Figure 6.15). Because the fertilizer is acidic, uncoated steel pipe would corrode quickly. Since zinc is a sacrificial metal, it soon begins to roughen inside, causing a turbulence that requires more energy to move the water through the pipe (Figures 6.16*a* and 6.16*b*). A higher head (pressure) is required to move the water through the galvanized sprinkler pipe than through a plastic pipe, for example, in order to maintain sufficient pressure at the outer sprinkler heads, requiring more electrical energy.

A search was made by the company engineers for a substitute pipe material to solve this energy loss problem. Although plastic had less friction than the other alternatives, it could not be substituted for two reasons: it tends to deteriorate from ultraviolet rays in sunlight, and it has insufficient tensile strength to support the weight of water in a suspended pipe. Aluminum, besides being expensive, would not substantially improve the energy loss problem. The Hazen–Williams formula of the coefficient of friction, gives a value of 100 for black steel pipe. Values for galvanized pipe and aluminum are both 120. Because weight was not a factor and higher strength was required, steel was still the preferred pipe material. The solution was to coat steel pipe with a suitable plastic material. Steel pipe was cleaned by shot-blasting and electrostatically sprayed inside and outside with an epoxy powder that was then heat dried. This epoxy coating is inert to the low acid in the fertilizer. It is very smooth and its Hazen-Williams number is 150, the same as that of plastic pipe. This investigation led to an amazing revelation: a substantial cost savings in electric power for the farmer's water pumps could be realized by reducing the friction in the pivot irrigation pipe. With the current cost of electricity in the Midwest being 6 to 9¢/kWh, a 10-machine operation (each machine covers 160 acres) over a 10-year period would realize a savings of $60,000 to $90,000 or more. Thus, the switch to epoxy-coated steel pipe gives the machine purchaser, the farmer, a competitive edge: it also will undoubtedly increase sales for that OEM company's product, the pivot irrigator.

Review Questions

1. Which is the more common classification system used in the manufacturing industry for machine and alloy steels, the SAE-AISI or the ASTM specifications?
2. What kind of steels are designated by the ASTM specifications?
3. Name the quenching medium used for W1 and O1 tool steels.
4. Which steel shafting is more expensive, cold rolled (CR) or ground and polished (G and P)?
5. Which steel bar having the same alloy content is more expensive, hot rolled (HR) or cold rolled?
6. State the greatest advantage of HSLA steels over plain carbon steels for use in modern automobiles.
7. One type of stainless steel that is often used for cutlery and surgical equipment can be hardened by heating and quenching. Under which number series is it classified?
8. As gray cast iron solidifies it is unable to hold more than 2 percent carbon in solution, yet it may contain 4 percent carbon. What happens to the remaining 2 percent?
9. Would you make a part from gray cast iron if it was to be subjected to high impact loads? Why?
10. What kind of cast iron might be acceptable for the part described in Question 9, particularly if the impact loading was moderate?
11. Why is a thin sheet of pure aluminum sometimes sandwiched (clad) on tough alloy aluminum?
12. In what two ways can aluminum be hardened?
13. What is the most widely used type of copper for electrical conductors?
14. Of all the means of identifying the alloy and carbon content of metals, which is the most accurate although the most expensive method?
15. If only low and high plain carbon steels are present, what simple method can be employed to distinguish between them?
16. A manufacturer of a consumer product wishes to make a material change from metal to plastic for the purpose of weight reduction and appearance. What should be done before the new product goes on the market?

Case Problems

Case 1: Separating Aluminum and Magnesium Bars

A manufacturing plant regularly received shipments of aluminum bars of various sizes on which machining and welding operations were performed. On one shipment the supplier discovered the bars had been sent out unmarked, and several magnesium bars had accidentally been mixed into the lot. There was no time to send for and receive a new shipment before running out of material. The two metals had to be separated, not only because magnesium was an unwanted material but because it is a pyrophoric metal, tending to catch fire and burn fiercely when finely divided, as in machining chips, and from welding, when proper precautions are not taken. Therefore, magnesium constituted an extreme hazard in this case, where it was not normally used. The metal bars all looked identical and they varied so much in size that weight differences could not be used readily to identify them by unskilled help. A quick, low-cost method was needed to separate these look-alike metals, one that an unskilled person could use. How would you solve this problem, considering what you have learned in this chapter?

Case 2: Cost of Steel Shipment

You receive a load of steel bars on a truck. The shipping order does not include the cost, which is on the invoice that will be mailed later. You need to know the approximate cost immediately in order to make a cost estimate for parts to be made from the steel bars. The load is made up of 20-ft-long steel bars: 10 flat bars (rectangular), ¾ × 2½ in., and 15 round bars 1¼ in. diameter. Metal shapes are sold by weight, and different metals have different weights per unit volume. Steel weighs 487 lb/ft^3. The price of the steel in this shipment is \$.27/lb for the flat bars and \$.31/lb for the round bars. Weight in cubic inches of a steel bar is found by multiplying the cross-sectional area by the length in inches, multiplied by the weight of 1 in.^{3}. There are 1728 in.^{3} in 1 ft^3. What is the total cost of the shipment?

PART II

SURVEY OF MANUFACTURING PROCESSES

Part II, Chapters 7 through 18, covers specific manufacturing processes. Conventional manufacturing processes including casting, cold and hot rolling, forming, forging, plastics molding, and joining of materials are discussed in detail. Newer processes and methods are presented where they apply, and the relationship between design tolerances and process capability is introduced.

Materials such as metals, plastics, and ceramics are made into useful articles and consumer products by many and varied means. Metals are cast into molds in a number of different ways to form intricate small shapes or massive machine parts. Metals are also rolled, pressed, and hammered into dies or forced through dies to make special shapes. For example, iron and steel are heated to a high temperature so they can be easily formed by forging (hammering and squeezing). Although forging was once only a hot-metal operation, today cold upset forging is practiced even on steel, and at intermediate temperatures a metallurgically superior product can be produced. For example, in warm forming, tough materials like SAE 52100 steel are routinely manufactured into high-quality parts with lower production costs than with cold forming.

A large segment of manufacturing is devoted to the processing of sheet metal in cold-working operations such as blanking and stamping. Home appliances, automobiles, and many other products depend on sheet metal processing.

Machinery of all kinds that requires precision parts depends on the machining and machine tool industry, which can be considered the foundation of modern manufacturing. The tool and die industry, though little known and understood, is the backbone of modern industry. Virtually every manufacturing process is in some way dependent on tool and die shops. Stamping, pressing, injection molding, and die casting are just a few of the processes for which tool and die services are absolutely necessary.

Manufacturing of plastic and composite materials is steadily growing and is replacing many products formerly made of metal, leather, and wood; however, many of these manufacturing industries are interdependent. For example, plastic injection molds require special tool and die machining processes, and manufacture of the molding machine itself is a machine tool process.

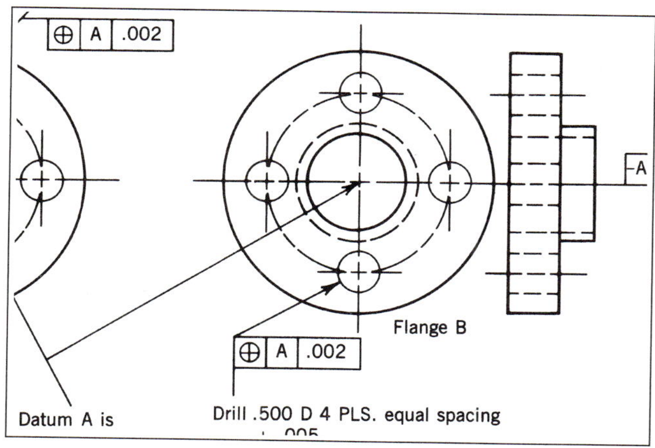

CHAPTER 7

Design Specifications and the Capability of the Manufacturing Process

Objectives

This chapter will enable you to:

1. Describe the information included in a manufacturing drawing.
2. Tell who is charged with the responsibility of creating manufacturing drawings.
3. Tell which employees directly utilize the information included in manufacturing drawings.
4. Describe how manufacturing engineers use manufacturing drawings in the design of manufacturing processes.
5. Explain the meaning of the capability of a manufacturing process.
6. Describe the relationship between blueprint dimensions and the capability of manufacturing processes.
7. Given the dimension and tolerance for a part feature on a blueprint, determine the required precision for the operation that creates the feature.
8. Given the dimension and tolerance for a part feature on a blueprint, determine the required precision of the measuring operation for the feature.

Key Words

assembly drawing
blueprint
capability (of a process)
detail drawing
dimension
feature
manufacturing drawing
precision (of a process)
Rule of Ten
tolerance
tolerance range

No comprehensive study of materials and manufacturing processes can be undertaken without considering the **manufacturing drawing.** This document, called a **blueprint,** records the physical specifications to be followed in the manufacture of the product. Complete conformance to the manufacturing specifications included in the blueprint is necessary for the creation of a quality product. In fact, to manufacturers, conformance to the blueprint specifications is the definition of product quality. Manufacturing engineers rely directly on the blueprint for guidance in designing the operations that compose a manufacturing process, and the process must be capable of creating a product that meets every specification included in the blueprint. Machine operators and other workers who are directly engaged in making the components and assembling the product are provided with blueprints that are followed to the letter in the execution of the manufacturing process.

This chapter will discuss the design specifications documented on the blueprint and their effect on manufacturing operations. A relationship will be established between tolerances and the capability of the forming, joining, machining, and measuring operations employed as steps in the manufacturing process used to make the product.

PRODUCT SPECIFICATIONS

Blueprints, including component **detail drawings** and **assembly drawings,** are created by the product designer to document the design of a product whether the product consists of one individual piece or an assembly of many components. Each individual part that is to be assembled into the product, except for purchased parts, requires the creation of a component detail drawing that defines the manufacturing specifications for that part. Each component of an assembly may also be included in an assembly drawing.

A typical component blueprint contains a detailed picture of the component in addition to various types of specifications including **dimensions,** surface roughness requirements, and a material specification. Each dimension on a blueprint must be accompanied by a tolerance, and when the manufacturing process is executed every dimension must be held within the specified tolerance.

Tolerances

An understanding of **tolerances** is vital to almost any successful manufacturing operation. A tolerance is a product specification that defines the acceptable range of sizes through which the part will fit and function as intended. Tolerance is required because it is not possible to manufacture any part to an exact size; however, it is possible to closely approach an exact size, depending on design requirements and the capability of the machinery used for manufacturing. It also is possible to measure parts reliably to assure that the product conforms to the tolerance. For each product component and assembly the designer must designate acceptable tolerances that will accommodate variations in the manufacturing process.

Tolerances generally fall into three important categories: size, form, and position. Several methods of specifying *size* tolerances may be used (Figure 7.1). Most manufacturing blueprints utilize *limits* tolerancing, in which the largest and smallest sizes are specified. This creates a range of permissible manufacturing error, or tolerance. Other common methods for designating tolerances state the theoretical exact size plus or minus a permissible amount of deviation. *Bilateral* tolerancing states the exact size with both plus and minus permitted deviation. *Unilateral* tolerancing states the exact size and includes only a plus or minus deviation, not both.

The designer, of course, would prefer that the parts be manufactured as near the exact theoretical size as possible; however, the tolerance is necessary to accommodate variations that exist in even the most carefully controlled manufacturing processes. For example, manufacturing precision parts to close (small) tolerances of a few millionths of an inch generally requires more care and attention than looser (larger) tolerance production. Close tolerance manufacturing requires higher technology in manufacturing equipment and inspection tools. This extra effort results in higher manufacturing costs when close tolerances are specified by the designer.

Manufacturing drawings contain form specifications and position dimensions and tolerances as well as size dimensions and tolerances. Specifications of form include squareness, parallelism, straightness, flatness, cylindricity, angularity, and concentricity. Position tolerances include the location of holes or other features (Figure 7.2). Size, form, and position specifications are all vital for the manufacture of a part that will fit and function as intended.

Consider the example illustrated in Figure 7.3. Flange A must bolt to mating part B. A size dimension is provided on the drawing showing the diameter of the bolt holes. A hole size sufficient to contain the bolt is naturally required; however, if the position of the holes is not maintained within tolerance the parts may not properly assemble, even if the diameters of the holes conform to the size tolerance. Therefore the manufacturer must maintain enough control of the manufacturing process to conform to both the hole position tolerance and the hole size tolerance.

THE CAPABILITY OF THE MANUFACTURING PROCESS

The error that occurs when any manufacturing operation is executed can be no greater than the error permitted by the tolerance. This requirement applies to each **feature** of

FIGURE 7.1
Types of tolerances and implied tolerances.

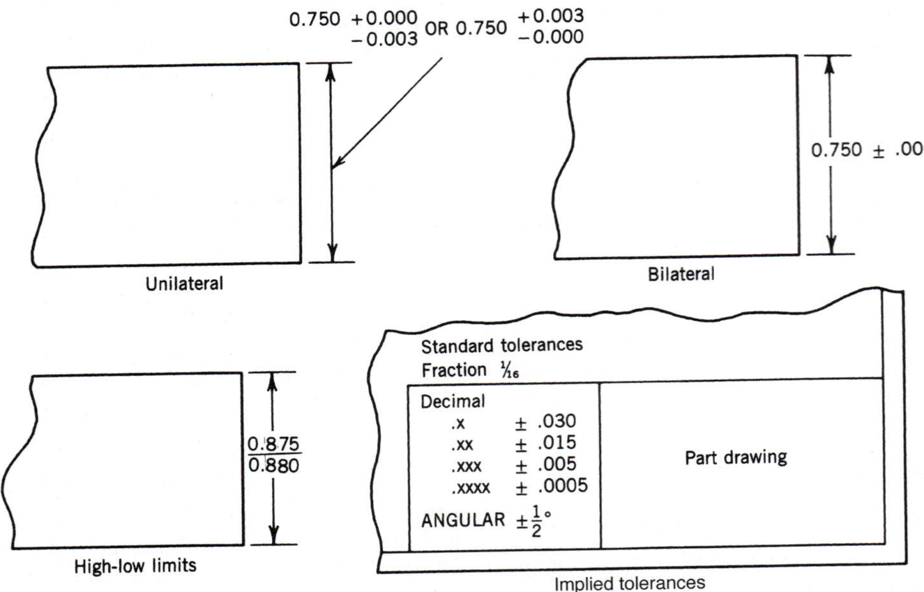

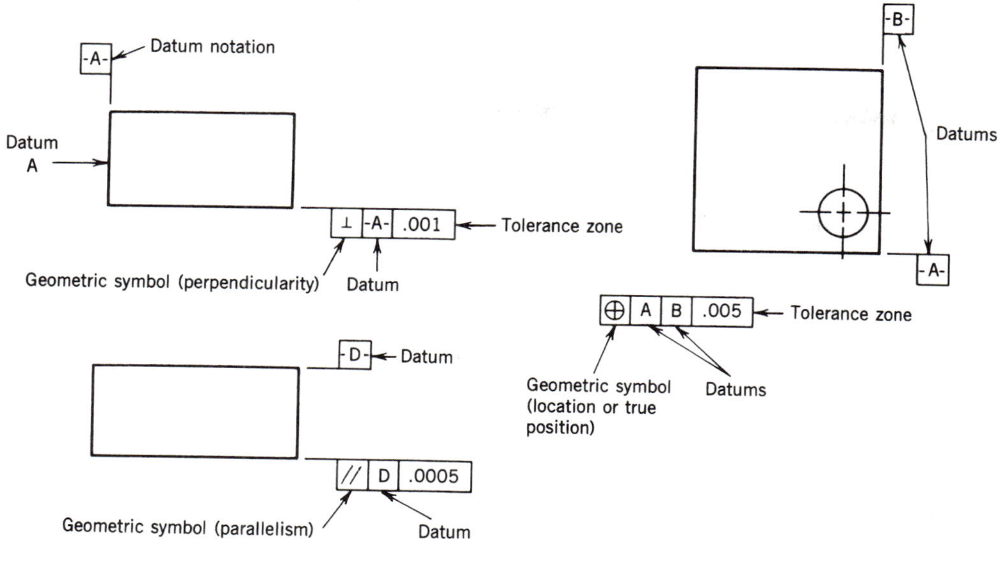

FIGURE 7.2
Examples of form and position tolerances.

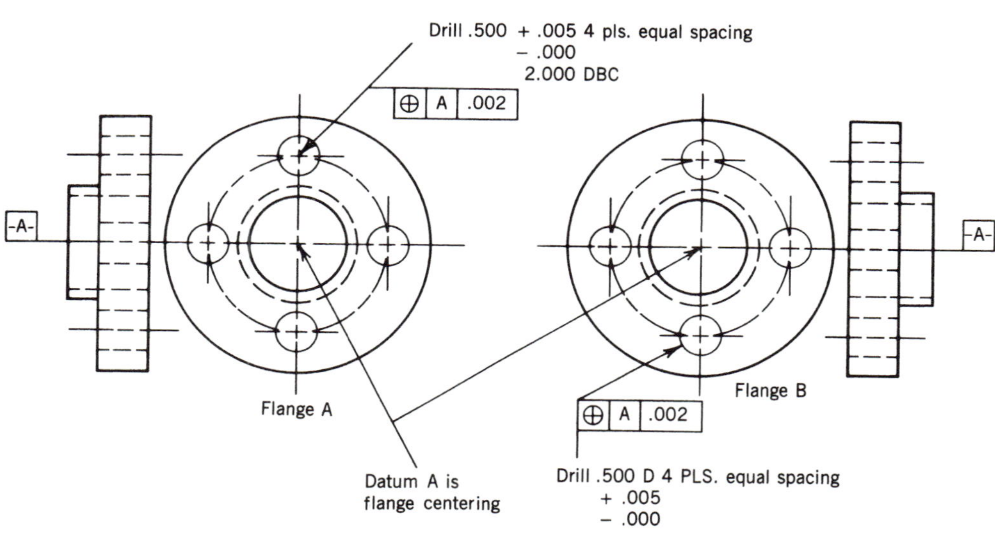

FIGURE 7.3
Accurate part feature locations are essential to successful assembly.

every component or assembly. The manufacturing process must be capable of creating any product feature, on a routine production basis, with no more error than is permitted by the tolerance. The total error permitted by a tolerance is called the **tolerance range.** For example, a ± .005-in. tolerance corresponds to a tolerance range of .010 in., which is the total amount of error permitted in the manufacture of the feature.

Precision

The tolerances on the manufacturing drawing dictate the minimum capability demanded of each operation of the manufacturing process. The **capability** of a manufacturing process is the **precision,** or repeatability, of the process, defined as the amount of error exhibited by the operation. For example, if a part has a length dimension of 1 in. with a ± 0.015 in. tolerance (.030 in. tolerance range), then the manufacturing process must exhibit a total error (precision) of 0.030 in. or less on a routine production basis.

Because the manufacture of a product feature includes both creating and measuring the feature, the amount of error permitted by a tolerance includes all error incurred in creating the feature plus all measuring error. For example, if the tolerance for a length dimension is specified as ± 0.005 in., then the tolerance range is 0.010 in., and the error in creating the feature plus the measuring error must be no greater than 0.010 in.

The Rule of Ten

Should equal amounts of error be allotted to making and measuring a product feature? It is reasonable that a manufacturing engineer would decide to reserve most of the permitted error for creating the feature and less error to measuring the feature. In fact, the common practice is to allot 10 percent of the total error to measuring the feature and to allot the remaining 90 percent of the total permitted error to making the feature. This common rule of thumb is called the **Rule of Ten.** The following example illustrates how the Rule of Ten is used.

Example: Rule of Ten

Blueprint specification: length = 1.394 ± .004 in.

- Tolerance = ± .004 in.
- Tolerance range = .008 in. = total permitted manufacturing error
- Precision ≤ .008 in. is demanded of the entire manufacturing process.

By the Rule of Ten the permitted measuring error = 10% of the tolerance range

- Precision ≤ .0008 in. is demanded of the measuring process

The error permitted in creating the feature = the remainder of the tolerance

- Precision ≤ .0072 in. is demanded of the operation that creates the feature

The Rule of Ten is employed, and 10 percent of the tolerance range (.0008 in. in this example) is allotted to measuring the length, and the remainder of the tolerance range (.0072 in. in this example) is allotted to cutting the part to length. In this example, .0008 in. of error is permitted in measuring the part, so the required measuring precision is .0008 in. Because .0072 in. of error is permitted in cutting the part, the precision of the cutting operation employed to create the feature is required to be .0072 in. or better. The measured length can vary no more than the tolerance range, .0080 in.

The Rule of Ten allocates more tolerance to the operation where more error is likely to occur, making the manufacturing process capable of creating a product that conforms to the tolerance specified on the blueprint.

The precision required of a measurement, cutting, forming, or joining operation is an important guide in selecting operations to be included in a manufacturing process that will routinely produce a quality product meeting all the blueprint specifications.

Review Questions

1. What are tolerances and why are they important in manufacturing?
2. Name three categories of tolerances and explain each.
3. What are form specifications and why are they important in manufacturing?
4. Is there any difference between tolerances of size and tolerances of position?
5. Cite several examples of form specifications.
6. Cite an example of a position specification.
7. Which is more important, form specifications or size specifications? Why?
8. A part is to be turned on a lathe. According to the blueprint dimension, the diameter of a part must be held to .500 ± .010 in.
   a. What is the tolerance range for the diameter? How much error is permitted in the process of turning and measuring this diameter?
   b. Use the Rule of Ten to determine the required precision for the measuring operation.
   c. Use the Rule of Ten to determine the required precision for creating the turned diameter.

Case Problems

Case 1: Tolerances

A manufacturer of electric motors acts as a supplier to different customers making use of several different motor models. The company sends a shipment of motors to a new customer with which it has not previously done business. A fit problem develops when the new customer attempts to attach the motors to its equipment. The customer company complains that the motor manufacturer's product is not within advertised tolerances; however, the motor manufacturer indicates that the motors in question when used by other customers have caused no assembly problems.

What could cause this problem? What steps might be taken to ensure that there will be no assembly problems in the future? Relate your answer to tolerances of size, position, and form.

Case 2: Translating Shop Terminology into Precise Dimensions

A millwright telephones your shop to request a number of roller chain sprockets, some to be bored to 2⅞ in. and some to 1$\frac{7}{16}$ in., "for a sliding fit." To do what the millwright wishes, you would have to make the bore 1½ to 2 thousandths of an inch larger than these standard shafting sizes. Convert these crude data first into precision decimal inch specifications and then to metric, because your shop is now on metric measure.

CHAPTER 8

Processing of Metals: Casting

Objectives

This chapter will enable you to:

1. Explain the use of patterns for sand casting.
2. Describe the green sand molding process.
3. Describe the shell molding and investment casting processes.
4. Describe centrifugal casting, permanent mold, and die casting processes.
5. Describe several common types of melting furnaces.
6. Show how good casting design can prevent many casting problems.

Key Words

casting	drag
draft	riser
cope	investment casting
sprue	pattern
lost wax	green sand
mold	core print
shrink allowance	core

Casting is one of the oldest methods of manufacturing metals. Prehistoric humans made tools by pouring molten metal into open molds made of stone or baked clay. Cast objects over 4000 years old have been found dating from ancient Assyrian, Egyptian, and Chinese cultures.

The process of **casting** metals is accomplished by pouring or forcing molten metal into a **mold** cavity having a desired shape. When the metal has solidified the casting is removed from the mold. Virtually any shape can be produced by this method, in some cases with such precision that subsequent machining is not required.

After a general discussion of the casting process and why it is so frequently used, this chapter sequentially examines the process. Because sand molding is so frequently used, that discussion is very thorough, but eight other casting processes are also covered. The melting of metal and some of the hazards represented by molten metal are discussed, including measures to ensure the safety of such operations. The chapter concludes with a discussion of the importance of casting design in avoiding process and product problems.

THE CASTING PROCESS

When designing a metal part to be manufactured an engineer must choose a method of production. The part can be made by one or more processes, including machining from solid metal, welding fabrication, powder metallurgy, pressing and cold forming, hot forging, or casting. The main advantage of casting over other manufacturing processes is that parts with intricate internal cavities/passages (e.g., faucets, exhaust manifolds) can be made. This geometric feature would be difficult or even impossible to produce with any other manufacturing process. Besides, the casting process results in parts with smooth, flowing designs, either for practical or decorative purposes. Also, the metal can be placed only where it is required. Thus, the economy of using less metal for a part (especially when a very expensive metal is used), the eye-pleasing appearance (such as in machinery housings), and the possibility of producing intricate shapes are the factors that set casting apart from other manufacturing processes.

Sand casting is typically not a rapid method of production, whereas die casting is a relatively rapid process. Other casting processes lend themselves to production, but none should be considered to have high production rates as compared with punch press work or powder metallurgy.

Casting processes involve a large segment of the metals industry. These range from the tiniest precision parts to huge castings for machinery sections weighing many tons (Figure 8.1). Some metals that are too difficult to machine, such as those used for aircraft turbine impeller blades, can be

FIGURE 8.1
Centrifugal casting, 131,000 lb, 5-ft. diameter, 35½ ft long, to be used for a suction roll shell in a paper machine. Here it is being machined to size (Copyright 1976, 1978, 1984, Sandusky Foundry & Machine Co., Sandusky, OH).

cast to a precision shape not requiring any subsequent machining. Other softer metals, such as aluminum, are used to form articles such as transmission cases and valve covers for automobiles. It would take many hours of machining to make a complicated carburetor part from solid metal in a machine shop, but it takes only seconds in a die casting machine.

Several general conditions must be met for the production of good castings, regardless of the method used:

1. A method of melting the metal to the correct temperature must be available.
2. A mold cavity of the desired shape must be formed, with sufficient strength to contain the metal without distorting or having too much restraint on the molten metal as it solidifies. Also, the mold must be designed to avoid porosity and cracking of the casting.
3. Molds must be arranged so that when molten metal is introduced into the mold, air and gases can escape so the casting will be free from gas-related defects.
4. As we will see, molds usually are split to facilitate the removal of the pattern and/or the casting. Depending on the cross-sectional area of the mold cavity, just the gravitational force of the liquid metal can be enough to separate the mold halves. Some casting methods apply additional pressure to the metal. In either case, a force adequate to hold the mold halves together must be provided.
5. Any mold material (or cores) in internal cavities must have a provision for its removal. Finishing operations are often required to remove any excess material from the casting.

FIGURE 8.2
Metal match-plate patterns (Eugene Aluminum & Brass Foundry).

PATTERNS

The first requirement in making a casting is to design and make a pattern (Figure 8.2). **Patterns** are usually made of wood when only a few castings are needed. For larger quantities and when wear (due to an abrasion by the molding material) becomes a problem, patterns are made of metal, such as aluminum or bronze. Hard, tough plastics are also used for patterns. Patterns include several types—single-piece, split-piece, loose-piece, match-plate, and cope and drag. The type of pattern used in the mold depends on the required production. For example, small production is economical with a single-piece pattern; however, large produc-

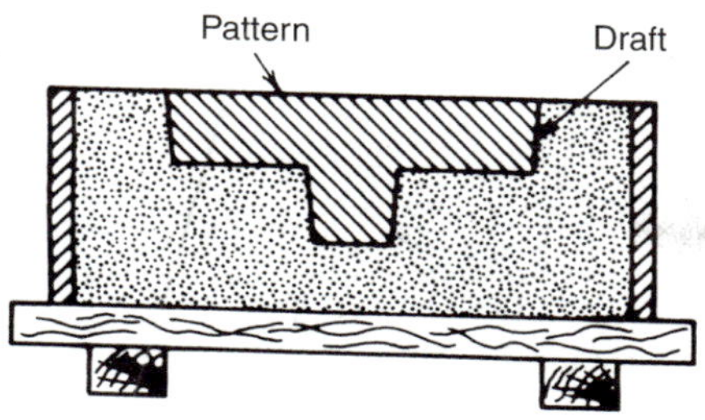

FIGURE 8.3
All sand patterns must have draft so they can be removed from the mold.

tion requires use of mechanized molding with application of a match-plate pattern. A previously made casting can also be used as a pattern if additional shrinkage is not a factor.

Shrinkage Allowance When metal solidifies from the liquid state, it shrinks. Thus, in order to reproduce casting of a desired dimension, a **shrinkage allowance** must be added to the pattern. Each metal has a different shrinkage. The following are shrinkages in in./ft for some metals:

Cast iron	1/8
Steel	1/4
Brass	3/16
Aluminum	5/32
Magnesium	5/32

Draft Patterns must have **draft** or taper to permit removal from the mold (Figure 8.3). Draft is usually about 2° to 3°. Without draft on the pattern, parts of the sand mold would be broken as the pattern is pulled out.

Other Allowances If there are areas on the casting that will have machined surfaces, the pattern needs to provide machining allowance that will leave additional metal for removal. The amount of machining allowance depends on the roughness and accuracy of the finished casting. Enough material must be left for subsequent machining in order to cut under sand inclusions on the surface of the casting.

SAND CASTING

In sand casting, a specially prepared sand that is mixed with different binders and additives is used as a mold material. The appropriately conditioned sand is then compacted around a pattern that has the shape of a desired casting. Sand has the advantage of being highly refractory (can resist high temperatures without melting), so metals like cast iron and steel can be cast easily in sand molds. Although there are many different molding materials, sand casting accounts for the greatest tonnage of all castings produced.

Molding Sands

The basic characteristics of sand are that it is (1) easily molded and capable of holding accurate detail, (2) reusable, and (3) inexpensive. Sand molding processes can be classified into the following groups: green sand molding, heat-cured resin binder processes, cold-box resin binder processes, no-bake resin binder processes, and silicate and phosphate bonds. The most common type of sand molding process is **green sand** molding. *Green* refers to the fact that the molding sand contains moisture. Other sand molding processes, because of their higher strength, are often used for cores that form holes and hollow spaces in green sand molds.

The most common molding sand is a natural silica sand (SiO_2); however, other sand types such as zircon ($ZrSiO_4$)and olivine are also used in special applications. The sands can be classified into two groups: natural and synthetic. Natural sand is the sand in its natural form. The synthetic sands are natural sands that have been washed, screened, classified, and blended to meet the requirements of a particular application. The synthetic sands are most commonly used in foundries.

General Characteristics of Molding Sands

Molding sand must have several characteristics:

1. **Cohesiveness** The ability to be packed (rammed) in a mold and retain its shape. This property is achieved by adding various sand additives (e.g., clay, water, resins).
2. **Refractoriness** The ability to withstand high temperatures.
3. **Permeability** Porosity that allows gases to escape through the mold.
4. **Collapsibility** The ability to allow freedom for the solidifying, shrinking metal to move without fracturing, and to allow the cast part to be removed easily from the mold. This property is also achieved by adding appropriate sand additives (e.g., corn flour, dextrin).

Preparation of Green Sand Mix

The green sand mixture (sand, clay, and water) is placed in a muller (Figure 8.4), where the ingredients are thoroughly mixed in order to obtain the proper consistency. A typical muller consists of a large tub in which an arm with rollers swings around, forcing the rollers over the sand.

To achieve required molding properties, the sand grains must be of the right size, and clay and water must be added in the right proportion. To obtain consistently

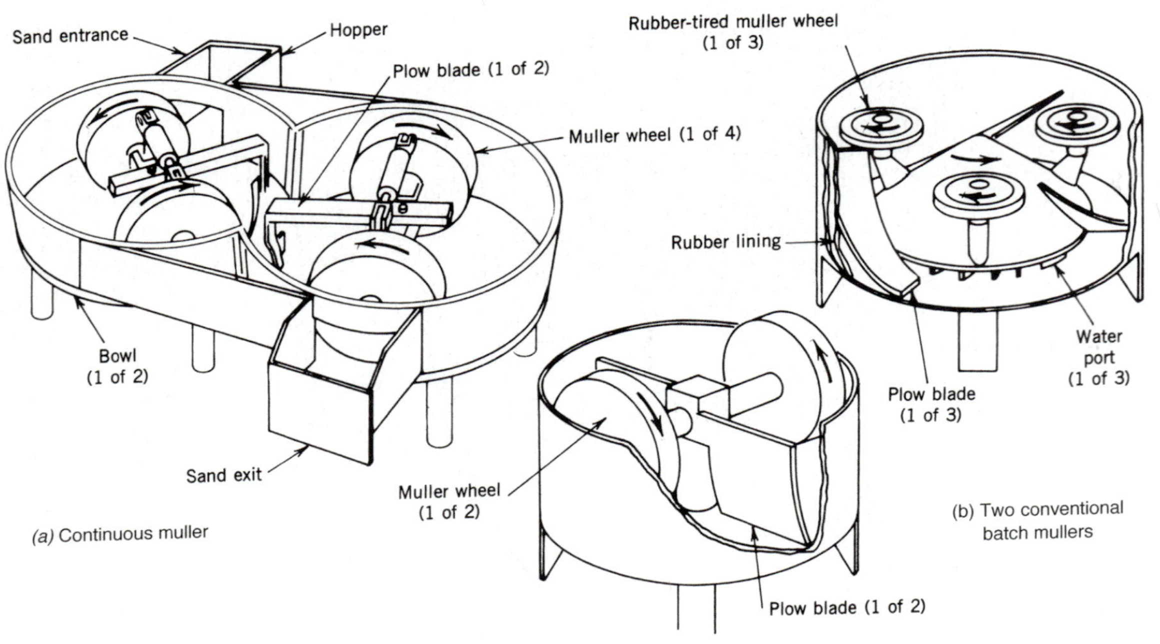

FIGURE 8.4
Three types of mulling machines used for the conditioning of molding sand. (*Metals Handbook,* Vol. 5, 8th ed., ASM International, 1970, p. 163. With permission.).

FIGURE 8.5
Casting sand is being pushed around on a jolt-squeeze machine (Eugene Aluminum & Brass Foundry).

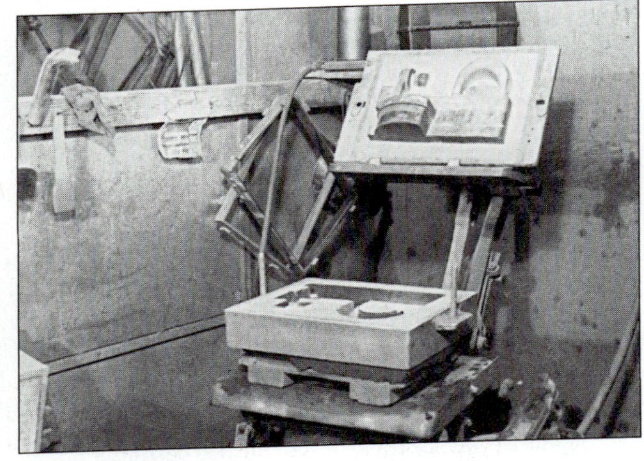

FIGURE 8.6
Drag (bottom half) of sand mold after it has been formed around the match-plate pattern (Eugene Aluminum & Brass Foundry).

good castings, the mold must be of appropriate hardness, strength, and permeability. Some standard tests are commonly applied in the foundries to control process variables (e.g., grain size and grain distribution, moisture content, green strength, hardness). These tests are performed both during sand preparation as well as during mold making in order to assure high-quality molds and castings.

Molding

A series of steps in manually producing castings can be seen in Figure 8.5 through 8.10. Although many kinds of patterns are used, such as split-piece, single-piece, and loose-piece types, patterns for green sand molding are usually made in two halves and are called *match-plate* patterns, or in the case of larger castings cope-and-drag patterns.

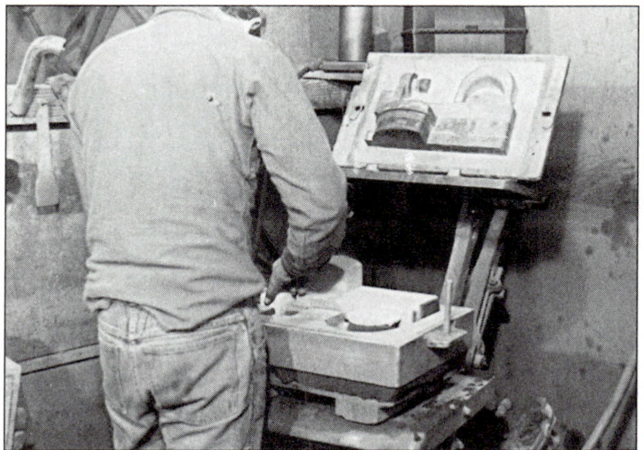

FIGURE 8.7
Core being placed in mold (Eugene Aluminum & Brass Foundry).

FIGURE 8.8
Cope (top half) being placed on drag flask (Eugene Aluminum & Brass Foundry).

FIGURE 8.9
Molten aluminum being poured into the mold (Eugene Aluminum & Brass Foundry).

FIGURE 8.10
Solidified casting broken out of molds (Eugene Aluminum & Brass Foundry).

Cope refers to the top half (Figure 8.11) and **drag** refers to the bottom half (Figure 8.12).

The cope-and-drag pattern is placed in a flask that is made in two halves (cope and drag) (Figure 8.13). First, the drag half of the flask is filled with sand. The sand is then rammed into place and afterward struck off (leveled off) even with the top of the flask with a straightedge (Figure 8.14). The drag half of the flask and pattern is also rammed in the same manner (Figure 8.15). The pattern also has an extension called a **core print** if a core is used. This space in the sand mold provides a support for the ends of the sand core. One or more vertical holes are provided for pouring the metal. This hole, called a **sprue,** has a pouring cup on the top. The role of the pouring cup is to make pouring easy and to keep the sprue constantly filled with metal. At its bottom the sprue is connected to the runner, which is connected to the mold cavity through the gate. When the molten metal is poured it fills the mold cavity and other holes in the top of the mold, called risers. The **risers** are used as reservoirs to feed liquid metal to the mold cavity as it shrinks while solidifying. They also provide for escaping gases and allow impurities to float to the top of the riser and out of the casting. The cope-and-drag pattern is made in segments that can be separated, such as the riser and downsprue, which are removed from the pouring side of the cope half of the flask, and the pattern is removed from the parting line side after it is turned over. Finally, the patterns are removed from the mold sections. Cores are installed, the two halves of the flask are assembled, and the mold is ready for casting. The resulting casting is similar to the pattern (Figure 8.16).

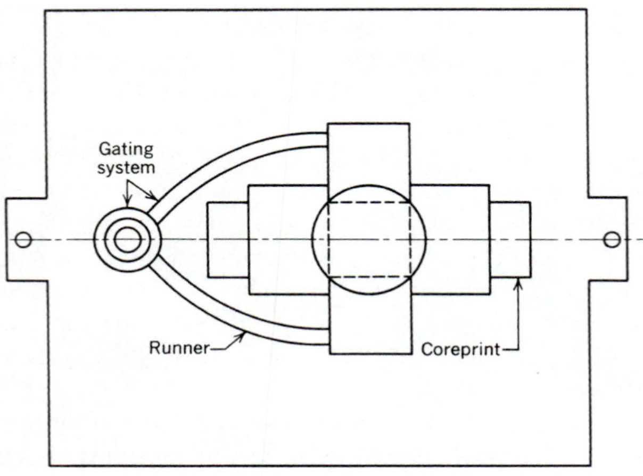

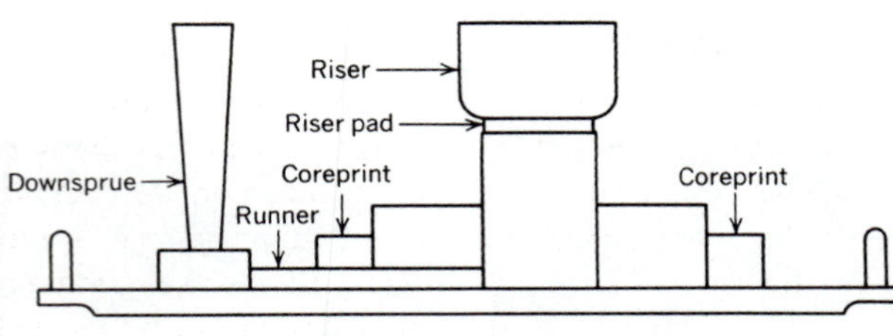

FIGURE 8.11
Top view and side view of cope match-plate pattern. The down sprue and riser patterns are removed from one side, and the cope pattern is removed from the other side (Neely and Bertone, *Practical Metallurgy and Materials of Industry,* 6th ed., © 2003 Prentice Hall, Inc.).

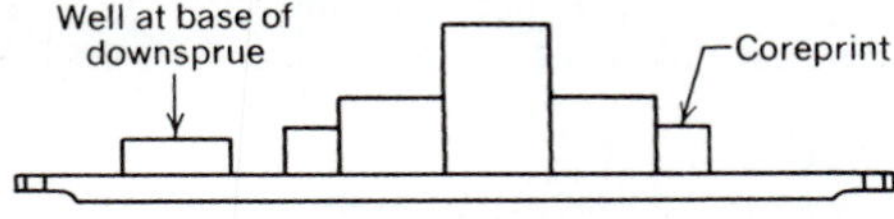

FIGURE 8.12
Drag match-plate pattern (Neely and Bertone, *Practical Metallurgy and Materials of Industry,* 6th ed., © 2003 Prentice Hall, Inc.).

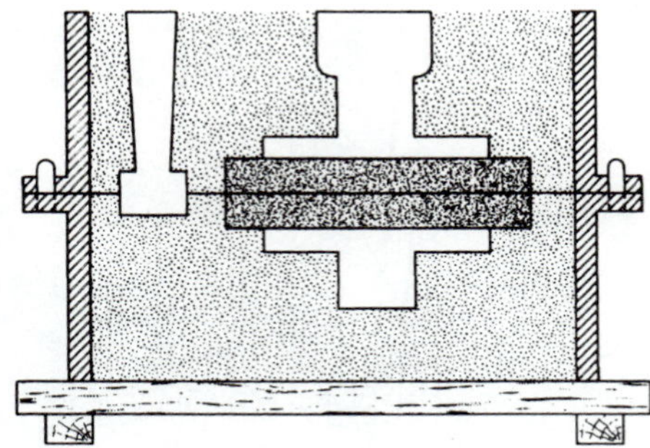

FIGURE 8.13
Sectional view of cope and drag with core in place on the mold board. At this point it is ready to make the casting (Neely and Bertone, *Practical Metallurgy and Materials of Industry,* 3rd ed. © 1989 Prentice Hall, Inc.).

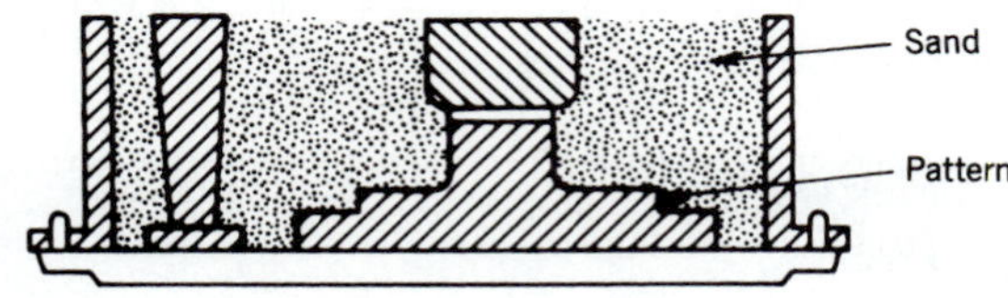

FIGURE 8.14
Sectional view of cope pattern in flask with the sand rammed in place and struck off.

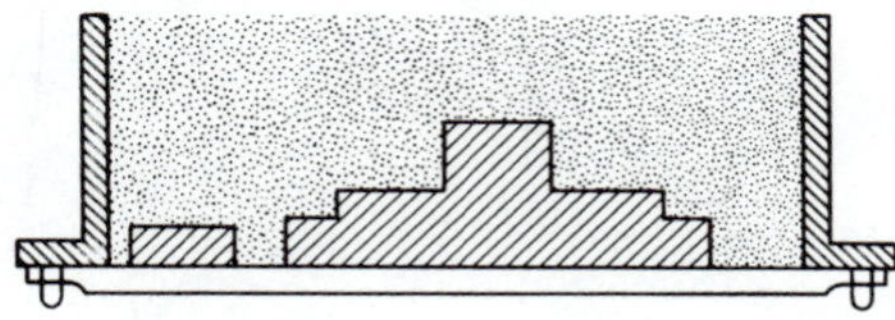

FIGURE 8.15
Sectional view of drag pattern in a flask with the sand rammed in place and struck off (Neely and Bertone, *Practical Metallurgy and Materials of Industry,* 3rd ed. © 1989 Prentice Hall, Inc.).

The sand mold and pattern, as well as the two halves of a sand mold are prevented from sticking together by dusting them with a commercially prepared parting powder. The inside of a sand mold is often coated with mold wash (e.g., graphite solution) to prevent the sand from sticking to the casting. It also results in a smooth surface of the casting.

Molten metal is poured into the mold, where it solidifies. After a specified cooling time the casting is removed from the mold. There are several ways to remove a casting. The most common method is vibration, in which the mold is broken by placing it in a shakeout, that is, a vibrating machine, that breaks the sand sticking to the casting, as well as the cores. The sand from the broken-up mold and from the shakeout is transferred to another machine that reconditions it and prepares it for reuse. The sand conditioning consists of crushing, screening, magnetically separating, and impinging of sand grains against a wear plate at high velocity. The fines (dust) are vacuumed away and the clean sand is transported to hoppers to be used again.

Cores

Cores are parts used to produce a desired cavity in a casting. To reduce the manufacturing cost, the cores should be made as a part of the mold whenever possible; however, due to the complexity of desired hollow sections, the cores are most commonly made separately. Cores are made by

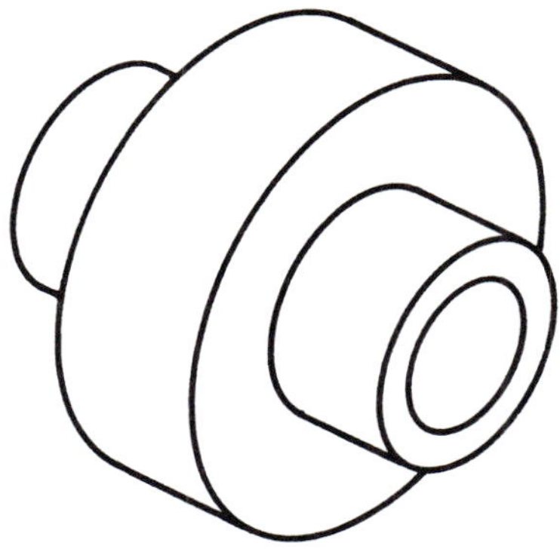

FIGURE 8.16
Completed casting (Neely and Bertone, *Practical Metallurgy and Materials of Industry,* 6th ed., © 2003 Prentice Hall, Inc.).

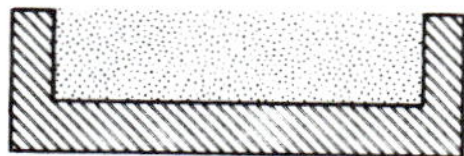

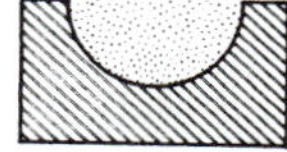

FIGURE 8.17
Sectional view of core box with the core (Neely and Bertone, *Practical Metallurgy and Materials of Industry,* 5th ed., © 2000 Prentice Hall, Inc.).

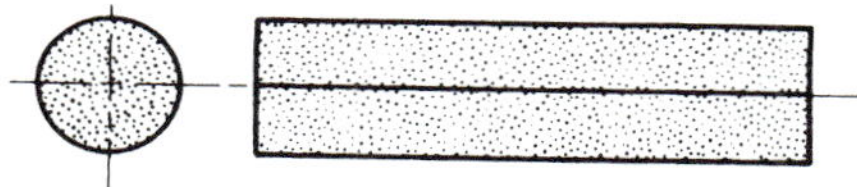

FIGURE 8.18
Two halves of core fastened together after being removed from core box and baked in oven (Neely and Bertone, *Practical Metallurgy and Materials of Industry,* 6th ed., © 2003 Prentice Hall, Inc.).

compacting a sand mixture into the core box (Figures 8.17 and 8.18), and subsequently curing the mixture. The compaction can be done manually or mechanically by using a core blowing machine. Today, most cores are made with one of many organic binder processes. These processes can be classified as no-bake binder processes, heat-cured binder processes, and cold-box binder processes. In the no-bake and cold-box processes the binder is cured at room temperature, whereas in the heat-cured binder process the binder is cured at elevated temperatures. Some cores are made hollow; thus, they use less sand and are also lighter to handle.

The core is placed in the sand mold and in the core print, and the mold is then ready for pouring (Figure 8.19).

The Advantages and Disadvantages of Sand Casting

The greatest advantages of sand casting are that almost any metal can be poured in the sand mold, and there is almost no limit on size, shape, or weight of the part. Sand casting provides the most direct route from pattern to casting. Tooling costs are low and the gravity-casting process is economical. Among the limitations involved in sand casting is the need for machining in order to finish the castings, especially large ones having rough surfaces. Other disadvantages are that it is not practical in the green sand process to cast parts with long and thin sections, and a new mold is required for every pour.

Sand castings are extensively used for machine tool housings, bases, slideways, and other parts, and they are also extensively used in the automotive industry for making engine blocks.

Castings must be defect free, and their quality must be consistent from casting to casting. Statistical process control

FIGURE 8.19
Large sand mold being gravity cast with molten cast iron (1983 © Pottstown Machine Company, Pottstown, PA).

(SPC) techniques are used often to improve quality and reduce rejection of parts.

EVAPORATIVE CASTING PROCESS

In the evaporative casting process, the pattern, sprue, and riser are made of foamed polystyrene. They can be made as a single piece, or they can be made separately and then glued together. The completed polystyrene pattern is then coated with a thin layer of refractory material. The coated pattern is placed in the mold and dry sand is compacted/vibrated around the pattern. When the metal is poured, the heat vaporizes the polystyrene pattern almost instantaneously, leaving the mold shape intact as it is being filled with metal (Figure 8.20). This process can be used for casting parts of any size and shape, since the patterns do not need to be removed prior to casting, thus eliminating the need for draft, on the pattern (Figures 8.21 and 8.22). Also, problems related to mold shift are eliminated, since the entire pattern is assembled before being covered with sand, allowing for corrections prior to pouring. Another advantage of the process is the low cost of sand preparation, since no additives or binders are used for sand conditioning. The process is economical only for prototyping or large production, that is, only if the polystyrene patterns themselves can be mass produced in a die.

FIGURE 8.20
Evaporative casting process. This photo depicts the way the intense heat of molten aluminum vaporizes a foam pattern, leaving behind a high-quality casting (Photo courtesy of Ford Motor Company).

FIGURE 8.21
Polystyrene (Styrofoam™) pattern for a bronze casting. The pattern is made in sections (Eugene Aluminum & Brass Foundry).

SHELL PROCESS

The **shell molding** process is a type of sand casting process that provides a finer detail and smoother finish

FIGURE 8.22
Bronze casting called *Coast Spirit,* permanently located in front of the library at the University of Victoria, British Columbia (Eugene Aluminum & Brass Foundry).

because the sand is finer and it is combined with plastic resin to make a smooth mold surface. The process can be automated for mass production. This method requires the use of metal patterns that are heated to temperatures of up to 450° F (232° C) and coated with a silicon release agent. The pattern is then placed on the dump box, which contains sand coated with phenolic resin (Figure 8.23). The assembly is then turned upside down, causing the sand to fall and cover the heated pattern. Some of the sand adheres to the pattern and hardens, forming a shell around it. The thickness of the shell depends on the length of time the pattern is in contact with the sand. After the prescribed curing time, the dump box is inverted to remove the loose sand. The pattern and adhering sand are placed in an oven and heated to a temperature of 600° F (316° C) for 1 or 2 minutes. The half-shell is then removed from the pattern and the two shells are glued or clamped together to form a complete mold. This thin mold is strong enough for casting small or thin parts, but heavier castings require the use of a backup material such as shot or sand in a pouring jacket.

The advantage of the shell process is the high quality of the surface finish of the casting. In addition, the mold is very strong, allowing casting of parts with fine details. Also, molds have long shelf life, thus allowing production flexibility. A disadvantage of the process is the relatively high cost of the core boxes; however, shell molding is an economical process for high-quantity production due to savings in the machining time of castings, low labor costs associated with the process, and high productivity. Many shapes can be made by this process. In the case of the complex shapes, individual segments are manufactured separately and assembled together into the mold. Shell mold casting has good repeatability, and most ferrous and nonferrous metals can be cast by this method.

PERMANENT MOLD CASTING

The greatest disadvantage of sand casting is that a new mold must be made for each casting. In addition, some inherent dimensional inaccuracies are present in the sand casting. These disadvantages gave rise to the development of a permanent mold. Despite the name, permanent molds can be reused at most for several thousand pours, after which they lose their true shape and must be scrapped. Most permanent molds are made of gray cast iron or steel. Graphite molds are often used for casting higher-temperature metals. The molds are made by machining processes and are hand finished or polished. A refractory wash is applied to the mold prior to casting in order to prolong its life. When cores are needed, they can be made from metal, in which case they are reused, or they can be made from sand, in which case they cannot be reused. The mold halves may be hinged (Figure 8.24) or mounted on a casting machine (Figure 8.25) so they can be opened and closed quickly and accurately.

In permanent mold casting, the metal is poured from a ladle, using gravity to fill the mold. Cast iron and nonferrous metals are cast in this manner. The molds are heated prior to a run, and the temperature is maintained by the molten metal. This is necessary to avoid chilling the metal too quickly, which can produce laps or cold-shuts in the casting (cracks where two adjacent portions of metal do not solidify together). Usually, only simple shapes are cast by this process because of the rigidity of the mold and shrinking of the metal when it solidifies. Because these molds are not permeable, they must be vented at the plane where the two mold halves come together. In addition to gravity pour, low-pressure and vacuum-pouring techniques are used in conjunction with the permanent mold process.

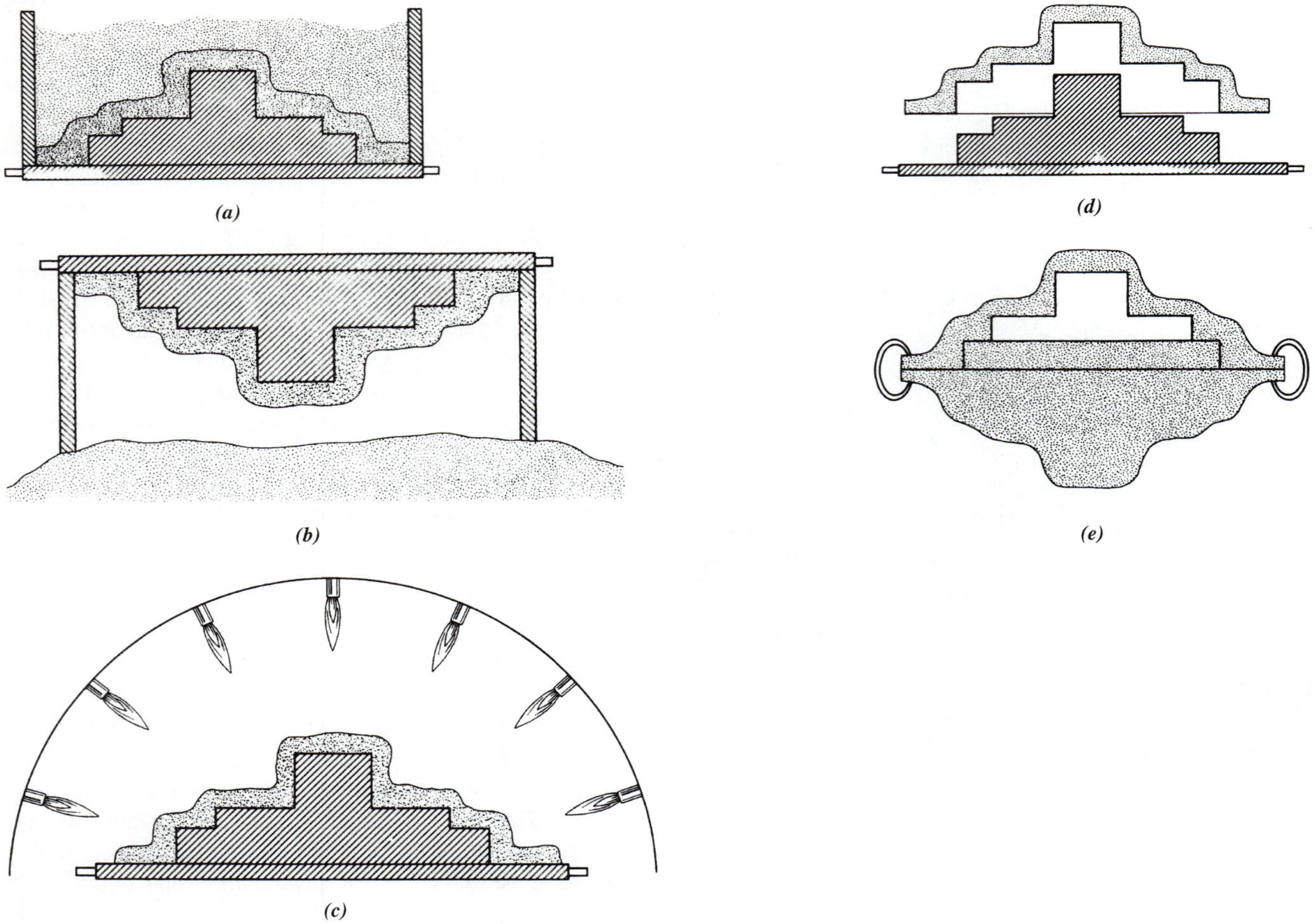

(a) *(d)*

(b) *(e)*

(c)

FIGURE 8.23
Shell molds. *(a)* Heated metal pattern in dump box with sand. *(b)* Dump box is turned, leaving some sand adhering to the pattern. *(c)* Pattern is placed in an oven for 1 or 2 minutes to harden it. *(d)* Half-shell mold is stripped from pattern. *(e)* Mold halves are clamped together with core. The top half-shell is shown as a sectional view (Neely and Bertone, *Practical Metallurgy and Materials of Industry,* 6th ed., © 2003 Prentice Hall, Inc.).

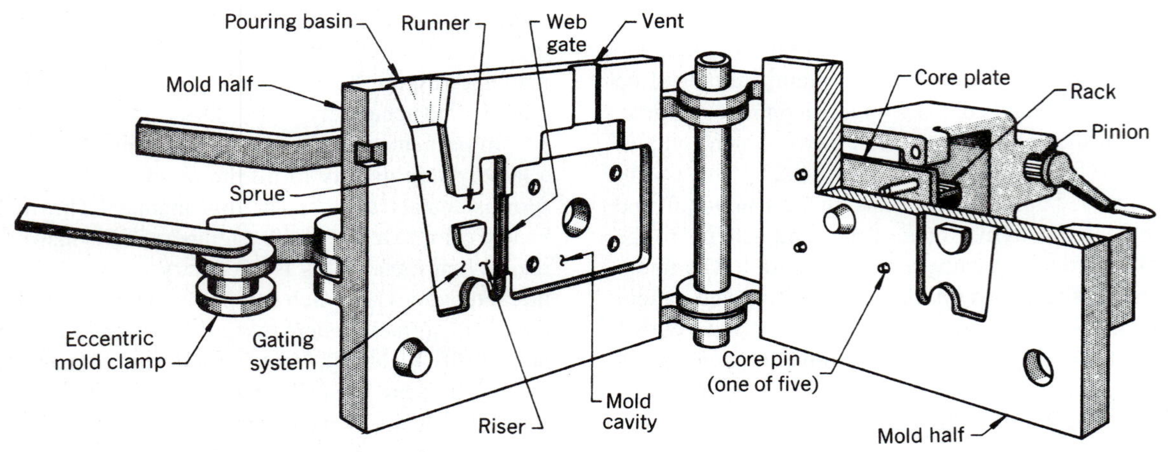

FIGURE 8.24
Book-type manually operated permanent mold-casting machine, used principally with molds having shallow cavities (*Metals Handbook,* Vol. 5, 8th ed., ASM International 1970, p. 266. With permission.).

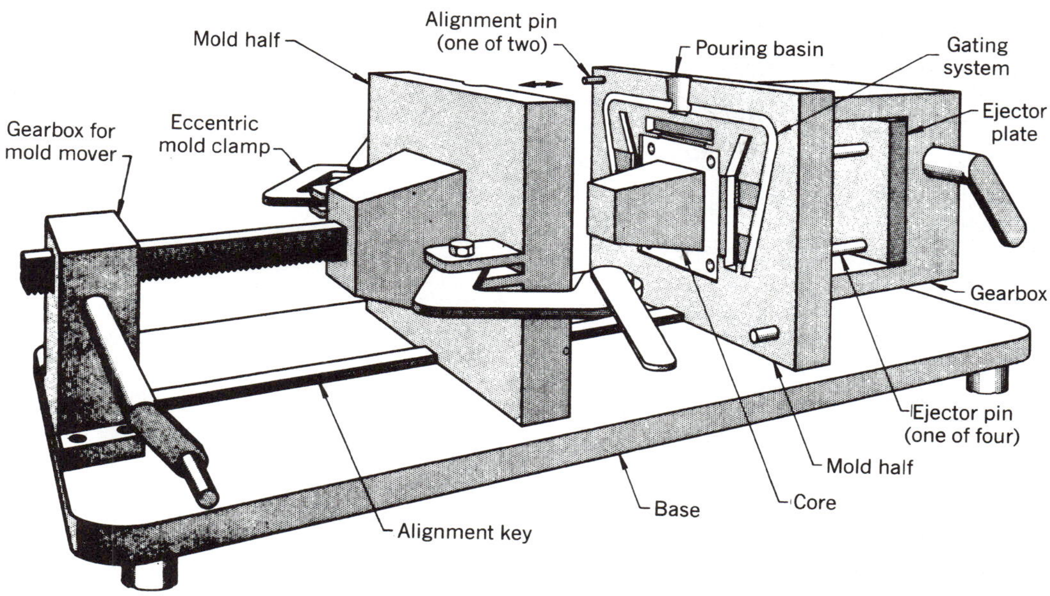

FIGURE 8.25
Manually operated permanent-mold-casting machine with straight-line retraction, required for deep-cavity molds (*Metals Handbook,* Vol. 5, 8th ed., ASM International, 1970, p. 266. With permission.).

Good surface finish, good dimensional accuracy, and low porosity can be obtained with permanent molds. Repeated use of molds and a rapid production rate with low scrap loss make this casting process ideal for moderate production runs of a few thousand pieces. Some disadvantages are high mold cost, limited intricacy of casting shape, and unsuitability for high-temperature metals such as steel.

SLUSH CASTING

Slush casting is used with permanent molds to make a shell of metal in the mold. The molten metal is poured into the mold and allowed to solidify to a certain wall thickness against the mold, and the remainder of the molten metal is dumped out, thus producing a shell. Toys, lamp bases, and ornamental objects are made with this process.

CENTRIFUGAL CASTING

Centrifugal casting is a process in which molten metal is poured into a rapidly revolving mold. The liquid metal is forced to conform to the shape of the mold by centrifugal forces many times the force of gravity. In this process the mold rotates about either a horizontal or a vertical axis. No core is needed to make an inner surface, since this process naturally produces a hollow shape such as pipe (Figures 8.26 and 8.27). The thickness of the mold controls the cooling rate and therefore the grain structure of the cast part. Also, if a hard outer wear surface is needed with an inner machinable soft metal, two dissimilar metals can be used—the outer one to harden when solidified and the inner one to remain relatively soft. For example, a thin wall of stainless steel can be poured first and then less expensive low-carbon steel can be poured, giving the pipe a corrosion-resistant jacket at low cost. Lighter elements, impurities, and slag collect on the inner wall of a centrifugal casting and can be removed later by machining.

Vertical centrifugal and semicentrifugal processes involve the rotation of a mold about its vertical axis. The metal is poured into a central reservoir that allows the liquid metal to flow outward into the spinning mold by centrifugal force. This process provides a denser structure in the metal than ordinary gravity casting does. It is ideal for steel wheels, brake drums, and cylinder barrels. A similar process, called *centrifuging,* rotates a group of molds or an entire mold located a certain distance from the center of rotation. A pouring basin allows the metal to flow outward into the molds. This method produces superior-quality castings.

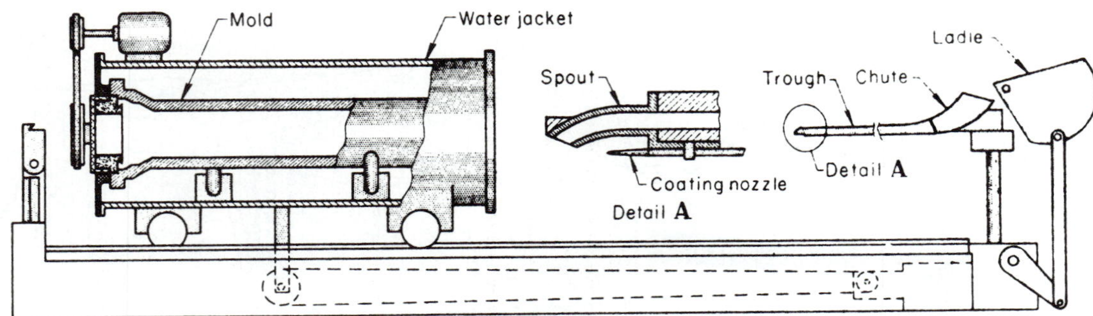

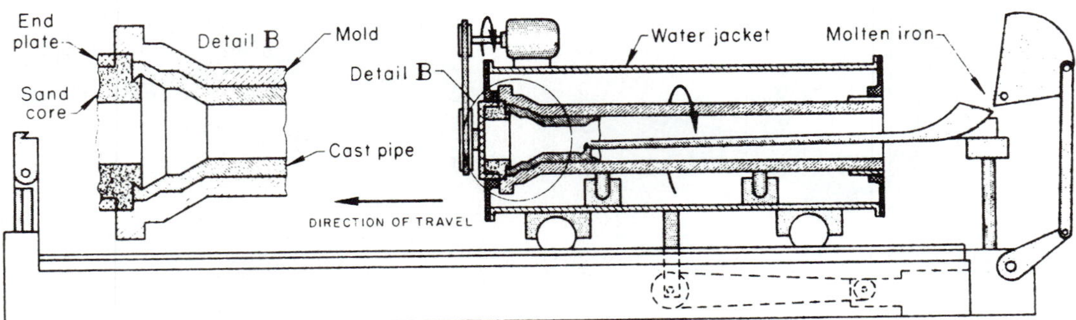

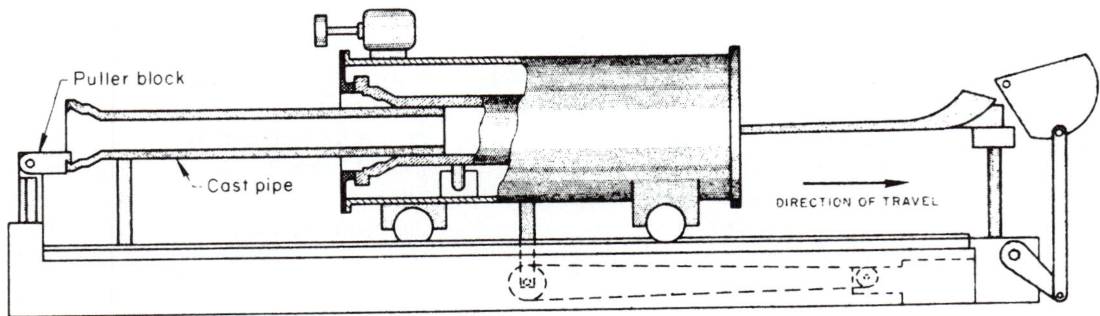

FIGURE 8.26
Machine for centrifugal casting of iron pipe in a rotating water-jacketed mold (*Metals Handbook,* Vol. 5, 8th ed., ASM International, 1970, p. 267. With permission.).

FIGURE 8.27
Large centrifugally cast tube (Copyright 1976, 1978, 1984, Sandusky Foundry & Machine Co., Sandusky, OH).

INVESTMENT CASTING

One of the oldest methods of casting metals is the **investment casting** (i.e., **lost wax**) process. The pattern is made of wax coated (invested) with a thick layer of refractory material (Figure 8.28). After the refractory material has hardened, the wax is melted and poured out of the mold and reused for another pattern. The mold is then preheated close to pouring temperature, and molten metal is then poured. This method is used for dentistry, arts and crafts, and for any type of very accurate near-net-shape casting. When it is used for industrial purposes, significant production rates are possible.

Plaster of paris can be used for the mold material in the lost wax process. The mold must be broken and destroyed

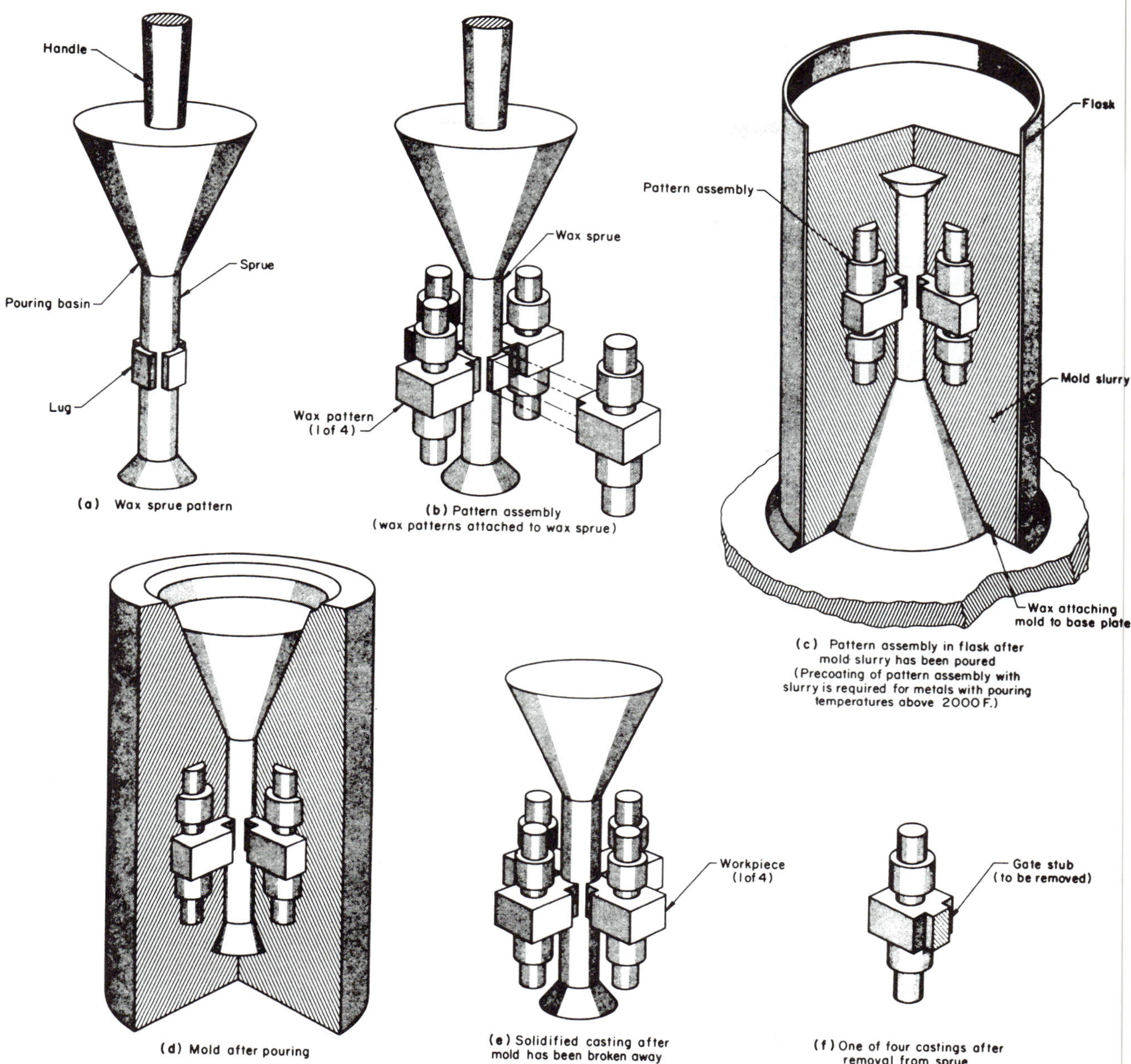

FIGURE 8.28
Steps in the production of a casting by the solid investment molding process, using a wax pattern. *(a)* Wax sprue pattern. *(b)* Pattern assembly (wax patterns attached to wax sprue). *(c)* Pattern assembly in flask after mold slurry has been poured. (Precoating of pattern assembly with slurry is required for metals with pouring temperatures above 2000° F.) *(d)* Mold after pouring. *(e)* Solidified casting after mold has been broken away. *(f)* One of four castings after removal from sprue (*Metals Handbook,* Vol. 5, 8th ed., ASM International, 1970, p. 239. With permission.).

to remove the cast part. Plaster of paris molds can be used only for the lower-temperature metals such as aluminum, zinc, tin, and some bronzes.

Metals having higher pouring temperatures, such as cast iron and steel, can also be mass precision cast. For a production casting process, a method of mass producing the wax patterns is needed. The patterns are mass produced by injecting the molten wax into a split mold in the same way plastic objects are rapidly made in injection molding machines (Figure 8.29). The wax patterns are assembled in groups on a sprue in clusters (Figure 8.30). The wax assembly is dipped into an investment, that is, slurry of refractory material. It is dipped several times to develop a fairly thick shell. The shell is then heated in an oven to melt out the wax, which is later

FIGURE 8.29
Automatic wax pattern molder. Large quantities of duplicate wax patterns can be produced on this machine (Courtesy of Mueller-Phipps International, Inc., Poughkeepsie, NY).

reused. Prior to casting, the mold is heated in a furnace close to pouring temperature and then may be placed in a flask where a heavy mold material is poured around it for additional support. The molten metal can be poured into the mold by gravity, pressure, or vacuum.

This method can be automated to some extent and is widely used in the automotive and aerospace industries. The following are some of the advantages of investment casting:

1. Unusual and nonsymmetrical shapes that would not allow withdrawal of an ordinary pattern are easily produced.
2. Smooth surfaces and high accuracy can be obtained. Machining can be reduced or eliminated.
3. Nonmachinable alloys can be cast with high precision.

Most parts made by this process are limited to a small size, and there are other limitations, such as in the use of holes and cavities. The process is more involved than some other casting processes, such as permanent molding and die casting, and is therefore more expensive per casting.

SHAW PROCESS

In this process, a rubbery jelling agent and a slurry of refractory aggregate is poured over the pattern. This rubbery mold hardens sufficiently to be easily stripped off the pattern and will return to the exact shape of the pattern. The mold is ignited to burn off the volatile elements, and it is then placed in a furnace and brought to a high temperature. The mold is then ready for pouring. The advantage of this process is good permeability and good collapsibility, which allow for production of delicate and intricate shapes with fine detail and higher-quality castings.

DIE CASTING

Die casting is similar to permanent molding in that a metal mold made in two halves is used. The difference is that the metal is not gravity poured into the mold (die), but instead the metal is injected under high pressures ranging from 1000 to 100,000 PSI. This requires massive machines that are generally operated hydraulically to exert the hundreds of tons of force necessary to hold the two halves of the die together when the molten metal is being injected at such high pressures (Figure 8.31).

The dies are usually made of alloy or tool steel and are quite expensive to make. Some have one or two identical mold cavities for larger parts, and others may have several different cavities (Figure 8.32). Some dies are more complicated and have sections that move in several directions. Grooves or overflows around the cavity on the parting face permit gases to escape. The overflows of excess metal must be trimmed off by a secondary operation after the casting is removed from the mold. This trimming is done with trimming dies that also remove the sprues and runners. The mold must also have provisions for water cooling so that a constant operating temperature can be maintained. Knock-out pins provide for ejection of the part when the die is opened (Figure 8.33). When cores are used they are made of metal and are usually drawn out before the die is opened. Cores are retracted either in a straight line or in a circular motion.

It is easy to see that these complex dies are quite expensive. Their costs can range from a few thousand dollars to $100,000 depending on the size of the die, its complexity, and the size of the casting machine used. Obviously, this

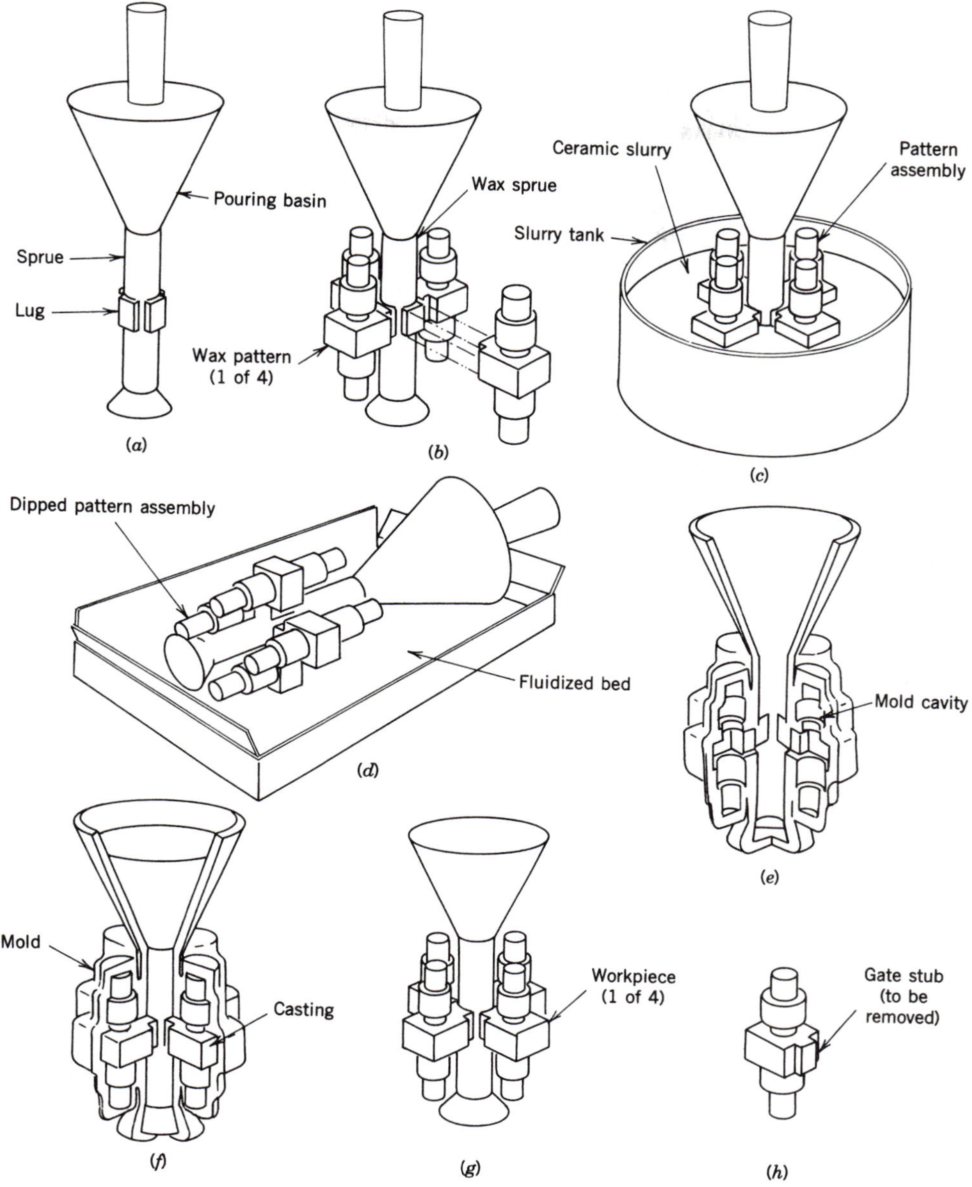

FIGURE 8.30
Steps in the production of a casting by the investment shell molding process, using a wax pattern. *(a)* Wax sprue pattern. *(b)* Pattern assembly (wax patterns attached to wax sprue). *(c)* Pattern assembly dipped in ceramic slurry. *(d)* Pattern assembly stuccoed in fluidized bed. Dipping *(c)* and stuccoing *(d)* are repeated until required wall thickness of mold is produced. *(e)* Completed mold after wax pattern has been melted out (mold shown in pouring position). *(f)* Mold after pouring. *(g)* Solidified casting after mold has been broken away. *(h)* One of four castings after removal from sprue. (*Metals Handbook,* Vol. 5, 8th ed., ASM International, 1970, p. 239. With permission.).

FIGURE 8.31
Die casting machine (TVT Die Casting & Mfg. Inc.).

process is not suitable for small-quantity production or for making very large parts. The large investment for dies and machines will pay off only when very large quantities (20,000 to millions of parts per year) are required.

Die casting is a means of producing castings of lower temperature alloys at a relatively high rate. These castings are usually thin walled, smooth, and highly accurate (Figures 8.34 and 8.35). The process is highly adaptable to the manufacture of small parts, such as automobile door handles, wiper motors, kitchen appliance parts, and thousands of small items we use every day.

The quality of die castings is high because of a rapid cooling rate that produces fine grains in the metal. The surfaces tend to be harder than the interior as a result of the chilling actions of the metal die. Porosity is sometimes a problem as a result of entrapped air, but with proper venting this can be overcome.

Zinc-based alloys are the most commonly used in die casting. Other metals used in die casting are alloys using copper, aluminum, and magnesium. Zinc alloys have the lowest melting point, about 700° F (371° C), and so have a less destructive effect on dies. Aluminum and magnesium alloys melt at about 1100° F (593° C), and copper alloys melt at about 1700° F (927° C). Therefore, dies using these alloys have a shorter life because of thermal shock, which causes crazing (microcracking) on the die surfaces.

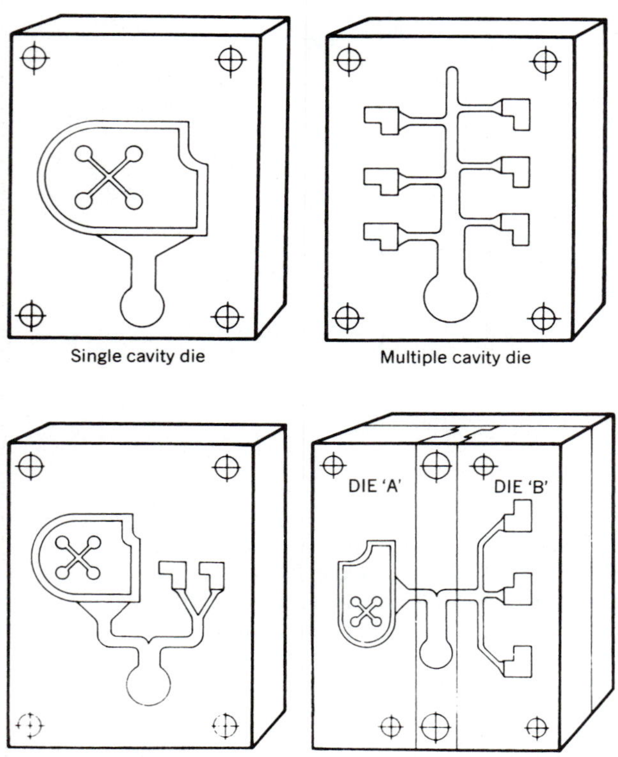

FIGURE 8.32
Die casting molds showing four different types of die cavities (Courtesy of American Die Casting Institute).

Die Casting Machines

The two basic systems and machines used for die casting are the cold-chamber machines (Figure 8.36) and the hot-chamber machines (Figure 8.37). Cold-chamber machines have a separate melting and holding furnace. For each

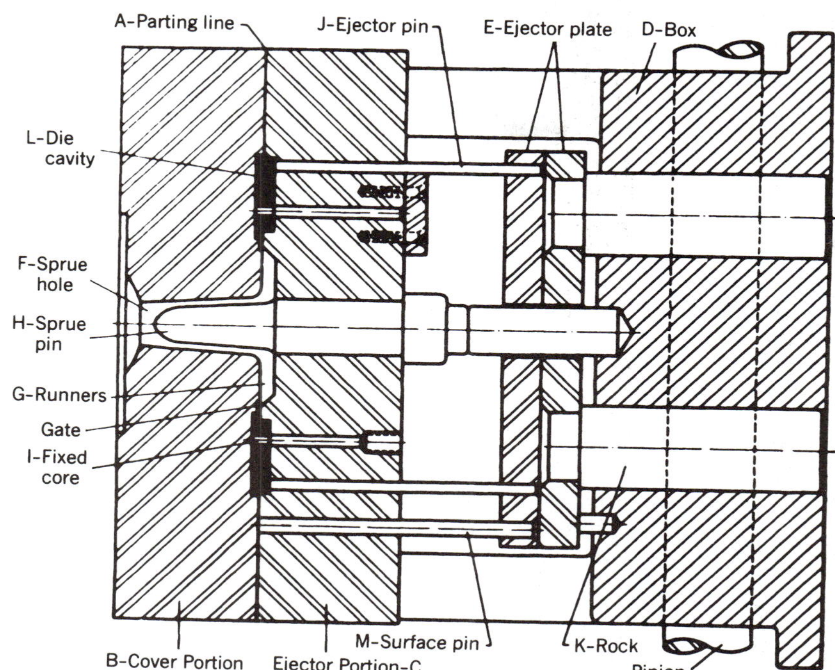

FIGURE 8.33
Typical die (Courtesy of American Die Casting Institute).

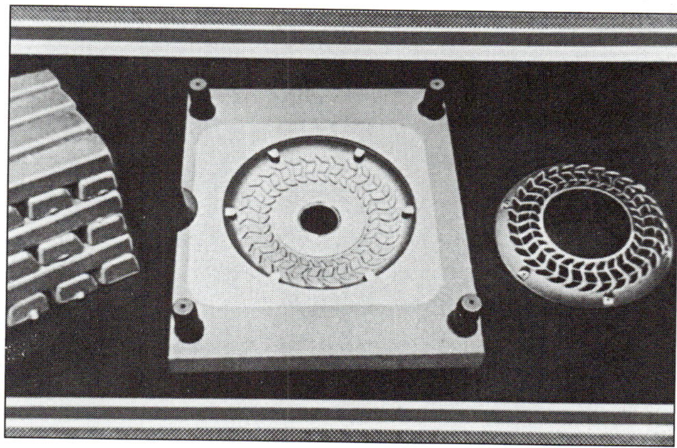

FIGURE 8.34
Part of a die showing mold cavity at center, raw material on the left, and product at the right (TVT Die Casting & Mfg., Inc.).

FIGURE 8.35
Some of the great variety of parts made by die casting (TVT Die Casting & Mfg., Inc.).

casting cycle, the molten metal is fed into the cold chamber and then is forced into the die by the plunger. These machines are often used for casting metals that melt at higher temperature because the chamber remains relatively cool and there is less tendency for spalling or cracking than in the hot-chamber machine. Although the cycle time is somewhat slower than that of the hot-chamber machine, the productivity is still high, as much as 100 cycles per hour.

A typical hot-chamber machine with a submerged gooseneck injector and plunger is illustrated in Figure 8.37. This type has an oil-fired or gas-fired furnace with a cast iron pot for melting and holding the metal. The plunger and cylinder are submerged in the molten metal that is forced

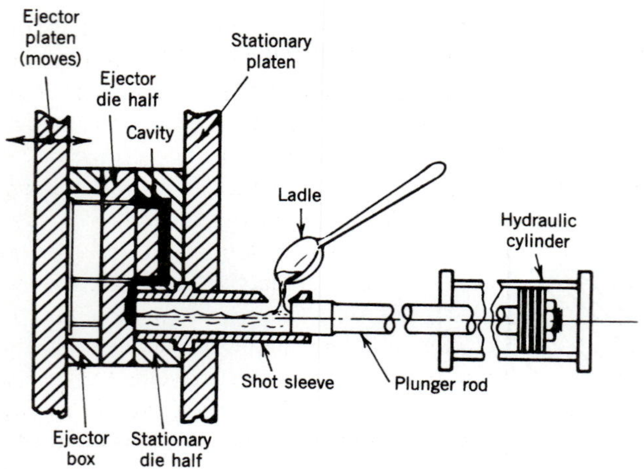

FIGURE 8.36
Cold-chamber machine. Diagram illustrates die, cold chamber, and horizontal ram or plunger in charging position (Courtesy of American Die Casting Institute).

through the gooseneck and nozzle into the die at each cycle. This type of die casting machine can be used only for the low-melting-point zinc-based and tin-based alloys. This machine is much faster in operation than the cold-chamber machine.

Several problems and difficulties have plagued the die casting process, especially when attempts were made to speed up or fully automate die casting systems. The major problems are flash (Figure 8.38), sticking, and soldering (molten metal adhering to the die). Many kinds of automatic unloading and loading devices have been developed, and die sprayers have been used to control these problems with mixed results. Other problems such as surface defects, dimensional inaccuracies, and internal porosity have been addressed also. Increasing the injection plunger velocity and reducing the total filling time reduced these problems.

FURNACES AND METAL HANDLING

The cupola furnace (Figure 8.39) has been of primary importance in the melting of cast iron for foundry work. Coke (produced from coal) is used for producing heat in the cupola furnace. A typical cupola furnace is a circular steel shell lined with a refractory material such as firebrick. It is equipped with a blower, air duct, and wind box with tuyeres for admitting air into the cupola. A sand bottom keeps the molten metal from burning through and is sloped so that the iron may flow out when the tap hole is opened. The process begins by starting a wood fire on the sand bottom and placing coke on top. When a 2- to-4-ft-thick bed of coke is thoroughly ignited, iron scrap or pigs are charged into the furnace, alternating with coke plus small amounts of limestone for a flux. A

FIGURE 8.37
Hot-chamber machine. Diagram illustrates plunger that is submerged in molten metal, gooseneck, nozzle, and dies. Modern machines are hydraulically operated and are equipped with automatic cycling controls and safety devices (Courtesy of American Die Casting Institute).

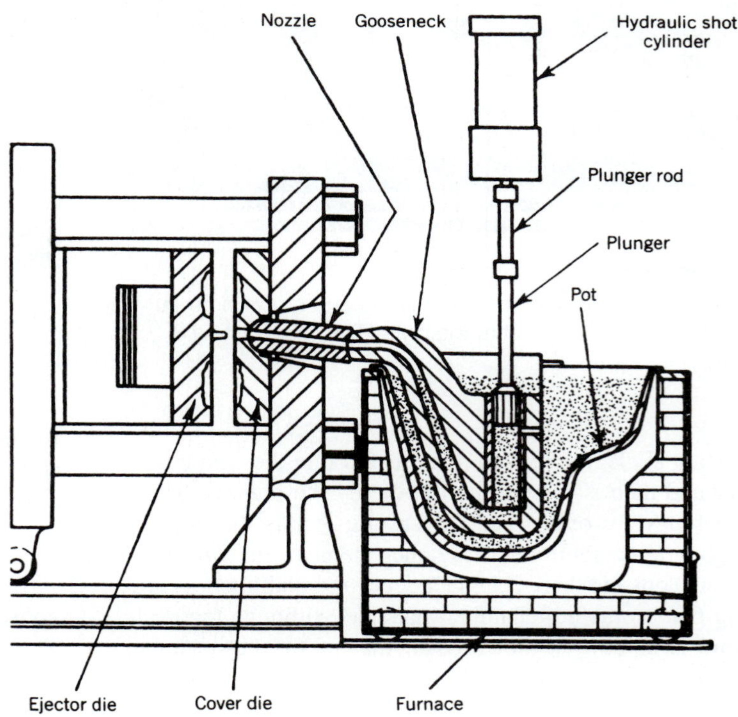

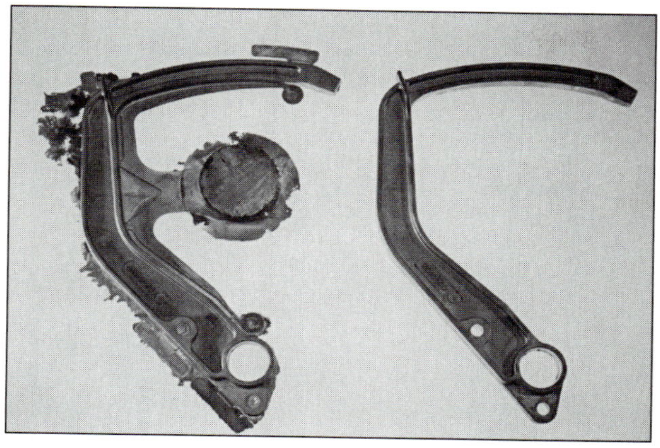

FIGURE 8.38
Die cast parts (clutch arm) showing flash on the left and flash removed on the right (Consolidated Metco, Inc.).

blower forces air up through the charge, and the iron begins to melt at the top of the coke bed. The molten iron is drawn off into a ladle, and suitable metallurgical tests are performed. The metallurgy of the cast iron is controlled at the furnace before the ladle is moved to the casting floor. These tests reveal the amount of carbon and other elements. Certain alloy-rich materials or silicon to control graphitization are added to the ladle. Also, if nodular iron is desired, elements such as magnesium or cerium are added to the ladle to cause the graphite to form in the shape of nodules (spheres) instead of flakes. (Nodular iron is much tougher and stronger than gray cast iron, which contains graphite in the form of flakes.) When the tests are completed and the additions are made in the ladle, the molten metal is ready to be poured into the prepared molds.

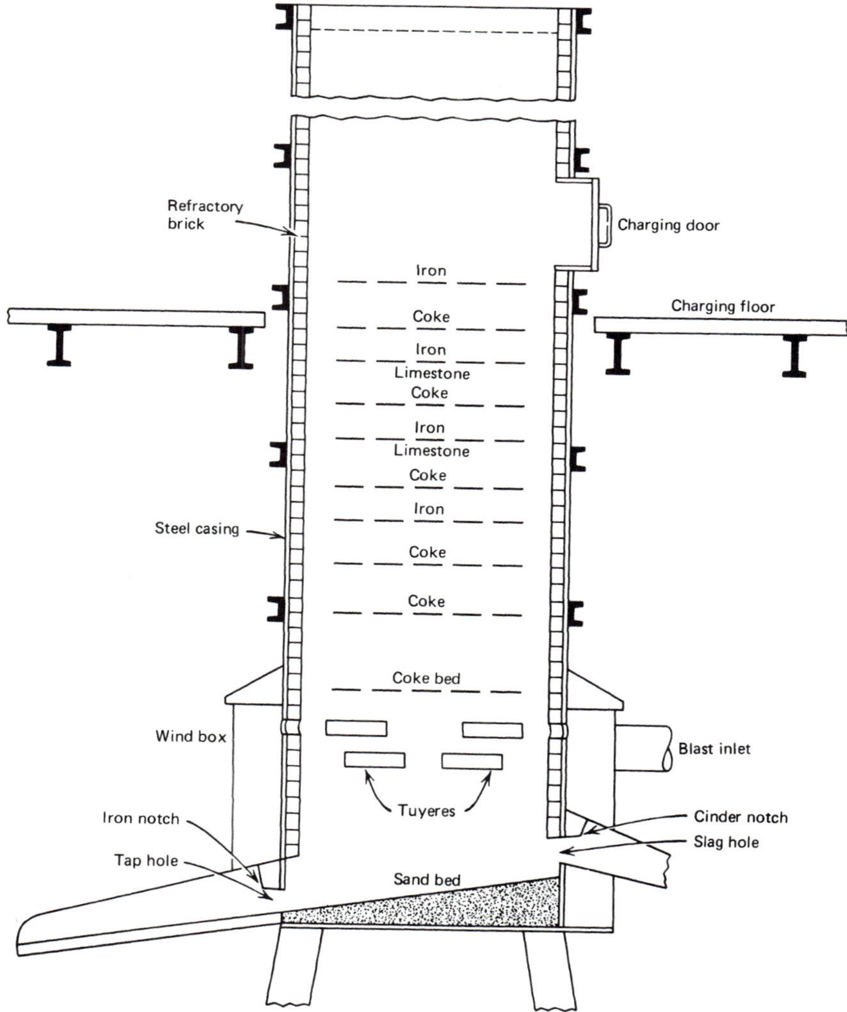

FIGURE 8.39
The cupola furnace (Neely and Bertone, *Practical Metallurgy and Materials of Industry,* 6th ed., © 2003 Prentice Hall, Inc.).

FIGURE 8.40
Molten metal being poured from an induction furnace into a ladle. The furnace is tilted to make the pour (Inductotherm Corp.).

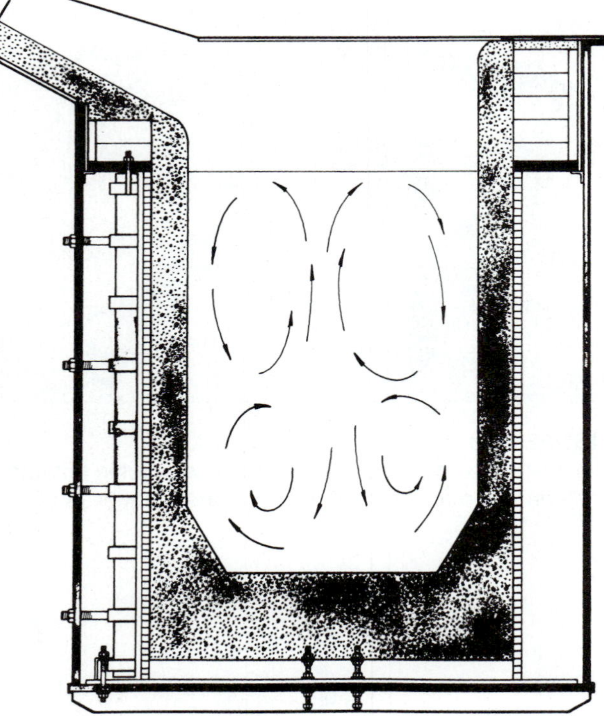

FIGURE 8.41
Cross section of a 60-cycle coreless furnace with a rammed lining (Courtesy of Ajax Magnethermic Corp., Warren, Ohio).

When fossil fuels such as coke are burned, considerable smoke and fumes are released into the atmosphere. In areas where air pollution is a critical problem, these furnaces are fitted with dust collectors and other air pollution equipment; where cheap electricity is available, the cupola furnaces are replaced with electric furnaces. Electric furnaces for melting cast iron can be either electric arc, electric induction, or electric resistance types (Figures 8.40, 8.41). Another advantage of the electric furnace over the cupola is that the former can be used to melt many other metals, for example, aluminum, copper, brass.

Gas-fired crucible or air furnaces are often used to melt nonferrous metals. This furnace consists of a heating chamber containing a crucible into which the metal charge is placed. The furnace may be small enough that the crucible of molten metal is lifted out of the furnace for pouring, or it may be so large that the whole furnace and the crucible tilt for pouring. In another adaptation the heating is done in a large brick-lined ladle by placing a gas-fired lid over it.

Larger, reverberatory-type gas-fired furnaces are also used, especially in melting nonferrous metals. Figure 8.42 shows such a furnace being charged with aluminum scrap.

MOLTEN METAL SAFETY

The hazards inherent in the handling of molten metal require thoughtful planning and engineering to ensure they are avoided. Molten metal is a threat to the safety of people and equipment if it is mishandled; therefore, the handling process, for example, cranes, ladles, fork trucks, and lifting devices, must be reliable and operate with a minimum of difficulty. The floor space or aisleways in which such equipment is used must be designed so they are free of obstructions and (if possible) people. Auxiliary equipment such as troughs, molds, and stirring and skimming equipment must be engineered to avoid spilling or splashing of metal.

Personal protective equipment varies with the metal involved but usually includes safety glasses, face shields, hard hat, molder's boots (or spats to cover shoelaces and shoe openings), heat-resistant leggings and aprons, and heat-resistant gloves; some plants require employees to wear flame-retardant cotton clothing and strictly forbid synthetic clothing. Some of these precautions are related to the heat involved in working around furnaces and with molten metals, but they all will also help protect the individual from coming in contact with molten metal in case there is an explosion.

A dangerous aspect of molten metal is the possible consequences of its coming in contact with water. The temperatures at which many metals are molten are many times the boiling point of water; if molten metal manages to trap or cover moisture or water, the water will rapidly transform to steam and explode as it expands. This explosion can hurl molten metal many feet, endangering people and equip-

FIGURE 8.42
Aluminum scrap being charged into a gas-fired melting furnace (Courtesy of Alcoa Inc.).

ment. The source of the moisture could be a cold cast iron mold on which moisture has condensed from the air, or it could be water spilled or leaked on the floor, or it could be moisture trapped inside scrap metal or the shrinkage crack of a pig that is then charged into a furnace. The correction of some of these hazards is obvious; others may not be. Scrap and pigs to be charged should be stored out of the weather. If that is not possible, they should be preheated before charging to drive off water. Cold molds should be preheated, or coated with a fuel such as kerosene.

Oddly enough, if molten metal is spilled into *enough* water there is no problem; the water does its job of cooling, and the metal solidifies before it can cause a problem, but a *little* water . . . and boom!

Molten metal can also explode with devastating force and consequences if it comes in contact with an oxide of another metal under just the right (or wrong) circumstances. This type of explosion is almost of an atomic nature and depends on the fact that energy is required to free metals from their oxides, therefore if the reverse happens, that is, if a molten metal suddenly converts back to its oxide, there is a tremendous release of energy. This can happen with all metals, but aluminum is possibly the worst; it is estimated that the conversion of 1 lb of aluminum to aluminum oxide releases three times the energy of a pound of nitroglycerine! Some of these types of explosions have been known to level whole plants.

The normal method of preventing these occurrences is to isolate the molten metal from the oxides of other metals. For example, if casting equipment is made of steel, it should be painted or covered with a material that will prevent it from rusting (turning into an oxide) and prevent the molten metal from coming in contact with it. The hazards will vary with the metals involved; just because one metal is safely cast does not mean that another will behave in the same manner. Industry standards can be very helpful in assessing the hazards involved with a particular metal.

POURING PRACTICE

Ladles are usually constructed of steel and lined with firebrick or other refractory material (Figure 8.43). The interior of the ladle is heated or kept hot while in use so as not to cool the molten metal. In small foundries, a handheld,

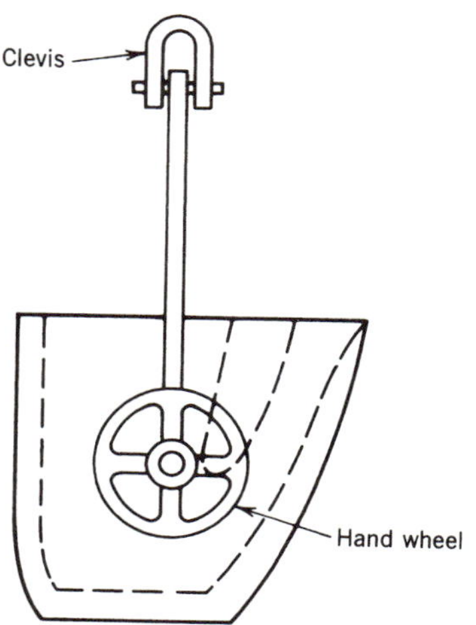

FIGURE 8.43
Teapot ladle.

FIGURE 8.44
Shank-type ladle.

shank-type ladle is often used to pour small quantities (Figure 8.44). This type requires two persons to use it. The teapot ladle can contain more metal and is supported on an overhead monorail or crane. The handwheel is turned to tilt the ladle and make the pour. This and the bottom-pour types keep the slag and oxidized metal from going into the mold. In some operations the molds are placed on a pouring floor, whereas in other, automatic or semiautomatic, operations the molds are carried along a conveyor to the ladles where a measured amount of molten metal is poured into the mold.

CASTING CLEANUP

After the solidified castings are removed from the mold, they are cleaned in a shakeout, and small parts are often put in a tumbler. The risers and gates must be knocked off (often this can be done on cast iron because of its brittleness) or cut off by means of sawing or use of an abrasive cutoff wheel.

CASTING DESIGN AND PROBLEMS

Because many factors influence the quality and performance of a casting, careful attention must be given to several design requirements if the best results are to be obtained. The more complex castings require more considerations than the simple ones. The draftsman or engineer should understand the problems involved or confer with the foundrymen for suggested alterations. Sufficient allowance of metal should always be provided when machining operations are to be performed. Size of the casting, and surface roughness, are involved in this decision. The location of the parting plane is also very important, since the pattern must be extracted at this plane without disrupting the sand mold. The following are other factors to be considered:

1. Weight of the casting and mold strength
2. Effective gating and sufficient riser
3. Number of cores and their placement
4. Required dimensional accuracy
5. Radii, thickness of sections, and amount of shrinkage

Cracking can occur at sharp corners and where thick sections join with thin sections (Figure 8.45). Proper radii can reduce this problem. Also, the cooling rates are greater for the thin section than for the thick one; this also causes cracking at the juncture. Cooling rates may be increased in a thick section by using "chills," metal sections embedded in the mold to absorb heat (Figure 8.46). The shape of a casting should be designed to have as uniform a thickness as possible, but in many products this is not possible. Parts that have ribs or spokes are subject to this kind of cracking, but often the ribs can be staggered (Figure 8.47). When parts intersect, such as in a 90° turn, interior corners (fillets) should be provided, but if they are too generous, hot spots may develop and cause shrinkage defects (Figure 8.48). Holes provided by cores can be used to eliminate these weak points

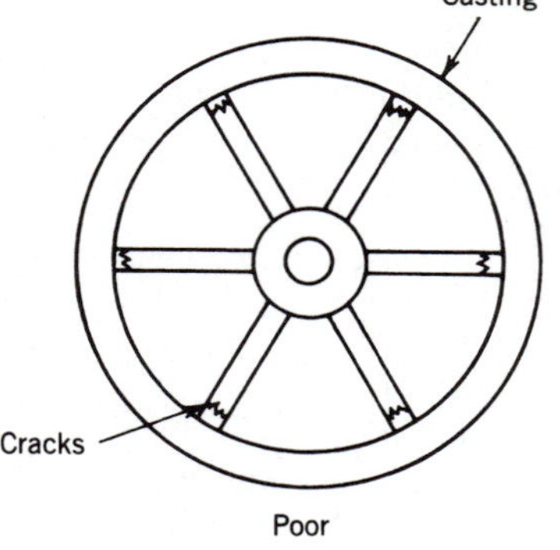

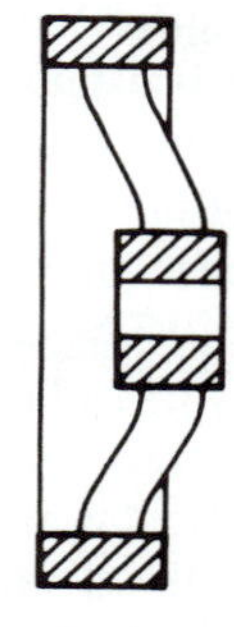

FIGURE 8.45
Casting problem: thin to thick section. The spokes, being smaller, cool more rapidly than the thick rim, causing them to break.

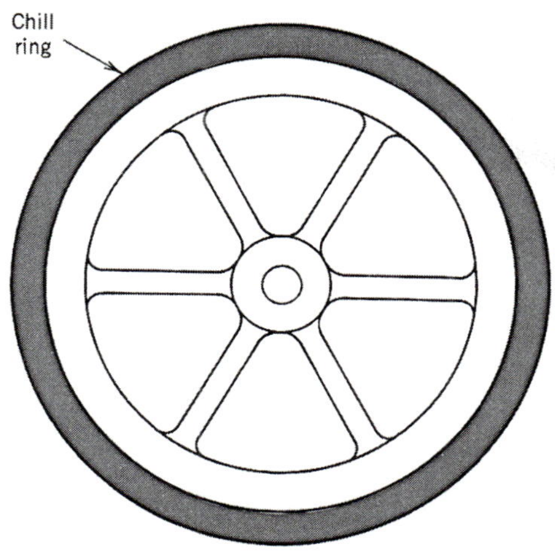

FIGURE 8.46
Chill in a mold around the rim of a wheel. This equalizes the cooling rate between the thick rim and the thin spoke. Odd numbers of spokes are often used to reduce cracking.

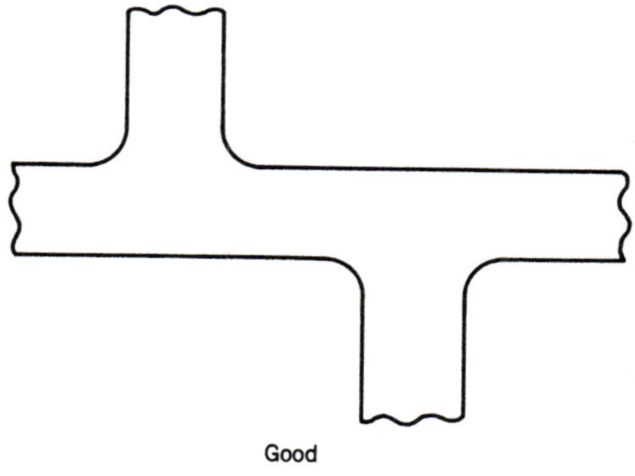

FIGURE 8.47
Staggered ribs.

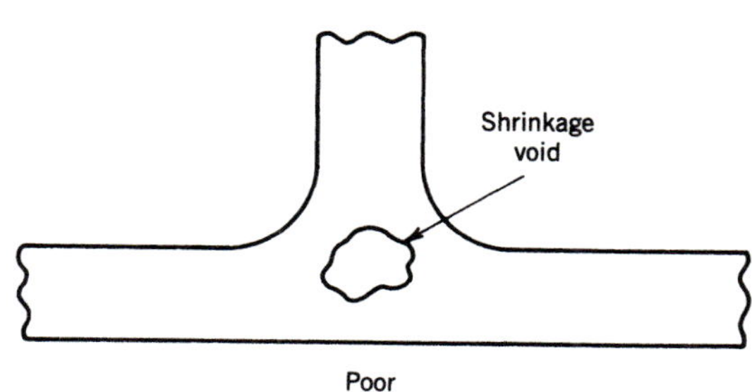

FIGURE 8.48
Shrinkage areas, hot spots cause voids.

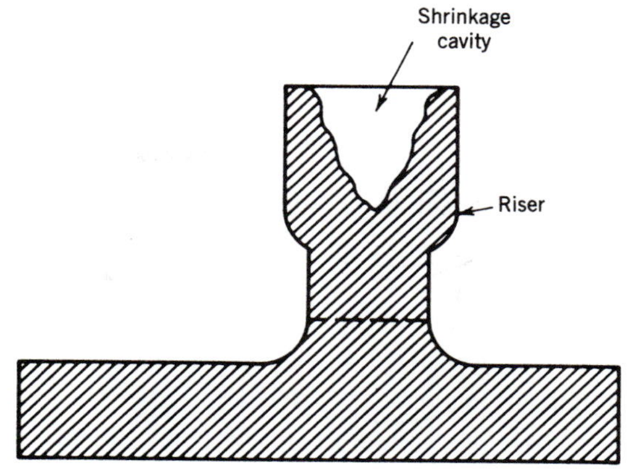

FIGURE 8.49
Shrinkage cavities.

in a casting. If risers are inadequate to supply molten metal when the casting is shrinking during solidification, shrinkage cavities can result (Figure 8.49).

Review Questions

1. When precision surfaces are required on a casting, which one of the several types of castings is most likely to need a machining operation subsequent to casting? Which two types are least likely to need subsequent machining?
2. What are patterns for sand casting made of? What material is used to make patterns for shell molds? Which kind of pattern volatilizes when the metal is poured?
3. What characteristic of sand casting patterns is most necessary for removal of patterns?
4. What is shrink rate and what does it have to do with pattern making?
5. What are cores and of what materials are they made?
6. If you had to choose between a steel casting and a gray cast iron casting for a crane hook, which would you choose? Why?
7. A gear housing for a tractor is to be sand cast. Which metal would you choose for this casting, gray cast iron or steel? Why?
8. You are required to manufacture 1000 simple, small parts that are to be cast in aluminum. Would using a permanent mold be less expensive per part than preparing 1000 sand molds? There is no time limit. Explain your reason.
9. What is slush casting and what is it used for?
10. Name a relatively rapid method of casting pipe and tubing.

11. In which type of casting is the pattern removed by heating and melting it? List three advantages in using this casting process.
12. Name the most rapid method of producing small cast parts. Is this system practical to use for only 20 parts? For 100? Explain the reasons for your answer.
13. Which fuel does the cupola furnace use? What metal is usually melted in it? Name one other type of furnace used for the same purpose.
14. What type or types of furnaces are usually used to melt nonferrous metals?
15. How does the molten metal get from the furnace to the mold?
16. What happens to the sand castings after they have solidified in the mold and cooled sufficiently to handle?
17. Why is it important to wear personal protective equipment (e.g., safety glasses, heat-resistant clothing) when working around molten metal?
18. A cast iron casting has failed in use by breaking at a T joint where one leg meets the other at the joint. On inspection, a large shrinkage cavity is found in the joint, which obviously caused the failure. How could this have been prevented?
19. In a sand casting, the top edge near the riser is full of impurities and porosity; also, a large shrinkage cavity is present. What could be the cause and how can it be corrected?
20. What is the major difference between a hot-chamber and a cold-chamber die casting machine? Which one is used mostly for low-temperature-melting alloys?

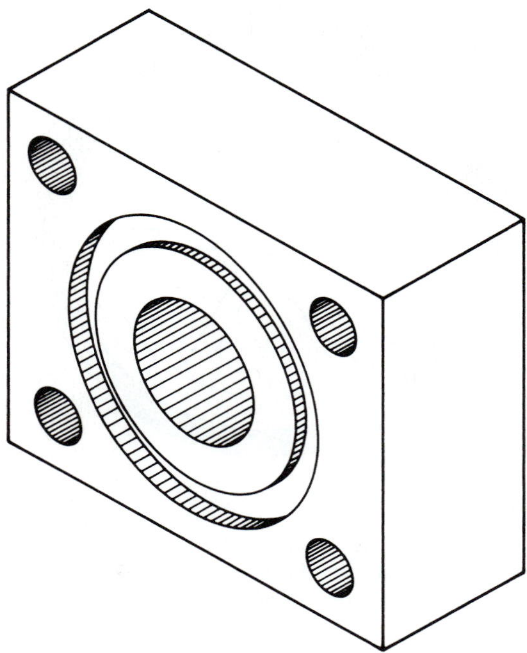

FIGURE 8.50
Metal cylinder head. See Case 1.

Case Problems

Case 1: Hydraulic Cylinder Head Problem

A design for a metal cylinder calls for a metal cylinder head (Figure 8.50) that must withstand considerable forces. Zinc-based alloys, aluminum, and magnesium cannot be used because they are not strong enough for the size limitations of the design. Cast iron, bronze, or steel may be used, since they all have sufficiently high tensile strengths. These metals vary somewhat in cost and in ease of casting. The least expensive material and process to be used to manufacture about 100 of these cylinder heads per 8 hours, for a total run of 1000 heads, must be selected. Only 20 days are given to produce them. The options are:

1. Machine all holes, cavities, and surfaces from solid bars of steel. These come in 20-ft-long bars that are sawn off in suitable lengths for machining.
2. Make sand castings of steel and finish by machining.
3. Make sand castings of cast iron and finish some surfaces by machining.
4. Cast the parts in cast iron by investment casting.
5. Use permanent molds and cast them in bronze.
6. Use the process of die casting with bronze and eliminate machining altogether.

Which method would produce the least costly parts and get the job done on time?

Case 2: Manufacturing a Small Bronze Gear

A small bronze gear must be made at a high production rate. Machining is ruled out as being too slow, and although extrusion (forcing metal through a die) of these gears could be a competitive option, it is also ruled out because no extrusion machine is available to the manufacturer. The order requires 10,000 gears to be delivered within 30 days. It will take about 18 to 20 days to make either sand casting patterns, permanent molds, or die casting dies. This leaves 10 to 12 days for production.

This means that about 1000 parts per day must be manufactured, allowing two days for packaging and shipping. The manufacturer will receive payment of $200,000 for the complete order. Which casting method should be used?

CHAPTER 9

Processing of Metals: Hot Working

Objectives

This chapter will enable you to:

1. Explain how molten steel is formed into industrial shapes.
2. Show the significance of recrystallization and grain structure of hot rolling and forging.
3. Describe several methods of hot forming metals and explain their advantages.
4. Describe two methods of hot forming pipe and tubing.

Key Words

hot working	cold working
recrystallization	anisotropy
direct chill (VDC, HDC)	swaging
forging	flash
direct extrusion	indirect extrusion

In earlier times, wrought iron and steel were forged by shaping a piece of red hot metal on an anvil by pounding on it with a hammer. The reason the blacksmith had heated the metal, of course, was to reduce its flow stress so each hammer blow could deform the metal a greater amount. The hammering process not only produced a desired shape, it also improved the grain structure of the metal, resulting in increased strength. When the blacksmith needed to make a round bar from a square one, a half-round die block, called a *swage,* was inserted in a square hole in the anvil, and an upper half-round die was held in the hand by a handle. The square bar was heated red hot in a forge and placed in the lower die while the upper one was hammered down on the bar. This process was repeated all along the bar, thus producing a length of round stock. This is roughly the process used for hot forging various shapes today, using open and closed dies and powerful presses. Today, steel bars of various shapes and dimensions are usually rolled in a steel mill; however, larger sections and other more complex and irregular shapes are often forged. Modern forging processes account for a large part of basic metalworking. Often a forged or rolled shape is subjected to a secondary process of metalworking, such as stamping or machining.

To many the term *hot working* means that the metal is deformed or worked at an elevated temperature; however, **hot working** technically means that the metal is worked at a high enough temperature that no plastic deformation, strain hardening, or **cold working** takes place. One way to check for hot working is to measure the hardness or strength of the metal before and after the working process; if no strengthening occurs, it is hot working. Another way to think of true hot working is that before the metal cools, the temperature is high enough that it anneals the metal and removes the cold work that would otherwise have occurred. The metals lead and tin provide unusual examples of this process. The temperature at which these metals will recrystallize is 25° F (−4° C). Because this temperature is below room temperature, if specimens of the metals are cold worked and left to sit on the lab bench, within hours all evidence of the cold working disappears. See Chapter 3 for a more thorough discussion of cold working and recrystallization.

It is often the case that it is less costly to perform operations such as rolling and extrusion at temperatures at which some cold working occurs, and this is a frequent practice. The consequences of this practice will be discussed in the section on recrystallization.

This chapter opens with an extensive look at the hot rolling process and its equipment, including a discussion of strand casting, which in modern mills is integrated with the hot rolling operation. An investigation of the phenomenon of recrystallization follows; in some products this process may be of no concern, but in others it may introduce unusual properties, and a certain amount of confusion.

Other hot working processes are then discussed: forging, the process, machines and dies; swaging; extrusion; drawing; spinning; the piercing process to produce seamless steel tubing; and finally, welded pipe.

HOT ROLLING

In Chapter 4 you learned how raw materials, iron ore, coke, and limestone are used to extract pig iron and how steel is produced in furnaces. The molten steel must then be formed into useful shapes. The process of shaping steel begins by teeming or pouring molten steel into ingot molds. The molds are then moved to the stripper where, after the metal solidifies, the ingot is removed. Before the ingot can be rolled, it must be heated uniformly to about 2200°F (1204°C) throughout in order to allow uniform flow of material during the rolling operation. The soaking is done in soaking pits (Figure 9.1).

During the ingot casting process the steel cools from its surfaces toward the center, forming large, columnar, dendritic grains; that is, the grains grow parallel to the direction of heat that flows through the walls of the mold, and then more **equiaxed grains** form in the center of the ingot where more of the heat has escaped through the top of the ingot (Figure 9.2 is a representation of the cross section of an ingot cast in a mold). These large grains are characterized by low strength and high elongation, which in most cases is considered undesirable. In addition, internal voids can be created owing to shrinkage of the material. The shrinkage voids result in reduced cross-sectional area, and even more importantly, they act as stress risers, which significantly reduce the strength of the structure. These undesirable grains and internal shrinkage voids can be removed or altered by the process of hot working: rolling, drawing, extruding, or forging.

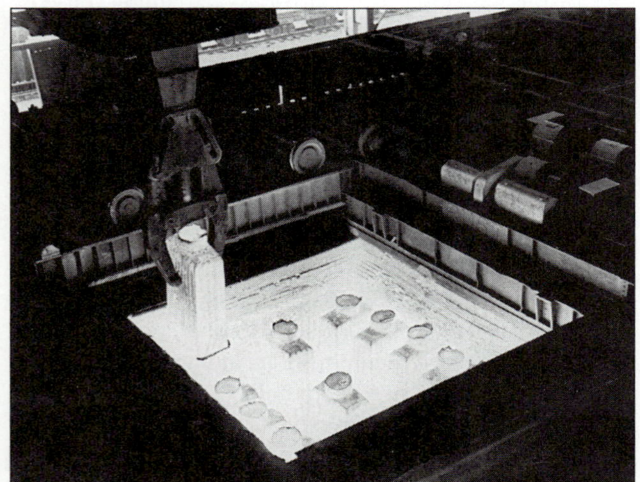

FIGURE 9.1
Powerful tongs lift an ingot from the soaking pit, where it was thoroughly heated to the rolling temperature (American Iron & Steel Institute).

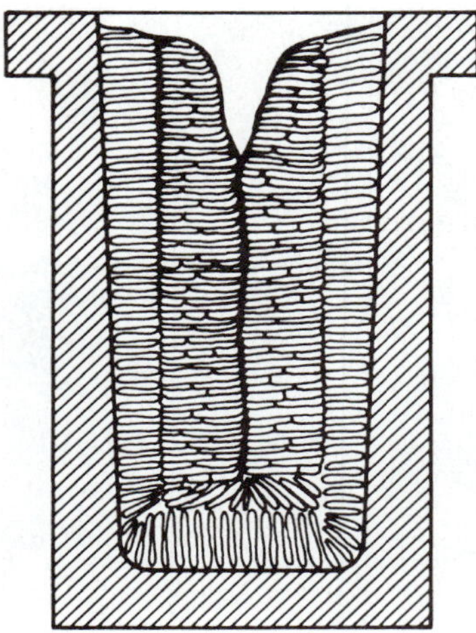

FIGURE 9.2
A representation of large dendritic crystals forming in a solidifying metal in a mold. These large grains will be reformed into smaller uniform grains by the rolling, or other hot-working processes.

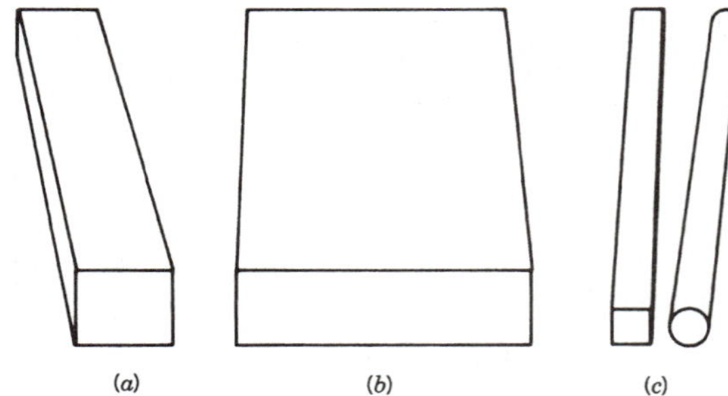

FIGURE 9.3
(a) Blooms, *(b)* slabs, and *(c)* billets are formed from ingots.

Rolling Mills

The soaked ingot is sufficiently plastic to be taken to the rolling mills to be shaped into one of the intermediate forms called blooms, slabs, and billets (Figure 9.3) or to directly produce structural shapes. *Blooms* are forms of semifinished steels that have a square cross section 6 × 6 in. or larger. A slab is rolled from an ingot or a bloom and has a rectangular cross section 10 in. or more wide and 1.5 in. or more thick. A billet is rolled from a bloom and is square, 1.5 in. on a side or larger. Blooms and slabs are most frequently rolled on a reversing two-high rolling mill (Figures 9.4 to 9.6).

Blooms (and often ingots) are further processed in structural rolling mills (Figure 9.7) into railroad rails or different shapes such as I beams, angles, wide-flange beams, zees, tees, and H-piles. They are made of plain

FIGURE 9.4
Reversing two-high roughing mill (American Iron & Steel Institute).

FIGURE 9.5
Operator at console of reversing two-high roughing mill (Courtesy of Bethlehem Steel Corporation).

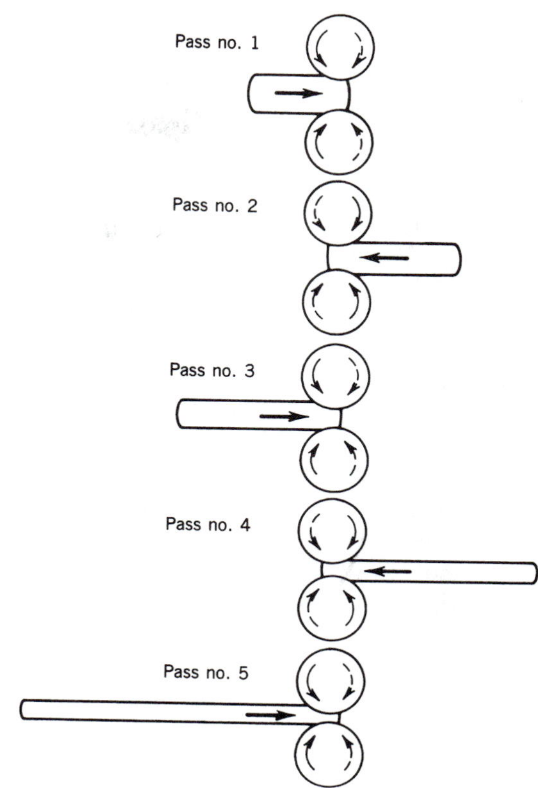

FIGURE 9.6
Diagram of sequences in rolling hot ingot into a slab on a reversing two-high rolling mill (American Iron & Steel Institute).

low carbon steel, which can be drilled, heated, or welded without such complications as work hardening, embrittlement, or cracking. Blooms are sometimes further reduced in size to billets on billet mills (Figure 9.8) before being processed into finished products such as bar and rod stock. Bar and rod (from which wire is made) can be made of either low-carbon steel or higher-carbon and alloy steel.

As the ingot is carried through the massive rolls the hot, plastic steel is squeezed to a smaller thickness. The rolls are brought closer together on each pass, causing the ingot to get longer and thinner as it passes between the rolls (Figure 9.9). Rolled shapes become quite long and must be cut to standard customer-length (Figure 9.10). Different configurations of the rolling mills include two-high, three-high, four-high, cluster mill, and tandem rolling mill. The two-high rolling mill can be used for reducing the stock by reverse action; however, because of the large inertia of the rolls it requires a complex mechanism for reversing direction of rotation. The three-high rolling mill has the advantage over the two-high mill that the direction of rotation of rolls does not

FIGURE 9.7
Structural steel and rails are rolled from blooms or ingots. Standard shapes are produced on mills equipped with grooved rolls. Wide-flange sections are rolled on mills that have ungrooved horizontal and vertical rolls (American Iron & Steel Institute).

FIGURE 9.8
Steel billet emerges from 35-in. blooming mill. The billet has been rolled from an ingot. The product of this mill will eventually be made into bars. Dials on the two-stand mill indicate size of billet being rolled. Alloy and tool steel ingots are processed in this particular mill (Courtesy of Bethlehem Steel Corporation).

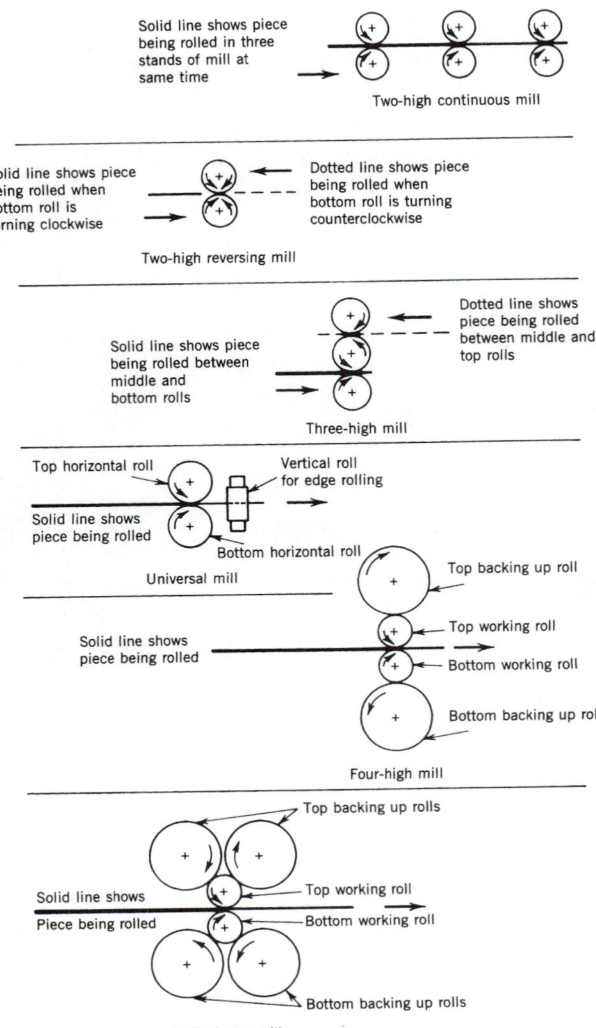

FIGURE 9.9
Diagrams demonstrating various arrangements of rolls and rolling procedures used in steel mills. Operating under extreme heat and pressure, rolls must frequently be changed, and a supply of new or refinished mill rolls is always maintained from the mill's own machine shop (American Iron & Steel Institute).

have to be reversed. In one pass, stock is passed between the top two rolls, then the stock is lowered and passed between the lower two rolls, after which it is lifted and the operation is repeated (Figure 9.9). The only disadvantage of this configuration is that it requires an expensive handling system. Four-high (Figures 9.9 and 9.11) and cluster mills (Figure 9.9) use rolls of small diameter to reduce the dimension of the stock, since smaller-diameter rolls require less energy and force for the same effect

FIGURE 9.10
Hot saw cuts finished rolled shapes to customer length after delivery from the 48-in. finishing mill (Courtesy of Bethlehem Steel Corporation).

FIGURE 9.11
Plate enters the four-high finishing stand in the 110-in. sheared plate mill. The plate is rolled to the desired length and thickness by being passed back and forth through the finishing stand rolls. The mill has AGC (automatic gage control) and is completely computerized (Courtesy of Bethlehem Steel Corporation).

as do large rolls; however, since small rolls are insufficiently rigid, they must be backed by larger rolls. Finishing mills for plates, strip, and sheets are often tandem rolling mills with a number of mill stands in series, each adding to the final shape. The tandem rolling mill requires a complex system for adjusting rotation speeds of the rolls in order to prevent uneven thickness or even fracture of the stock.

All hot-rolled steel forms a black or gray mill scale on the surface that is caused by oxidation of the red-hot steel in the presence of oxygen in the air. The oxide scale helps protect the steel from corrosion after it has cooled.

STRAND CASTING

A second procedure of steel processing that bypasses ingot teeming, stripping, soaking, and rolling in roughing mills is called strand casting (Figure 9.12). It is also called *continuous casting,* because the molten metal is continuously supplied to a tundish or molten metal reservoir (Figure 9.13), where it is fed through a water-cooled copper mold from which it emerges as a continuous strand of steel in a desired cross section as slabs or billets. Sprays of water under high pressure cool and harden the metal still further. The strand of metal is cut into suitable lengths by a traveling torch as the steel moves along on rolls toward the mill stands (Figure 9.14). In some strand casters, the descending column of metal is cut to the desired length while it is still in the vertical position, then the cut length falls forward onto the rolls and is carried away. At this time the metal is at a rolling or forging heat, just under the temperature of solidification. Both ingot and strand casting procedures are primarily designed for high-volume production. The continuous strand casting unit at Bethlehem's Burns Harbor, Indiana, plant can convert 300 tons of molten steel to solid slabs in about 45 minutes.

In the strand or continuous casting process the metal cools from its surface to the center, just as a mold-cast ingot would, so there is still the possibility of forming large dendritic grains, as previously described. In strand casting, however, there is a constant supply of molten metal to feed the cavity, so a shrinkage cavity is of less concern. The large columnar grains will be worked by subsequent hot working.

The discussion so far in this chapter has been on the processing of steel ingots, blooms, slabs, and billets. In completing this section on strand casting we now turn to nonferrous metals. For the common nonferrous metals that are produced in large volume, such as aluminum, copper, brass, and magnesium, the melting process is usually just that, a melting process. Commercially pure metals and scrap are placed in reverberatory, gas-fired furnaces (usually), melted and cast, usually by some process that directly produces an ingot that can be easily further processed, similar to the strand casting process just discussed.

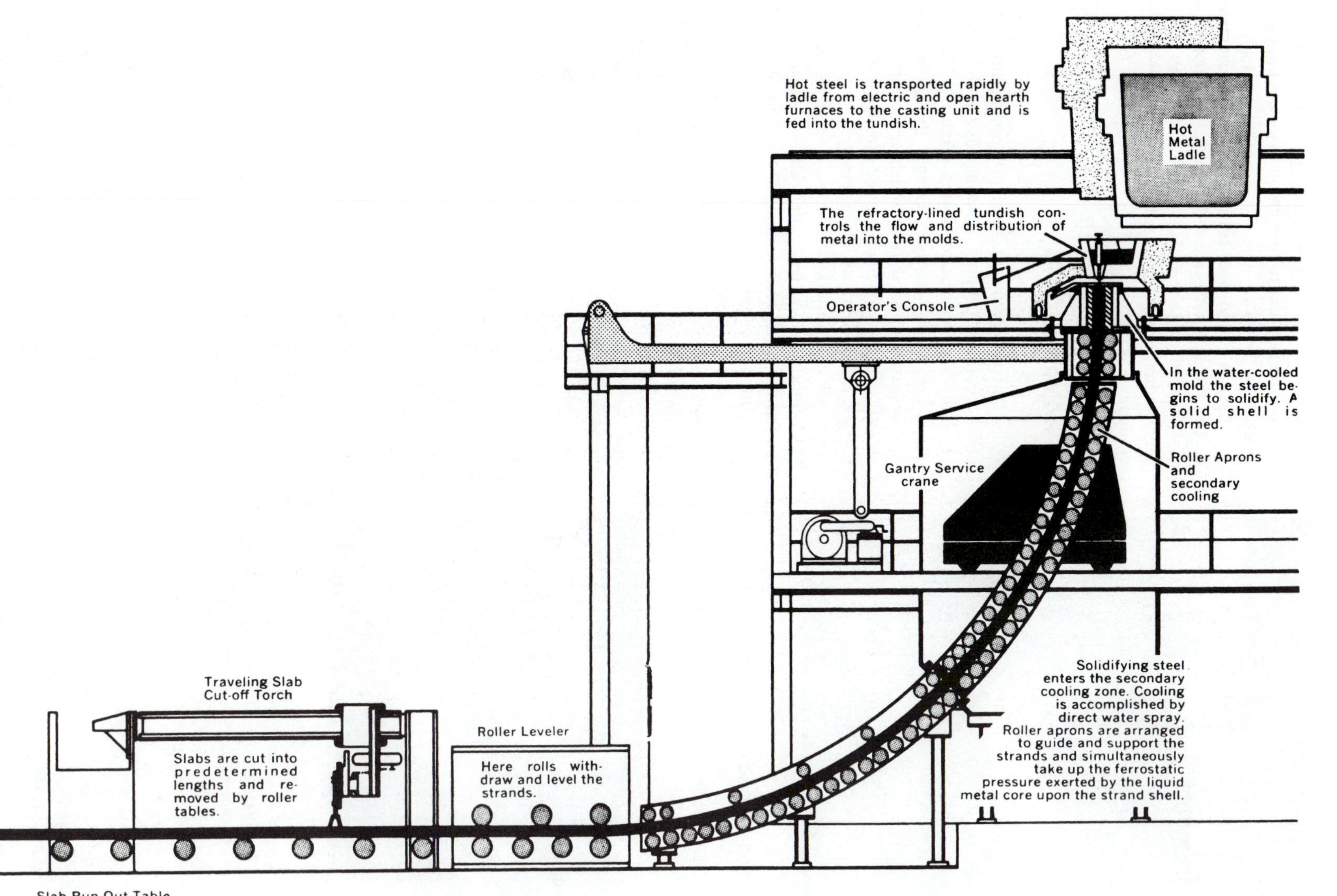

FIGURE 9.12
Diagram of continuous (strand) casting (American Iron & Steel Institute).

FIGURE 9.13
Molten steel is poured from a ladle into the tundish, which regulates the flow of liquid steel into two water-cooled copper molds for the slabs to be continuously cast (Courtesy of Bethlehem Steel Corporation).

FIGURE 9.14
Perfectly formed slabs of steel are cut to required lengths by an automatic flame torch after passing through the continuous slab caster (Courtesy of Bethlehem Steel Corporation).

Many of these metals are cast by a vertical **direct chill** (VDC) process. In this process metal is delivered, by gravity, to a round, square, or rectangular mold. Water is sprayed on the outside of the mold (that is where the direct chill begins). At the start of the process a "bottom block" is placed in the mold that fits tightly enough that the metal will solidify before it leaks out. After a shell of metal freezes on the block and in the mold a lowering device (e.g., a hydraulically controlled piston) lowers the bottom block out of the mold, and a long ingot (or ingots; more than one can be cast at a time) is cast into a pit. The depth of the pit and length of the lowering equipment are the factors limiting the length of the ingot cast. The water, which starts by flowing down the outside of the mold, also flows down the ingot when it emerges from the mold; this adds to the "direct chill," so cooling is usually very effective.

With some added equipment this vertical process can effectively be turned on its side so it becomes horizontal direct chill, or HDC. Now it functions much like the strand caster described earlier, except saws, rather than a cutting torch, cut the ingot to lengths appropriate for the next process.

RECRYSTALLIZATION

Steel ingot, with its typically coarse columnar grain structure, is quite unsatisfactory for applications where high strength is required. A part made directly from steel ingot can easily fail under impact load. The columnar grains in a cast ingot must go through recrystallization to give steel the required strength. Recrystallization can occur during an elevated-temperature process, such as forging or rolling (Figure 9.15).

When metal is worked at or below the recrystallization temperature the metal is effectively "cold worked." During such a process, grains are deformed and their internal energy is increased; that is, a small percentage of the cold working done is stored in the atom lattice. Heating this worked metal to an appropriate temperature will cause recrystallization of the grains. During **recrystallization** numerous nuclei are formed within the larger deformed grains. Keeping the structure at a high temperature will cause the nuclei to spread until they run into each other, thus defining new, much smaller grains. In order to initiate this process it is required that at least minimum critical work be imparted to the material, usually about 3 to 5 percent. The rate at which transformation occurs depends on the temperature to which the part is being heated. Factors that influence the grain size resulting from hot deformation are (1) initial grain size, (2) amount of deformation, (3) finishing temperature, and (4) rate of cooling.

In some metal-working operations, such as forging and direct extrusion, the percent cold work imparted may not be uniform across the part, or along its length. This can be detrimental to the use of the part, however, if the part is later heated to its recrystallization temperature (say, for heat treating); the grains *will not* be the same size after they recrystallize.

If the structure is left for an extended period of time at high temperature, the recrystallization process will be followed by a grain-growth stage at which smaller grains will continue to grow and eventually start merging into larger grains. The process of recrystallization and grain growth can occur simultaneously with a hot-working operation, and sometimes it is intentionally performed, as an intermediate annealing process. This allows the material to be worked to a great degree without fracturing.

The recrystallization temperature for a metal depends on the percent cold work done to the metal, but some general values are as follows:

Copper (99.999% pure), 250°F (121°C)
Copper with 5% zinc, 600°F (316°C)
Aluminum (99.999% pure), 175°F (79°C)
Aluminum (99.0% and alloys, 600°F (316°C)
Low-carbon steel, 1000°F (538°C)

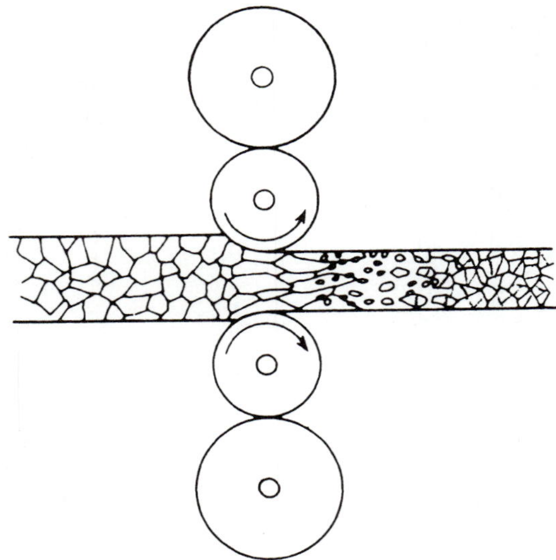

FIGURE 9.15
Recrystallization in hot rolling (Neely and Bertone, *Practical Metallurgy and Materials of Industry,* 6th ed., © 2003 Prentice Hall, Inc.).

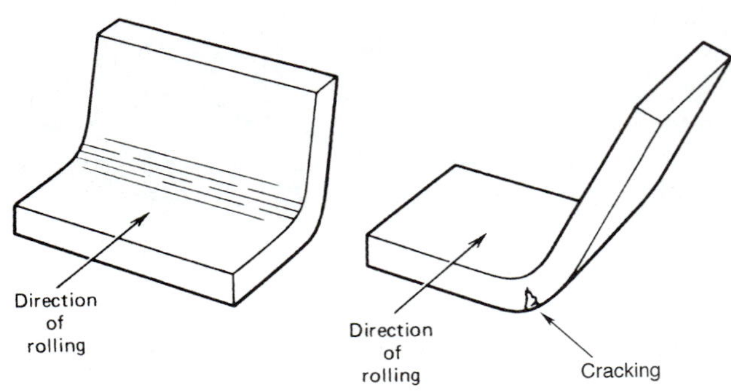

FIGURE 9.16
This diagram illustrates the fibrous quality of rolled steel, called anisotropy. Since the fibers are in the direction of rolling, the metal is stronger along one axis than along the other (Neely and Bertone, *Practical Metallurgy and Materials of Industry,* 6th ed., © 2003 Prentice Hall, Inc.).

FIGURE 9.17
Woodcut showing a large wrought iron ship's anchor being forged in eighteenth century France. The arm and the shaft of the anchor are being forge welded together in this view (Dover Publications).

Note that adding impurities or alloying elements tends to *raise* the recrystallization temperature. For additional data see Table 3.4.

The forming processes result in a condition called **anisotropy**, that is, a metal has mechanical properties (e.g., strength, elongation) that vary with direction. This is due to the grain deformation and their longitudinal orientation. For example, the material is more ductile, that is, can be deformed without cracking or splitting, to a greater extent along one direction than the other. As seen in Figure 9.16, the metal can be bent to a smaller radius in the direction of rolling than it can at 90° from direction of rolling. The anisotropy of cold- or hot-formed metals is very important in the welding and cold-forming processes.

FORGING PROCESSES

In ancient times, copper and other nonferrous metals that are sufficiently soft and ductile were shaped while cold by hammering. Iron and steel, which are very strong but not ductile enough to be hammered into a shape while cold unless the piece is very small or thin, were heated before forming. Heating metals to forging temperatures greatly increases their **plasticity** and workability. Therefore, hot **forging** was and still is a useful method of forming steel articles, both large and small.

In ancient Egypt, Greece, Persia, India, China, and Japan forging was used to produce armor, swords, and agricultural tools. In the thirteenth century, the tilt hammer came into use in which water power was used to raise or tilt a lever arm to which a weight was attached. When it fell, a heavy blow was delivered to the hot iron. In the mid–seventeenth century at the Saugus Iron Works in Massachusetts, a 500-lb weight on the end of a wooden beam was used as a forging hammer in which a cogged wheel driven by a waterwheel periodically raised and dropped the weight onto an anvil. Thus it was possible to make large forgings such as heavy ship's anchors (Figure 9.17). In 1783, Henry Cort of England built a rolling mill that had grooved rolls, making shaped bars of uniform cross section possible.

Forging gives metal good static, fatigue, and impact strength. This is why most good tools are forged, and critical parts such as automobile wheel spindles and axles are made as forgings. Forged parts are often machined to provide bearing surfaces, splines, key seats, and so forth, but the essential strength of the forging remains. When parts are made by machining from the solid material, the grain flow is not always in the most advantageous direction and does not flow around shoulders and corners the way forgings do (Figure 9.18). Thus, machined parts are not as strong or resistant to failure as forgings.

In some applications the ingot is given an *ABC upset* to remove the cast structure. The process works the ingot in all 3 directions before the forged piece is created; it is very similar to kneading bread dough.

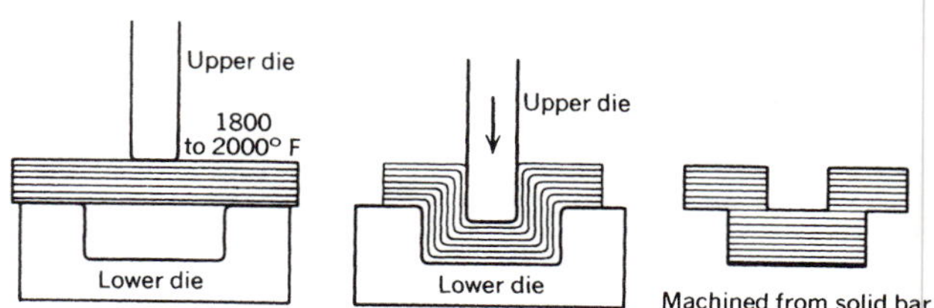

FIGURE 9.18
Grain flow in a solid bar as it is being forged, compared with a machined solid bar (Neely and Bertone, *Practical Metallurgy and Materials of Industry,* 6th ed., © 2003 Prentice Hall, Inc.).

FIGURE 9.19
A large forging to be used in a large aircraft. Subsequent machining operations will be performed on it.

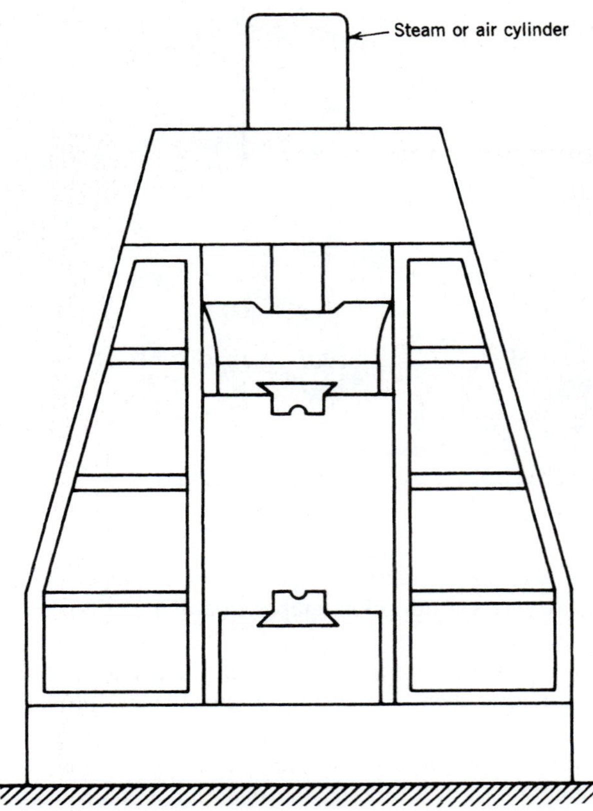

FIGURE 9.20
Drop-forging hammer. Steam or air pressure is used to raise the hammer and force it down on the heated metal.

Forging Machines

Today, large forgings (Figure 9.19) are produced in massive hydraulic presses. Smaller articles are forged by a variety of methods and machines. **Open-die drop forging,** impression-die drop forging, press forging, **swaging,** and **upset forging** are the most commonly used types of forging operations in manufacturing. When choosing a particular forging operation, several things must be considered: force and energy required, size of the forging, repeatability, speed and workability, and metallurgical structure. In addition, the amount of time hot metal is in contact with the die is important to die life and is also a consideration when choosing a forging machine for a particular forging operation.

Drop Hammer The drop hammer (Figure 9.20) is a development based on the old tilt hammer. Today, drop hammers are usually steam or air assisted to increase the force of the hammer blow and to raise the weight after each blow. Large articles such as machinery shafts that are to be finished by machining are usually open-die forged, using repeated blows while the hot metal is turned and moved with tongs or a manipulator.

During this process the metal is squeezed or hammered to increase its length and decrease its cross section, or it is upset, that is, pressed to increase its diameter and shorten the length. Drop hammers perform

FIGURE 9.21
Progressive dies used with a mechanical press. Here a round bar at forging heat will be moved into the die by means of mechanical fingers. At each stroke kickouts move the partially formed forging to the next die. The bar is converted to two connecting rod forgings, and the last die operation is that of removing the flash (Courtesy of National Machinery LLC).

FIGURE 9.22
Automatic mechanical presses are high-production forging machines (Courtesy of National Machinery Co. Maxipres® is a registered trademark of National Machinery LLC).

repetitive blows to bring the workpiece to size, whereas mechanical and hydraulic presses can often make a forging in one or two strokes, however, both types of forging machines are adaptable to automated forging processes.

In open-die forging, drop-hammer operators require a skill in giving the hammer just the right amount of force. When shaped upper and lower dies are used, the process is called impression die forging. This method increases production rates and ensures repeatability. Often, two or more progressive dies are used, each contributing to the final shape (Figure 9.21).

Mechanical Presses These machines make use of a heavy flywheel that stores energy for the forging stroke. Mechanical presses (Figures 9.22 and 9.23) of this type have an eccentric shaft or crank that moves the ram down and back to the starting position when a clutch is

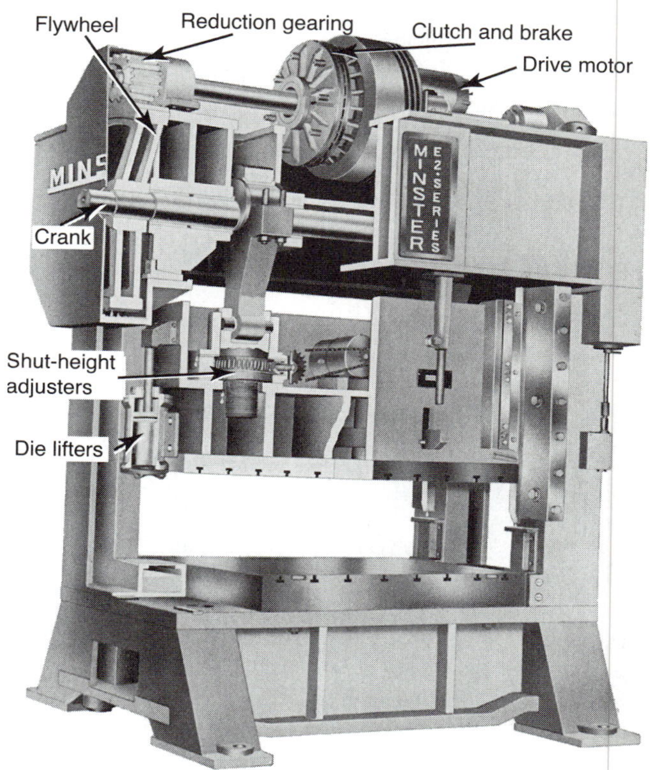

FIGURE 9.23
Mechanical press with a cutaway view showing the driving mechanism. The maximum die used in a press depends on shut height, which is the vertical opening with the ram down and adjustment up (Courtesy of the Minister Machine Company).

actuated. The greatest pressure is exerted at the bottom of the stroke. An inherent problem with these mechanisms is the possibility of jamming the ram near the bottom dead center, but most mechanical presses have a stall release that frees the stuck ram and upper die when this happens. Compared with drop forging machines, these machines operate more quietly, allow more accurate control, and produce higher force and energy. For this reason, the mechanical press is often chosen when a part must be made to closer tolerances. During the past decades, there has been a gradual trend toward mechanical press forging. Fewer blows are required per forging and less operator skill is needed with these machines. Mechanical presses squeeze rather than strike the hot metal with a shorter forging-to-die contact time, thus extending die life.

A modern version of an old principle (Figure 9.24) is a high-energy type of mechanical forging press (Figure 9.25) that utilizes a vertical screw (Figure 9.26) instead of a crank or eccentric shaft to transfer the flywheel energy to the ram. This arrangement provides uniform force throughout the stroke, unlike crank presses in which the major tonnage is confined to the end of the stroke. Like the drop hammer, the screw press does not have a bottom dead center. A continuously rotating flywheel imparts motion to a screw when a clutch is engaged. It appears that the screw press combines two of the best features of the drop hammer and the crank press, high impact and a squeezing action. Even with hard-blow forging, the press frame and mechanical components are not subject to high stresses, and the die is in contact with the hot metal a very short time. Often only one or two blows are needed to complete a single forging (Figure 9.27). High-energy forging presses provide high flexibility, since they can be used for either open or closed die forging operations.

FIGURE 9.25
High-energy forging press (Courtesy of Siempelkamp Corporation).

FIGURE 9.24
Many years ago parts were made one at a time by an operator, often under dangerous conditions, unlike automatic press systems today (Graebener Press Systems, Inc., Providence, RI).

Hydraulic Presses Hydraulic presses (Figure 9.28) are the slowest types, but their advantage is that they can exert very high forces required for large forgings. Instead of mechanical linkages and cranks, the press ram is powered with one or more large hydraulic cylinders. Oil is forced at high pressure into the cylinders, and the ram moves at a constant rate. Smaller cylinders return the ram to its upper position. Large hydraulic presses can exert thousands of tons of pressure, making them ideal for some of the largest forgings (Figure 9.29).

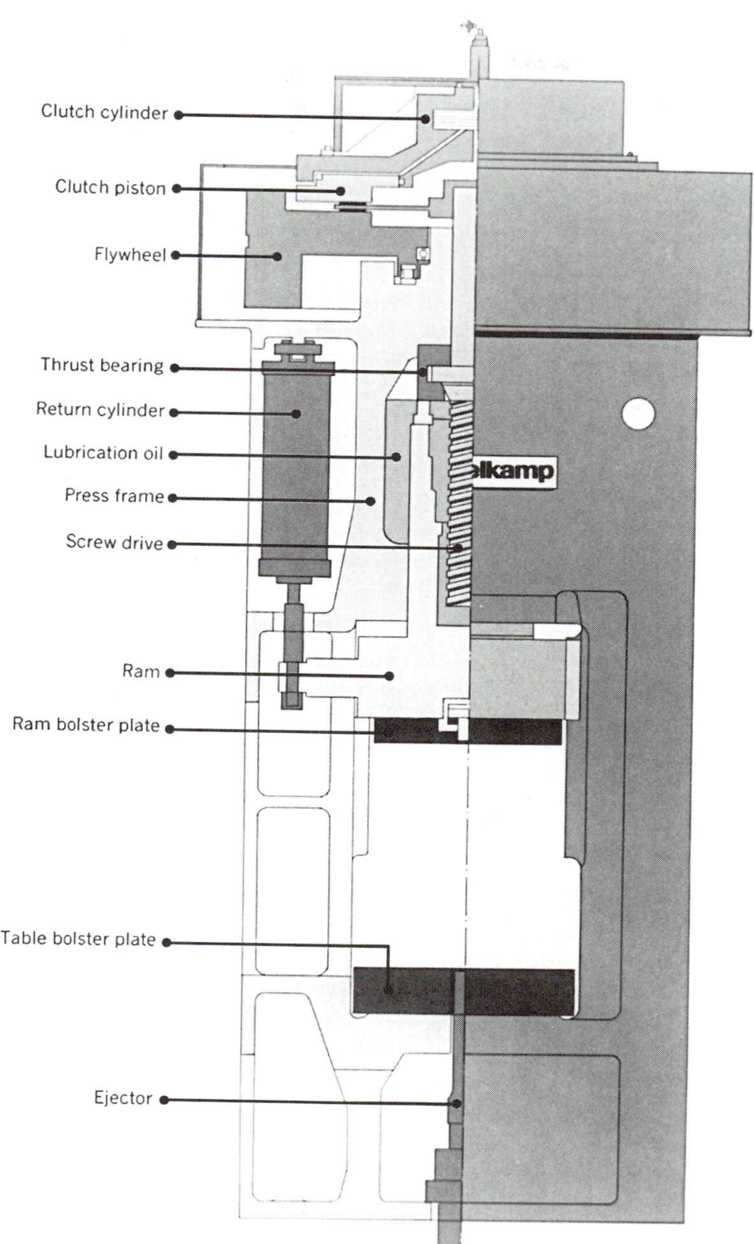

FIGURE 9.26
Drawing of the parts of a high-energy screw press (Courtesy of Siempelkamp Corporation).

FORGING DIES

Forging dies can be classified into three general groups: (1) open dies, (2) impression dies, and (3) flashless dies. In open-die forging, the workpiece is compressed between two flat or almost flat dies that do not hinder lateral flow of the material. In impression-die forging, the material is squeezed between two dies where lateral motion is significantly restrained. A small amount of metal is allowed to flow out of the die. This excess material is called *flash,* and it is removed after forging. Flashless die forging involves compression of material in which the lateral flow of material is completely restricted, and no flash is formed during operation. The process requires that the exact amount of material be placed into a die. If too much material is placed, forces exerted on

FIGURE 9.27
Preform in one-blow forging on a high-energy press. *(a)* Preforging or descaling. *(b)* One-blow finish forging (Courtesy of Siempelkamp Corporation).

FIGURE 9.28
Large open-die hydraulic forging press with manipulator positioning the forging ("Forging Presses," Schirmer-Plate-Siempelkamp, Hydraulische Pressen GmbH, Krefeld, Germany).

FIGURE 9.29
This huge forging press is used to dish heavy plate for pressure tank heads ("Pressing for All Applications in Metalforming," Schirmer-Plate-Siempelkamp, Hydraulische Pressen GmbH, Krefeld, Germany).

the die can be so high that the die can be damaged. Conversely, if not enough material is placed, then an incomplete filling of the die will result.

Design of forging dies can be very simple (open dies) or very complex (impression dies), and it is similar to that used for die casting. In the case of the impression dies, there is a parting line where the two halves of the die meet. To allow flow of excess material (flash) out of the die, a gap between the two dies (gutter) is provided, which is connected to the die cavity through a channel (land). The excess metal—**flash** (Figure 9.30)—must be removed in a secondary operation, usually in a small shear press. There must be adequate draft so the part can be removed from the upper and lower die halves without sticking. Generous corner fillets and corner radii are necessary to allow for a smooth flow of material (Figure 9.31).

Normally, only simple designs can be made in forging dies (Figure 9.32). Sometimes a series of dies is necessary to preform the material so it will conform to the shape of the next set of dies until the final shape is made (Figure 9.33 to 9.35). Computer-aided design (CAD) is quite useful in designing these multiple dies. From the starting shape of used stock, each step altering the shape of the die prior to the finished shape can be displayed on the screen. The determination can be made

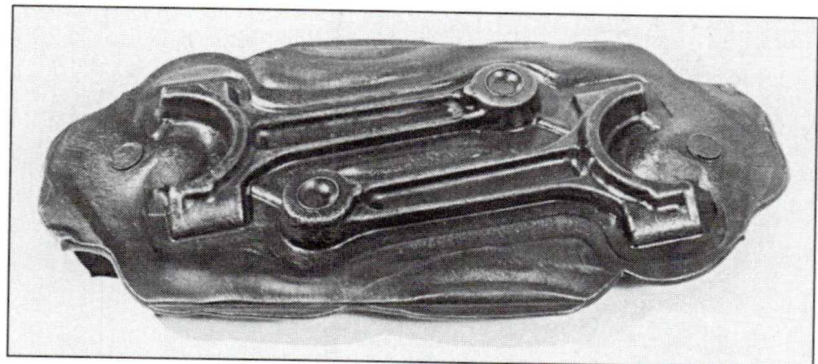

FIGURE 9.30
Forged connecting rods showing flash that must be removed, either in the automatic die set or as a secondary operation in a separate die (Courtesy of National Machinery LLC).

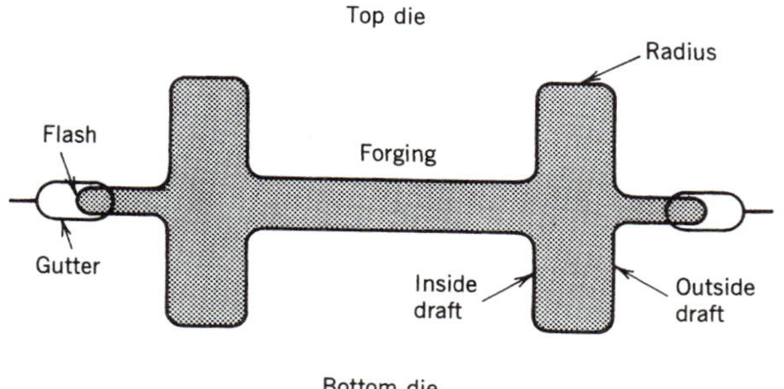

FIGURE 9.31
Die showing parting line and gutters for flash overflow.

FIGURE 9.32
Typical automatic progressive die set for use with a 2000-ton press. Guide pins at the rear align the upper half of the die. Parts at the front show the progressions in making the part; the last one to the right shows the sheared-off flash (Courtesy of National Machinery LLC).

as to the amount of deformation and its required force for each step so it will be within the capacity of the equipment with the least strain on the dies. Cast and powder metal preforms are often used in automated forging operations.

A forging die must be able to withstand extremely high temperatures of the work piece. Special hot-work tool steels are used for this purpose to prevent spalling and cracking of the dies.

UPSET FORGING

The term *hot upset forging* refers to a deformation operation in which a cylindrical stock is increased in diameter and reduced in length. Its prime application is in the fastener industry, for manufacturing nails, bolts, and similar hardware products (Figure 9.36).

Hot-forming machines are massive and complex automatic feeding, shearing, forging, and punching systems capable of very high output (Figure 9.37). The process of nut making takes four steps on a three-die hot former (Figure 9.38). The heated metal stock is fed into a stop where it is sheared off (Figure 9.39) (p. 186). The first die upsets the blank; the second die forms the blank to a hexagonal shape and partly forms the hole; and the third die punches out the hole and the blank nut is ejected. One of the advantages of hot forming is that a large percentage of the hot blank can be displaced to make various shapes. Bolt heads, flanged axles, and similar shapes are made in a process called **heading,** which is similar to the nut-making process. Heading can be done by either a hot forging or a cold upset process. The trend seems to be toward the cold upset forging process and away from hot-work upsetting; however, shapes such as bearing caps and front wheel hubs are hot formed, whereas more symmetrical shapes such as wrist pins and gear blanks are cold formed. Heading and

FIGURE 9.33
Hand-fed crankshaft progressive die used on a 3000-ton press (Courtesy of National Machinery LLC).

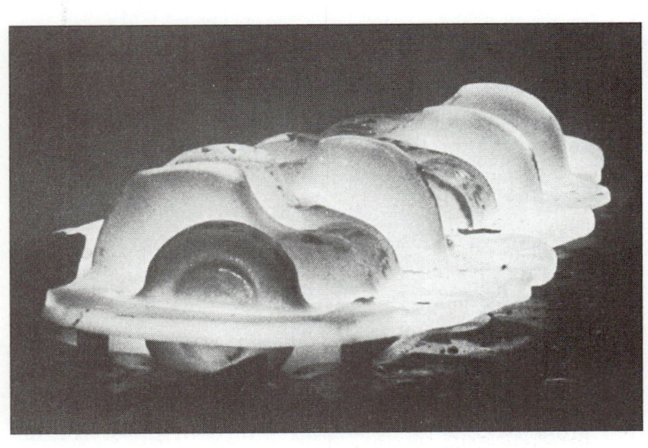

FIGURE 9.34
Hot crankshaft forging in a die after the last blow. It will now be removed from the die and the flash will be sheared off (Courtesy of National Machinery LLC).

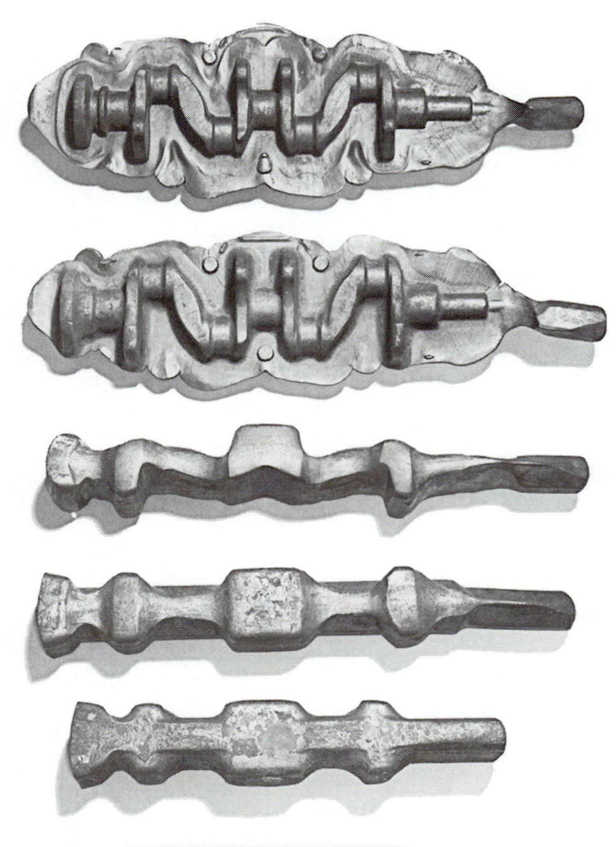

FIGURE 9.35
Stages in the formation of a crankshaft by hot forging from bottom to top (Courtesy of National Machinery LLC).

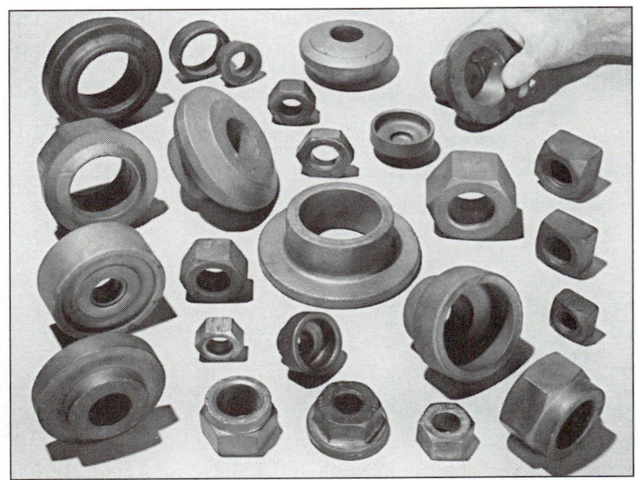

FIGURE 9.36
These are typical shapes made in hot formers—short, round, symmetrical, and with a hole (Courtesy of National Machinery LLC).

FIGURE 9.37
A hot former is an automatic feeding, shearing, forging, and punching system capable of very high output (Courtesy of National Machinery LLC).

the cold upset process are fully covered in Chapter 10, Processing of Metals: Cold Working.

Another type of very high production forming system that is in sharp contrast with the older methods (as shown in Figure 9.24) is the knuckle-joint press principle (Figures 9.40 and 9.41). Machines of this sort use progressive dies in an advanced automated system. Small parts such as socket wrenches and bicycle bearing races (Figure 9.42) are produced with such presses. Hot, cold, or warm forming can be done on knuckle-joint presses, depending on the requirements of a design and toughness of the metal. Open-end and box-end wrenches are perfect examples of parts that require a forging process. For example, an end wrench made from steel strip by blanking would be a very poor product, tending to bend or break in use. Forgings are much stronger. Combination box-end and open-end wrenches are produced on two presses that are linked by a conveyor (Figure 9.43). Together they produce about 1800 pieces per hour. Since the material is a chrome–vanadium alloy, it can be only hot

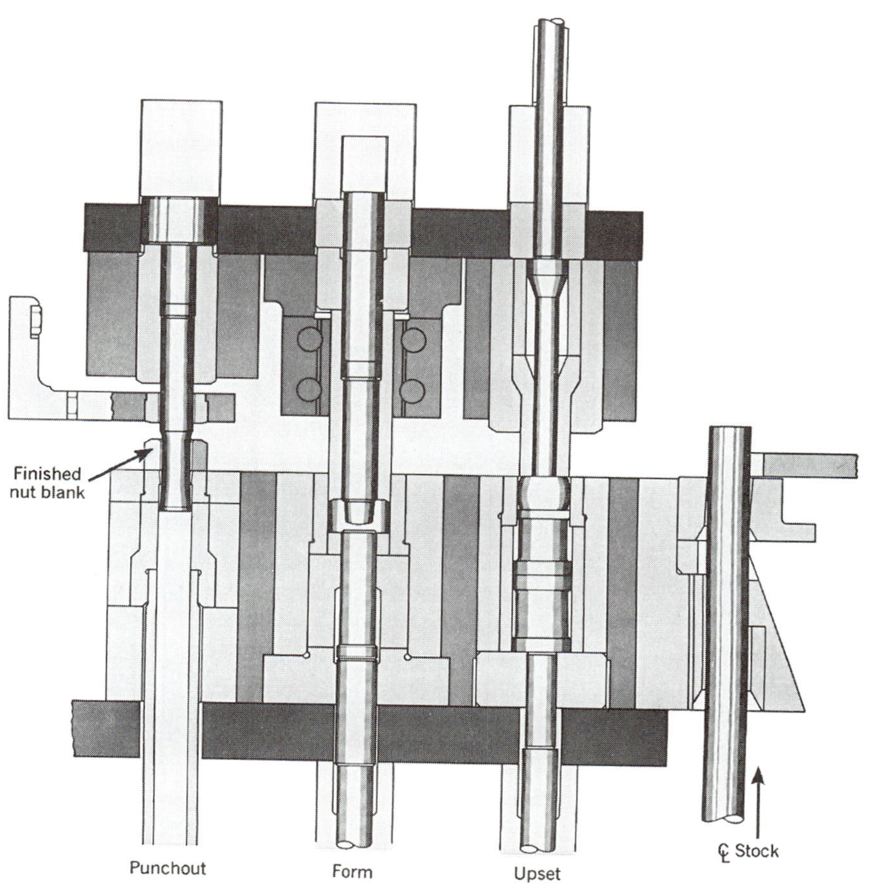

FIGURE 9.38
The tooling arrangement for a three-die hot former progressing from right to left shows the steps in nut making—upset, form, and punch out the hole (Courtesy of National Machinery LLC).

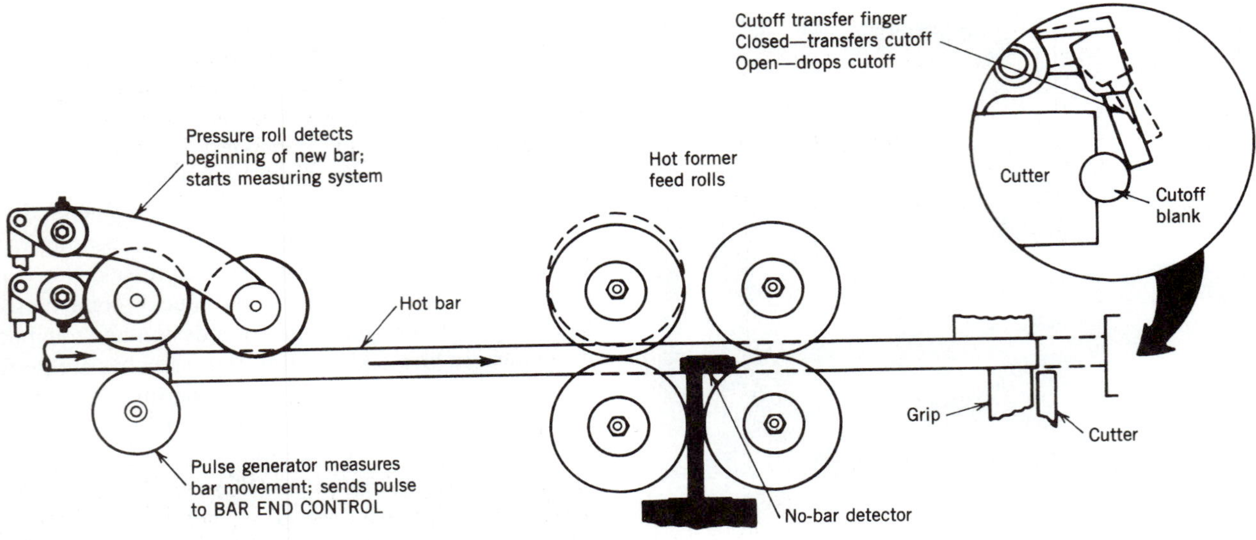

FIGURE 9.39
Hot shearing. A computer measures bar end movement into the hot former, then signals feed rolls and cutoff transfer to cycle at preset times (Courtesy of National Machinery LLC).

FIGURE 9.40
Press system for the coining of bicycle bearing races. Parts are oriented and conveyed up the incline on the left, then fed down the tube escapement device that admits them individually into the open die. The blanks continue through a five-station progressive die with the operation steps shown in Figure 9.42 (Graebener Press Systems, Inc., Providence, RI).

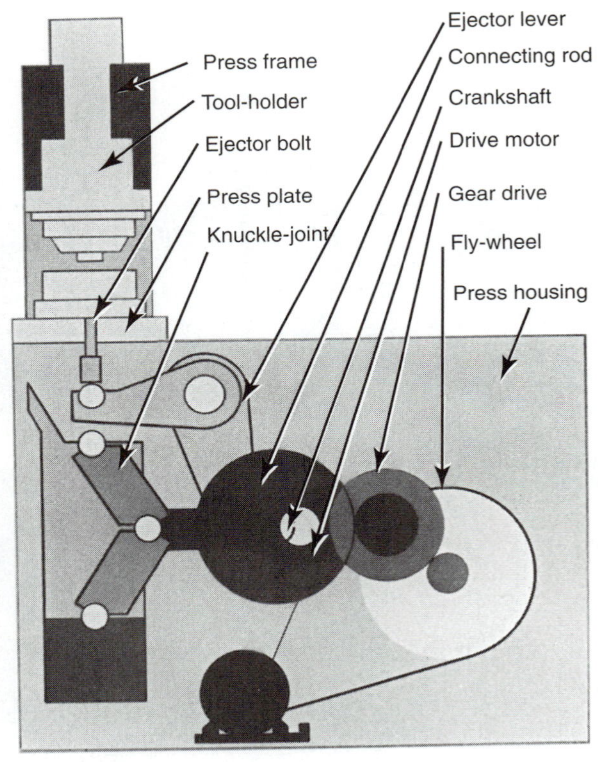

FIGURE 9.41
Principle of the Graebener knuckle-joint press system (Graebener Press Systems, Inc., Providence, RI).

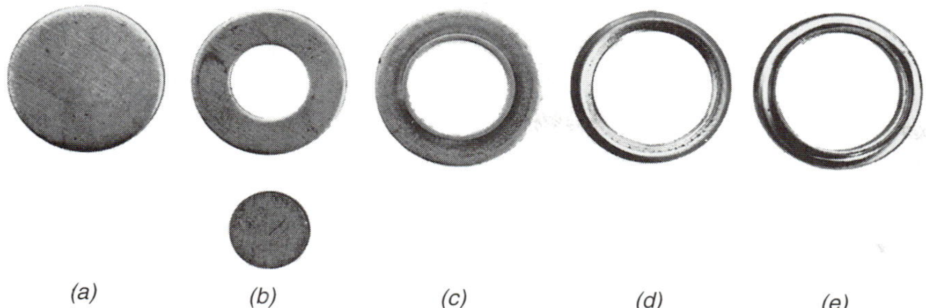

FIGURE 9.42
Sequence in producing bicycle bearing races: *(a)* Material. *(b)* Blanking. *(c)* Upsetting. *(d)* Forming. *(e)* Finishing (Graebener Press Systems, Inc., Providence, RI).

forged. Combination wrenches are produced in a sequence of operations (Figure 9.44). Wrenches are carried out of the machine after the final finishing operation (Figure 9.45). The wrenches collect in containers (Figure 9.46) for transfer to other operations.

FIGURE 9.43
Interlinked press line for the production of ⅞-in. combination wrenches: (left) Line 1—GK 360 and (right) Line 2—GK 800 (Graebener Press Systems, Inc., Providence, RI).

SWAGING

Swaging, like nut making and bolt heading, can be either a hot- or a cold-forming operation. Metals with properties similar to the aluminum alloys are usually able to be swaged cold; however, to swage stronger metals considerably more force and more massive machines are needed for cold than for hot forming. Therefore, larger, heavier parts are usually hot formed.

Swaging is normally done by rotating a spindle between a set of cylindrical rollers mounted in a ring around the spindle (Figure 9.47). In the end of the spindle is a slot into which is fitted a die holder, with backers, and a set of dies. The diametrical distance between the rollers is a little less than the diameter of the tool set (dies and backers) so that as the spindle rotates the dies are forced together with great force as they encounter each opposing pair of rollers; the blows delivered by this process are quite rapid—1800 to 4000 per minute. A workpiece, usually cylindrical, is fed into tapered dies that gradually work the metal down to the finish size of the dies. In a modification of this process, called *radial forging,* the striking rolls rotate, and the dies and backers remain fixed. The workpiece is then rotated in the dies, but the result is essentially the same.

Typical swaged products are shown in Figures 9.48 to 9.50. The aluminum softball and baseball bats so popular today are manufactured using the cold swaging process.

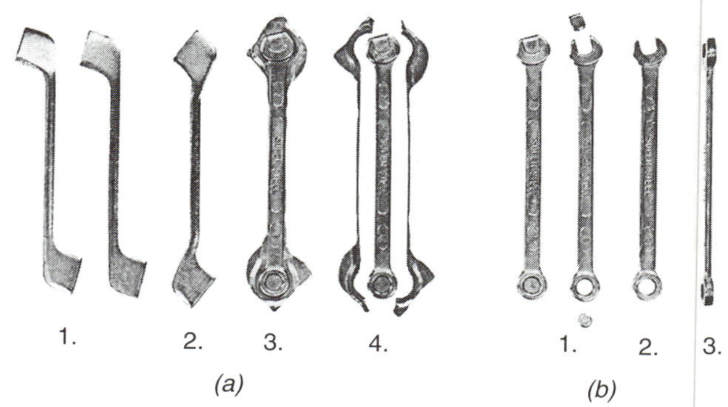

FIGURE 9.44
Sequence of operation for ⅞-in. combination wrenches: *(a)* On a GK 800: (1) Setting. (2) Straightening of head. (3) Forming. (4) Trimming. *(b)* On a GK 360: (1) Piercing. (2) Flat coining. (3) Edge coining (Graebener Press Systems, Inc., Providence, RI).

FIGURE 9.45
Wrenches emerging from automatic machine (Graebener Press Systems, Inc., Providence, RI).

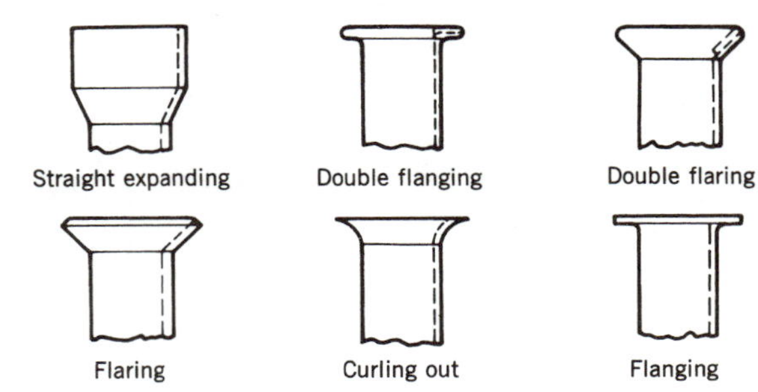

FIGURE 9.48
Types of expanded tube ends (Courtesy of Alcoa Inc.).

FIGURE 9.46
The finished product—⅞-in. combination wrenches (Graebener Press Systems, Inc., Providence, RI).

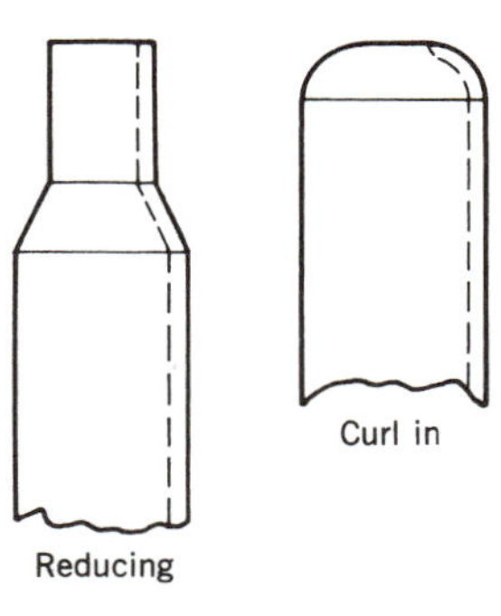

FIGURE 9.49
Types of reduced tube ends (Courtesy of Alcoa Inc.).

FIGURE 9.47
Tubing being necked down by the swaging process (Courtesy of Alcoa Inc.).

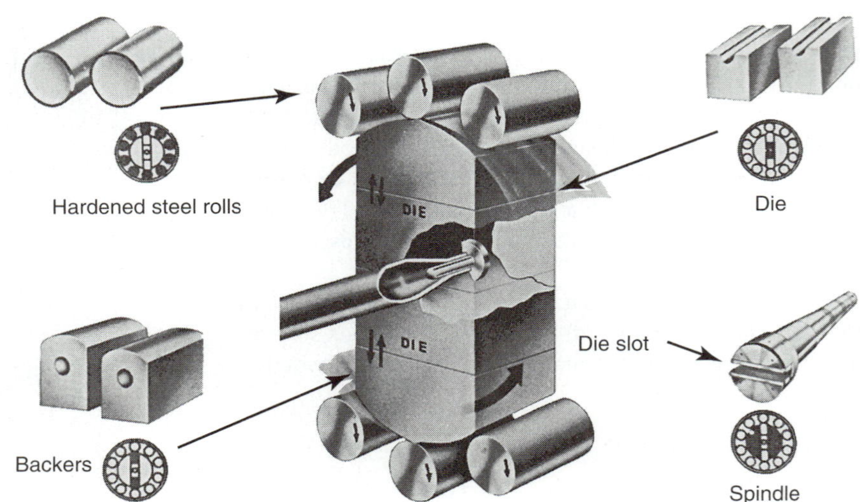

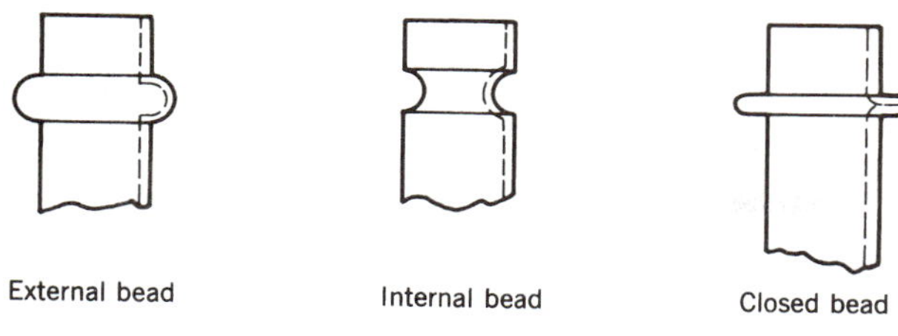

FIGURE 9.50
Types of beaded tubes (Courtesy of Alcoa Inc.).

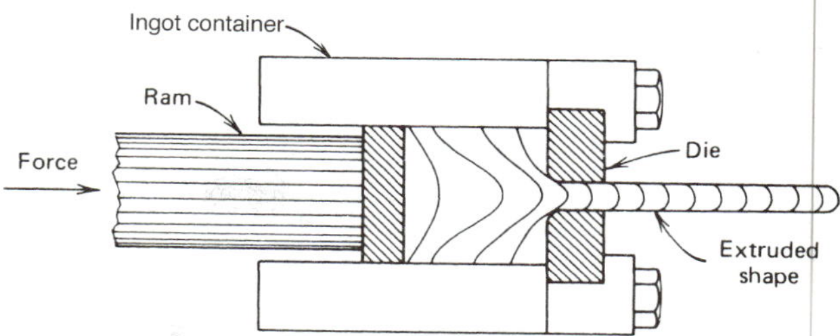

FIGURE 9.51
Direct or forward extrusion. The flow lines show how the metal is pushed out of the die by the use of great forces. Softer metals are usually extruded cold, whereas harder metals are brought to a forging heat before being extruded (Neely and Bertone, *Practical Metallurgy and Materials of Industry,* 6th ed., © 2003 Prentice Hall, Inc.).

HOT EXTRUSION

Extrusion is a process of forcing metal through a die, similar to squeezing toothpaste out of a tube. The very soft nonferrous metals (lead, tin, zinc) may be extruded cold or warm, whereas the stronger nonferrous metals, for example, aluminum, copper, brass, magnesium, and ferrous metals usually need to be at a forging temperature to make them plastic enough to be extruded. Square and round tubular products, structural shapes, and round, square, or hexagonal solid shapes are some examples of extruded pieces. There is almost no limit to the kinds of intricate cross-sectional shapes that can be made by this process. Extrusion produces surfaces that are clean and smooth with accurate shapes that can be held within close tolerances. The great advantage of the extrusion process is that it allows manufacture of intricate shapes. For example, inexpensive bronze gears of various size and shapes can be extruded in lengths of 20 to 40 ft, and later cut off to desired lengths; however, in cases where a product can be rolled instead of extruded, the choice should be a rolled product if large amounts are needed, because rolling is a less expensive method for shaping large quantities of metal. For short runs, extrusions may be more economical, since tooling costs are much lower.

The two basic types of extrusion process are (1) **direct extrusion** and (2) **indirect extrusion,** illustrated in Figures 9.51 and 9.52. In direct extrusion metal is placed into a container, a ram is pushed, and the work billet is forced to flow through one or more openings in a die. Alternatively, in indirect extrusion a die is mounted on the ram. As the ram is pushed against the work billet material flows through the die opening and through the ram.

The major disadvantage of the *indirect* method is that whatever is extruded has to be small enough to go through the opening in the ram, so product size has a definite limitation. Conversely, in the *direct* method the extrusions can be almost equal in size to the billet container, and with special tooling, sections wider than the container can be produced.

One disadvantage of the *direct* method is that the billet is pushed through the ingot container during the

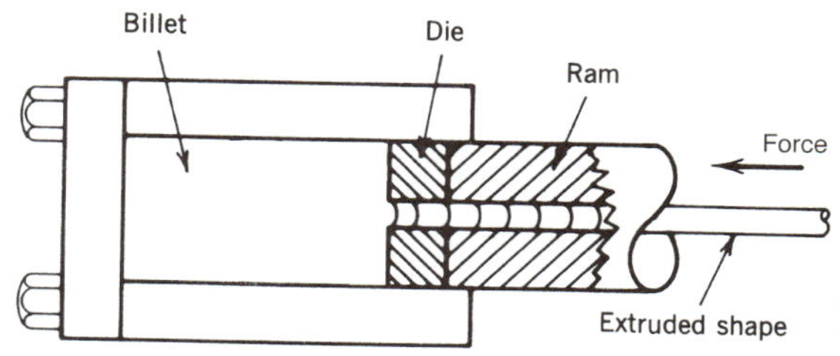

FIGURE 9.52
Indirect or backward extrusion.

process; that is, there is relative motion between the billet and the container. This causes two problems: (1) The frictional force between the metal and the container consumes a large percentage of the press capacity. (2) The friction between the billet and the container causes working of the metal before it ever reaches the die; thus, the working done on the product is not uniform but increases as extrusion proceeds. If the metal temperature is below the recrystallization temperature, then the metal is cold worked, and the amount of cold working done increases toward the rear of the product; if the product is later reheated, as in heat treating, the recrystallized grain size will vary from very large in the front end to very small at the rear—usually not a desirable situation.

By comparison, the major advantage of indirect extrusion is that the work imparted to the metal is uniform along its length, since there is no relative motion between the ingot and ingot container. This also means the pressure of the metal on the die is more uniform, so dimensional control is usually better.

Figure 9.53 illustrates the two methods of making hollow shapes or tubing by direct extrusion. To extrude

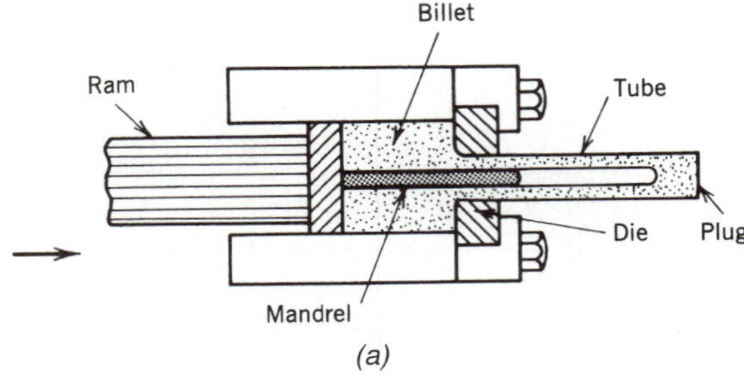

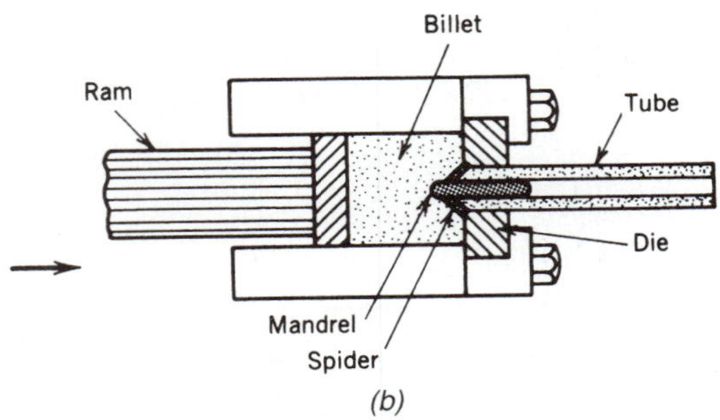

FIGURE 9.53
Forward extrusion. *(a)* Mandrel fastened on the ram producing a hollow shape. *(b)* Use of the spider mandrel to produce hollow extruded shapes.

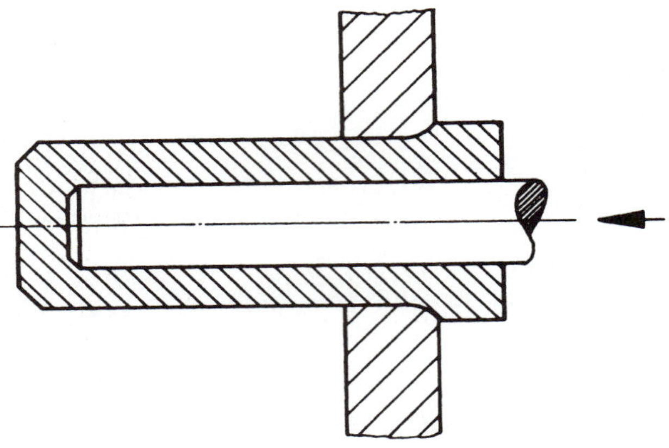

FIGURE 9.54
Diagram showing the principle of producing thick-walled steel bodies ("Presses for All Applications of Metalforming," Schirmer-Plate-Siempelkamp, Hydraulische Pressen GmbH, Krefeld, Germany).

something with a hole in it requires a die to form the outside, and a mandrel to form the inside. In view (*a*) a mandrel is fastened to the ram; the ingot either had a hole drilled in it before being put in the container, or the mandrel pierced the hole. On application of the press force the metal extrudes between the die and mandrel and produces a tube or hollow shape. In view (*b*) there is a die to produce the outside surface and a mandrel to generate the inside surface, but the mandrel is not fastened to the ram. Instead, it is fastened to a bridge or spider that goes *across* the die opening. This requires that the metal split at the bridge and weld back together again as it goes through the die.

HOT DRAWING

Drawing is a process in which the cross section of a bar, rod, or wire is reduced by pulling it through a die opening. (A mandrel pushes on the work piece and thus pulls the work through the die; see Figure 9.54.) The process is similar to extrusion, except that in this process the material is pulled through a die, whereas in extrusion the material is pushed through. Cold drawing,

FIGURE 9.55
After the mandrel has forced the hot metal through the draw ring, a stripping device removes the forging as the mandrel is drawn back. The arrow indicates the workpiece ("Presses for All Applications of Metalforming," Schirmer-Plate-Siempelkamp, Hydraulische Pressen GmbH, Krefeld, Germany).

covered in the next chapter, uses relatively thin metal and changes its thickness only slightly, whereas hot drawing deforms the metal to a very great extent. Hot drawing is used to make thick-walled parts of simple cylindrical shapes (Figures 9.54 to 9.56). Heavy-duty

FIGURE 9.56
Hot-drawing press, called a *push bench,* for the production of thick-walled hollow bodies and pressure vessels. The arrow indicates the workpiece ("Presses for All Applications of Metalforming," Schirmer-Plate-Siempelkamp, Hydraulische Pressen GmbH, Krefeld, Germany).

hydraulic cylinders, artillery shells, and oxygen tanks are made by the hot drawing process.

HOT SPINNING

Although most metal spinning operations are carried out on cold metal (discussed in the next chapter), very large, tough metals are spun while hot. Domed heads for pressure vessels are often made by hot spinning. This process consists of shaping flat or preformed metal disks over a rotating form. Pressure applied by the spinning tool causes the hot metal to flow over the form. Because simple tooling is needed for spinning, it is a less expensive method of forming circular and cylindrical shapes than using drawing or stamping dies.

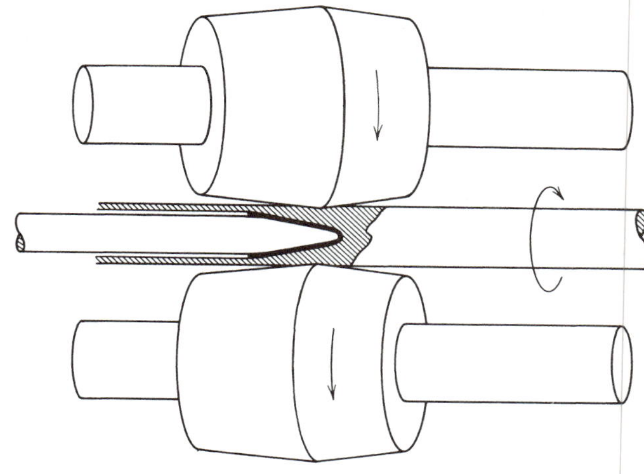

FIGURE 9.57
Piercing of solid billet to make seamless tubing (Neely and Bertone, *Practical Metallurgy and Materials of Industry,* 6th ed., © 2003 Prentice Hall, Inc.).

SEAMLESS TUBING

Seamless tube is hot formed by the Mannesmann process of roll piercing (Figure 9.57). A heated, cylindrical billet is passed through rotating conical rolls whose axes are in different planes. This action creates high tensile stresses that form a crack at the center of the billet that flows over a conical mandrel that shapes and sizes the thick-walled tube. A following operation in a rolling mill with grooved rolls and a mandrel further elongates the tube and reduces its wall thickness. Seamless tube is often cold drawn as a finishing operation.

PIPE WELDING (ROLL FORMING)

Hot-formed pipe is made of **skelp,** a low-carbon semi-finished steel strip. Essentially, the hot strip is drawn through a series of rolls that gradually curl the flat strip into a cylinder (Figure 9.58). As the pipe is formed the edges are forced together, and at a white heat they make a weld bond. This process is called *butt welding.* Since the skelp is unwound from a continuous coil, the pipe must be cut to lengths with a flying saw, one that follows the moving pipe as it cuts it to length. Small diameter pipe, up to 3 in., is butt welded. For larger diameters it is lap welded to give it greater strength. Hot-formed pressure-welded pipe is much less expensive than seamless tubing and is widely used for underground water systems, household plumbing, and for many industrial uses; however, because it is pressure welded (butt welded), it is not acceptable for high-pressure applications such as in hydraulic machinery, where seamless pipe is used.

FIGURE 9.58
Small-diameter pipe formed by continuous butt welding. Skelp is formed and welded into pipe as it passes (from right to left) through sets of rolls. A flying saw cuts the pipe into lengths. The arrow indicates the workpiece (Courtesy of Bethlehem Steel Corporation).

Review Questions

1. By what means are white-hot ingots formed into slabs, blooms, and billets?
2. What modern method of steel manufacture bypasses the ingot stage and produces slabs or blooms directly?
3. Large dendritic crystals that weaken the metal form in cast metals such as ingots and strand casting. How are these removed from the metal?
4. Why is a cast ingot (one that has had no working) not able to recrystallize even if heated for a long time?
5. What would be the consequences of finishing a series of metal-working operations at too high a temperature?
6. What is anisotropy in metals?
7. Why are very large forgings heated although smaller forming operations are sometimes carried out cold? Explain.
8. What kind of forging machine would you choose to use for a very large open-die operation? Which would you choose for making a run of 10,000 high-precision parts?
9. Hot metal tends to stick to forging dies. What design characteristic helps to prevent this from happening?
10. What is "flash" and how is it removed?
11. If you were required to design a machine part, such as an automobile axle, that would be subjected to high shock loads, would you choose a forging? A casting? Or would you machine it from solid bar stock? Why?
12. How are bolt and nut blanks formed?
13. What kinds of shapes can be made by hot extrusion?
14. A tank end having a hemispherical shape must be made of ¾-in.-thick flat steel. Name two methods by which this operation can be accomplished.
15. Briefly explain how seamless tubing is made and how common steel water pipe is made.
16. Would extrusion or cold rolling be more practical for producing various odd-shaped cross sections from soft metals such as aluminum in short runs? Why?

Case Problems

Case 1: Failure of a Tractor Spindle

A front wheel axle spindle on a tractor became damaged from a bearing failure. A replacement part would take 6 weeks to get if it was ordered from the factory, and the farmer needed the tractor for his spring planting. A new spindle was made at a local machine shop from solid steel of the same alloy as the original. After 2 weeks of use, it broke in half and the wheel fell off. Considering what you have learned in this chapter, what do you think caused the failure of the machined spindle?

Case 2: Hot Pressing

Open-die forging on flat anvils usually requires a pressure of 3 to 5 tons per square inch on the tool faces for mild steel at a forging temperature. Harder alloy steels may require two or three times that amount, and up to 15 tons per square inch is needed for the very hardest steels. Pressures may need to be doubled if swages or dies are used.

A 3-in.-square billet, 10 ft long, of hot-rolled mild steel is to be open-die forged into a lifting arm for a hoist. Upper and lower swaging dies will be used to form half the length of the bar into an elliptical shape from the square. Each of the lower and upper swages has a surface area of 20 $in.^2$. A 300-ton hydraulic forging press is available. Will it be sufficient for the job? Approximately how much pressure will this forging operation require?

CHAPTER 10

Processing of Metals: Cold Working

Objectives

This chapter will enable you to:

1. Explain the effects of cold working on metals.
2. Describe how hot-rolled steel is prepared and cold finished in steel mills.
3. List a number of cold-forming operations and explain their principles, advantages, and uses.
4. Account for the difference between cold-rolled and machined parts and select one process over another for a particular product.
5. Make a choice among cold-forming operations for a particular product.

Key Words

plasticity	residual stresses
ductility	recovery anneal
malleability	process anneal
elastic recovery	annealing
cold working	stress relief
work hardening	brittle

In ancient times, **cold working** was accomplished by hammering on soft metals such as gold, silver, and copper for jewelry and other ornamentation. Iron wire was made in the Middle Ages by pulling iron rods through progressively smaller holes in a steel plate. Later, the improved properties of cold-drawn products were discovered and a new industry was born. After being hot formed in a steel mill, steel shapes such as bars, sheets, and tubes could be descaled and further shaped by the process of cold forming. Today, cold-formed products range from very fine hypodermic needles to huge pipeline tubes and from tiny hair-size filaments to propeller shafts for ships. Almost any conceivable shape can now be made by one or more processes of cold forming.

Some of the advantages of cold working as compared with hot working are better surface finish, closer dimensional tolerances, better machinability, superior mechanical properties (and consequently better strength-to-weight ratio), and enhanced directional properties. Some of the negative consequences of cold working are that the metal becomes more **brittle** and less workable, so **annealing** is often required to continue the process; the metal may contain residual stresses that can cause warping or distortion when machined or welded; more massive and powerful equipment is required; and subsequent heating (e.g., welding) will change the cold-worked structure and reduce its strength.

This chapter begins with a discussion of the basic properties of metals that enable them to be deformed "cold," that is, at room temperature. These properties are studied by relating them to the stress–strain diagram of a tension test. The remainder of the chapter is an investigation of the common methods of cold working metals. Some of these are cold rolling, blanking, pressing, drawing, forming, extruding, bending, straightening, roll forming, and spinning.

FACTORS IN COLD WORKING

Plasticity

One of the most valuable characteristics of metals is known as **plasticity,** the ability of metals to be deformed permanently in any direction without cracking or splitting. The higher temperatures used in hot working make metals more plastic, but recrystallization at the high temperatures prevents them from increasing in strength. To achieve higher strengths and hardnesses the metal must be deformed at normal temperatures, that is, below the recrystallization temperature. Also, plastic deformation can occur only at stresses above the elastic limit or yield point, as illustrated in Figure 10.1 with tensile test curves for three types of steels. The tensile test is itself a cold

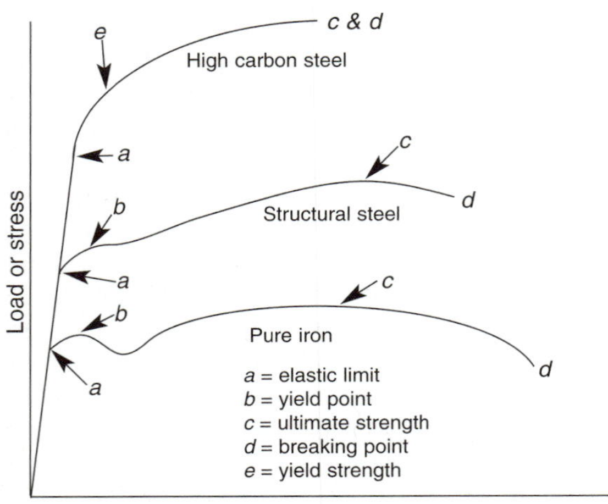

FIGURE 10.1
Above the yield point more stress is required as the cold work continues. The *d* on the diagram incidates where failure occurs (Adapted from Neely and Bertone, *Practical Metallurgy and Materials of Industry*, 6th ed., © 2003 Prentice Hall, Inc.).

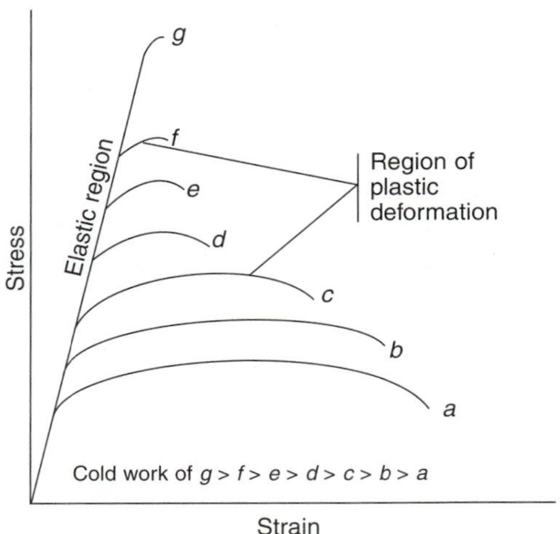

FIGURE 10.2
A representation of test results for a ductile metal that has been cold worked by rolling and then tensile tested, in which specimen *g* received the greatest percent of cold work, and *a* the least. As the amount of cold work increases from *a* to *g* the yield and tensile strengths increase, but ductility, as measured by the elongation, decreases. There is so much cold work in *g* that shortly after achieving the maximum load it breaks; very little ductility remains.

deformation process, so for each test shown when the stress is above the elastic region (point *b* or *e*) the strain (metal deformation) increases substantially. Eventually the metal cannot strengthen itself sufficiently to carry more stress (the curve peaks at point *c*), and failure occurs very quickly at point *d*.

When metals are cold worked to a certain point, the next operation requires forces greater than those previously applied to deform the metal further. Each operation brings the particular metal closer to its ultimate strength and point of rupture, as shown in Figure 10.2, which depicts the results of tensile tests of a metal that had increasing amounts of cold work done by rolling. Note that as the amount of cold work increases the yield strength (where each curve departs from the elastic region) gets closer to the maximum stress or tensile strength, and the strain at failure (called the *elongation*) decreases. The object in cold working metals, therefore, is to stop well short of failure. Also note the slope of the elastic region, which is the modulus of elasticity, *E*, does not change with the amount of strain or cold work. *E* is a constant for a given composition of a metal or alloy.

The degree of deformation (amount of cold working) determines the level of toughness, strength, hardness, and remaining ductility. Different specifications for a manufactured product can thus be obtained. In the case of sheet steel, for example, it is possible to produce from one-quarter-hard steel to full-hard steel. Quarter-hard steel can be bent back 180° without breaking, whereas half-hard steel can be bent only 90°, and full-hard steel can be bent only 45° on a radius of about the material thickness. If more deformation is needed, then a **process anneal** is used to restore plasticity, which at the same time reduces the strength of the metal and lowers the force or stress required to deform it further. Process anneal on the cold-worked steel is often carried out in a closed container of inert gas to avoid scaling problems, a technique called a **bright anneal.** See Chapter 3, Changing the Properties, for a discussion of annealing processes.

Ductility and Malleability

Ductility is the property of a metal that allows it to deform permanently or to exhibit plasticity without rupture while under *tension.* Any metal that can be drawn into a wire is ductile. The ability of a metal to deform permanently when loaded in *compression* is called **malleability.** Metals that can be rolled into sheets or upset cold forged are malleable. Most ductile metals are malleable, but some very malleable metals such as lead are not very ductile and cannot be easily drawn into wire. Some ductile metals are steel, aluminum, gold, silver, and nickel. A few nonferrous alloys such as brass and Monel are also quite ductile, but most alloys of steel are less ductile than plain carbon steel. Some metals and alloys, such as stainless steel, high-manganese steel, titanium, zirconium, and Inconel, tend to **work harden,** that is, to quickly increase hardness as cold working progresses; however, most metals tend to work harden to some extent. Austenitic stainless steels will tend to remain ductile until a very high hardness is achieved. This is an engineering property that is valuable when higher hardness as well as toughness is a requirement for a cold-formed product. Annealing will restore a softer condition as well as the ductility of these work-hardened metals. Work hardening is often a troublesome difficulty in machining operations.

Elastic Behavior

When a metal is placed under stress within its elastic range it will return to its former shape when the load is removed. If the metal takes on a permanent set by loading it beyond the elastic limit (into its plastic range), it will be permanently deformed but will bounce back to some extent because of its elastic properties. This characteristic of metals, called **springback** or **elastic recovery** (Figure 10.3), is a design consideration for forming and bending dies and fixtures. Parts are thus overbent several degrees, ironed, or bottom beaded to counteract springback.

The elastic property of metals also can cause other problems. The metals used in the vast majority of applications are polycrystalline; that is, they consist of many crystals or grains that were initially formed when the metals were solidified. Although cold working tends to stretch out the grain structure in the direction of working, the grains have atom lattices that are not aligned. Thus, the grains react to the plastic deformation differently, and just because the working forces have been removed and the overall stress is zero, each individual grain is not at a zero stress. Instead, **residual stresses** are left behind. Some of these are separating or tensile stresses, and others are compressive, but the result is they are alive and still able to exert a force. An analogy is squeezing a handful of rubber erasers; they push back.

These residual stresses are the type of stresses that can cause a correctly bored hole to be oblong, or a straight piece of bar to become a banana when it is machined on one side, or an otherwise aligned weldment to become warped. These stresses also can influence where a corrosive environment will attack.

The solution to such problems is to use a **stress relief** or **recovery anneal**. This is an anneal at a temperature *below* the metal's recrystallization temperature (because the objective is to remove elastic not plastic deformation or stresses). For pure copper the temperature is about 400°F (200°C), for aluminum alloys it is about 600°F (315°C), and for steel it can be upward of 800°F (425°C).

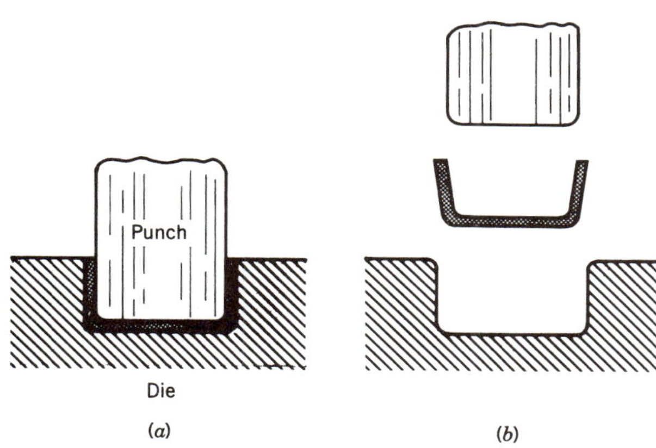

FIGURE 10.3
Cold-working operation showing springback, When sheet metal is formed as in *(a)* and is removed from the die as in *(b)* it will partially revert to its original shape.

COLD ROLLING IN THE STEEL MILL

Preparing Hot-Rolled Steel for Cold Rolling

Hot-rolled steel bars and plates must be sufficiently oversize because the cold-finishing process reduces them in cross section. These are also sometimes straightened in a set of straightening rolls (Figure 10.4). The surface must be clean and free from scale. The hot-rolled bars or sheets are placed in a hydrochloric or sulfuric acid dip that removes the scale; this is called **pickling.** The acid is washed off and the steel is dipped in lime water to remove any acid residue. Sometimes the hot-rolled steel requires full annealing to make it as soft and ductile as possible prior to the cold-working operation.

If the metal has already had some cold rolling its grain or crystal structure is permanently altered, flattened, and lengthened by the process of cold rolling (Figure 10.5). After a process anneal, the recrystallized metal can be further cold worked, since the soft condition of the grains allows for more deformation to take place.

The Cold Strip Mill

The coils of pickled hot-rolled steel pass through a series of rolling mills at a high rate of speed; 1 to 2 mil/min is not uncommon. The strip begins with a thickness that is a little less than ⅛ in. and a length of about ¾ miles. Within 2 minutes it is squeezed to the thickness of two playing cards, and it is more than 2 miles long (Figure 10.6). After annealing, the strip passes through a temper mill. There it is given the desired thickness, flatness, and surface quality, and the desired temper or hardness. Sometimes the full coil is shipped to customers; other times the coiled strip is cut into lengths for shipping flat.

FIGURE 10.4
Straightening rolls. The rolls alternately bend the metal slightly beyond its elastic limit in two axes so that when it emerges it is straight.

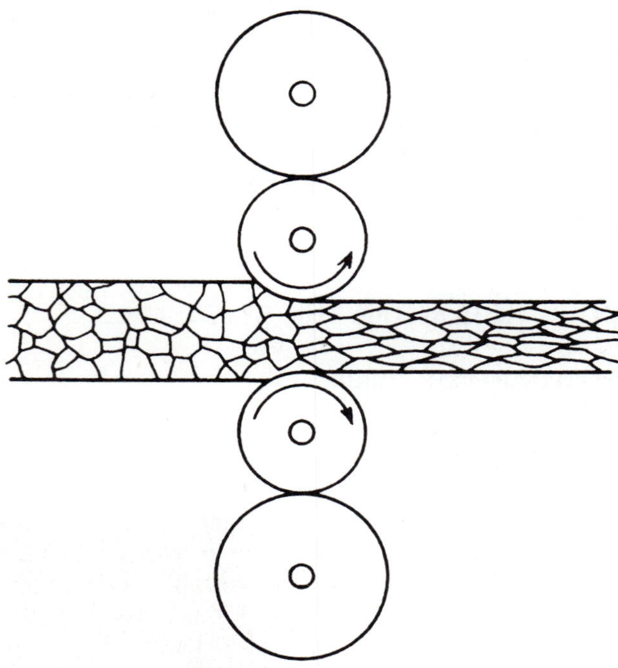

FIGURE 10.5
Cold rolling flattens and lengthens the grains in the direction of rolling (Neely and Bertone, *Practical Metallurgy and Materials of Industry,* 6th ed. © 2003 Prentice Hall, Inc.).

FIGURE 10.6
Cold strip mill showing the operator at his console in the foreground and the rolling mill stands in the background (American Iron & Steel Institute).

Thin-gage, high-precision steel sheets were made possible by the Sendzimer mill (Figure 10.7), which was first developed to produce thin sheets for lightweight radar equipment used in aircraft. Today these thin sheets are needed for space programs. This mill also produces steel foil of 0.003 in. thickness or less for packaging and other uses. The work rolls on these mills are quite small in diameter, from 1 to 2½ in., and they are sometimes several feet in length. These are supported by a cluster of backup rolls. High surface quality of steel strip can be obtained by planishing and polishing with abrasives. Stainless steel sheet and strip are sometimes given a mirror finish before being shipped to the customer.

Metals other than steel, such as aluminum, copper, and titanium, are also rolled into bars, plates, and sheet stock, using similar methods. Both hot and cold rolling processes are used for most nonferrous metals.

Surface Coatings on Sheet Steel

Because iron and steel have a tendency to rust in the presence of oxygen and moisture, many protective coatings for them have been developed. One of the first of these was the coating of iron with tin by hammering it onto steel sheets. This method can be traced back to the thirteenth century. Tin does not corrode in the presence of moisture and will readily bond to clean iron by hot dipping or by plating by the electrolytic method—the way most tin plate is produced today (Figure 10.8). The tin plating on steel may be less than 0.001 in. thick. When available supplies of tin began dwindling and the demand for it became great worldwide, steel companies began to search for substitutes for tin-coated steel. Chromium plate is sometimes used as a substitute, and other coatings are now being used for some purposes. Galvanized sheet steel (Figure 10.9) is used for corrugated roofing material, culvert pipe, and many small articles such as steel pails. Manufactured steel parts for boats, street lighting fixtures, and other exposed metal objects are often galvanized. Although zinc can be deposited by the electrolytic process, by far the most prevalent plating method is the hot-dip process. In the sheet steel mill this process is continuous and often includes a final coating of

FIGURE 10.7
Sendzimer mill for rolling steel strip to close tolerances. Only the two small central rolls touch the metal; the outer cluster of rolls is for backup and bearings. These mills are used to make thin strip and foil (Courtesy of American Iron & Steel Institute).

FIGURE 10.8
Steel coated with tin by the electrolytic process leaves the plating line (Courtesy of Bethlehem Steel Corporation).

FIGURE 10.9
Minimized spangle steel coils await shipment. These coils are produced on the plant's modernized galvanized line (Courtesy of Bethlehem Steel Corporation).

paint that is dried in bake ovens as the steel moves through them. These galvanized and painted sheets are capable of being formed or bent without destroying their protective coatings. Many other coatings are also put on steel to protect it from deterioration or to enhance its appearance. These will be covered in Chapter 18, Corrosion and Protection of Materials.

COLD ROLLING OF NONFERROUS METALS

The malleability of many of the nonferrous metals makes them ideal materials to be cold rolled. The very soft metals such as tin and lead, with recrystallization temperatures below room temperature, can be cold worked extensively without annealing. The stronger nonferrous metals and their alloys, such as aluminum and copper, are also capable of extensive reductions in thickness by cold rolling. The FCC (face-centered cubic) atom lattice of these metals (with its large number of slip planes) makes them especially capable of cold rolling. Magnesium, with its CPH (close-packed hexagonal) lattice is more difficult to cold work.

The volume of aluminum alloy sheet is large enough that the rolling process is very similar in appearance to the large-volume steel sheet mills, that is, hot lines of 10 stands of rolling mills half a mile long, followed by multiple-stand, high-speed cold mills. For the thinnest of gages almost pure aluminum is rolled to foil. The "household foil" purchased in the supermarket is typically 0.0007 in. thick, and the "heavy-duty" foil is 0.001 in. thick. The foil used in cigarette packages is often only 0.00022 in. thick. For these lighter gages the last rolling pass is done by "pack rolling": two thicknesses are rolled together and then later separated.

BLANKING AND PRESSING

One of the most versatile forms of metal working is that of converting flat sheet and strip metals into useful articles. Sheet metal can be pierced, punched, or blanked to produce flat shapes. It can be drawn or cupped to form such items as pots and pans. It can be rolled into cones or cylinders or pressed into automobile body shapes; it also can be shaped by many other forming operations.

Although the cutting of a blank is actually a shearing operation, it is included here because it is the first step in forming a product from sheet stock. **Blanking** is the operation of cutting a flat shape out of a strip of sheet metal in which the hole material is saved for further operations. If the hole material is scrap, it is called **punching.** By this process any shape hole can be made. **Piercing** is the operation of cutting (usually small) round holes in sheet metal; if the holes are small and close together, it is known as **perforating.** Blanking with a conventional punch and die (Figure 10.10) is usually employed where large quantities are required. It usually is used only for thin sheet, but thicker material can be punched or blanked, depending on such factors as press size, blank size, and material thickness (Figure 10.11). Some blanks may be more economically cut by circular shears (Figure 10.12) or nibbling machines (Figure 10.13). Where square or rectangular shapes are needed blanks may be cut with sheet metal shears (Figure 10.14). Blanks may be sawed in a vertical band saw, several pieces stacked together (Figure 10.15). Handheld power shears can also be used to cut blanks (Figure 10.16). It is obvious that using any of these methods to make blanks is much slower than using a die set, and they are useful only when a few parts are to be produced.

FIGURE 10.10
Blanking die set in a press showing a punch making round blanks (Courtesy of Alcoa Inc.).

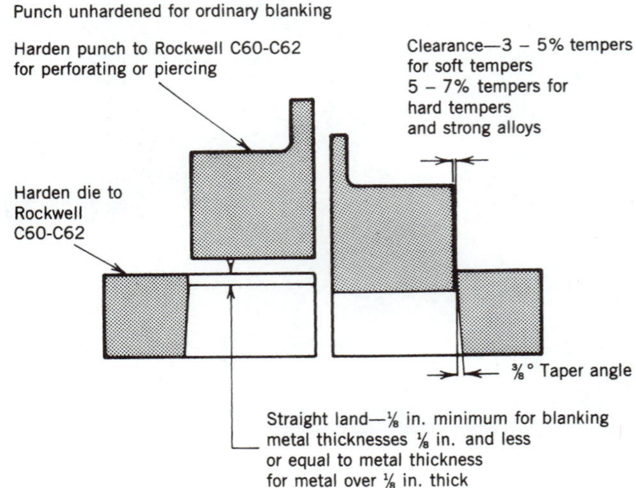

FIGURE 10.11
Hardness and clearance of blanking tools may vary with the type of sheet metal used (Courtesy of Alcoa Inc.).

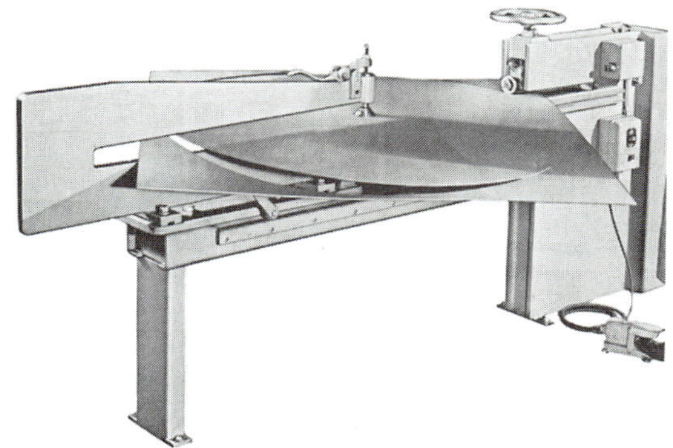

FIGURE 10.12
Circular shears cutting a large disk (Courtesy of Niagara Machine & Tool Works).

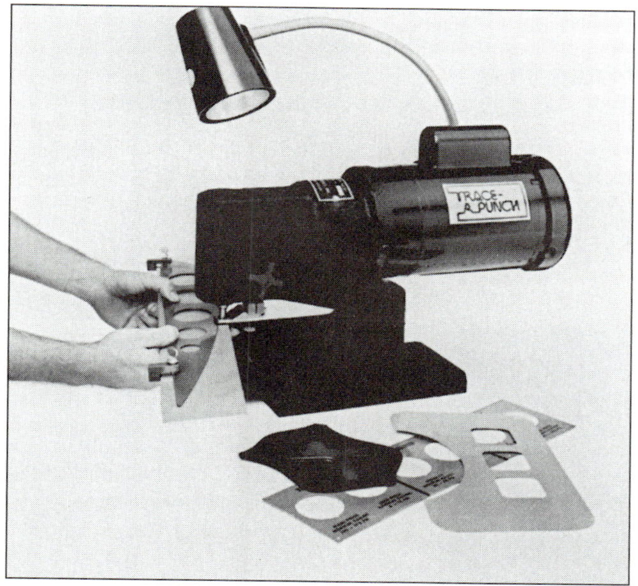

FIGURE 10.13
Duplicating nibbler. This machine cuts sheet metal by means of a small oscillating punch. Here it is being moved along a template to duplicate that shape (Courtesy of Heck Industries, Inc.).

The basic components of piercing and blanking die sets, as shown in Figure 10.10, are a punch, a die, and a stripper plate. The punch holder (top plate) of the die is attached to the press ram, which moves the punch in and out of the die. The die holder of the die set (bottom plate) is attached to the bolster plate of the press. Guideposts in the die set on which bushings slide maintain precise alignment of the cutting members of the die. The stripper plate removes the material strip from around blanking and piercing punches. The material strip is advanced after each punching stroke by a feeding mechanism.

FIGURE 10.14
Guillotine shear for shearing sheet and plate.

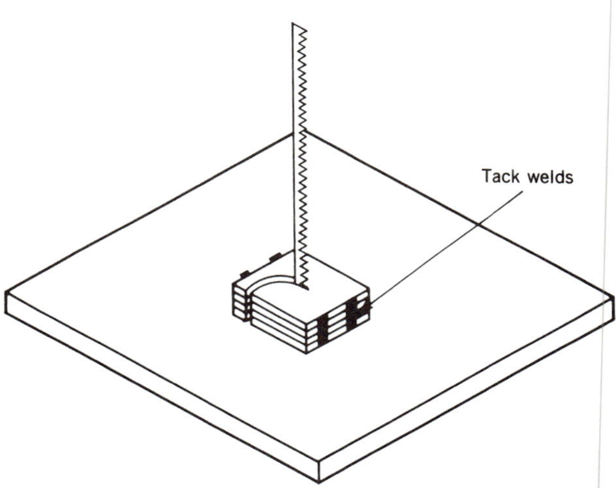

FIGURE 10.15
Band sawing a stack of blanks.

An alternative method of blanking uses steel-rule dies (Figure 10.17). This method was first used for cutting softer materials such as cardboard, cloth, and plywood, but the process is now used for metal. A steel band is fastened on edge in a groove in the upper die plate. Neoprene rubber pads take the place of the usual stripper plate. Steel-rule dies have an expected life of about 100,000 parts. Conventional blanking dies may produce three times that many parts before resharpening is required. An open lower die can be used, or a solid maple or urethane slab can be used to cut against, lessening the cost.

When mechanical presses (Figure 10.18) are fitted with progressive punching and forming dies, a continuous manufacturing of products is possible. Parts for electrical outlet boxes are formed from steel strip (Figure 10.19).

FIGURE 10.16
A portable handheld power shears for cutting sheet metal blanks (Courtesy of Robert Bosch Power Tool Corp.).

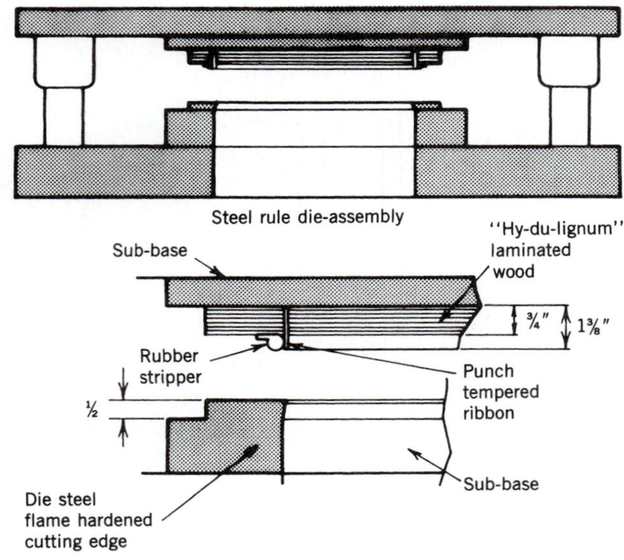

FIGURE 10.17
Steel-rule die blanking technique (Courtesy of Alcoa Inc.).

FIGURE 10.18
Mechanical presses such as this one are extensively used for punching, blanking, and forming of sheet metal strip. In this picture coiled steel is being fed into progressive dies from the right (Courtesy of The Minster Machine Company).

DRAWING, FORMING, AND EXTRUDING METAL

It is easy to imagine molten metal flowing by gravity into molds, and steel softened by heating to a white heat being hammered or forced into shape. Also, the hardness and toughness of cold metals are familiar characteristics. It is more difficult to understand how cold metal can be made to flow when sufficient pressure is applied. Virtually every operation performed on hot metal can also be done on cold metal, but the limit to the size of these parts is in the massiveness and power of the machinery. Because the cost of heating metal is eliminated in cold forming, it is a less expensive method, especially for the smaller piece parts, and there is no thermal damage to dies.

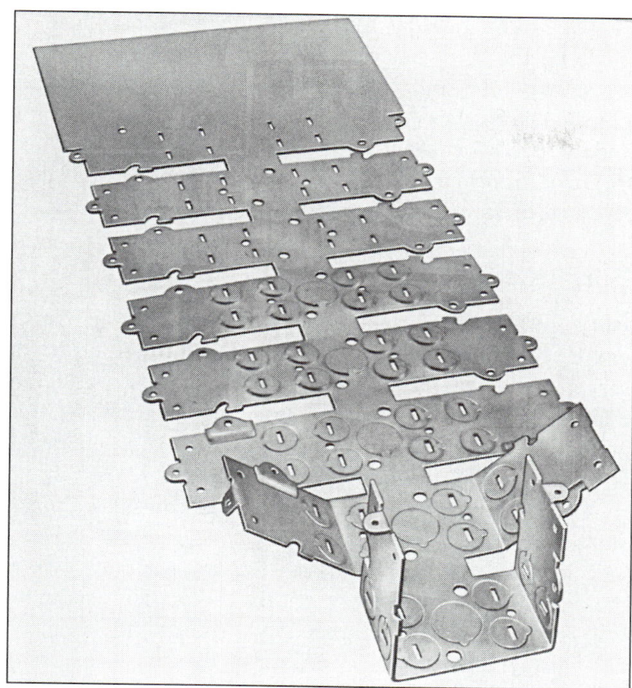

FIGURE 10.19
Electrical outlet boxes are punched and formed with a progressive die (Courtesy of The Minster Machine Company).

Combination hot, cold, and/or warm forming has the advantage of reducing equipment costs where several operations are involved and often reduces the need for secondary machining and operating personnel. Larger parts can be hot formed and later finish formed by cold forming methods. There is a new trend toward warm forming in which metals that are difficult to cold work, such as titanium, stainless steel, and some alloy steels, can be formed without the great pressures and stresses of cold forming. For example, bearing races are now being warm formed of tough alloy steel that is subsequently hardened and ground. Steels that contain more than 0.40 percent carbon are difficult to cold form but can easily be formed while warm, that is, below the transformation temperature but at or above the recrystallization temperature, about 900° F (482° C) for steel. Warm forming can substantially increase production rates for some materials.

Coining and **embossing** are stamping operations that form the surface of metal. Impressions of letters, figures, and patterns are formed by pressing them onto the metal. Usually, knuckle-joint presses are used for coining and embossing because these machines are of rigid construction, have rapid short strokes, and exert intense pressure during the last part of the stroke as a result of the knuckle-joint action. See Chapter 9 and Figure 9.41 for a description of knuckle-joint presses.

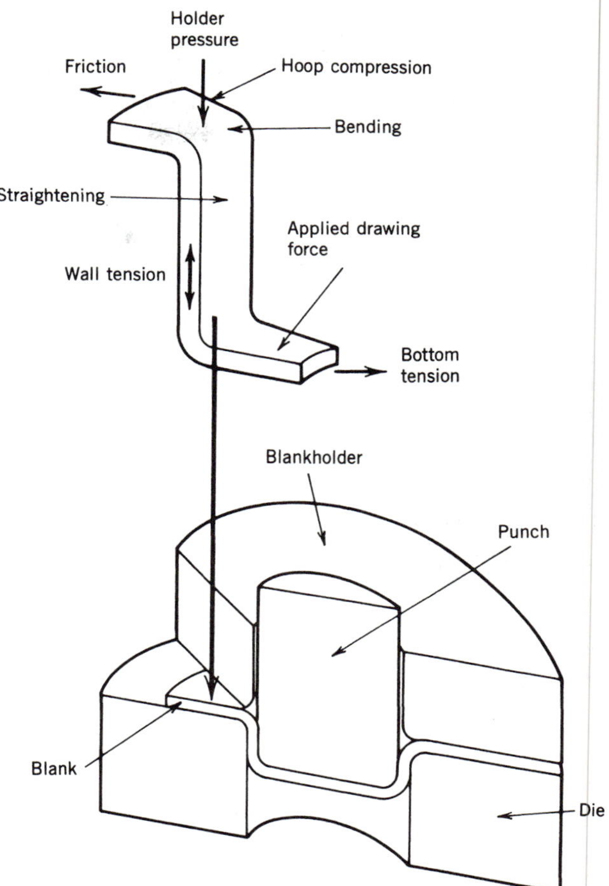

FIGURE 10.20
Forces applied to a blank during cold drawing operation (Courtesy of Alcoa Inc.).

Drawing Plate, Sheet, and Foil

This kind of drawing process is one in which a flat piece of metal is formed into a hollow shape by applying force with a punch to the center portion of the metal. The punch draws the metal into a die cavity (Figure 10.20). In this operation, the metal is stretched on the side walls and made to assume an exact thickness. Sheet metal drawing is usually performed in a vertical press.

Blanking and piercing are sometimes combined in one operation for the purpose of continuous manufacture of small parts such as metal cases and bottle caps. This is usually done with progressive dies (Figure 10.21). Deep drawing is often done with several presses, each having a different die that contributes to the final shape (Figure 10.22). When the wall thickness must be further reduced and the parts lengthened, as in the manufacture of cartridge cases, an ironing process is used (Figures 10.23 and 10.24). Sometimes an anneal is required between pressings.

FIGURE 10.21
Blanked and drawn parts showing progression of drawing operation (Courtesy of Alcoa Inc.).

The very popular aluminum beverage container is made by the *draw and iron* process, the basics of which are shown in Figures 10.23 and 10.24. An aluminum slug, punched from sheet produced by hot rolling of cast ingot and then cold rolling, is drawn to a cup, ironed, and trimmed to form the body of the can. A stamped lid is then fastened to the trimmed end to close the can.

Bar, Tube, and Wire Drawing

Hot formed ferrous and nonferrous metal bars and tubes can be reduced in size by being drawn through a die (or series of dies) slightly smaller in size than the original or entering stock. Cleaning and descaling may be required before the drawing operation. This *bar drawing* operation is performed on a draw bench (Figure 10.25). The drawing process hardens the metal and gives it a smooth finish suitable for many applications.

Seamless tubing is also cold drawn on a draw bench; the only difference in the operation is that there is a mandrel inside the tube to thin the walls and provide an internal finish (Figure 10.26). Because of this finish operation

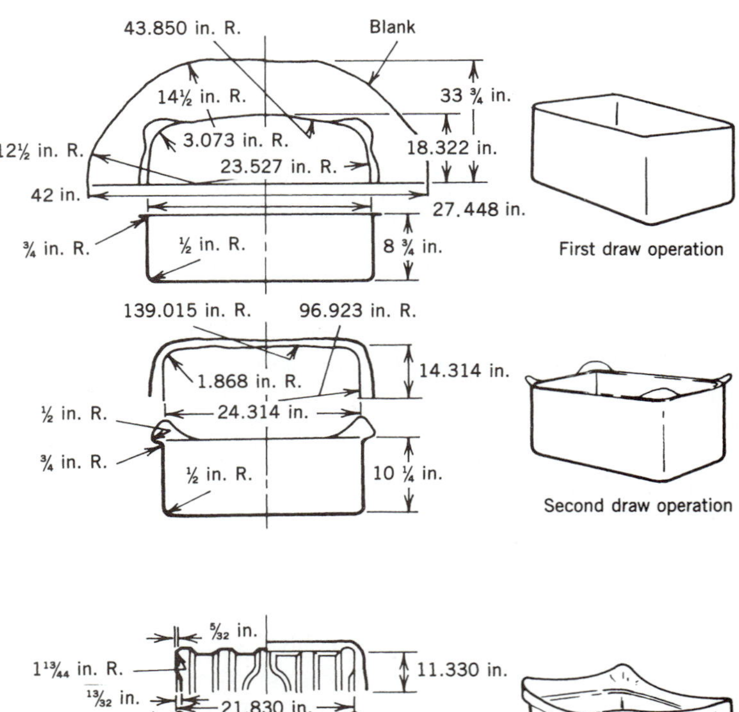

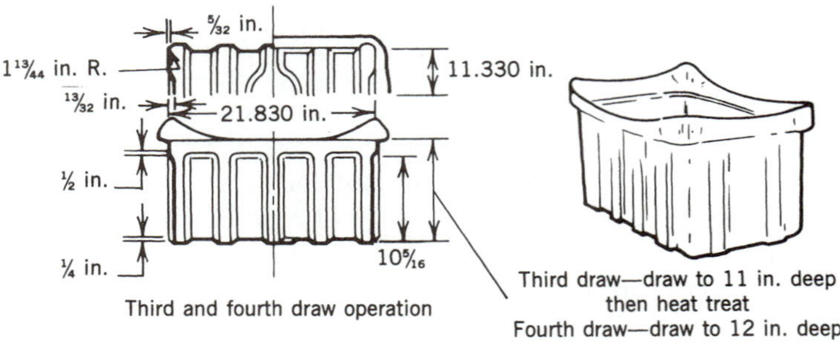

FIGURE 10.22
Sequence of draw reductions for 6061 aluminum alloy rectangular case (Courtesy of Alcoa Inc.).

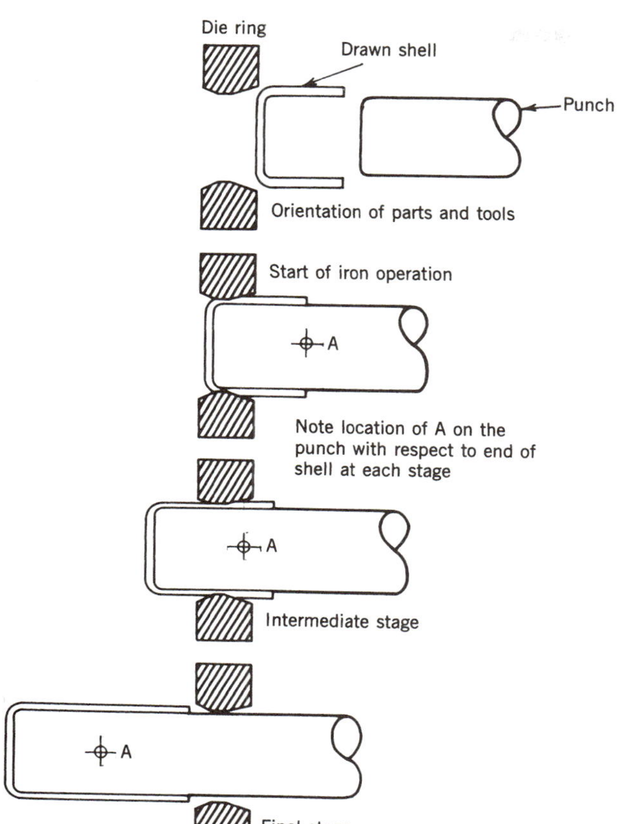

FIGURE 10.23
Ironing a drawn shell (Courtesy of Alcoa Inc.).

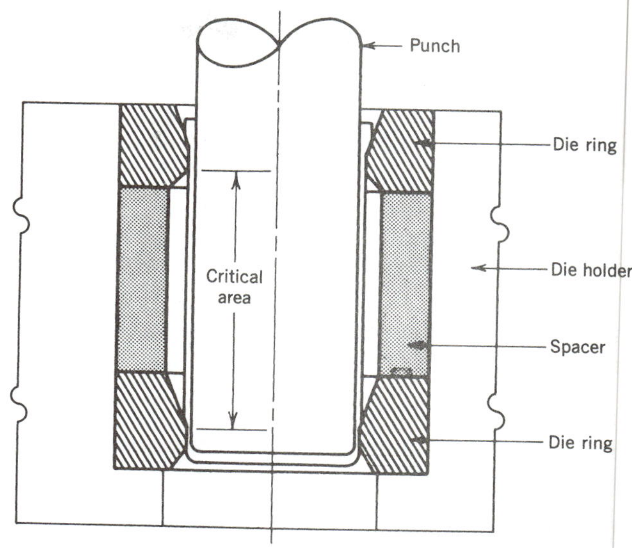

FIGURE 10.24
Multiple reduction with stacked ironing dies (Courtesy of Alcoa Inc.).

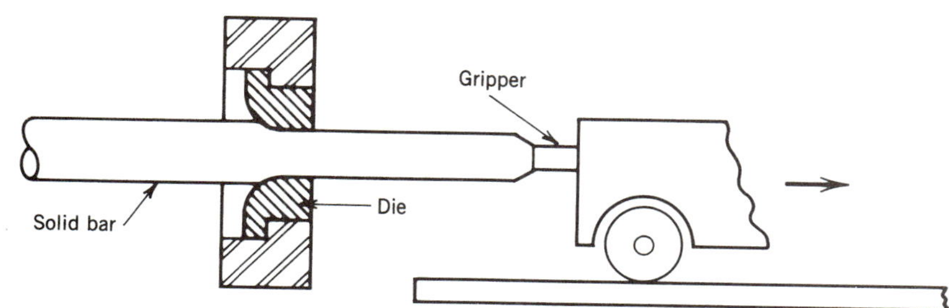

FIGURE 10.25
Draw bench for solid-bar stock.

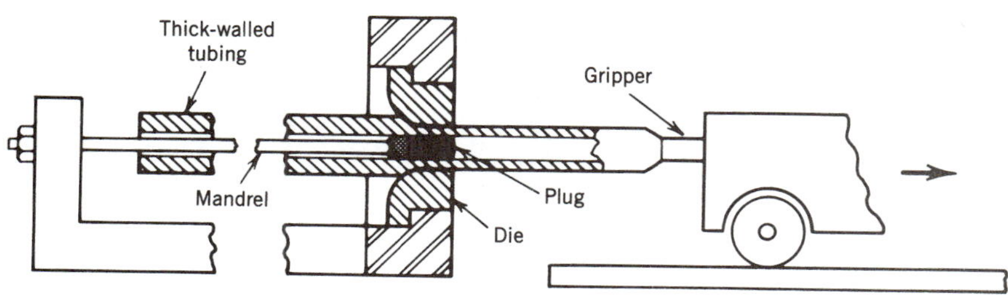

FIGURE 10.26
Draw bench for seamless tubing.

and the higher cost of pierced seamless tubing, cold-drawn seamless tubing is considerably more expensive than butt-welded black pipe. Seamless steel tubing, after a honing operation, is widely used for hydraulic cylinder manufacture, and smaller diameters are used for high-pressure pipes that carry oil or other fluids.

Wire drawing is basically the same process as bar drawing except that it involves much smaller diameters of metal and it is a continuous process done on rotating equipment (Figures 10.27 and 10.28). The wire is drawn through a series of dies, each one slightly smaller in diameter than the previous one. The wire is pulled by a rotating capstan or drum that is located between each set of dies. Coolant, acting as a lubricant and to cool the die, is flooded over the dies, since this is a cold-forming operation and unwanted heat is developed in the operation. The finished wire is wound on a reel.

FIGURE 10.27
Battery of modern wire drawing machines. Rod enters at the left, is reduced in size as it passes through successive dies, and is coiled at the right (Courtesy of Bethlehem Steel Corporation).

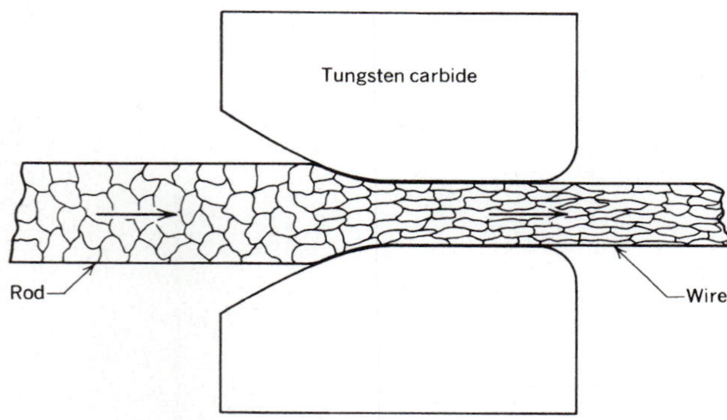

FIGURE 10.28
Enlarged cross section of wire drawing through a die. (Neely and Bertone, *Practical Metallurgy and Materials of Industry,* 6th ed., © 2003 Prentice Hall, Inc.).

In a similar fashion, smaller diameter coiled tubing (e.g., aluminum, copper) is able to be produced on rotary drawing equipment. Starting with starting stock produced by hot extrusion, the tube is passed through a series of draws using dies of a decreased diameter. The inside diameter (ID) is controlled by a floating plug or mandrel; for each draw in the sequence, the plug is designed with two diameters and a tapered section between them. The smaller diameter is the desired finished ID for that draw, and the larger diameter is just smaller than the ID of the tube entering that draw. The friction forces that are present force the plug into position in the die, but its larger diameter prevents it from passing through the die; thus it "floats" in position, controlling the ID.

The limitation of the drawing process, whether on a bench or rotary equipment, is the ability of the metal to become stronger as it is drawn through the die. Since the tube or bar is in tension if the pulling force is too large, it will just pull the metal apart. Referring back to Figures 10.1 and 10.2, as a metal is cold worked it becomes stronger, but if the forces required to cold work the metal and overcome the friction in the die become greater than the strength of the metal (which, remember, is getting smaller and smaller as it is drawn through the dies), the metal will break. In order to continue the drawing sequence the metal may have to be annealed.

FIGURE 10.29
Wire upsetting machine (Courtesy of National Machinery LLC).

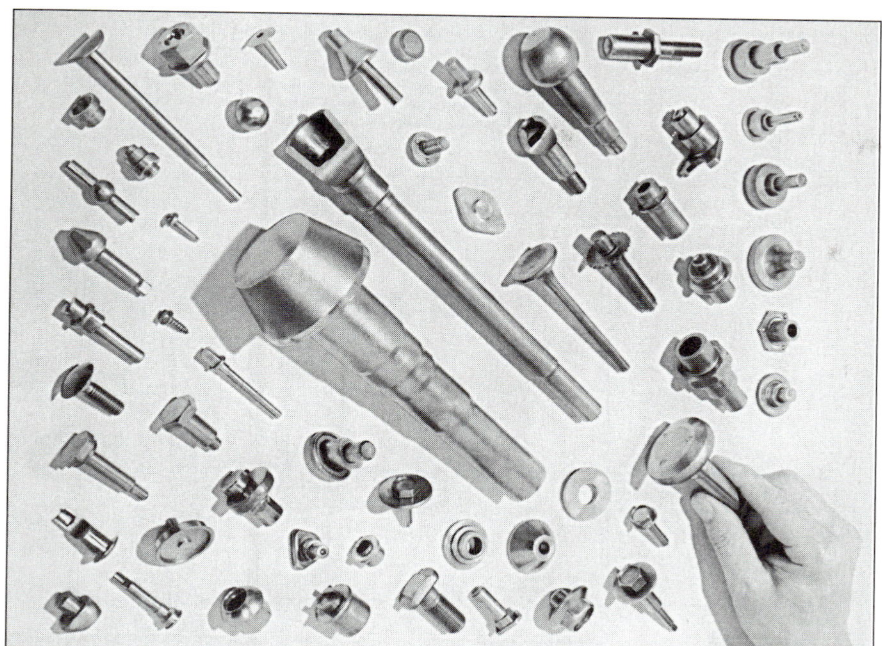

FIGURE 10.30
An array of cold upset parts (Courtesy of National Machinery LLC).

Cold Forming

Cold forming, also called *cold heading* or *cold forging*, is a metal upsetting process carried out on machines designed for rapid production of small parts from wire stock (Figure 10.29). Upsetting machines are frequently rated by maximum diameter cutoff capacity. For example, a ⁵⁄₁₆-in. cold header can shear alloy steel up to ⁵⁄₁₆ in. in diameter. Generally, only small parts such as screw and bolt blanks, rivets, and ball bearing blanks are produced on cold upset machines, but some machines are capable of upsetting much larger parts (Figure 10.30). The largest single use of upsetting machines is in forming heads on fasteners such as rivets and screws. Threads are subsequently rolled on the blanks, giving them a much greater strength and fatigue resistance than cut screws (Figure 10.31).

In the past, most threaded fasteners were made on automatic screw machines that turned the part to size from bar stock and then threaded it with a die, followed by a cutoff operation. Use of automatic screw machines is necessary when high precision and complex shapes are involved, but they are much slower than cold-heading machines; a small part made in 10 seconds is acceptable on an automatic machine; however, production rates on upsetting machines can be as high as 36,000 per hour for small unpierced rivets, and No. 8 size screw blanks can be made at a rate of 27,000 per hour. Bolts ⅜ in. in diameter can be headed, pointed, and threaded at a rate of 15,000 per hour. Another advantage of forming parts is

FIGURE 10.31
Grain structure of a cold-formed bolt blank (Courtesy of National Machinery LLC).

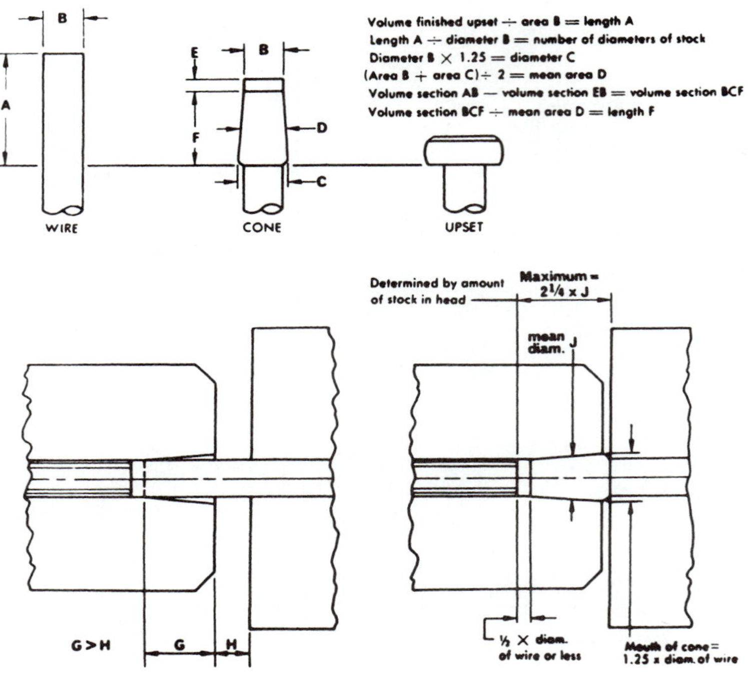

FIGURE 10.32
Cone upsetting. Sequence of operations in upsetting a bolt head (Courtesy of National Machinery LLC).

that nothing is wasted; almost no scrap is cut away, as in machining. There are also few rejects with the upsetting process.

The process and sequence of operation for a cold-heading operation is shown in Figure 10.32. The wire or bar is first formed into a cone before the head is made, because the unsupported length tends to buckle if the heading operation is done first. From the figure you can see that the cone dimensions are very carefully determined to provide the correct volume of metal to form the head. Several modes of upsetting are shown in Figure 10.33, some parts are upset in the punch, some in the die, some in both punch and die, and others between the punch and die. The five basic operations performed in cold-forming machines are shown in Figure 10.34. Combinations of these give cold forming great versatility. An example is the spark plug shell (Figure 10.35). Figure 10.36 shows the progression of upsets and extrusions used to make the spark plug center post. This sequence of operations and the tooling involved is shown in Figure 10.37.

Nut blanks are made in four- or six-die nut formers (Figure 10.38). The operation is similar to upsetting and to the nut-blank hot formers discussed in Chapter 9. As in other cold upset operations, the advantage over hot forming is the higher production rate and lower cost. A sequence of four-die nut forming is shown in Figure 10.39.

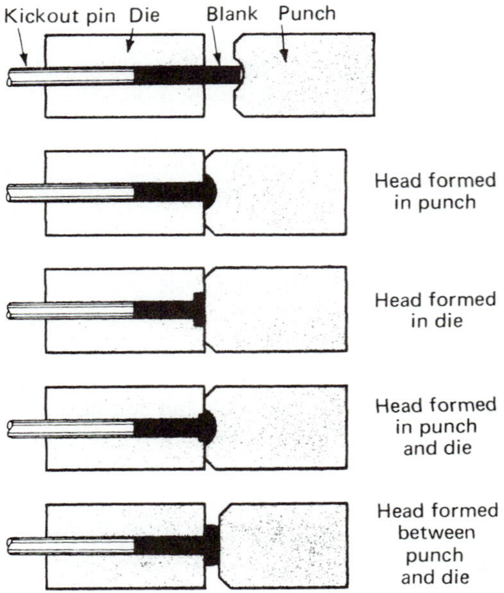

FIGURE 10.33
Modes of upsetting (Courtesy of National Machinery LLC).

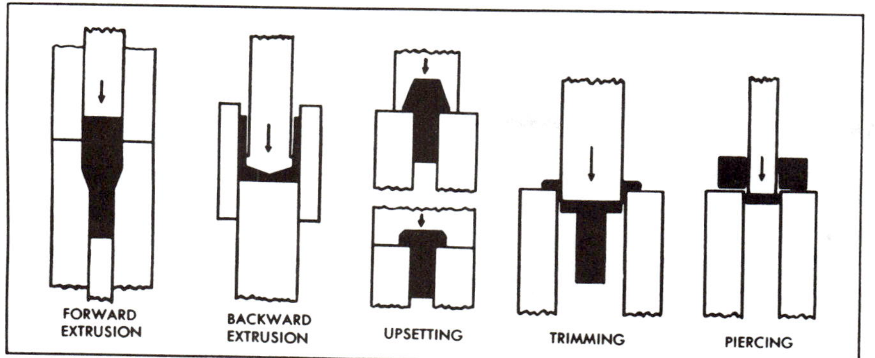

FIGURE 10.34
Five basic operations performed in cold-forming machines. Combinations and variations of these operations give the cold former a wide range of application (Courtesy of National Machinery LLC).

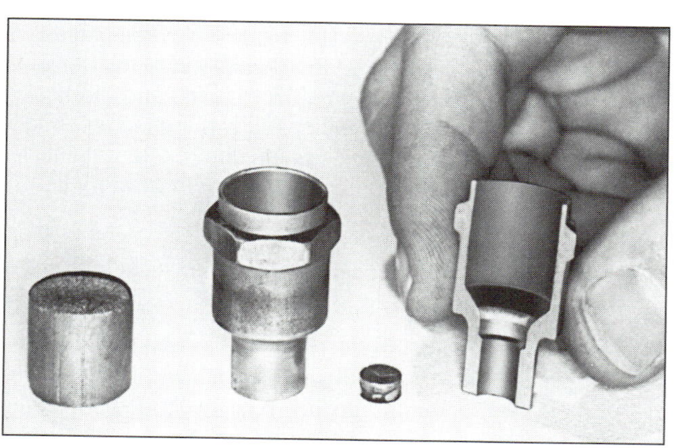

FIGURE 10.35
From blank to finished spark plug shell in a six-forming die machine (Courtesy of National Machinery LLC).

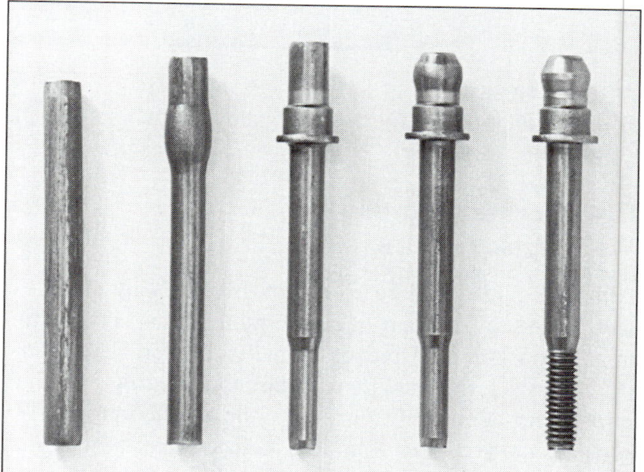

FIGURE 10.36
Progression of upsets and extrusions used to make spark plug center post (Courtesy of National Machinery LLC).

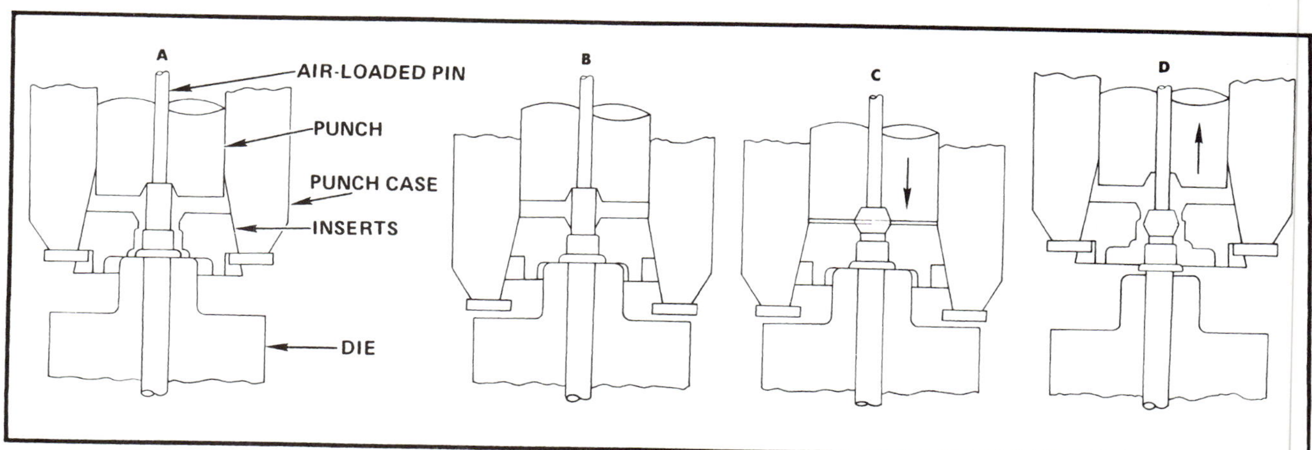

FIGURE 10.37
(a) Air-loaded pin punches the spark plug center post blank into die. *(b)* Die forces inserts into closed positions. *(c)* Continued advance of pin upsets metal into cavity formed by inserts. *(d)* As punch case withdraws, inserts open to allow them to clear largest upset diameter (Courtesy of National Machinery LLC).

FIGURE 10.38
Four-die nut former (Courtesy of National Machinery LLC).

FIGURE 10.39
Four-die nut forming. The top row of operation shows forming done by the punches; the bottom row shows forming done in the dies. The first blow upsets the blank and starts the hex formation. The second blow displaces metal away from the center of the blank toward the corners of the hex. The third blow finishes both faces and marks the hole for punching out. The final blow punches the slug, forming the hole. This method retains controlled metal flow for high-quality nuts (Courtesy of National Machinery LLC).

Cold Forming Threads, Worms, and Gears

Although thread and gear forming is not a new idea, it is well suited to be a part of a new trend toward the continuous processing of metals. Ideally, the process would progress from the raw ore to continuous casting to rolling mills to upset and roll forming to the finished product. In such a system, waste would be reduced, and transferring and shipping of materials would be eliminated.

External thread forming by the process of cold rolling screw threads is a follow-up operation after cold heading screw blanks; however, threads may also be cut as a machining operation on cold-formed blanks. The decision whether to machine threads or to roll them can be based on several factors.

1. **Plasticity** The amount of deformation required is directly related to the plasticity of the metal. Some grades of aluminum, gray cast iron, and die-cast metals cannot be rolled because of their insufficient elongation and reduction of area.
2. **Work hardening** Most stainless steels and high-manganese steels rapidly work harden when cold worked. These metals do not readily lend themselves to cold rolling.
3. **Workpiece design** Some parts are not adaptable for rolling, for example those with threads that are too close to a shoulder (Figure 10.40). Some tapered threads and threads on inaccurate blanks should be cut instead of rolled.
4. **Finish** Rolling is far superior to cutting if the finish is a consideration.
5. **Surface hardness** The rolling process produces a greater surface hardness (Figure 10.41) that wears well and helps resist metal fatigue (Figure 10.42) at thread roots.

FIGURE 10.40
If threads must be made close to a shoulder, cutting is better than rolling.

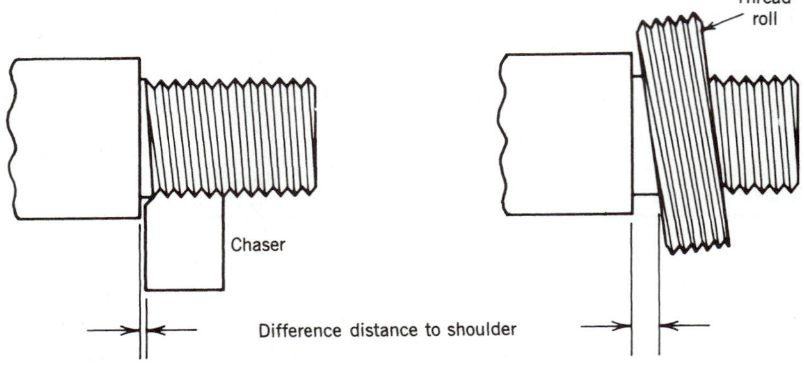

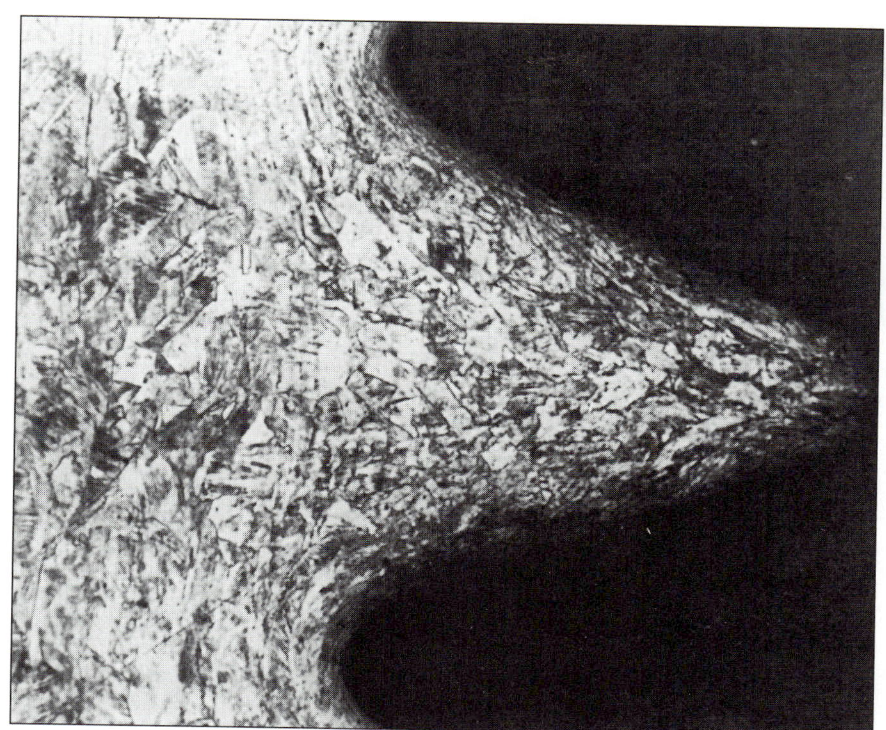

FIGURE 10.41
Longitudinal section of a type 304 stainless steel fastener that was cold worked by being thread rolled in the annealed condition. Cold-worked austenite (and probably some martensite) is in the area immediately below the surface of the rolled thread (50 ×) (*Metals Handbook,* Vol. 7, 8th ed., ASM International, 1972, p. 134. With permission.).

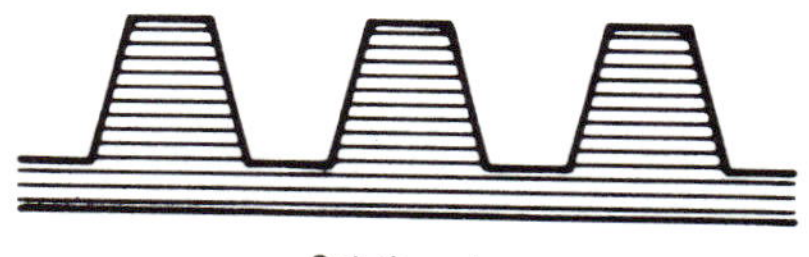

FIGURE 10.42
Rolled threads have a harder surface than cut threads and are generally stronger.

6. **Higher speeds** Generally, thread rolling is a somewhat faster operation than cutting threads with dies.
7. **Tensile strength** Rolled threads are stronger than cut threads because of grain flow (Figure 10.42).

Thread rolling machines have two or three hardened, threaded rolls between which threads are formed on blanks. Several methods are used for forming threads, splines, worms, and gears: infeed forming, through-feed forming, the reciprocal method, the infeed–through-feed–out method, and the automatic continuous method.

Internal threads are often cut using taps, but they may also be formed using thread-forming taps (Figures 10.43 and 10.44). As in thread rolling, metal is displaced to form the threads, work hardening and smoothing it. Both cutting and forming methods for internal threads are very rapid processes, and the choice of method can be based on the same factors as those for external roll forming of threads.

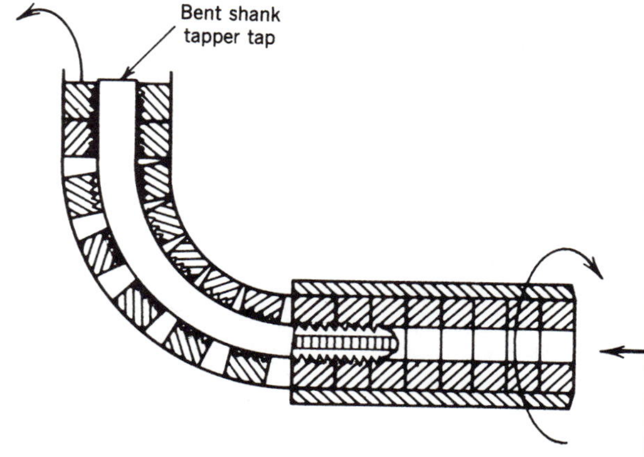

FIGURE 10.43
Production tapping of nuts is usually done with bent shank taps to make it a continuous process.

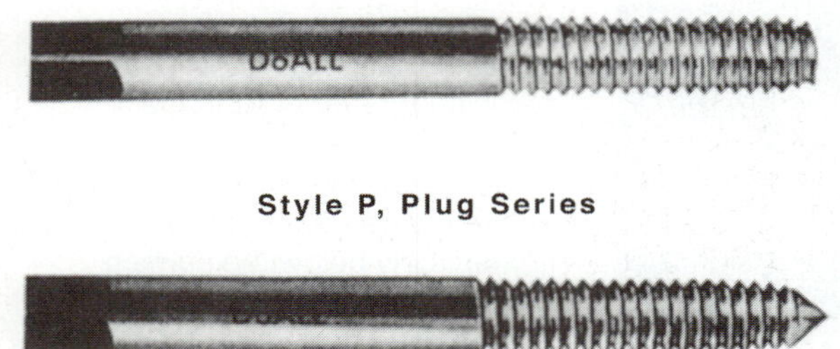

FIGURE 10.44
Thread-forming taps (Courtesy of DoAll Company).

A *worm* is a spiral or helical groove, similar to a screw thread that engages with a worm wheel or worm gear to impart motion. Worms have traditionally been machined on lathes or milling machines, since they are highly precision parts; however, it is possible to achieve the required precision by the cold-rolling process with the added advantage that the part receive a very smooth finish and work-hardened surface. This reduces friction and provides a longer service life. In this process, a blank is placed in a rolling machine and moved between rolls in the infeed process, where hundreds of tons of pressure is applied as the rolls and blank turn and form the shape of the worm.

Helical and spur gears are formed exactly as worms are formed. Smaller gears are cold formed, but larger gears with coarser teeth are hot formed. For high-precision gears, a follow-up operation is necessary. Even machine-cut gears require a finishing operation of gear shaving or grinding. This topic will be further discussed in Chapter 13.

Cold Extrusion

Direct and Indirect Extrusion The basics of the direct and indirect extrusion processes were described in Chapter 9 and illustrated in Figures 9.51 through 9.53. Those techniques are used to extrude steel, and nonferrous metals such as aluminum, copper, brass, magnesium, and their alloys, and require the metal be heated to reduce its yield strength.

For cold extrusion only the very soft metals can be used, for example, tin and lead. With that exception the cold extrusion process proceeds in the same manner as the hot extrusion process, as shown in Figures 9.51 through 9.53.

Impact Extrusion This process (Figure 10.45) is used to produce products such as collapsible tubes for toothpaste. In this process a thick slug of metal is placed in the cavity of a die. The slug is struck by a punch having the size and shape of the inside of the part. The impact of the punch is sufficient to cause the metal in the blank to flow very quickly backward through the space between the punch and die cavity. The part adheres to the punch and is removed by a stationary stripper as the punch is withdrawn. Impact extrusion is a very rapid production method of forming cold metal.

MISCELLANEOUS METAL-WORKING PROCESSES

Bending, Straightening, and Roll Forming

Bending refers to a simple bend in one axis. When two or more bends are made simultaneously, as with a die, the process is usually called forming or drawing. Bending

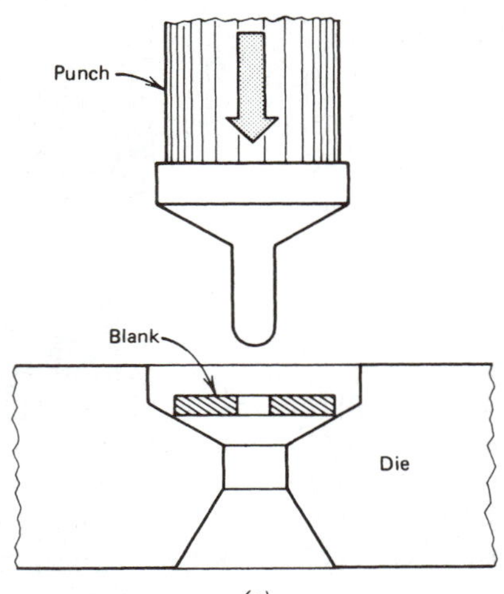

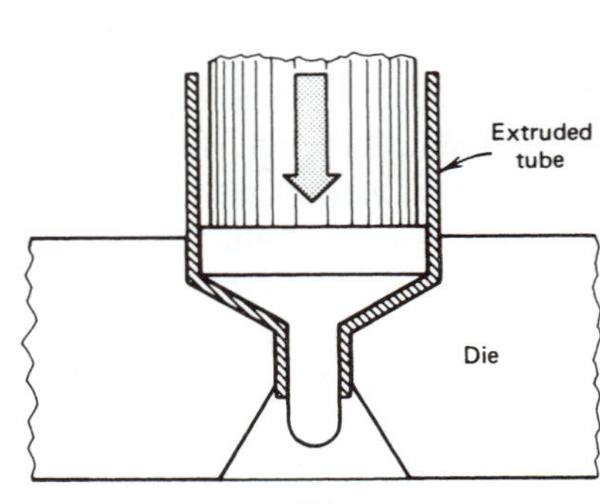

FIGURE 10.45
The method of forming thin-walled tubes by impact extrusion. A flat blank is placed in the die *(a)* and the punch is brought down rapidly with a single blow *(b)*. The material in the blank then "squirts" upward around the punch. When the punch is withdrawn, a stripper plate removes the tube (Neely and Bertone, *Practical Metallurgy and Materials of Industry,* 6th ed. © 2003 Prentice Hall, Inc.).

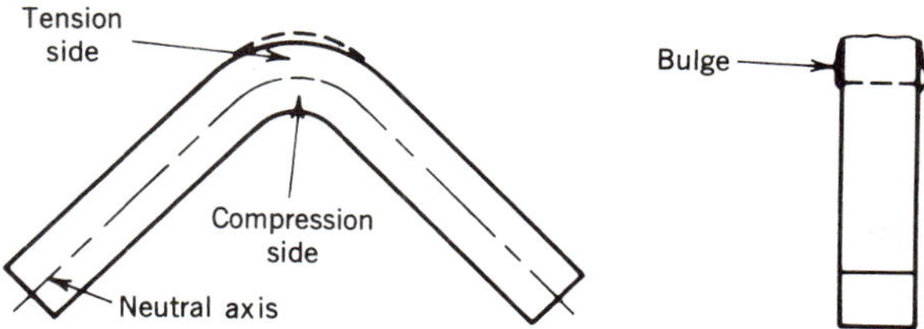

FIGURE 10.46
Simple bend in one axis.

FIGURE 10.47
Bar folder. Such a device is used for bending light sheet metal.

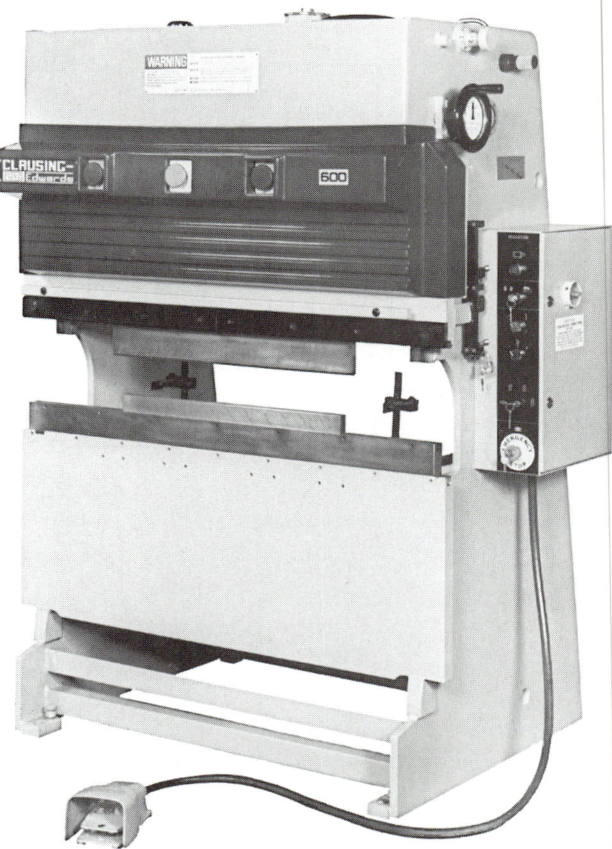

FIGURE 10.48
Vertical press brake (Courtesy of Clausing Machine Tools).

metals causes a plastic deformation about a linear axis, making little or no change in the surface area of the bend. The neutral axis of the bend (Figure 10.46) is not located at the center, equidistant between the outer and inner surfaces of the bend, because the yield strength of metals in compression is somewhat higher than in tension and this causes the inner compression sides of the bend to bulge, whereas the outer tension side is thinned and reduced in width. Also, as in other cold-forming operations, there is a tendency for the metal to spring back or unbend. This tendency varies directly with the modulus of elasticity of the metal. For example, lead has almost no springback, aluminum has very little, but stainless steel has considerable springback.

Machines that make relatively sharp bends or small radii in sheet metal are called *bar folders* (Figure 10.47). These tools are used to bend sheet metal for architectural shapes, furnace and air conditioning pipes, and sheet metal products. When bends in heavier sheet metal are required or more complex bends are needed, press brakes are used (Figures 10.48 to 10.53). These machines are designed to make short strokes to a predetermined depth, and they are powered either with hydraulic cylinders or in much the same way as a crank punch press.

Roll Bending This process, shown in Figures 10.54 to 10.56, is used to form curved shapes, cylinders, or rings. The metal used can range from thin sheet metal to massive bar or plate stock. Plates, bars, and structural shapes are formed by this method.

Cold-Roll Forming Cold-roll forming (Figure 10.57) consists of bending a flat strip into a complex shape by a series of rolls, each contributing to the final shape. Channels, moldings, building trim and rain gutters are formed by this method (Figure 10.58) (p. 216).

Straightening or Flattening This is a process designed to remove unwanted bends from metal. Sheet or bar stock can be straightened by passing it through a series of rolls,

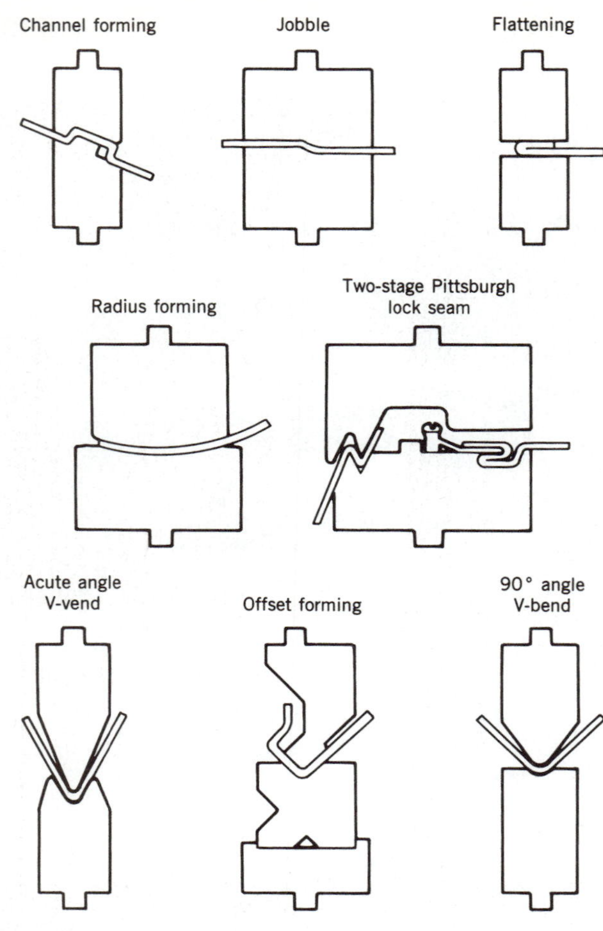

FIGURE 10.49
Typical press brake tools (Courtesy of Alcoa Inc.).

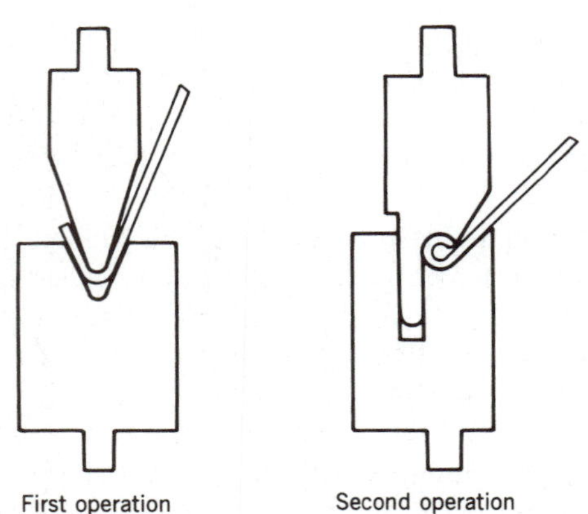

FIGURE 10.50
Press-brake forming of bead in two operations (Courtesy of Alcoa Inc.).

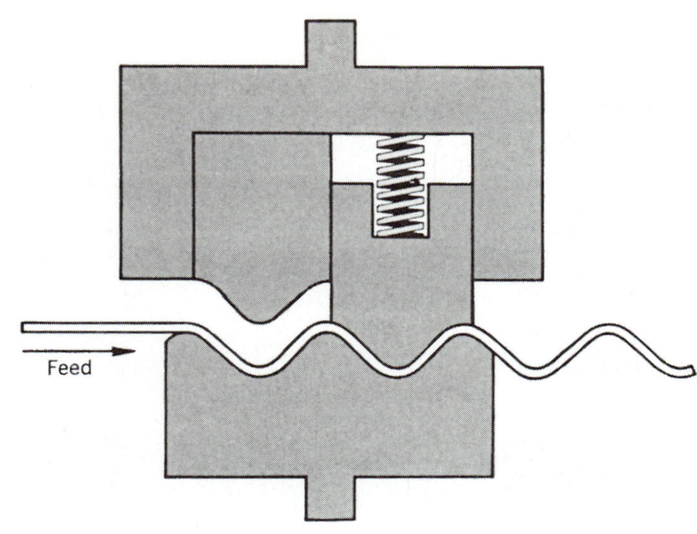

FIGURE 10.51
Press-brake forming of a corrugated sheet (Courtesy of Alcoa Inc.).

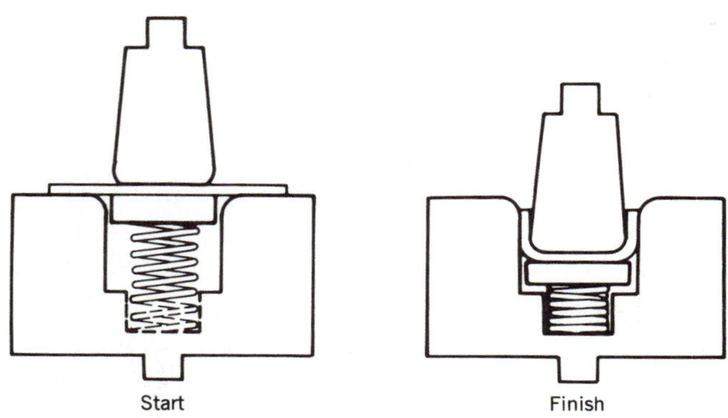

FIGURE 10.52
Brake tool with spring-loaded pressure pad in bottom die (Courtesy of Alcoa Inc.).

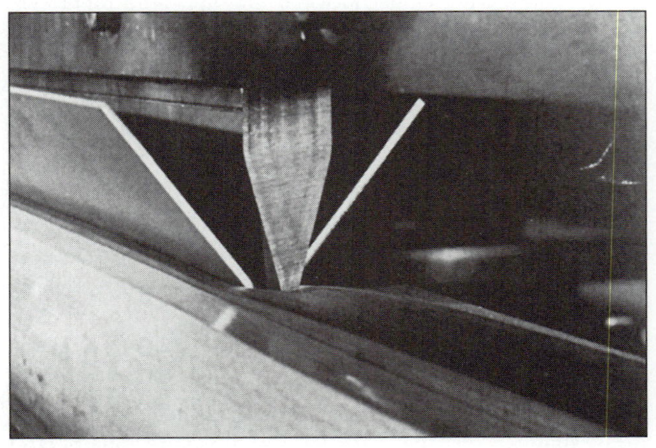

FIGURE 10.53
Brake tools used with urethane die (Courtesy of Alcoa Inc.).

FIGURE 10.54
Roll bending is used to form cylindrical parts (Courtesy of Alcoa Inc.).

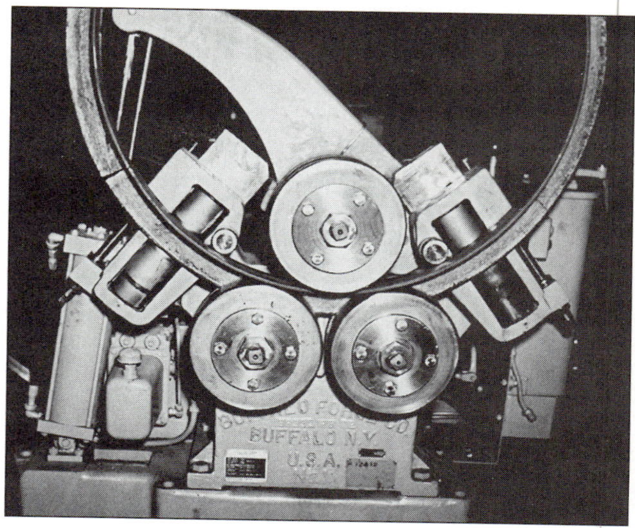

FIGURE 10.55
Vertical roll-bending machine. Heavy bars may be rolled into rings with this machine (Courtesy of Buffalo Forge Company).

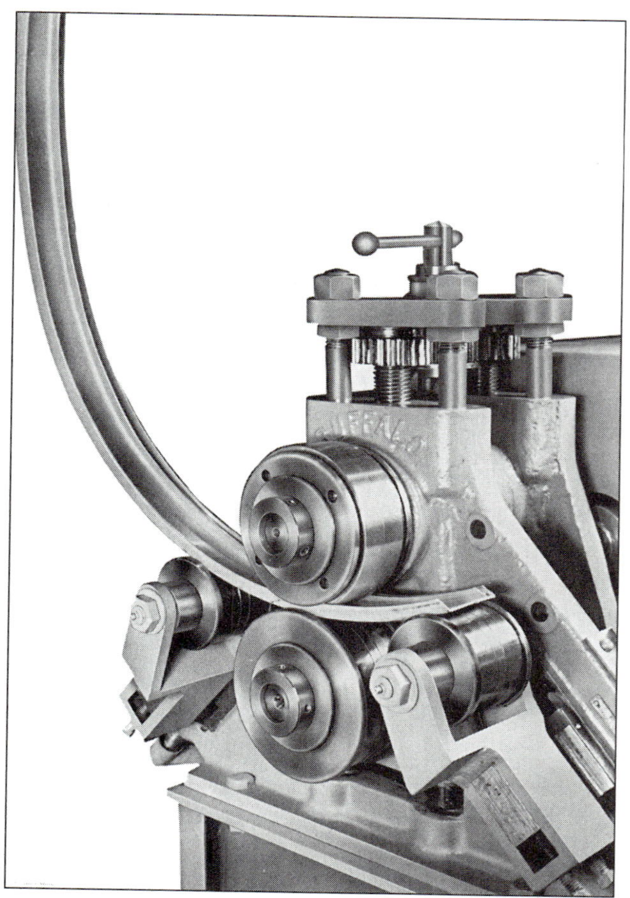

FIGURE 10.56
The pinch-type roll bender is used to form shapes, such as channels, into rings (Courtesy of Buffalo Forge Company).

FIGURE 10.57
Roll-forming machine (Courtesy of Alcoa Inc.).

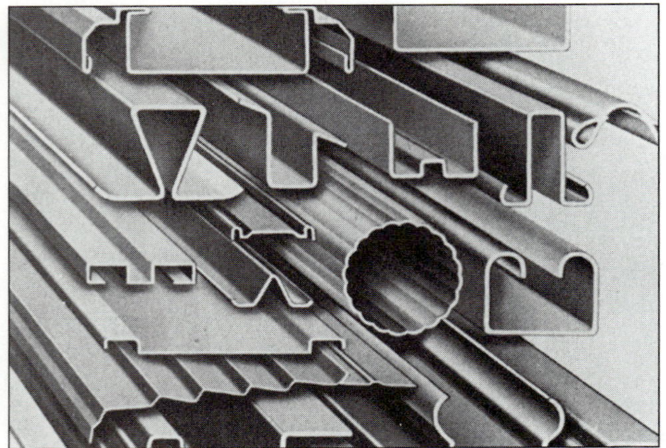

FIGURE 10.58
Typical roll-formed shapes (Courtesy of Alcoa Inc.).

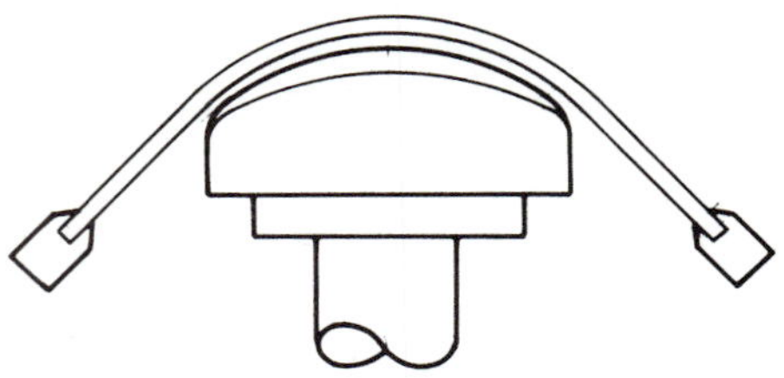

FIGURE 10.59
Stretch forming (Courtesy of Alcoa Inc.).

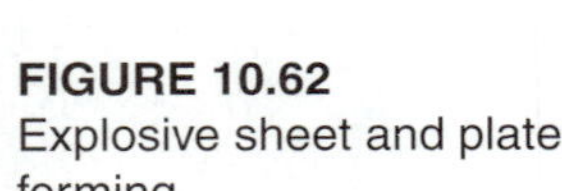

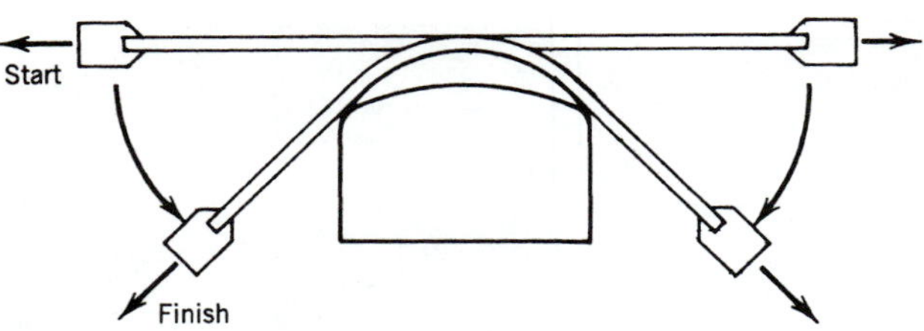

FIGURE 10.60
Stretch-wrap forming (Courtesy of Alcoa Inc.).

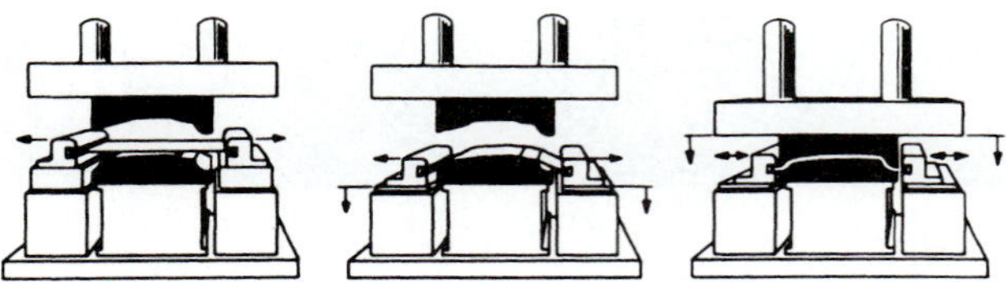

FIGURE 10.61
Stretch-draw forming (Courtesy of Alcoa Inc.).

as shown in Figure 10.4. The metal is bent back and forth to a point slightly beyond its elastic limit, thus removing previous permanent deformations. Heavy sections are often straightened in a press by applying pressure on the bend while the part is supported at two points.

Stretch Forming Light-gage sheet metal can be done by stretching it over a contoured form or die block. This is done on a mechanically or hydraulically powered machine that has gripping jaws (Figure 10.59). The wing spans for some large aircraft are formed by stretch forming.

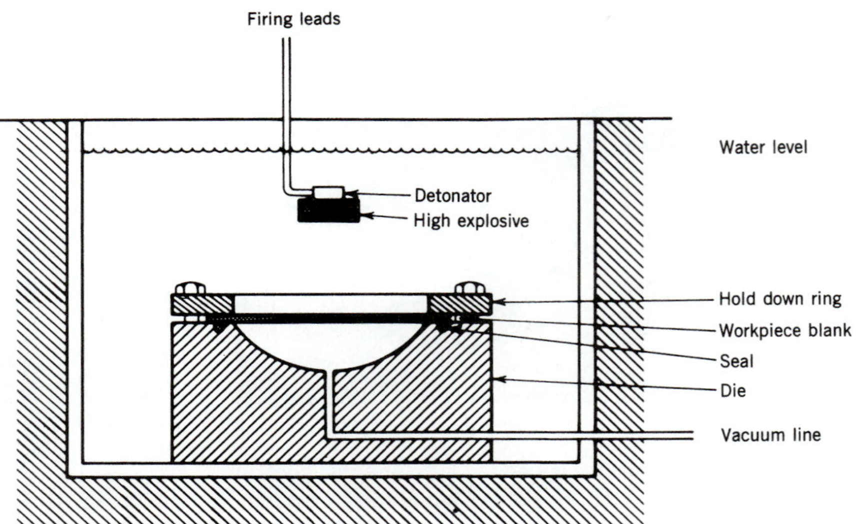

FIGURE 10.62
Explosive sheet and plate forming.

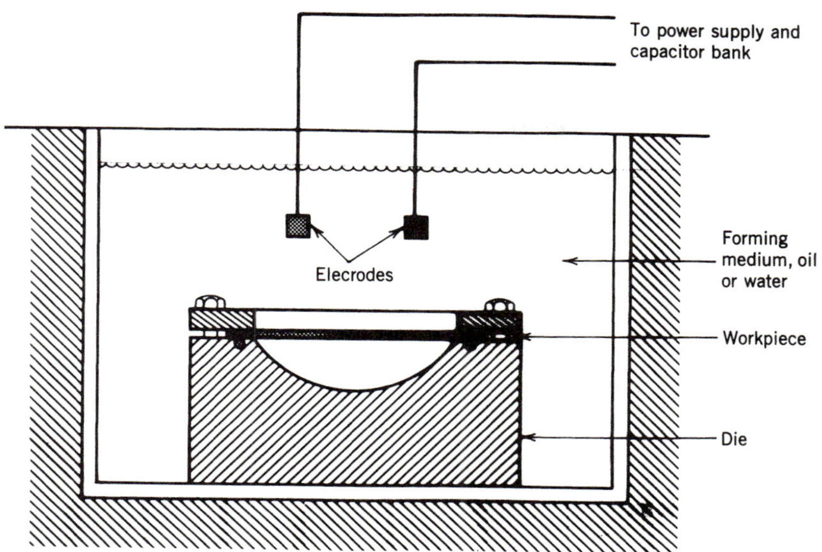

FIGURE 10.63
Components in electrohydraulic metalworking.

Stretch-Wrap Forming This process (Figure 10.60) is similar to stretch forming. The blank is first stretched beyond the yield point and then wrapped around a form block. Stretch-draw forming (Figure 10.61) is a shallow stamping process in which bottom and top dies are brought together to form a stretched sheet. The principal advantages of stretch forming are that it greatly reduces springback, virtually eliminates wrinkles, and reduces tooling cost.

High-Energy-Rate Forming Processes

High-energy-rate forming processes (HERF) involve a high rate of workpiece deformation. The most important of these techniques are chemical (explosive), electrohydraulic, electromagnetic, and high-energy mechanical forming.

Explosive Forming Explosive forming (Figure 10.62) is used for operations such as shaping metal parts, cladding, joining, and forming powdered metal parts.

Electrohydraulic Metalworking This process (Figure 10.63) is similar to explosive forming. Instead of a chemical explosion to produce shock waves, an electric arc along a bridgewire immersed in a liquid provides the high energy needed to form the metal.

With explosive forming there is virtually no limit to the size or thickness of plate that can be formed. This is a low-volume production process, but the original investment is also low. Springback is reduced. Large dome shapes and curved gore segments are produced by this means.

Coil around
tubing and insert

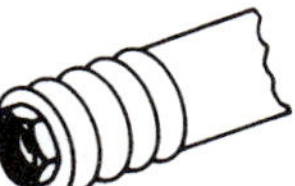

Tubing is compressed
into grooves of insert

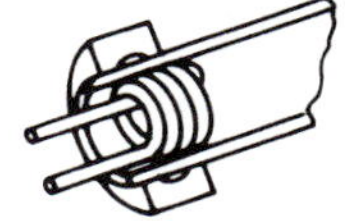

Coil inserted
inside of tubing
surrounded by die

Tubing is expanded
into die
to form beading

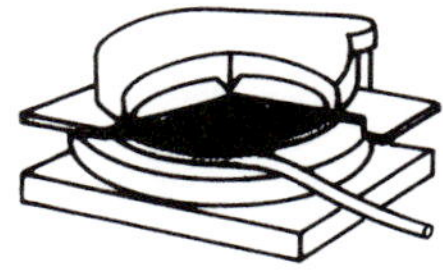

Coil placed under
flat sheet
with die on top

Flat sheet is
formed to
contours of die

FIGURE 10.64
Typical applications of magnetic pulse compression, expansion, and flat forming coils (Courtesy of Alcoa Inc.).

Magnetic Pulse Compression and Expansion This technique (Figure 10.64) is used for metals that are good electrical conductors. A magnetic field acts like a compressed gas to deform the metal. It is used to blank, bulge, compress, perforate, dimple, swage, flange, and assemble parts.

FIGURE 10.65
Products formed by metal spinning (Courtesy of Alcoa Inc.).

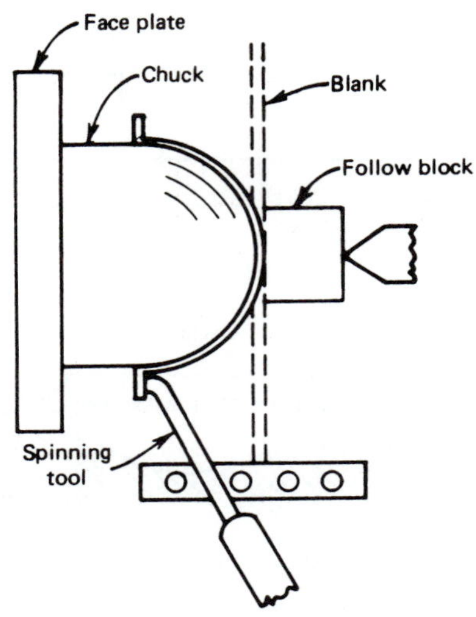

FIGURE 10.66
Diagram of metal spinning. Flat circular blanks are often formed into hollow shapes such as photographic reflectors. In a spinning lathe, a tool is forced against a rotating disk, gradually forcing the metal over the chuck to conform to its shape. Chucks and follow blocks are usually made of wood for this kind of metal spinning (Neely and Bertone, *Practical Metallurgy and Materials of Industry*, 6th ed. © 2003 Prentice Hall, Inc.).

Metal Spinning and Flow Forming

Some metals such as titanium alloys do not draw well in press forming; however, they can be formed into cylindrical, tapered, or curved shapes by metal spinning (Figure 10.65). *Metal spinning* is a process in which a disk of metal is rotated and forced against a form, called a chuck (Figure 10.66), with no intended reduction of wall thickness. In contrast, in *flow forming* (Figure 10.67) the blank is intentionally reduced in thickness while being lengthened. Objects such as seamless tubes and oxygen tanks and many industrial products and consumer items can be cold formed with this method. Shear spinning (Figure 10.68) is similar to flow forming in that the cross section of the blank is reduced. In shear spinning the blank is usually a disk instead of a tube.

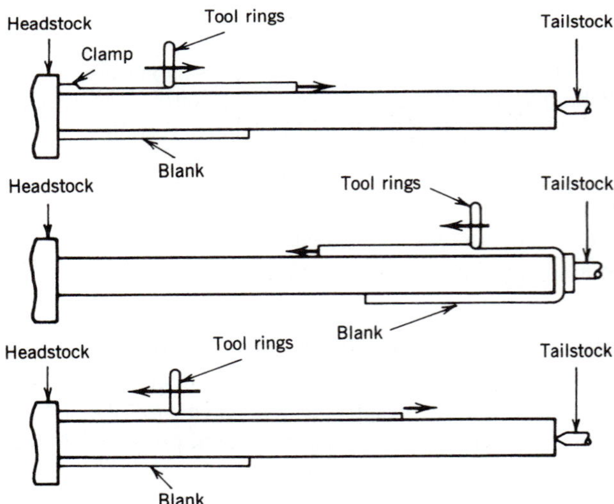

FIGURE 10.67
Three methods for using the flow forming process to spin tubular parts (Courtesy of Alcoa Inc.).

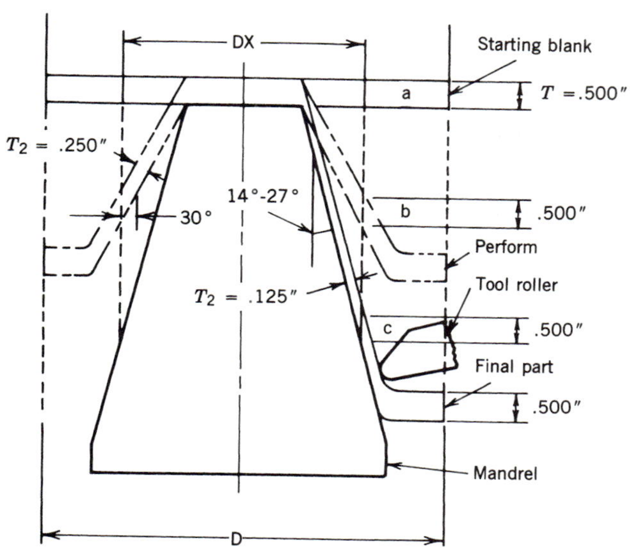

FIGURE 10.68
The principle of the shear spinning process (Courtesy of Alcoa Inc.).

FIGURE 10.69
In the electric resistance welding process, skelp coming through the forming rolls is welded automatically (Courtesy of Bethlehem Steel Corporation).

Shearing Operations

Shearing is a sheet metal cutting operation that makes no chips. It consists of two straight- or square-edged hardened blades, the bottom one stationary and the top one fastened to a ram. Other forms of metalworking that also shear metal are blanking, trimming, piercing, notching, and shaving. These shearing operations differ from simple shearing in that the cutting edges are curved, whereas shear blades are always straight.

Slitting A shearing process in which rolls of sheet metal are cut into narrower widths is termed **slitting.** The rotary shears of the larger-diameter slitting blades on one cylindrical roll fit into grooves on the mating roll.

Forming Pipe and Tubing

Pipe and tubing are formed with rolls in a manner similar to that of cold-roll forming. Smaller sizes of square, rectangular, and round tubing are rolled to shape and the edges are electric-resistance welded together (Figure 10.69). This process makes use of the natural resistance of the metal when an electric current is applied close to the edges to be joined to generate sufficient heat for welding. Larger-diameter pipes, such as those used for gas and oil pipelines are made in special hydraulic presses (Figures 10.70 to 10.72). A final O-forming

FIGURE 10.70
Pipe-forming press-crimping tool ("Presses for All Applications of Metalforming" Schirmer-Plate-Siempelkamp, Hydraulische Pressen GmbH, Krefeld, Germany).

FIGURE 10.71
Second stage in large pipe forming—U-canning transport ("Presses for All Applications of Metalforming" Schirmer-Plate-Siempelkamp, Hydraulische Pressen GmbH, Krefeld, Germany).

press makes the cylindrical shape (Figure 10.73). The edges are joined by arc welding processes (Figure 10.74) and the pipe is tested hydrostatically by capping the ends and pumping in water under pressure (Figure 10.75). For some uses the pipe is galvanized or covered with a corrosion resistant coating.

Rubber Pad Presses

Certain products that do not require deep draws can be produced on rubber pad presses (Figure 10.76). Forming is done with an elastic pad that presses sheet metal against forms made of metal or of dense laminated wood (Figure 10.77). The elastic pad is usually made of poly-

FIGURE 10.73
O-forming press gives the pipe its final shape ("Presses for All Applications of Metalforming" Schirmer-Plate-Siempelkamp, Hydraulische Pressen GmbH, Krefeld, Germany).

FIGURE 10.74
The final step in making a large-diameter steel pipe is welding the curved plates together by the submerged arc process (Courtesy of Bethlehem Steel Corporation).

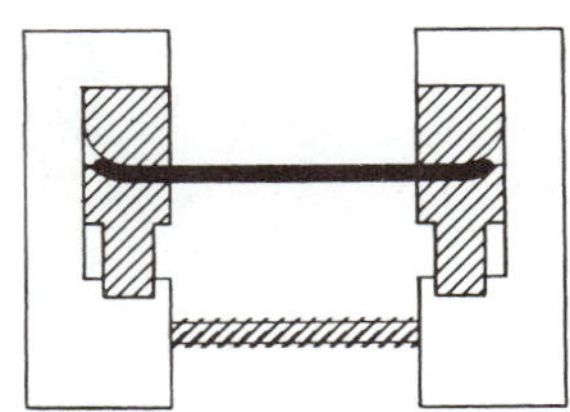
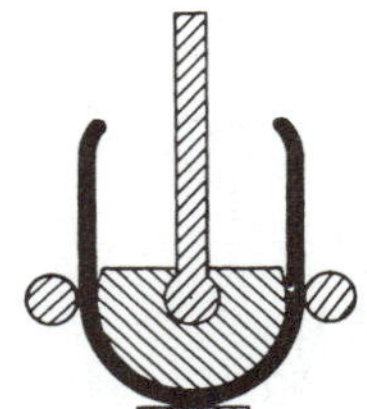
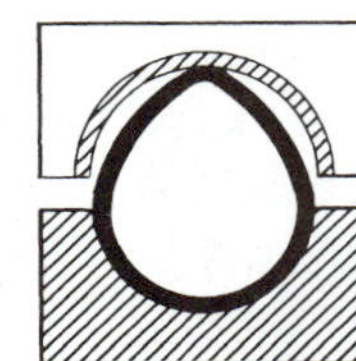
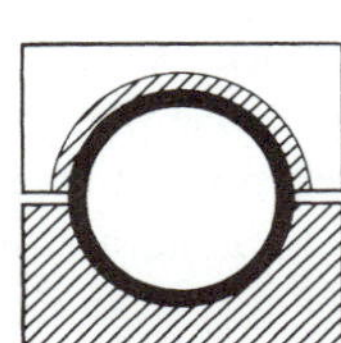

FIGURE 10.72
Steps in formation of large steel pipe ("Presses for All Applications of Metalforming" Schirmer-Plate-Siempelkamp, Hydraulische Pressen GmbH, Krefeld, Germany).

FIGURE 10.75
Hydrostatic pipe tester ("Presses for All Applications of Metalforming" Schirmer-Plate-Siempelkamp, Hydraulische Pressen GmbH, Krefeld, Germany).

FIGURE 10.76
Rubber pad press showing forming tools on the press table ("Rubber Pad Presses," Schirmer-Plate-Siempelkamp, Hydraulische Pressen GmbH, Krefeld, Germany).

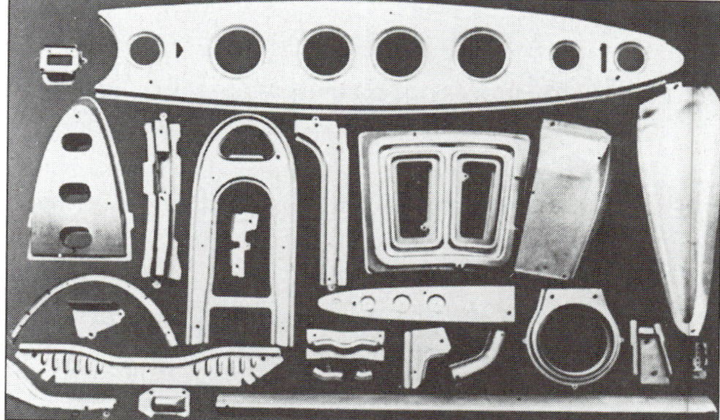

FIGURE 10.77
A large number of different components can be made simultaneously during one press cycle with rubber pad presses ("Rubber Pad Presses," Schirmer-Plate-Siempelkamp, Hydraulische Pressen GmbH, Krefeld, Germany).

urethane or similar material. The forming principle is shown in Figure 10.78. The major advantages of this method of sheet metal forming are that manufacturing costs for either short or long runs are low, dies are inexpensive, and large objects can be made, up to 5- by 10-ft plan area. This process is particularly useful for aircraft component manufacturers. Materials such as titanium and stainless steel can also be formed at slightly elevated temperatures by this method.

Review Questions

1. What is the property of metals that makes possible the cold deformation of metals?
2. When considerable deformation is required in a metal, what can be done to avoid brittleness or rupture?
3. Are closer dimensional tolerances and better finish advantages of hot forming, or of cold forming?
4. In which type, cold forming or hot forming, is more massive machinery required to form a given size part?
5. What is springback?
6. What must be done to hot-rolled steel in the mill before it can be cold finished?
7. Cold-rolled steel strip is the material used for many products such as auto bodies and kitchen appliances. How is it made?
8. Aluminum foil is used every day in our homes. How is it made?
9. What is the difference between blanking and punching?
10. How can a sheet metal blank be formed into a cup shape?

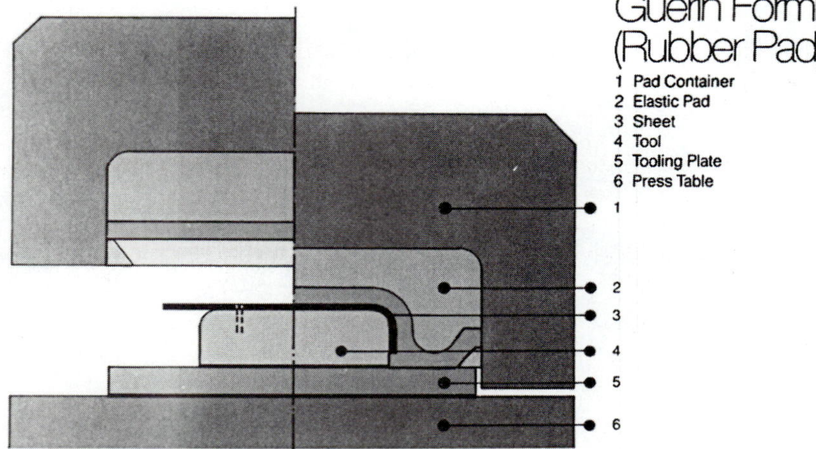

FIGURE 10.78
Principle of rubber pad forming process ("Rubber Pad Presses," Schirmer-Plate-Siempelkamp, Hydraulische Pressen GmbH, Krefeld, Germany).

11. How is the process of making wire similar to that of cold drawing bars?
12. Cold-forming upsetting machines can make parts from wire stock at a high rate of production, as much as 36,000 parts per hour. What kind of parts are usually made on these machines?
13. Larger, more complex products than those of Question 12 are made from bar stock on cold-forming machines that do combined operations in a sequence of operations. Name two products made on these machines.
14. How are threads, worms, and gears cold formed?
15. Threads cannot be rolled on gray cast iron, die cast metals, and 301 stainless steel. Why not?
16. In what ways are rolled threads superior to cut threads?
17. What are residual stresses, how are they caused, and what is a possible solution to them?
18. When press brakes or bar folders are used, what is the process called: forming, shaping, bending, or drawing?
19. What is the difference between metal spinning and flow forming?
20. Large steel pipes are electric arc welded after being formed. How are small round, square, or rectangular tubes joined at the seams after being formed?
21. If thin aluminum panels, 4 by 8 ft, needed a shallow, ribbed form to give them rigidity and 300 were required, which would be more economical to use, standard drawing dies, or a rubber pad press?

Case Problems

Case 1: Cold Working versus Hot Forming

Automotive wrist pins were once made only on automatic screw machines, an operation that turned 40 percent of the starting stock into scrap. Wrist pins are usually less than 1 in. in diameter, about 3 or 4 in. long, and hollow. These have been formed of steel for the past 30 years. They must have a good finish and accurate dimensions. Which method, hot forming or cold forming, do you think the manufacturers chose for making this part?

Case 2: Stainless Steel Nuts

A manufacturer of fasteners uses a four-die nut former to produce blanks. It received an order for 80,000 stainless steel Type 305 hex nuts. A trial run was made with the stainless steel bar stock, and the piercer punch began to break down after only a few hundred nuts were produced. What change in the operation can be made so the manufacture of these nuts can be made profitable with no breakdowns?

Case 3 (Part 1): Punching Pressure

The rule for finding the approximate pressure for punching circular holes in sheet steel is

$$P = D \times t \times 80$$

where

D = diameter of punched hole

t = thickness of sheet steel

P = pressure (force in tons)

You are given a die set having four punches: two are 1¼ in. in diameter and two are ¾ in. in diameter. You are asked to determine the press tonnage required to punch holes in ½-in. sheet steel. What size press will be required for this operation?

Case 3 (Part 2): Punching Pressure

When a punched part is not circular, the same formula as in Part 1 may be used, but instead of the diameter of the hole, one-third the length of the perimeter or outline of the hole is used. You are asked to determine the pressure required to punch a rectangular hole in ¼-in. steel plate that is 3½ in. long and 2½ in. wide. What size press will be needed?

Case 4: Force Required to Shear Plate

Squaring shears have an upper and a lower blade or knife (Figure 10.79). The lower one is in a level position and is stationary, whereas the upper one moves up and down and has rake to allow for a gradual cutting action. Most plate shears have ½ in./ft rake, and scrap shears have up to 2 in./ft rake because distortion of the cut piece is not important and larger rakes require less force. One formula to determine the force required to shear mild steel plate is

$$\text{maximum } KP = \frac{174{,}000 \times T^{1.6}}{R^{0.79}}$$

where

KP = knife pressure, lb

R = rake, in./ft

T = plate thickness, in.

Using this formula, calculate the pressure required to shear 1¼-in. steel plate with ½ in./ft rake, and with 2 in./ft rake. Also determine KP for ½-in. plate with ½ in./ft rake.

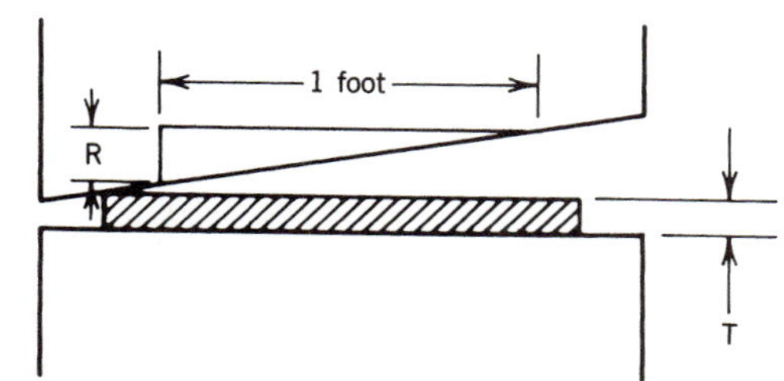

FIGURE 10.79
See Case 4.

CHAPTER 11

Powder Metallurgy

Objectives

This chapter will enable you to:

1. Explain the methods by which metal powders are produced.
2. Understand and explain the processes involved in simple die compaction.
3. Describe the metallurgical principles involved in bonding of powders during the sintering process.
4. State the advantages of several advanced powder metal processes.
5. Determine which parts could and should be made using powder metal processes.

Key Words

atomization	diffusion
cold isostatic pressing (CIP)	hot isostatic pressing (HIP)
compaction	sintering
densification	

Powder metallurgy (P/M) is one of the four major methods of shaping metals (machining, hot and cold plastic deformation, casting, and P/M). The P/M process is essentially the compression of finely divided metal powder into a briquette of the desired shape that is then heated but not melted to form a metallurgical bond between the particles.

Although the P/M manufacturing method dates to the nineteenth century, it was not until recent decades that this field gained wide acceptance and use, and technological advances in P/M continue to grow very rapidly. Products that are difficult if not impossible to produce by other means are being manufactured with P/M at high production rates at very competitive cost.

Parts manufactured by the P/M process have found a widespread use in a variety of applications. P/M products are used in the transportation industry (automobiles and trucks), in farm and garden equipment, and in household appliances. Many new applications will be found in the future for this unique method of forming metals.

This chapter discusses the many uses and advantages of P/M parts as well as the manufacturing processes commonly employed to produce them.

HOW P/M PARTS ARE MADE

The basic conventional process of making P/M parts consists of two basic steps—compacting (molding) and sintering (Figure 11.1). In addition to these two basic manufacturing steps involved in the P/M process, secondary operations are commonly performed to impart final desired properties to the P/M product (e.g., coining, sizing, repressing).

In the first step (**compacting**), loose powder (or a blend of different powders) is placed in a die and is then compacted between punches. This operation is commonly performed at room temperature. The compacted part (Figure 11.2), called a **briquette** or green compact, is now a solid shape; however, green compact can easily be broken or chipped and requires careful handling. In the second step (**sintering**) the briquette is heated in an appropriate atmosphere to a temperature high enough to cause the powder particles to bond together by solid-state diffusion (Figure 11.3) and to homogenize any alloy constituents in the powder. Melting does not normally occur. The P/M part is now ready for use unless other finishing operations are needed.

Secondary operations may include sizing, machining, heat treating, tumble finishing, plating, or impregnating with oil, plastics, or liquids. Secondary operations can significantly increase the cost of the finished part, therefore designers should limit the use of secondary operations and, if possible, complete the product in the first two basic steps. However, the sintering process tends to deform and shrink the shaped briquette slightly, so some parts (e.g., precision gears) require a finishing operation to obtain the desired tolerances.

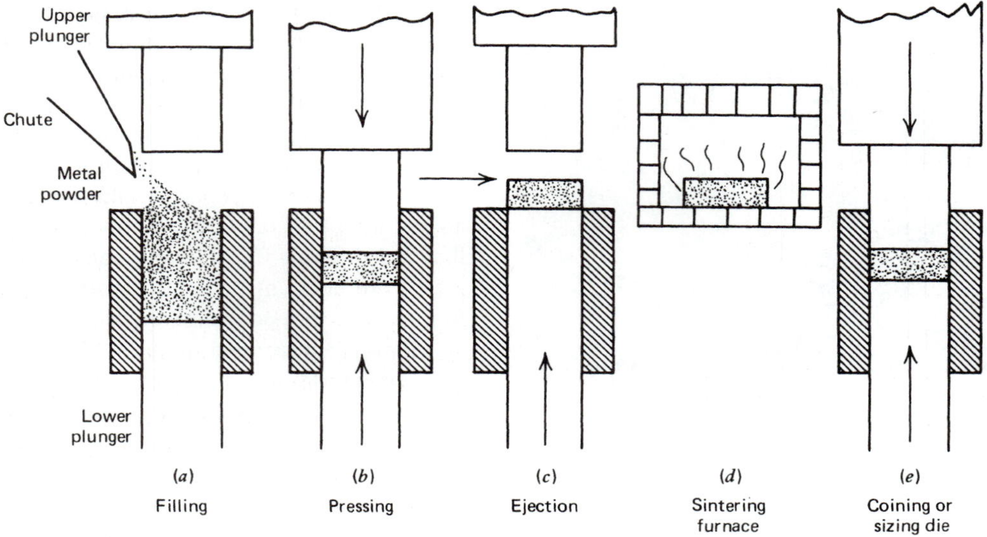

FIGURE 11.1
Sequence of operations for powder metallurgy. The depth of the die cavity and the length of the plunger stroke are determined according to the density required. *(a)* A measured amount of metal powder is placed in the die cavity. *(b)* Pressure is applied. *(c)* The briquette is ejected from the die cavity. *(d)* The parts are sintered at a specified temperature for a given length of time. The parts are now ready for use. *(e)* If more precision is needed, they can be sized in a coining die (Neely and Bertone, *Practical Metallurgy and Materials of Industry,* 6th ed., © 2003 Prentice Hall, Inc.).

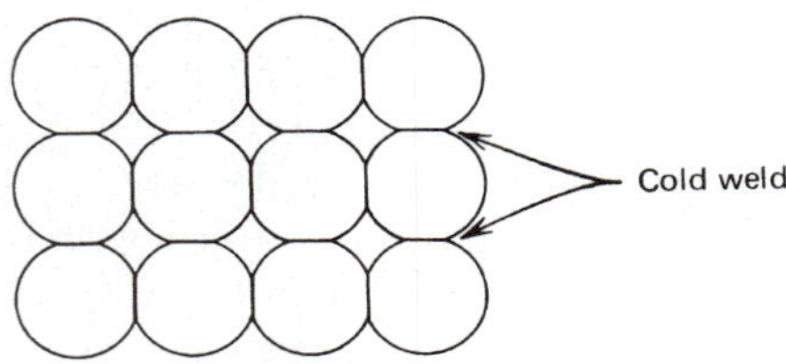

FIGURE 11.2
The compressed particles in the briquette are cold welded together at this stage with very weak bonding (Neely and Bertone, *Practical Metallurgy and Materials of Industry,* 6th ed., © 2003 Prentice Hall, Inc.).

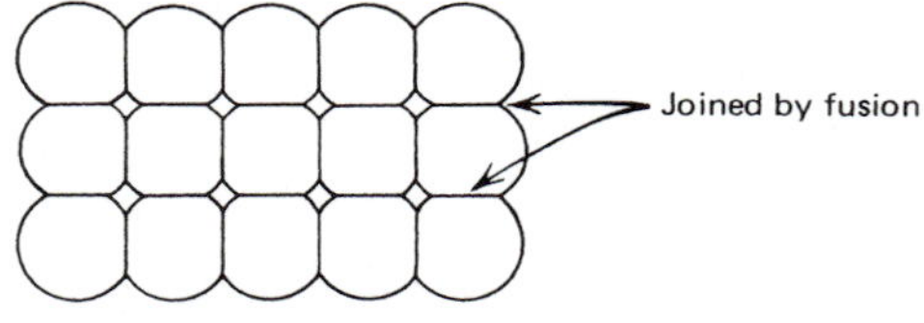

FIGURE 11.3
After sintering, the particles in the briquette become fused together (Neely and Bertone, *Practical Metallurgy and Materials of Industry,* 6th ed., © 2003 Prentice Hall, Inc.).

METAL POWDERS

A number of different metals and their alloys are used in P/M (e.g., iron, alloy steel, stainless steel, copper, tin, lead). The three most important methods of producing metal powders are (1) atomization, (2) chemical methods, and (3) electrolytic processes.

Atomization is a process in which a stream of molten metal is transformed into a spray of droplets that solidify into powder. Molten metal spray can be produced in several ways. The most common method is to use a stream of high-velocity gas to atomize the molten metal. This method has several variations. In one method, the gas stream is expanded through a venturi tube, which siphons the molten metal from the crucible located below the tube. The gas breaks the stream of molten metal into small droplets that then solidify as they are carried by the gas stream. In another variation, the crucible with bottom gate is located above the gas tubes. The metal flows under the influence of gravity and passes through the gas stream, which breaks the molten metal stream. The solidified metal droplets are then collected in a collection chamber. In addition to gas, water and synthetic oils can also be used in the atomization process.

Several chemical methods can be used to make metal powders, including reduction and precipitation. Chemical

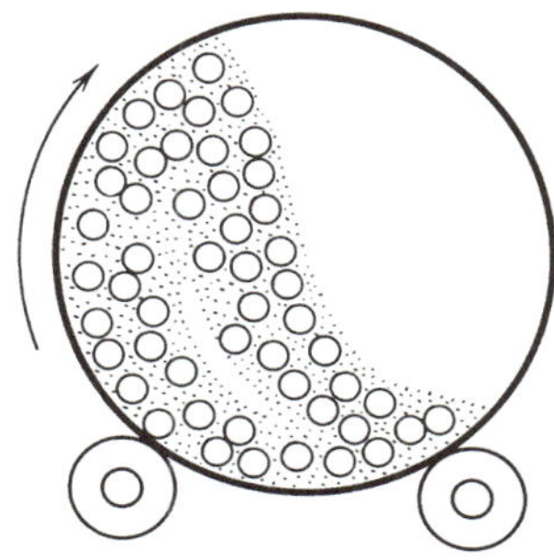

FIGURE 11.4
Action of the ball mill is shown as a continuous grinding as the drum rotates (Neely and Bertone, *Practical Metallurgy and Materials of Industry,* 6th ed., © 2003 Prentice Hall, Inc.).

reduction is a process in which metal powders are formed by chemical reaction between metal oxides and reducing agents (e.g., hydrogen or carbon monoxide). Hydrogen or carbon monoxide reacts with oxygen in the metal oxide, producing pure metal.

The electrolytic process that is utilized to precipitate metal powders begins in the electrolytic cell where the source of desired metal is the anode. As the anode is dissolved the desired metal is deposited on the cathode. After this step is complete, the metal deposit is removed from the cathode and is washed and dried.

In each process, the powders may be ground further to a desired fineness, usually in a ball mill (Figure 11.4). Metal powders are screened, and larger particles are returned for further crushing or grinding. The powders are classified according to particle size and shape in addition to other considerations such as chemical composition, impurity, density, and metallurgical condition of the grains. Particle diameters range from about 0.002 in. to less than 0.0001 in. Test sieves are used to determine particle size. This method of testing has been standardized throughout the industry.

Powders are often blended by tumbling or mixing. Lubricants (e.g., graphite) are added to improve flowability of material during feeding and pressing cycles. Deflocculants are also added to inhibit clumping and to improve powder flow during feeding.

POWDER COMPACTION

Compacting or pressing gives powder products their shape. Pressing and sintering techniques can be separated into two types: conventional and alternative. The method most commonly used today is the conventional approach, which consists of the pressing operation first, followed by sintering. The alternative techniques can be classified into (1) alternative compaction methods, (2) combined compaction and sintering, and (3) alternative sintering methods.

In the conventional compacting process, the powder is pressed unidirectionally in a single- or a double-acting press. Unlike liquids, which flow in all directions under pressure, powders tend to flow mainly in the direction of the applied pressure (Figures 11.5 and 11.6). Engineering properties such as tensile and compressive strength depend to a great extent on the density of compacted material. Hot pressing, in which the powder is pressed in the die at a high temperature, produces a density approaching that of rolled metal. Die compaction can be done either hot or cold.

Mechanical presses (Figure 11.7) are favored when the required load is not too high. Eccentric (crank) presses rarely exceed 30 tons (60 tons would be considered quite large). Toggle-type presses may reach a 500-ton force.

FIGURE 11.5
Custom-blended powder is precision metered into a die cavity and compacted under high pressure (Pennsylvania Pressed Metals Inc.).

FIGURE 11.6
Green compacts of gears are produced and stacked ready to be taken to sintering furnaces (Pennsylvania Pressed Metals Inc.).

FIGURE 11.7
Automatic presses have controlled molding speeds up to 50,000 parts per hour (Pennsylvania Pressed Metals Inc.).

FIGURE 11.8
Large hydraulic press used for briquetting metal powders or sponge for experimental purposes (Bureau of Mines).

The advantage of mechanical presses is that they are high-production machines. Hydraulic presses (Figure 11.8) can exert forces in excess of 5,000 tons, and they have the advantage of a long stroke and easily adjustable stroke length; however, these machines provide low productivity compared with mechanical presses.

Compaction of powders with various presses has the advantage of speed, simplicity, economy, and reproducibility. Such compaction produces a strong, dimensionally accurate, and relatively inexpensive product; however, it does have limitations. The aspect ratio (length to diameter) must be relatively small. Parts with a large aspect ratio will have uneven densities, being denser nearest the punches. These parts may have nonuniform and uncertain properties and should not be made by die compaction. Grooves or undercuts or parts with thin sections cannot be made by simple die compaction, so not every part is a good candidate for powder metallurgy; however, some of these limitations are overcome by alternative forming techniques such as split-die techniques to provide undercuts, isostatic pressing, and densification methods.

Advanced Processes

Because conventional presses can compact powder along only one axis, such presses cannot make some shapes, including hollow hemispheres, long parts, and internal threads; however, one method allows pressure to be applied from all directions: isostatic pressing. In **cold isostatic pressing (CIP),** the powder is loaded into molds made of rubber or other elastomeric material and subjected to high pressures at room temperature (Figure 11.9). Pressure is transmitted to the flexible container by

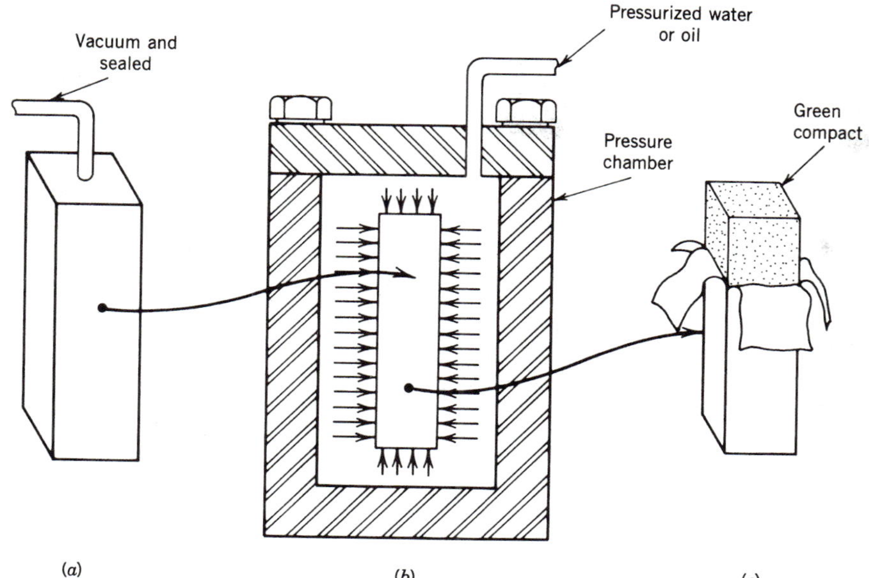

FIGURE 11.9
Isostatic pressing. *(a)* Prepared powder is placed inside a flexible container or mold. A vacuum is drawn in the mold and it is then sealed. *(b)* The powder and mold are then placed in a pressure chamber into which water or oil is pumped under pressures of 15,000 psi or more. *(c)* The green compact is removed from the pressure chamber and the container is stripped off.

water or oil. The compacted parts are removed and sintered, followed by secondary operations if needed. With **hot isostatic pressing (HIP),** an inert gas such as argon or helium is used in a pressure chamber to provide the squeeze. This gas is reclaimed between each batch of pressings. Hot isostatic pressing provides more density and achieves a finer microstructure than the cold process. Powders are often preformed to an oversize shape prior to being placed in the isostatic chamber. Heat is applied to the preform by induction for a short time while the gas pressure compacts the preform. Temperatures may be as high as 1600 to 2000°F (871 to 1093°C) with pressures in excess of 15,000 psi. Isostatic pressing is useful only for certain special applications.

CIP is a comparatively slow process, and HIP is even slower. Parts made by either CIP or HIP are not limited by the shape constraints of rigid tooling.

Powder Forging

Fully dense P/M parts equaling or surpassing the mechanical properties of wrought products are being produced in commercial quantities by powder forging (P/F). The green compact or preform is made in a conventional press and then sintered. These operations are then followed by a restrike (forge) that brings the part to the final density. Mechanical properties may sometimes exceed those of wrought metals because a more uniform composition is achieved in P/M processes. Fatigue strength and impact strength are particularly high in powder forgings compared with conventional P/M parts. P/M bearing races have been shown to outlast wrought steel races by a factor of 5 to 1.

Metal Powder Injection Molding

A P/M technology that borrows a plastic injection molding process shows great promise for production of small precision parts. In fact, some variations of this process can use plastic molding machinery. In order to inject powders into molds, the particle size must be much finer than that used for conventional P/M processes. This "dust" is combined with a thermoplastic binder. The molding step is performed at injecting pressures of about 900 psi and about 325°F (163°C). The result is a green compact that is sintered in the conventional fashion after the thermoplastic binder has been removed in an oven at about 400°F (204°C). Thin walls, high densities, unsymmetrical shapes, and accurate dimensions are possible with this method.

Metal Powder-to-Strip Technology

Direct rolling of metal strip from a powder slurry (powder-to-strip process) is a process in which thin strips are directly produced, avoiding numerous hot or cold rolling operations. In this process, an appropriate powder mix is blended with water and a cellulose binder to form a fine slurry. The slurry is deposited on a moving band as a continuous film (Figure 11.10). After drying, the moving strip is compacted between rolls and then sintered, first to remove the binder and then to bind the particles. It is rolled a second time and resintered to remove porosity. As in all these advanced P/M processes, metals or alloys that cannot be formed in any other way can be produced with powdered metals. Bimetal alloys can be produced in a strip, and high-strength titanium strip is being produced for the aircraft industry.

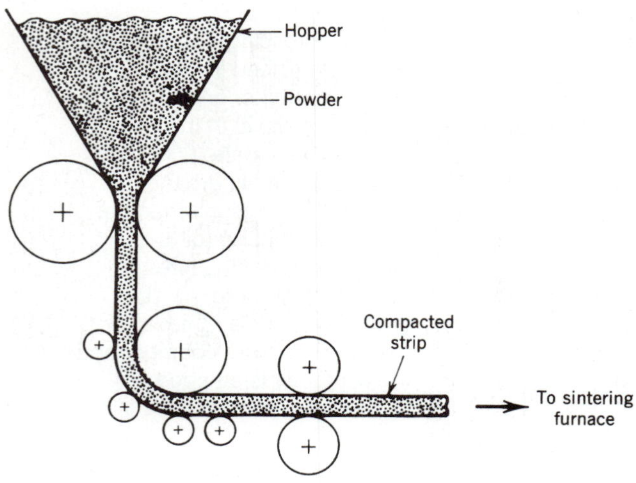

FIGURE 11.10
Powder rolling can produce a compacted strip of difficult-to-work, refractory, or reactive metals.

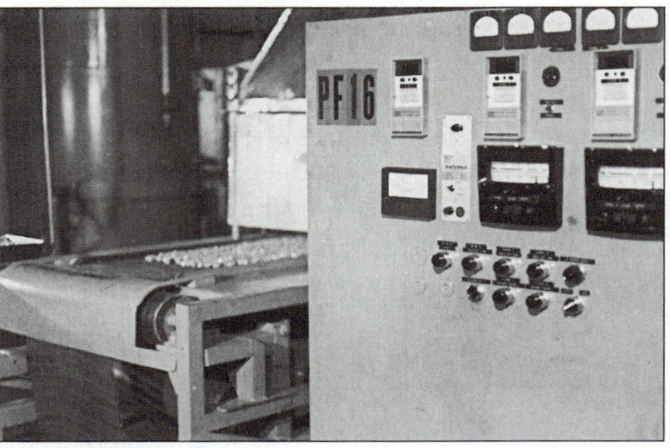

FIGURE 11.11
Molded briquettes are sintered in controlled atmosphere furnaces to bond metal powder particles at temperatures up to 2200°F (1204°C) (Pennsylvania Pressed Metals Inc.).

Powder Extrusion

Metal powders can be hot extruded with or without presintering. Metal powders are placed inside a can that is then evacuated and sealed. The unit is then heated and extruded. Metal billets and tubing are made from powder by this process.

SINTERING

In *solid-phase sintering* the green compact part must be heated to 60 to 80 percent of the melting point of the constituent with the lowest melting point. This usually requires from 30 minutes to 2 hours in a sintering furnace to produce metallurgical bonds. The following are the important changes that take place during the process of solid phase sintering:

1. **Diffusion** This takes place on or near the surface of the particles as the temperature rises. For example, any carbon present in the voids between the particles will diffuse (penetrate) into the metal particles.
2. **Densification** Particle contact areas increase considerably. Voids decrease in size, therefore lowering porosity. As a result, there is an overall decrease in the size of the part during the sintering process. The green compact must be made larger to allow for this shrinkage.
3. **Recrystallization and Grain Growth** Because sintering is usually carried out well above the recrystallization temperature of metals, grain growth can occur within and in between particles. Methods of inhibiting excessive grain growth are often used, since large grains tend to weaken metals.

Alternatively, *liquid-phase sintering* is carried out above the melting point of one of the constituents. When one of the blended metal powders has a melting point below the sintering temperature, a liquid phase of that metal fills the voids between the particles that do not melt. Infiltration is a process in which the pores or voids of a sintered or unsintered compact are filled with a metal or alloy of a lower melting point. For example, a steel–copper compact is heated at a temperature lower than the melting point of the steel compact and higher than the melting temperature of copper. The molten copper is drawn into the pores of the compact and fills the voids. This process increases densities and tensile strengths considerably.

High density and low porosity are not always desirable. Porous filters or prelubricated bearings are produced by loose sintering or by combining the powder with a combustible or volatile substance, which is later removed by sintering after the green compact is made. Very large parts are sometimes formed with very low pressures or none at all, and then loose sintered (called *pressureless sintering*) followed by a cold forging operation.

Sintering furnaces on production lines are typically of the continuous type (Figure 11.11). Furnace atmospheres usually consist of a hydrocarbon gas; however, with certain metals or alloys other gases may be used. Some manufacturers use a nitrogen gas atmosphere for both ferrous and nonferrous metals.

SECONDARY OPERATIONS

For many products the slight variations in dimension that occur during sintering are acceptable; however, where close dimensional tolerances must be maintained the product must be finished after sintering. Common secondary operations are densification, sizing, impregnation, infiltration, heat treatment, surface treatment and machining. Some of these secondary operations are described next.

Sizing is similar to forging: the part is forced into a sizing die of the correct shape by a sizing punch. This operation is performed to improve dimensional accuracy.

Small electric motors and many other mechanisms have sintered, porous, prelubricated bearings that last the lifetime of the motor. These bearings are made by impregnation of the sintered P/M part with oil, which is stored in the pores of the metal. A thin film of oil from the impregnated bearing covers the shaft, and when the temperature rises from rotation of the shaft and friction, more lubricant is supplied from the bearing. When the shaft stops and the temperature drops, the oil is reabsorbed by capillary action, so that none is lost by dripping or leakage.

P/M articles can be finished by plating; however, because of the porosity of sintered products, precautions should be taken to avoid trapping the plating electrolyte in the pores, which could cause corrosion and ultimate failure of the part. Sintered parts that are infiltrated with lower-melting-point alloys are less likely to have this problem. These articles can be painted or coated using the same processes used for rolled or cast metals.

Machining and precision grinding are common secondary operations. Drilling and threading holes and reaming operations are often necessary for plain die compacted products because some features cannot be formed in the die. Grinding is commonly performed on cylindrical or flat surfaces, because many P/M parts are relatively hard.

Depending on the content, P/M parts can be heat treated by the same methods as parts formed by other processes. Steel parts can be hardened by quenching and then tempered. Annealing is sometimes necessary before machining can be performed.

P/M PRODUCTS AND THEIR USES

A wide array of small parts made by the P/M process can be seen in Figure 11.12. Manufactures often choose P/M over other manufacturing methods because of the following characteristics of P/M parts:

1. Superior engineered microstructures and properties with precise control
2. Consistent properties and quality
3. Controlled porosity for filters and self-lubrication
4. Very low scrap loss
5. Wide variety of shape designs
6. Unlimited choice of alloys and composites

FIGURE 11.12
These are some of the thousands of intricate shapes and designs that are produced by the powder metallurgy process (Pennsylvania Pressed Metals Inc.).

7. Low-cost, high-volume production
8. Good surface finishes
9. Close dimensional tolerances
10. Little or no machining required.

A single part can be made having different properties in different areas of the part. A part made by the P/M process can take the place of several pieces made by other methods. Internal and external gear teeth with an adjacent shoulder can easily be formed by P/M methods.

Surprisingly, P/M steels are almost as strong as wrought steels, and powder forging processes increase the tensile strength. The versatility of the P/M process allows parts to be made lighter than with other manufacturing processes. P/M processes allow parts to be made of very hard metals, such as tungsten carbide cutting tools for machine tools. The P/M process can also be used to make friction materials (in the form of bimetal powder materials that are bonded to a steel base) and aluminum-based antifriction materials containing graphite, iron, and copper. Copper–nickel powders are often formed as a layer on steel strip and then sintered. The sintered strip is impregnated with babbitt metal and formed into bearings for automobile and aircraft engines.

Some disadvantages of P/M are found in the conventional cold die compacting and sintering process. Since P/M products are somewhat porous and present a larger internal surface to any corrosive atmosphere, they have lower corrosion resistance than solid metals. P/M products also tend to have reduced plastic properties (ductility and impact strength) compared with conventionally produced metals.

FACTORS IN DESIGN OF P/M PRODUCTS

When parts are made by the conventional powder metal process, several elements of design should be observed. Thin sections and feathered edges should be avoided (Figure 11.13). Generous fillet radii should always be provided in a die (Figure 11.14), and internal holes should have rounded corners (Figure 11.15). External corners should be chamfered (Figure 11.16), and narrow deep slots should be avoided (Figure 11.17). Splines or keyseats should have rounded roots.

Secondary machining processes are common practice for precision P/M parts. Holes, tapers and drafts, countersinks, threads, knurls, and undercuts usually must be machined after the parts are sintered; however, holes in the direction of pressing are readily produced in P/M parts. Round holes are easiest to produce, but shaped holes, keys, splines, hexagonals, squares, and any blind holes can also be made.

Holes that are not in the direction of pressing generally have to be machined later. Some redesign often can eliminate secondary machining. In Figures 11.18*a* and 11.18*b*, the purpose of the undercut must be kept in mind. In this case, it is simply a relief for a sleeve whose internal edge should not contact any radius on the flange. A P/M design that will still do the job and be even stronger is shown in Figure 11.18*b*. Vertical (straight) knurls and splines can be made in the direction of pressing, but diamond or angled knurls cannot be pressed because they will interfere with part ejection.

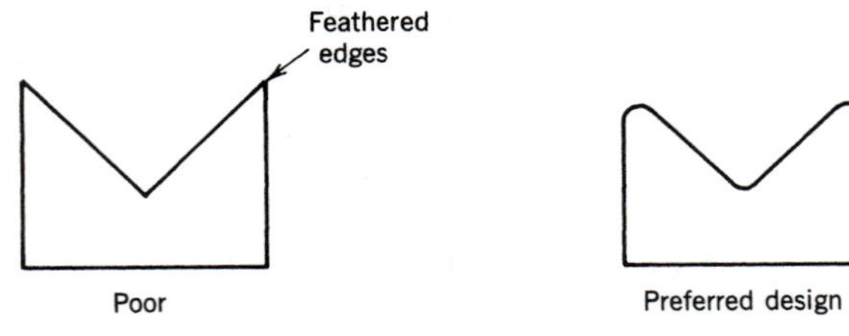

FIGURE 11.13
Metal powders do not fill dies with sharp feathered edges well. A preferred design is one with rounded corners.

FIGURE 11.14
Sharp corners on shoulders should be avoided. A fillet radius should be used.

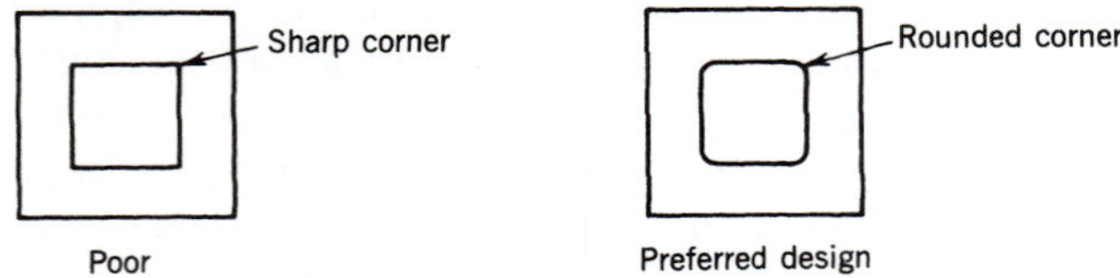

FIGURE 11.15
Avoid sharp corners. Internal shapes should have rounded corners.

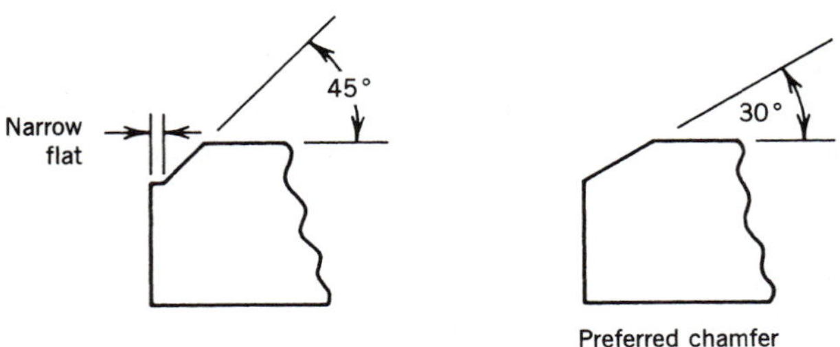

FIGURE 11.16
Both of these chamfers are used, but the 30° angle is the better form.

FIGURE 11.17
Deep slots are not easily made by the P/M process. Shallow grooves with rounded bottoms are better.

FIGURE 11.18
Undercuts as in *(a)* are difficult to make, requiring special expensive dies. The design for an undercut at *(b)* serves the same purpose and is easily done with simple pressing dies.

P/M is the ideal way to make cams and gears. Clusters of gears or gears and cams together are easily produced by die compaction. Gears with flanges or other adjacent surfaces, both internal and external, are very difficult to make by other processes such as machining or forming, but are easily produced by P/M processes. Helical gears with angles up to 45° can theoretically be made, and gears with helix angles up to 30° are manufactured in quantity. Hubs for gears and sprockets should be located as far as possible from the root diameters of the gear; that is, the hub should be as small as possible while retaining sufficient strength.

Probably the greatest advantage in adopting the P/M process over other methods of manufacture is that it allows for redesign that makes use of the great versatility of the P/M process. Many small mechanisms such as those found in pneumatic drills, electronic printers and sequencers, door locks, firearms, and sewing machines have a number of small parts that can often be combined into one piece by redesigning for powder metallurgy. Savings in production can be realized and the mechanism can be simplified as well. P/M is not suitable for every metal product, but its use should always be considered when designing a part to be manufactured.

Review Questions

1. In simple die compacting, what are the three steps in producing a precision finished part?
2. Name the three most important methods of producing metal powders.
3. A heat-resistant alloy part that has a melting point higher than that of steel is to be used to make 20,000 small symmetrically shaped parts. Both P/M and die casting processes have high production rates, and making the dies would cost about the same for each. Which method would you choose? Why?
4. Why would a bushing or sleeve of 1-in. diameter and 4-in. length have uncertain mechanical properties and probably fail in use if it were produced by simple die compaction?
5. Name two methods of increasing the density of P/M products.
6. Simple die compaction presses metal powders in only one axis, making undercuts, threads, and cross holes impossible. By what two methods of compaction can these operations be performed?
7. Name three processes of metalworking other than die compacting of powders that have been adopted for P/M.
8. In liquid-phase sintering, does all the metal powder reach the melting point? Explain.
9. How does infiltration increase density?
10. What method is used to make metal filtration devices and prelubricated porous bearings?
11. Is it possible to harden and temper a P/M part to make a wear-resistant mechanical part?
12. A manufacturer must tool up to produce 20,000 metal handwheels with a knurl on the outer edge of the wheel. The parts will be produced with P/M simple die compaction. The type of knurl is not specified. What kind of knurl, diamond or straight, should be used? Why?

Case Problem

Case: Improving a Good Product

A one-piece drive hub and flange was produced successfully by the P/M process from iron powder; however, under some shock load conditions, the hub failed. It was found to have considerable porosity that contributed to its somewhat low impact strength. How can this product be made more dense with increased impact strength without changing the process from plain die compaction and without changing present die size?

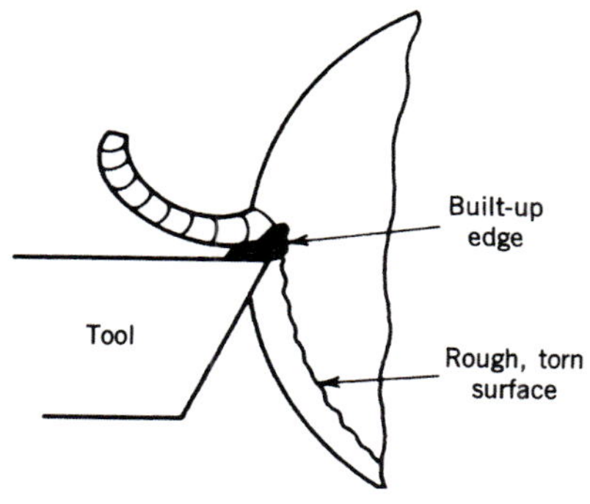

CHAPTER 12

Principles of Machining Processes

Objectives

This chapter will enable you to:

1. Understand the principles of material removal and the basic mechanisms of machining.
2. Explain the basic machining motions and the purpose of speeds and feeds.
3. Understand the effect of heat and pressure on tool wear and their influence upon cutting tool design.
4. Describe the properties of commonly used cutting tool materials including the economic justification for the use of expensive cutting tool materials.
5. State the importance of rake, clearance, cutting edge, and lead angle for cutting tools.
6. Show why cutting fluids are used.

Key Words

built-up edge (BUE)
carbide
ceramics
chip
cutting fluids
cutting speed
depth of cut
feed rate
high-speed steel
machinability
rake angle
shearing
spindle speed

Machining is essentially the process of removing unwanted material from wrought (rolled) stock, forgings, or castings to produce a desired shape, surface finish, and dimension. It is one of the four major types of manufacturing processes used to create product components. Machining is done by shaving away the material in small pieces, called *chips*, using very hard cutting tools and powerful, rigid machine tools. The cutting tool may be held stationary and moved across a rotating workpiece as on a lathe, or a rigidly held workpiece may move into a rotating cutting tool as on a milling machine.

This chapter presents the basic mechanisms involved in material removal, including the process parameters speed, feed rate, and depth of cut. Machinability, the relative ease or difficulty encountered in machining differing engineering materials, is discussed, and the impact of cutting tool materials and cutting tool geometry are explored.

Historically, machining processes probably began with Besson in 1569 when he invented the screw-cutting lathe and later when a practical version was built by Henry Maudslay in 1800. Horizontal milling machines first appeared in 1820 when Eli Whitney used them for the manufacture of firearms, and the vertical milling machine first appeared in 1860. Through continuous improvement these primitive machine tools have developed into the variety of manufacturing machinery that is used today. Many modern machine tools are controlled by a computer (numerical control) and can perform complex machining operations without the guidance or constant attention of a machinist. Some specialized machines are designed primarily for manufacturing at high production rates. The more versatile machines are designed to make one-of-a-kind prototype parts for research and development (R&D), tool and die making, or to repair existing machinery. These versatile machines are usually found in tool rooms or local machine shops and are operated by general machinists or tool and die makers. The less highly skilled workers who operate production machinery are called *machine operators.*

Few manufacturing technologies can achieve the precision of the various machining processes. The common 0.001 in. and 0.0001 in. precision that is possible for modern machining processes is far superior to the precision of common casting, molding, and forming operations. Large, heavy sections such as those used for dies are machined to precise dimensions, a virtual impossibility for many forming processes that handle only thin sheet materials. The basic machining processes are shown in Table 12.1

Machining processes remove material in the form of chips that are disposed or recycled. Machining is more costly than casting, molding, and forming processes, which are generally quicker and waste less material, but machining is often justified when precision is needed.

TABLE 12.1
Basic machining processes

Operation	*Diagram*	*Characteristics*	*Type of Machine*
Turning	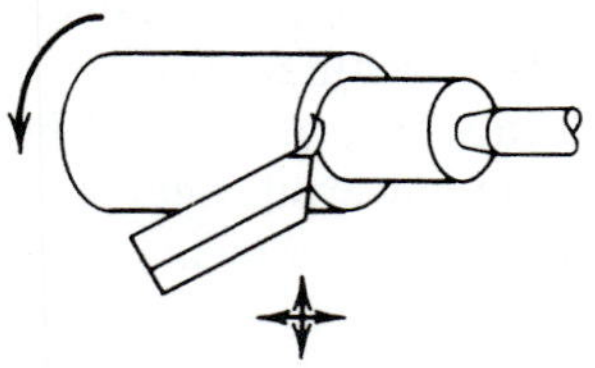	Work rotates, tool moves for feed	Lathe and vertical boring mill
Milling (horizontal)	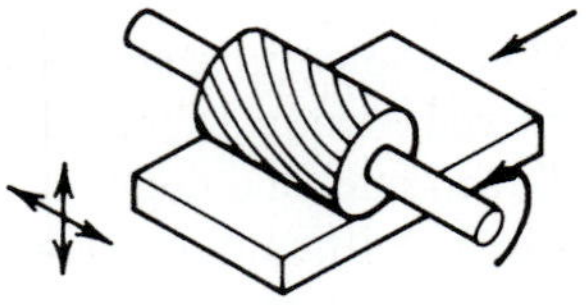	Cutter rotates and cuts on periphery. Work feeds into cutter and can be moved in three axes.	Horizontal milling machine
Face milling	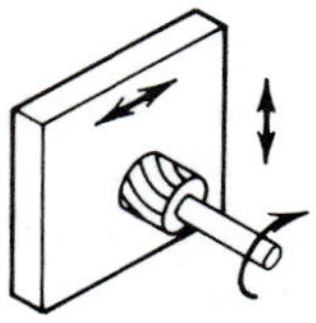	Cutter rotates to cut on its end and periphery of vertical workpiece.	Horizontal mill, profile mill, machining center
Vertical (end) milling	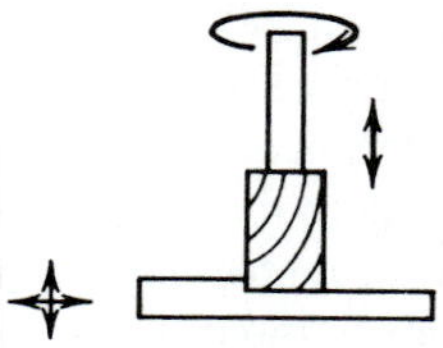	Cutter rotates to cut on its end and periphery, work moves on three axes for feed or position. Spindle also moves up or down.	Vertical milling machine, die sinker, machining center
Shaping	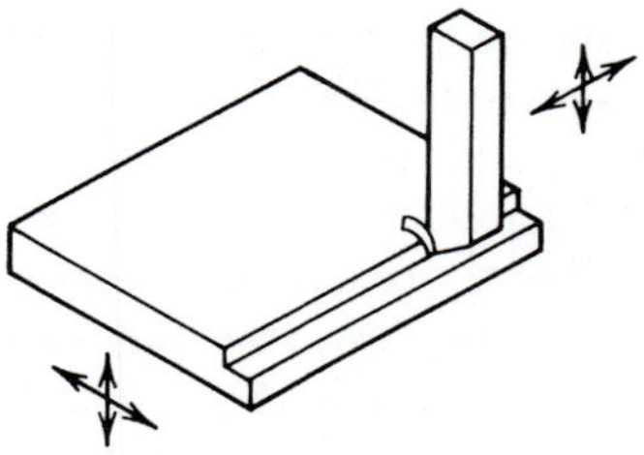	Work is held stationary and tool reciprocates. Work can move in two axes. Toolhead can be moved up or down.	Horizontal and vertical shapers
Planing	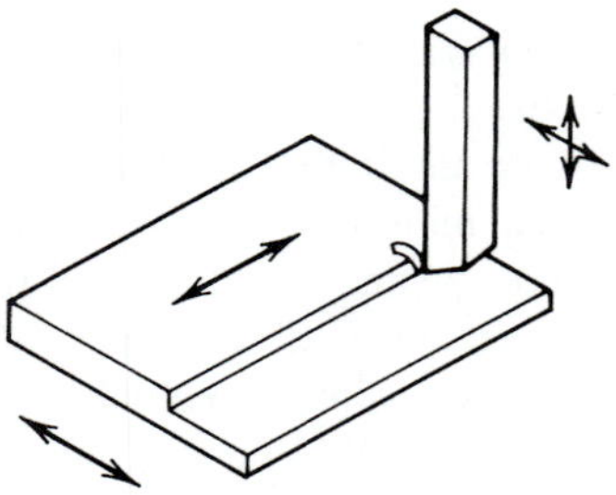	Work reciprocates while tool is stationary. Tool can be moved up, down, or crosswise. Worktable cannot be moved up or down.	Planer

TABLE 12.1 ***(continued)***

Operation	*Diagram*	*Characteristics*	*Type of Machine*
Horizontal sawing (cutoff)	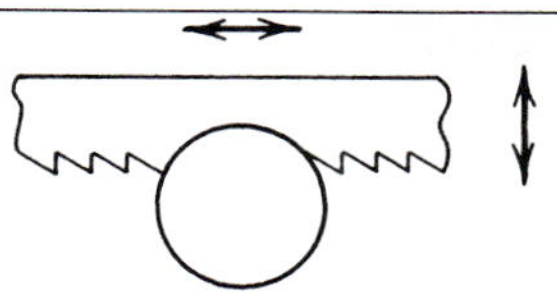	Work is held stationary while the saw either cuts in one direction as in bandsawing or reciprocates while being fed downward into the work.	Horizontal bandsaw, reciprocating cutoff saw
Vertical bandsawing (contour sawing)	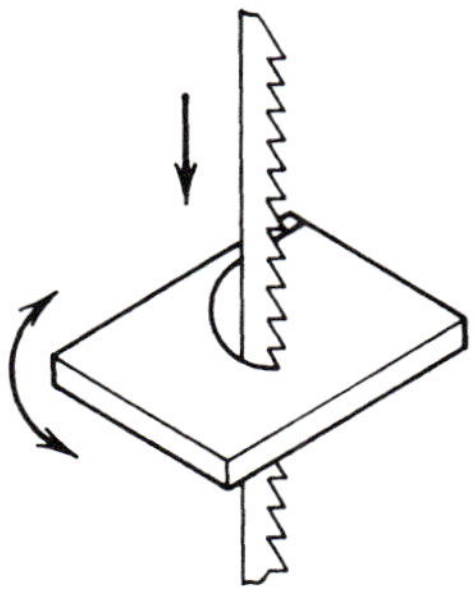	Endless band moves downward, cutting a kerf in the workpiece, which can be fed into the saw on one plane at any direction.	Vertical bandsaw
Broaching	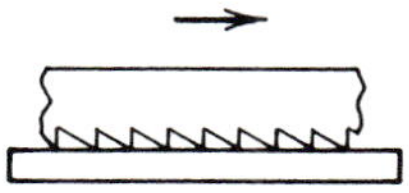	Workpiece is held stationary while a multitooth cutter is moved across the surface. Each tooth in the cutter cuts progressively deeper than the previous one.	Vertical broaching machine, horizontal broaching machine
Horizontal spindle surface grinding	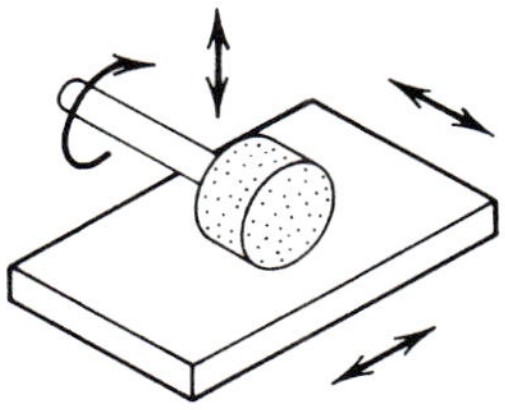	The rotating grinding wheel can be moved up or down to feed into the workpiece. The table, which is made to reciprocate, holds the work and can also be moved crosswise.	Surface grinders, specialized industrial grinding machines
Vertical spindle surface grinding	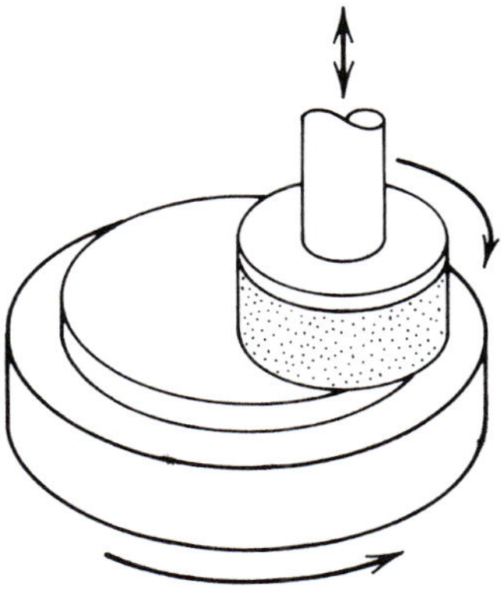	The rotating grinding wheel can be moved up or down to feed into the workpiece. The circular table rotates.	Blanchard-type surface grinders

(continued)

TABLE 12.1 ***(continued)***
Basic machining processes

Operation	*Diagram*	*Characteristics*	*Type of Machine*
Cylindrical grinding	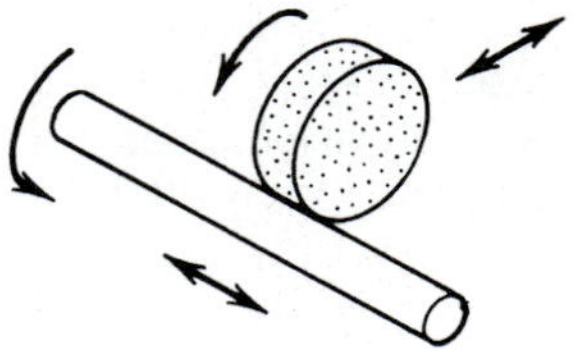	The rotating grinding wheel contacts a turning workpiece that can reciprocate from end to end. The wheelhead can be moved into the work or away from it.	Cylindrical grinders, specialized industrial grinding machines
Centerless grinding	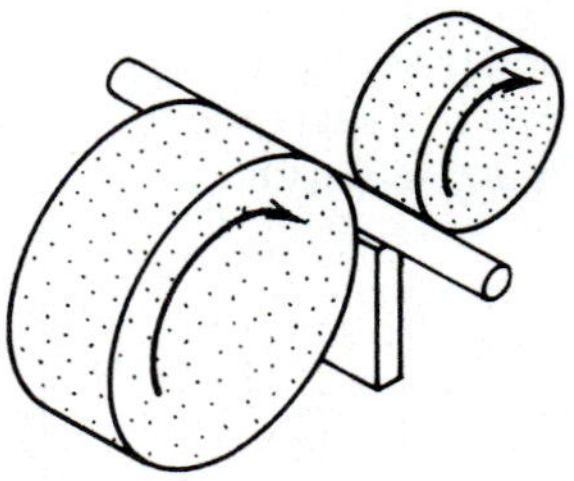	Work is supported by a workrest between a large grinding wheel and a smaller feed wheel.	Centerless grinder
Drilling and reaming	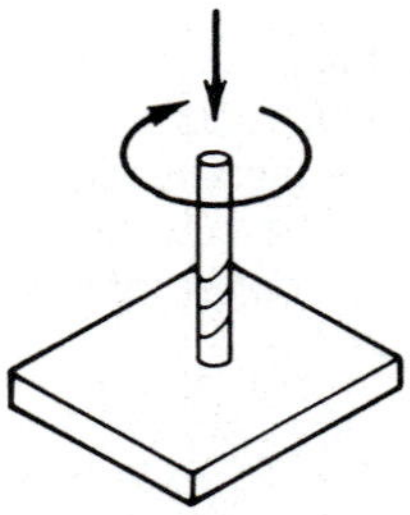	Drill or reamer rotates while work is stationary.	Drill presses, vertical milling machine
Drilling and reaming	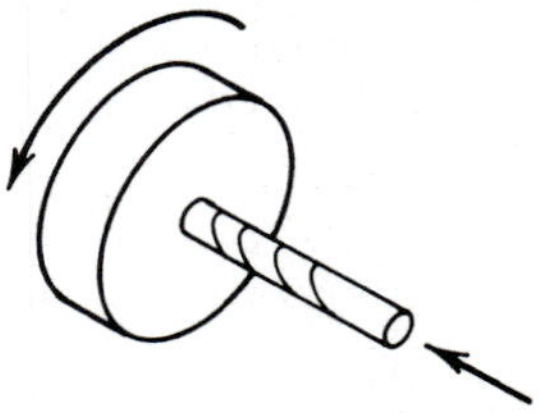	Work turns while drill or reamer is stationary.	Engine lathes, turret lathes, automatic screw machines
Boring	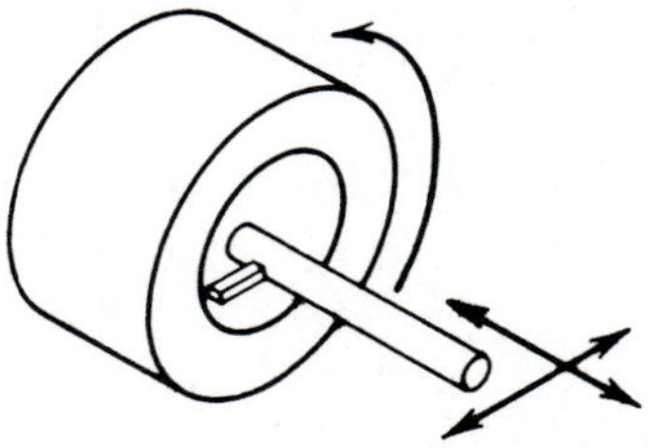	Work rotates, tool moves for feed on internal surfaces. (On some horizontal and vertical boring machines, the tool rotates and the work does not.)	Engine lathes, horizontal and vertical turret lathes, and vertical boring mills

For example, gears for an alarm clock may be entirely acceptable if stamped out of sheet metal or produced by plastic injection molding, which is far less expensive than machining, but gears for an automobile transmission must be hard, strong, and precisely made so they will function quietly and efficiently when subjected to variable forces and heavy shock loads. These gears must be made by machining processes and then heat treated and precision ground.

The performance of the cutting tool used to remove workpiece material determines the efficiency and cost of a machining operation. The geometry of the cutting edge controls the shearing action as a chip is torn away from the part. The cutting tool material determines how fast the operation may progress, and since time is money in manufacturing activities this is an important factor in the cost of the operation.

MOTION AND PARAMETERS: SPEED, FEED, AND DEPTH OF CUT

Machining operations require two basic simultaneous motions; one motion creates cutting speed, and the other is the feed motion. *Cutting speed* is the rate at which the workpiece moves past the tool or the rate at which the rotating surface of the cutting edge of the tool moves past the workpiece. Regardless of whether the tool rotates or the workpiece rotates, the relative motion between the two creates the cutting speed. English units for cutting speed are feet per minute (fpm), which is often called surface feet per minute (sfpm or sfm). Metric units are meters per minute (m/min). (Table 12.2 suggests cutting speeds for machining several metals using high-speed steel cutting tools.) Higher cutting speed shortens the time required to complete the machining cut but can greatly shorten the useful life of the cutting tool. Cutting speeds that are too low tend to tear instead of cut, produce rough finishes, and distort the grain structure at the surface of the workpiece, all of which can cause early failure of a machined part (Figures 12.1 and 12.2). Speeds should be as high as can be maintained without causing the tool to wear out too quickly. Recommended cutting speeds for machining operations can be found in commonly available tables. These suggested speeds vary based on the workpiece material, cutting tool material, and type of machining operation.

Most machine tools have rotating spindles, so the rate of rotation (the spindle speed) for the machining operation must be established. This rotation of the machine spindle generates the cutting speed necessary for the cutting tool to efficiently shear material from the workpiece. The spindle speed is calculated based on the cutting speed to be used in the operation, and units are revolutions per minute (rpm). (See the appendix at the end of this chapter for spindle speed calculations.)

FIGURE 12.1
This micrograph shows the surface of the specimen that was turned at 100 sfm. The surface is irregular and torn, and the grains are distorted to a depth of approximately 0.005 to 0.006 in. (250 ×).

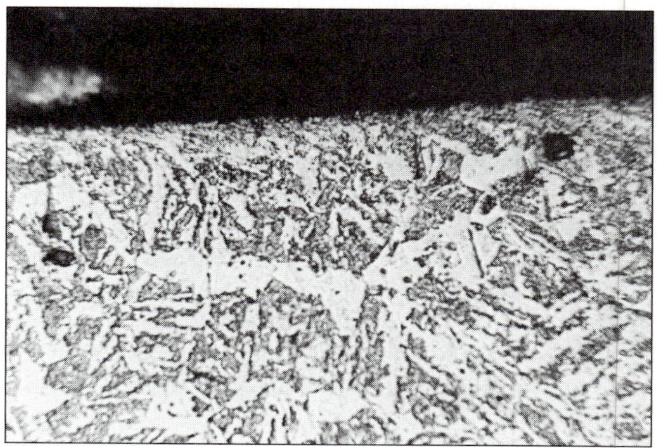

FIGURE 12.2
At 400 sfm this micrograph reveals that the surface is fairly smooth and the grains are only slightly distorted to a depth of 0.001 in. (250 ×).

TABLE 12.2
Cutting fluid and speed table for machining various metals with high-speed tools

Material	*Cutting Fluid*	*Speed (fpm)*
Aluminum and alloys	Soluble oil, kerosene, light oil	200 to 300
Brass	Dry, soluble oil, synthetic coolant solution, light mineral oil	150 to 300
Bronze	Dry, soluble oil, mineral oil, synthetic coolant solution	200 to 250
Cast iron, soft	Dry, air, synthetic coolant solution	100 to 150
Cast iron, medium	Dry, air, synthetic coolant solution	70 to 120
Cast iron, hard	Dry, air, synthetic coolant solution	30 to 100
Magnesium and alloys	60-second mineral oil	300 to 600
Steel, alloy	Soluble oil, synthetic coolant solution	30 to 90
Steel, low carbon	Soluble oil, synthetic coolant solution	90 to 120

TABLE 12.3
Lathe feeds, inches per revolution

Material	*Roughing Feed*	*Finishing Feed*
Cast iron	.010 to .020	.003 to .006
Low-carbon steel	.010 to .020	.003 to .005
High-carbon steel, annealed	.008 to .020	.003 to .005
Alloy steel, normalized	.005 to .020	.003 to .005
Aluminum alloys	.015 to .030	.005 to .010
Bronze and brass	.010 to .020	.003 to .010

The *feed motion* is the advancement of the cutting tool along the workpiece (for lathes) or the advancement of the workpiece past the tool (for milling machines). The rate at which the tool or workpiece moves is the *feed rate.* The basic measure of the feed rate is the distance advanced per revolution of the machine spindle. Units for this feed per revolution are inches per revolution (ipr) or millimeters per revolution (mm/rev). (See the appendix at the end of this chapter for feed rate calculations.) Table 12.3 suggests feed rates for machining several different metals. Faster feeds shorten the time required to complete the machining cut but create rougher workpiece surfaces. Feed rates for roughing (rapid material removal) should be as heavy as the tool material and the machine tool can withstand without failure, and finishing feeds should be fine enough to produce the desired smooth surface finish on the workpiece.

The rate at which the workpiece advances (measured in inches or millimeters) per minute is an alternative way to specify the feed rate. The units are inches per minute (ipm) or millimeters per minute (mm/min). The feed per minute is calculated based on the spindle speed and the feed per revolution. (See the appendix at the end of this chapter for feed rate calculations.)

The tool must be engaged in the workpiece to remove material. The amount of engagement, called the *depth of cut,* is equal to the thickness of the layer of material removed from the workpiece. The depth of cut also determines the width of each chip removed from the part as it is machined.

MACHINING: SHEARING CHIPS FROM THE WORKPIECE

Pressure, Heat, and Friction in Machining

Machining is similar to whittling wood with a sharp knife. The cutting tool exerts enough force on the workpiece to smoothly cut away material in the form of chips. This action results in the continuous shear failure of the work material at the cutting edge of the tool (Figures 12.3 and 12.4). Imagine how much pressure is required to shear chips

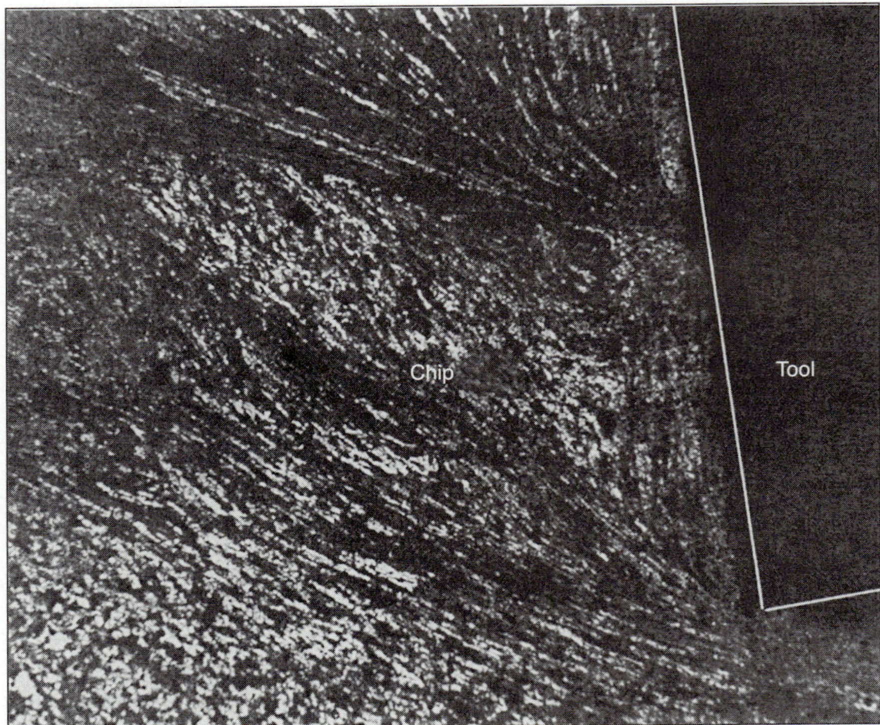

FIGURE 12.3
Micrograph of chip formation showing shear plane. Point of negative rake tool producing a discontinuous chip (100 ×) (Neely and Bertone, *Practical Metallurgy and Materials of Industry,* 6th ed., © 2003 Prentice Hall, Inc.).

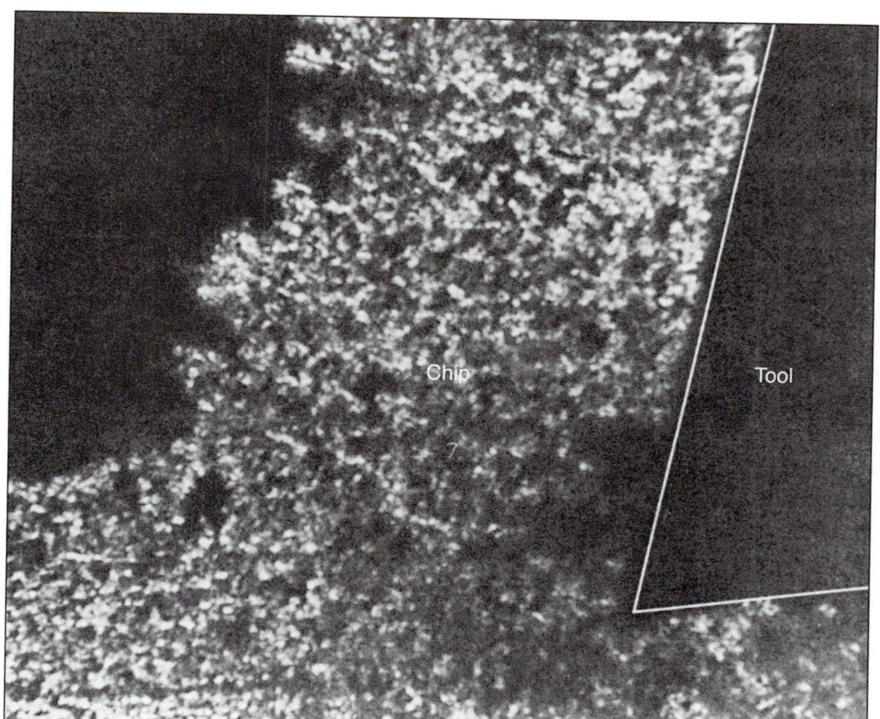

FIGURE 12.4
Positive rake tool showing less distortion ahead of the tool than produced by a negative rake tool (100 ×) (Neely and Bertone, *Practical Metallurgy and Materials of Industry,* 6th ed., © 2003 Prentice Hall, Inc.).

from a steel workpiece and make it appear to be as easy as cutting butter. This tremendously high pressure causes the tool and workpiece to flex or deflect away from each other. This movement affects precision, surface finish, and tool life, so deflection must be opposed by rigidity. The machining system is only as strong as its weakest link, so the workpiece, workholding tooling, machine tool, toolholder, and cutting tool must all be extremely stiff to oppose deflection.

The high forces in machining create a considerable amount of heat near the cutting edge. Most of this heat is generated within the shearing process, and some heat is created by friction between the tool and the workpiece. Most of the heat is carried off in the chips, but the remainder stays in the tool and workpiece, creating a large amount of thermal stress and softening the tool. Cutting fluid (coolant) is often used to bathe the tool and workpiece to remove much of the heat and minimize damage to the tool and part.

Workpiece Materials and Their Effects

The workpiece material plays an important part in chip formation. Greater cutting forces are required to shear harder and higher-strength materials, causing more deflection between the cutting tool and workpiece. More heat is also generated, which creates higher operating temperatures in the tool and part. The increased friction and tool temperature encountered when machining hard or abrasive materials can accelerate tool wear, shortening the tool life, the amount of time that the tool will survive while cutting. Softer, lower-strength materials shear more easily and may be machined faster while causing less tool wear, but very soft materials such as soft pure aluminum or hot-rolled (HR) low-carbon steel are somewhat "gummy" when cut on machines and tend to cause **built-up edge (BUE)** on the cutting tool. Built-up edge (Figure 12.5) is a common machining problem in which workpiece material becomes welded to the cutting edge of the tool, changing the geometry of the cutting edge. BUE causes chips to be torn away rather than cleanly cut, resulting in rough part surfaces, and it may damage the tool.

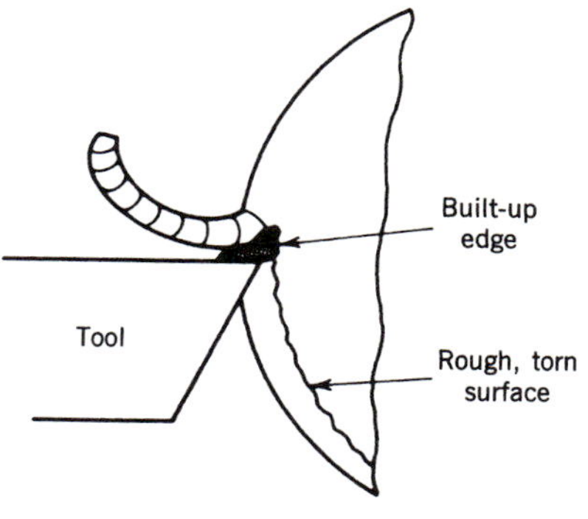

FIGURE 12.5
Tool with built-up edge.

Correct cutting speed and tool geometry helps eliminate built-up edge and produce a clean cutting action with improved surface finish.

Fragmented chips called *discontinuous chips* are produced when brittle materials are machined. Brittle gray cast iron chips break immediately when sheared from the part and have the appearance of a powder. When ductile materials are machined the chip is formed as a coil or helix, a continuous chip. Because the tool is in constant contact with the chip, added friction causes operating temperatures to be higher when continuous chips are produced.

When ductile materials are machined, heavier (larger) feed rates produce chips that curl tightly and break readily, and lighter (smaller) feeds tend to produce long stringy chips. Chips that break after one curl, forming a 9-shape, are the most desirable form of continuous chips for ease of handling and for safety's sake.

The **machinability** of a workpiece material is the ease with which it can be machined. Machinability ratings compare various materials in terms of cutting ease or difficulty resulting from the forces required and heat generated in the shearing of a material. The relative machining difficulty of various materials may be compared by measuring the power required and cutting tool life for each workpiece material. Commonly machined materials have been machinability rated using a scale based on low-carbon soft steel (AISI B-1112), which was arbitrarily given a rating of 100 percent on the machinability scale (Table 12.4). Metals that are more easily machined have a higher rating than 100 percent, and materials that are more difficult to machine are rated lower.

TABLE 12.4
Machinability ratings and approximate hardness of annealed steels

AISI Classification	*Machinability Rating, %*	*Approximate Hardness—HB (Brinell)*
B-1113	135	200
B-1112	100	205
C-1118	80	160
C-1020	65	150
C-1040	60	200
A-8620	50	220
A-3140	55	200
A-5120	50	200
C-4140	50	200
Cast iron	40 to 80	160 to 220
302 stainless	25	190

(John E. Neely, *Practical Metallurgy and Materials of Industry,* 2d ed., Wiley, New York, © 1984.)

CUTTING TOOL MATERIALS

Commonly machined materials include plastics, aluminum, many varieties of steels including heat-treated tool steels that are as hard as a knife blade, ceramics, and many others. Cutting tools must be capable of retaining their hardness at high temperatures (hot hardness). Better hot hardness permits tools to operate at higher cutting speeds, thereby improving productivity.

A variety of cutting tool materials are needed. Some must be very hard for long tool life and to machine hard workpiece materials. Others must be very shock resistant (tough) to withstand interrupted cuts (intermittent cutting action). An example of an interrupted cut would be turning a bar that has a hexagonal cross section. The cutting tool will, on the first cut, contact only the corners of the rotating bar, creating six impacts each time the bar rotates. The mechanical shock of this intermittent cutting action tends to fracture brittle cutting tools.

Cutting speed has a pronounced effect on the useful life of a cutting tool. Higher speeds increase both friction and the shear rate, generating added heat that makes cutting tools wear faster or break down. This increases tool costs and tool changing delays, but higher speed also decreases the time required to perform a machining operation, and that improves productivity. The chosen cutting speed strikes the balance between tool wear and productivity. Any cutting tool materials that can operate at higher cutting speeds with no increase in the rate of tool wear enable manufacturers to increase productivity without paying a penalty in tooling costs and delays. Millions of dollars are spent each year to develop new cutting tool materials that can produce such rewards.

High-Speed Steel Cutting Tools

One hundred years ago hardened plain carbon tool steel was used for tools to cut metals. Cutting speeds were very slow because carbon steel would permanently lose its hardness if it got very hot. Today, plain carbon tool steel is used only for low-speed, low-temperature applications such as hand tools, including files, knives, and chisels. A type of tool steel developed around 1900 was called high-speed steel (HSS) because it could get hot, cool down, and retain its hardness. High-speed steel is less expensive than most cutting tool materials, and steel tools can be machined to complex shapes in their soft state, heat treated for hardness, and then ground to a sharp cutting edge. Because of its versatility and low cost, HSS is today the most commonly used cutting tool material in machining applications. High-speed steel drills, milling cutters, and lathe tools are widely used. After the cutting edge dulls, HSS tools are sharpened using a grinder, greatly increasing the useful life of the tool.

Carbide Cutting Tools

Most cutting tools used for production machining are made of tungsten carbide. Compared with high-speed steel tools, carbide cutting tools have much better hot hardness, so they can machine at higher temperatures without softening and destroying the cutting edge. Cutting speeds are three to four times faster for carbides than for HSS tools. In the making of a carbide tool, tungsten carbide particles are mixed with a cobalt powder, compressed into a briquette of the required tool shape, and then sintered in a furnace, causing the cobalt to bind the tungsten carbide particles into a very hard, strong, solid material. Carbide is made in grades of varying hardness and toughness, and titanium carbide and tantalum carbide are sometimes added to the mixture to provide greater hardness for wear resistance. Most carbide tools used today in manufacturing operations are throw-away inserts that have several indexable cutting edges (Figure 12.6). Carbide inserts are clamped into different toolholders for external and internal machining (Figures 12.7 and 12.8). Milling cutters, turning tools, boring bars, and many other types of toolholders equipped with carbide inserts have long life, and the inserts can be quickly rotated (indexed) to expose a fresh cutting edge or replaced when they wear out.

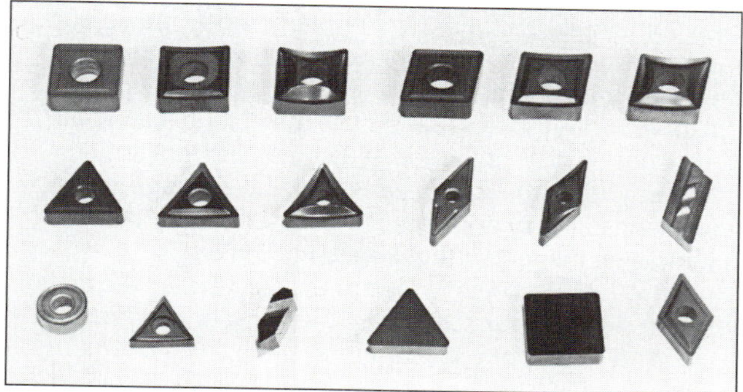

FIGURE 12.6
Carbide inserts (Courtesy of Kennametal Inc.).

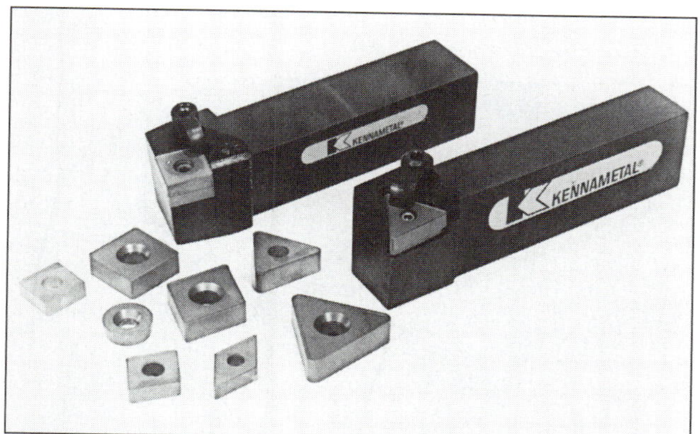

FIGURE 12.7
Carbide toolholders for turning machines (Courtesy of Kennametal Inc.).

FIGURE 12.8
Boring bar holder with interchangeable heads for internal turning operations (Courtesy of Kennametal Inc.).

Ceramic Cutting Tools

Ceramic tools are often used to machine hard workpiece materials and have better hot hardness than carbide. These operate at much higher speeds than carbide inserts but require a very rigid setup (machine, cutting tool, workpiece, work-holding system) and will not endure the impact of heavy interrupted cuts. Ceramic cutting tools are manufactured in insert shapes similar to carbide inserts.

Coated Carbide and Cermet Cutting Tools

Tungsten carbide inserts coated with titanium nitride (TiN), ceramic (aluminum oxide), and/or titanium carbide combine the qualities of carbide and the wear resistance of the coating material to permit increased cutting speeds and added productivity. Cermet, a mixture of carbide and ceramic that is sintered into inserts, competes closely with the productivity of coated carbide tools.

Diamond and CBN Cutting Tools

Diamond cutting tools can produce exceedingly smooth surface finishes and hold very close tolerances. Diamond tools are manufactured from polycrystalline powder (PCP), and it retains a sharp, stable cutting edge, but it is prohibitively expensive for many applications. Cubic boron nitride (CBN), a manmade material, is second in hardness to diamond. CBN is used to "hard turn" steel workpieces that are too hard to be machined by less exotic cutting tool materials, but it is also very expensive.

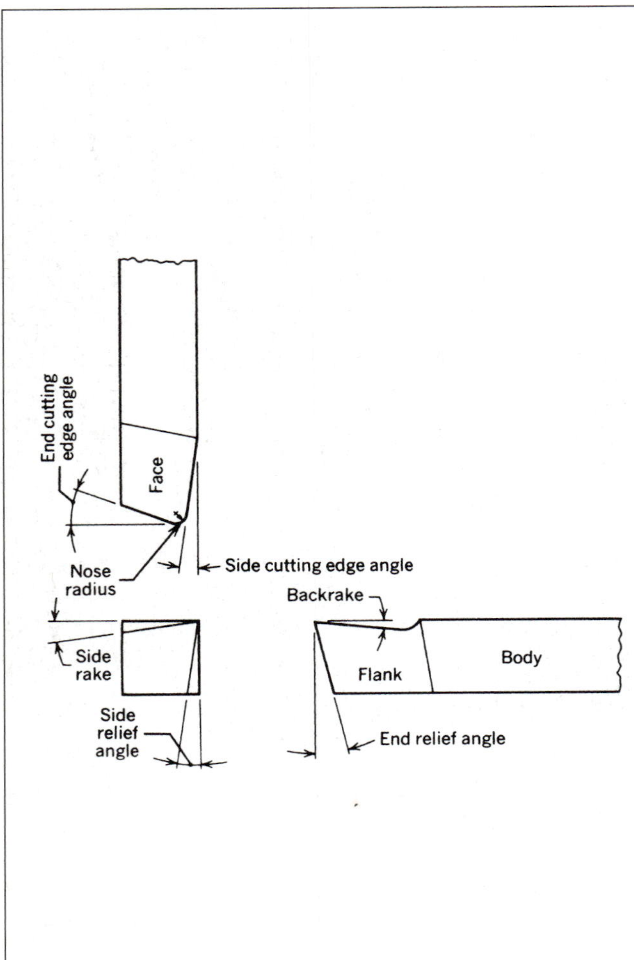

Important Elements of Cutting Tool Geometry

1. The tool shank is the area where the tool is held and clamped by the toolholder.
2. Back rake, a component of the rake angle, is very important to the smooth chip flow that is needed to have a uniform cut and a smooth surface finish.
3. The side rake, the other component of the rake angle, directs the chip flow away from the cutting edge. Large back and side rakes are used for machining soft workpiece materials to minimize built-up edge.
4. The end relief angle prevents the front flank of the tool from rubbing on the workpiece.
5. The side relief angle prevents the side flank of the tool from rubbing on the workpiece and allows the tool to feed into the work material.
6. The side cutting edge angle (SCEA) may vary considerably depending upon the application. For roughing it should be approximately 15 to 30° as shown in Figure 12.9, but tools used for cutting square shoulders or for light machining could have reversed angles from 5 to 30° depending on the application. The side cutting edge angle may be provided by angling the toolholder (creating a lead angle), by grinding the desired angle on the tool, or by a combination of both. When a large nose radius and a small depth of cut are employed in finishing operations the SCEA has no effect on the cut. The side cutting edge angle shown in Figure 12.9 directs the cutting forces into a stronger section of the top. It also helps direct the chip flow away from the workpiece (avoiding damage to the workpiece surface) and reduces the thickness of the chip (decreasing the forces required to cut the chip).
7. The nose radius of cutting tools is established to help provide the required surface finish. A larger tool nose radius provides a smoother surface.

FIGURE 12.9
The parts and angles of a tool. Angles given are only examples. The listing to the right describes elements or tool geometry (White, Neely, Kibbe, Meyer, *Machine Tools and Machining Practices,* Vol. 2, © 1977 John Wiley & Sons, Inc. This material is used by permission of John Wiley & Sons, Inc.).

CUTTING TOOL GEOMETRY

Tool geometry (shape) varies considerably depending on the machining application. The shape and angles of the cutting tool must be accurate in order to have sufficient strength at the cutting edge, to ensure good chip formation and flow, and to provide sufficient relief (clearance) so the tool will cut into the work material and not rub. An example of a right-hand cutting tool is illustrated in Figure 12.9. There are many shapes of cutting tools, including left-hand, threading, and form tools, but all of them must have correct angles for relief and rake. Chip breakers are sometimes needed on tools to control chip formation (Figures 12.10 and 12.11).

At what angle would you hold a knife blade when peeling an apple? If you hold the knife blade perpendicular to the surface of the apple (a neutral or zero rake

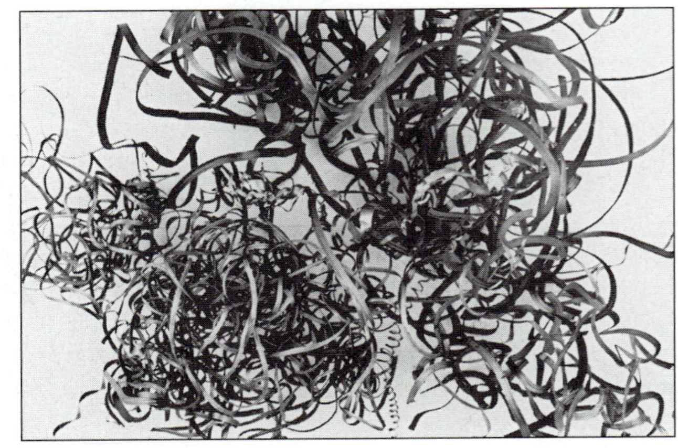

FIGURE 12.10
Stringy chips.

FIGURE 12.11
Chips that are 9-shaped are best.

angle) the knife will scrape the skin off the apple, requiring a large amount of force and cutting very poorly. If you hold the knife blade at an angle greater than 90° (a negative rake angle), which would seem to be backward, the blade will cut even less effectively, requiring a very large force to push the skin off, possibly crushing the apple. The most effective knife blade angle, of course, is the acute angle that one would normally use when peeling an apple (a positive rake angle). A positive rake angle will shear material smoothly and efficiently while requiring less force. This rake angle relationship also holds for machining operations, although the angles become more complex. Positive rake angles are always preferred, but the poor tensile strength of many cutting tool materials makes the use of negative or neutral rake angles necessary in many cases.

All cutting tools must have clearance to prevent the flank of the tool, which is located next to and beneath the cutting edge, from rubbing the workpiece. If the flank of the tool rubs the workpiece it prevents the cutting edge from reaching the workpiece. Grinding a positive rake tool in the flank area provides a relief angle for the needed clearance but weakens the cutting edge. The operating orientation of negative and neutral rake tools provides natural clearance. It is not necessary to grind the flank of these tools, so negative and neutral rake tools are stronger than positive rake tools. This extra strength can be extremely important, especially in heavy machining operations where large forces are required to remove large amounts of material or in interrupted cuts where impact resistance is needed.

Chip Control

Breaking chips is very important. Chips that do not break readily and form strings longer than the ideal 9-shaped chip create handling and safety problems. Chips are as sharp as knives or saws and can cause deep cuts. Chips should not be handled with bare hands, and long chips are especially difficult to handle and dispose of, creating a workplace hazard. Many automated manufacturing machines carry chips away for disposal on a conveyor. Short, broken chips are a necessity for this automatic machinery because stringy, wiry chips would entangle machine parts and eventually jam the conveyor. Chips tend to fly from the work at high speed, sometimes as far as 20 to 30 feet away. They can easily injure the eye if safety glasses are not worn, and hot chips can penetrate clothing or soft-soled shoes.

Many cutting tools have special chip-breaking geometries called chip breakers (Figures 12.12 and 12.13) that break the chips by sharply curling them (Figure 12.14). The shape of chip breakers is vital to chip control.

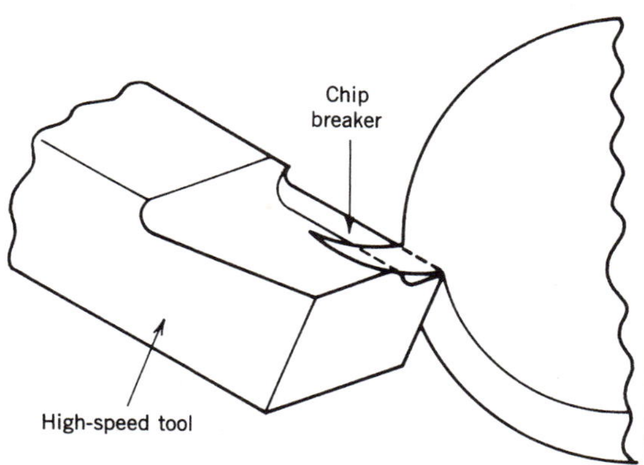

FIGURE 12.12
High-speed steel chip breakers.

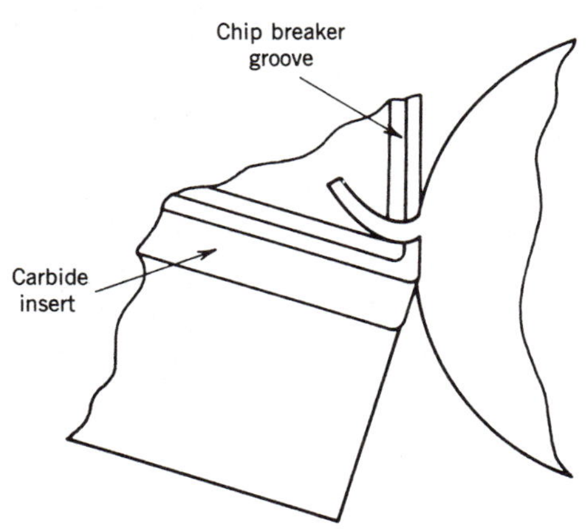

FIGURE 12.13
Carbide chip breakers.

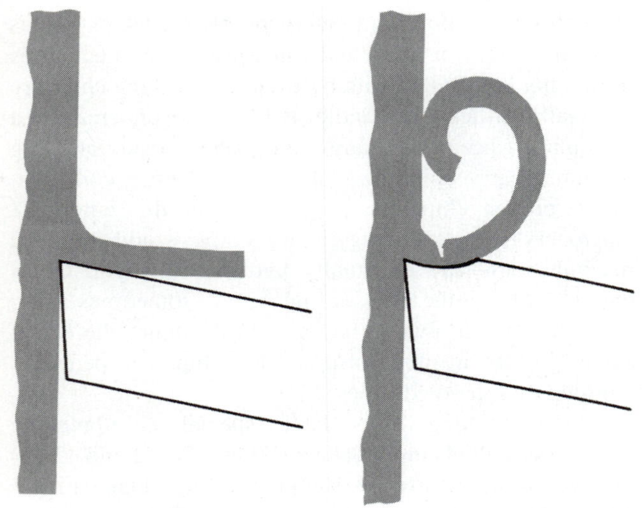

FIGURE 12.14
How chips are broken. Chip flow with plain tool and with a chip breaker (Kibbe, Neely, Meyer, and White, *Machine Tool Practices,* 7th ed., © 2002 Pearson Education, Inc.).

Chip-breaking geometry is most needed when machining soft workpiece materials or when finish machining with a light (small) feed rate.

CUTTING FLUIDS

Cutting fluids are usually liquid, but air is sometimes used to avoid contamination for certain workpiece materials. In general, cutting fluids serve to dissipate heat, to lubricate between the tool and workpiece for lower friction, and to carry away chips from the cutting area. There are essentially two types of cutting fluids, cutting oils and coolants. Some cutting operations, such as using a tap to cut threads, create a considerable amount of friction and need a cutting oil for good lubricating qualities. Cutting oils can be either animal fats or petroleum-based. Higher-speed cutting operations require relatively less lubrication and more of the heat removal qualities that are found in a coolant. Coolants are usually water-based soluble oils that have excellent heat transfer qualities. Petroleum oils or waxes are treated so that they emulsify when mixed with water to produce a milky white solution. Other chemicals such as wetting agents, rust inhibitors, antibacterial agents, and polarizing agents are added to improve the coolant. Many synthetic coolants have been developed for use as emulsions or cutting oils for specific and general uses.

Some machine tools used in tool rooms or for low-volume production are operated without coolant, but most production machines are fitted with coolant pumps and tanks. In some production machining operations, however, coolant is not used. For example, in an intermittent cutting setup with a carbide tool where there is alternate heating and cooling of the carbide, coolants would add to the thermal shock, which can crack the carbide insert, so no cutting fluid is used. Also, due to the risk of thermal shock and the superior hot hardness of ceramic, cutting fluids are not normally required with ceramic tools. This practice of "dry machining" is becoming more commonplace owing to environmental concerns involving the disposal of spent cutting fluid.

APPENDIX: CALCULATION OF SPEEDS AND FEEDS FOR MACHINING OPERATIONS

Let: V = cutting speed
N = spindle speed
f_r = feed per revolution
f_m = feed per minute

Recommended cutting speeds may be obtained from commonly available tables or by utilizing machinability ratings. Recommended feeds per revolution may also be obtained from charts. These recommended speeds and feeds represent only a beginning point, so they must be tempered considering the conditions of each machining operation and should be optimized in production.

Speed Calculation

The spindle speed may be approximated given the cutting speed and the diameter of the rotating tool or workpiece.

$$\text{English units: } N \cong \frac{4V}{D}$$

where: N is spindle speed in rpm
V is cutting speed in feet per minute
D is diameter in inches

$$\text{Metric units: } N \cong \frac{300V}{D}$$

where: N is spindle speed in rpm
V is cutting speed in meters per minute
D is diameter in millimeters

Feed Rate Calculation

The feed per minute may be calculated given the feed per revolution and the spindle speed.

English units: $f_m = f_r \times N$

where: f_m is inches per minute feed rate
f_r is inches per revolution feed rate

Metric units: $f_m = f_r \times N$

where: f_m is millimeters per minute feed rate
f_r is millimeters per revolution feed rate

Review Questions

1. In terms of interchangeability of manufactured parts, what characteristic of machining processes is unique and most important?
2. Define the process of machining metal.
3. Describe the process by which metal is removed by a cutting tool in the machining process. Is the metal removed in chips by splitting the metal ahead of the tool, or by plastic deformation?
4. The most common cutting tools used in machining operations of all kinds are high-speed steel, ceramic inserts, diamond tools, and varieties of carbide tools. Of these, which type of tool is by far the most commonly used in production machining operations?
5. Stringy, wiry chip formations are hazardous and cannot be handled or disposed of easily. How can this kind of chip be avoided? What is a more desirable chip form?
6. Holes made by twist drills are usually rough and often oversized. How can a drilled hole be made to have a more accurate dimension with a smoother finish?
7. How can a built-up edge cause a reamer or other cutting tool to produce a rough finish and inaccurate dimensions?
8. Cutting speeds (sfm) for manufacturing should be as high as possible without burning the tool or workpiece. If the speed for a lathe operation has been determined to be 300 rpm for high-speed tools, how fast should the machine turn for carbide tools?
9. Cutting fluids come in two types. Name each and describe its special qualities.
10. When are lathe operations sometimes performed without cutting fluid?

Case Problem

Case: Chip Control Problem

High-speed steel cutting tools were used for production turning of stainless steel fittings on an automatic screw machine. The setup person originally sharpened the cutting tools by offhand grinding on a pedestal grinder. A problem developed in which wiry chips entangled the mechanism, causing lost machining time in shutdowns for cleaning out the entangled chips. What would you recommend to correct this problem?

Key Words

lathes
machining centers
milling machines—horizontal milling machines
milling machine—vertical milling machine
grinding machine—surface grinder
grinding machines—cylindrical grinder
gear shaping
gear hobbing
shapers
honing
lapping
milling cutters—end mills
milling cutters—face mills
milling cutters—plain milling cutters
sawing machines
drilling machines
threading

CHAPTER 13

Machine Tool Operations

Objectives

This chapter will enable you to:

1. Decide when to use toolroom machinery and when to use production machinery for making a machined part.
2. Decide which type of drilling or sawing machine is needed for a given operation.
3. Explain the special uses of vertical and horizontal spindle turning, milling, and boring machines.
4. State how machine threads are cut and how they are designated.
5. Describe the processes of making many kinds of internal and external gears and splines.
6. State the principles and uses of abrasive machining.
7. Determine how grinding machines and processes are used in manufacturing.

The following factors are obviously considered when choosing the best manufacturing process to perform a needed machining operation: finished part configuration, tolerances, surface finishes, cost, and available equipment. Less obvious considerations that must not be overlooked are production volume and the skill level of the work force. Low-volume and one-of-a-kind toolroom machining are very labor intensive and utilize manually operated machine tools such as drill presses, engine lathes, and milling machines. Medium-volume and high-volume (mass production) machining are called production machining. Production machining is repetitive, and frequent repetition of machining operations justifies more sophisticated machines and tools that produce parts more quickly and save labor. CNC (computer numerical control) machine tools were originally developed for use in low to medium volume manufacturing, but today they are commonly used in one-of-a-kind and high volume manufacturing as well.

Low-volume or one-of-a-kind operations require the constant attention of a skilled machinist. General machinists require much more training than production machine operators because of the many skills needed to produce complete, precision-machined parts from mechanical drawings. An even more highly skilled trade is that of a tool and die maker who is a specialized machinist, trained in toolmaking. Usually the most highly regarded and highest paid of all machinists in the trade, toolroom machinists are always in demand. Besides making new tooling these machinists often repair existing tools and dies, repair machine parts for local industries, and make prototypes of new products (Figure 13.1). When these new prototypes have been developed sufficiently in the toolroom machine shop, a manufacturing plant can then put the machining process into production using numerically controlled machines, automatic machines, or flexible manufacturing systems (FMS) to produce the parts in the desired volumes.

FIGURE 13.1
Robot positioner. Robot arms and components such as this one must first be made by conventional machining processes before being manufactured in volume (Barrington Automation Ltd.).

FIGURE 13.2
Vertical band saw (DoALL Company).

Many large manufacturing plants have machine shops for toolmaking, machinery repair, prototype work, and low-volume machining. Small privately owned job shops can be found in virtually every city. They do one-of-a-kind work and produce short runs of products as suppliers to larger companies that assemble and market the end product.

This chapter discusses the various material removal processes commonly encountered in low-, medium-, and high-volume manufacturing. Beginning with cutoff and hole-making processes, the text continues through lathe and milling machine operations and concludes with coverage of shaping, broaching, gear cutting, and abrasive machining. The presentation of each process includes a basic description of the procedure, machine tools, cutting tools, and work-holding tooling employed for production at various volume levels.

BASIC MACHINE TOOLS

Basic machine tools such as metal cutting saws, drill presses, lathes, and milling machines are used in conventional machining. They are not normally used for production purposes unless they are automated using computer control.

Metal Sawing Operations

The vertical band saw (Figure 13.2) removes unwanted metal without turning all of it into chips. For some operations the saw cut can be made and no subsequent machining is required. In other cases the blueprint specifies precision-machined surfaces, so finishing operations are required after the part is sawed. A band saw blade has a carbon steel body, and high-speed steel teeth are welded into most blades to create cutting edges. Blades are cut to length and welded on a resistance butt welder (Figures 13.3 and 13.4), sometimes called a "flash" welder, to form endless bands. Internal features can be cut by inserting the blade through a drilled hole in the workpiece and then welding the blade together. Contours, small or large radii, and many other shapes can be cut with vertical band saws.

Cutoff Saws

Metal bar stock for manufacturing is usually shipped in 12- to 40-ft lengths and range from 1/4-in. diameter round

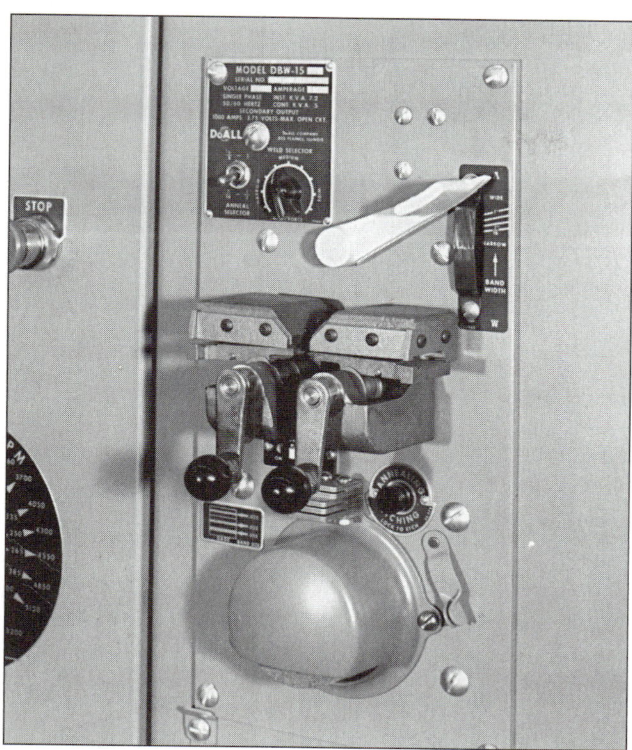

FIGURE 13.3
Flush-mounted blade and grinder enable the operator to weld saw blades for external and internal sawing (DoALL Company).

FIGURE 13.4
Ends of band being welded in a resistance "flash" butt welder.

bars to those 12 in. in diameter and over. Universal milled flat bar stock is rolled in sizes up to 14 in. wide, and there is a great variety of round, square, and rectangular bar stock and a wide variety of round, square, and rectangular tubing, plus many other shapes. All these must be cut to the length needed for specific part manufacturing. The most common type of cutoff saw used for this purpose is the horizontal band saw (Figure 13.5). Automatic band saws, often controlled by microprocessors (Figure 13.6), cut off long bars to programmed lengths, feeding the bar after each cut. Most of these saws use cutting fluids to prolong saw blade life. These machines are mainly used for production.

The reciprocating cutoff saw uses a straight blade and is often called a power hacksaw because its action resembles the operation of a hand hacksaw with its back-and-forth motion. It has a slower cutting action than the band saw, but it is still used in many shops because it can cut harder material than the band saw with less blade breakdown. It is useful for cutting off tough, hard metals such as titanium, zirconium, stainless steel, and tool steels.

Drilling Machines and Equipment

Drilling machines are designed primarily for making holes in metals and other materials; however, other operations such as counterboring, spotfacing, reaming, and tapping are often performed on them. Since hole drilling operations range from a few thousandths of an inch in diameter (size of a hair) to over a foot in diameter, several kinds of drilling machines are needed. Some drilling machines are suited to low-volume work and others are production machine tools. The three basic types of drill presses are sensitive, upright heavy duty, and radial arm drills. Other variations used are multiple spindle drill presses, turret-type drills, and deep hole drilling machines.

The sensitive drill press (Figure 13.7) is so named because the operator "senses" or "feels" the cutting action of the drill while holding the handle that feeds the drill into the work. These machines have no power feed and are usually limited to holes of up to 1/2-in. diameter. Sensitive and upright heavy duty drill presses are sized by the largest diameter of a circular piece in which a centered hole can be drilled. Thus, if the distance from the column to the center of the spindle is 10 in., then it is called a 20-in. drill press.

Ordinary sensitive drill presses used in machine shops can drill holes no smaller than about 1/16-in. diameter owing to the higher spindle speed (rpm) required to maintain the cutting speed for smaller drill sizes. For example, with a cutting speed of 100 sfm, a 10-in. diameter drill would need to rotate at 40 rpm, which would be realistic for this diameter drill. If the drill were only 0.010 in. in diameter, the spindle speed would have to be 40,000 rpm; however, it has been found that these high rotational speeds are not realistic. Even

FIGURE 13.5
Horizontal band saw (DoALL Company).

FIGURE 13.7
Sensitive drill presses are used for light-duty drilling (Clausing Machine Tools).

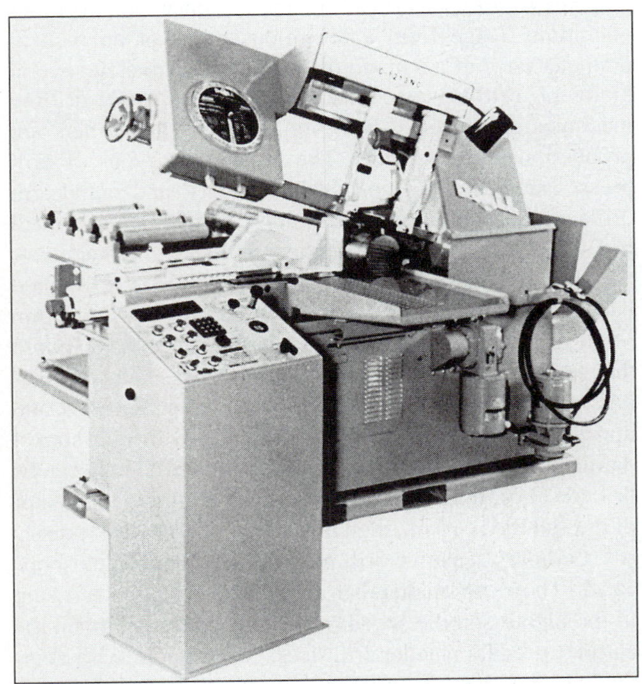

FIGURE 13.6
Automatic band cutoff saw (DoALL Company).

at 10,000 rpm, which would be conservative for a 0.010-in. drill, fine, powdery chips often develop, which can quickly dull the drill. Holes smaller than approximately 1/16 in. must be drilled on a microdrill press, and most microdrilling takes place at speeds between 5000 and 12,000 rpm.

Heavy-duty upright drilling machines (Figure 13.8) have power feeds and reversing capability. On some machines tapping accessories are provided that feed the tap with the correct thread pitch and then rapidly reverse to remove the tap from the workpiece. A coolant pump is

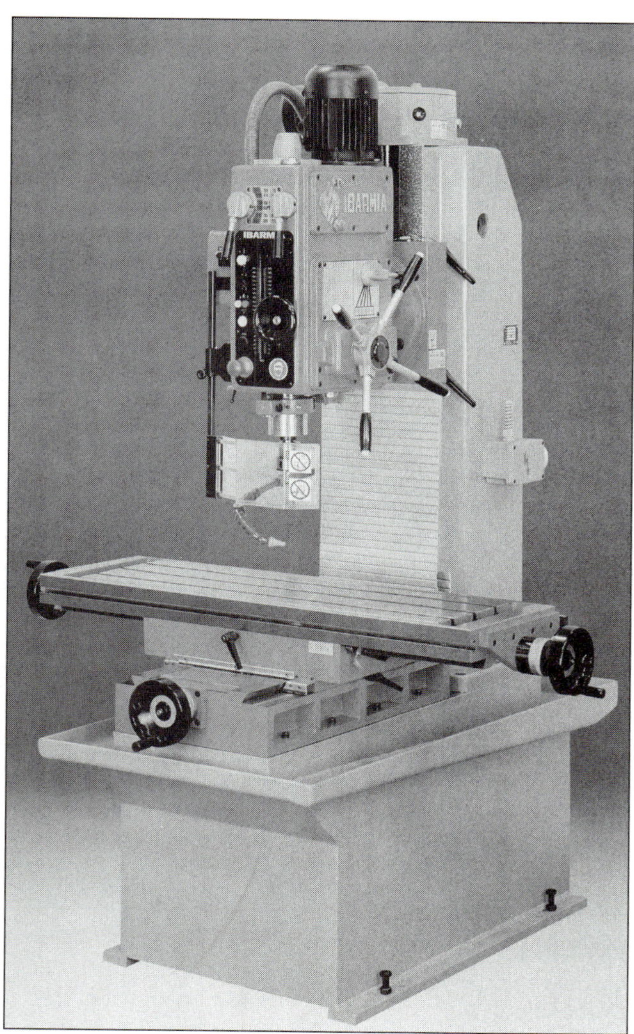

FIGURE 13.8
Upright heavy-duty drilling machine (Ibarmia Drilling Machine Division).

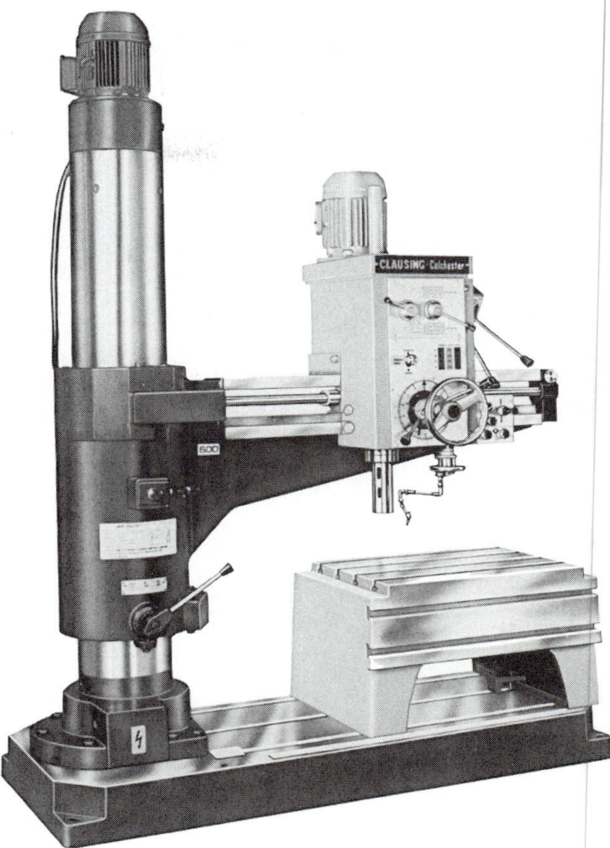

FIGURE 13.9
Radial arm drill press (Clausing Machine Tools).

often utilized with these heavy-duty drill presses to provide an ample supply of pressurized cutting fluid. A wide range of hole sizes can be drilled on these machines, often from 1/4 to 3 1/2 in. in diameter, the largest twist drill size. Larger holes are normally made on radial arm drilling machines with flat spade drills.

Radial arm drill presses (Figure 13.9) are made in a variety of sizes. They are sized by the diameter of the column and the length of the arm as measured from the center of the spindle to the outer edge of the column. Small workpieces can be shifted around by the operator until the desired hole location is directly under the twist drill, as is done on sensitive drill presses; however, when the workpiece is a large casting that weighs many tons it is not convenient to reposition the part for each drilled hole. The drill head of the radial arm drill press, which contains the spindle and the drilling tool, is easily moved along a rail on the arm to locate the tool over the workpiece. When it is located at the drilling position, the radial arm can be clamped to the column and the drill head can be clamped to the arm. Like the upright machine, the radial arm drill press has a power feed mechanism in addition to a hand-feed lever. It also has a coolant tank, usually located in the base, with a pump to filter and recirculate cutting fluid.

Gang drills (Figure 13.10) are used where a series of drilling and reaming operations must be performed in a sequence. Each drill head has its own special tooling and a fixture for holding the workpiece. Time-consuming tool

FIGURE 13.10
Gang drills (Ibarmia Drilling Machine Division).

changing and setups are eliminated by using gang drills. Drilling machines with multiple spindles (Figure 13.11) allow many drilled holes to be made in one pass. Drill jigs that guide the drills to precise locations are often used with multiple spindle drills. Modern CNC machining centers (Figure 13.12) are often used for precision drilling and reaming operations because of their capability to position the tool precisely under computer control.

Cutting Tools: Drills, Taps, and Reamers

High-speed steel twist drills (Figure 13.13) are the most common tools used for general purpose drilling in metals. The twisted flutes provide for chip clearance and removal. The helix of the flute also provides positive rake (Figure 13.14), which is needed for most work materials. Straight flute drills (Figure 13.15) provide zero rake and are used for plastics, brass, and bronze, which have the tendency to "grab" when drills have positive rake. Zero rake can also

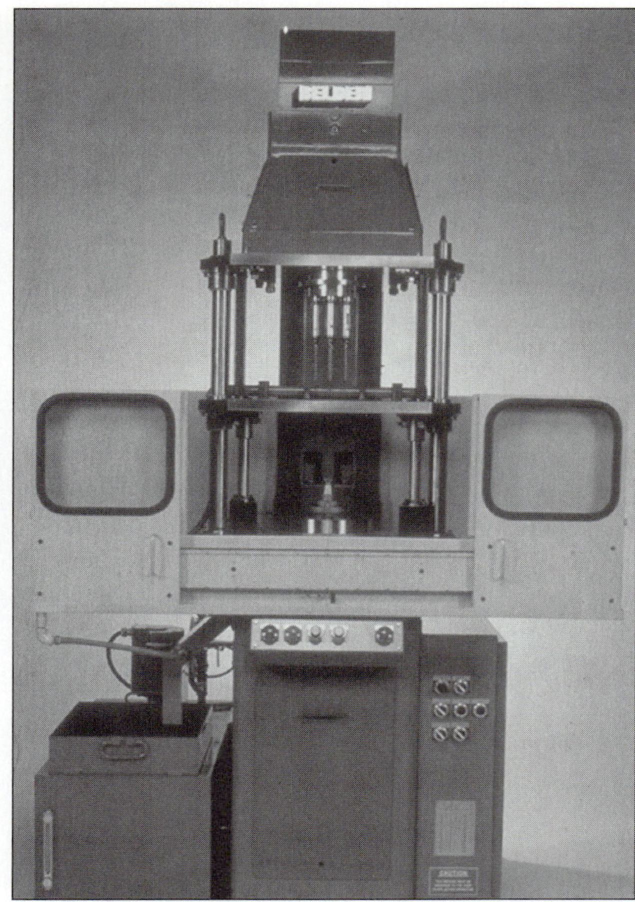

FIGURE 13.11
Multiple spindle drilling machine (Belden Machine Corporation).

be provided on a standard twist drill by grinding a flat on its cutting edge (Figure 13.16).

Spade drills (Figure 13.17) are made in sizes from very tiny to 12 in. in diameter and are widely used for drilling holes 1 in. and larger. A spade drill has the advantage of a very rigid shank, enabling heavier cuts to be taken, and the main body is often provided with holes to pump coolant through to the blade. Blades are easily sharpened and replaced. Spade drills are frequently equipped with a stepped blade to produce a bevel or chamfer on the edge of the hole to facilitate tapping and assembly.

FIGURE 13.12
Machining centers such as this one perform many different cutting operations automatically (Cincinnati Milacron, Inc.).

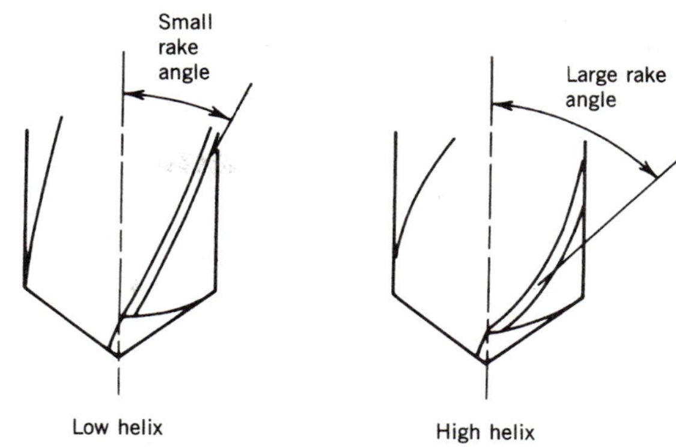

FIGURE 13.14
Helix of twist drill provides low or high rake angle.

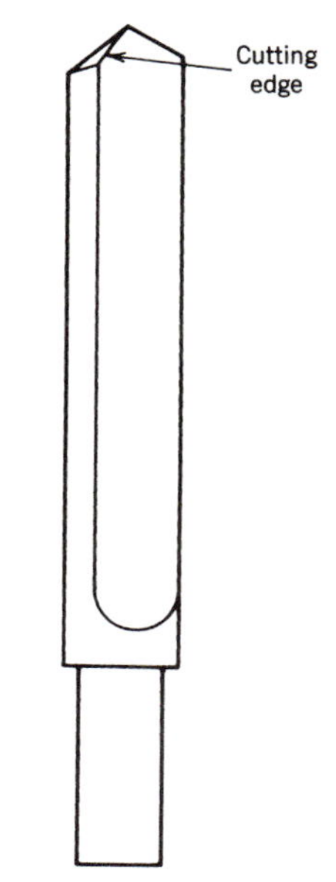

FIGURE 13.15
Straight flute drill.

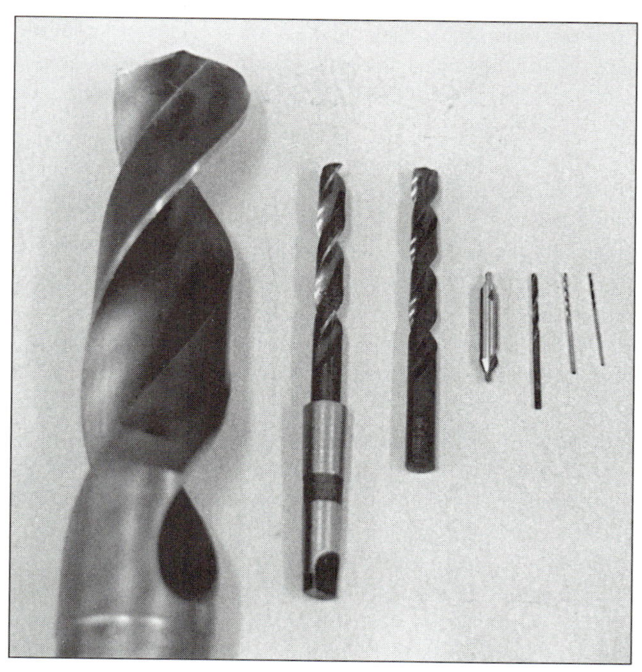

FIGURE 13.13
A variety of high-speed steel twist drills.

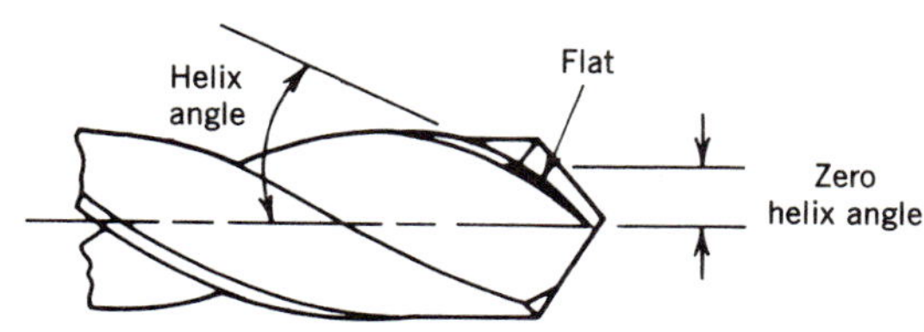

FIGURE 13.16
A flat ground on the cutting edge of the drill is often necessary when drilling brass, bronze, and plastics.

FIGURE 13.17
Taper shank spade drill (Courtesy of DoALL Company).

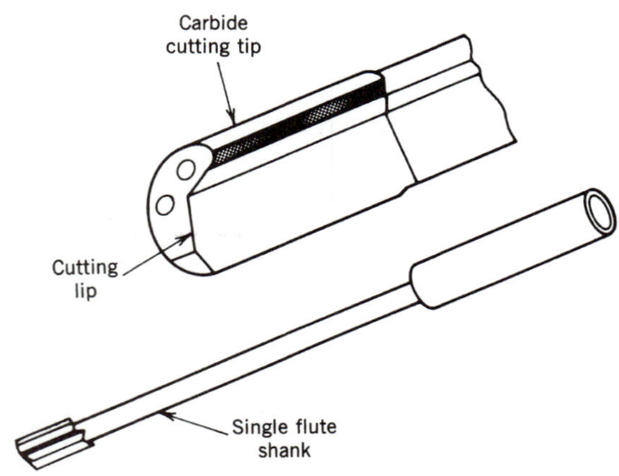

FIGURE 13.18
Gun drill.

Gun drills (Figure 13.18) are capable of making very precise holes many feet deep. Frequently used for drilling holes in automotive engine blocks, among other deep hole applications, they are used with special machinery that guides the drill and pumps coolant through it at high pressure.

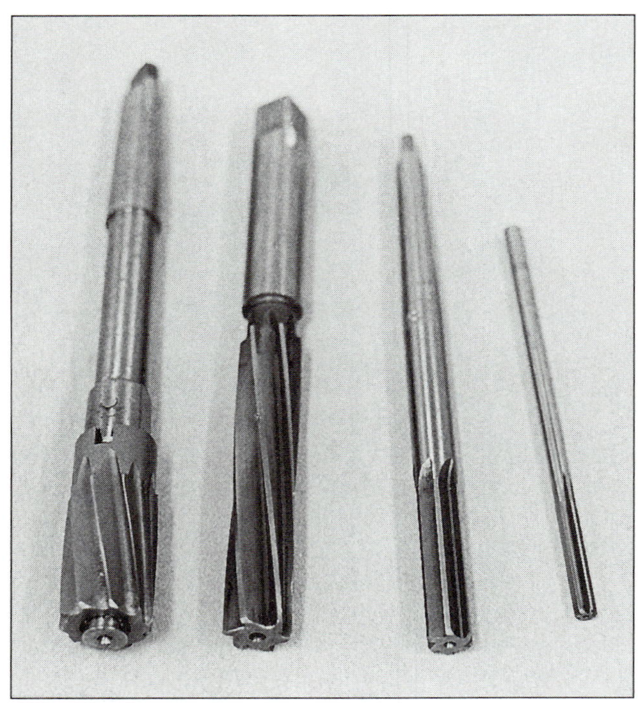

FIGURE 13.19
A variety of reamers.

A center drill is used for making a 60° conical hole in the end of a shaft for turning in a lathe. This hole accepts a center mounted in the tailstock of the lathe to provide support for the end of the long, slender workpiece. Center drills are also used for spotting, providing a starting hole for larger drills.

Most drilling operations lack precision. Control of hole diameter, location, roundness, straightness, and surface finish is difficult. Reamers (Figure 13.19) are used to enlarge an existing hole and give it a more accurate diameter with an acceptable finish; however, if the drilled hole is off center, rather than relocating it the reamer will follow the existing hole. Boring operations, unlike reaming, do not have a tendency to follow the existing hole and can restore an off-center hole to true position. Reamers are used on drill presses, lathes, and vertical milling machines.

Special tools called counterbores are used to enlarge existing holes (Figure 13.20). Countersinks provide a bevel to receive certain bolt heads, and spotfacing is done to provide a machined seat for bolt heads or nuts.

Taps are hardened thread cutting tools with flutes that collect chips. Taps are brittle, and breakage is a serious problem in manufacturing processes. Standard hand taps (Figure 13.21) include taper, plug, and bottoming taps. Threaded internal holes can be made by hand-twisting the

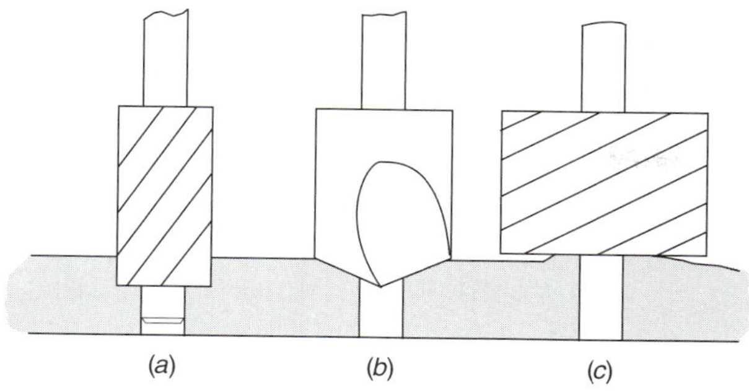

FIGURE 13.20
(a) Counterbore. *(b)* Countersink. *(c)* Spotfacing.

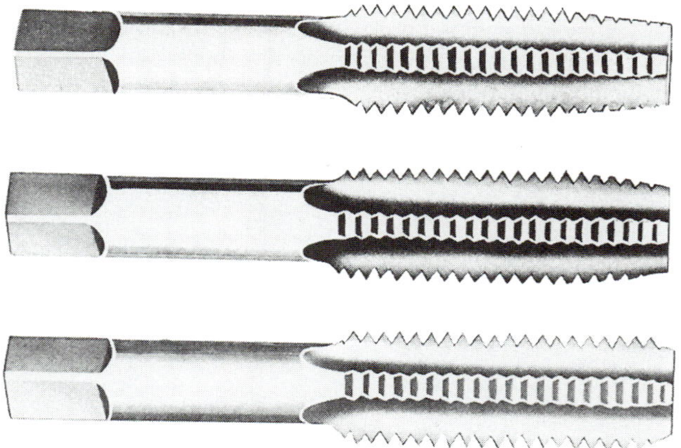

FIGURE 13.21
Standard set of three taps. Top to bottom: taper, plug, and bottoming.

FIGURE 13.22
Spiral point tap (DoALL Company).

tap into a properly sized drilled hole by means of a tap handle. Machine tapping is a higher-speed operation in which plug taps are most often used. In machine tapping, spiral point taps (Figure 13.22) are far more efficient and are less likely to break, but they are effective only for the tapping of through holes. Spiral point taps push the chips forward through the hole and would cause jamming at the bottom of blind (not through) holes.

Tapping a blind hole presents problems owing to incomplete threads at the bottom of the hole and because chips collect at the bottom of the hole. Bottoming taps are designed with only one or two tapered teeth to cut complete threads near the bottom of blind holes. Spiral flute taps (Figure 13.23) are also helpful when tapping blind holes because they direct the chips upward along the flutes and out of the hole.

FIGURE 13.23
Tapping with a spiral flute tap (DoALL Company).

Threading dies for making external threads are used in a diestock, a handle that holds the die, for hand threading. Dies for high-production threading on machine tools are covered later in this chapter.

TURNING MACHINES AND EQUIPMENT

Turning machines rotate a workpiece against a tool that gradually moves along the workpiece, producing an accurate cylindrical or tapered (conical) surface. Tracing attachments make possible almost unlimited curved shapes on turned cylindrical parts. Computer numerical control–operated turning machines (Figure 13.24) can produce a great variety of cylindrical, curved, and

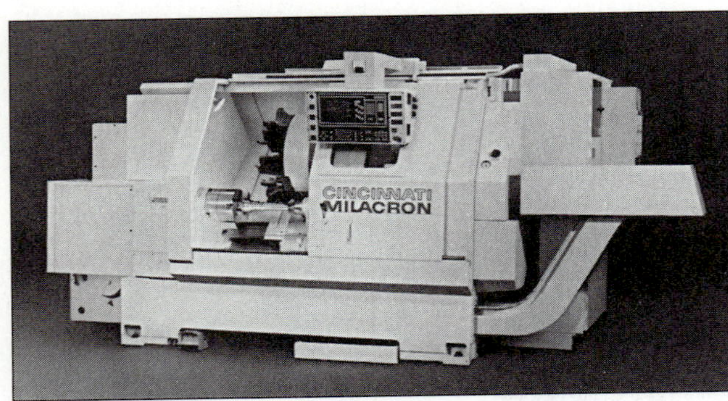

FIGURE 13.24
NC lathe (Cincinnati Milacron, Inc.).

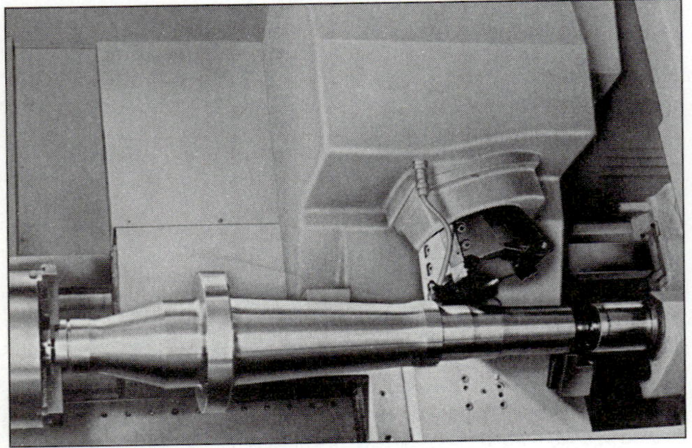

FIGURE 13.25
Curved and tapered turning operation on a shaft in an NC lathe (Cincinnati Milacron, Inc.).

tapered shapes (Figure 13.25) under computer control. These NC machines are fully automatic and require an operator to only load and unload the workpiece. Turning machines include the engine lathe with its horizontal spindle (Figure 13.26) and the vertical spindle machines, the vertical boring mill (Figure 13.27), and vertical turret lathe (Figure 13.28).

The Engine Lathe

The principal function of the lathe is to remove unwanted material to form cylindrical and conical shapes. This is done in a turning operation that utilizes a single-point tool (as contrasted with a multiple-point tool such as a milling cutter). Other tool shapes are used to make grooves, to cut off round stock, and to make various kinds of threads by chasing (taking a series of cuts in the same groove until a full thread is made).

FIGURE 13.26
A 15-in. swing engine lathe. Swing (largest diameter that can be turned) is one measurement of lathe size. (Clausing Machine Tools).

FIGURE 13.27
Vertical boring mill.

FIGURE 13.28
Vertical turret lathe (Ductr Mfg.).

The engine lathe acquired its name from the steam engine that first powered it, an improvement over the earlier foot- or hand-powered lathes. It has been the most important tool in machine shops and manufacturing plants for over 150 years. Today these valuable machines are still very much in use in tool and die shops, job machine shops, and machine repair shops, although they have been replaced by CNC lathes and screw machines for production purposes. Lathes range in size from tiny jeweler's lathes to huge machines that turn massive forgings (Figures 13.29 and 13.30).

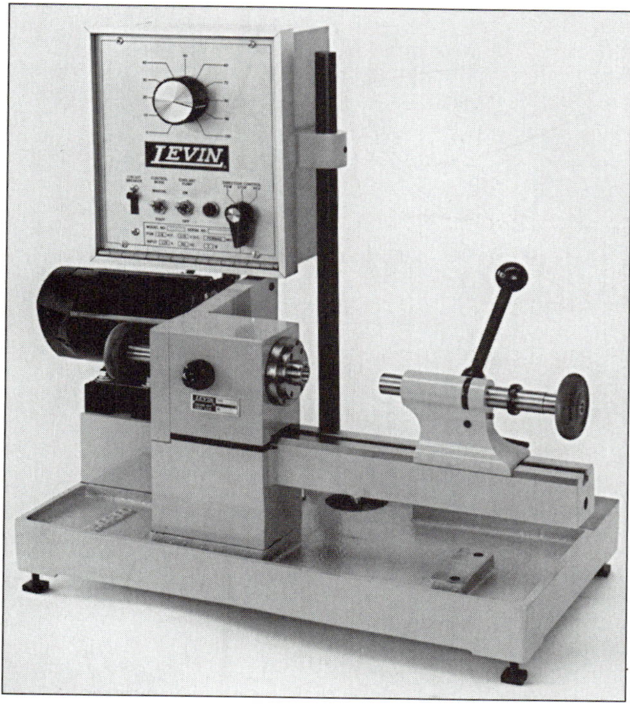

FIGURE 13.29
Jeweler's lathes are used to make very tiny precision parts. Many attachments for toolholding can be used with these machines (Courtesy of Louis Levin & Son, Inc.).

FIGURE 13.30
A large heavy-duty engine lathe (Clausing Machine Tools).

Threads

Although cutting threads is one of the important functions of the lathe (Figure 13.31), most threads are cut on production machines. There are a number of thread forms that have been standardized, and some of the more commonly used ones can be seen in Figures 13.32 through 13.43 (pp. 259–262). Screw thread sizes are expressed in terms of major (outside) and minor (inside) diameter, pitch diameter, number of threads per inch, and pitch (distance between threads) (Figure 13.44).

Even though the American National Form as a thread standard with its common thread series National Coarse (NC) and National Fine (NF) is obsolete, it is still in use in many machine shops. Even some tap drill charts are still being produced based on this popular system. The U.S. American standard for Unified threads is the official thread

FIGURE 13.31
Single-point threading on a lathe.

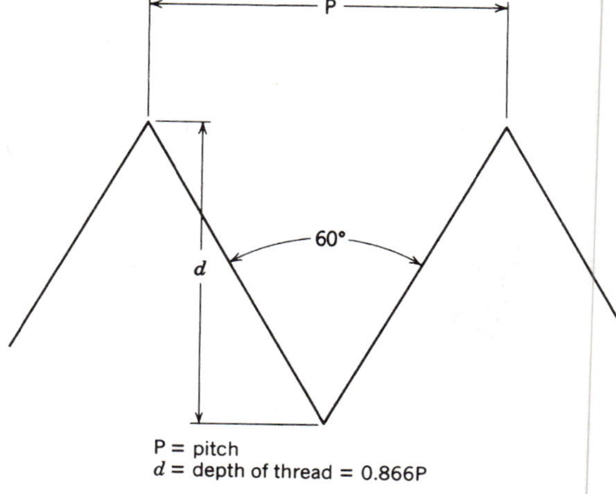

FIGURE 13.32
The 60° sharp V-thread form (Kibbe, Neely, Meyer, and White, *Machine Tool Practices,* 7th ed., ©2002 Pearson Education, Inc.).

FIGURE 13.33
The American National form of thread (Kibbe, Neely, Meyer, and White, *Machine Tool Practices,* 7th ed., ©2002, Pearson Education, Inc.).

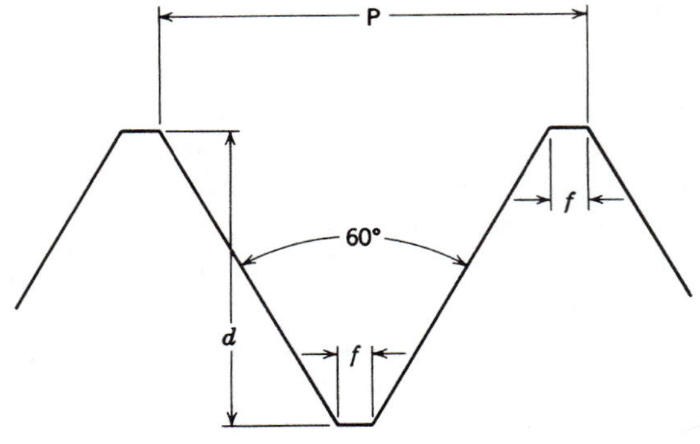

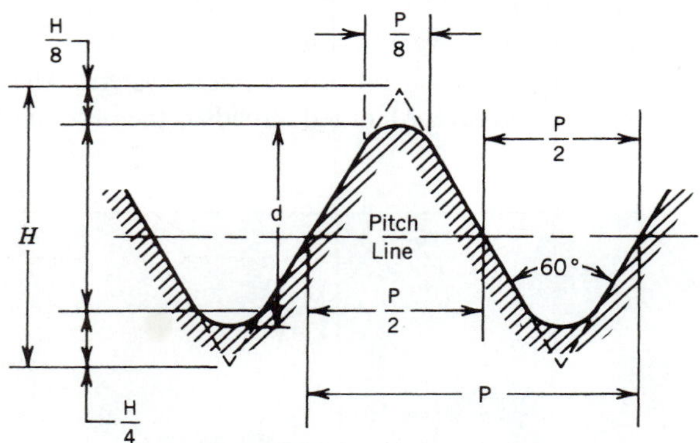
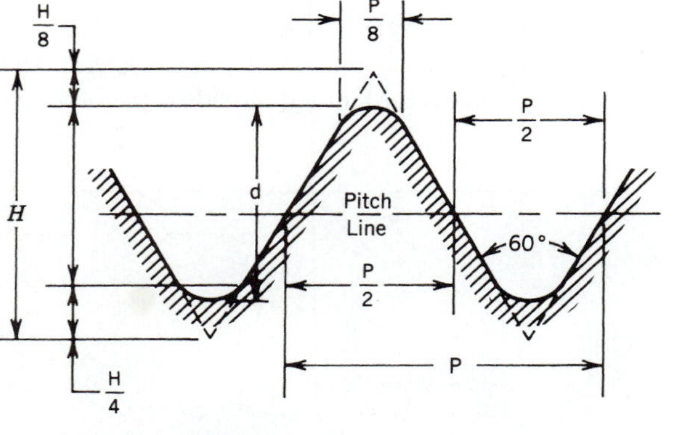

FIGURE 13.34
Unified thread forms.

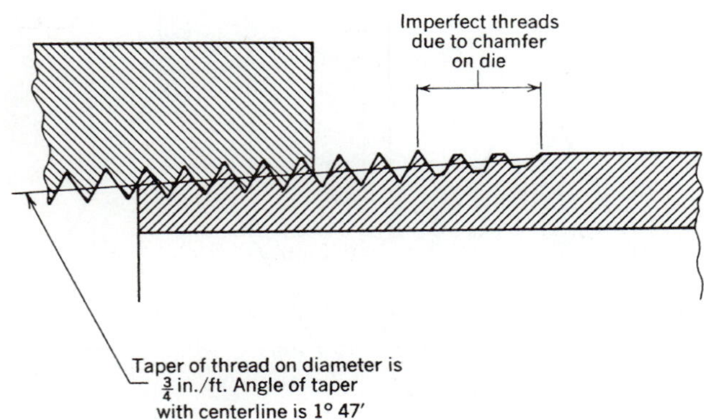

FIGURE 13.35
American National standard taper pipe thread form (Kibbe, Neely, Meyer, and White, *Machine Tool Practices,* 7th ed., ©2002, Pearson Education, Inc.).

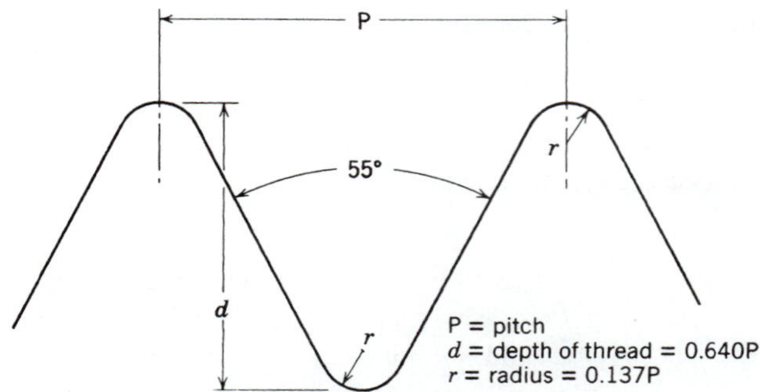

FIGURE 13.36
Whitworth thread form (Kibbe, Neely, Meyer, and White, *Machine Tool Practices,* 7th ed., ©2002, Pearson Education, Inc.).

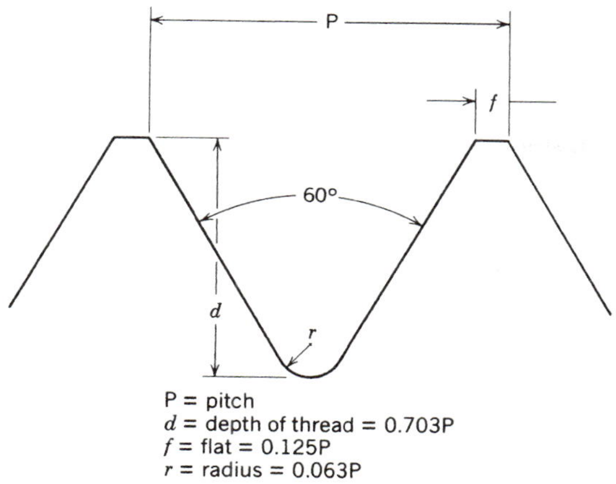

FIGURE 13.37
The SI metric thread form (Kibbe, Neely, Meyer, and White, *Machine Tool Practices,* 7th ed., ©2002, Pearson Education, Inc.).

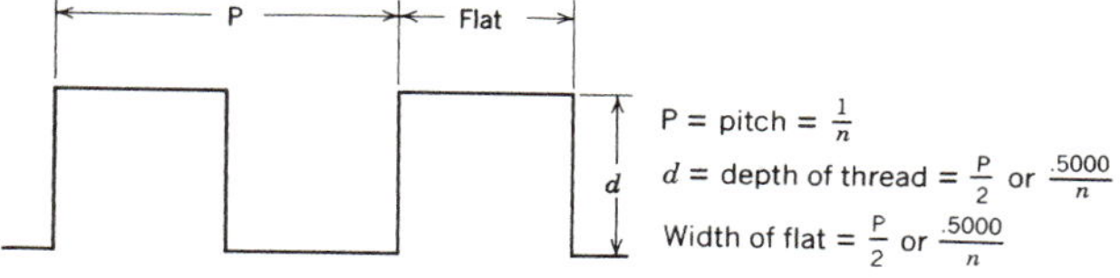

FIGURE 13.38
The square thread form (White, Neely, Kibbe, Meyer, *Machine Tools and Machining Practices,* Vol. II, ©1977 John Wiley & Sons, Inc. This material is used by permission of John Wiley & Sons, Inc.).

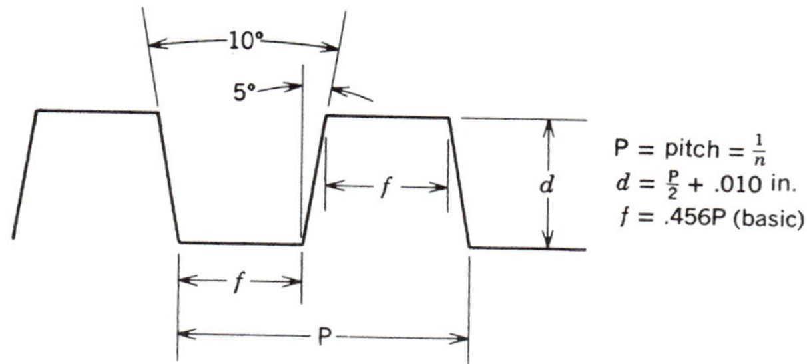

FIGURE 13.39
Modified square thread form (White, Neely, Kibbe, Meyer, *Machine Tools and Machining Practices,* Vol. II, ©1977 John Wiley & Sons, Inc.This material is used by permission of John Wiley & Sons, Inc.).

standard at the present time with its basic Unified Coarse thread (UNC) and Unified Fine thread (UNF). American and Unified standards are interchangeable for the most part; fits and tolerances make up the principal difference.

Screw thread manufacturing in the United States utilizes either the Unified or the Metric series. Both have similar thread forms but different pitches, pitch diameters, and major diameters. For metric thread forms the pitches

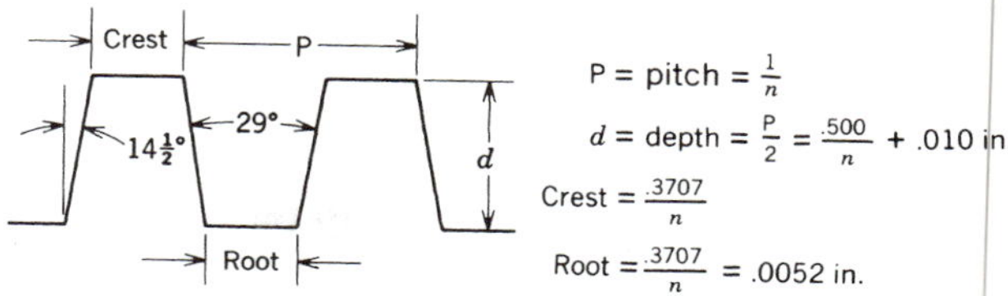

FIGURE 13.40
Acme thread form (White, Neely, Kibbe, Meyer, *Machine Tools and Machining Practices,* Vol. II, ©1977 John Wiley & Sons, Inc. This material is used by permission of John Wiley & Sons, Inc.).

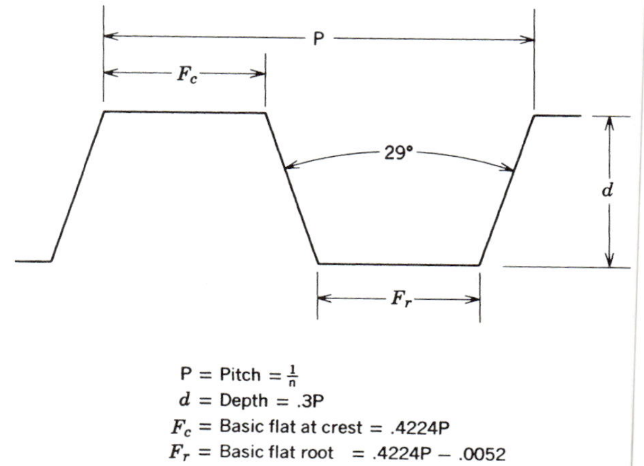

FIGURE 13.41
American Standard Stub Acme thread form (White, Neely, Kibbe, Meyer, *Machine Tools and Machining Practices,* Vol. II, ©1977 John Wiley & Sons, Inc. This material is used by permission of John Wiley & Sons, Inc.).

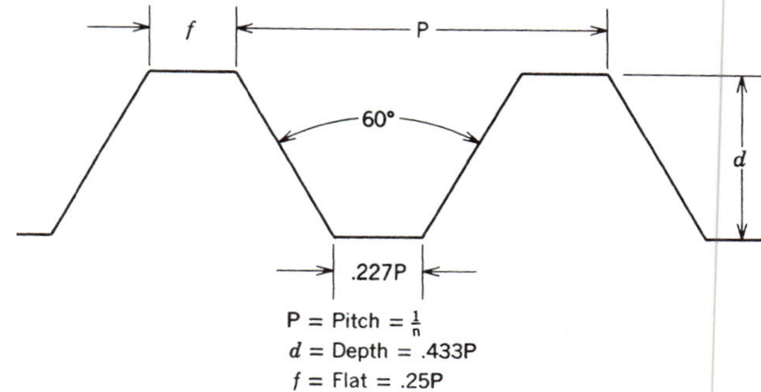

FIGURE 13.42
Stub Acme thread form (White, Neely, Kibbe, Meyer, *Machine Tools and Machining Practices,* Vol. II, ©1977 John Wiley & Sons, Inc. This material is used by permission of John Wiley & Sons, Inc.).

and pitch diameters may vary by country and often are not interchangeable. Besides the British ISO, there is the Système International (SI), the Löwenherz metric thread used in Germany, SAE spark plug threads, the British

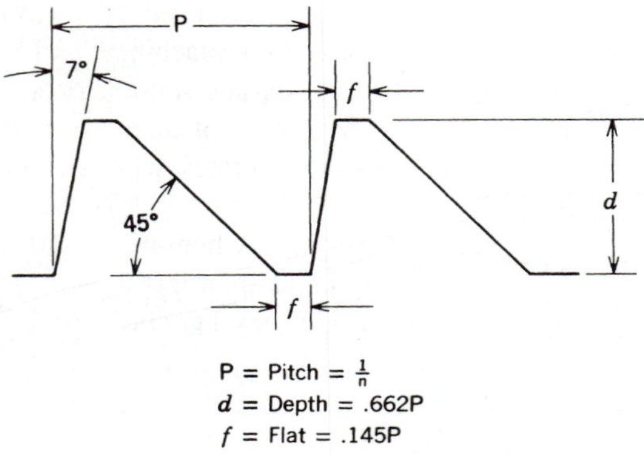

FIGURE 13.43
Buttress thread form (White, Neely, Kibbe, Meyer, *Machine Tools and Machining Practices,* Vol. II, ©1977 John Wiley & Sons, Inc. This material is used by permission of John Wiley & Sons, Inc.).

standard for spark plugs, and numerous differences among manufacturers worldwide.

In the Unified and the ISO metric thread systems there are coarse and fine pitches, and the pitch varies with the diameter. The angle of the thread teeth is 60° in both standards. Threads are designated in the Unified standard beginning with the nominal size first, followed by the number of threads per inch, the thread series, and class of thread. If the letters LH are added, the thread is left-hand. An example is 3/4-10 UNC-2A. The symbols 1A, 2A, and 3A indicate an external thread tolerance; 1A is a loose fit, and 3A is a tight fit. Three other tolerances are used for internal threads, 1B, 2B, and 3B. These can be adjusted to get a desired fit; for example, a 2A bolt can be matched with a 3B nut. Metric threads are likewise designated with the major diameter first, followed by the pitch, and then the tolerance grade. An example is M20 × 2 6H/5g6g.

Internal threads that are tapped must be predrilled (tap drilled) to a certain diameter to produce a correct percentage of thread for ease of tapping in a given material. For example, a 75 percent thread is correct for soft steel, but a 60 percent thread (larger tap drill) may be better for 301 stainless steel. The tap drill size dictates the *actual minor diameter* and can be determined with the following formula:

$$\text{percentage of thread} = \frac{\text{major diameter} - \text{actual minor diameter}}{2 \times \text{basic thread height}} \times 100$$

or,

$$\text{actual minor diameter} = \text{major diameter} - 2 \times \text{basic thread height} \times \text{percentage of thread}$$

The height of the tooth (basic thread height) for Unified threads equals the pitch × 0.541.

Tapered threads such as those used to join pipes can be chased on an engine lathe by using a taper attachment. The symbol for American standard taper pipe thread is NPT.

Work-Holding Devices on Lathes

It is essential that lathe workpieces be held securely to the machine spindle to prevent the possibility that the rapidly spinning part could be thrown from the machine spindle, causing damage or injury. It is also important that the desired center of the workpiece be aligned to the center of the machine spindle. Poor workpiece centering causes

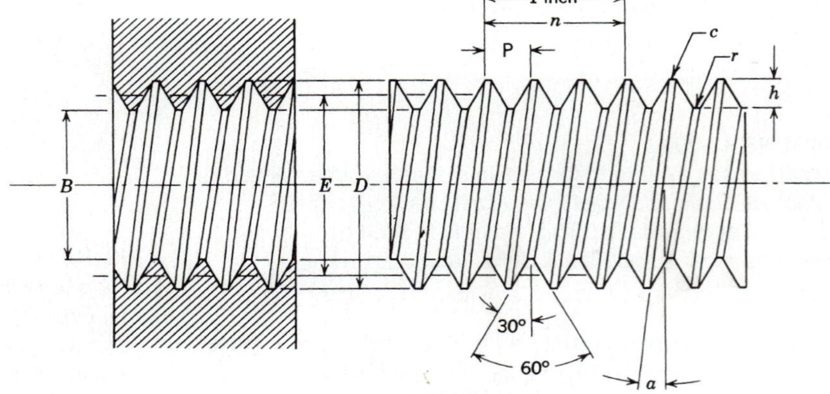

FIGURE 13.44
Thread nomenclature and dimensions (Kibbe, Neely, Meyer, and White, *Machine Tool Practices,* 7th ed., ©2002, Pearson Education, Inc.).

D Major Diameter
E Pitch Diameter
B Minor Diameter
n Number of threads per inch (TPI)
P Pitch
a Helix (or Lead) Angle
c Crest of Thread
r Root of Thread
h Basic Thread Height (or depth)

inaccurate location of machined diameters and creates an imbalance condition that interferes with smooth operation of the lathe.

Tooling used to locate and hold a workpiece on a lathe includes independent (usually four-jaw) chucks (Figure 13.45), universal (two-, three-, and six-jaw) chucks (Figure 13.46), and collets (Figure 13.47). Independent chucks have jaws that must be adjusted separately to grip and locate the workpiece. Square, rectangular, and odd-shaped parts are usually held in four-jaw independent chucks. Most universal chucks have jaws that all move together or apart the same amount when a chuck wrench is turned. Round and hexagonal parts or bar stock can be quickly gripped and centered with these chucks, so they are often called *self-centering chucks.*

Even though most universal chucks have three jaws, six-jaw universal chucks are able to grip round materials more securely than three-jaw chucks. Three-jaw chucks for engine lathes often have two interchangeable sets of jaws, one set for internally gripping hollow parts such as rings and one set for holding workpieces externally. Most chucks have removable jaws (called *top jaws*), each attached to a master jaw. These can be quickly removed and reversed to change from external to internal gripping (Figures 13.48 and 13.49). Soft jaws are not hardened and can be machined while in place on the lathe. These are used to hold odd-shaped parts.

Power chucks (Figure 13.50) are used on turret lathes, CNC lathes, and other production machines. The power chuck is commonly a three- or six-jaw self-centering chuck that is opened and closed by means of an air or hydraulic cylinder that is fastened to and rotates with the machine spindle.

Most self-centering chucks, when new, will center and hold round parts within less than 0.001 in. of runout (eccentricity). However, when they become worn, runout

FIGURE 13.45
Four-jaw independent chuck (Buck Chuck Company).

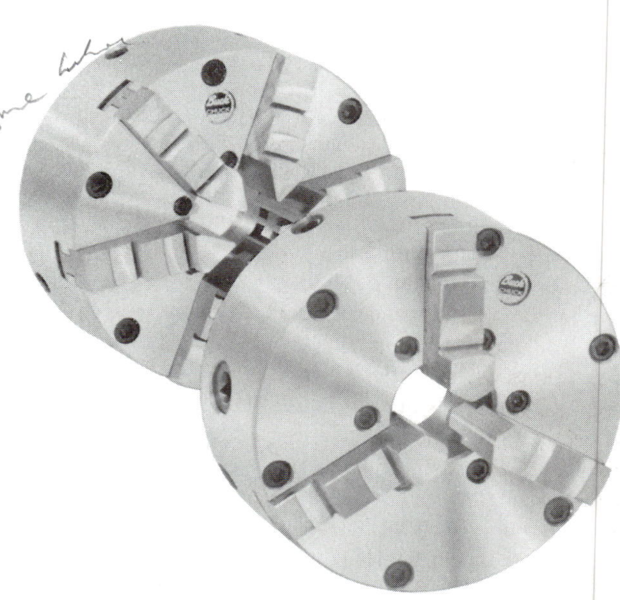

FIGURE 13.46
Three- and six-jaw chucks. These standard chucks require an extra set of jaws for gripping large diameters (Buck Chuck Company).

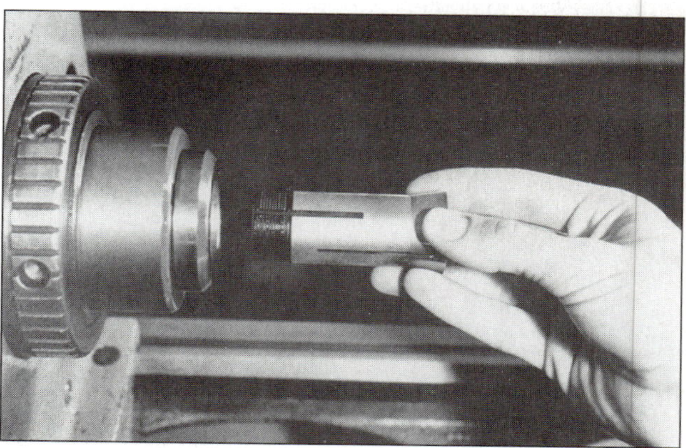

FIGURE 13.47
Collets are held in the spindle nose with an adapter and a draw bar.

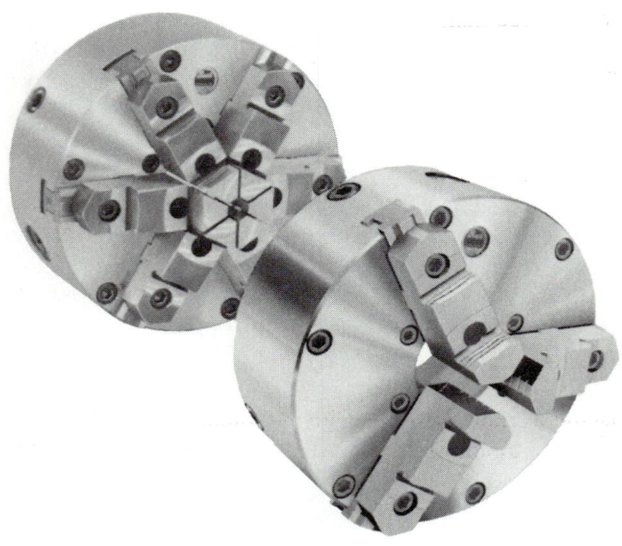

FIGURE 13.48
Three- and six-jaw chucks with top jaws (Buck Chuck Company).

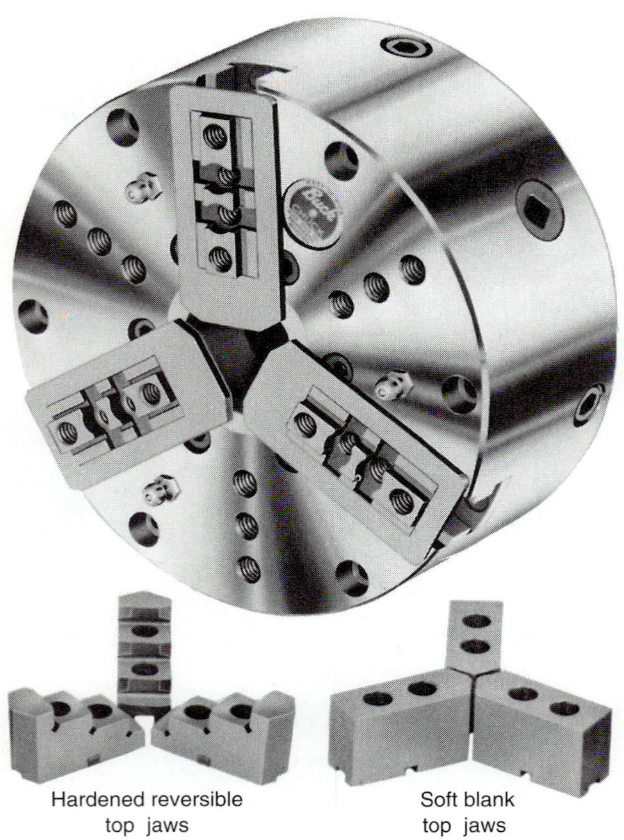

FIGURE 13.49
Three-jaw chuck showing hardened reversible top jaws and soft blank top jaws (Buck Chuck Company).

FIGURE 13.50
Power chucks are often used on production machinery (Buck Chuck Company).

FIGURE 13.51
Adjust-Tru® chuck. Chuck is adjusted at *G* to eliminate runout using a dial indicator. Jaws are tightened at *C* (Buck Chuck Company).

may exceed 0.010 in. Some chucks have adjustment screws to compensate for wear; these can be adjusted to hold runout within 0.0005 in. (Figure 13.51).

A standard collet is a spring steel sleeve with slits along its length that permit it to spring open a small amount and accept the workpiece. To close the collet, an external taper on the collet is pulled or pushed into a mating taper, squeezing the collet securely closed around the workpiece. Collets are capable of consistently holding round precision workpieces to less than 0.001 in. of runout. Steel collets expand very little, so they cannot hold parts that vary from the nominal size of the collet by more than a few thousandths of an

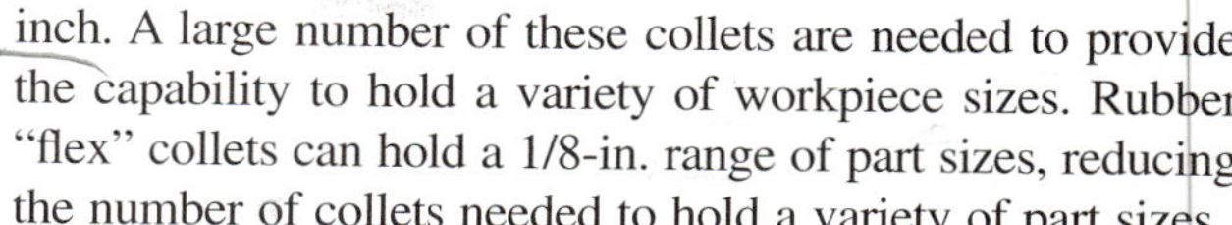

inch. A large number of these collets are needed to provide the capability to hold a variety of workpiece sizes. Rubber "flex" collets can hold a 1/8-in. range of part sizes, reducing the number of collets needed to hold a variety of part sizes.

Chucks and collets are fastened to the spindle by several methods—flange nose, camlock, long taper key drive, and threaded nose (Figure 13.52).

When long, slender workpieces are held between centers on a lathe, conical tools called *centers* are inserted into matching depressions on each end of the part. One center is held in the lathe spindle, and the other center is held in the lathe tailstock, providing support for the outer end of the workpiece. A tool called a *lathe dog* is required to assure that the workpiece rotates with the spindle. The dog clamps onto the part and seats against the spindle to drive the part.

Mandrels (Figure 13.53) are a form of work-holding device for parts that have a central hole or bore. The mandrel holds the work concentric to the bore, so that any machined features are also concentric. Workpieces are held by friction, either by means of a slight taper (0.006 in./ft) on a mandrel that is pressed into the bore, or by an expanding mandrel. The mandrel is commonly held between centers. Stub arbors, as they are sometimes called, are a form of mandrel that is held in a chuck (Figure 13.54). The term *arbor* correctly refers to a rotating member that holds a cutting tool such as the arbor of a milling machine, and *mandrel* correctly refers to a rotating member that holds a workpiece; however, these terms are often interchanged in common shop terminology and in product literature.

Am. St. flange nose

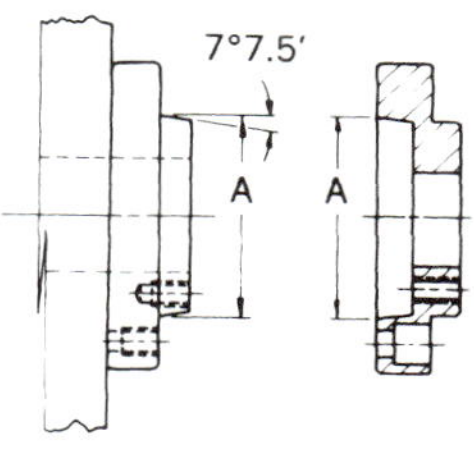

Camlock
Camlock stud length is adjustable to suit spindle.

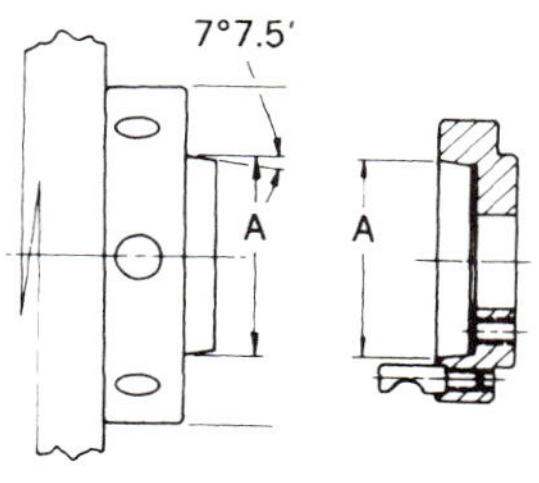

Long taper key drive

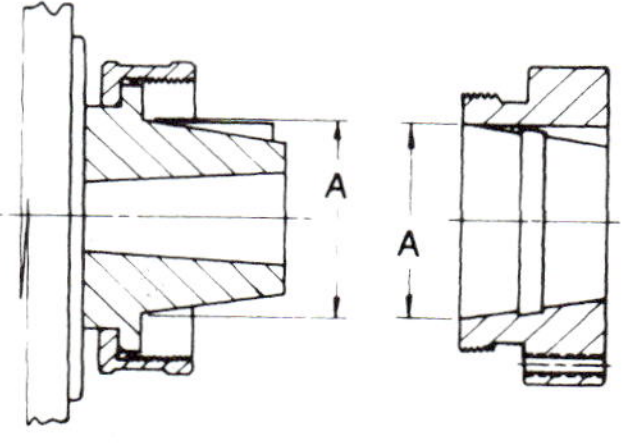

Threaded mounting plates

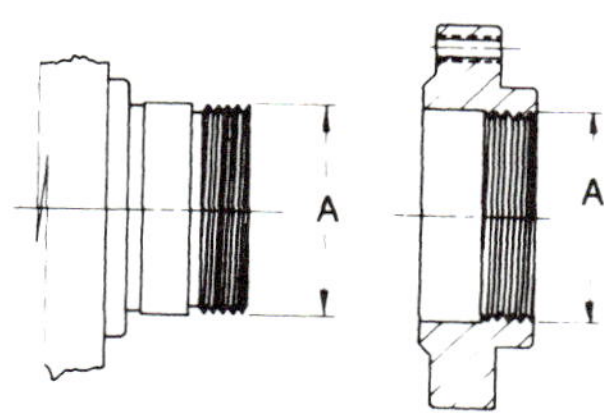

FIGURE 13.52
Four basic types of spindle nose arrangements for mounting chucks and other work-holding devices (Buck Chuck Company).

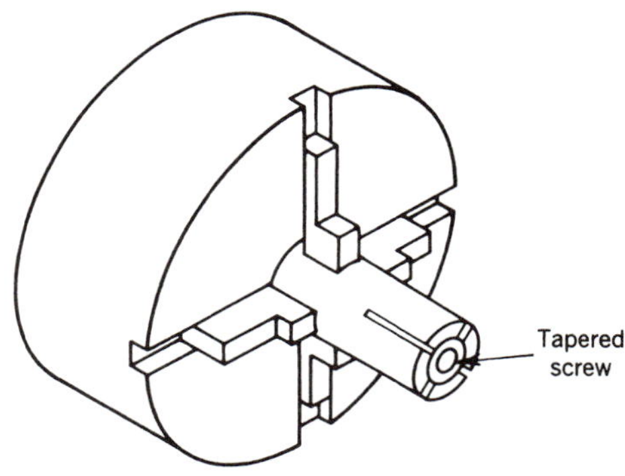

FIGURE 13.54
Stub mandrel. Part to be machined is gripped in the bore by expanding the mandrel with a tapered screw.

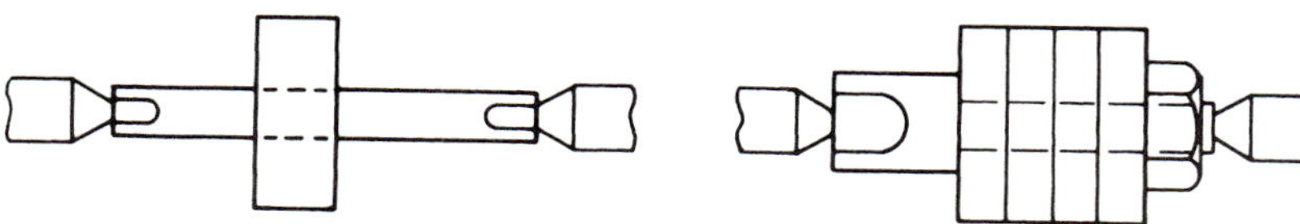

FIGURE 13.53
Left, tapered mandrel (0.006 in./ft) holds part to be turned by friction in the bore. Right, gang mandrel holds several parts for machining on a straight threaded screw.

Turning Operations

In *turning cuts,* the carriage guides the cutting tool as it feeds parallel to the centerline of rotation of the workpiece to create cylindrical shapes (Figures 13.55 and 13.56) or at an angle to create conical shapes (tapers). When a cut is taken across the end of a piece by moving the cross-slide instead of the carriage, the procedure is called *facing.* If a bore is required in a solid workpiece, a hole is drilled by holding the drill in the tailstock and feeding it in with the handwheel (Figure 13.57). The hole can then be reamed (Figure 13.58) or bored with a single-point tool (a boring bar) that is fastened to the carriage (Figure 13.59). Counterbores, internal and external grooves, and threads can be cut on the lathe.

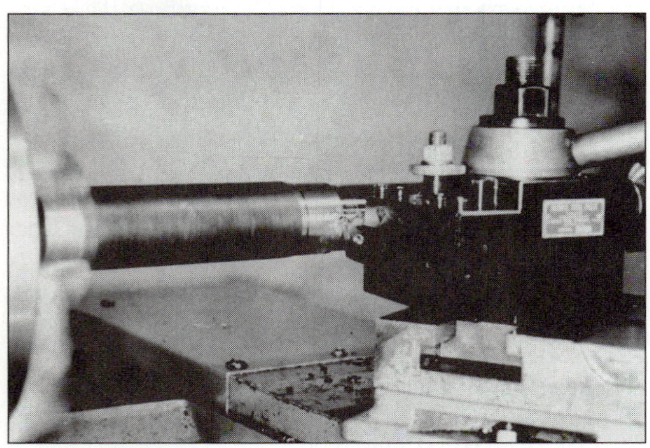

FIGURE 13.55
Turning operation. Long pieces should be supported between the chuck and the center.

FIGURE 13.56
Short workpieces can be turned while supported in the chuck only.

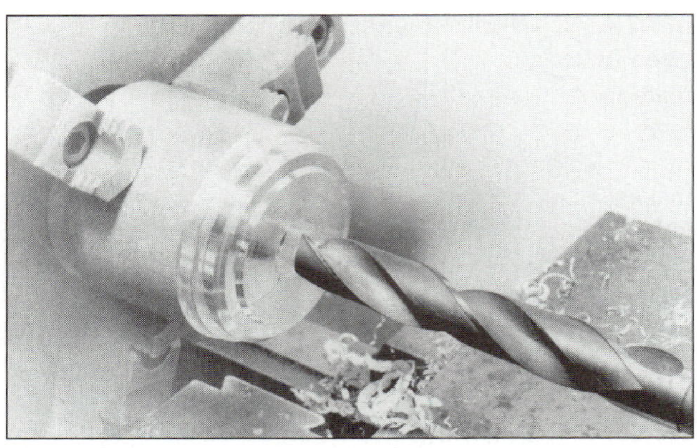

FIGURE 13.57
Drilling in the lathe. The drill is supported in the tailstock and is forced into the metal by means of a handwheel.

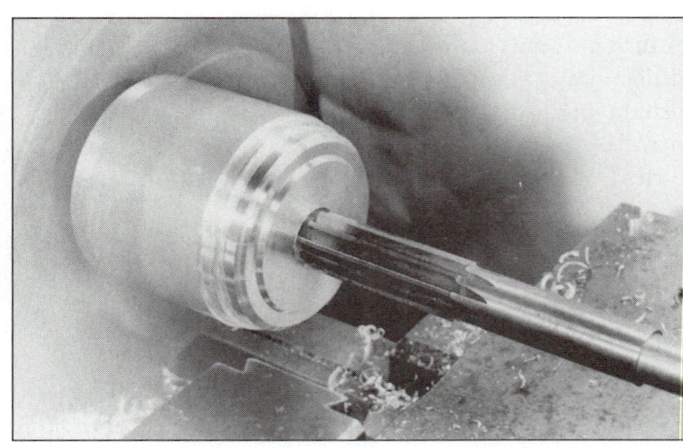

FIGURE 13.58
A drilled hole can be finished to size by reaming.

FIGURE 13.59
Accurate holes may be made by boring with single-point tools.

The diameter of a reamed hole is limited to the exact size of the reamer, and if the drilled hole is eccentric, the reamed hole will also be poorly located, as the reamer follows the existing hole. Boring has the capability to remove more material from one side of a hole than the other to make the hole concentric to the centerline of rotation of the lathe spindle. Hole diameters can be varied by merely repositioning the boring bar; however, boring precision holes requires much more time, care, and skill than reaming.

Turret Lathes

For the most part, skilled machinists use engine lathes, and semiskilled machine operators use turret lathes and other production machines. Turret lathes (Figure 13.60) are similar in many ways to engine lathes. The difference is that turret lathes are arranged for repetitive, rapid production and often perform several operations in quick succession. The workpiece is held in a chuck that is often air operated. A turret, usually six sided, contains different tools such as drills, taps, and turning and knurling tools. The cross-slide usually holds a parting or cutoff tool to sever the finished part from the bar stock. The CNC lathe evolved from the turret lathe and often has a similar configuration.

FIGURE 13.60
Turret lathes are semiautomatic production machines (Clausing Machine Tools).

Turret lathes require an operator at all times. The operator simply moves levers to perform machining operations in the proper sequence to make the part and checks finished dimensions. Tooling is all preset by a setup machinist.

External threading on turret lathes is done by die heads. There are two basic types of die heads, radial (Figure 13.61) and tangential (Figure 13.62). Both types cut the thread full depth in one pass and can be quickly opened to remove the die from the part by means of a handle or trip lever. The cutters (thread chasers) in radial die heads have conventional cutting edges, as

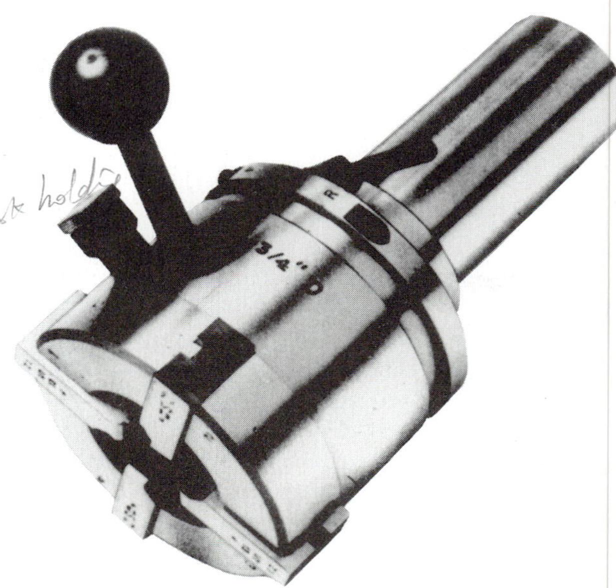

FIGURE 13.61
Radial die head (TRW Cutting Tools Division).

FIGURE 13.62
Tangent die head (TRW Cutting Tools Division).

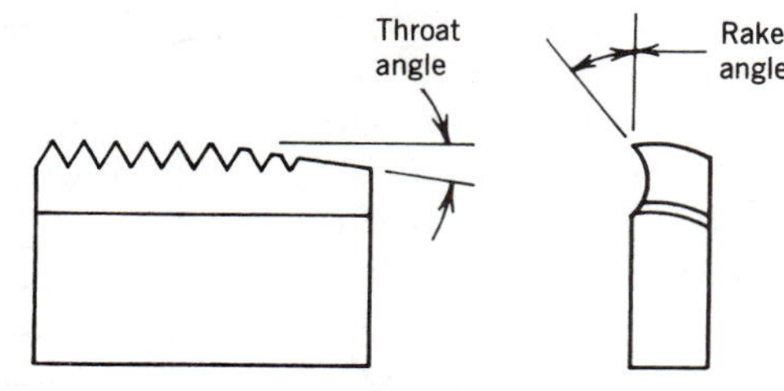

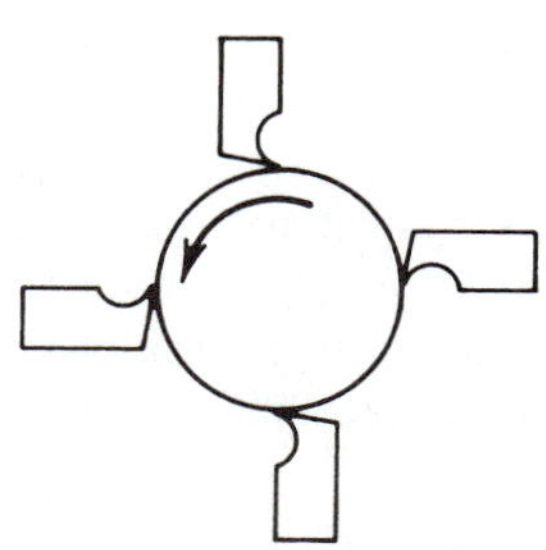

FIGURE 13.63
Principle of radial thread cutting.

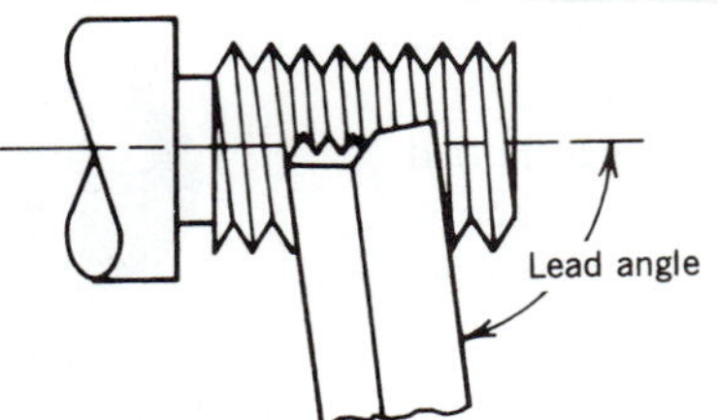

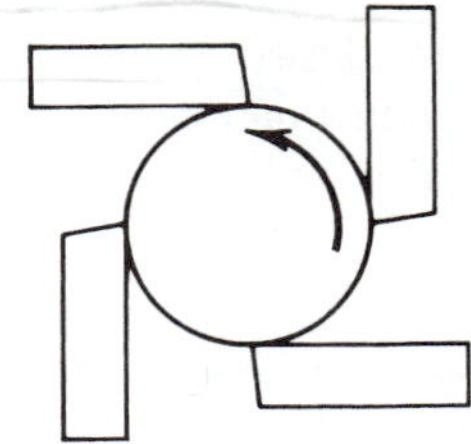

FIGURE 13.64
Tangential principle of thread cutting.

shown in Figure 13.63. Tangential chasers are oriented tangent to the workpiece surface and cut on their ends (Figure 13.64). Tangential die chasers are easier to sharpen and last longer than radial chasers. Internal threads are quickly produced on turret lathes by collapsing taps (Figure 13.65). Ordinary taps, after completing the thread, must be screwed out of the work by reversing the spindle rotation. Collapsing taps (Figure 13.66) simply collapse to reduce their diameter when the thread is completed and can be removed quickly. These quick-opening die heads and collapsing taps can be adapted for use in engine lathes, drill presses, and other machine tools.

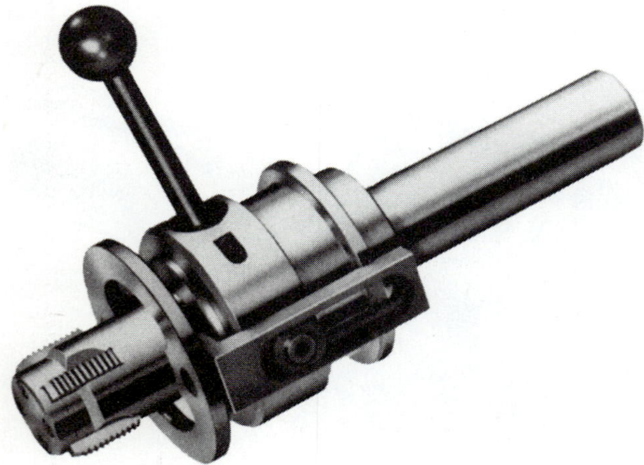

FIGURE 13.65
Collapsing taps make possible the removal of the tap without stopping or reversing the spindle rotation (TRW Cutting Tools Division).

FIGURE 13.66
Here a collapsing tap is mounted on the turret of a turret lathe so it can be quickly swung into position for making an internal thread (TRW Cutting Tools Division).

Screw Machines

Screw machines (Figure 13.67) are similar to turret lathes in that they have preset, indexing tooling and follow a preset sequence of operations to produce a part. Most screw machines are cam-driven and produce small

FIGURE 13.67
Screw machine.

FIGURE 13.68
Horizontal boring mill (Ductr Mfg.).

parts at a rapid rate without needing constant attention by an operator. A stock feed mechanism feeds metal bars into the machine as needed, and stock feeders are kept full by an operator who is in charge of a number of machines. In addition to small threaded fasteners, screw machines can produce a large variety of small cylindrical and tubular parts.

Vertical Turning Machines

Vertical turret lathes (see Figure 13.28) have the advantage of supporting large heavy parts on a horizontal chuck surface that rotates on a vertical spindle. These machines are capable of taking heavy cuts in tough materials with a high material removal rate.

Vertical boring mills (see Figure 13.27) are similar machines that are built on a much larger scale, some having rotating tables over 50 ft in diameter. Massive steel castings weighing many tons can be placed on their tables, clamped in place, and then bored, counterbored, faced, or turned. Many vertical boring mills are equipped with indexable turrets similar to turret lathes, so various tools can be brought into use quickly.

Horizontal Boring Mills

Horizontal boring mills (Figure 13.68) make use of a single-point tool for machining precision bores, especially in large or awkward parts. Like the lathe, they have a horizontal spindle to which boring bars may be attached, including line-type bars with an outboard bearing support. Unlike the lathe, the horizontal boring mill cuts using a tool that is attached to the rotating spindle. The workpiece is mounted on a sliding table that can be moved in two horizontal axes, and the spindle and outboard bearing may be raised or lowered. These useful machines can also be utilized for hole-making and milling operations, but they are being replaced rapidly by CNC machines.

Cutting Tools: Milling Cutters

A milling machine makes use of a rotating cutter with two or more teeth to remove material from a workpiece. Milling cutters must have correct relief and rake angles (Figure 13.69). Cutters used on horizontal milling machines for slab milling and slotting are called *plain milling cutters* (Figure 13.70). Plain milling cutters have cutting edges on their periphery. *Form-relieved cutters*

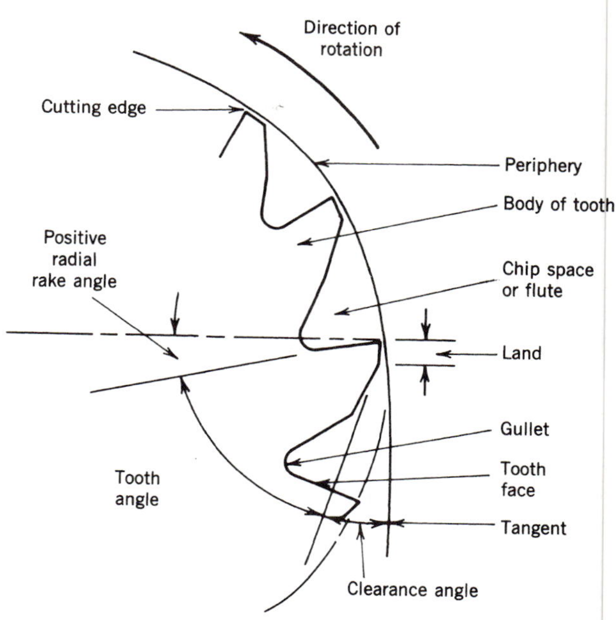

FIGURE 13.69
Milling cutter angles.

FIGURE 13.70
Plain milling cutters.

FIGURE 13.71
Form-relieved gear and radius cutters.

have a custom form such as a radius or gear shape and must be sharpened on the tooth face instead of on the periphery of the cutter. Form-relieved cutters are used on horizontal spindle machines for cutting gears, producing fillets and corner radii, and making other special shapes (Figure 13.71).

Large vertical spindle milling machines use *face mills* for machining flat surfaces and utilize smaller-diameter end mills having two or more flutes. (Figure 13.72). *End mills* cut on the sides (periphery) and on the end (face), so vertical plunge cuts can be made as in drilling, followed by horizontal milling cuts using the same tool. A blind keyseat (Figure 13.73) can be machined into a shaft, or an internal pocket can be milled on a part with an end mill.

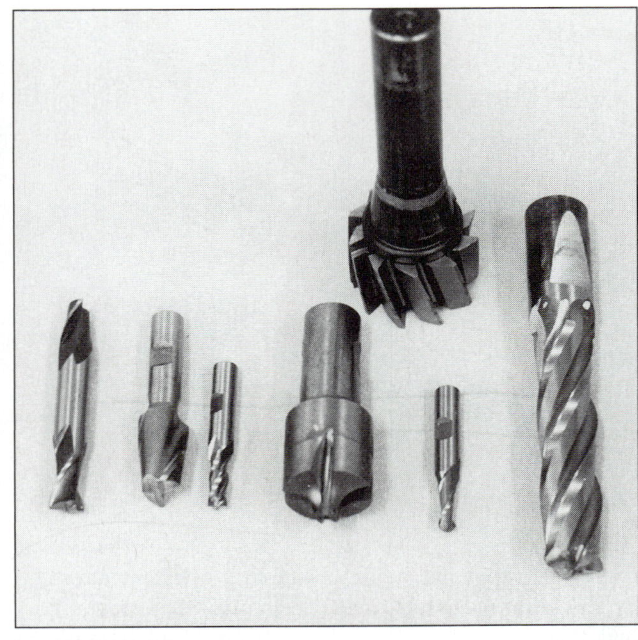

FIGURE 13.72
A variety of end mills.

FIGURE 13.73
Blind keyseat.

MILLING MACHINES AND EQUIPMENT

There are basically two types of milling machines, those with horizontal spindles (Figure 13.74) and those with vertical spindles (Figure 13.75). There are many combinations of these two types, and some machines permit their spindles to be swiveled into either orientation. Sizes range from small bench-top models to huge planer-type mills that can machine castings weighing many tons. CNC machining centers with horizontal spindles. (Figure 13.76) are extensively used to perform multiple machining operations in one setup.

FIGURE 13.74
Horizontal milling machine (Cincinnati Milacron, Inc.).

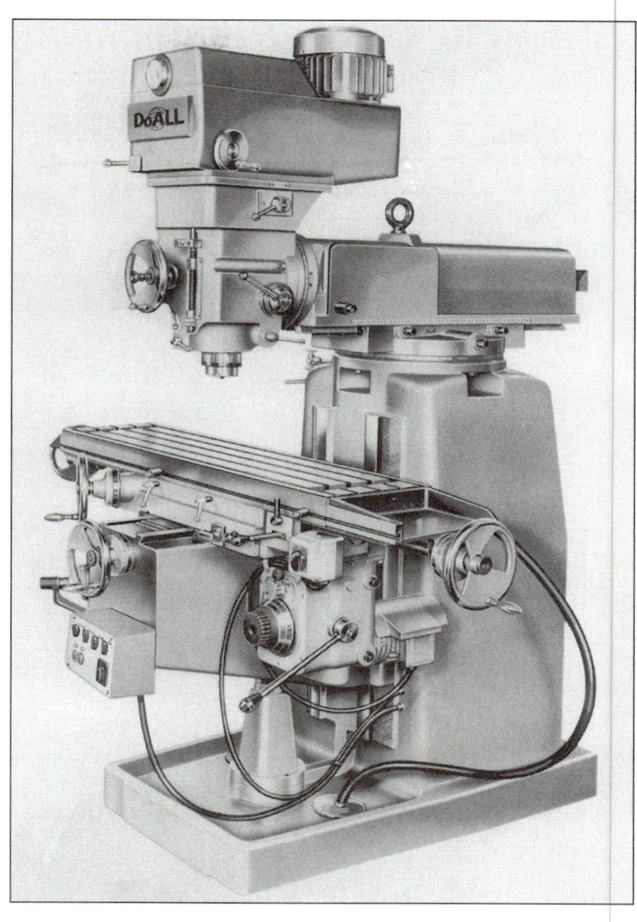

FIGURE 13.75
Vertical milling machine (DoALL Company).

FIGURE 13.76
Profiling mill. Aircraft plants use these NC machines for milling parts.

FIGURE 13.77
Horizontal milling machine, gear-cutting operation.

Conventional horizontal spindle milling machines used in machine shops and tool rooms employ various types of milling cutters: plain milling cutters for slab milling, side milling or staggered-tooth cutters for deep slots, saws or slitting cutters for grooves, and face mills for milling flat surfaces. These cutters either are made of high-speed steel, or they have carbide cutting edges. Horizontal mills are sometimes used in machine shops for making splines and gears (Figure 13.77).

Vertical spindle milling machines are very useful machines in toolrooms and machine shops, and they utilize tooling ranging from end mill cutters to face mills and fly cutters. These versatile machines can be used for drilling, reaming, and boring operations as well as for milling operations. Vertical mills are often operated by numerical control for drilling hole patterns or for milling operations.

FIGURE 13.78
Shaper.

SHAPERS AND PLANERS

Shapers (Figure 13.78) are designed to cut metal by forcing a single point tool across a workpiece in a reciprocating action. These machines are primarily used for low-volume machining and toolroom work. Shaping is also used in gear manufacture, where it is very useful for low-to medium-volume machining of gear teeth. Planers (Figure 13.79) are virtually obsolete as machine tools, but they are still useful for machining some extremely large machine parts. On the shaper, the ram moves back and forth holding the tool that cuts the stationary workpiece. In contrast, on the planer, the workpiece is clamped to the table and moves back and forth past a stationary tool. Some large planers have been retrofitted

FIGURE 13.79
Planer shop floor. Planers are useful for machining extremely large heavy parts, such as castings for presses (National Machinery LLC).

with multiple milling heads, and the table speed is altered to convert them into planer mills, which are much faster and more versatile machine tools.

BROACHING

Broaching is the precision cutting of a material by a tool incorporating a series of progressively stepped teeth. Flat surfaces, contours, internal splines, and external splines are some of the shapes produced by broaching. Broaching is usually a one-stroke operation in which roughing, semi-finish, and finish cuts are all done in a single machine stroke; however, when excessive stock prevents a one-stroke application, multiple strokes can be used, either using multiple internal broaches or by adjusting the position of an external broach.

Broaching is ideally suited for high-production machining (Figures 13.80 to 13.82). For example, forged diesel engine connecting rods require a machining operation

FIGURE 13.80
Twin-ram with in-line transfer progressive broaching machine that cuts blind external splines in two different parts. One part is loaded while the other is broached (Apex Broach & Machine Co.).

FIGURE 13.81
Part being loaded under the front ram while the rear ram is stroking down (Apex Broach & Machine Co.).

FIGURE 13.82
Progressive tooling for two different parts (Apex Broach & Machine Co.).

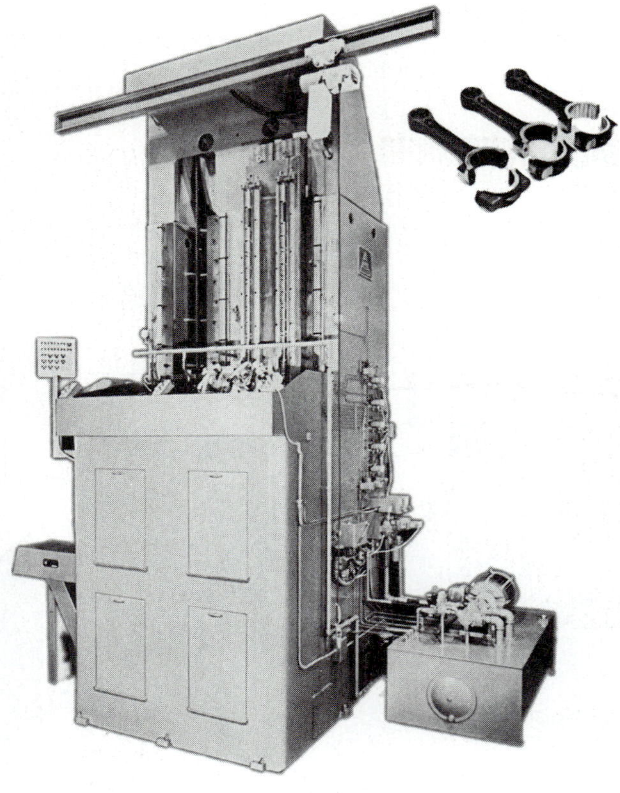

FIGURE 13.83
Broaching machine making connecting rods and caps (Apex Broach & Machine Co.).

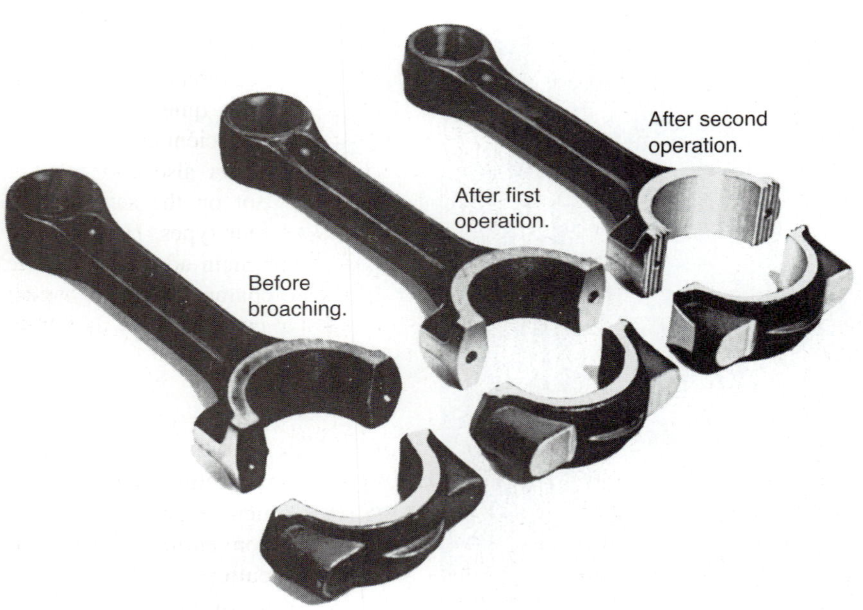

FIGURE 13.84
Connecting rods and caps (Apex Broach & Machine Co.).

FIGURE 13.85
Blind spline broaching machine. Inset shows blank and internal broached part. These automatic transmission converter-clutch hubs are splined at the rate of 240 hubs per hour, 12 times faster than conventional gear shaping methods (Apex Broach & Machine Co.).

(Figures 13.83 and 13.84) on 12 surfaces that are flat, round, and serrated. These connecting rods are finish machined by broaching at the rate of 112 per hour. Another high-production broaching operation on a vertical machine is seen in Figures 13.85 and 13.86. Where quite long broaches are needed, horizontal broaching machines (Figures 13.87 and 13.88) are used. External broaching of large flat surfaces is an economical means of machining when a large volume of piece parts is involved.

GEARS

One of the most efficient methods of transferring rotary motion from one shaft to another is through gearing. Large amounts of torque can be transmitted in a small space at the desired ratio by using gears. When they are properly made and kept lubricated, they are quiet, dependable, and long lasting.

Two friction disks or crossed belts can be used as examples to explain the involute curve employed in the operation of gears (Figure 13.89). The involute curve has been proven by long experience to be the most satisfactory profile for spur and helical gear teeth. This gear tooth form transmits motion smoothly and uniformly. The involute can be described as a curve traced by a point on a crossed belt as it moves from one pulley to another without slippage (Figure 13.90). If the crossed belts are superimposed on the friction disks (Figure 13.91), it can be seen that the outer diameters of the disks represent the pitch circles of the gears; the pulleys are the base circles, and where the belt crosses the line of centers is the pitch or tangent point of the friction disks. Figure 13.92 shows the terms used to identify parts of spur gears.

Many types of gears are made for various purposes. *Spur gears* are used to transmit motion between parallel shafts that turn in opposite directions. *Helical gears* are used for the same purpose, but they are quieter-running gears. *Bevel gears* connect intersecting shafts on the same plane but at an angle. *Miter gears* are bevel gears that connect shafts at a 90° angle. *Worm gears* connect shafts at 90° angles but not on the same plane. *Hypoid* gears such as those used in automotive differentials are similar to worm and bevel gears. They are quiet running like worm gears and have some of the efficient characteristics of bevel gears. *Crossed helical gears* also connect two shafts that are at an angle but not on the same plane. Internal gears can be helical or spur types. *Herringbone gears* have V-shaped teeth. Their main advantage is that they eliminate the end thrust characteristic of helical gears. *Gear racks* are designed to convert rotary motion to linear motion using a rack and pinion.

Gear-Cutting Operations

Several methods are used to manufacture gears. Where high production is not a requirement, external gears can be cut one tooth at a time on conventional horizontal milling machines with form cutters. A single gear milling cutter mounted on an arbor of a milling machine

FIGURE 13.86
Close-up of turntable on broaching machine. Two parts are broached at the same time under a single ram. Automated pick-and-place arms unload the machine in just 10 seconds (Apex Broach & Machine Co.).

FIGURE 13.87
Horizontal broaching machine and broach (Apex Broach & Machine Co.).

FIGURE 13.88
Surface-to-air missile casings. Internal grooves are made with the horizontal broach in Figure 13.87 (Apex Broach & Machine Co.).

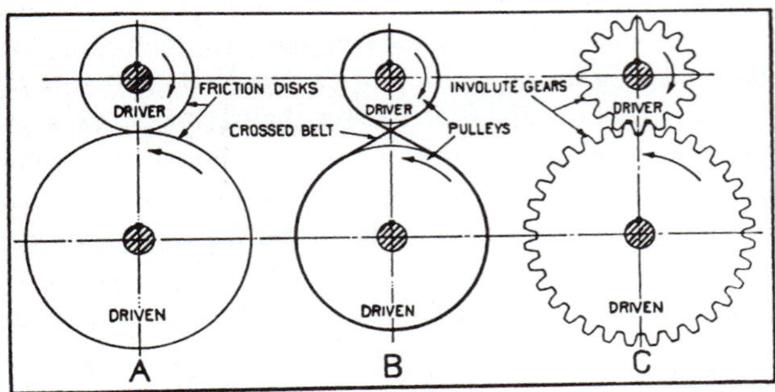

FIGURE 13.89
Three means for transmitting rotary motion from one shaft to another (Fellows Corporation).

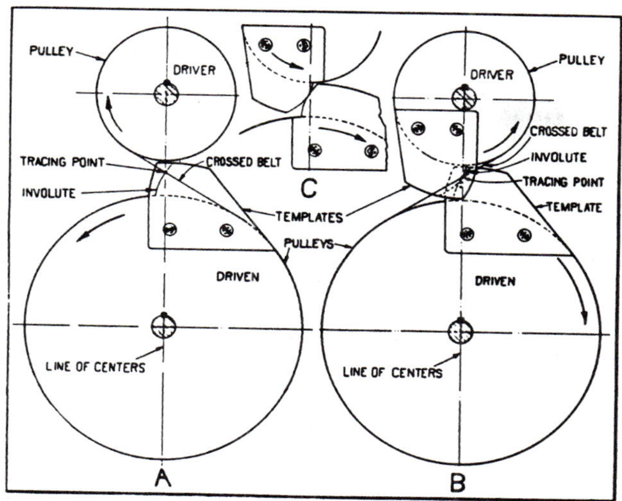

FIGURE 13.90
Diagram illustrating that the involute is that curve traced by a point on a crossed belt as it moves from one pulley to another (Fellows Corporation).

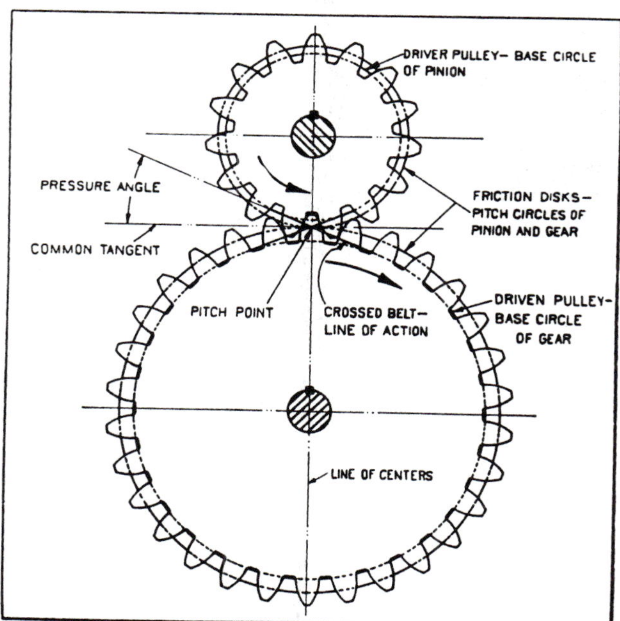

FIGURE 13.91
Diagram showing gears, pulleys, and crossed belt superimposed on friction disks, illustrating basic gear elements (Fellows Corporation).

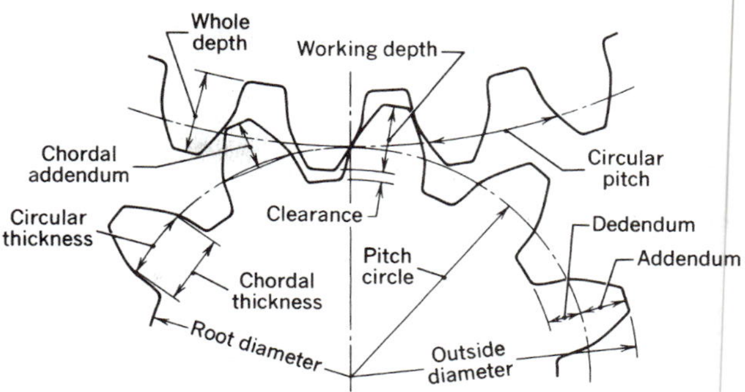

FIGURE 13.92
Spur gear terms (White, Neely, Kibbe, Meyer, *Machine Tools and Machining Practices,* Vol. II, ©1977 John Wiley & Sons, Inc.).

FIGURE 13.93
Hobbing machine producing gears (Foote-Jones/Illinois Gear).

removes material on the periphery of a gear blank to form the gear teeth. This is usually done in two cuts, first a heavy roughing cut at each tooth location all around the blank and then a light finishing cut. These precise tooth divisions are positioned by means of a dividing (indexing) head. This is a relatively slow gear-making process that is useful when a special gear is needed for a prototype or a repair operation.

High production rates for external gears can be realized in the process of hobbing. Hobbing (Figures 13.93 and 13.94) is a generating process in which a rotating cutter is slowly fed into a rotating workpiece (called a gear blank). Extremely large gears can be generated by the hobbing process (Figures 13.95 and 13.96). Racks

FIGURE 13.94
CNC gear hobbing machine (The Gleason Machine Division).

FIGURE 13.96
Hobbing a large helical gear for a hot strip mill (Foote-Jones/Illinois Gear).

(Figure 13.97) can be produced by hobbing, and the hobbing process is also used for making small gears, splines, and roller chain sprockets. Often, when the blanks are thin, as with plate sprockets, many blanks are stacked together on a mandrel and cut in one operation. Modified herringbone gears (Figure 13.98) and large bevel gears (Figure 13.99) can be produced on special gear hobbing machines.

Gear shaping is a common gear-cutting process that is used in medium- and high-volume production. The principle of gear shaping is based on the fact that the teeth on two meshing gears rotating together will follow an invo-

FIGURE 13.95
Extremely large gears can be produced by the hobbing process (Foote-Jones/Illinois Gear).

FIGURE 13.97
Cutting large heavy-duty racks for steel mill equipment (Foote-Jones/Illinois Gear).

FIGURE 13.98
Herringbone gears are made on special machines (Foote-Jones/Illinois Gear).

FIGURE 13.99
Large bevel gears are also produced on special machines (Foote-Jones/Illinois Gear).

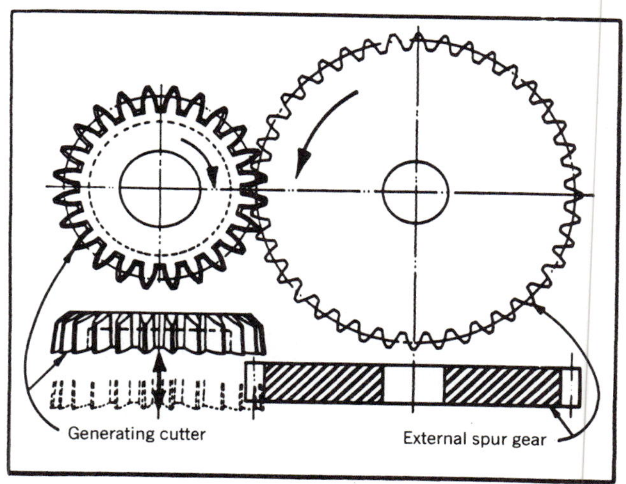

FIGURE 13.100
Illustration showing how the gear shaper employs the "molding-generating" process for cutting gears (Fellows Corporation).

lute curve. If one gear is hardened and provided with suitable cutting edges and clearances, it can be used as a generating cutter for a mating gear blank (Figure 13.100). The cutter is made to reciprocate like the tool in a vertical shaper while the tool and blank slowly rotate together. The cutter is fed into the blank until the gear teeth are machined to the correct size and shape (Figure 13.101 and 13.102). These machines are often computer controlled (CNC). Both internal and external gears can be produced using gear shapers (Figures 13.103 to 13.106). Very large and very small gears as well as helical and special shapes can be produced on gear-shaping machines. Internal splines and some external gears that are easily made on gear shapers cannot be made by other conventional gear cutting processes such as hobbing.

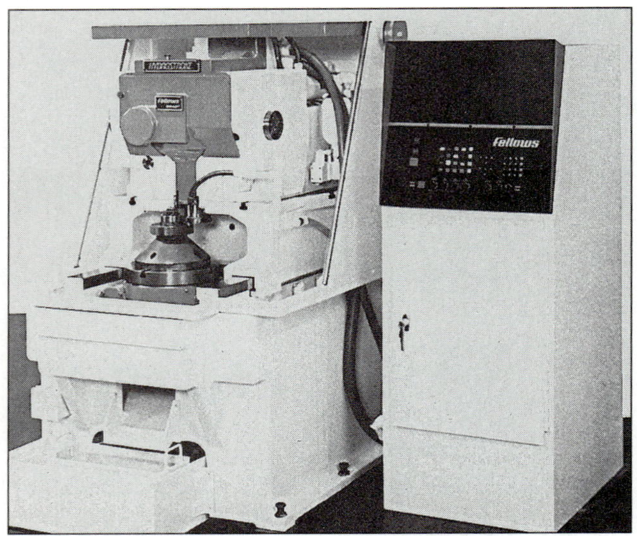

FIGURE 13.101
CNC gear shaper (Fellows Corporation).

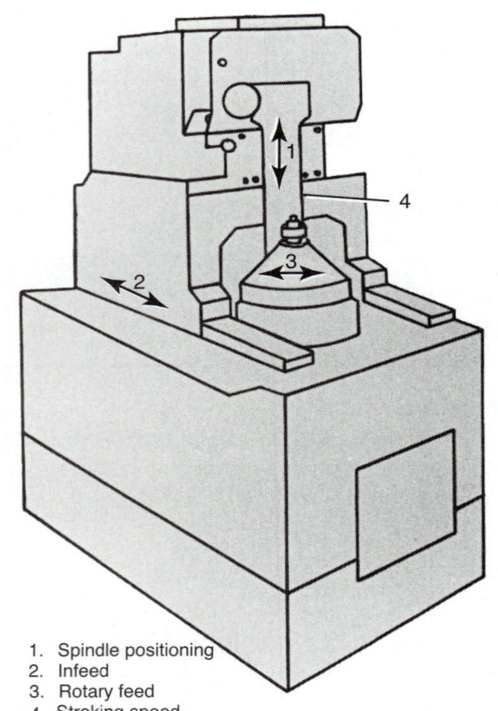

FIGURE 13.102
Schematic of gear shaper (Fellows Corporation).

FIGURE 13.103
Internal gear shaping (Fellows Corporation).

FIGURE 13.104
External gear shaping. Here a crowned gear (higher in the center) is being generated on a gear shaper (Fellows Corporation).

Hypoid gears are generated on special machines (Figures 13.107 and 13.108). Like other automotive gears, they must be finished by other processes after they are machined (Figure 13.109).

Gear finishing operations include gear shaving, flame hardening, and gear grinding. A rotating cutter shaped like a gear is made to mesh and turn with a machine cut gear before it is heat treated. This eliminates most indexing errors and roughness.

FIGURE 13.105
Quite large external gears may be produced by the gear-shaping process (Fellows Corporation).

FIGURE 13.106
A large internal ring gear being generated by the gear-shaping process (Fellows Corporation).

A flame or induction hardening process (Figure 13.110) is often performed on the teeth of a gear or the splines of a shaft, followed by a finish grind to restore accuracy. This assures strong, shock-resistant and wear-resistant gears that mesh together accurately and run quietly.

ABRASIVE MACHINING

Grinding processes remove metal by using abrasive grains as cutting tools.

FIGURE 13.107
Spiral bevel and hypoid gear generator (Foote-Jones/Illinois Gear).

FIGURE 13.108
Close-up of a hypoid gear being cut from a solid blank on a gear generator (The Gleason Machine Division).

FIGURE 13.109
Lapping of intermediate-size spiral bevel and hypoid gears (Foote-Jones/Illinois Gear).

FIGURE 13.110
Flame hardening gear teeth with pyrometer-controlled machine (Foote-Jones/Illinois Gear).

Common industrial abrasive materials used in grinding are silicon carbide for cutting cast iron workpieces and aluminum oxide for cutting other ferrous materials. Cubic boron nitride and artificial diamond are called *superabrasives.* Their higher cost is offset by their ability to cut very hard materials such as tungsten carbide and by their capacity for higher-speed, high-productivity grinding. Since these abrasives are harder than hardened metals, the grinding process is ideally suited to the removal of material from hardened workpieces. Also, since there are many small cutting edges traveling at a high cutting speed, extremely good surface finishes can be obtained by abrasive machining. Demanding tolerances can be maintained on a production basis with these precision machines. Precision grinding operations often follow a conventional machining process or heat treatment process to provide the final dimension and surface finish to the workpiece.

Many variables exist in the production of uniform precision parts by grinding. The complexity in this area of machining is greater than that of single-point tooling for metal cutting. Grinding wheels tend to break down and become smaller as they are used. This must be compensated for by adjusting downfeed or infeed. Grinding wheels tend to be self-sharpening. When the grains dull they fracture, exposing new, sharp grains.

Wheels must be trued and dressed occasionally to maintain their shape and remove any grains that fail to self-sharpen. This is usually done on production machines with a diamond dresser.

Feed rate, wheel speed, sparkout time, and wheel dress can be set and adjusted on the grinding machine. Quality and productivity factors in grinding include surface finish, surface integrity, cycle time, problems of burning the workpiece, and cost per piece. A large variety of grinding wheels are available, from soft to hard and from open to dense bond, and various grit sizes are used. Choosing the correct wheel for the job requires extensive study of abrasive wheels and their uses.

Grinding Machines

Grinding machines range from shop pedestal grinders for sharpening small tools to large specialty machines such as crankshaft grinders. Surface grinders, used to grind flat faces, include low-volume horizontal spindle tool room grinders (Figure 13.111) and higher-volume vertical spindle "Blanchard"-type grinders that can grind multiple parts at once. Workpieces are usually held on a magnetic base or table if they are ferromagnetic. The table holding the workpiece passes under the rotating abrasive wheel, enabling the wheel to make repeated cuts.

Cylindrical grinders (Figure 13.112) are designed to grind surfaces on rotating parts that are cylindrical in shape. Many of these machines are equipped to do internal grinding of parts that are held in a chuck. Special production machines are available for internal grinding in large volumes.

FIGURE 13.111
Surface grinders are used to produce flat surfaces (Clausing Machine Tools).

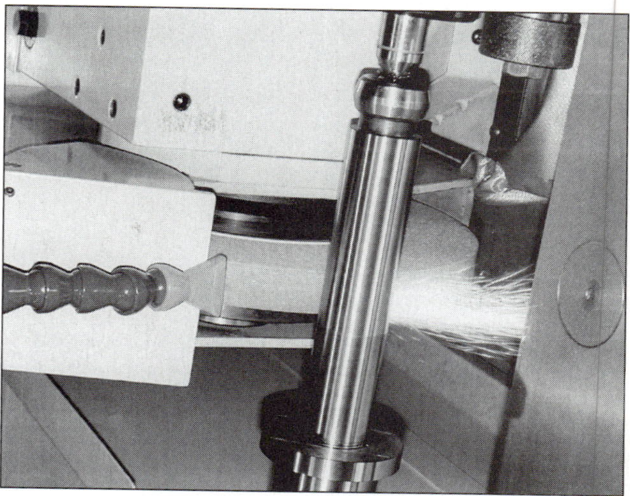

FIGURE 13.113
Infeed grinding in which the workpiece traverses across the wheel to a preset stop (Okamoto Corporation).

FIGURE 13.112
Cylindrical grinder.

FIGURE 13.114
Straight plunge grinding on a shaft. In this operation the wheel is fed straight in (Landis Tool).

Production Grinding Processes

Three methods of grinding the circumferance of cylindrical parts are used in manufacture: infeed grinding, plunge feed grinding, and centerless grinding. *Infeed grinding,* in which the grinding wheel moves along the work axis similar to a turning cut (Figure 13.113), is usually used where long cylinders or shafts must be ground to a uniform diameter. This method is most often used in small machine shops.

Plunge grinding (Figures 13.114 and 13.115) is often used in production grinding where large numbers of parts are manufactured to exacting specifications and where short lengths are involved. In high-volume production grinding, automatic workpiece loading on the fly (without stopping shaft rotation) greatly reduces cycle time. Wheels are dressed with diamond tools or wheels at predetermined intervals under computer control, and automatic compensation is used to correct for wheel imbalance. In-process gaging (Figure 13.116) allows workpiece diameters to be checked as the grinding process proceeds. With NC programmable machine controls (Figure 13.117) dimensions can be controlled and measured to 0.00004 in., 0.001 mm, or less.

FIGURE 13.115
Plunge grinding on a crankshaft (Landis Tool).

FIGURE 13.116
In-process gaging. Dimensions can be checked while the machine is in operation with this setup (Landis Tool).

Cylindrical grinders with swiveling grinding heads are called *angle grinders* (Figure 13.118). On these machines both diameters and shoulders can be finished in one plunge grinding operation. Grooves in wheels or rings can be plunge ground by fastening them onto an arbor (Figure 13.119). Many dimensions on one shaft can be finish ground in one operation with multiwheel production grinders (Figures 13.120 and 13.121).

Centerless grinding is done by means of a large grinding wheel, a regulating wheel (Figures 13.122 and 13.123), and a work rest to support the rotating workpiece.

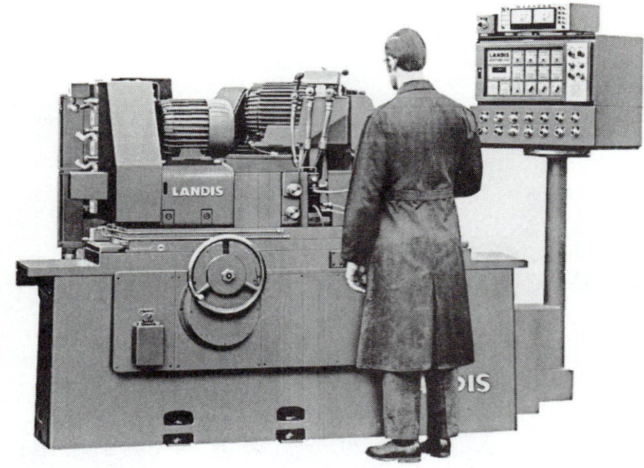

FIGURE 13.117
Small crankshaft grinder. Grinding process can be programmed on this machine for a precision automatic operation (Landis Tool).

Centerless grinding is a high-production method of machining precision rods, shafting, and small parts. Commercial ground and polished (G and P) shafting is produced on centerless grinders. Centerless internal grinding is accomplished with three rolls supporting and rotating the workpiece while the inside surface is ground (Figure 13.124).

HONING, LAPPING, AND SUPERFINISHING

When smooth internal finishes are required, such as in tubes for industrial hydraulic cylinders, a preferred method of finishing is *honing* (Figure 13.125). The honing head follows the existing hole or work surface and rotates while it oscillates in the bore. If the hole needs to be realigned, a boring operation must first be performed before honing. An example of multiple spindle honing machines used in manufacturing is engine block honing where all cylinder bores in a block are honed at one time.

Lapping is an abrasive finishing process in which fine abrasives are charged (embedded) into a soft material called a *lap*. Lapping can be done by hand (Figure 13.126) on a cast iron surface, or it can be done on a special machine (Figures 13.127 to 13.129).

Superfinishing (Figure 13.130) is similar to honing, and it can be done on a flat or cylindrical surface. In superfinishing, a lubricant is used to maintain separation

(Text continues on p. 288)

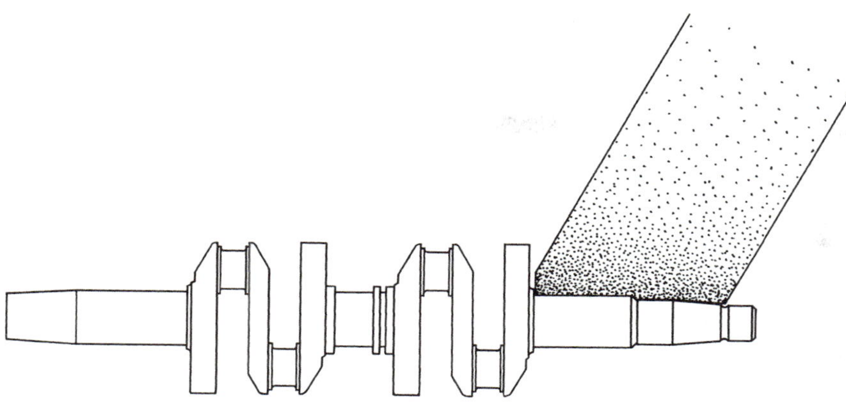

FIGURE 13.118
Angular plunge grinding on a crankshaft. Two diameters, fillets, faces, radii, and a tapered diameter are made in one plunge grind (Landis Tool).

FIGURE 13.119
Grooves can be made by the plunge grinding process. Grinding the diameter of a large bearing race mounted on an arbor (Landis Tool).

FIGURE 13.120
Multiwheel grinding machine increases productivity (Landis Tool).

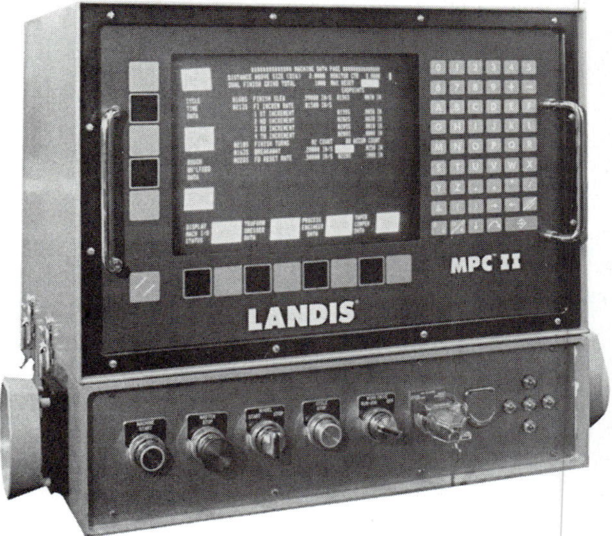

FIGURE 13.121
Programmable machine control for the multispindle grinder allows the operator to control wheel feed, in-process gaging, wheel balance, and monitoring dimensions (Landis Tool).

FIGURE 13.122
Centerless grinding. Through-feed grinding of bearing rings (Lidkopings Mekaniska Verkstads AB, Sweden).

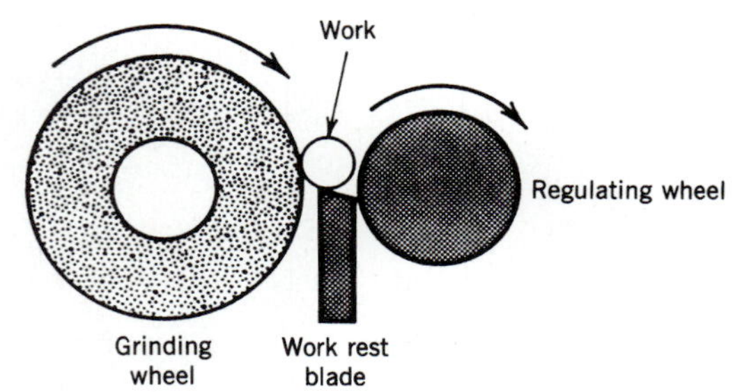

FIGURE 13.123
Principle of centerless grinding.

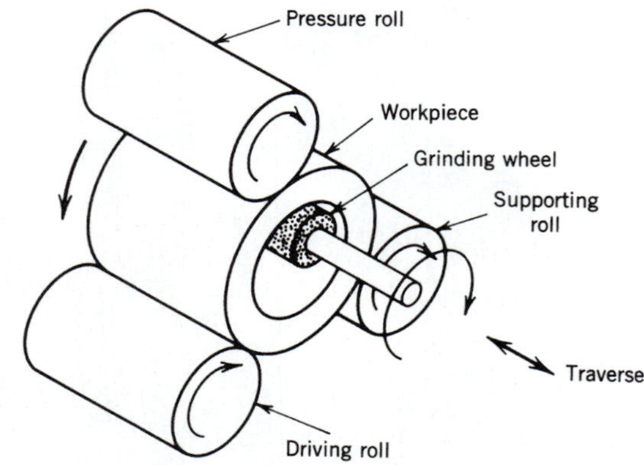

FIGURE 13.124
Centerless internal grinding.

FIGURE 13.125
Honing a cylinder in an engine block (JG Engine Dynamics).

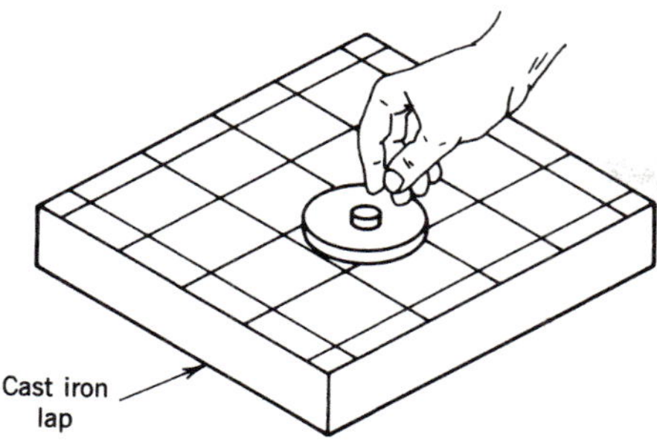

FIGURE 13.126
Lapping by hand. Grooves hold the abrasive. A circular motion is used to ensure flatness.

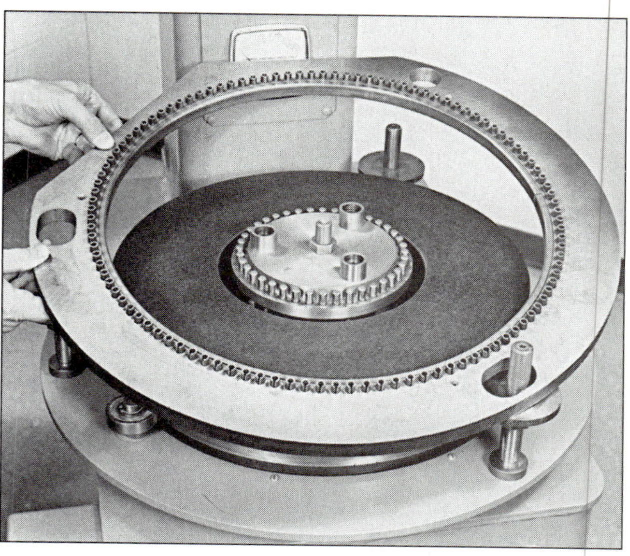

FIGURE 13.128
The planetary lapping attachment provides for complete coverage of laps during operation (DoALL Company).

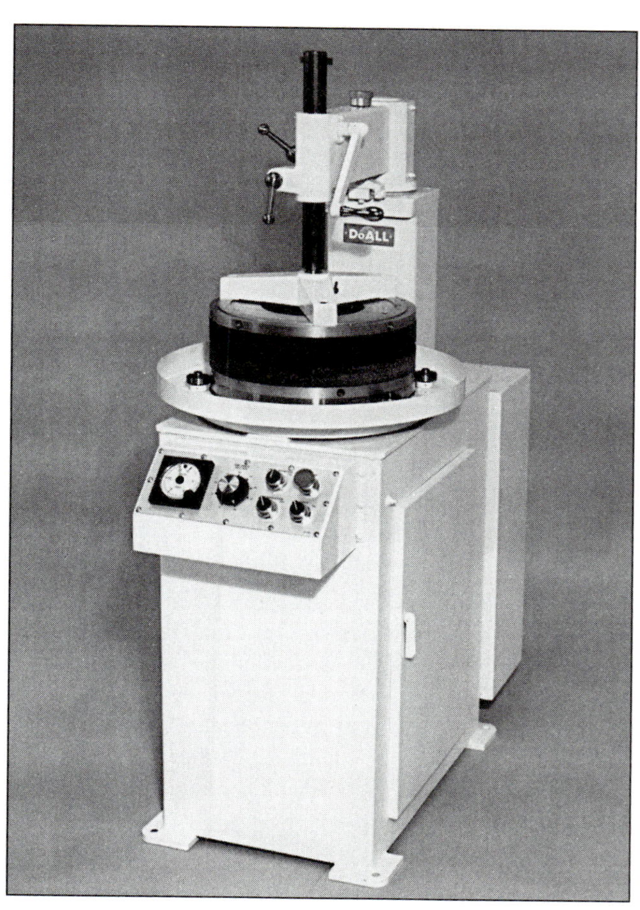

FIGURE 13.127
Lapping machine (DoALL Company).

FIGURE 13.129
These typical lapping workholders are custom designed for each individual flat or cylindrical workpiece application (DoALL Company).

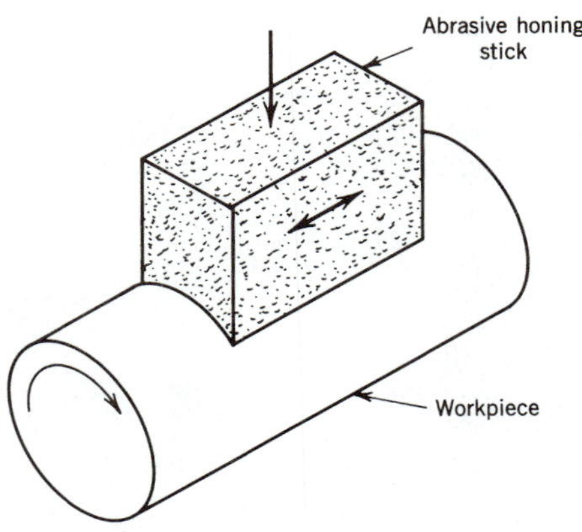

FIGURE 13.130
Superfinishing.

of a honing stone and the workpiece. Tiny projections causing roughness on the workpiece are cut off by the honing stone because they extend through the lubricant. By maintaining a controlled pressure, a desired degree of smoothness or surface finish is obtained, and the controlled-viscosity lubricant prevents further cutting action.

Review Questions

1. Saw blades (bands) for vertical and horizontal band saws are made in long strips that are coiled for packaging and shipment. How are these strips made into endless bands?
2. What is the greatest advantage of vertical band sawing over other machining processes?
3. Two drilled holes are required on a part, one 0.010 in. diameter and one 0.100 in. diameter. Which drilling machine or machines should you use to drill these holes?
4. Using the following formula, calculate how fast the drill press should rotate a 1/2-in. diameter drill if the cutting speed is 90 sfm:

$$N \cong \frac{4V}{D}$$

5. A massive steel casting weighing 3 tons requires that several 3-in. diameter holes be drilled into it. Which type of drilling machine would be best suited for this job?
6. An NC lathe produces accurate shafts having several diameters without the need for an operator to be in constant attendance. Would you call this an engine lathe or a production machine? Explain.
7. Threads are often designated omitting the tolerance symbols. Explain the meaning of the following: 3/8 − 16 UNC, and M16 × 2.
8. Internal and external threads can be made with hand taps, dies, or on an engine lathe with a single-point tool. How are threads cut at rapid production rates on turning machines?
9. How can a round or hexagonal bar be securely and accurately gripped for turning operations in a lathe? How can a square or rectangular piece be gripped for turning?
10. On which type of turning machines are air-operated chucks commonly used?
11. What principal advantage does a turret lathe have over an engine lathe?
12. A blind keyseat is a short groove that begins and ends anywhere along a shaft but not on the very ends. Which machine, a vertical or a horizontal spindle milling machine, would be most appropriate for cutting a blind keyseat?
13. Name two methods by which internal splines can be cut.
14. Hobbing is used almost exclusively for what machining operation?
15. How can hardened workpieces that cannot be machined with ordinary cutting tools be finish machined to a precise dimension with a very smooth finish?
16. How are steel workpieces usually clamped to the reciprocating table on a small surface grinder?
17. What is the difference between plunge grinding and plain infeed grinding on cylindrical grinders?
18. A precision part must be held to a tolerance of ±0.0001 (±0.0025 mm). Would you attempt to finish such a part on a turning machine? Explain your answer.
19. Fifty thousand hardened capscrews are made a few thousandths of an inch oversize and must be ground to a diameter of ±0.0001 in. (cylindrical surface). Select which grinding process and machine should be used to get the job done most economically: (a) center-type cylindrical grinder, (b) multiwheel grinder, or (c) centerless grinder.
20. Where is honing most likely to be performed, on flat surfaces, outside surfaces of cylinders, or in bores? What is the principal purpose of honing?

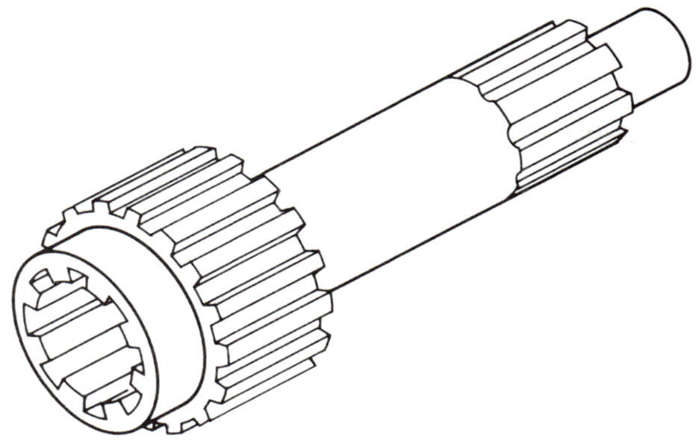

FIGURE 13.131
Transmission shaft. See Case 1.

Case Problems

Case 1: The Manufacture of a Transmission Gear and Shaft

A manufacturer is asked to design and build a small tractor with a four-speed manual transmission. You are asked to decide on the means of manufacture, on a cost-effective production basis, of a transmission shaft (Figure 13.131). One end of the shaft has a spline and the other end has a spur gear. The gear end is hollow and has a blind end internal spline. The gear, the internal and external splines, and the two bearing surfaces must be hardened. The gear must also be a precision one with close tolerances. The blank will be forged from a high-chromium carbon steel that can be hardened. What processes would you use to produce and finish this geared transmission shaft?

Case 2: The Robotic Wrist Component

A prototype (one only) of a robotic wrist component similar to the one shown in Figure 13.132 is required before it is put into production. It must be tested and perhaps revised and retested before it is ready to be manufactured in quantity. The prototype is mostly made of aluminum alloy. Where and how should it be made? Should the parts be:

a. Cast in a foundry?

b. Machined from aluminum alloy bar stock on production machines?

c. Machined from aluminum alloy bar stock by broaching the holes and surfaces?

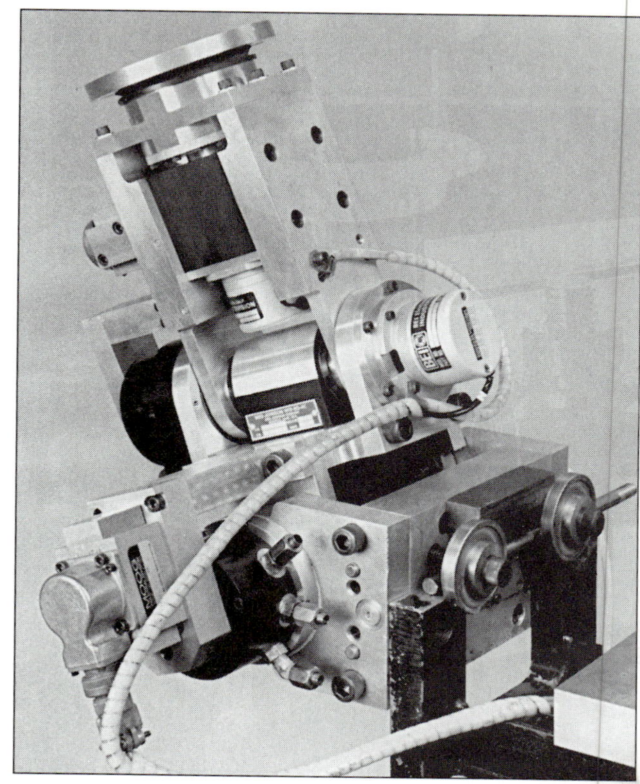

FIGURE 13.132
Robotic wrist component. See Case 2 (Bird-Johnson Company).

d. Machined from aluminum alloy bar stock in a tool room (machine shop) one piece at a time?

Explain why you chose your answer.

Case 3: RPM for Drilling Operations

A production drilling machine needs to be adjusted to the correct rpm for several drill sizes, 1/4-, 1/2-, and 1 1/2-inch diameter. The drilling operations are all in steel, which requires a cutting speed of 90 sfm. Using the formula

$$N \cong \frac{4V}{D}$$

where:

N is spindle speed in rpm
V is cutting speed in sfm (feet per minute)
D is diameter of the drill in inches

list the correct rpm settings for each drill size.

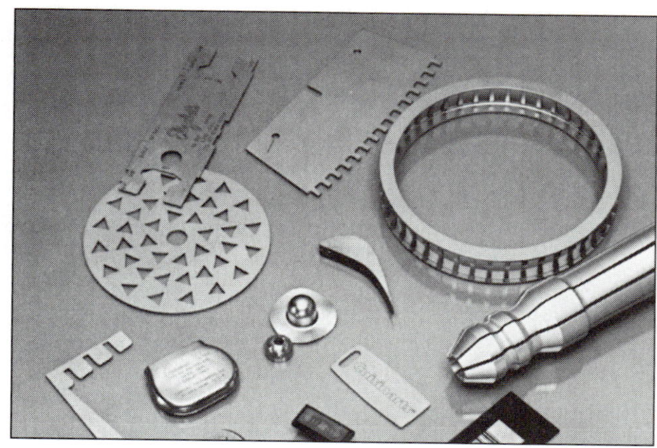

CHAPTER 14

Nontraditional Manufacturing Processes

Objectives

This chapter will enable you to:

1. List common nontraditional machining processes.
2. Describe in general terms how these processes work.
3. Describe applications of the processes.

Key Words

dielectric oil	deplating
electrolyte	coherent light
lasers	plasma

The previous chapters described machining processes in which workpiece material is removed in the form of chips through direct contact with a cutting tool. These processes are often called *traditional* machining processes, and they account for most materials processing.

The design and manufacturing engineer, in the continuing search for design improvements, stimulates development of alternative materials processing methods and takes advantage of new technologies as they evolve. Machining processes have developed in recent years that make use of electricity, laser light, electrochemical methods, and extremely high temperature gas plasma. These processes, because of their unique capabilities, have opened many new avenues of design to the engineer. The processes described in this chapter are often referred to as *nontraditional* machining processes because they do not involve the chip-making process of conventional machining. They are nonetheless well established as standard manufacturing processes.

This chapter describes nontraditional machining processes, including the following:

1. Electrodischarge machining (EDM)
2. Electrochemical machining (ECM)
3. Electrolytic grinding (ELG)
4. Laser machining
5. Ultrasonic machining
6. Water jet machining and abrasive water jet machining (AWJ)
7. Electron beam machining (EBM)
8. Plasma cutting

ELECTRODISCHARGE MACHINING

Electrodischarge machining, commonly known as **EDM,** removes workpiece material by an electrical spark erosion process. This process is accomplished by establishing a large potential (voltage) difference between the workpiece to be machined and an electrode. A large burst (spark) of electrons travels from the electrode to the workpiece. When the electrons impinge on the workpiece some of the workpiece material is eroded away.

The EDM process (Figure 14.1) takes place in a dielectric (nonconducting) oil bath. The dielectric bath concentrates the spark and also flushes away the spark-eroded workpiece material. A typical EDM system consists of a power supply, dielectric reservoir, electrode, and workpiece. The EDM machine tool has many of the same features as do conventional machine tools including worktable positioning mechanisms and measurement devices. Many EDM machine tools (Figure 14.2) are equipped with numerical control systems, providing the versatility of CNC for workpiece positioning and tool control functions.

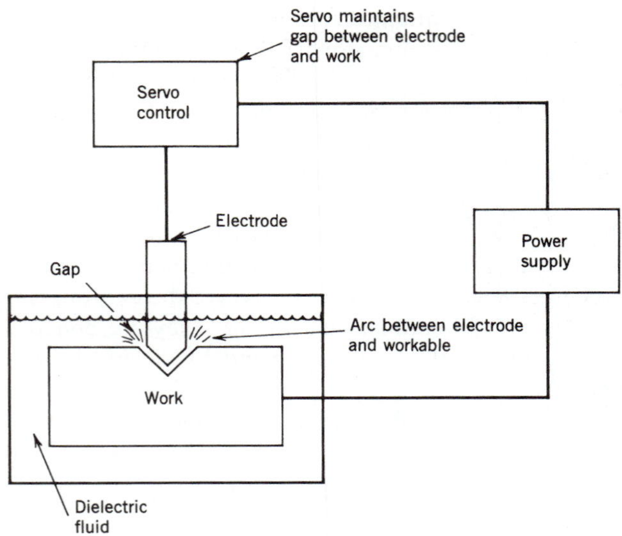

FIGURE 14.1
Basic EDM system components.

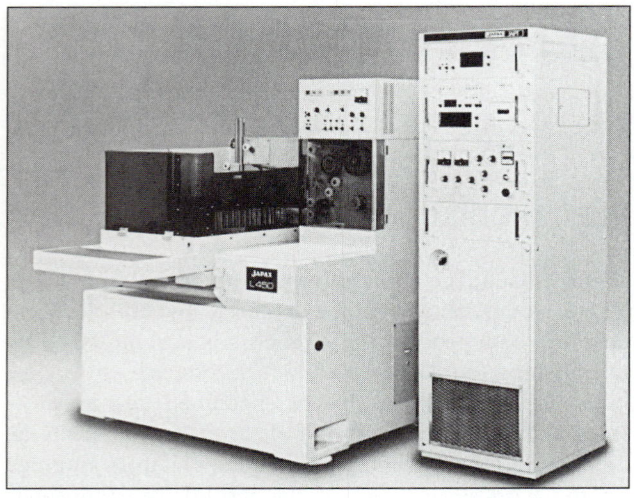

FIGURE 14.2
CNC wire EDM machining center (Sodick Inc.).

EDM Electrodes

In the EDM process, the shape of the electrode controls the shape of the machined feature on the workpiece. EDM electrodes may be made from copper or carbon (graphite) and shaped to the desired geometry by molding or machining. Erosion of the electrode slowly takes place during the EDM process, and in time the electrode becomes unusable. To circumvent this problem, a

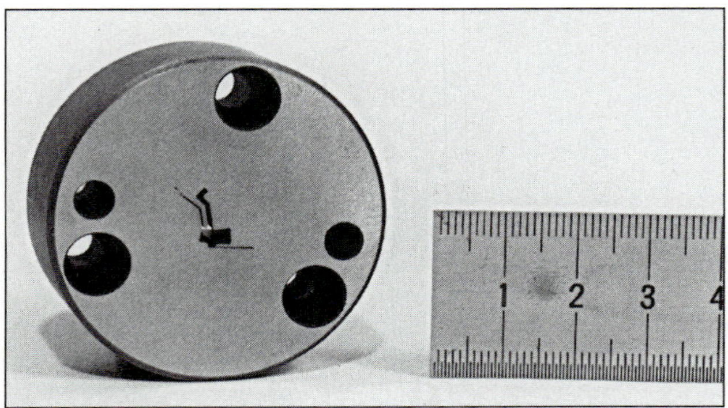

FIGURE 14.3
Blanking die for watch parts made with wire EDM (Sodick Inc.).

roughing electrode may be used to generally shape the workpiece, and then a finish electrode may be utilized to complete the process and establish final dimensions and geometry.

Wire EDM

Wire EDM uses a slender wire as an electrode and is extremely useful for making narrow slots and cutting detailed internal features in the workpiece (Figure 14.3) for detailed tool and die making. Wire EDM is similar in some ways to band sawing in conventional machining, but the saw teeth do the cutting in band sawing, whereas in wire EDM a spark occurs between the wire electrode and the workpiece. Wire feed is continuous during the process, and the wire cannot be reused.

Different metals are used for wire electrodes. Brass is one popular material. EDM wire can be very small in diameter, on the order of 0.005 in. Sapphire guides maintain the alignment of the wire as it passes through the workpiece. Wire tension is maintained by weighted pulleys or other mechanical tension schemes.

Wire EDM using small diameter wires permits extremely narrow slots to be machined in the workpiece, and the kerf (the width of the cut) is only slightly wider than the wire diameter. This process is highly suited to cut intricate slotted features in tools and dies.

Advantages and Applications of EDM

EDM processes can accomplish machining that would be impossible by traditional methods. Odd electrode shapes

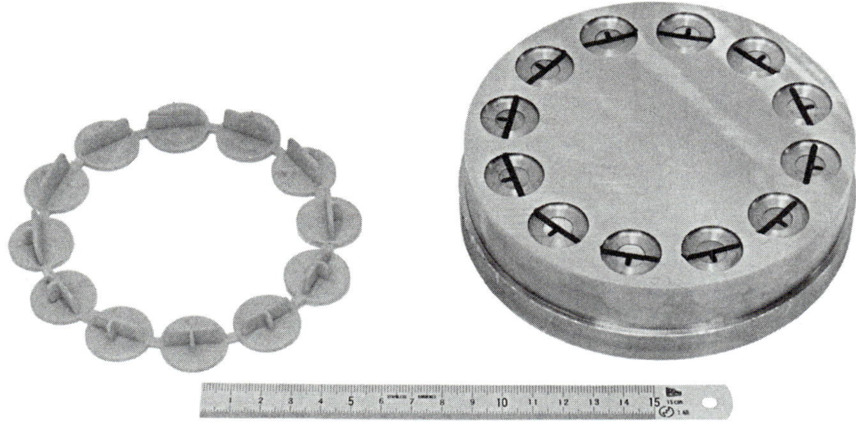

FIGURE 14.4
Mold machined with ram-type EDM machine (Sodick Inc.).

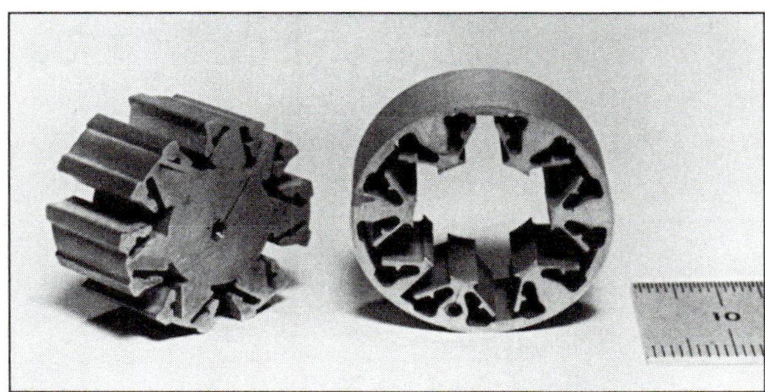

FIGURE 14.5
Intricate internal geometry machined with wirecut EDM (Sodick Inc.).

are reproduced in the workpiece, and fine details can be created, such as those needed in tool, die, and mold work (Figures 14.4 and 14.5). When EDM is coupled to numerical control, an excellent machining system is made available to the tool and die maker.

Because electrodischarge machining is accomplished by spark erosion, the electrode does not touch the workpiece and is therefore not dulled by hard workpiece materials that would dull conventional cutters. The EDM process is commonly used to spark erode very hard metals, and this process has found wide application in removing broken taps without damaging the workpiece.

Disadvantages of EDM

The EDM process is much slower, from a metal removal standpoint, than conventional machining processes involving direct cutter contact. There is also a possibility of softening the workpiece surface in the area of the cut in the parts being machined; however, EDM can accomplish many machining tasks that could not be done by conventional machining. Thus, the EDM process has become well established as a versatile manufacturing process in modern industrial applications.

ELECTROCHEMICAL MACHINING (ECM) AND ELECTROCHEMICAL DEBURRING (ECDB)

Electrochemical machining (ECM) is essentially a reverse metal-plating process (Figure 14.6). The process takes place in a conducting fluid (electrolyte) that is pumped under pressure between the electrode and the workpiece. As workpiece material is deplated it is flushed away by the flow of electrolyte, and the workpiece material is removed from the electrolyte by a filtration system.

The ECM electrode does not wear, and sufficient flow of the electrolyte prevents workpiece material from plating onto the electrode.

Advantages and Applications of ECM

Like EDM, ECM can accomplish machining of intricate shapes in hard-to-machine material. The process is also burr free and does not subject the workpiece to distortion and stress, as do conventional machining processes. This makes ECM useful for work on thin or fragile workpieces. ECM is also used for part deburring in the process called electrochemical deburring. This process is useful for deburring internal workpiece features that are inaccessible to traditional mechanical deburring.

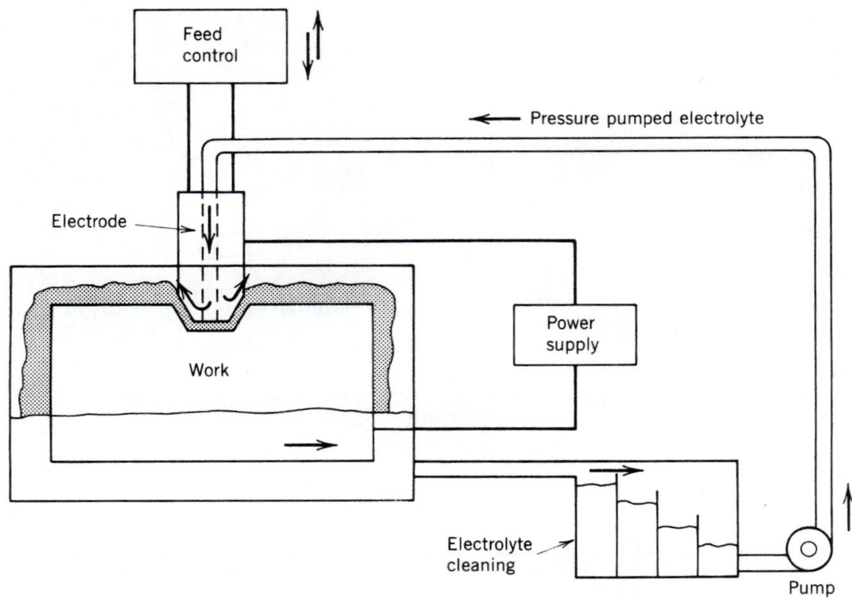

FIGURE 14.6
Basic ECM system components.

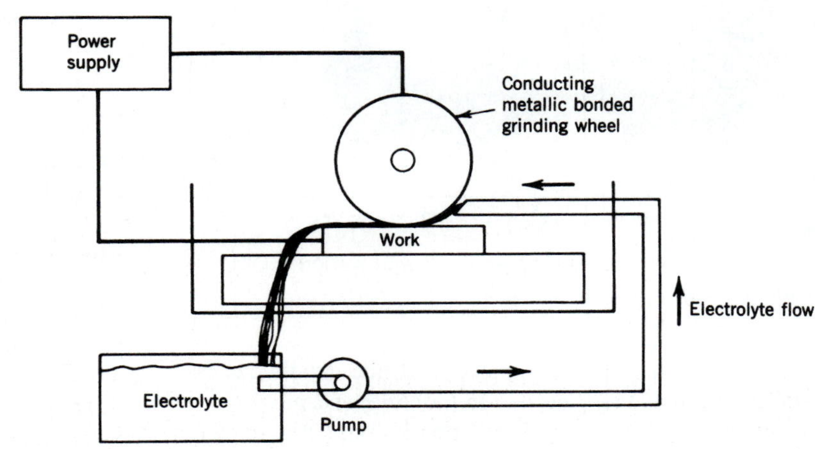

FIGURE 14.7
Basic ELG grinding system.

ELECTROLYTIC GRINDING (ELG)

In the process of **electrolytic grinding (ELG)** an abrasive wheel much like a standard grinding wheel is used. ELG, like ECM, is a deplating process, and workpiece material is carried away by the circulating electrolyte. Also, like ECM, ECG is effective in removing material from very hard workpieces. As the ECG process deplates a portion of the workpiece material it also transforms the surface layer of material into an oxide layer that is more readily removed by the abrasive wheel. The abrasive wheel bond is metal, thus making it a conductor of electricity. The abrasive grains in the grinding wheel are nonconducting and help maintain the gap between wheel and work while removing the oxide layer from the workpiece.

The ELG System

The basic ELG system (Figure 14.7) consists of the appropriate power supply, the electrode (metal-bonded grinding wheel), work-holding equipment, and the electrolyte supply and filtration system. Workpiece material is carried away by the electrolyte solution.

Advantages and Applications

Because this process is electrochemical and mechanical, unlike conventional grinding, the abrasive wheel in ELG wears little in the process. ELG is burr free and will not distort or overheat the workpiece. The process is therefore very useful for small precision parts and thin or fragile workpieces, and the ECG process can remove material from very hard, conducting workpieces at a higher material removal rate than can conventional abrasive machining processes.

LASERS AND LASER MACHINING

Lasers

Laser is an acronym for *l*ight *a*mplification by *s*timulated *e*mission of *r*adiation. By electrically stimulating the atoms of certain materials, such as various crystals and certain gases, the electrons of these atoms can be temporarily displaced to higher electron shell energy level positions within the atomic structure. When the electrons fall back to their originally stable levels photons of light energy are released (Figure 14.8). This light energy can be enhanced and focused into a coherent beam and then used in manufacturing, medical, measurement, and other useful applications. The applications of lasers in manufacturing are widespread, and laser energy finds many applications other than use as a cutting tool.

Laser Machining

In laser machining, the coherent laser light becomes the cutting tool. Integrating the laser with CNC machine tool positioning (Figures 14.9 and 14.10), provides extremely versatile systems with many capabilities.

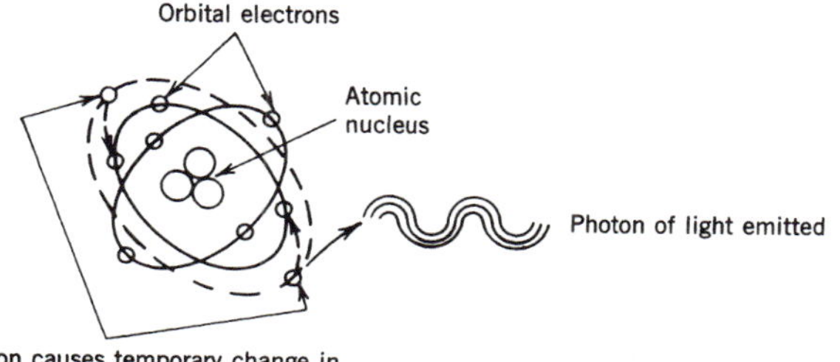

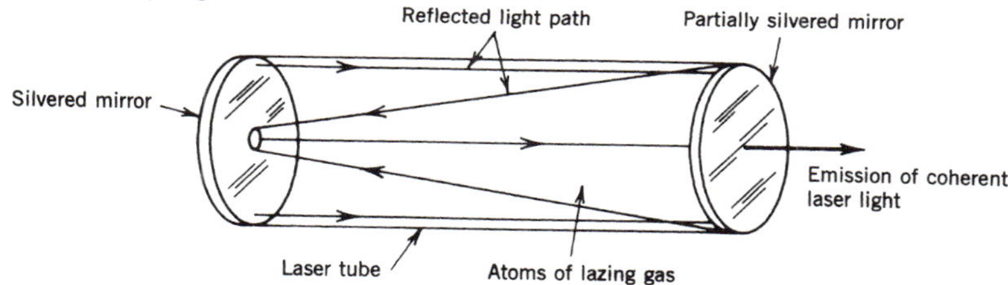

FIGURE 14.8
Photon emissions and laser generation.

FIGURE 14.9
CNC laser machining center (Coherent, Inc., Palo Alto, California).

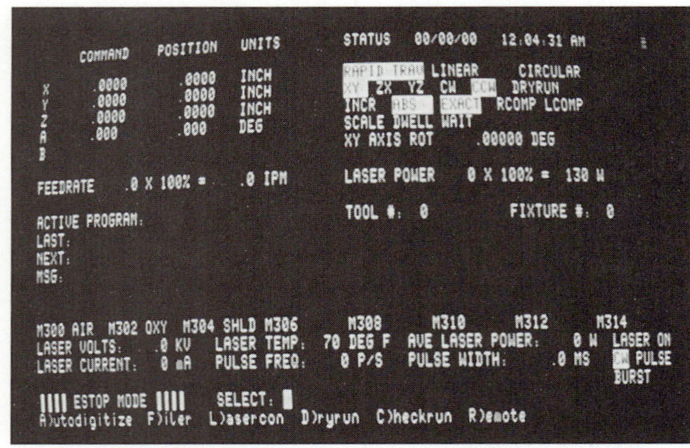

FIGURE 14.10
Video display on CNC laser machining center showing process status during production (Laserdyne).

Applications of laser machining tools include cutting plates (Figure 14.11), cutting shapes from thin materials (Figure 14.12), slotting stainless tubing (Figure 14.13), marking (Figure 14.14), and engraving. Lasers are also effectively applied in welding processes, as discussed in Chapter 15.

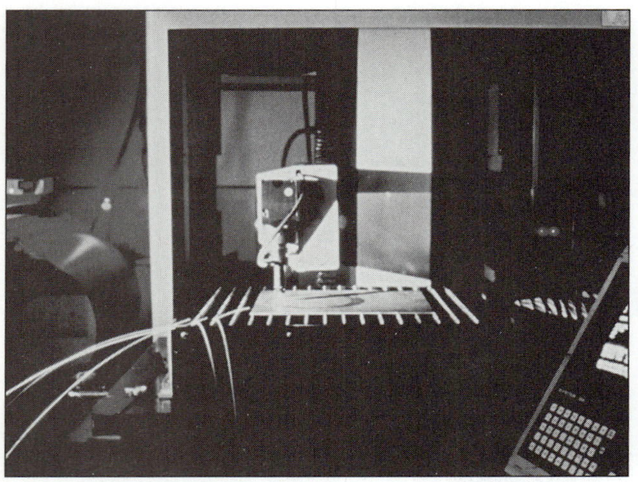

FIGURE 14.11
Laser plate cutting (Laserdyne).

FIGURE 14.13
Stainless steel tube slotting using a laser (Laserdyne).

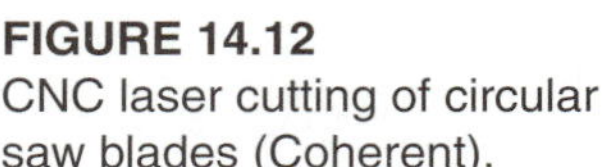

FIGURE 14.12
CNC laser cutting of circular saw blades (Coherent).

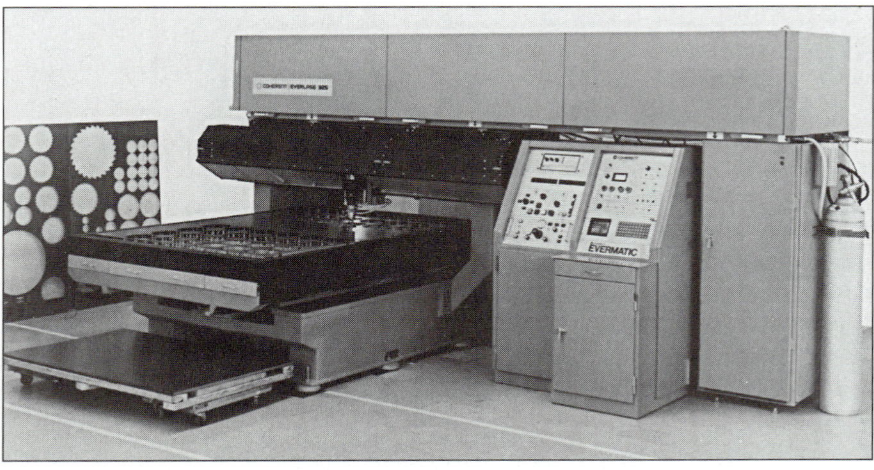

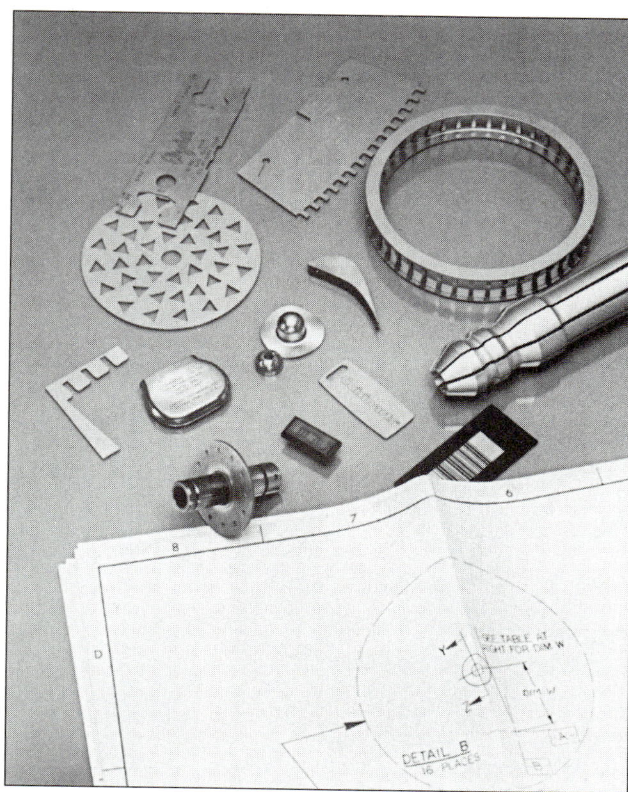

FIGURE 14.14
Laser marking and other laser-machined products (Laser Fare).

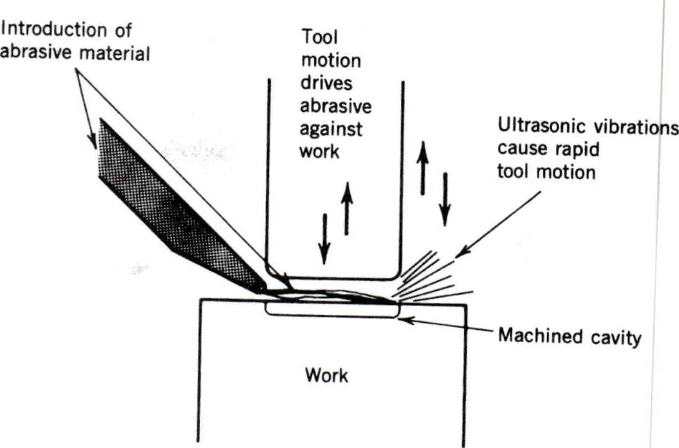

FIGURE 14.15
Ultrasonic machining.

ULTRASONIC MACHINING

Ultrasonic machining is an abrasive process similar to sand blasting. High-frequency vibration (Figure 14.15) is used as the motive force that propels abrasive particles against the workpiece. Advantages of this process include the ability to machine very hard material with little distortion. Good surface finishes may be obtained, and part features of many different shapes can be machined.

WATER JET MACHINING AND ABRASIVE WATER JET MACHINING

The cutting tool in the **water jet machining** process is an extremely high pressure water jet. Pressures of several thousand pounds per square inch are used to pump water through a small nozzle that directs a narrow stream at the workpiece. This process is effective in cutting fabric, foam rubber, and similar low-strength materials.

In **abrasive water jet machining (AWJ)**, the efficiency of the process is enhanced by adding an abrasive material to the water. Both pressure and mechanical abrasion join forces to accomplish cutting and machining tasks. The process may be used for hard metallic materials, but there are limitations on the thickness of the materials being machined.

ELECTRON BEAM MACHINING

Electron beam machining (EBM) is related to EDM and to electron beam welding. In EDM, however, a burst of electrons (spark) impinges on the workpiece, whereas electrons in a continuous beam are used in the EBM process. In EBM the workpiece material is heated and vaporized by the intense electron beam. This process, like electron beam welding, must occur in a vacuum chamber, and appropriate shielding must be employed to protect personnel from X-ray radiation.

PLASMA TECHNOLOGY

Plasma cutting is an extremely fast process for cutting, welding, and machining nonferrous metals and stainless steel. Plasma is created by passing a gas through an electric arc. The gas is ionized by the arc, and an extremely high temperature results. Temperatures in plasma arcs can exceed 40,000°F, many times hotter than any arc or flame temperature produced by other processes (Figure 14.16). When the superheated gas is forced through a venturi, a high-velocity jet is created that instantly melts metals on contact and blows the molten material away from the cut.

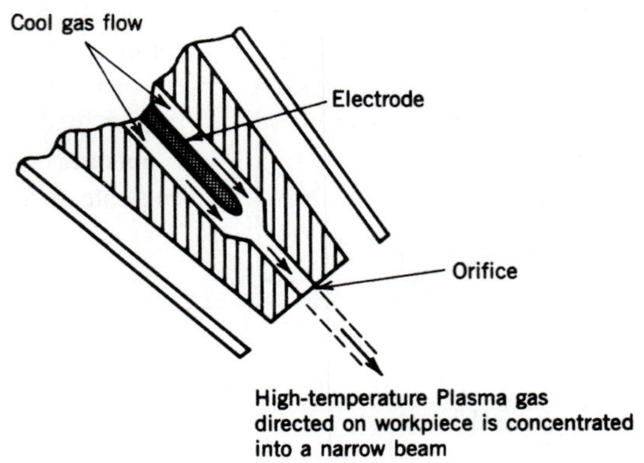

FIGURE 14.16
Plasma arc.

Plasma arcs of such high temperatures have a multitude of uses, not only in machining but for other applications such as ore smelting and waste metal recovery operations. There is also ongoing research in the applications of high-temperature plasmas for incinerating industrial wastes.

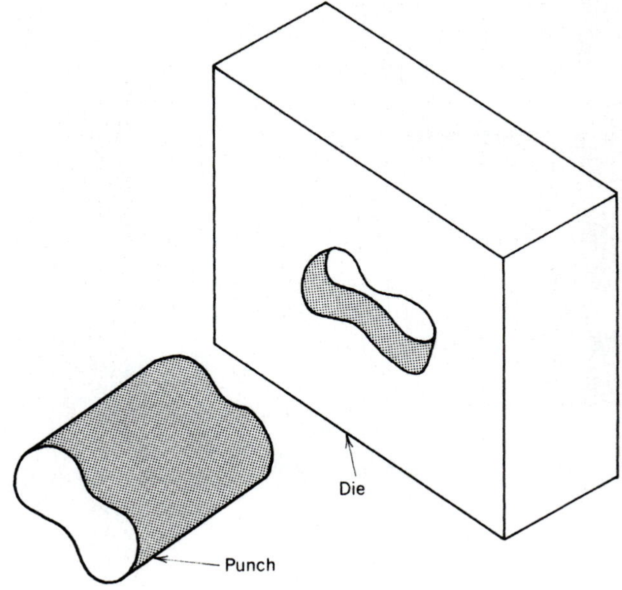

FIGURE 14.17
Punch and die set. See Case 1.

Review Questions

1. Describe in general terms how the following processes work: EDM, ECM, ELG, laser machining, ultrasonic machining, abrasive water jet machining, and plasma cutting.
2. Which processes might be used for deburring applications?
3. Which process would be used for detail die slotting?
4. What temperatures can be reached in a plasma system?
5. Which process is essentially deplating?
6. What is the primary function of the grinding wheel in ELG?
7. What provides the energy in ultrasonic machining and what does the cutting?
8. Which process is accomplished in a dielectric fluid?
9. Which process uses a conducting fluid (electrolyte)?
10. Describe in general terms how a laser functions.

Case Problems

Case 1: Punch and Die Tooling

A punch and die set is required as shown in Figure 14.17. Which process discussed in this chapter would be most applicable to the manufacture of this tooling?

Case 2: Internal Deburring

A complex part with several intersecting drilled holes must be as burr free as possible. Because many of the internal features are inaccessible for mechanical deburring, which process discussed in this chapter might be used to accomplish the task?

CHAPTER 15

Joining Processes

Objectives

This chapter will enable you to:

1. Choose the joining technology that is best suited to a given application.
2. Describe common methods of joining materials and cite their advantages and disadvantages.
3. Describe mechanical fastening methods and their applications.
4. Describe common welding processes and their applications.

Keywords

peel load
shear load
flux
shield gas
metal-inert gas (MIG)
tungsten-inert gas (TIG)
threaded fasteners
plasma arc
brazing
soldering
riveting

Manufacturing processes commonly require assembly involving several different pieces made of the same or different materials. In manufacturing, it is almost always necessary to join the materials at some point in the process. Plastics and wood material are glued, fabrics are stitched, and metal components are welded, bolted, or riveted. The purpose of this chapter is to survey the major methods of joining materials in manufacturing. In general, joining processes can be divided into the following categories:

1. Mechanical fasteners
2. Adhesive bonding
3. Welding
4. Brazing and soldering

MECHANICAL FASTENERS

Many mechanical fastening systems and components are employed to join materials. These include threaded fasteners, nails, staples, rivets, stitching, tying, snaps, pins, retaining rings, pressing, crimping, and specialty fastening systems.

Threaded Fasteners

Threaded fasteners include screws, bolts, and nuts. These fasteners are very common in all mechanical assemblies. All threaded fasteners make use of the inclined-plane principle from mechanics to put pressure on the parts that are to be assembled. Threaded fasteners will withstand large loads, are available in a wide variety of sizes and materials, and can be designed to meet many specialty fastening applications.

Bolts and nuts are found in almost every mechanical assembly. The bolt is designed to work in conjunction with a nut, with the material to be joined compressed between the nut and bolt head (Figure 15.1). Bolts with threads all the way to the head, called *cap bolts* or *cap screws,* are designed to screw into tapped holes, capturing the parts to be joined between bolt head and base material.

Machine screws are similar to bolts, but are usually smaller (Figure 15.2). They have many of the same fastening applications as bolts and nuts. Sheet metal screws are made with coarse threads and are designed for joining thin sheet materials. All threaded fasteners have features that permit them to be secured with various tools such as wrenches and screwdrivers.

The two major advantages of using threaded fasteners for joining are that they can be disassembled, and they come in a wide variety of types and sizes. These advantages make the system extremely versatile and likely to remain in wide use.

Disadvantages include labor-intensive assembly, requirements to stock a large number of types and sizes, and the possible loss of the hardware during assembly,

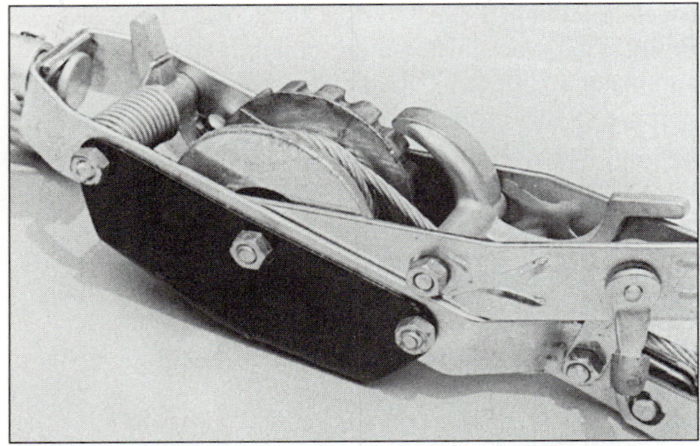

FIGURE 15.1
Common bolt and nut assembly.

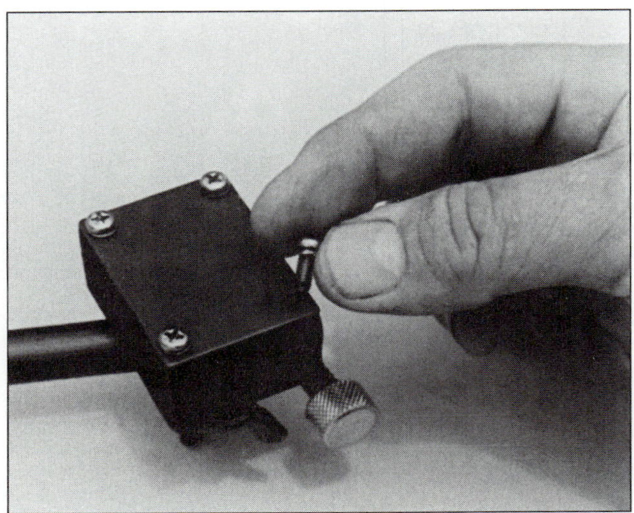

FIGURE 15.2
Machine screw assembly.

disassembly, or service. Easily damaged threads on fasteners can make assembly or disassembly difficult or impossible. Any threaded fastener will require a predrilled or punched hole in the workpiece before the materials can be assembled or joined.

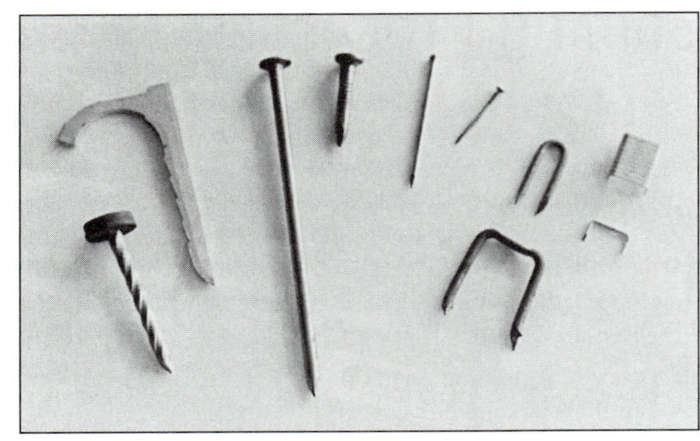

FIGURE 15.3
Common types of nails.

Nails and Staples

The foremost application of fastening by nails is in wood building construction. Threaded fasteners may also be used in some wood construction applications; however, the additional time required to drill holes for bolts and to make up assemblies usually limits the use of threaded fasteners to large wood beam constructions that are designed to support large loads.

A typical nail consists of a stiff wire with a head on one end for driving a point on the other end to penetrate the material to be joined. Nails come in a variety of sizes, styles, and materials including steel, brass, and aluminum (Figure 15.3). As the nail is driven into wood it pushes aside the grain of the material. The frictional forces exerted by the wood recompressing holds the fastener in place. Nailing at an angle to the forces trying to separate the material can add some additional mechanical holding capacity. More holding capacity can be obtained by using fast helix nails. These have some of the characteristics of threaded fasteners and are used in crate and pallet manufacturing.

Nails are extremely well suited for static assemblies such as houses. They do require considerable energy for driving, which is often done by hand; however, nailing by machine and by explosive charges can expedite this joining process.

Nailing is not suitable for dynamic loads in an assembly. The fastener will pull out if the structure is sufficiently distorted. A material assembled with nails must also be compressive by nature so that the fastener will not fall out.

Stapling is related to nailing in that a staple is basically a two-pointed nail. Stapling is a versatile mechanical fastening system, highly adaptable to high volume production of furniture and paper boxes, plus installation of many materials such as insulation and ceiling tiles. Stapling is best used where loads on the structure are moderate. It is not suitable for heavy loads because of the usually short penetration of the fastener into the material being joined. Because stapling can be most unsatisfactory in this respect, this system should be selected only after a careful engineering evaluation of the loads the assembly will encounter.

Threaded fasteners, including sheet metal screws, wood screws, and many types of specialty hollow wall fasteners, are available to the building industry (Figure 15.4).

Rivets

Rivets are a very common and extremely versatile mechanical joining system. They are used to join materials ranging from heavy steel structurals to light-gage sheet metal, plastics, and composites. An example is the popular pop rivet used in many common fastening applications. Rivets are widely used in the aircraft industry, where it is desirable to have as little projection of the fastener head as possible above the surface of the materials being joined. This greatly reduces air friction on skin and cowl surfaces subjected to airflow during flight (Figure 15.5); however, since supersonic aircraft require almost totally smooth skins, riveting is giving way to adhesive bonding in this application.

Riveted structures are used extensively as internal aircraft structural components (Figure 15.6). Extremely strong but very light ribbed structures can be fabricated by

FIGURE 15.5
Riveting flush with the material surface is used extensively in aircraft cowl and skin applications in order to reduce airflow turbulence.

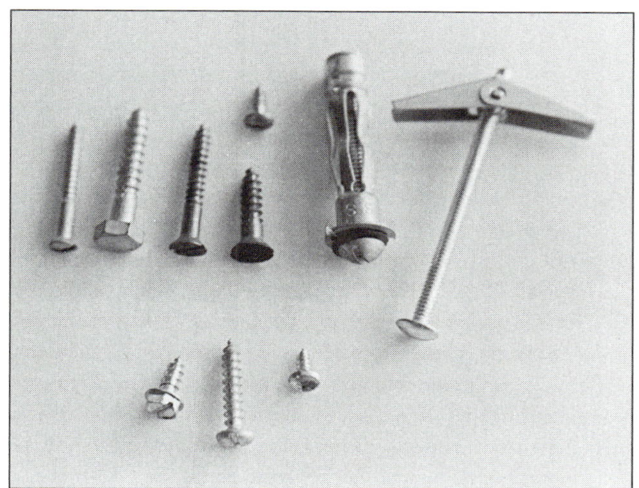

FIGURE 15.4
Wood and sheet metal screws and special hollow wall fasteners.

FIGURE 15.6
Riveting internal aircraft structures makes them very strong but also very light.

FIGURE 15.7
Strong but lightweight internal riveted rib structure in the fuselage of the F/A22 fighter jet (Lockheed Martin Co.).

riveting (Figure 15.7). Machine riveting is widely used in aircraft production. Because rivets are retained on the back side, as are many threaded fasteners, they are well suited to dynamic loads and will not loosen through distortion of the assembly.

Much riveting is accomplished with power machinery. The equipment used may be portable or may be bench type. The equipment illustrated in Figure 15.8 swages rivets using a kneading action from the center of the rivet outward. Multipoint rivet-

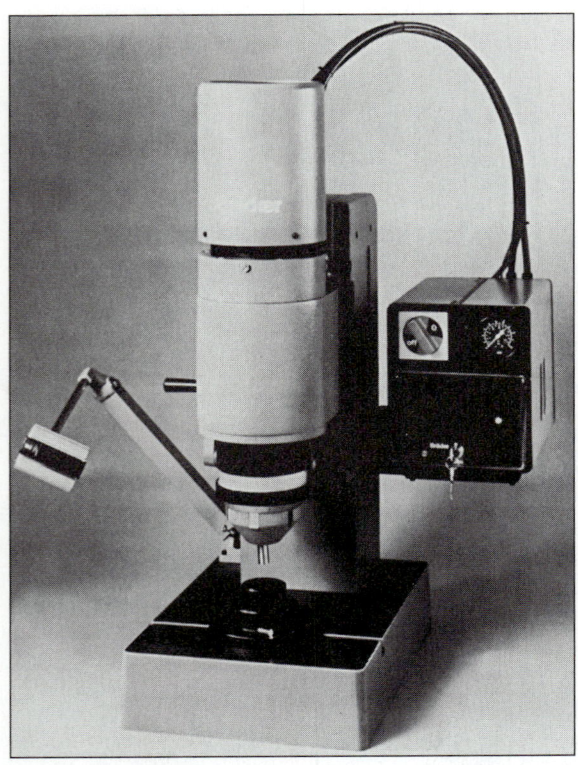

FIGURE 15.8
Bench rivet machine forms rivets with a center-outward (radial) action causing little distortion of the parts being joined (Baltec Corporation).

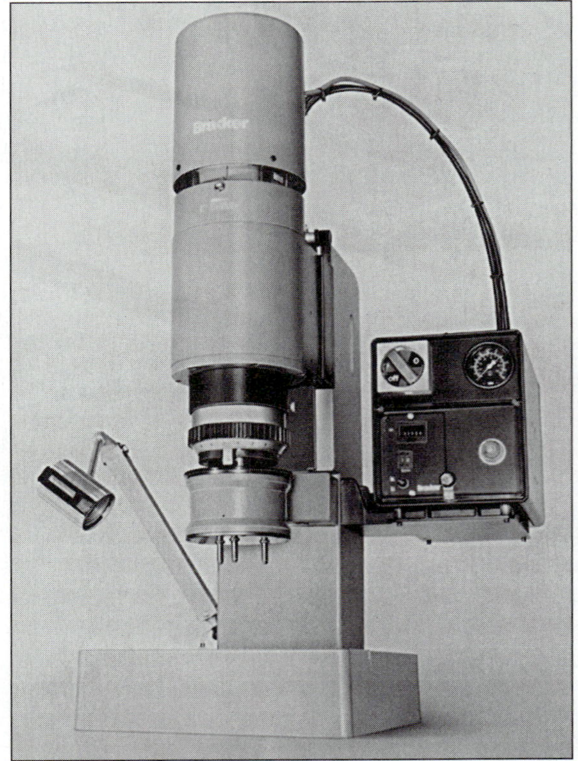

FIGURE 15.9
Multipoint radial riveting machine for die cast metals and plastics (Baltec Corporation).

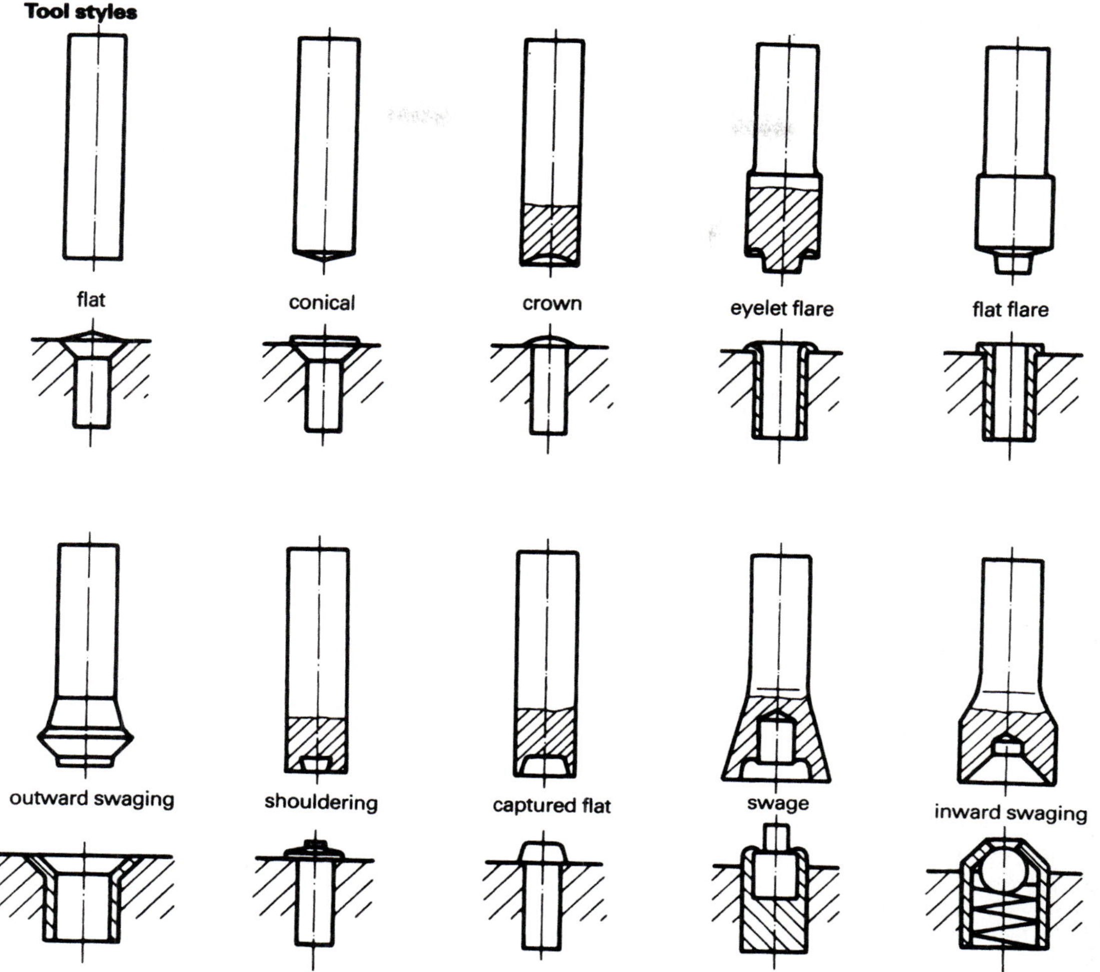

FIGURE 15.10
Rivet types and applications (Baltec Corporation).

ing machines can swage several rivets at one time (Figure 15.9). Many riveting tools and rivet styles are available to meet a broad range of riveting applications (Figure 15.10).

Advantages of riveting are strength with a minimum-sized fastener and the ability to join material by insertion and attachment of the fastener from only one side of the assembly. Disadvantages include labor-intensive installation, requirements for predrilled or punched holes in the material, and difficulty in disassembly. Most rivets must be drilled out in order to take an assembly apart. Reassembly requires replacement of the fastener with new hardware.

Stitching, Tying, and Snaps

Stitching and tying are found primarily in the fabric industry. Machine lock stitch is an essential process in the making of clothing. The modern sewing machine has perfected this method of material joining. Other applications include sack sewing with an easily unraveled chain stitch. Tying with rope, string, or cord has many temporary applications. Mechanical snaps are frequently used for temporary joining of fabrics and canvas.

Pins and Retaining Rings, Pressing, and Crimping

Assembly of mechanical components often makes use of pins and retaining rings. These fasteners make disassembly possible, but precision machining of the parts is often required. Retaining rings may be used to hold mechanical parts together (Figure 15.11). Pins are often found where parts must be joined and disassembled frequently (Figure 15.12).

Many parts in mechanical assemblies can be joined by pressing (press fitting). In this joining method, the

FIGURE 15.11
Mechanical assembly using retaining rings.

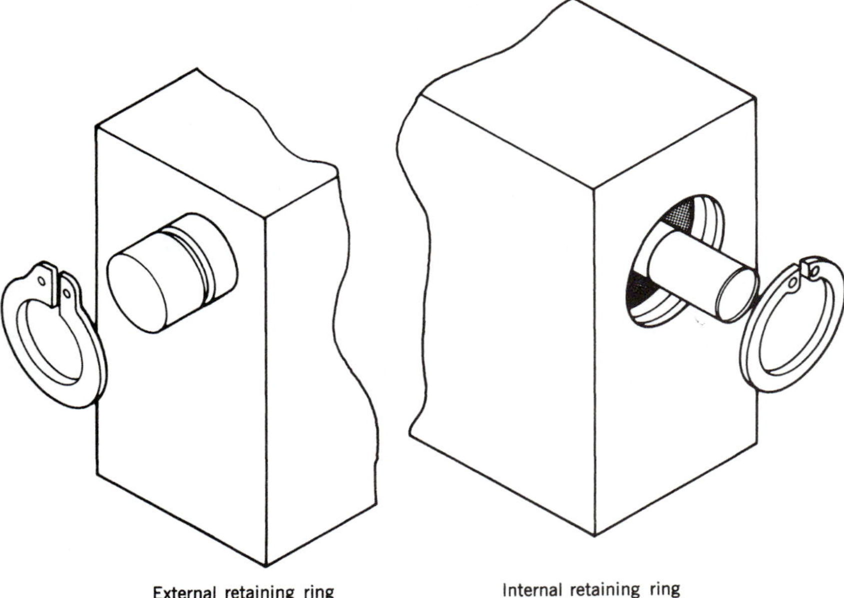

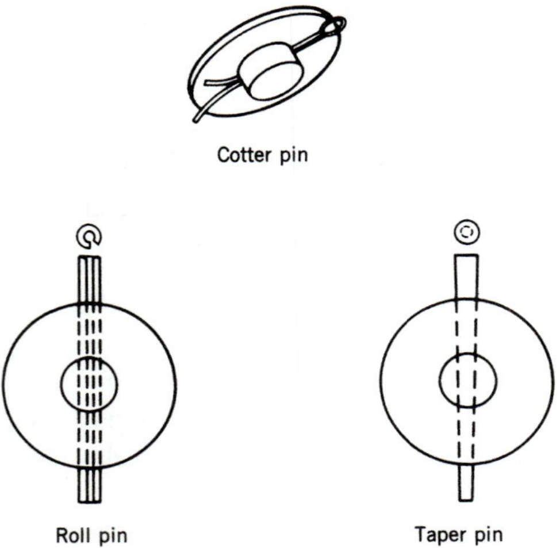

FIGURE 15.12
Mechanical assembly using pins.

FIGURE 15.13
Mechanical assembly using press fitting.

mating part dimensions overlap slightly and the parts are forced together under mechanical pressure. The method has many applications, especially in assemblies where bearings and bushings are to be retained without the need for additional mechanical hardware (Figure 15.13).

Also related to mechanical pressing are shrink and expansion fitting, in which parts in an assembly are joined by heating or shrinking one or the other, fitting the parts, and then returning them to normal temperature. This process makes use of the normal expansion and contraction of material as temperature is varied.

Crimping is another method of material joining, in which mechanical force is applied to compress or expand one piece of material against another piece to join the parts. Generally, crimping requires a groove or other feature to provide a place where the material of the mating part can grip (Figure 15.14).

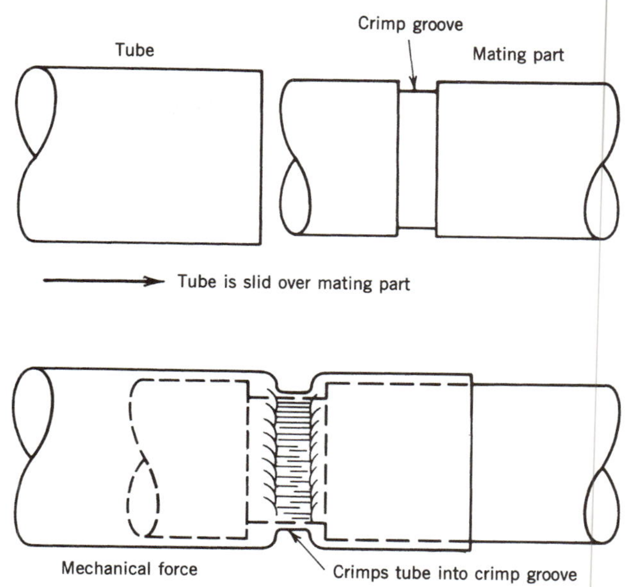

FIGURE 15.14
Mechanical assembly with crimping.

Seaming is related to crimping. This joining process is widely used in the sheet metal industry for closing duct work and other sheet metal fabrications. Common sheet metal seams include those illustrated in Figure 15.15. The industry makes use of a number of hand- and power-operated machines to roll, form, and crimp the various types of joint seams.

Specialty Fastening Systems

Specialty fastening systems have found wide applications. One example, plastic barbs commonly known as Velcro, consists of a pad of plastic barbs designed to grip into mating materials (Figure 15.16). This fastening system has many uses where temporary joining of materials is desirable. Plastic barbs are not suitable for joining materials that will be subjected to heavy pull-apart loads; however, this system is extremely useful for temporary joining of fabrics and other light-duty holding applications.

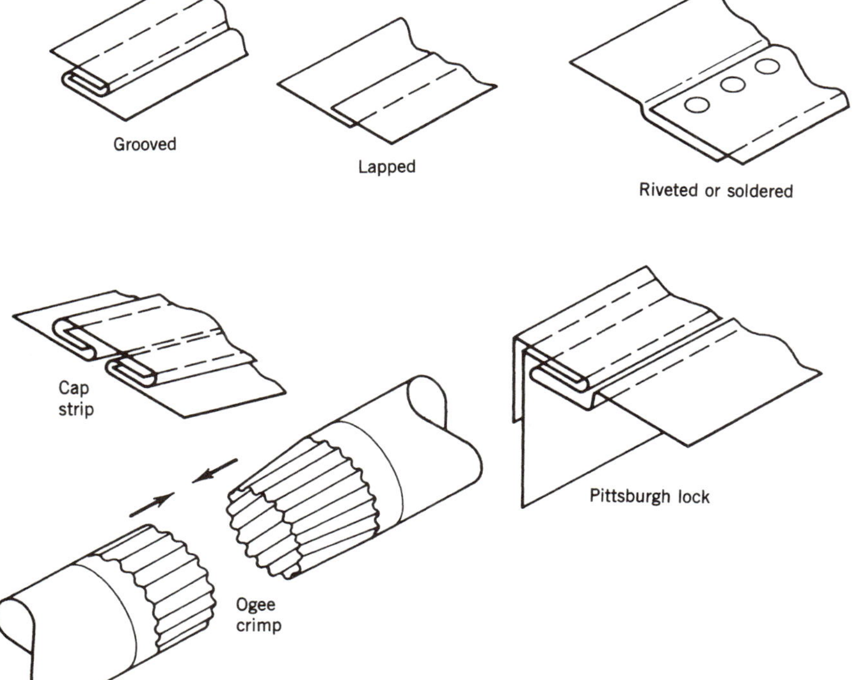

FIGURE 15.15
Joining sheet metal by various seams.

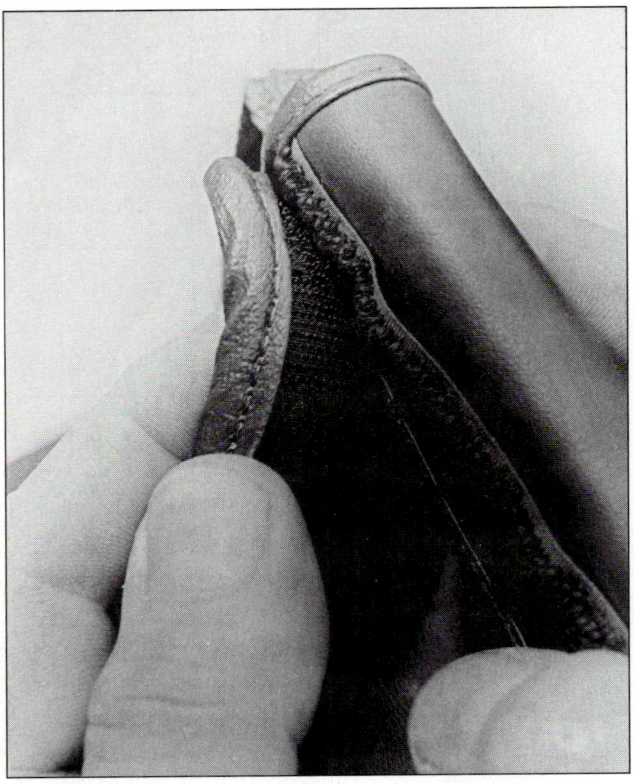

FIGURE 15.16
Plastic barbs, a unique material-joining method.

ADHESIVE BONDING

It is naturally desirable to join material without labor-intensive preparation such as hole drilling or punching and the insertion of mechanical hardware. Processes involving adhesives have developed rapidly in recent years and are now widely used. Research and development into adhesive joining of metals, plastics, and composite materials is under continual development. Adhesives are divided into two major groups: (1) natural adhesives and (2) plastic resin adhesives. Advantages gained by adhesive bonding include the following:

1. Dissimilar materials can be joined that cannot be joined by other processes.
2. Mechanical preparation such as hole drilling is eliminated.
3. Adhesive bonding joins materials with no gaps, as there would be between bolts or rivets in a mechanical assembly. Overall strength of the joint can therefore be enhanced.
4. Many adhesives are not subject to deterioration from service environments and therefore can be used in a wide range of temperature and corrosive environments.
5. Joined materials are not subjected to mechanical or heat distortion, as they would be if joined with mechanical or welding processes. Hence, precision parts can be joined without risk of damage or distortion.

Disadvantages of adhesive bonding include the following:

1. Specific joint preparation may be required.
2. Adhesive material may be expensive and require application in controlled environments.
3. Curing time may be long and may require a special environment.
4. Mechanical strength of the adhesive-bonded joint may depend greatly on how the joint is loaded. Shear loading is best followed by tension. Where loading of the joint tends to peel materials apart, adhesive bonding may prove unsatisfactory (Figure 15.17).

Natural Adhesives

Natural adhesives include (1) vegetable and animal glues and casein, (2) sodium silicate, and (3) natural gums.

Vegetable glues are used in the paper industry and have applications in woodworking. Examples include wallpaper paste made from starch, and the gum arabic used on envelopes and postage stamps. Animal glues made from bone and hide are adequate for some applications, but they tend to deteriorate in moist environments. Casein, made from milk, is a water-based adhesive.

Sodium silicate shows good temperature characteristics, and it can be used as a binder to make some types of composite material.

Natural gums have long been used as adhesive materials. These include natural rubbers, sealing waxes, and asphalt.

Plastic Resin Adhesives

Synthetic resins (plastics) used for adhesives have undergone extensive development and have found an important place in material joining processes. In many cases plastic resin adhesives have proved to be superior to mechanical joining systems.

Resins used for adhesives come from either of the major plastic classifications: thermosets and thermoplastics. The thermosetting plastics undergo fundamental chemical change while hardening and cannot be reused or recycled. Thermoplastics, on the other hand, can be heated, resoftened, and reused.

Thermoplastic resins used for adhesives include cellulose nitrate, acetate, acrylic, cyanoacrylate, and vinyl.

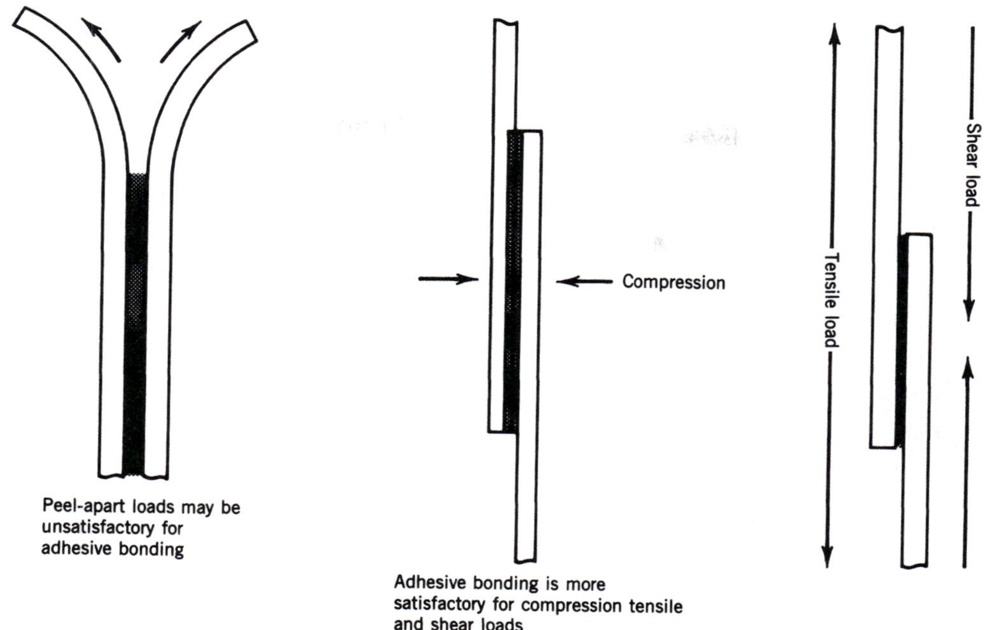

FIGURE 15.17
Types of loads applied to adhesive-bonded joints.

Thermosetting resins used for adhesives include polyesters, epoxies, phenolics, ureas, and silicones.

Adhesive Bonding in Aircraft Production

Adhesive bonding in aircraft assembly has undergone much development in recent years. The bonding of aircraft skin panels has numerous advantages over joining them with screws or rivets, such as:

1. Bonding resists corrosion (Figure 15.18).
2. Adhesive bonds are as strong as or stronger than riveted joints.
3. Joint loading is more uniform with adhesive bonding (Figure 15.19).

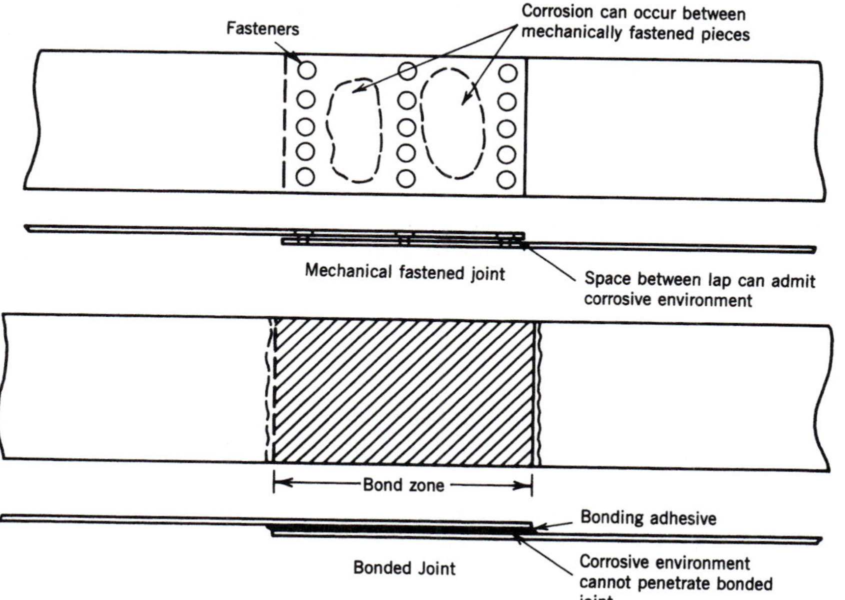

FIGURE 15.18
Bonded joints are resistant to corrosion.

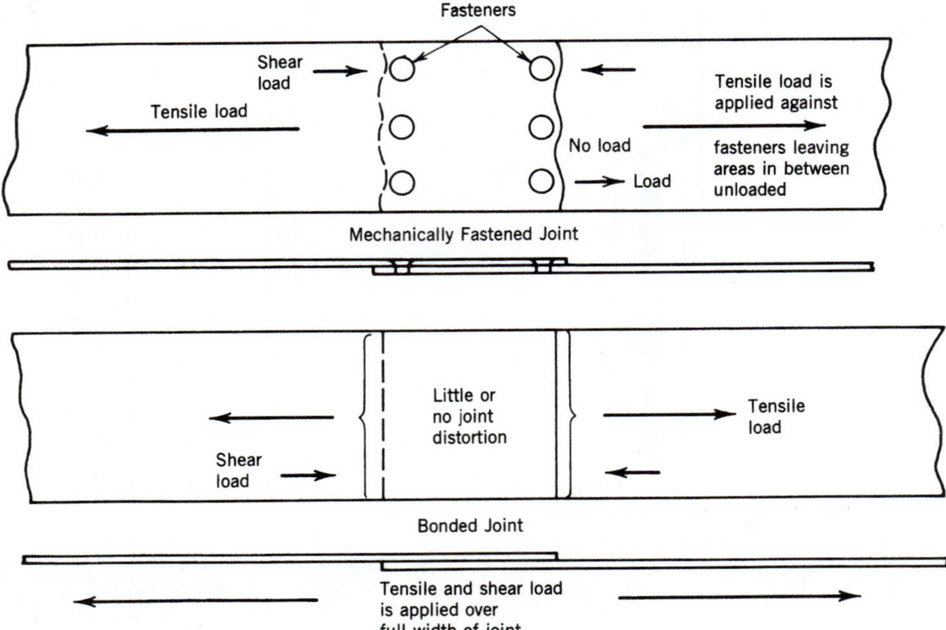

FIGURE 15.19
Bonded joints distribute load better than mechanical joints.

4. Joint fatigue and crack development are reduced using bonded joints.
5. Bonded joints are more effective in sound and vibration dampening in aircraft structures.
6. Bonded joints eliminate the need for many fasteners and thus may result in a substantial weight saving, essential to modern fuel-efficient aircraft.

The requirements of aircraft skin panel bonding are stringent. The adhesive used must have high resistance to peel-apart and shear loads. The bond joint must be able to resist penetration to corrosive fluids and must also be able to withstand high temperatures in its service environment.

Another requirement of bonded joints is consistency of the bond throughout the joint area. In addition, the adhesive material must contain a corrosion inhibitor and must not crack or crumble when drilled for attachment to any mechanically fastened components. The most versatile adhesives that demonstrate these properties necessary for aircraft bonding requirements come from the epoxy group.

FIGURE 15.20
Aircraft skin panels being cleaned prior to bonding.

Aircraft Bonding Technology

Skin panels are first sprayed with a protective coating. They are then formed by pressing to the contour shape of the aircraft fuselage. The protective coating previously applied is not affected by the forming process. The formed skin panels are then cleaned (Figure 15.20). Cleaning, which is essential to a proper bond, is done with an alkaline solution and water rinse. This is followed by an acid rinse to slightly etch the surface of the skin panel. Acid etching provides an effective bonding surface. The panel then goes to a temperature- and humidity-controlled adhesive bonding room where the epoxy-based adhesive primer is applied. Air spraying is used, with the spray air having been carefully cleaned of all oil and moisture. A full and uniform coating, required for proper bonding, is applied at this stage in the process. Primed panels are then cured and stored in a clean room environment. Bonding occurs in a clean-room environment where the atmosphere is maintained at a positive pressure. The epoxy-based adhesive is applied in a layer 0.01 in. thick. Parts are joined and held in position by nylon tape (Figure 15.21). The joined assemblies are then placed on bonding fixtures in plastic bags and a vacuum is drawn, creating a downward pressure on the parts. The next process phase, bonding and curing, is accomplished in a large-capacity autoclave (Figure 15.22). The autoclave is closed, heated, and purged with an inert gas under pressure. After a suitable period, pressure and temperature are reduced and the bonded assembly is removed from the autoclave.

FIGURE 15.21
Panels are held in position with nylon tape prior to bonding.

FIGURE 15.22
Bonded panels enter the large autoclave for curing.

FIGURE 15.23
Ultrasonic testing of panel bond joints.

After completion of the bonding process, assemblies are cleaned and prepared for inspection. Samples of the bonded assemblies undergo peel and shear testing. Bonded joints undergo ultrasonic inspections (Figure 15.23) to determine joint integrity. The final phase is the painting and assembly of the skin panel on the aircraft (Figure 15.24). Large aircraft skin panels are precision components and can be damaged unless handled carefully. It is interesting to note the methods by which this is accomplished. An array of vacuum cups is used to pick up and handle the skin panel. The method is very secure, and it will not damage the precision panel component.

WELDING PROCESSES

It would be difficult to overstate the importance of welding in modern material-joining processes. These processes make fabrication of complex metal assemblies possible. In welding, the materials to be joined are heated to the fusion temperature, at which point they are joined and allowed to cool. Filler material, generally of the same chemical composition as the parent metal, may be added to fill the weld

FIGURE 15.24
Suction cups hold the panel while it is positioned for installation.

joint during the process. Common industrial welding processes include oxy–fuel welding, arc welding, resistance welding, solid state, and other welding processes.

Oxyfuel Welding and Cutting

The oxyfuel welding process uses an oxygen-supported hydrocarbon (e.g., acetylene, methane, hydrogen, propane) gas flame to heat the parts to welding temperature. The most commonly used gas is acetylene because of the resulting high flame temperature and heat content. The oxyacetylene process, often known as *gas welding,* is used mostly for repair welds on light to medium applications or where electric welding might not be available. Gas welding requires considerable technique and skill on the part of the operator.

In the gas welding process, gas and oxygen are separately delivered under regulated pressure to the welding torch and are mixed in the mixer within the torch. The fuel–oxygen mixture is ignited as it exits the torch nozzle, and the operator adjusts the mix by valves on the torch assembly. The flame mixture may be adjusted to be neutral, oxidizing, or reducing. The neutral flame is a result of the complete combustion of acetylene, the oxidizing flame has an excess of oxygen, and the reducing or carburizing flame has an excess of fuel gas. The type of flame needed depends on the particular type of application. For example, the carburizing flame can be used to inject carbon into the surface of the metal being heated, thus affecting the process of carburizing for surface heat treatment; however, the most commonly used type of flame is neutral.

Oxyacetylene flame processes are also used extensively for cutting applications (Figure 15.25). This process plays an important part in manufacturing. Oxyacetylene cutting torches can cut through metal that is several inches thick, making them indispensable for man-

FIGURE 15.25
Flame cutting with an oxyacetylene cutting torch is a widely used process.

ufacturing structural fabrications such as bridges, ships, and oil field and farming equipment.

During this process, preheat jets (mixture of oxygen and acetylene) in the torch tip heat the metal to a red-hot state and maintain this temperature. When the proper cutting temperature is reached, a pure stream of oxygen is directed through a hole in the center of the torch tip. When the oxygen makes contact with the red-hot pre-heated metal, rapid oxidation takes place almost instantly, and the metal is cut at that point. An operator skilled in flame cutting can produce a very smooth torch-cut surface.

Flame cutting has also been computer automated for production purposes. Cutting torches that are solidly mounted and have their movement controlled by mechanical feeding mechanisms can produce very smooth flame-cut surfaces that may require little or no finishing machining operations.

Electric Arc Welding

Electric welding processes are the most common method of joining metals during manufacturing processes. The bulk of steel structural fabrication and construction is done by electric arc welding processes (Figure 15.26) or electric resistance welding processes.

Although the individual processes differ to some extent, they all convert the flow of electric current to heat so that the material to be welded may be heated to proper fusion temperatures. The welding system consists of a power supply capable of sustaining large currents at relatively low voltages and appropriate cables connected to the workpiece, thus completing the electrical circuit. A typical arc welding system configuration is shown in Figure 15.27.

FIGURE 15.26
Large structural steel fabrication joined by electric arc welding processes (Miller Electric Manufacturing Company).

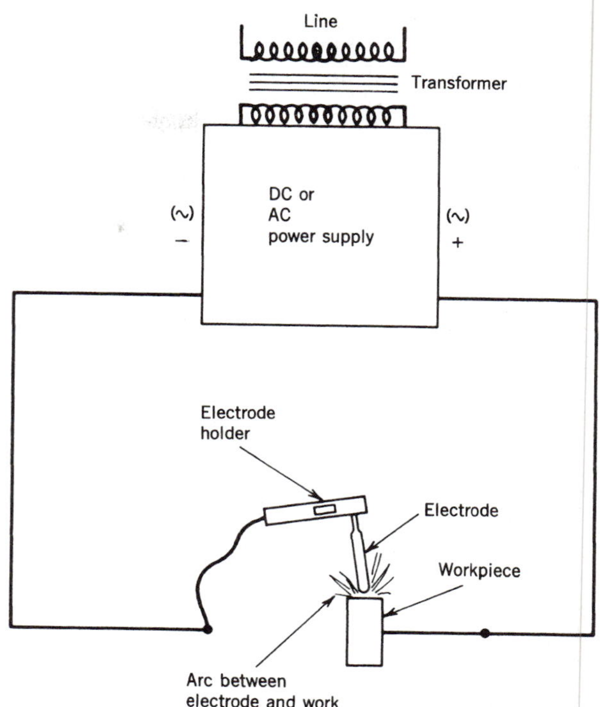

FIGURE 15.27
Electric arc welding system components.

Several electric welding processes are in common use. They have wide capability and can be used on a variety of metals in configurations ranging from heavy sheets, structural shapes, and plates to very thin sheets. Common electric welding processes include shielded metal arc welding, inert shield gas arc welding, gas metal arc welding, gas tungsten arc welding, plasma arc welding, resistance welding, stud welding, and induction welding.

Shielded Metal Arc Welding In the shielded metal arc welding (SMAW) process, current flows from the power supply through copper wire cables to the electrode holder, which is usually held in the welder's hand. The welding power supply may be powered from the utility line, or it may be engine driven. Engine-driven power supplies permit electric welding to be accomplished at locations where no line power is available. This makes the welding system extremely useful for field repairs such as those often made on farm equipment.

FIGURE 15.28
The engine-driven power supply for electric welding is indispensable equipment for on-site machinery repairs. (Miller Electric Manufacturing Company).

FIGURE 15.29
Electric arc weld.

A wire electrode is clamped in the electrode holder (Figure 15.28). The electrode, often called a *welding rod,* is about 12 in. long and is usually coated with a hard dry coating material. The coating serves several functions during the welding process: (1) fluxing, (2) arc shielding, and (3) arc stabilization. The fluxing material cleanses and deoxidizes the molten metal. It also forms slag, which slows down the freezing rate of the weld and controls the shape and appearance of the weld bead. The coating also provides a protective atmosphere that serves to shield the weld zone from atmospheric oxygen and nitrogen. The arc-stabilizing elements in the coating maintain a steady arc during operation.

The wire core of the welding rod acts as the filler metal in the joint between the materials being welded. As the operator touches the rod to the work current flows across the junction, and an electric arc is created. Heat develops from the large current being dissipated at the workpiece and in the electrode, which causes the filler rod to melt and deposit into the joint (Figure 15.29).

Considerable skill is required on the part of the operator to properly maintain the arc length and to make correct movements so as to deposit filler metal neatly into the welded joint. Heat is controlled by adjusting the current flow from the power supply. Current flow is based on the type of rod used and the thickness of the metals being joined.

The operator (welder) must be protected from the intense heat and ultraviolet light coming from the welding arc by wearing appropriate clothing and a face hood containing a darkened-glass view port. This enables the welder to observe the welding being done while the eyes are protected from the intense light emitted by the welding arc.

The shielded metal arc welding process is used widely in all types of manufacturing applications. It is especially popular in fabrication industries such as ship, bridge, pipeline, and tank manufacturing and in many other industries that fabricate products from structural metals, plates, and heavy-gage sheet steel materials.

Inert Gas Shielded Arc Welding In the shielded metal arc welding process, the gases resulting from burning the coating on the welding rod provide a certain degree of protection from external air contamination of the weld zone. Since the weld takes place at elevated temperatures, contamination from the air surrounding the weld zone poses a significant problem. For example, during the welding of thin steel sheet metal, the heat from the welding process cannot be dissipated fast enough to prevent oxidation of the welded material without shielding. In addition, nonferrous metals with an even higher affinity for oxygen than steel will be consumed before they can be joined by the welding process.

An excellent method of controlling the gas contamination problem is inert gas shielded arc welding. In this type of welding, the entire weld zone is surrounded by

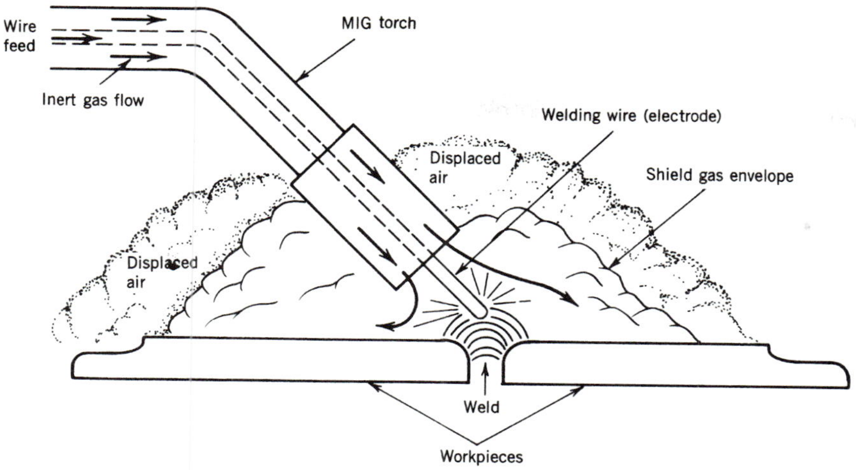

FIGURE 15.30
GMAW or MIG welding shields the weld zone with inert gas.

an inert gas such as argon, helium, or carbon dioxide (Figure 15.30) or by various mixtures of these gases. The inert gas displaces the air in the immediate vicinity of the weld zone, and welding takes place in an uncontaminated environment. The major shielded gas welding processes include gas metal arc welding and gas tungsten arc welding. There are also several other arc welding processes.

Gas Metal Arc Welding The **gas metal arc welding (GMAW)** process, also known as **MIG (metal inert gas)** welding, is a widely used and efficient process that can be applied over a wide range of metal thickness and types of materials.

The MIG system consists of a constant voltage power supply, wire feed unit, and a shield gas cylinder with pressure and flow regulators (Figure 15.31). The MIG electrode is a continuous wire, preferably of the same material that is being welded. Electrode wire is fed continuously from a reel while the welding process is taking place (Figure 15.32). At the same time shield gas flows from the welding gun (Figure 15.33), sur-

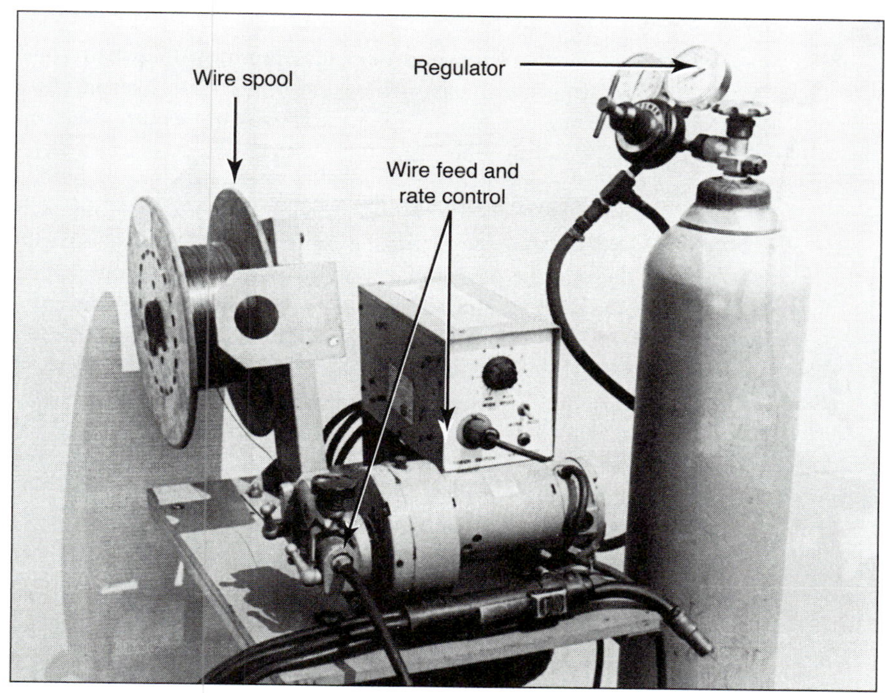

FIGURE 15.31
MIG welding equipment includes wire feeder and shield gas cylinder.

FIGURE 15.32
The MIG electrode and shield gas are fed to the weld in a continuous flow (Miller Electric Manufacturing Company).

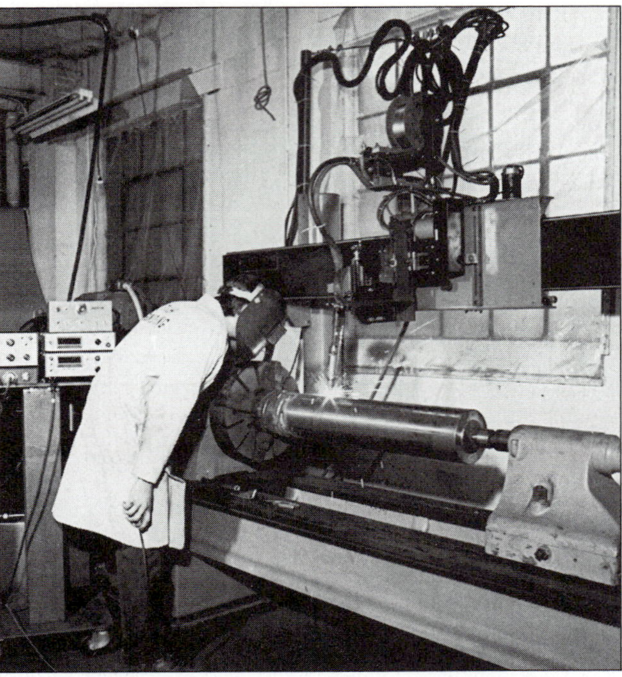

FIGURE 15.34
Lathe-mounted workpiece undergoing weld buildup by the MIG process (Miller Electric Manufacturing Company).

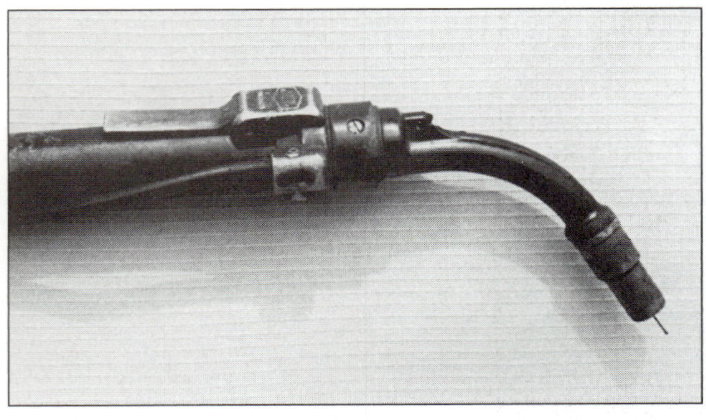

FIGURE 15.33
The MIG torch.

rounds the wire, and displaces air from the weld zone. Welding current, shielding gas flow, and electrode feed rate may be varied to meet the exact requirements of the welding job.

Variations of the MIG process include the flux cored arc welding process in which the flux is contained inside a hollow electrode wire. Fluxed electrode welding may be performed with or without shielding gases.

Because the process takes place in the inert gas environment, the weld quality of the MIG process is unsurpassed. Applications include the rehabilitation of worn parts by weld buildup (Figure 15.34). In the illustrations, a shaft has been mounted in a lathe where it can be rotated, and a MIG bead can be deposited evenly and accurately on the workpiece (Figure 15.35). After the welding is complete, the part may be machined to restore its original dimensions.

With the convenience of the continuously fed electrode and the inert gas shielding, the MIG process is

FIGURE 15.35
Weld built up by the MIG process is virtually free of slag pockets and inclusions. After weld buildup is completed, the part may be remachined to restore original dimensions (Miller Electric Manufacturing Company).

highly suited for production welding of ferrous and nonferrous metals, particularly steel and aluminum (Figure 15.36). The MIG process is highly suited to both semiautomation (Figure 15.37) and full automation using industrial robots (Figure 15.38).

Gas Tungsten Arc Welding **Gas tungsten arc welding (GTAW),** commonly known as **TIG (tungsten inert gas)** welding, is a nonconsumable arc welding process that uses a tungsten electrode that does not melt during the welding process. Filler material is fed externally into the welded joint during welding. The process is somewhat like the oxyacetylene process in this respect, the difference being that the heat is supplied by an electric arc established between the tungsten electrode and the workpiece. TIG also uses a shielding gas to surround the weld zone and to displace atmospheric contaminants (hydrogen, oxygen, and nitrogen).

The TIG system (Figure 15.39) consists of a power supply, cables, and a welding torch. In manual operation, the welder manually feeds the filler rod into the arc, where it melts and is deposited into the welded joint (Figure 15.40). Shielding gas flow and welding current are controlled by a foot pedal. Some TIG torches that are used for higher welding currents are water cooled. This helps dissipate heat, thus preserving the electrode and making the torch more comfortable to hold.

FIGURE 15.36
The operator can reach the work easily using an efficient portable MIG welder with overhead suspended wire feeder (Miller Electric Manufacturing Company).

FIGURE 15.37
Automated MIG welding equipment (Miller Electric Manufacturing Company).

FIGURE 15.38
The MIG welding process is highly suited to industrial robot automation (DaimlerChrysler Corporation).

TIG process is extremely useful for welding applications where detail welding is required. TIG can be used to weld thin sheets of metal a few thousandths of an inch thick. Since the process takes place in the inert gas envelope, it is possible to achieve superior quality welds. For this reason, TIG may be used where a weld of outstanding quality is required. An example would be the root pass in joining nuclear system piping. After a TIG root pass is made at the bottom of the weld joint, the remaining weld filling of the joint may be done by other processes such as MIG. TIG is also superior for welding nonferrous metals, especially aluminum.

Submerged Arc Welding Submerged arc welding (Figure 15.41) is a shielded electric arc process. The welding

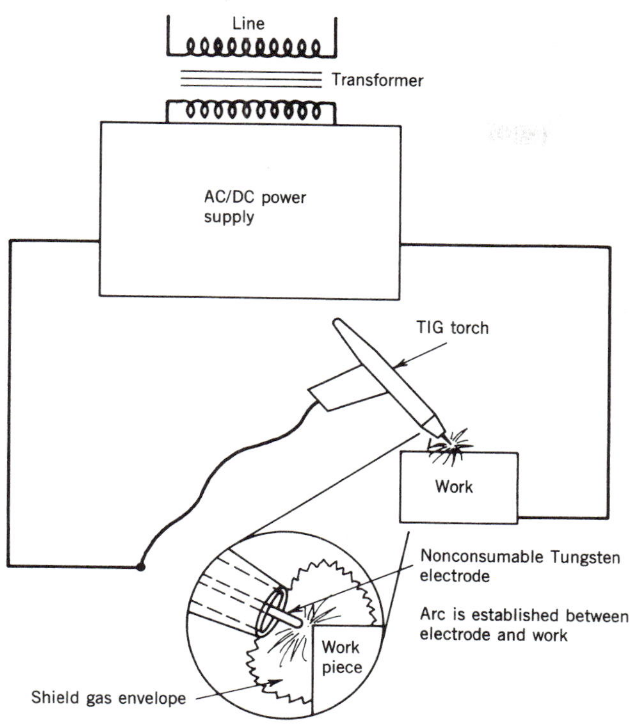

FIGURE 15.39
GTAW or TIG welding system.

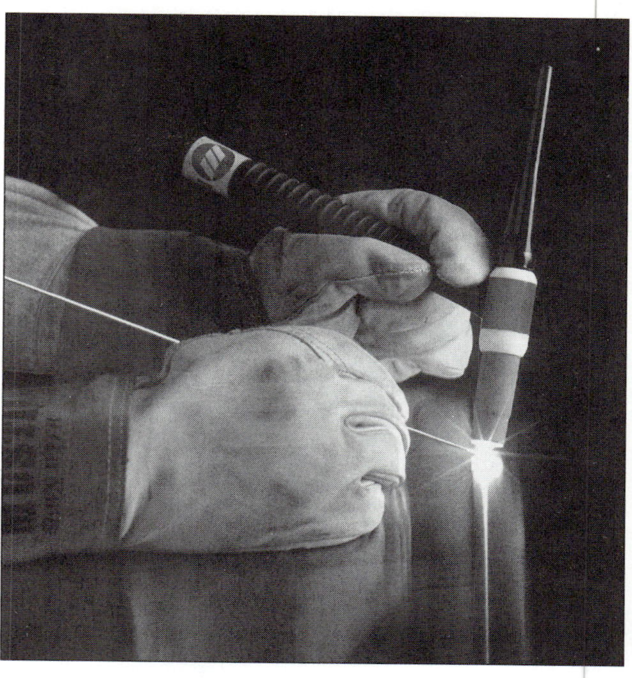

FIGURE 15.40
In this close-up view of TIG welding the TIG torch and filler rod are visible in the welder's hands (Miller Electric Manufacturing Company).

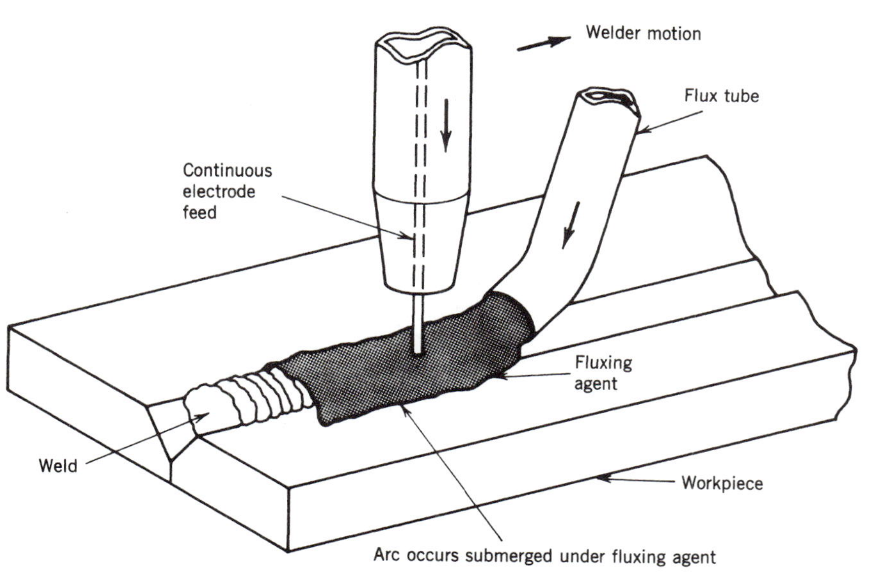

FIGURE 15.41
Submerged arc welding.

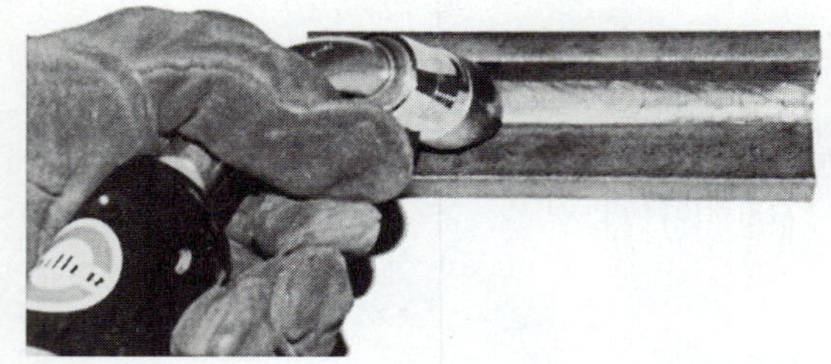

FIGURE 15.42
Submerged arc results in a weld of exceptional quality (Miller Electric Manufacturing Company).

FIGURE 15.43
Submerged arc welding equipment (Miller Electric Manufacturing Company).

process takes place buried under a fluxing agent that is being deposited in the joint ahead of the electrode. The fluxing compound acts in much the same way as a shielding gas in the MIG process by preventing atmospheric gases from contacing the molten metal. The fluxing compound also acts in much the same way as a coating in the SMAW process by cleansing impurities from molten metal and slowing down the cooling rate of the weld. In addition, flux prevents spatter, and thus results in a better quality weld (Figure 15.42).

Submerged arc welding equipment consists of the wire feed mechanism, the flux hopper, and the power supply (Figure 15.43). The flux hopper is air pressurized to provide the flux feed. The process is suitable for large welds and for automatic welding of pipe and major structurals such as I beams, channel beams, and H beams, but it is limited to a flat or horizontal welding position so that the fluxing agent will remain in the weld area.

Plasma Arc Welding Heat for the plasma arc process (Figure 15.44) is derived by ionizing (creating an imbalanced electric charge) a plasma gas by an electric arc. The resulting concentrated plasma arc can deliver a large amount of heat at very high temperatures to the weld zone. Plasma arcs can generate temperatures of more than 40,000°F, much higher than can be achieved by any fuel combustion process. Plasma arc welding is fast and clean and its arc concentration properties permit high-quality welds with minimal effect on the surrounding welding area. The process is suitable for welding many materials and is commonly used for cutting. Additional uses of extremely high temperature plasma arcs are being investigated and applied in many other areas of industrial processing.

FIGURE 15.44
Plasma arc welding.

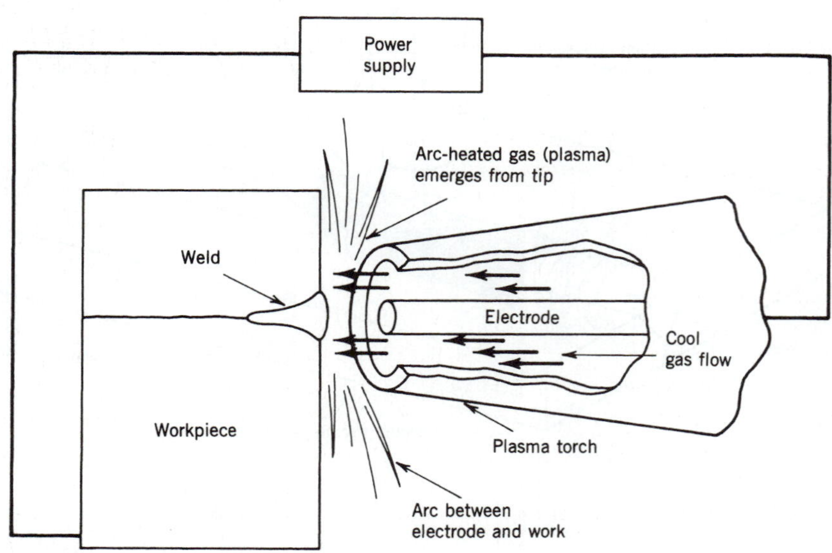

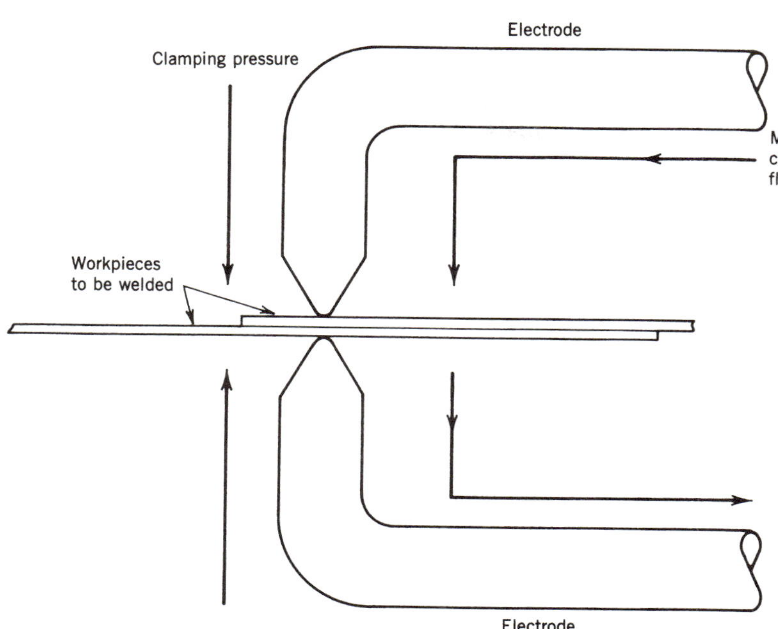

FIGURE 15.45
Spot welding.

Resistance Welding

Most metals, particularly steels, have resistance to electrical current flow. This characteristic can be used to heat the metal to welding temperatures, at which point the parts to be joined are forced together, thus completing the weld. No externally supplied filler material is required and the welding process does not require any shielding gas. One popular and versatile resistance process is spot welding (Figure 15.45). Workpieces are clamped between two electrodes, the current flow heats the metal to a fusion temperature, and the mechanical pressure of the electrodes forces the material to fuse together. Since the welding process results in a relatively small weld size, several spot welds are usually necessary to ensure adequate strength in large fabrications.

The spot welding process is very fast and is therefore highly suitable for volume production. Spot welding is widely used in the automobile industry, and the nature of the process is such that it is readily automated using industrial robots (Figure 15.46). Microprocessor controls are used on spot welding equipment to control welding variables (e.g., current, contact time).

Resistance welding is also used to join the ends of band saw blades (Figure 15.47). A band saw blade weld will require annealing (softening) after the welding process so that the band will have the flexibility required in service. Resistance welding can be applied to seams and long joints by using roller electrodes. The process called *resistance seam welding* is used in the manufacture of pipe and tube (Figure 15.48).

FIGURE 15.46
Automated spot welding in automobile production (Daimler Chrysler Corporation).

Stud Welding In the stud welding process, an arc is established between the stud and the mounting surface. The stud is secured in a stud gun (Figure 15.49) and a ceramic ferrule surrounds the base where the weld will occur. The ceramic ferrule serves two purposes: (1) it concentrates the heat, and (2) it serves as the mold that shapes the weld in the fillet form. The ferrule is broken off after the weld is completed. When the stud gun is energized, the solenoid in

FIGURE 15.47
Resistance flash butt welding a band saw blade.

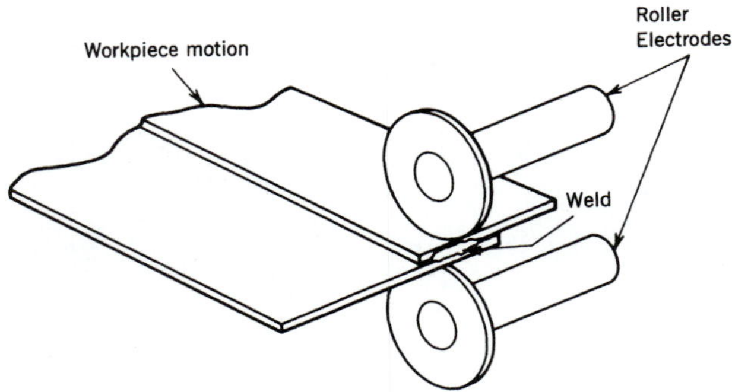

FIGURE 15.48
Resistance seam welding using roller electrodes.

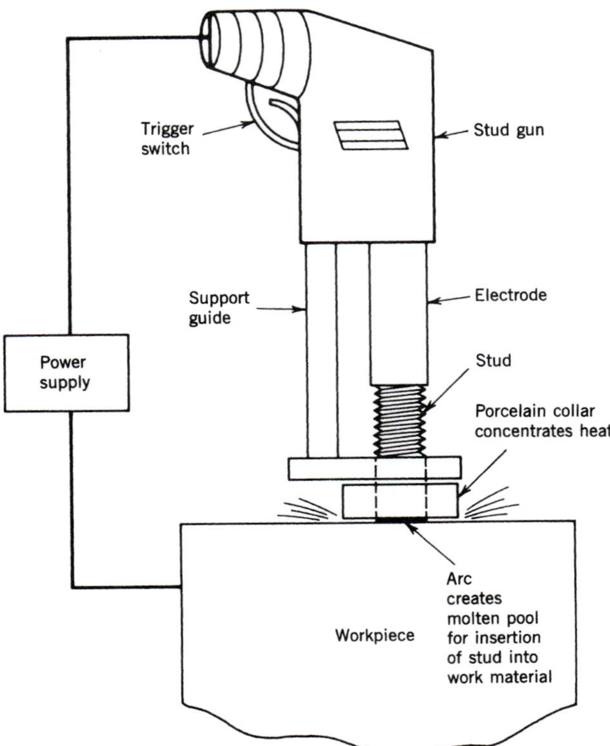

FIGURE 15.49
Stud welding.

the gun withdraws the stud from the base metal, at which moment an arc is established. The arc heats and melts the base metal and stud. After a specified time interval, a mechanical spring is released, which forces the stud into the molten pool of metal, accomplishing the weld. The process takes place almost instantly.

Studs of many different sizes and styles may be installed by this fast and efficient process. Stud welding is a very convenient method of attaching a threaded fastener to a flat surface. Nonthreaded fasteners including attachment hooks or clips may also be installed by stud welding.

Induction Welding Heat for induction welding (Figure 15.50) is generated by inducing a rapidly oscillating electromagnetic field into the parts to be welded. Parts may be placed in an induction coil field, and current induced into the workpiece as the field rises and falls in the induction

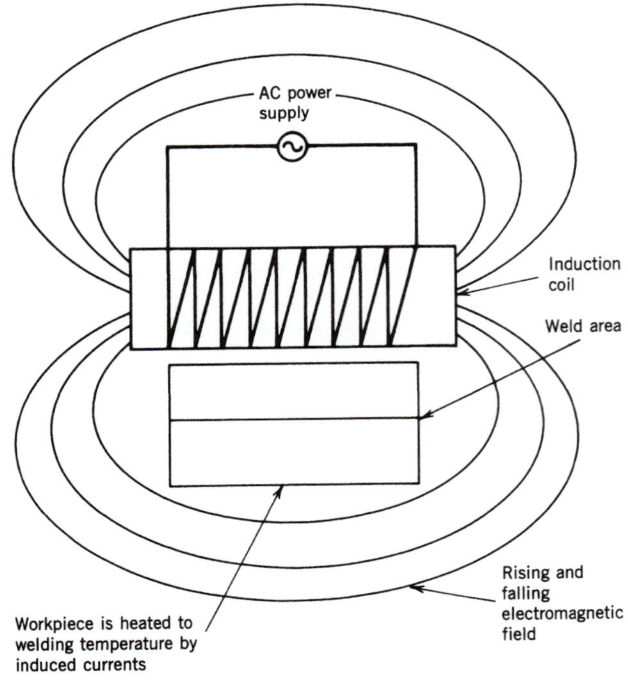

FIGURE 15.50
Induction welding.

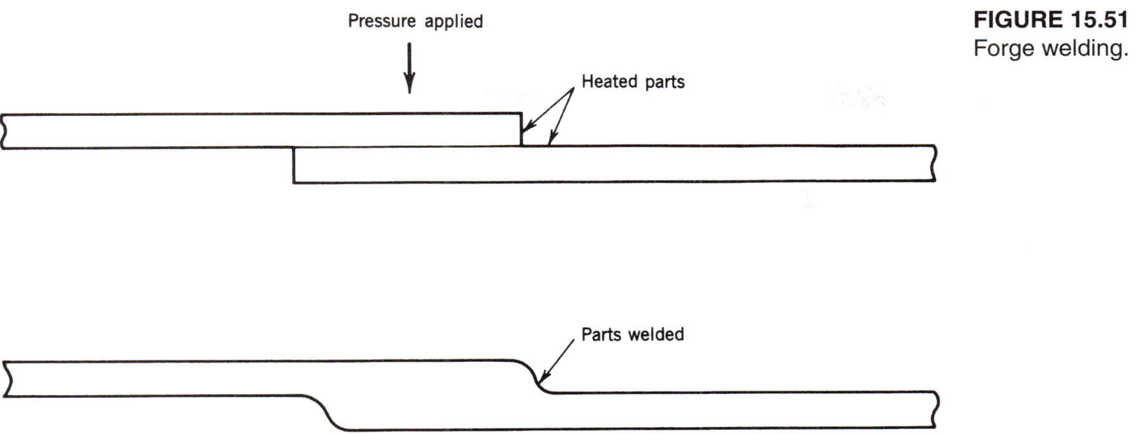

FIGURE 15.51
Forge welding.

coil causes rapid heating. Induction heating is useful for production welding. It also is used as a heating method for other processes.

Solid-State Welding Processes

Forge Welding In forge welding (Figure 15.51), probably best known from the days of the village blacksmith, parts to be welded are heated in a forge fired with charcoal or coal. The fire is forced-air fed by a bellows or a motor-driven fan under the coal bed to increase the temperature. Parts are placed in the fire and are heated to a point slightly below melting. The parts are then removed from the fire and placed in contact to be joined. Pressure is applied by hammering, in the case of the blacksmith, or by mechanical pressing or rolling in more modern techniques. Assuming correct heating and scale-free surfaces, a weld occurs.

Ultrasonic Welding Heat for ultrasonic welding (Figure 15.52) is generated by subjecting the parts to be welded to high-frequency vibrations in the ultrasonic range. The mechanical energy of the vibration causes the parts to heat as they are forced together by mechanical pressure. The process has applications for precision welding in a variety of materials and where the electromagnetic forces and intense heat of other welding processes would damage the product. The process can be used for only very thin materials.

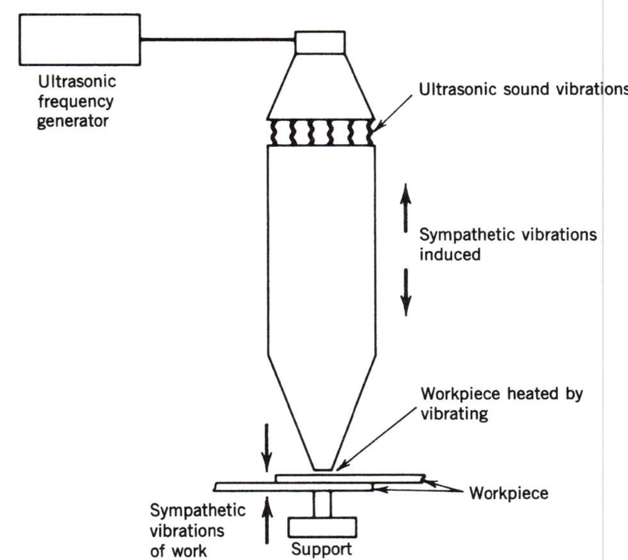

FIGURE 15.52
Ultrasonic welding.

Other Welding Processes

Other welding processes cannot easily be classified into one of the previous groups. Some of these processes include thermit welding, electron beam welding, and laser welding.

Thermit Welding In the thermit process, parts to be joined are submerged in a bath of molten metal that is sufficiently hot to heat the parts to be welded to a fusion temperature. The thermit mixture consists of metallic oxide of the metal to be welded, reducing metal, and another metal that will achieve a very high temperature when ignited. A typical mixture consists of iron oxide, aluminum, and magnesium. Magnesium is ignited, and as it burns it develops the necessary heat to start an exothermic chemical reaction between aluminum and iron oxide. As a result, aluminum oxide and pure iron are formed. Because the aluminum oxide is lighter, it will float to the surface, where it is removed. The molten iron is used to create the weld.

The thermit weld takes place in a mold, so that the molten metal is retained as it solidifies. In this respect thermit welding is similar to casting (Figure 15.53). Different

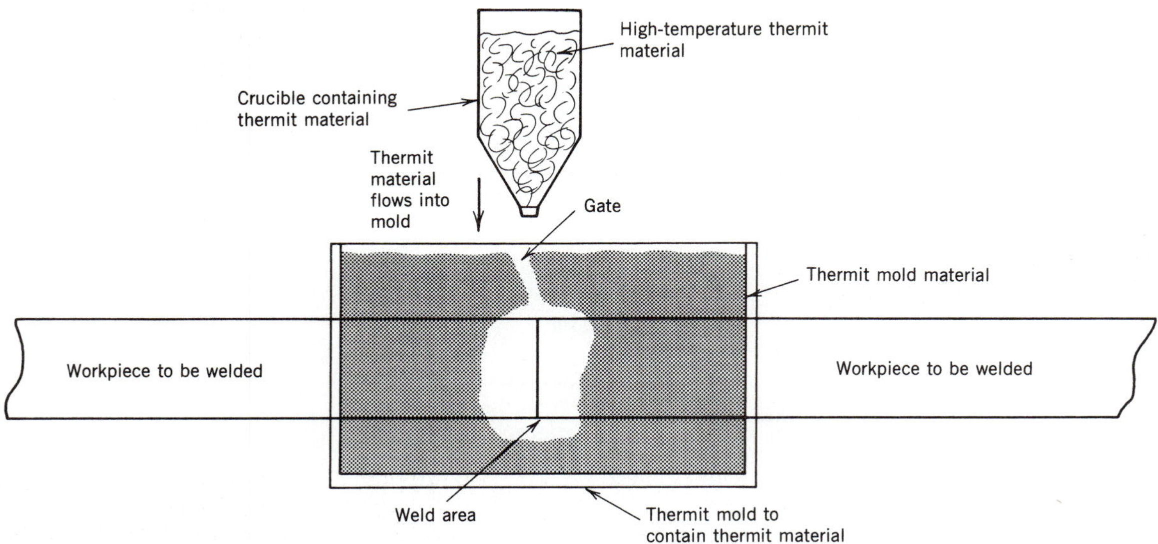

FIGURE 15.53
Thermit welding.

refractory and high-melting-temperature materials can be used to make the mold. Parts to be welded must be cleaned and correctly aligned. Preheating is often required. An example of a thermit welding application is the joining of railway rail ends so that the rail becomes continuous, which makes for a much smoother roadbed.

Electron Beam Welding Electron beam welding (Figure 15.54) is a high-energy welding process that has found wide application in modern industry. Heat for the process is achieved by accelerating and focusing a stream of electrons in an electromagnetic field and allowing the beam to impinge onto a concentrated area on a workpiece. The kinetic energy of an electron is small. However, when that energy is multiplied by the number of electrons in the beam it is sufficient to generate heat that can melt a workpiece. To reduce or even eliminate the scatter of the electrons that occurs when an electron beam passes through air, the process is accomplished in a vacuum. The process is highly suitable for precision welding on parts that have already been machined. Because the weld penetration and location can be controlled precisely, this process has wide application in joining precision assemblies with a minimum of heat distortion. The electron beam process is also used for hard facing teeth on band saw blades, thus permitting the manufacture of superior cutting bands for sawing difficult-to-cut materials.

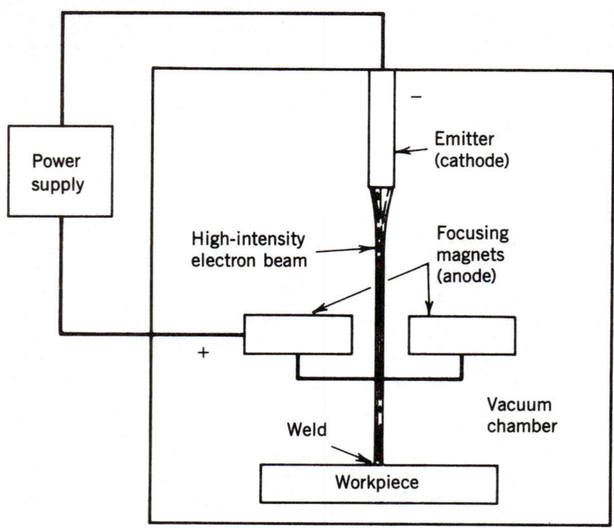

FIGURE 15.54
Electron beam welding.

Laser Welding. The laser, discussed in the previous chapter, is an intense beam of coherent light. Laser light is generated by particular agitation of atoms or molecules in gases such as carbon dioxide and neon or crystals such as ruby. The light emitted from the gas or crystal laser is a single color and is an extremely tight or coherent beam. The coherent beam of a high-powered laser can be extremely intense, which makes the laser quite useful for industrial applications. The laser has found many applications in modern industrial processes, including joining processes.

In laser welding (Figure 15.55), heat is generated by directing the laser beam onto the parts to be welded. The beam is concentrated and also can be applied in areas of extremely small diameter. Precision welds (Figure 15.56)

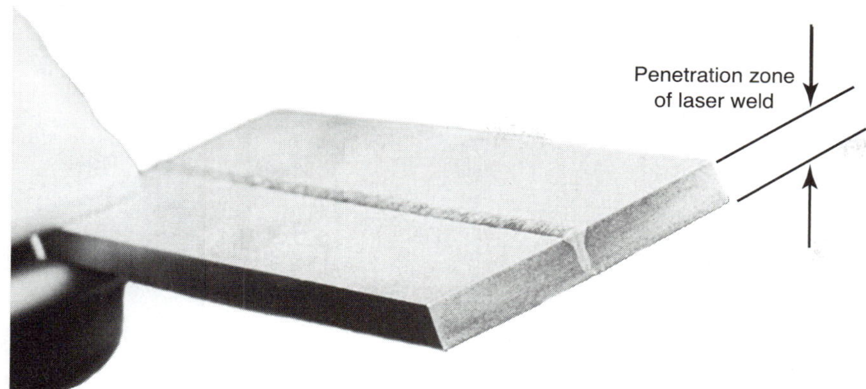

FIGURE 15.56
Laser welding results in excellent penetration and little distortion of surrounding material (Coherent Inc., Palo Alto, California).

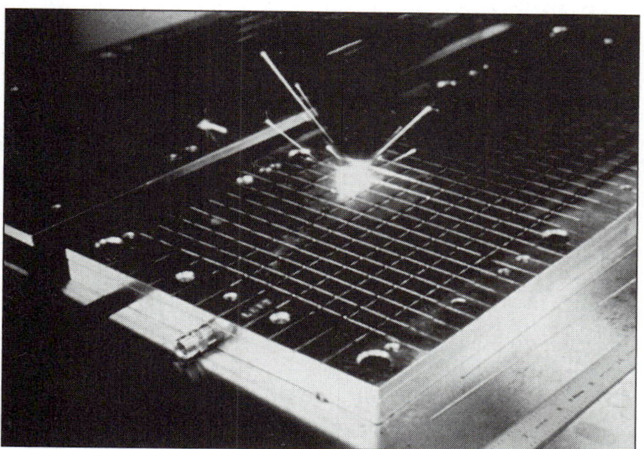

FIGURE 15.55
Laser welding (Laserdyne).

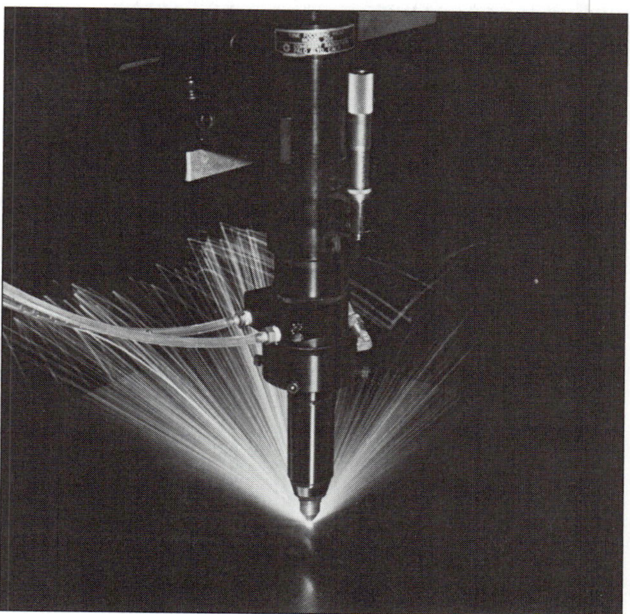

FIGURE 15.57
Laser cutting (Coherent Inc., Palo Alto, California).

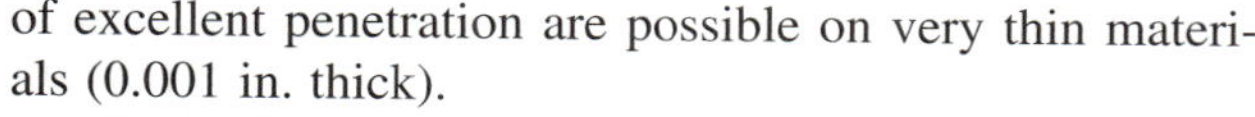

of excellent penetration are possible on very thin materials (0.001 in. thick).

Related in many ways to the welding of metals, the ability of the laser to seal other materials such as human tissue has led to many amazing applications. Lasers can also be used to cut (Figure 15.57), drill, plate, heat for selective hard facing, and etch markings on parts.

BRAZING AND SOLDERING

Welding requires heating parts above their melting temperature, where they fuse into a common mass across the weld zone. If filler material is used, it too becomes part of the common mass. Alternatively, in brazing and soldering, the filler metal melts at a lower temperature than the melting temperature of the parts to be joined, and the parts themselves are not melted. The filler metal does not actually become part of the common material in the weld. The joining of the material is purely mechanical; that is, filler metal grips the parent metal by adhesion forces. The major difference between brazing and soldering is that brazing processes use filler metals with higher melting temperatures than soldering processes.

Soldering and brazing do not provide as strong a mechanical joint as do the fusion welding processes.

Nonetheless, they have wide application in manufacturing as efficient material joining processes. It is important that these processes are applied only where soldering or brazing is a suitable joining method. They must not be used in applications where strength requirements necessitate the complete fusion bond accomplished in the welding processes.

Brazing

Brazing processes use filler material with a melting temperature above 840°F and below the melting temperature of the base metal. The process requires properly cleaned surfaces, which can be accomplished by mechanical (e.g., grinding, sand blasting, wire brushing, filing) or chemical means (e.g., solvents, alkaline baths, acid baths, salt bath pickling). In addition, the parts to be joined must be properly positioned with respect to each other. That is, the gap between the parts must be such that capillary forces can pull the filler material into a joint. A larger or smaller distance between the parts may adversely affect bonding. Heat sources commonly used in this process are the same as those used in welding: chemical, mechanical, and electrical.

Brazing filler metals are classified according to their chemical composition: aluminum–silicon, copper, copper–zinc, copper–phosphorus, nickel–gold, heat-resisting materials, magnesium, and silver. The selection of braze material is based on the metal being brazed.

Some of the common brazing processes include dip brazing, furnace brazing, induction brazing, infrared brazing, resistance brazing, and torch brazing. Joint strength is much higher with the brazing and braze welding processes than with soft soldering processes.

Soldering

Soldering processes use filler material with a melting temperature below 840°F and below the melting temperature of the base metal. Typical soldering metals are alloys of nonferrous metals such as lead and tin. The most commonly used solder is 50 percent lead, 50 percent tin alloy. Soldering is commonly used in joining sheet metal when low to medium strength is required, and in leak-proof joints. Solders are used extensively in the electronics industry for electronic circuit card soldering (Figure 15.58). The solder alloys have excellent electrical conductivity and lend themselves to automated soldering applications.

Soldering, similar to brazing, requires proper joint preparation and material cleaning to ensure properly soldered joints. Fluxes are almost always used to aid in cleaning during the soldering process. Soldering can be applied to many materials; however, there are some metals for which soldering is unsatisfactory.

The same processes and heating methods used for brazing are also used for soldering: e.g. dip soldering, furnace soldering, induction soldering, and infrared soldering.

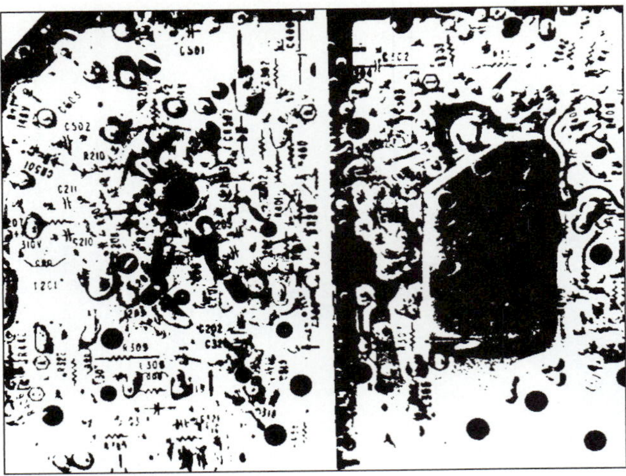

FIGURE 15.58
Electronic printed circuit card soldering using soft lead–tin alloy solders.

THERMAL SPRAYING

Thermal spraying is a process of depositing a layer of metal or ceramic material onto the surface of a part. The process is used to provide surface coatings of different characteristics, such as coatings to improve abrasive wear or corrosion resistance, to change thermal conductivity, or simply to build up the surface. For example, thermal spraying can be used to apply a layer of aluminum on a steel structure (such as a bridge) to slow down or eliminate corrosion and to prolong the life of a structure. Also, disk teeth on farm equipment can be coated with wear-resistant material to reduce the wear caused by rock and dirt abrasion.

Several processes have been developed for thermal spraying: flame spraying, high-velocity oxyfuel (HVOF) spraying, arc spraying, plasma spraying.

In the processes of thermal spraying (Figure 15.59), small droplets of molten metal or ceramics are carried by hot gases and are deposited on the workpiece, where they adhere. The strength of the mechanical bonding at the interface between the part and the sprayed-material buildup is increased by creating a rough surface on the

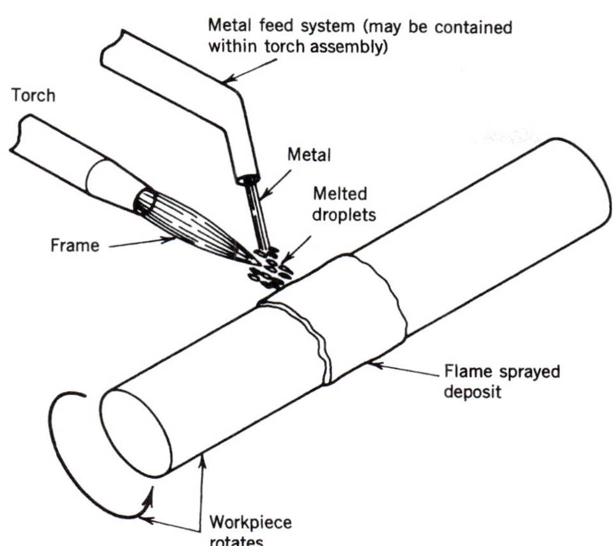

FIGURE 15.59
Thermal spraying.

part using machining for round parts and shot blasting for flat surfaces. Thermal spraying requires that the substrate be properly cleaned prior to depositing, and spraying should be performed immediately after the cleaning operation in order to avoid oxidation. The part to be sprayed should be preheated to approximately the same temperature that will be attained during spraying in order to prevent cracking of the sprayed surface. The resulting spray deposit is porous. In applications where porosity is undesirable, the part can undergo a fusing or recasting operation. In this operation the surface is heated to slightly above the melting temperature of the sprayed layer. This is most commonly performed with an oxyacetylene torch; however, other processes can be used such as laser heating.

WELDING PLASTICS

Welding of plastics is an increasingly important process, since plastics are widely used materials, especially where corrosion resistance and weight constraints are important. All plastics can be classified into two groups based on their chemical composition and their elevated-temperature characteristics. The two groups are thermosetting polymers and thermoplastics. Thermosetting materials cannot be welded by any method that involves heating, whereas thermoplastic materials can be readily welded.

Thermoplastics may be welded by several thermal or chemical joining processes. The various thermal joining processes are as follows: electromagnetic, friction, heated surface, high frequency, hot gas, implant, radiant, ultrasonic, and vibration.

Several steps are involved in thermal welding of thermoplastics—surface preparation, heating, application of pressure, diffusion, and cooling. Surface preparation is very important, since most molded plastics are contaminated with mold release compounds. Heating can be accomplished by one of the previously mentioned methods. Pressure can be applied manually, in presses, or in automatic fixtures. Diffusion occurs when the interface between the parts becomes liquid, after which the solidification occurs during the cooling stage.

Solvents may be used in place of heat in plastic joining. The plastic surface in the joint to be welded is rendered liquid by the application of appropriate solvents. After the parts have been joined and the solvents have evaporated, the joint becomes as homogeneous as the parent material.

Review Questions

1. Name three mechanical fastening methods.
2. Discuss the relative advantages and disadvantages of common mechanical fasteners.
3. An assembly must be taken apart often for maintenance. Which mechanical fastening system would be the best for this application?
4. Where would you most likely find stitching, tying, and snaps used as a material joining process?
5. What are the advantages of adhesive bonding?
6. What are the disadvantages of adhesive bonding?
7. In what industry would you likely find adhesive bonding and why?
8. List three plastic resins used in adhesive bonding.
9. Describe the inspection method used in the aircraft industry to check for defects in bonded joints.
10. List four common electric welding processes.
11. What is GMAW or MIG?
12. What are the advantages of GMAW?
13. What is GTAW or TIG?
14. How does GTAW differ from MIG?
15. Describe shielded metal arc welding and discuss its applications.
16. What is stud welding?
17. What is the function of the shield gas in shielded arc welding?

18. In the MIG process, how is the electrode delivered to the weld area?
19. What material is used for TIG electrodes?
20. If welding is needed where no access to utility company electric service is available, what welding processes could be used?
21. What is the resistance in resistance welding and what does it have to do with the process?
22. Describe laser welding and discuss its advantages.
23. Describe the thermit welding process.
24. What conditions are necessary for electron beam welding?
25. Describe the submerged arc process.
26. How do soldering and brazing differ from welding?
27. What are the most common soldering filler materials?
28. What are the most common brazing filler materials?
29. Discuss where thermal spraying, metallizing, and hard facing might be used.
30. How can plastic be welded?

Case Problems

Case 1: Welding Thin Materials

A new business begins to manufacture patio furniture made from thin-wall steel tubing. Shielded metal arc welding is initially employed, but metal burn-through occurs immediately. The production supervisor, an experienced welder, recommends that shielded metal arc welding be replaced with TIG. The vice president of manufacturing suggests that this would be expensive and the problem could just as easily be solved by further training of the production welders in order to improve their welding skills. In your opinion, who is right and what would you recommend in this case?

Case 2: Collapsing Furniture

The Super Cut Rate Wood Furniture Manufacturing Co. is receiving customer complaints that some of its furniture is falling apart soon after purchase. Super Cut Rate sends its director of quality control to check on the problem. The QC director finds that on the furniture assembled by high-production stapling, the fasteners are pulling out when the furniture is subjected to normal-use loads. A new method of fastening must be used. Considering the primary factors of high production of a low-cost product, screws, bolts and nuts, and gluing are ruled out as being too expensive and labor intensive. What other methods might be employed to solve this production problem?

CHAPTER 16

Processing of Plastics and Composites

Objectives

This chapter will enable you to:

1. Identify common methods of processing plastics and composites.
2. Describe in general terms how the processes work.

Key Words

composite	matrix
reinforcement	PMC
MMC	CMC
prepregs	pultrusion
filament winding	lamination
injection molding	extrusion

Product performance and manufacturing costs are major factors in motivating the designer to seek new product materials and new manufacturing methods to process them. For example, fuel efficiency in automobiles as well as in aircraft has become an important consideration in recent years. To meet these requirements, the designer has turned to nonmetallic materials that demonstrate both light weight and strength. Numerous other advantages in using nonmetallic material such as plastics and advanced composites include corrosion resistance; resistance to denting, chipping, and changes in temperature; and ease of manufacturing.

The purpose of this chapter is to survey the major methods of processing these materials. The chapter begins with a discussion of the changes that are occurring in the world of polymer and composite materials. Next, composites are defined and described. Then, the major molding and shape-producing methods used to process polymers into plastic products are reviewed. The methods used to produce the three types of composites are described, and some applications of composites are discussed briefly. The chapter concludes with a short discussion of the tool and die requirements for producing plastics and composites.

THE CHANGING WORLD OF PLASTIC AND COMPOSITE MATERIALS

It would be difficult to envision the modern world without plastics and composites. The product line of these materials is vast and expanding every day. Although metals will remain in widespread use, plastics and composites are fast replacing metals in a vast array of manufactured products. Why? Because of the often high cost of metals and the large energy requirements in metal processing, and the high strength and light weight of plastic and composite materials, which now permits their use in applications where previous designs required metals.

Composites

Previously, we discussed metals, polymers, and ceramics separately, as individual materials; now we are going to discuss them being used together in a relatively new class of materials called **composites.**

Depending somewhat on the definition, the use of composites began early in the twentieth century with a phenolic–paper laminate used for electrical insulation. Fiberglass boats appeared in the mid-1940s. Until fairly recently, composites were thought of as polymer material surrounding or encasing various strengthening reinforcements. Today the material doing the encasing is termed a **matrix,** so these types of composites are now termed **PMCs,** or polymer matrix composites; they take their place alongside two more recently developed composites—**MMCs,** metal matrix composites, and **CMCs,** ceramic matrix composites.

Although the three types of composites have some significant differences, they are similar in their general makeup. Each has a polymer, metallic, or ceramic matrix. The **reinforcements** used inside the matrix can vary, but the same material may be used with each of the matrices. Composites differ from alloys, polymers, and ceramic compounds in that the matrix and reinforcement are separate from each other. A material may be added to a metal, polymer, or ceramic for strengthening purposes, but the material added becomes a part of the original material; reinforcements used in composites do not.

The reinforcements used vary from short or chopped fibers, flakes, and particles to filaments and wires to continuous woven fibers and honeycombs. The short, discontinuous reinforcements will increase mechanical strength, but they are not as effective as the continuous reinforcements that have the ability to transfer or redistribute loads throughout the composite.

Because there are two types of polymers, thermosets and thermoplastics, there are two types of PMCs, and of course there are a number of polymers within each of the two types.

The metal matrix composites use metal alloys for their matrices with reinforcement provided by particles or filaments of high-performance materials. Examples of discontinuous materials are glass fibers, silicon carbide whiskers, and alumina particles or short polymer fibers; continuous fibers may be of carbon, boron, alumina, or silicon carbide. The metals used in the matrix may be aluminum and magnesium, for their low density; titanium, for strength at higher temperatures; and copper, for electrical and thermal conductivity. Other metals are used for matrices depending on requirements.

The ceramic matrix composites are so far very niche oriented. For example, CMCs are being applied where very special circumstances can take advantage of their ability to withstand high temperatures. Many of these applications are in aerospace. In these composites the matrix is a ceramic, and the reinforcing materials can be any of the materials already discussed, both continuous and discontinuous.

Meeting Design Requirements

In recent years, engineering development of plastic and composite materials has yielded products that are equal to or better than their metal counterparts in many ways. One significant factor is weight. Lighter-weight vehicles in both the automotive and aerospace fields have generally increased fuel efficiency. Plastic and composite materials therefore have generated much interest as materials for these applications.

Composite engineering has yielded structures that meet design strength requirements, thus permitting extensive use of these materials in modern jet aircraft (Figure 16.1). The illustration shows an aircraft in which the fuselage and other major structural components are made completely from composite materials.

FIGURE 16.1
The Beechcraft Premier I aircraft is an example of the use of composite material for structural aircraft components. The fuselage and control surfaces are of composite materials (Raytheon Aircraft Company).

Advanced composite as well as plastic materials have also been used for major structural components on full sized military and jet aircraft (Figures 16.2 and 16.3). The advanced composite empennage of the Boeing 777 weighs 1500 lb less than if it were an all-aluminum assembly. In addition, composite materials are used for floor beams, spoilers, flaps, and other parts of the aircraft. Several aircraft manufacturers are building entire major structural components such as fuselages and wing structures from these advanced materials. The weight saving in

FIGURE 16.2
High-strength but light-weight composite materials greatly enhance the performance and fuel efficiency of the F16 (Lockheed Martin Aeronautics Company).

FIGURE 16.3
The tail (empennage) of the 777 Boeing jetliner makes extensive use of composite materials, which weigh 1500 lbs less than the all-aluminum version (Copyright The Boeing Company).

FIGURE 16.4
Plastic and composite materials are highly suited to the manufacturing of one-piece complex-shaped products (Goodyear Tire and Rubber Company).

aircraft and auto structures will play an important part in the fuel efficiency ratings of these products.

Another common example is the use of plastic material such as laminated products as a very durable substitute for wood in a broad range of applications. These materials are easily fabricated, easily cleaned and cared for, and can be made very thin and therefore very light while still retaining the beauty of wood-grain designs.

Because plastics and composites are often in a liquid or softened state prior to their formation into products, these materials are highly suited to the manufacture of one-piece complex shapes such as automotive body components. The truck and bus hood and fender components seen in Figure 16.4 are made from plastic composites and weigh less than 100 lb. They are also resistant to corrosion and impact damage and are unaffected by variations in temperature.

PROCESSING METHODS FOR PLASTICS

So far we have discussed many processes in which metals are processed into useful products. Many of the same manufacturing-process principles are applied in the processing of plastic materials. These materials may be cast, molded, extruded, machined, fabricated, laminated, and rolled. The equipment used for processing plastic material is specifically designed to process these materials.

Three major methods make up a large portion of plastic and composite processing techniques. These include blow molding, injection molding, and extrusion. Several other processes are also used.

Blow Molding

In **blow molding**, air is used to force a mass of molten plastic against the sides of a mold shaped in the form of the desired end product (Figure 16.5), such as a milk bottle. Blow molding may be used when the end product must have an enclosed hollow internal shape. An example

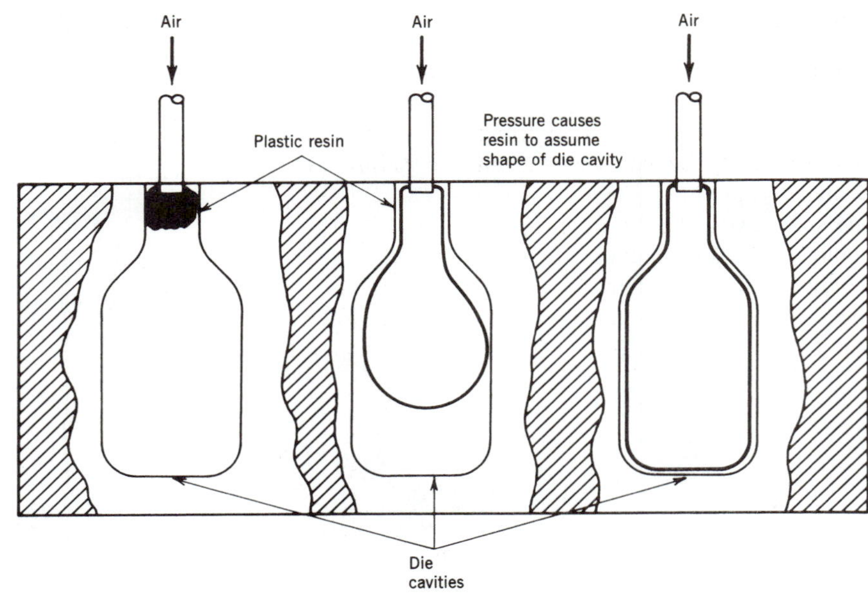

FIGURE 16.5
Blow molding process.

FIGURE 16.6
Injection stretch blow molding machine used to produce bottles ranging from 150 ml to 3 liters in size and round square, or oval in shape. The bottles can be seen exiting the machine to the left (Magic North America).

of this application is containers such as the plastic jugs illustrated in Figure 16.6. Machinery used in this process will produce units at a high rate of production.

This method can also be used for sheet forming in which a sheet of plastic is clamped between a form and die. Air pressure forces the heated plastic against the die, where it assumes the die shape. A common and versatile application of the blow molding process is the production of refrigerator liners. This process is particularly well suited to the forming of such large fabrications.

Related to this process is vacuum forming, in which a vacuum is drawn on one side of the material and air pressure on the opposite side forces the material against the mold or form.

Injection Molding

Injection molding is another versatile and widely used process. Examples of injection-molded products are found everywhere. In this process, molten plastic is forced into a metal die cavity that has been machined to the shape of the desired end product. When the plastic material has solidified sufficiently, the die is opened and the part removed. Figure 16.7 shows a cutaway view of a plastic injection molding machine (Figure 16.8). Raw plastic material in small pellets (Figure 16.9) is placed in the machine hopper. The raw material then enters the heater, where it is melted. Molten plastic is then injected into the die cavity by hydraulic or direct mechanical pressure (Figure 16.10).

The pressure at which the metal is injected, when spread over the projected area of the die cavity, creates a force that must be offset by the mechanism used to close the die. This force, plus a safety factor, determines the rating of the machine, usually in tons.

Large-capacity injection molding machines may exert several hundred tons of pressure and can be used to fabricate large one-piece plastic parts. Examples include automotive body components, including fender/hood assemblies, bumpers, and grilles (Figure 16.11).

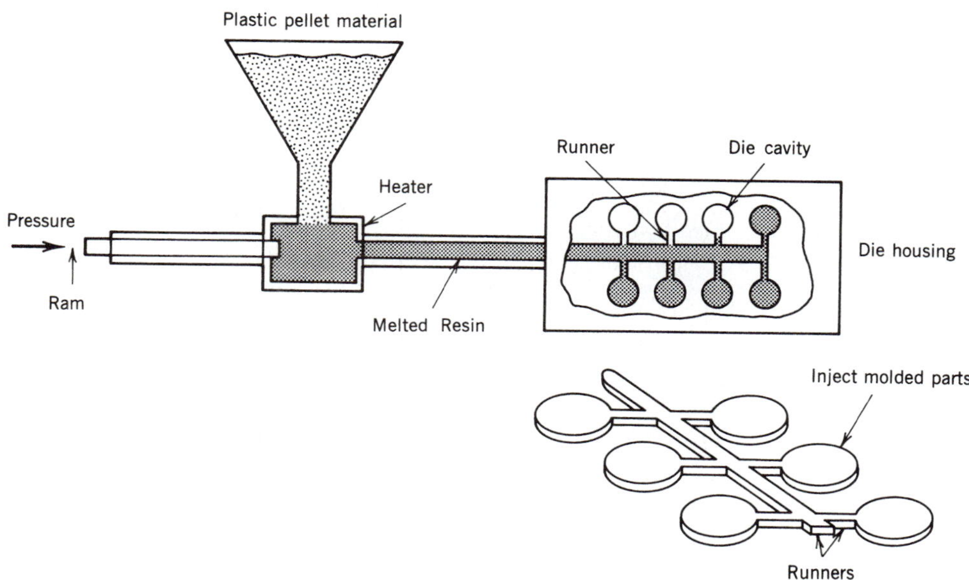

FIGURE 16.7
Injection molding process.

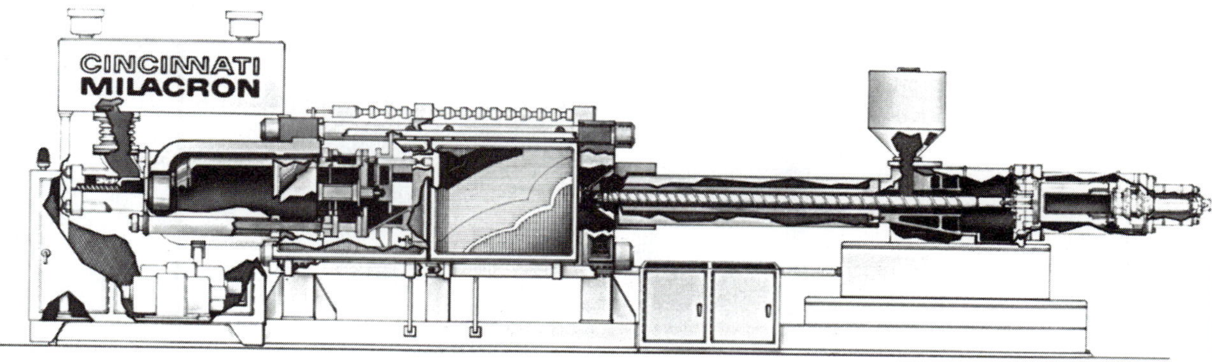

FIGURE 16.8
Cutaway view of an injection molding machine (Cincinnati Milacron Plastics Machinery Division).

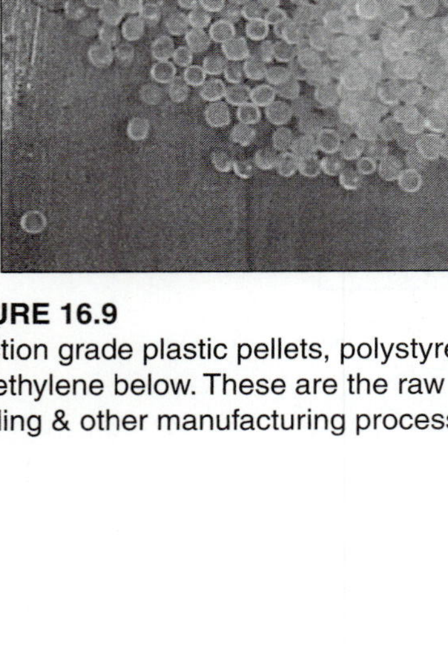

FIGURE 16.9
Injection grade plastic pellets, polystyrene above and polyethylene below. These are the raw materials for molding & other manufacturing processes.

FIGURE 16.11
One-piece plastic fender and cowel components made by injection molding (Goodyear Tire and Rubber Company).

The injection molding process is extremely versatile for making parts with fine detail and complex shapes. Examples include high production of small precision-detailed parts such as those for plastic models.

Reaction Injection Molding Reaction injection molding (RIM) is the latest injection molding technique. In this process, two base thermoset resins (monomers) are mixed together just as they enter the mold (Figure 16.12). A chemical reaction occurring at low heat takes place, and the plastic material (polymer) of the end product is formed at that instant. The RIM process does not require the plastic materials to be heated before molding. Thus, the process is both energy efficient and very fast. Large parts weighing many pounds can be made in a few seconds, and the process minimizes scrap.

FIGURE 16.10
Injection molding machine, 250-ton capacity, with computer microprocessor controls (Cincinnati Milacron Plastics Machinery Division).

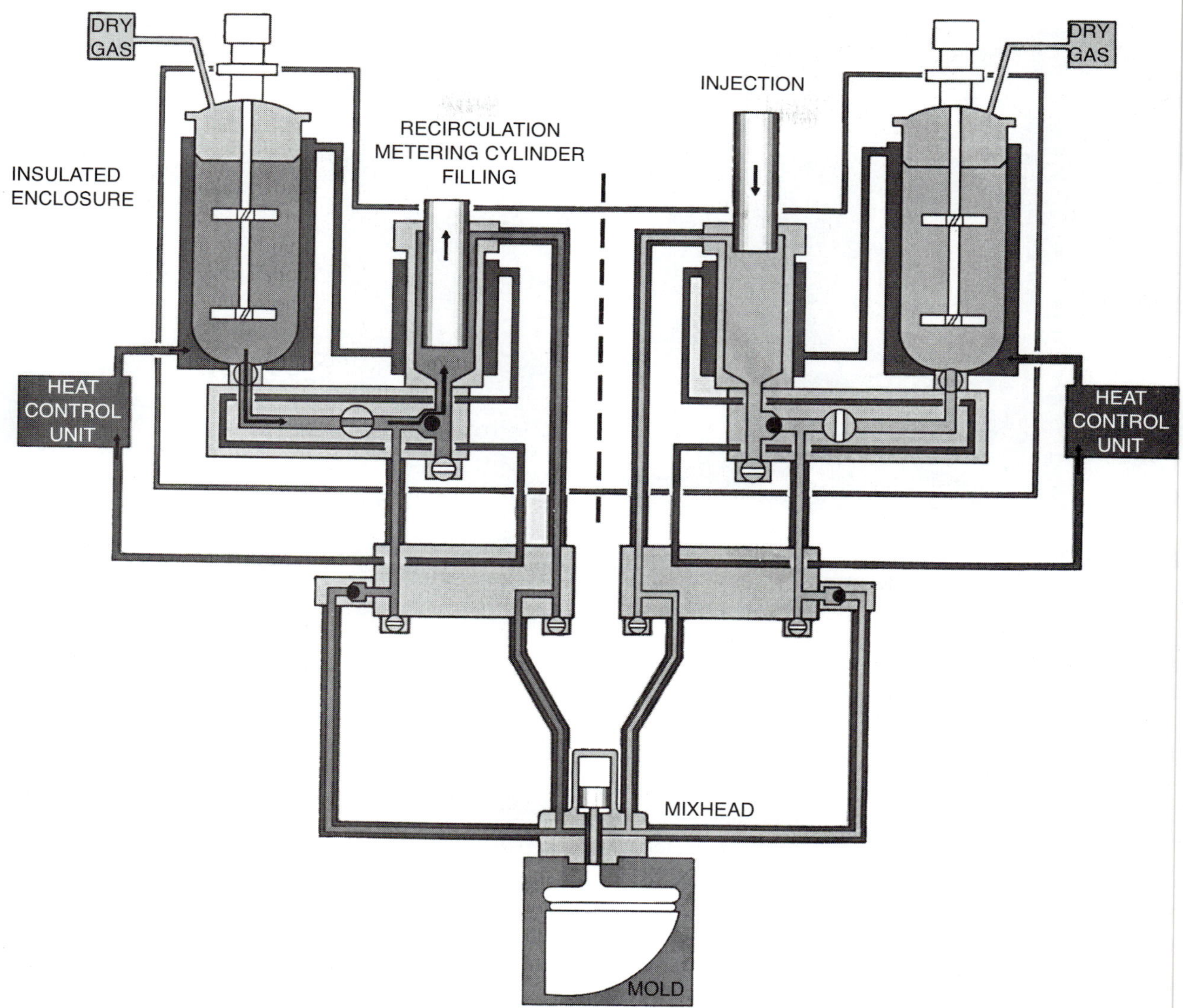

FIGURE 16.12
Reaction injection molding process (RIM) (Cincinnati Milacron Plastics Machinery Division).

Reaction injection molding machines (RIMM) may be of the platen design (Figure 16.13). The machine illustrated is seen with essential safety guards removed so that details of construction may be seen. This machine is highly suited for production of large parts, such as auto bumpers and fenders.

Extrusion

Extrusion is another versatile and widely used plastic and composite processing method. In this process, the material is forced through an extrusion die that forms the end product (Figure 16.14). The major component of a plastic extruder is the extruder screw (Figure 16.15). The extruder may be of the twin screw design, in which two screws with variable lead sections are in mesh. Parallel screw extruders also are used. This system provides the mechanical force to push the plastic material through the extrusion die. Extrusion is the popular manufacturing process for products requiring long dimensions in one axis, such as plastic pipe and plastic molding.

FIGURE 16.13
Platen-type reaction injection molding machine (RIMM) uses the latest technique (RIM) in injection molding processes. This illustration shows the machine with essential safety guards removed so as to better show machine parts (Cincinnati Milacron Plastics Machinery Division).

The illustration (Figure 16.16) shows a plastic extruder for manufacturing plastic pipe. The specific tooling for this product is attached to the machine on the left side.

The process of **calendering** or rolling plastic material is used to form sheets and film. Plastic is considered to be film if its thickness is less than 0.010 in.; if it is thicker, the material is considered to be sheet. Sheet and film are formed

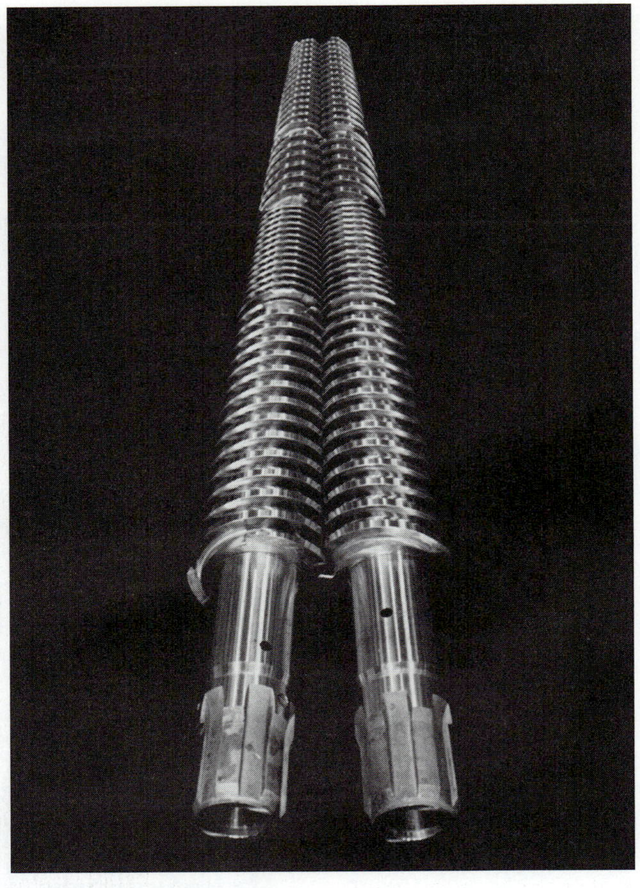

FIGURE 16.15
Twin-tapered extruder screws, uniquely engineered major components of a plastic extruder (Cincinnati Milacron Plastics Machinery Division).

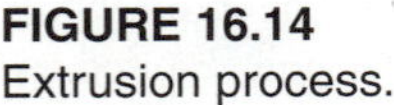

FIGURE 16.14
Extrusion process.

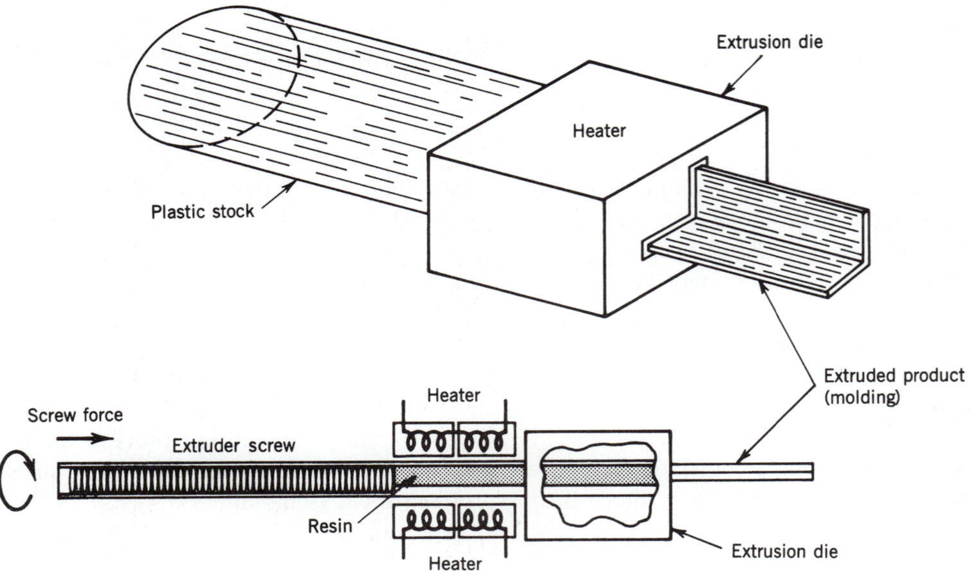

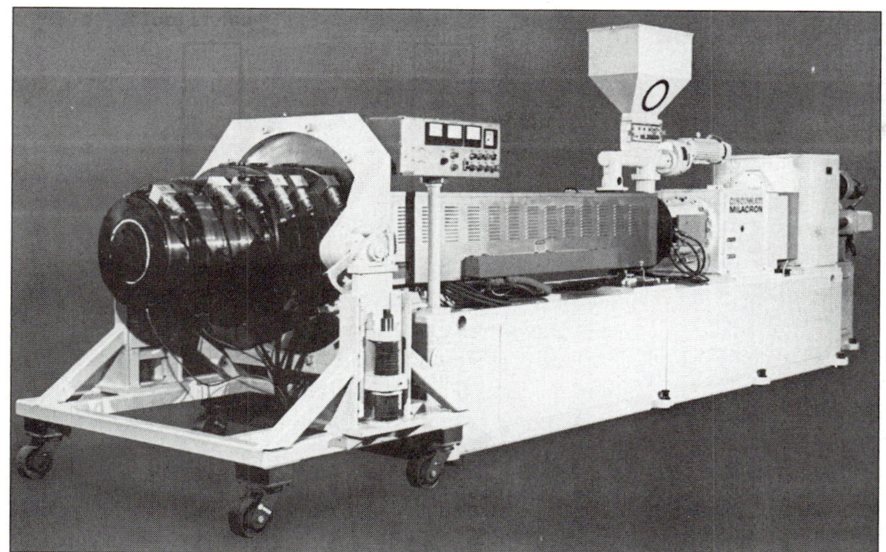

FIGURE 16.16
Plastic extrusion machine for plastic pipe. Tooling is seen attached to the machine on the left side (Cincinnati Milacron Plastics Machinery Division).

by pressing the material between heated rollers. Thickness is determined by the spacing between the calendering rolls. Plastic coatings are also applied to cloth by this method, forming the many useful plastic-coated fabrics appearing in a large number of consumer and industrial products.

Other Molding Processes

Other processes that make use of molding principles include compression molding, transfer molding, rotational molding, and solvent molding.

Compression Molding The compression molding process is similar to injection molding, but it is primarily used with thermosetting resins. In compression molding, the resin is in pellet or dry powder form and may be mixed with a binder agent to provide strength. In this respect the resin can become a plastic composite material. The dry resin is mechanically compacted into a mold shaped to produce the desired end product. Heat and pressure are then applied to melt the resin and cause it to flow throughout the mold. A thermosetting plastic material undergoes a fundamental chemical change in this process, and a permanently hardened part is produced. The process is good for making handles and knobs for kitchen utensils, since thermosetting resins can be subjected to heat without becoming soft.

Transfer Molding Transfer molding is similar to compression molding except that the resin is heated to a liquid state before being forced into the mold. This process may be used where intricately shaped parts are required or where metal inserts are molded into plastic. Metal parts that are positioned in the mold will remain in location while the liquid resin flows around them. After the plastic has solidified, inserted metal parts are securely held in place.

Rotational Molding In rotational molding, the heated mold cavity is spun about 2 axes, and centrifugal force causes the resin to be distributed in a thin layer over the inside of the mold. Heating is required to liquefy the resin. The process is capable of producing hollow products (tanks, floats) with little internal stress and very uniform wall thickness. The tooling costs are unusually low compared with those of processes such as injection molding (see Figure 16.17).

Solvent Molding In solvent molding, a mold of the desired end-product shape is dipped into or filled with a liquid resin material. When the mold is withdrawn or emptied out, a thin layer or coating of resin adheres to the sides. After the solvent has evaporated, the formed part may be removed. Additional layers may be built up to increase the thickness of the end product and the process may also be used as a method to plastic-coat another part.

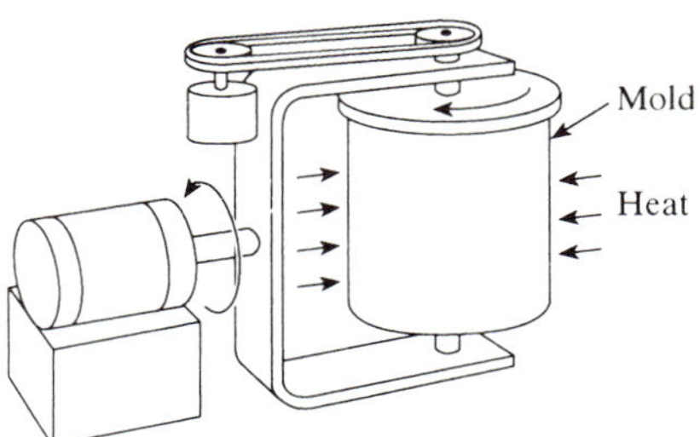

FIGURE 16.17
The mold is rotated about two axes in rotational molding (Budinski & Budinski, *Modern Materials and Manufacturing Processes*, 7th ed., ©2002 Prentice Hall, Inc.).

Other Processing Methods for Plastics

Other plastics processing methods include casting, in which liquid resins are poured into molds. The resin contains a catalyst that causes the resin to harden after a short period of time. Resin casting is popular in the home-craft area and yields many decorative products. The process also can be used to encase an object in a transparent block of resin. Color may be added to clear casting resins. The process must be done quickly or the catalyst will harden the resin before it can be poured. Casting can be used to make sheets, rods, tubes, special shapes, film, and sheeting from both thermoplastic and thermosetting resins.

Rigid plastic and composite materials can be shaped effectively by most of the common machining processes. The material can be milled, turned, sanded and ground, broached, reamed, drilled, and tapped. A number of plastics are readily available in bar stock form for processing by machining methods. Plastics suitable for machined products include nylon, acetyl (delrin), fluorocarbon (Teflon), ABS, and PVC. Common machined plastic products include gears, bearings, threaded fasteners, and bushings.

Plastic coatings may also be brushed or sprayed on metals or woods to protect them from corrosion. Clear plastic coatings are used as wood and metal finishing media. Plastics are also used in paints, where color as well as durability properties of the plastic can make a superior coating medium. Examples include epoxy- and polyurethane-based paints.

In the process of high-pressure **lamination,** which generally makes use of thermoplastic resins, sheets of the material are placed between heated steel platens, which are then compressed with hydraulic pressure. The process bonds the layers of the laminate into a rigid sheet.

Laminated sheets can also be formed around corners and then bonded to wood bases. The products resulting from this process are used extensively for kitchen countertops. Very attractive products that simulate woods, but have the advantages of plastics, are created by this process.

Processes used to join pieces of plastic in assemblies include welding by heat and ultrasound, gluing with solvents and other resin adhesives, sewing (sheet and film), and fastening with mechanical fasteners such as bolts and nuts, screws, and rivets.

COMPOSITE PROCESSING METHODS

Many composites require two processes before they are able to meet a material need—the first one to create the composite itself and the second to shape and/or fit the composite material to the final application. In other cases a composite may be prepared and shaped to meet final requirements in one step.

In this section we will briefly explore the methods used to create and/or form the three types of composites outlined earlier—polymer matrix composites (PMCs), metal matrix composites (MMCs), and ceramic matrix composites (CMCs).

Polymer Matrix Composites (PMCs)

The mechanical properties of plastics can be enhanced greatly by adding a reinforcing agent to the resin. Plastic resins may be reinforced with cloth, paper, and fibers such as glass and graphite. The reinforcing fibers can be in short pieces randomly oriented within the base resin, or fibers running unbroken through the resin providing the major structural component for the material.

Although plastics with the shorter reinforcements may not have quite the same strength and load-carrying capability as those with the longer fibers, they are nonetheless excellent structural material for a variety of products.

PMCs consist of the plastic material reinforced with some type of stranded fiber material. The fiber reinforcement material usually makes up about half of the total material weight. Examples of the fiber materials used in composites are graphite and glass. The fiber functions as a structural component of the composite and is designed to take the load stresses applied to the composite structure. Were it not for the fiber portion of the composite material, only the resin portion of the material would be subjected to the applied loads. This structure alone would not be able to withstand the forces applied and would fail in service (Figure 16.18).

Both thermoplastics and thermosetting plastics are used in PMCs. Research has shown that the thermosetting plastic will probably be more popular for use in advanced composites. The reason for this is that many applications of advanced composites required the product to be subjected to considerable heat. The thermosetting plastics are more suitable to these applications.

Previously in this chapter, we discussed methods of processing polymers. A number of these methods can also be used to process PMCs, especially where the shorter reinforcements are used. It should be easily recognized that if the reinforcing fibers or particles can be kept in suspension in the polymer, the resulting viscous material can be injection molded, reaction injection molded (RIM), transfer molded, or used in any other process that can handle a viscous fluid. Therefore, in the remainder of this section we will deal with those processes that are somewhat different, especially in their use of longer or larger reinforcements. We begin, however, with a discussion of prepregs and sheet-molding compound, both materials from which PMC products might be made.

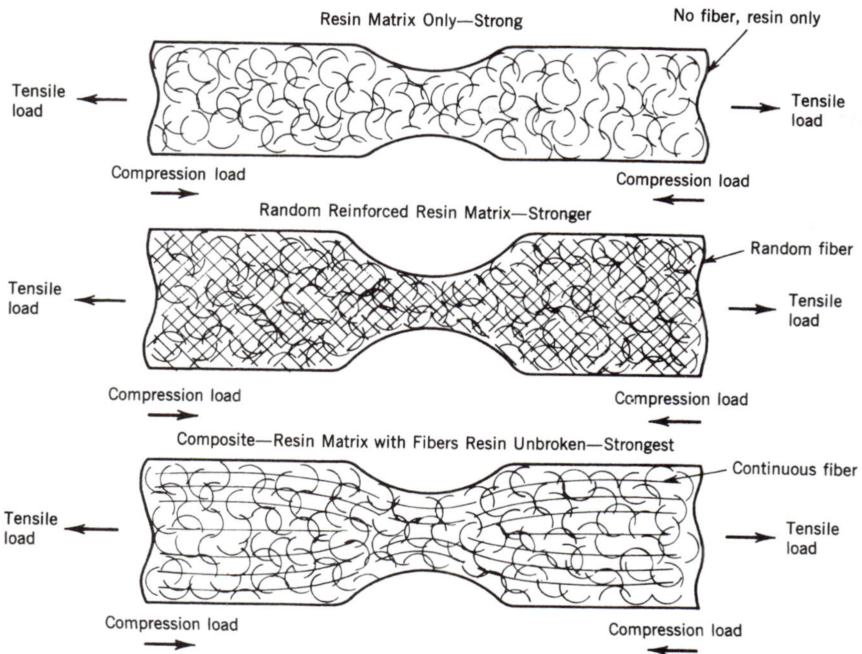

FIGURE 16.18
Load applications to resins, reinforced plastics, and composites.

Three other processes for manufacturing PMC products will then be described—pultrusion, filament winding, and lamination.

Prepregs In producing **prepregs**, an intermediate (*pre*impregnated) product, many continuous strands of fiber are carefully aligned and then coated with the appropriate thermoplastic or thermoset resin and made into a sheet or tape. Individual pieces of the tape or sheet are used to form the laminated product and then heated for curing.

Sheet-Molding Compound Instead of the continuous fibers used in the prepregs just described, sheet-molding compound is made from chopped fibers that are deposited, with random orientation, on a resin-covered carrier film, such as polyethylene. Another layer of resin covers the fibers, and then a top sheet or film (e.g., polyethylene) covers the resin. The resulting sandwich is pressed between rollers and coiled up or kept as flat sheets. The molding compound is stored in a controlled environment through a maturation process that lasts about 24 hours. It must then be kept at a temperature low enough to delay curing; the shelf life is about 30 days.

Molding Some of the molding methods previously described that can take advantage of the prepregs and molding compounds are compression and transfer molding. In these cases the reinforcements are placed into the mold cavity before or during the application of the polymer material.

Hand Layup A process with wide application that provides a good illustration of the hand layup method is the use of fiberglass. In this process alternating layers of glass fiber fabric and resin are coated over a mold or form built in the shape of the desired end product. Short pieces of glass fiber may also be mixed with the resin while it is still in a liquid state. A popular application of this method is in manufacturing for boats (Figure 16.19) and other large hollow fabrications such as swimming pools.

FIGURE 16.19
Reinforcing resin with glass fibers in the process of fiberglassing yields extremely durable structures such as boat hulls (Wellcraft Marine Corporation).

FIGURE 16.20
Glass fiber application (Wellcraft Marine Corporation).

FIGURE 16.22
Finishing a fiberglass reinforced boat hull (Wellcraft Marine Corporation).

FIGURE 16.21
Smoothing fiberglass-reinforced resin in boat manufacturing (Wellcraft Marine Corporation).

The process is accomplished by building up layers of fiber strands or sheet fabric (Figure 16.20) and resin. Pieces of glass fiber are laid over the desired form, and the resin is applied by spray application or hand spreading (Figure 16.21). The liquid resin and fiber mixture may be troweled and smoothed while it is still in a liquid state. After hardening, the fiberglass fabrication may be refinished by mechanical abrasive processes such as sanding and buffing (Figure 16.22).

The end product is extremely strong, durable, and lightweight. It is also impact resistant and not subject to corrosion, and almost any shape can be fabricated given the proper tooling.

Pultrusion Pultrusion is illustrated in Figure 16.23; fibers are pulled or drawn through a liquid resin and then through a heated die that forms the desired shape. Pultrusion is much like extrusion except for the pulling rather than pushing of the material through the extrusion die. Composite products manufactured by pultrusion methods include structural members and tube.

Filament Winding In the technique of filament winding (Figure 16.24), the fiber is wound back and forth on a cylindrical form. This method is used to produce cylindrically shaped products such as tanks or other pressure vessels. After curing, the form is removed, leaving the hollow composite product.

Lamination Laminating alternating layers of resin containing the structural fiber is the third major method of composite manufacturing (Figure 16.25). This process is similar to the technique used with fiberglass; however, the composite fibers are continuous throughout the material, whereas in fiberglass short pieces of glass fiber are randomly distributed throughout the resin structure.

Metal Matrix Composites (MMCs)

Liquid Matrix The most straightforward of the MMC methods involves casting the molten matrix metal around solid reinforcements, using either conventional casting techniques or a pressurized gas on the liquid matrix to force it into and around a preformed reinforcement, which is often made of metal sheet or wire, or ceramic fiber.

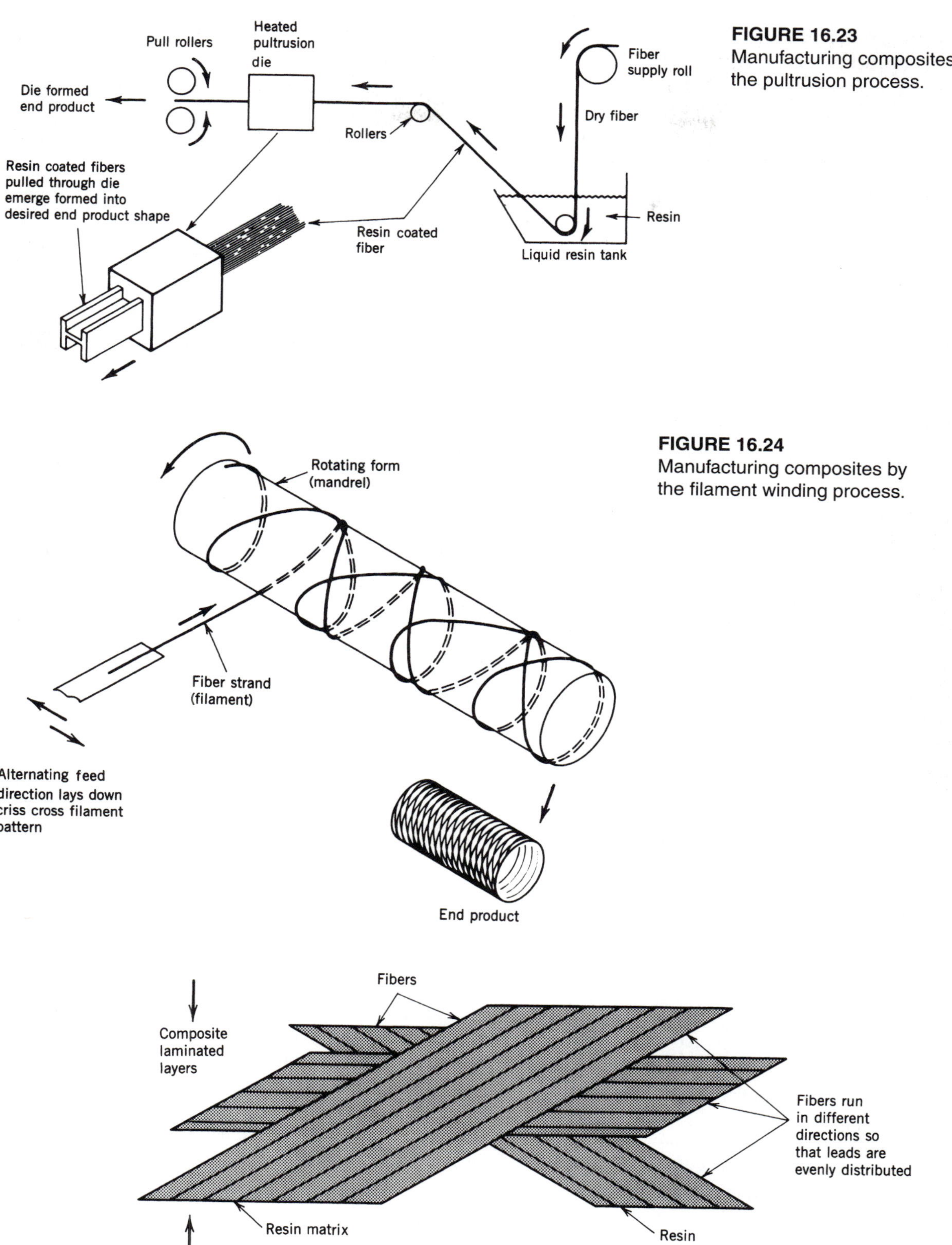

FIGURE 16.23
Manufacturing composites by the pultrusion process.

FIGURE 16.24
Manufacturing composites by the filament winding process.

FIGURE 16.25
Manufacturing composites by the lamination process.

Powder-Metallurgy Techniques The reinforcement fibers, whiskers, or particles are carefully mixed with the powdered metallic matrix so they are uniformly distributed in the mixture. This mixture of powdered metal and reinforcements is then compacted to form the desired part. An initial compaction may be done cold, followed by sintering, or all the compacting may be done at the sintering temperature. The sintering fuses the various particles together into a shape and size that should be very close to the desired dimensions.

Liquid–Solid Processing This is a casting technology in which the reinforcement being added to the metallic matrix is in its mushy stage: that is, it is partly frozen, partly liquid.

Ceramic Matrix Composites (CMCs)

The most common process used in producing CMCs is slurry infiltration, in which the slurry contains the ceramic matrix powder. A fiber preform of the desired product is hot pressed and impregnated with the slurry, then sintered.

FIGURE 16.26
Aluminum honeycomb sandwiched and bonded between graphite composite skins makes an extremely strong and lightweight structure for the F-18 fighter aircraft (Northrop Grumman Corporation).

COMPOSITE APPLICATIONS

As composite materials undergo further development their uses will extend to a larger line of products. In the beginning of this chapter, the applications in aircraft were mentioned briefly. In this application, the aircraft design engineer seeks materials with favorable strength-to-weight ratios. Strength coupled with light weight can result in products that require less fuel to propel them. This consideration is extremely important in the production of aircraft, spacecraft, and automobiles of today and tomorrow.

The honeycomb composite structure creates a product that demonstrates these highly desirable characteristics of light weight and strength. In Figure 16.26 aerospace technicians lay an aluminum honeycomb core on a graphite composite skin. The metal honeycomb, which is in itself a high-strength lightweight material, will be sandwiched between two layers of the graphite composite. The entire assembly will then be cured by heat and pressure, thus making a durable aircraft structural component.

Large components can more easily be made from composite materials than from metals. Figure 16.27 shows a large graphite composite component for a jet aircraft. The technician is seen applying graphite-epoxy prepegs to a precision bonding fixture. When 50 to 60 plys have been built up, a vacuum will be applied to the bonding fixture so the part will closely adopt the shape and size of the fixture. The fixture and part will then be placed in an autoclave (a controlled heating chamber), curing the prepeg.

The Beechcraft Premier I we saw in Figure 16.1 at the beginning of the chapter provides an excellent example of a large composite structure. Figure 16.28 shows its one-piece carbon-fiber epoxy fuselage being manufactured. This structure is produced by precise placement of prepeg strips that ensures maximum strength and durability. The carbon-fiber composite material is also used in the horizontal stabilizers, vertical stabilizer, flaps, ailerons, and spoilers.

Air and land vehicles are not the only applications that benefit from an improved strength-to-weight ratio. Overhead electrical conductors must be strong enough to support the weight of wire strung between transmission towers that are spaced many feet apart. If the normal methods of strengthening are used on the conductor itself (heat treating or cold working) its conductivity suffers greatly. Instead, a typical method of strengthening is to put steel strands in the core of the cable. Figure 16.29 shows strands of an aluminum conductor with a core of strands of an MMC—an aluminum matrix with alumina reinforcement fibers. This combination can yield higher conductivity at the same weight per foot.

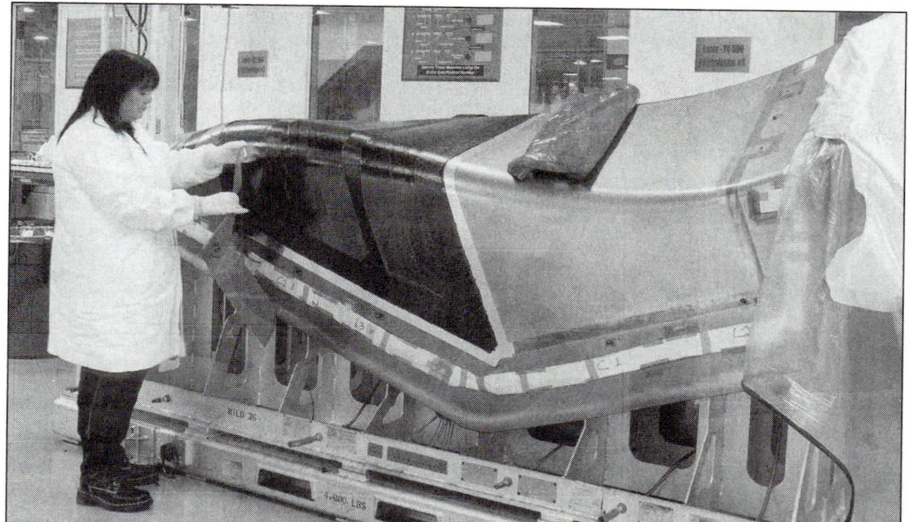

FIGURE 16.27
Tooling is all-important to the formation of plastic and composite products. A technician lays up graphite composite material on a large precision bonding fixture (Lockheed Martin Co.).

FIGURE 16.28
The forward fuselage for the Beechcraft Premier I (see Figure 16.1). The carbon fiber, sandwich-type construction does away with internal stiffeners and ribs used in conventional construction, and provides a cabin with integral door frames, windows, etc., that can be pressurized (Raytheon Aircraft Company).

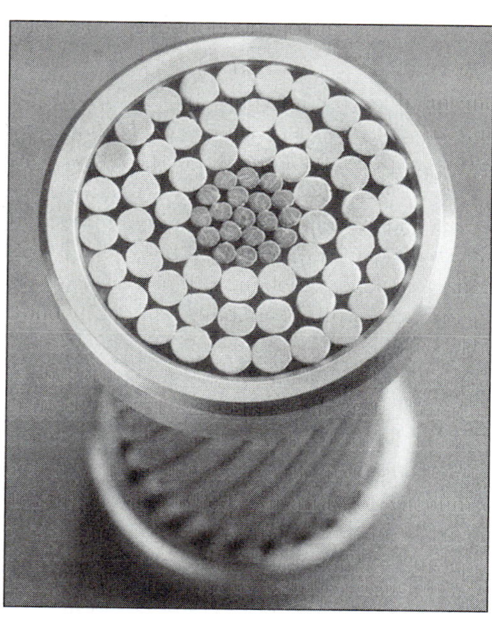

FIGURE 16.29
Weight per foot is very important when selecting materials for overhead electrical conductors. This picture shows the cross section of a conductor being developed by 3M and a foreign partner. The aluminum conducting strands can be seen on the outside, surrounding a core of aluminum matrix composite (AMC) strands. The AMC provides the strength typically provided by much heavier steel wires. The rings that can be seen around the conductor are there only to keep this section of cable from unwinding (3M Company, St. Paul, MN).

Figure 16.30 is a photo of an improved overhead-valve pushrod for a high-performance automobile engine. It is made of an MMC with aluminum as the matrix and alumina (Al_2O_3) fibers as the reinforcement. Combining a ductile, light metal with a strong reinforcement that is almost as light yields a light, strong composite. The alumina reinforcement, which has a very high modulus of elasticity (see Chapter 3), improves the stiffness of the composite. When the performance of the composite pushrod is compared with that of a 4340 heat-treated steel pushrod it is found that the composite is stronger, stiffer, and has better damping capability.

FIGURE 16.30
A photo of newly developed overhead-valve pushrods made of an aluminum-matrix, continuous alumina fiber–reinforced composite material. The tensile and compressive strengths, stiffness in bending, and damping capability of the composite are a significant improvement over a 4340 steel pushrod and can result in improved engine performance (3M Company, St. Paul, MN).

FIGURE 16.32
Rotors (or flywheels) for high-speed (~100,000 rpm) applications are subjected to very high stresses. This is a good application for MMCs because they can be fabricated with reinforcements to withstand both hoop and radial stresses (3M Company, St. Paul, MN).

Figures 16.31 and 16.32 illustrate applications that take advantage of the high strength and low weight of the ceramic-reinforced aluminum-matrix composite material. The alumina used in these composites is rods or fibers of 10 to 12 μm (microns), or 0.010 to 0.012 mm in diameter—very small. Although we think of alumina as being very brittle, it is very strong (tensile strength about six times that of heat-treated aluminum alloys) and stiff (modulus of elasticity about five times that of aluminum). When the composite material is 40 percent aluminum and 60 percent alumina the resulting tensile

FIGURE 16.31
Using an aluminum matrix composite for an auto's brake calipers (seen here) can cut their weight in half but gives them equal strength with less vibration (3M Company, St. Paul, MN).

strength and modulus of elasticity are about three and a half times that of heat-treated aluminum, or the composite material has properties that are higher than those of steel, but at 45 percent the weight!

For further information on 3M metal matrix composites see http://www.3m.com/market/industrial/mmc/.

TOOL AND DIE MAKING FOR PLASTICS AND COMPOSITE PROCESSING

As is the case in processing most raw materials into useful products, special tooling is required in the plastics and composite processing industry. Here the raw material is often in a liquid or flexible state prior to its formation into the end product. Therefore, considerable tooling is required to hold or form the material until steps are taken in the process that alter the state of the material to make it rigid.

In processing plastics and composites, much of the tooling appears in the form of molds or forms. Most of this tooling is made from metal and is manufactured to exact dimensional specifications.

Tool and die making is therefore an integral and indispensable part of any manufacturing industry. The tool and die maker, who represents the upper end of the skilled machinist trade, will always be needed to build the tooling that makes mass production possible. The toolmakers of today and tomorrow have at their disposal all the modern improvements that an advancing technology makes possible. Among these are computer-aided design (CAD) and the versatility of computer numerical control machining (CNC) for making the high-precision tooling required by space-age manufacturing.

Modern methods of high-production plastic and composite part production will yield parts at very low cost; however, the tool and die cost must be included in the product price. Each plastic part produced includes a portion of the tooling price. The first part produced represents the full cost of tooling plus the material consumed in making the part. As each successive part is produced the cost of tooling is spread out further and further. Eventually, enough parts are produced to underwrite the tooling cost, and from then on, the only cost is for material and production overhead expenses. If a large number of production parts are further produced, a good return is received on the tooling investment.

Production tooling, once manufactured, can be stored and reused at a later date. If a manufacturer receives an additional order for the same product, it is not necessary to remanufacture the tooling, and the customer may be able to purchase the manufactured product at a more reasonable price.

Review Questions

1. What factors are involved in selecting a plastic or composite material over metals?
2. What are the advantages of using these materials rather than metals?
3. List three major methods of processing plastics and describe how they work.
4. Give an example of a blow-molded product.
5. Cite an example of an injected-molded product.
6. Cite an example of an extruded product.
7. What is RIM and how does it differ from conventional injection molding?
8. Briefly describe molding processes other than RIM.
9. What types of plastic products can be made by machining?
10. Describe reinforced plastic material and cite a common example.
11. What is the purpose of reinforcing plastic and what materials may be used in this process?
12. Name a use of high-pressure laminates.
13. How does a composite material differ from a reinforced plastic?
14. What medium provides the structural component in a composite?
15. List and describe three major methods of composite manufacturing.
16. What process would be best for manufacturing a composite water tank? A composite structural beam? An aircraft composite deck walkway?
17. What is tool and die making and where does it fit into the plastics and composite manufacturing industry?

Case Problem

Case: Choosing Different Material

To conserve weight in an aircraft structure, an analysis is made of the aluminum structural beams that support the internal platform decks. Because these components are not flight-critical items, their strength-to-weight ratios can be maximized. In other words, they need only be strong enough to support required loads but need not have the safety factors required of such flight-critical components as wing and tail spars.

One proposal advanced by structural engineering is to cut lightening holes in all deck beams to conserve weight; however, CAD analysis suggests that if enough material were to be cut from the beams to make a significant weight saving, the beams would be borderline as to their load-carrying capabilities. What would you suggest to solve this problem?

CHAPTER 17

Processing of Other Industrial Materials

Objectives

This chapter will enable you to:

1. Be generally familiar with common industrial materials other than metal and plastic.
2. Be familiar with their manufacturing processes.

Key Words

tension load	reinforced concrete
calendering	refractory materials
plate glass	laminated wood

Although metals and metalworking make up a large portion of materials and manufacturing processes, many other materials and processes are important in manufacturing. Ceramics and wood, for example, have long been in use. These materials are undergoing new developments, and many new applications are being found for them in the manufacture of industrial products.

Nonmetallic materials are used in conjunction with metals in all types of products. In many cases these materials, because they are more economical, stronger, lighter, and easier to process, are replacing metals. Foremost among these are plastics and composites, as previously discussed. Important industrial materials discussed in this chapter include the following:

1. Glass
2. Ceramics
3. Wood, wood products, and paper
4. Fabrics
5. Rubber
6. Natural materials
7. Construction materials

GLASS

Glass consists primarily of fused silica (SiO_2), a major constituent of beach sand. The uses of glass are widespread and include many common products such as windows, containers, and lenses. The glass-window-wall commercial building has become very popular in modern building construction (Figure 17.1).

Many of the advantages of glass containers are unsurpassed. Glass will withstand chemical attack and can also withstand large variations in temperature. Lead crystal (Figure 17.2) is noted for beautiful designs in tableware. Lead is added to glass for windows that provide shielding from ionizing radiation while permitting radioactive equipment to be observed safely. Another outstanding property of glass is its ability to be fashioned into precision optical lenses for eyeglasses, telescopes, microscopes, and many other optical instruments.

A relatively new and developing technology in the uses of glass is fiber optics. Here, the glass is drawn into a thin strand, much like a glass wire. Strands are grouped into a fiber-optic bundle and then sheathed in plastic for durability and protection. Fiber-optic technology is a rapidly developing field for the transmission of information in telecommunications and computers. Figure 17.3 shows a fiber-optic bundle designed to transmit digitized data. The fiber bundle is much smaller and lighter than the standard electrical cable, and it can transmit more channels of data than its wire cable counterpart.

Sheet glass is manufactured on a production basis by the latest computer-controlled automated production

FIGURE 17.1
Plate glass used in building construction turns entire walls into windows.

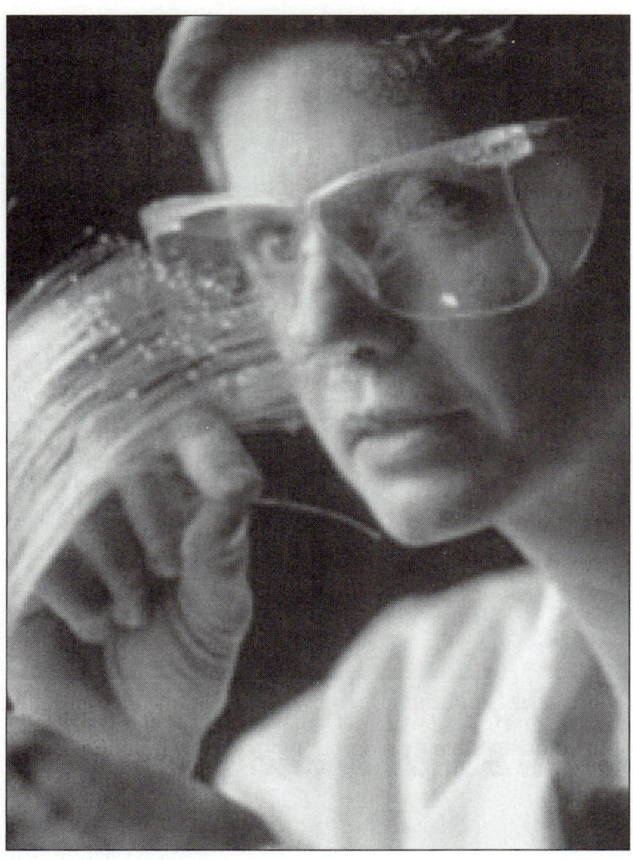

FIGURE 17.3
Glass fiber-optics uses strands of glass to transmit light signals in communications and computer systems. (Photo Researchers).

FIGURE 17.2
Lead crystal makes highly artistic and beautiful glass products.

methods. The float method (Figure 17.4) produces a continuous ribbon of glass, creating sheet glass products such as windows, mirrors, and automotive items.

Other glass product manufacturing processes include pressing, blowing, and combination press and blow operations (Figure 17.5). These processes are most often performed on production equipment in a highly automated manner (Figure 17.6) and are used primarily in glass container production.

Other glass manufacturing processes include cutting sheets into various shapes, edge beveling, and grinding. Hole-drilling enables the assembly of glass pieces using mechanical fasteners.

Tempering turns standard glass into plate glass. The glass is heated in a large furnace to a semiplastic state and then is removed from the furnace and quickly air quenched. This process creates closely spaced swirl-

FIGURE 17.4
Float glass manufacturing produces a continuous ribbon of glass (Pilkington).

shaped stress patterns in the glass. Tempering is done to provide glass with mechanical strength for such applications as refrigerator shelves, store fixtures, doors, and windows that might be subject to breakage or impact. Should plate glass be broken, many small and relatively harmless pieces result, as opposed to large very hazardous slivers.

Glass in its multitude of product forms is an extremely useful and versatile industrial material. Annual glass production in the United States amounts to several million tons per year.

CERAMICS

Modern product design has motivated great interest in and development of ceramic engineering materials. Ceramics are made from the metallic oxides of such metals as

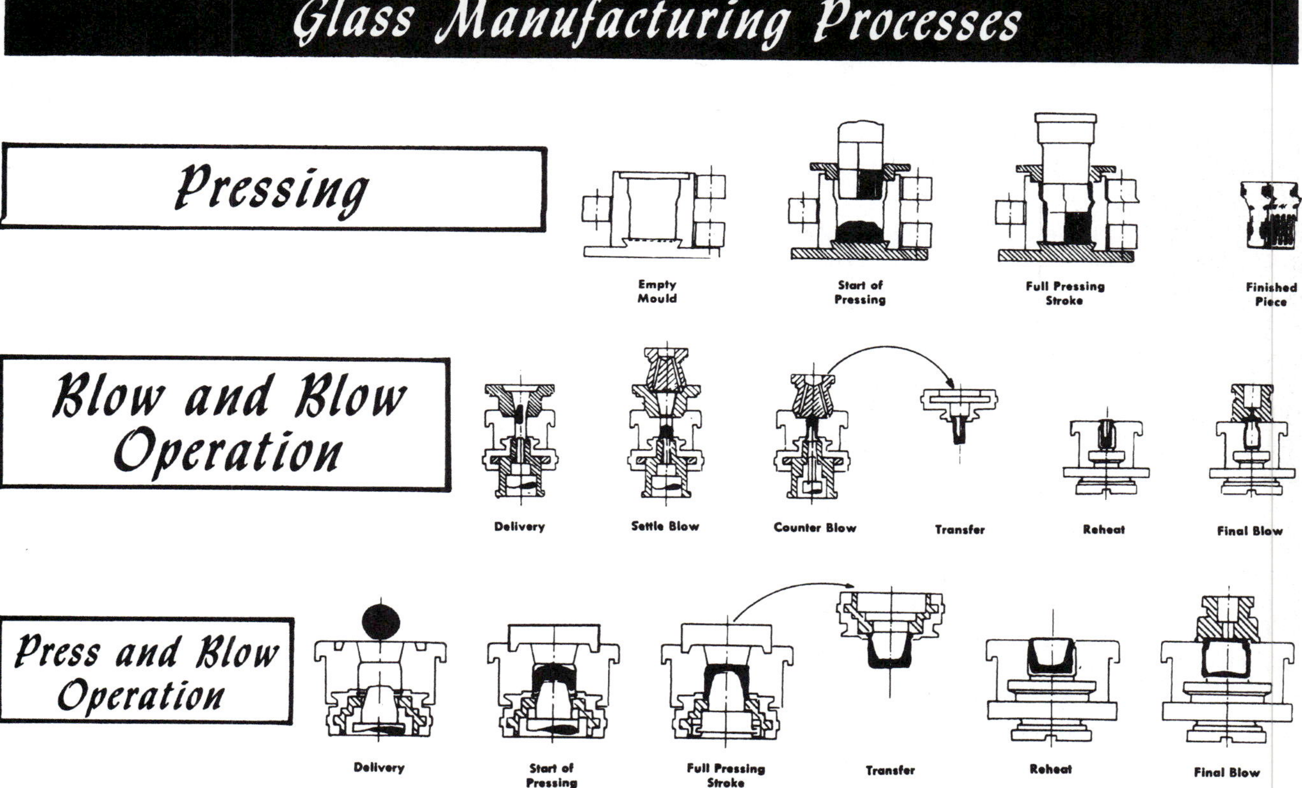

FIGURE 17.5
Glass manufacturing process for containers (The Glass Group, Inc.).

FIGURE 17.6
Automated glass container manufacturing (The Glass Group, Inc.).

silicon, aluminum, and magnesium. Clay-based ceramic materials are also in wide use and are extremely popular in arts and crafts.

Materials used for industrial ceramic products include carbides, borides, nitrides, and silicides. Ceramic materials have unique properties that make their applications for industrial products unique. These advantageous characteristics include having high melting points (up to 7000°F), being chemically inert, being refractory (heat reflective), and withstanding large compressive (crush or pressure) loads. Certain ceramics such as those from the silicon nitride group are practically as hard as diamond and as light as aluminum. Disadvantages of ceramic materials include brittleness and a tendency toward weakness when subjected to tension (pull-apart) loads.

From the viewpoint of the industrial product designer, the ceramic property of resistance to high heat levels is one of the most interesting. Ceramic materials can withstand temperatures of several thousand degrees and still maintain their shape and strength. It is this characteristic that makes them useful for making parts that will be subjected to extreme temperatures including jet engine turbine components, reciprocating engine valves, and exhaust gas-driven turbines in automobiles.

The plasma arc with its extremely high temperature characteristics has been used to produce high-quality ceramic powders, the raw material from which ceramic products are produced. After the product is shaped using pliable ceramic materials, it must then be furnace fired to achieve its final strength and form characteristics. Manufacturing processes for ceramics include extrusion, pressing, casting, machining, injection molding, flame sprayed coating, and sintering.

Uses of ceramics include cutting tools for machining, abrasives for grinding, and coatings on cutting tools to improve their performance characteristics. Industrial products include turbine rotors, engine valves, ball and roller bearings, electrical insulators, bricks, tiles, dishes, and containers.

WOOD, WOOD PRODUCTS, AND PAPER

Throughout the history of manufacturing, wood has been an important and useful material for a multitude of products. Because wood is a natural and often readily available material, it is logical that its use is widespread. Wood is a renewable resource, and modern forest conservation practices are critical in ensuring that we have a continuing supply.

It is interesting to note that wood was a fundamental fuel source well into the twentieth century. In fact, one factor that contributed to the Industrial Revolution, after which manufacturing shifted toward coal and petroleum-based fuel supplies, was a shortage of firewood accompanied by high wood prices. Today there is a renewed interest in wood as a fuel source for domestic heating and even for industrial and power generation purposes; however, the return to wood fuel on a large scale is unlikely because of cost, availability, air pollution control, and its value as a construction material.

Wood and Wood Product Processing

Many types of woods are used in constructing wood products. The primary use of wood is in structural lumber for homes and small building construction (Figure 17.7). Trees that grow straight and tall are most useful for lumber. These include pine, fir, spruce, and redwood. Trees are selected, felled, and transported to high-production sawmill operations where standard structural lumber is produced. The modern sawmill makes use of manufacturing automation, and standard lumber is produced at a high rate.

Waste from sawing, which includes sawdust, slabs, and trimmings, was at one time burned at the mill site. Today, sawdust and wood chips are too valuable to be burned and are used to produce many other wood products, such as plywood and chip-wood products. These materials play an important part in supplementing structural lumber as a building material.

Wood Products In addition to structural lumber, wood products include plywood in which thin layers of wood are glued into panels so that the grain of each layer runs perpendicular to the next (Figure 17.8). Plywood is an indispensable building material in modern wood building construction. Plywood panels are strong and are available in standard sized sheets in several thicknesses.

FIGURE 17.7
Structural lumber is a versatile building material for house construction.

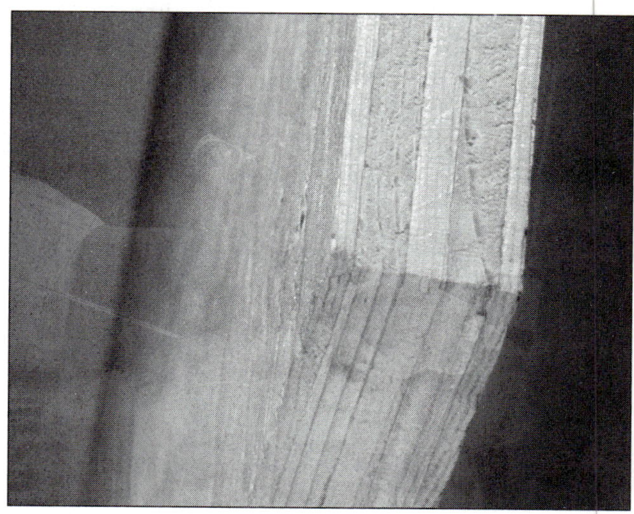

FIGURE 17.8
Plywood consisting of veneer layers glued together in sheets is an exceptional building material (Color-Pic, Inc.).

Other wood products include wood-chip products such as flake board (Figure 17.9). This material allows wood waste to be used in a profitable manner. The material is suitable for a variety of products such as cabinets and furniture.

Wood should not be discounted as a structural building material. Glue-laminated beams (Figure 17.10) are well suited to structural building requirements. They will support large loads and are often lighter and less expensive than metal beams.

FIGURE 17.9
Flake board, a wood-chip product, makes an effective building material and makes use of wood chips that would otherwise become waste products.

FIGURE 17.10
Glue-laminated wood beams are an excellent structural building material for buildings of large size. This material demonstrates high strength-to-weight ratios while minimizing cost of the structural components.

Hardwoods The list of hardwoods used for furniture and flooring is long and distinguished. Common hardwoods used in furniture manufacturing include mahogany, rosewood, oak, teak, birch, maple, hickory, and walnut. Several other woods from tropical areas are also used in furniture manufacturing.

Furniture manufacturing is a large industry employing many of the same manufacturing techniques and fastening systems used in metalworking. Wood must be cut, shaped, and smoothed. Woodworking machine tools and processes resemble their metalworking counterparts in many ways. The most popular process is assembly with modern glues and adhesives as well as mechanical fasteners including nails, staples, bolts, and screws. Typical wood manufacturing equipment includes lathes, shapers, routers, sanding machines, and various types of saws. Woodworking is one area of manufacturing in which hand crafting (Figure 17.11) and hand carving still have an important place (Figure 17.12)

FIGURE 17.11
The construction of fine furniture will always ensure employment of the artist and highly skilled craftsperson (Ethan Allen).

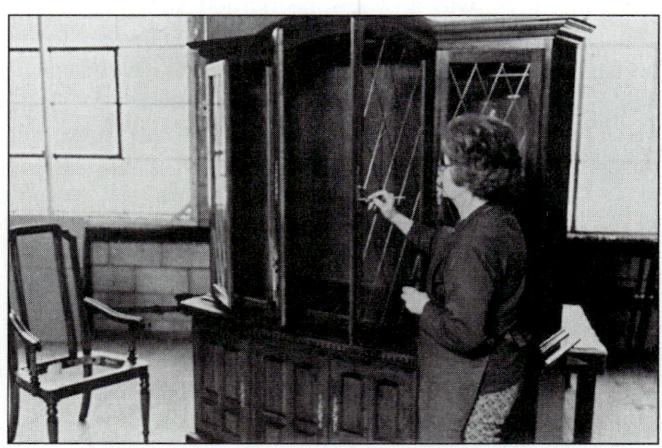

FIGURE 17.12
Construction of fine wood furniture often involves delicate hand work during the manufacturing process (Ethan Allen).

Paper

The importance of paper as a modern industrial material should not be underestimated. Paper is an indispensable material for the support of modern life. Major uses of paper include printing and packaging. Modern paper manufacturing makes available a large number of specialty paper types to meet the many specialized needs of the paper user. Heavy-gage paper is used as an inexpensive construction material for many consumer products, and paper has even been used for clothing. Rising energy costs and an emphasis on conservation of natural resources has prompted the growth of a large wastepaper recycling industry in recent years. Today, wastepaper is a valuable commodity that is reused effectively rather than dumped or burned.

The primary raw material for paper is cellulose fibers from wood or other plants. Various forest trees are used for wood pulp, and at present much paper is made from recycled wastepaper products. Various fabrics in rag form including felt, linen, and wool are also used in papermaking to achieve desirable paper characteristics such as strength and texture.

The papermaking process is somewhat complex and, like many other continuous industrial processes, is highly automated, carefully monitored, and computer controlled. In the papermaking machine, a liquid mass of wood pulp

FIGURE 17.13
Air-jet weaving machines produce high-quality woven fabrics at high production rates (Burlington Industries).

and rag pulp is fed onto a traveling wire screen conveyor. As the wire screen moves, water drains from the wet pulp fiber mass that is transported by the screen. Suction systems remove more water, and the pulp mass begins to solidify as paper. Further operations remove additional water by pressing and rolling. Final finishing is accomplished by calendering rolls that exert many tons of pressure on the paper to provide the desired thickness and surface finish.

FABRICS

Fabrics are important for clothing and for upholstery both in furniture and in the auto industry. Fabrics may be divided into two major classifications, synthetic and natural. Synthetic fabrics such as polyester are primarily derived from plastic polymers. Synthetic fabrics are in widespread use for clothing, carpeting, and upholstery applications. Natural fabrics made from natural fibers include cotton, wool, and silk.

Many types of fabrics, both synthetic and natural, are made by weaving threads of the material into cloth of sufficient size for manufacturing fabric products. Modern fabric mills make use of highly automated and computer-controlled water- and air-jet weaving machines (Figure 17.13).

Fabric processing involves cutting and joining primarily by stitching, and modern methods are applied including computer control of cutting and sewing. Mass-production piece-part cutting of clothing pieces may be done by band sawing or with laser cutting systems.

Modern fabric manufacturing is a continuous and highly automated process in which computer process control is widely used (Figure 17.14). Fabric mills make use

FIGURE 17.14
State-of-the-art fabric production making use of modern CAM results in products of high quality (Burlington Industries).

of the latest factory automation such as the wire-guided computer-controlled material-handling system shown in Figure 17.15. Fabric finishing and dyeing are maintained at a high level of consistency using computer-controlled processes.

RUBBER

Natural rubber or latex is derived from the natural gum of the rubber tree. Prior to World War II, this material was important in the manufacture of many products, and

FIGURE 17.15
Automation in fabric manufacturing material handling involves driverless computer-controlled transports (Burlington Industries).

sources of supply were considered to be valuable. The geography of the war placed sources of natural rubber out of reach, and the needs of the military exceeded the available supply. The situation stimulated the development of synthetic rubber that was chemically derived from the basic chemical constituents of natural rubber. Once the chemistry of rubber was understood, a virtually unlimited supply of the synthetic material was assured.

The applications of rubber are widespread. The material is used for coatings, tubing, hose, and tires. Rubber processing methods include many of the same processes as used in plastics. Among these are various molding processes for products such as tires. Extrusion is used for tube, moldings, and hose. Calendering (rolling) is used to make sheet products and to coat fabrics and metal with rubber.

NATURAL MATERIALS

Natural materials include clay and adobe from which bricks are made. These materials have long and distinguished histories as fundamental building materials. Stone, including granite, marble, slate, and sandstone have also been used as material for buildings and decorative facings. Although these materials are giving way to concrete, steel, glass, plastic, and aluminum in modern buildings, they are still used to some degree as construction materials and are still very much in evidence in distinguished architecture throughout the world.

Leather, derived from animal hides, is another natural material with a long history of use, and it is still widespread in use today. Shoe manufacturing makes use of leather, and although many modern shoes make use of plastic and other synthetic materials, leather still remains a superior material for this application.

Natural fabrics such as wool, cotton, linen, and silk still have wide appeal for clothing and other uses. Even though synthetic fabrics are in wide use, natural fabrics will always have a prominent place in fabric product manufacturing, and their product applications are widespread and well known.

CONSTRUCTION MATERIALS

Construction materials are those from which structures are built, whereas *engineering* materials are those from which products are made. Construction materials include many natural materials such as stone and wood, or material made from natural components such as bricks and rammed earth (compacted dirt). These materials have long and distinguished histories and will continue in use for building structures.

In modern construction of large structures such as buildings, bridges, and highways, steel and concrete stand out as the most versatile and widely used materials. Structural steel is an essential material for modern large-scale construction. Modern steel production along with advanced metallurgical technology has developed high-strength structural steel that permits the safe construction of extremely tall buildings. Plans are presently being drawn for skyscrapers that will exceed 2000 ft in height (200 stories) but will occupy a minimum of space at their base. It is likely that future building construction in urban areas will favor high-rise design in order to save ground space. Structural steels will play an ever increasing part in this type of building construction.

When steel and concrete are used jointly an unsurpassed construction material is created. Concrete (portland cement) is primarily a mixture of limestone and clay that has been pulverized to an extremely fine powder. This material is mixed with proper portions of sand, graded gravel (aggregate), and water, making a thick fluid mass. Large mixing machines that are often truck-mounted are used to prepare the mix so that the concrete may be easily delivered to a job site.

In most cases where concrete is the primary construction material, it is reinforced with steel reinforcing rods often known as rebar. Reinforced concrete has many advantages as a construction material. It can be cast into almost any shape using simply constructed wood forms, and it can bear large loads. The material is exceptional for dams, bridges runways, highways, and buildings. Steel reinforcement is often arranged in a lattice and laid into the form before the concrete is poured. The lattice may be welded or wired together to hold it in position. After the concrete is poured, the steel lattice is permanently encased and will increase the load-bearing capacity of the concrete structure. High-strength concretes containing polymers will withstand loads of over 10,000 psi.

Precasting is a popular and versatile method of manufacturing structural components for buildings (Figure 17.16). In this method, building walls are precast in molds and then removed and transported to the building site where they are bolted together. This method is popular for commercial building, and erection time is very fast. Concrete walls for buildings can also be poured in place and when properly reinforced are suitable for high-rise buildings.

FIGURE 17.16
Buildings constructed from precast concrete walls are erected quickly in a large variety of sizes and floor plans.

Review Questions

1. Name some advantages nonmetallic materials have over metallic materials.
2. What are ceramics and what are they used for?
3. What single property makes ceramic products so useful?
4. List some of the major manufacturing processes used to make ceramic products.
5. What are the major applications of wood and paper?
6. What processes are used in woodworking? Do these resemble machining processes in any way?
7. List some common natural materials.
8. What are the advantages, if any, to using natural material over synthetic materials? What are the disadvantages, if any?
9. What is concrete and what are its primary applications?
10. What can be done to concrete to enhance its strength?

google.-
corrosion protect of ship's fins

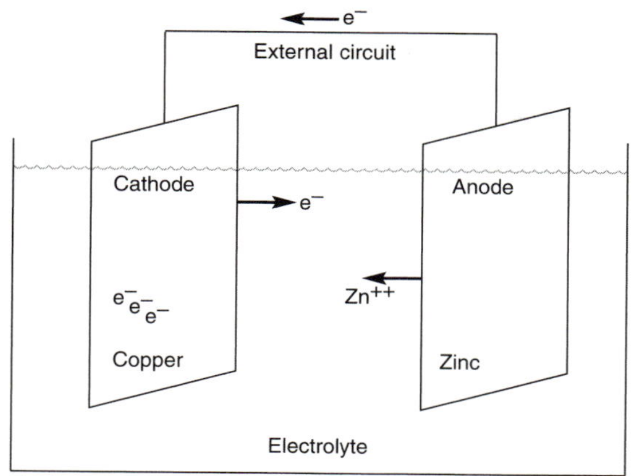

CHAPTER 18

Corrosion and Protection of Materials

Objectives

This chapter will enable you to:

1. Discuss the causes of corrosion.
2. Explain how corrosion may be slowed or prevented.

Because many materials, especially metals, react chemically with their service environments, it is often necessary to provide protective coatings and coverings for them to prevent or slow surface deterioration. Failure to provide surface protection can result in rapid destruction of many types of materials, thus requiring expensive replacement. Other materials, such as wood and plastics, are subject to physical deterioration from weather, sunlight, and water and must also be protected from the environment. The purpose of this chapter is to survey the various means by which materials may be protected from environmental degradation.

We begin by looking at the two major processes by which metals are corroded—direct oxidation and galvanic corrosion. This investigation will be done at the fundamental level of the combination of elements and the exchange of ions and electrons. Once these fundamentals are mastered, developing strategies to retard or combat corrosion are rather easily identified. The remainder of the chapter is devoted to investigating some of the more effective ways of corrosion prevention and control.

METAL CORROSION 

Most common metals exist in nature chemically combined with other elements as various oxides (ores). Examples include iron ores, from which metallic iron is extracted, and bauxite, from which aluminum is derived. These ores are mined and refined by various processes to extract the metallic components that in turn become the common familiar metals. There is a natural tendency for many refined metals to revert to their natural state after the metallic component has been extracted. Because both air and water are common service environments for many metals and because oxygen is a large percentage of both, many metals react with the oxygen present and to revert in varying degrees to their original oxide (ore) state.

Iron is a good example. Existing in nature as an oxide (iron oxide, e.g., Fe_2O_3), the metallic element iron is chemically combined with oxygen. The refining process separates the metallic iron from the oxygen and from then on, if the iron is used in the presence of oxygen, it is a struggle to keep it from turning back into iron oxide. This process, a slow oxidation of the iron, is known as rusting. In the case of iron and steel, the layer of rust (iron oxide) does not adhere to the metal surface but flakes off, exposing the surface to further rusting. Metals such as aluminum and chromium, however, form adherent oxides and are thus protected. The chromium present in stainless steels provides this protection and is discussed shortly. (Gold, on the other hand, does not readily react with oxygen at all and is therefore found in nature in its pure form.)

The oxidizing process, termed *direct oxidation* corrosion, can occur slowly as in normal rusting, or quickly, as in the case of steel heated to high temperatures. This oxidation occurs during heat treating and hot-rolling operations, and the high temperature accelerates the process, and a thick layer of black oxide forms. This oxide is called mill scale and is noticeable as the black coloring seen on most structural steel.

Metal corrosion can be classified according to the two major processes by which it occurs. One is the process just described, direct oxidation corrosion; the second process is galvanic corrosion, in which corrosion takes place at normal temperatures in a moist or electrolytic service environment.

Galvanic corrosion gets its name from the galvanic effect, which is the electric potential or voltage that results when two different metals are connected electrically in the presence of an **electrolyte**—a medium, usually liquid, capable of conducting electric current, that is, a flow of electrons. The voltage thus generated causes a flow of electrons between the metals and into the electrolyte. Under these conditions metallic **ions** of one of the metals will dissolve in the electrolyte; that is, the metal experiences **galvanic corrosion**.

By experimentation, the tendency of different metals to corrode relative to each other in the electrolyte seawater has been determined. The resultant ranking is termed the **galvanic series**. The more easily corroded metals are at the top of the list and are said to be **anodic**; the less corrodible metals are at the bottom of the list and are said to be **cathodic**. Some common metals in the list, from anodic to cathodic, are magnesium, zinc, aluminum, carbon steel, alloy steel, cast iron, tin, lead, brasses and bronzes, copper, nickel, and stainless steels.

For example, if zinc is in electrical contact with copper (see Figure 18.1) in the presence of an electrolyte, the zinc is more anodic than the copper and will be corroded; the copper will be cathodic and will not corrode. In summary, four conditions are required for galvanic corrosion: (1) two different metals, (2) in an electrolyte, (3) with an electron flow *between* the metals through an external circuit or connection, and (4) electron flow *into* the *electrolyte*. As stated here, these four requirements are clear, but in some situations they are not. For example, the two different metals may be different areas of the same piece of metal that act anodic–cathodic to each other, and the electrolyte may be moisture left behind by a fingerprint.

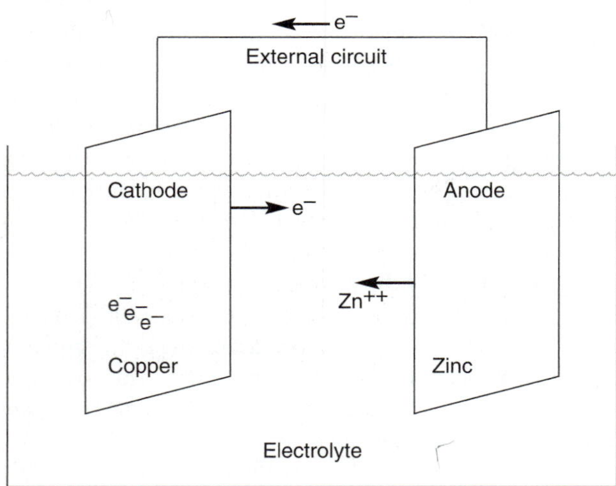

FIGURE 18.1
Galvanic corrosion: two metals in an electrolyte connected electrically cause electron flow through that connection and removal of metallic ions (corrosion) from the anodic metal.

Two other general principles are important: (1) The farther apart the metals are in the galvanic series the larger the voltage that will exist between the two metals, and (by Ohm's law) the larger the voltage, the larger the electron flow and therefore the more rapid the rate of corrosion. (2) The larger the areas of the two metals, the larger the electron flow in the external circuit and into the electrolyte.

PROTECTION METHODS

In developing strategies to protect a metal from its environment the designer or engineer needs to take into account the two corrosion processes, direct oxidation and galvanic corrosion, and the four conditions required for galvanic corrosion.

The method used to protect a given material from influences of the environment will depend on several factors, such as cost, how corrosive the environment is, and how directly the need to protect the material is tied to structural design requirements.

There are two principal strategies for protecting metals subjected to direct oxidation:

1. Shielding the material from its service environment by coating it with another material. Examples include painting, plating, cladding, and treating with various oxides. As long as the integrity of the coating is maintained, the base material will be preserved.
2. Establishing the chemical makeup of the parent material such that service environments do not cause it to degrade. For example, the alloying of metals with noncorrosive alloys makes them less subject to oxidation forces. This is frequently done for many consumer products, but alloying is generally much more expensive than coating a less expensive material with surface protection; however, the buckets of a gas turbine blade must function at a high temperature in a corrosive environment. The choice of sophisticated, expensive alloys is the solution currently employed.

To combat galvanic corrosion one or more of the four previously mentioned conditions must be avoided or interrupted. For example:

1. *two different metals:* Obviously, using only one metal is a solution, but if that is not an option, then metals should be used that are as close to each other as possible in the galvanic series.

2. *in an electrolyte:* The presence of an electrolyte should be avoided if possible, and if contact occurs only occasionally (such as rainwater), then provision should be made for the electrolyte to drain away. Shielding the metals from the electrolyte by covering their surfaces, by painting, for instance, is also a good technique.
3. *with electron flow between the two metals:* To reduce, and perhaps eliminate, the flow of electrons between the different metals, an insulator can be placed between them. For example, aluminum stadium seats bolted to a steel grandstand should have insulating washers inserted so the two metals are not in electrical contact.
4. *and electron flow into the electrolyte:* In Figure 18.1 electrons (e^-) are shown collecting on the more cathodic metal (the copper). If these electrons are not removed by the electrolyte, the galvanic process stops! Thus, if the electrolyte is a moving fluid, it should be slowed down; if the area of the cathodic metal (relative to the anodic metal) is kept small, there is less opportunity for electron removal, and the rate of corrosion is reduced. (Note that this suggests it is better to use *less* of the cathodic metal in relation to the anodic. In a steel rain gutter copper rivets should be used because steel is anodic to the cathodic copper; steel rivets in a copper gutter will not last long!)

A special application of the last concept is to purposely add a more anodic metal to a structure so that the anodic metal corrodes in place of the more cathodic structure. The anodic metal then corrodes and protects the more cathodic structure. In a like fashion steel or cast iron pipe lines are converted from being anodic to being cathodic by the attachment of a more anodic metal, like zinc, and are thus protected. The zinc is a **sacrificial anode** for the other metal, which is now more cathodic. A very common application of this technology is **galvanized steel**—steel covered with more anodic zinc, which protects the steel from corrosion. Because this method turns the metal into a noncorroded cathode it is termed **cathodic protection**.

An examination of Figure 18.1 shows that the role of the cathode is to supply electrons to the electrolyte. If the steel pipeline just discussed is supplied with electrons from a direct current source it is converted from being anodic and corroded, to being cathodic and protected. Natural gas companies use this technique, termed an **impressed current**, to protect their miles of steel and cast iron pipelines.

Cladding

Cladding is adding a layer of one metal to another metal to protect against corrosion, to improve mechanical characteristics such as hard facing, or in some cases to improve appearance. Cladding may be done by weld buildup of a layer of metal, or by adhesive bonding the metal to another. Cladding may also be applied by bonding the metals during a rolling operation. See Figure 6.6 and the accompanying discussion.

In the clad-welding process, weld beads are laid down on the base metal side by side and layer by layer. The process is tedious if done by hand and usually requires a shield gas welding process to eliminate slag pockets and other inclusions in the cladding. After clad welding, parts may be machined to achieve desired dimensions.

Alloying and Oxidizing

Alloying Alloying is an effective but often expensive method of material protection. Certain metals, such as chromium, have excellent anticorrosion properties. Plain carbon steel, on the other hand, has an affinity for the oxygen in air and will rust (oxidize) if exposed to a normal outdoor environment. It is also subject to galvanic corrosion; however, if the steel is alloyed with chromium, creating stainless steel, corrosion by rusting may be reduced or totally eliminated. Figure 18.2 shows pieces of stainless and plain carbon steel that have been exposed to a normal outdoor environment for the same period of time. The plain carbon steel shows considerable rust, whereas the stainless steel shows very little evidence of corrosion; however, even stainless steel will corrode in some environments, so it is important to select anticorrosive alloys when using this method.

Other metals that will provide anticorrosion characteristics include copper–nickel alloys (Monel), molybdenum, lead, nickel, and pewter. Lead, used in lead–acid batteries for plates, serves in an extremely corrosive acid

FIGURE 18.2
A heavily rusted piece of unprotected carbon steel is shown resting on top of a piece of stainless steel, which shows little corrosion. Both were exposed to a normal outdoor environment for the same length of time.

environment. Gold is also an excellent anticorrosive metal and is often used to plate other metals for corrosion protection.

Another possible advantage of alloying is that the anticorrosion properties exist throughout the thickness of the metal. Thus, even if the metal surface is damaged by abrasion, denting, or scraping, the anticorrosive properties will be preserved. This is not the case with plated, painted, or otherwise coated materials. Because the anticorrosion medium is only on the surface and most often is extremely thin, scratch or scrape damage can expose base material to the environment, permitting corrosion to begin. Figure 18.3 shows painted steel on which the integrity of the paint has been breached, thus exposing the bare metal to the environment. This exposes a place where corrosion can begin.

Oxidizing A metallic material that tends to react with oxygen and oxidize may be protected by carrying out this very process on the surface of the metal if the oxide formed adheres; this happens naturally with aluminum and chromium. Common oxidizing processes include black oxidizing and anodizing.

In *black oxidizing,* parts are heated in a carbon or oxidizing salt environment and then quenched. Black oxidized parts take on a black color and are protected from rusting by their oxide surface.

Anodizing is a popular surface protection process, especially for aluminum. The process is performed electrically and results in a layer of oxide on the surface of the part that is thicker than the layer that forms naturally. This layer prevents further oxidation by the service environment (Figure 18.4). Once again, the anodizing layer must be maintained intact. This layer, which is less than 0.0005 in. thick, can be damaged to the point where the base metal is exposed to the corrosive environment.

FIGURE 18.3
Damaged surface of painted steel exposes base metal and permits corrosion to occur.

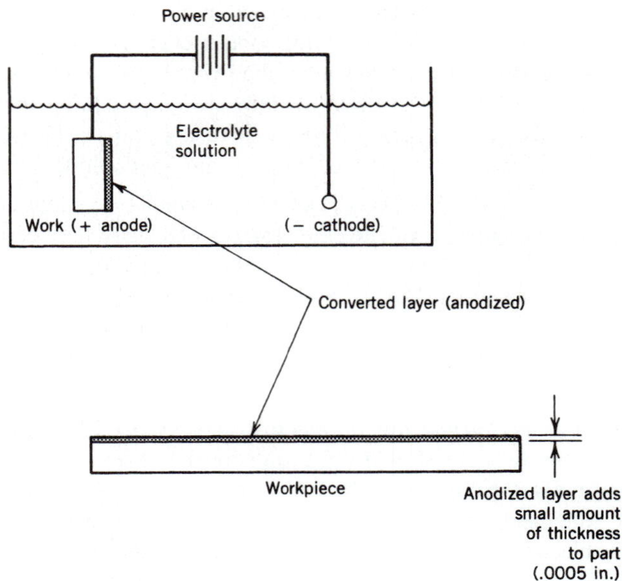

FIGURE 18.4
Anodizing process.

Plating The various plating processes make up a large portion of methods used to protect metals from their service environments. The theory behind plating is that if a thin layer of anticorrosive material can be plated to a base metal, then the material can be protected from corrosion. Figure 18.5 clearly demonstrates the application and advantage of plating as an anticorrosion measure. Note how the unprotected steel on the bracket is heavily rusted; however, the zinc-plated fastener shows little or no degradation from corrosion.

Plating processes are used widely and have numerous advantages. For example, sheet steel is used for a vast number of products. It is easily formed, can be effectively assembled by high-production methods such as spot welding, and is relatively inexpensive. Its primary drawback is that it is subject to rapid rusting in typical service environments. One solution is to use stainless sheet steel; however, this product is much more expensive and less easily fabricated.

Common low-carbon sheet steel is one product that can be protected effectively from rusting by plating it with an anticorrosive metal (such as zinc) or a noncorroding metal (such as tin). If zinc is used, the steel is **galvanized** and protected against galvanic or electro-

FIGURE 18.5
Unprotected steel is heavily rusted, whereas the plated fastener shows no evidence of corrosion.

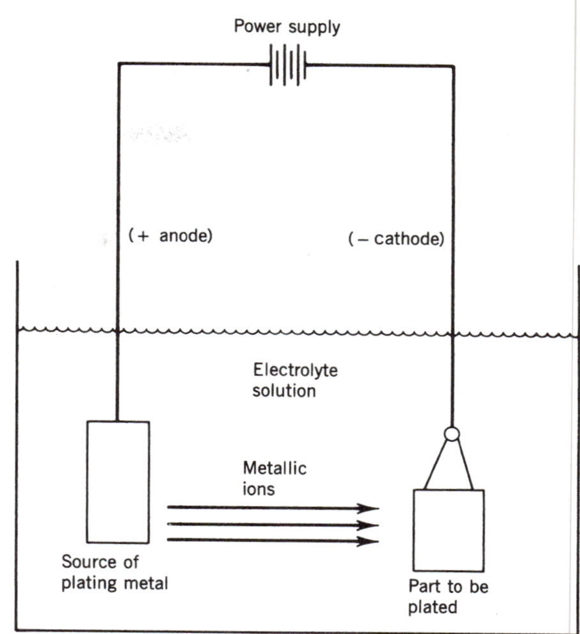

FIGURE 18.6
Electroplating process.

chemical corrosion. If tin is used, the steel is **tin plated**; its surface is covered with a layer that protects it from the environment. Because the tin is cathodic or more corrosion resistant than the steel, the steel is *not* protected by the tin against galvanic attack but is protected as if by a layer of paint. Thus, the useful production properties of sheet metal are maintained while the material is protected against corrosion from its service environment.

Electroplating Processes Many metals are electroplated onto other base metals by **electroplating** processes. Parts to be plated are placed in a tank containing a salt solution (electrolyte) of the plating metal. A direct current (DC) voltage is applied across the part to be plated (− cathode) and the source metal (+ anode). The current flow causes metal ions to move from the source to the parts, where they are deposited as plated metal (Figure 18.6). Galvanizing is one common process where a zinc coating is electroplated onto steel (Figure 18.7).

Other metals that are electroplated include tin, cadmium, nickel, chrome, brass, silver, and copper. Gold has many industrial plating applications in electronics because of its superior electrical conductivity properties. Silver and brass plate add beauty and serve to resist corrosion (Figure 18.8). Electroplating of a hard metal such as chromium is also used to improve wear resistance or to build up a part that has been worn undersize.

Electroless Plating Processes Plating without electricity is also a widely applied process. Nickel is an example. The process of electroless nickel plating yields a more

FIGURE 18.7
Examples of common galvanized products.

FIGURE 18.8
Silver plating (left) and brass plating (right) add both beauty and resistance to corrosion.

even thickness on complex parts than will electroplating processes.

Electropolishing

In the electropolishing process, a small amount of surface material is deplated from the part. The process is accomplished by placing the parts in an acidic electrolyte where a DC potential exists between parts (anode) and cathode (the other side of the power supply).

The process leaves an extremely smooth surface while not removing enough workpiece material to affect precision dimensional characteristics. It is also useful for smoothing out surfaces on irregularly shaped objects or where the surfaces of parts are not visible or accessible.

Painting

Painting is probably the most widely used process for surface protection of metals and other materials. Its primary advantages include ease of application, wide color variety, and reasonable cost. Its primary disadvantage is that because painting is a surface treatment, damage can expose the base material to the corrosive environment. However, the process is in wide use and will probably remain a mainstay of surface protection.

Although there are many types of paints, and paint technology undergoes constant development, paints are still basically a liquid that dries or cures to form a film that holds a pigment; the pigment is a solid particle that provides the color. Solvents are added as required to improve the application of the paint. Traditionally, paints used natural resins with oils such as linseed, tung, and soybean; but paints are increasingly becoming polymeric coatings using solvents such as water that minimize the environmental problems inherent in disposing of petroleum solvent fumes.

Some of the newer formulations are based on common polymers: vinyl, epoxy, urethane, alkyd, phenolic. Two of the most widely used polymers are the water-based polyvinyl acetates and acrylics. Each of these is a particular polymer with its own advantages and disadvantages.

The thickness of the paint varies with the anticipated environment: 0.001 in. for decorative purposes, 0.003 to 0.005 in. for outdoor and general industrial environments, 0.015 in. for chemical plants, 0.090 in. for asphalt coating of pipes to be buried.

Paint Applications Paint is typically applied by dipping, brushing, rolling, and spraying. Spray technology is most popular in mass-production applications and is the only way to achieve the truly smooth coat necessary in most manufactured product applications, such as on automobiles.

FIGURE 18.9
Spray application of paint.

Spray painting can be done by air-driven sprayers (Figure 18.9), but overspray must be captured and retained in order to meet most current air pollution requirements, especially if nonwater solvents are used. Personnel must also be protected during spray painting operations.

Electrostatic spray painting eliminates overspray and the resulting pollution. This process is extremely effective from two important standpoints—better coverage of product and less consumption of paint. In the electrostatic process, a high DC voltage difference is maintained between the paint sprayer and the part to be painted. Paint particles flowing from the sprayer acquire the charge opposite that of the part and are therefore attracted to it. The process is extremely fast, and since the part has an equal charge on all sides, the paint is attracted to the back as well as to the front. Thus, parts are literally painted all over at once (Figure 18.10).

High-pressure paint pump sprayers have become very popular in recent years. In this equipment, the paint is atomized by pumping it through a small nozzle at very high pressure. High-pressure pumping is very fast in application, and the technology is readily available.

Paint spraying, regardless of the application equipment, requires considerable skill on the part of the operator to achieve a smooth and even coat on the workpiece.

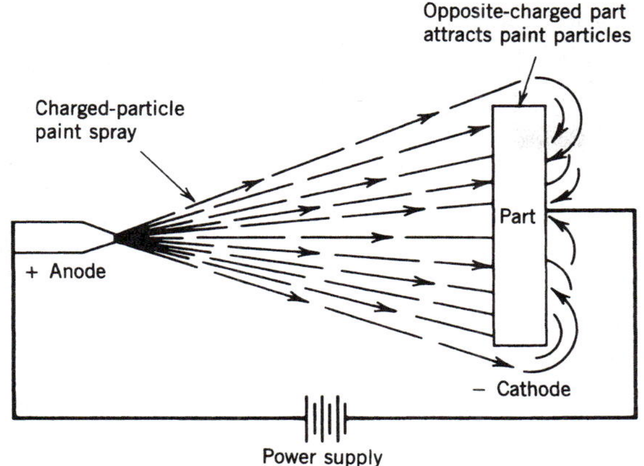

FIGURE 18.10
Electrostatic spray-painting process.

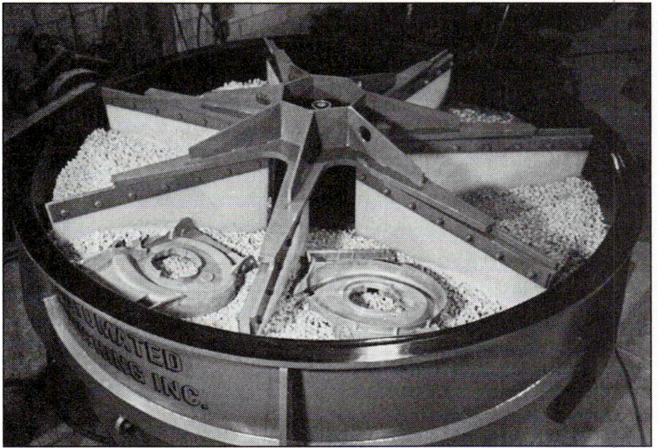

FIGURE 18.11
Vibratory deburring machine with a capacity of 160 ft^3 showing compartments, workpieces, and abrasive pellets (Automated Finishing Inc.).

Like many industrial processes, spray painting can be accomplished effectively with computer-driven industrial robots. Since robots cannot make the intuitive adjustments a human is capable of, some products may have to be specially located for the robot to reach surfaces with the spray gun. For example, in an automobile production line the bodies may have to be positioned diagonally on the conveyor so that the spray-paint robots may access them for the computer-controlled production painting operation. Since painting must be accomplished on underside surfaces and overhead locations, the jointed-wrist type industrial robot is popular for this application.

Paints must dry after application. This can be a time-consuming operation and can be speeded up by the application of heat and circulating air. Parts are conveyed to a heated drying facility where heaters expedite the drying process.

PREPARING MATERIALS FOR SURFACE PROTECTION

Before material can be surface protected, it must be properly prepared. Failure to do this will result in a poor application of the particular surface protection process. Several methods are used to prepare materials.

Mechanical Methods

Commonly used mechanical methods physically abrade the material to remove mill scale, dirt, and corrosion. For example, in sand- or shot blasting, high-velocity abrasive particles are propelled against the workpiece. Abrasives in the form of sandpapers, abrasive wheels, or wire wheels (brushes) may also be used.

Tumble cleaning and deburring is another popular method for both cleaning and removing burrs from machined workpieces. In this process, parts to be cleaned or deburred are placed in a vat or vibratory bowl with small pellets of abrasive material (Figure 18.11). The bowl is vibrated rapidly, causing the parts to be tumbled along with the abrasive material. Equipment used for this process ranges in size from small to quite large. Vibratory cleaning and deburring machinery also can be equipped with washing, drying, and material-handling conveyor systems (Figure 18.12).

These methods are satisfactory but may be labor intensive, dirty, and sometimes dangerous. It is also difficult to clean all surfaces on a complex part by brushing or grinding. Sandblasting is a very abrasive process and will erode metal and wood surfaces rapidly. Using glass beads for the abrasive eliminates most of this effect, but this process must be done in a confined area where the fine dust can be collected and removed from circulation. Sandblasting must also be done under controlled conditions, and the personnel must be protected from the flying sand.

Ultrasonic cleaning of parts is accomplished in a container with a cleaning medium. The bath is vibrated by ultrasonic sound waves. The process is a combination of both chemical and mechanical cleaning.

In the process of buffing, parts are brought into contact with a rapidly rotating wheel, usually made from layers of cloth sewn together. Various grades of abrasive materials in cake form are hand applied to the wheel. These materials include various metallic oxides and rouge. The buffing process provides highly polished mirrorlike finishes on metals; however, it is labor intensive and often requires that the part be previously abraded with

FIGURE 18.12
Through-flow vibratory machines equipped with part washers, dryers, and material-handling conveyors (Automated Finishing Inc.).

coarser materials so that final buffing will yield a desirable result. This process is popular in art metal and jewelry metal work.

Chemical Processes for Cleaning and Material Preparation

Chemical preparations are usually necessary prior to painting. In pickling, material is dipped in an acid bath to remove oxides. This process must be followed by a rinse to remove the pickling acid.

Solvents can be used to remove grease and oil. The parts can be placed in the vapor of the solvent. When vapor condenses it washes the parts. This process is called *vapor degreasing.* Alkaline solutions are also used as a washing agent.

To provide for a better bond between paint and material, two phosphate coating processes are used. In *bonderizing* the parts are dipped in a phosphating solution. The resultant coating permits paint to adhere more readily. The phosphate process of *parkerizing* also provides a corrosion-resistant primer coating under a painted surface.

Review Questions

1. Name the primary causes of corrosion.
2. Describe the galvanic series and how it can be used to select metals to reduce corrosion.
3. What four conditions are required for galvanic corrosion to occur? Describe how this information can be used to avoid galvanic corrosion.
4. Why not manufacture everything out of solid anticorrosive materials?
5. Describe plating processes and discuss their advantages and disadvantages.
6. What is anodizing?
7. What is galvanizing? How does the protection it provides differ from tinplating?
8. How can paint coatings be applied?
9. Are there any disadvantages to painting as a corrosion-prevention measure?
10. Describe electrostatic painting and discuss its advantages over air spraying.
11. How can materials be mechanically cleaned prior to painting or plating?
12. What chemical preparations are used to prepare materials for plating or painting?
13. Compare Figures 18.1 and 18.6. What purposes are served by the anodes and cathodes in each figure?

Case Problems

Case 1: Abrasion Resistance

An escalator side plate is subjected to light but constant abrasive forces. In your opinion, which of the following would be the best method of protecting this material, with an emphasis toward minimum maintenance and long-term preservation of the product?

- Plating
- Painting
- Anodizing
- Making from solid material
- Galvanizing

Case 2: Rust Prevention

Large steel fabrications are designed to be used in a brackish estuary. The products are painted with the best available anticorrosive marine paints. What other methods may be used to further prevent corrosion of these items in their service environments?

PART III

MEASUREMENT AND QUALITY ASSURANCE

Part III introduces measurement and quality assurance with particular emphasis on tolerances, measuring processes, and calibration. Quality assurance is an integral part of today's manufacturing environment. No manufacturing process can be successfully designed and implemented without full consideration of how the quality of the product will be assured. This includes determining when, where, and how the product will be inspected as discussed in Chapter 19, Quality Assurance and Control.

Quality must be maintained at every step in the manufacturing process, and quality is not assured unless all product specifications are met at each of those steps. Tolerances establish the precision required of the manufacturing process, and in order to assure that the product meets all specifications existing tolerances must receive full consideration. In designing a process to achieve this level of precision, the precision of both the measuring process and the manufacturing operation itself must be considered. The precision required of the measuring process affects the design of the measuring operation as well as the choice of inspection instrument. These issues are addressed in Chapter 20, Inspection and Measurement.

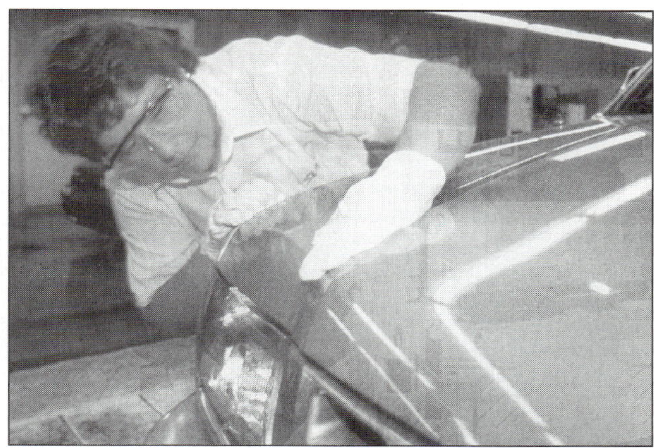

CHAPTER 19

Quality Assurance and Control

Objectives

This chapter will enable you to:

1. Define the purpose and intent of quality control and quality assurance.
2. List the major responsibilities of quality control and quality assurance personnel in manufacturing.
3. Discuss aspects of nondimensional quality control.
4. Discuss quality control and assurance after manufacturing.

Key Words

in-process inspection	nondestructive testing
source inspection	dimensional inspection
visual inspection	final inspection
performance inspection	warranty

Product quality is of paramount importance in manufacturing. If quality is allowed to deteriorate, then a manufacturer will soon find sales dropping off followed by a possible business failure. Customers expect quality in the products they buy, and if a manufacturer expects to establish and maintain a name in the business, quality control and assurance functions must be established and maintained before, throughout, and after production. Generally speaking, **quality assurance** encompasses all activities aimed at maintaining quality, including quality control. This chapter presents an overview of quality assurance activities. Quality assurance can be divided into three major areas:

1. Source inspection before manufacturing.
2. In-process quality control during manufacturing.
3. Product service and warranties after manufacturing.

SOURCE INSPECTION BEFORE MANUFACTURING

Quality assurance begins long before any actual manufacturing takes place, starting with **source inspections** conducted at the plants that supply materials, discrete parts, or subassemblies to manufacturer. The manufacturer's source inspector travels to the supplier factory and inspects raw material or premanufactured parts and assemblies. Source inspections present an opportunity for the manufacturer to sort out and reject poor quality materials or parts before they are shipped to the manufacturer's production facility.

The responsibility of the source inspector is to check materials and parts against design specifications and to reject the items if specifications are not met. Source inspections may include many of the same inspections that will be used during production:

1. Visual inspection
2. Metallurgical testing
3. Dimensional inspection
4. Destructive and nondestructive testing
5. Performance inspection

Visual Inspection

Visual inspections examine a product or material for such specifications as color, texture, surface finish, or overall appearance of a component or assembly to determine if there are any obvious defects or deletions of major parts or hardware (Figure 19.1).

FIGURE 19.1
Making a product visual inspection (Subaru-Isuzu Automotive, Lafayette, Indiana).

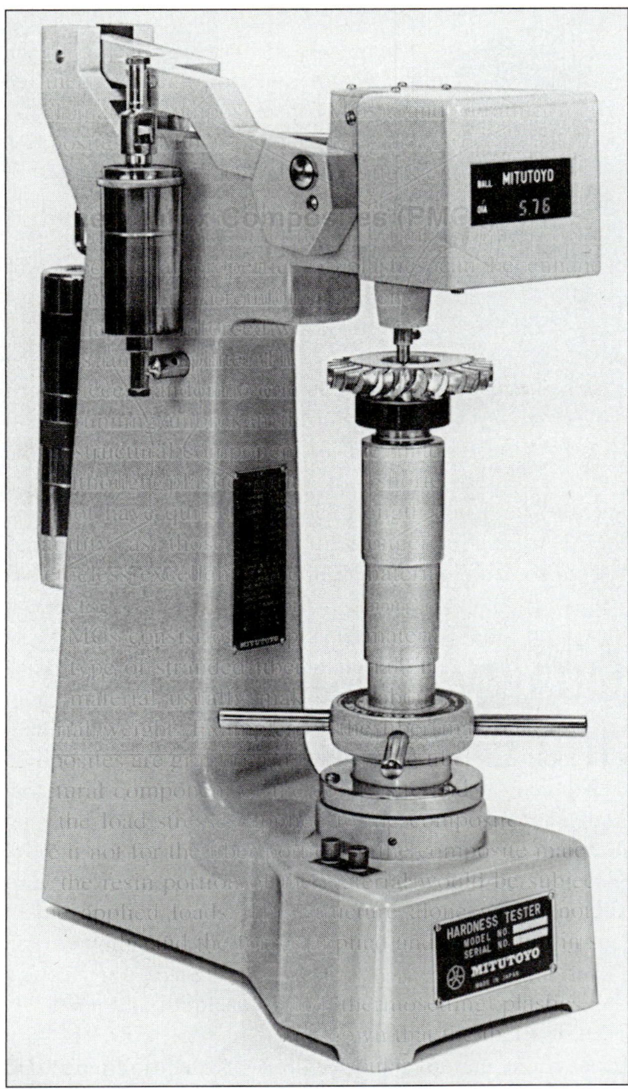

FIGURE 19.2
Hardness testing is an example of metallurgical inspection (Mitutoyo American Corporation).

Metallurgical Testing

Metallurgical testing is often an important part of source inspection, especially if the primary raw material for manufacturing is stock metal such as bar stock or structural materials. Metals testing can involve all the major types of inspection including visual, chemical, spectrographic, and mechanical, which include hardness (Figure 19.2), tensile (Figure 19.3), shear compression, and spectrographic analysis for alloy content. Metallurgical testing can be either destructive or nondestructive.

Dimensional Inspection

Few areas of quality control are as important in manufactured products as dimensional requirements. Dimensions are as important in source inspecting as they are in the manufacturing process. This is especially critical if the source supplies parts for an assembly. Dimensions are inspected at the source factory using standard measuring tools plus special fit, form, and function gages that may be required. Meeting dimensional specifications is critical to interchangeability of manufactured parts and to the successful assembly of many parts into complex assemblies such as autos, ships, aircraft, and other multipart products.

Destructive and Nondestructive Testing

In some cases it may be necessary for the source inspections to call for destructive or **nondestructive tests** on raw materials or parts and assemblies. This is particularly true when large amounts of stock raw materials are involved. For example, it may be necessary to inspect castings for flaws by radiographic, magnetic particle, or dye penetrant techniques before they are shipped to the manufacturer for final machining. Specifications calling for burn-in time for electronics or endurance run tests for mechanical components are further examples of nondestructive tests.

It is sometimes necessary to test material and parts to destruction, but because of the costs and time involved destructive testing is avoided whenever possi-

FIGURE 19.3
Tensile testing a sample.

ble. Examples include pressure tests to determine if safety factors are adequate in the design. Destructive tests are probably more frequent in the testing of prototype designs than in routine inspection of raw material or parts. Once design specifications are known to be met in regard to the strength of materials, it is not often necessary to test parts to destruction unless they are genuinely suspect.

Performance Testing

Performance testing involves checking the function of assemblies, especially those of complex mechanical systems, prior to installation in other products. Examples include electronic equipment subcomponents, aircraft and auto engines, pumps, valves, and other critical systems.

QUALITY CONTROL DURING MANUFACTURING

Quality control during manufacturing involves many of the same checks as in source inspection. A great deal of time and effort is invested in in-process quality control by the manufacturing industry. Major considerations for in-process quality control include the following:

1. Receiving inspection
2. First-piece inspection
3. In-process inspection
4. One hundred percent inspection
5. Final inspection

Receiving Inspection

In receiving inspection, parts, material, and assemblies are checked against design specifications as they arrive at the manufacturer's production or assembly facility. If the items received are premanufactured parts requiring further assembly or processing, receiving inspection will inspect for correct dimensions and any other specifications required by the design. Thorough source inspection will reduce or eliminate the need for receiving inspection.

First-Piece Inspection

Any time a new part or product is made for the first time or when a production machine is set up for repeat production after having been used for another job, a first-piece inspection is required. When the first item completes the manufacturing process it is inspected, and if it conforms to specifications further production is allowed to continue. If the first piece is not acceptable, the machine or the entire production line must be stopped and reset to correct any problems. First-piece inspection is necessary before production can begin.

In-process Inspection

In-process inspection must be applied continuously during production. Each production station must have the necessary tools to accomplish the required inspection operations. Past practice of employing roving quality inspectors to randomly inspect parts throughout the factory has largely been replaced by in-process inspection performed by the machine operator or assembler at his or her workstation. In this manner the worker, often aided by statistical process control (SPC) measures, is made responsible for product quality.

One Hundred Percent Inspection

One hundred percent inspection refers to the inspection of every part or assembly produced. Though this might appear

FIGURE 19.4
Automotive dynamometer testing (James D. Halderman)

to be a way to solve all quality control problems, it is often impractical and altogether too expensive to accomplish in routine manufacturing. This is especially true in high-volume production where it would not be feasible to inspect every item produced. Also, 100 percent inspection does not always assure perfect quality. Inspection operations can be fallible and fail to detect defective products, especially in the case of highly repetitive manual inspection.

Regardless of its faults, 100 percent inspection is often used for critical components. Critical aircraft parts may be 100 percent inspected, since the possible acceptance of an out-of-specification part could have serious consequences. Another situation where such inspection may be applied is where nonconforming parts begin to show up continually in the production process, and the problem must be tracked down and immediately corrected. One hundred percent inspection may be necessary until production quality is restored, when normal in-process quality control may resume.

Final Inspection

Final inspection presents another opportunity to detect parts or assemblies that do not meet specifications. Nonconforming parts are rejected to be scrapped or recycled for rework if possible. It is naturally desirable to avoid scrapping as many parts as possible, since they often represent a large production expense.

Nondimensional Quality Control

Quality control and assurance practices must be observed for both dimensional and nondimensional specifications with equal dedication. This includes a wide variety of tests and inspections on all types of products, especially complex operating assemblies.

FIGURE 19.5
Vaporizing an engine crank case oil sample for spectrographic analysis (AGCO Corporation).

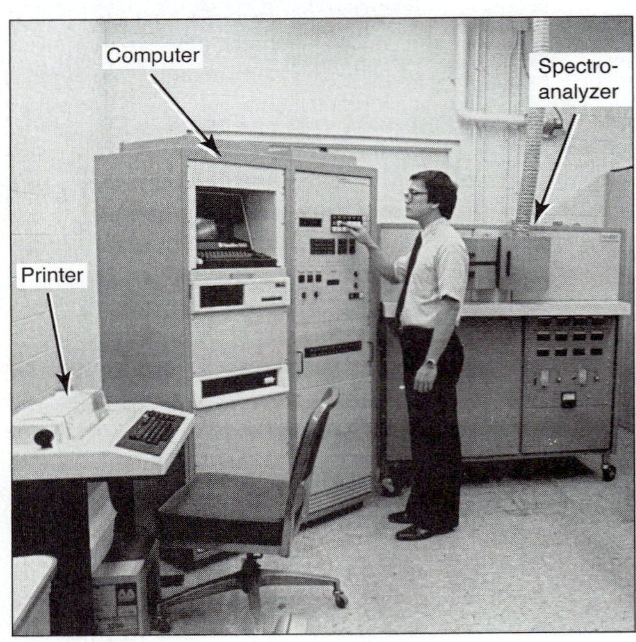

FIGURE 19.6
Computerized spectrographic analysis of the oil sample (AGCO Corporation).

Nondimensional quality control inspection and testing are far-reaching and often involve not only state-of-the-art technology but many engineering innovations. One example is the automotive dynamometer test that measures power output and emission levels of engines (Figure 19.4). Computerized equipment with

FIGURE 19.7
Space shuttle flight stress simulator subjects air frame to forces that it will encounter in actual flight.

the ability to analyze test data with great speed plays a vital part in the overall quality control and assurance effort. Oil samples from lift truck engines are spectrographically analyzed to determine chemical and molecular content. The samples are first vaporized (Figure 19.5), and then the light energy emissions from this process are computer analyzed (Figure 19.6) to determine the chemical makeup and content of foreign matter in the oil sample. The process can determine the type and amounts of metal particles suspended in the oil, and the data can be used to determine where metallic abrasion is taking place within the engine during operation.

Aircraft and spacecraft airframes are tested to the stress levels that will be experienced in service. Spacecraft are enclosed in a large steel framework where hydraulic jacks exert forces of up to 1 million pounds to simulate what the vehicle will experience during launch, spaceflight, and reentry into the earth's atmosphere (Figure 19.7). Another example of nondimensional quality control is analysis of microwave oven emissions (Figure 19.8). This type of inspection is vital to consumer protection.

PRODUCT SERVICE AND WARRANTIES AFTER MANUFACTURING

Quality control and assurance do not end at the manufacturer's shipping dock. They continue well after manufacture in the form of warranties and product support providing factory or factory-authorized service, spare parts, field service, service contracts, reliability studies, and failure reporting.

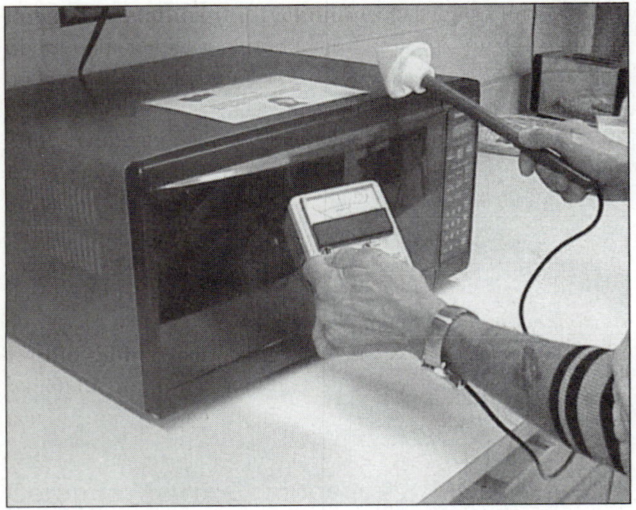

FIGURE 19.8
Analysis of microwave oven emissions.

Purchasers of almost any product expect and are entitled to a **warranty** or guarantee that will be in force for a period of time after the product arrives in the hands of the end user. Warranties vary in length ranging from a few months to lifetime full replacement. Some warranties cover full repair and replacement for a short period of time and cover only specific components for an extended time. Extended warranties may be purchased at extra cost in many cases. Related to this are service contracts for which the customer pays a flat price, sometimes a percentage of the purchase price, in exchange for repair service for the duration of the contract.

Critical equipment that fails in service must often be restored to functional capability as quickly as possible, and efficient service can be critical to the continuing function of businesses. Many manufacturers provide factory service on their products. This capability may be extended over wide geographical areas by using factory-authorized service through other businesses. Spare parts can often be purchased from either the original maker or a factory-authorized service center.

Many manufacturers, especially those of complex products, are concerned with reliability and failure rates of their products in service. Product failure and reliability analysis is useful in determining where improvements or correction of deficiencies in the production or design need to be made. Reliability analysis can greatly improve a manufacturer's product quality. The compilation of reliability statistics can be used effectively in product advertising, thus enhancing the manufacturer's standing in the business.

PRODUCT TESTING

Product testing is vital to successful manufacturing, and manufacturing industries often place a large emphasis on product testing. Testing may take place at the prototype stage before an actual product is manufactured for sale. Product testing also may be carried out in controlled markets where a few preproduction models are placed in selected hands for the purpose of testing and evaluation. In the case of aircraft, for example, a license to manufacture will not be granted until performance and safety specifications have been met. The manufacturer must assume the costs of manufacturing and testing preproduction prototypes.

Endurance testing of automobiles is another significant product testing activity. Production models are test driven through many different environments to simulate what the vehicles will encounter in actual service. By continuous testing, several years of typical service of the vehicle can be compressed into a few months. The data collected is valuable in determining design problems and in predicting the reliability and effective service life of the product.

Review Questions

1. What is the difference between quality control and quality assurance?
2. What are the three major areas of quality assurance?
3. What are the responsibilities of the source inspector?
4. What items does the source inspector look for?
5. Give two examples of metallurgical testing.
6. How would a given metal be checked for alloy content?
7. What is involved in a dimensional inspection?
8. Why are dimensional inspections important in manufacturing?
9. What is the difference between destructive and nondestructive testing?
10. Where might destructive testing be applied?
11. What is a performance inspection?
12. What are the major points of in-process quality control?
13. How are statistics used in quality control?
14. What is the purpose of a first-piece or first-article inspection?
15. What are some examples of nondimensional quality control?
16. What quality assurance factors occur after manufacturing and shipping?
17. Why is product testing important to a manufacturer?

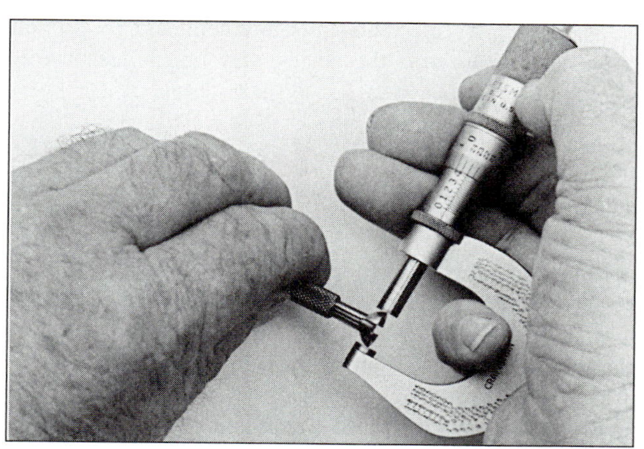

CHAPTER 20

Inspection and Measurement

Objectives

This chapter will enable you to:

1. Describe why, how, and with what tools a product is inspected.
2. Describe a gage that can be used for comparison inspection.
3. Identify common tools for dimensional measurement.
4. Select an effective measuring instrument for a given application.
5. Describe the relationship among tolerance, precision, and resolution and use this to select inspection methods and equipment.
6. Define calibration, relate the importance of measurement standards, and describe how instrument traceability is assured in measurement operations.
7. Explain how the effect of temperature changes differs for steel and aluminum workpieces.

Key Words

measurement standards
fixed gages
transfer measurements
calibration
micrometers
vernier scale
precision
deviation gages
coordinate measuring machine (CMM)
surface finish (surface roughness)
interchangeability
direct-reading instrument
calipers
resolution

Manufacturing involves the modification of raw materials for the purpose of meeting the product specifications that are established when the product is designed. The manufactured product must conform to the intended specifications so that it may perform its function effectively. Many types of specifications exist. The weight, length, area, or volume of an object may be specified. For any surface of an object the shape, smoothness, or ability to reflect light may be specified. Sound pitch or volume level, speed, acceleration, and many other factors also may be specified. This chapter deals only with specifications of size that are critical to manufactured products, such as length, width, height, diameter, depth, and thickness, location specifications, and form specifications involving angles, shape, or surface qualities.

Any product must be inspected to judge compliance with the established specifications. This step is necessary because a product that does not conform to the design specifications may fail to perform its intended function and therefore be of little value. Should this occur the manufacturer will have wasted the material, employee time, use of machinery, and the efforts of the company to support the manufacturing operation. More importantly, a nonconforming product may fail in use, leading to customer dissatisfaction and possible injuries for which the manufacturer may be liable.

Products are inspected in many different ways. A product may be visually inspected to ensure general conformance to specifications. A feature of the product such as length may be compared with an inspection tool to judge whether it meets a specification, or a feature may be measured with an instrument and the resulting reading compared with the specification to judge conformance. All of these methods—visual inspection, comparison inspection, and measurement—are commonly used in manufacturing.

Computers and other electronic devices have found wide application in precision measurement. They have made high-precision measurement both routine and reliable, permitting precision manufacturing to flourish and produce many of the high-quality products that consumers enjoy. Examples of electronic measuring devices include the three-axis coordinate measuring machine (Figure 20.1) and many types of digital electronic instruments and gages used for precision measurement (Figure 20.2).

FIGURE 20.1
The three-axis coordinate measuring machine used in auto body dimensional inspection (DaimlerChrysler Corporation).

FIGURE 20.2
Electronic comparators and digital height master (The L.S. Starrett Company).

This chapter discusses measurement and inspection processes along with the gages and other instruments employed.

INSPECTION

Because manufacturing operations involve varying degrees of precision and create products of many different sizes and shapes with both internal and external features, many different inspection instruments are needed. These range from simple plugs, rings, and templates to semi-precision tools such as steel rules and protractors and to very precise measuring instruments such as micrometers, height gages, and coordinate measuring machines.

Measurement in manufacturing processes is commonly confined to size specifications and locations. Frequently used expressions of size include width, depth, outside and inside diameter, center-to-center distance of part features such as holes, and wall thickness of tubes and pipes (Figure 20.3). In addition, form specifications (Figure 20.4) such as angles, perpendicularity (squareness), flatness, straightness, concentricity (or eccentricity), and surface roughness are encountered in manufacturing operations, although these require special measuring devices.

COMPARISON INSPECTION

Many inspection operations are performed in production merely for the purpose of determining conformance to tolerances, and an actual reading of the part size is not needed. Here the techniques of **comparison inspection** can be applied. In comparison inspection, dimensions are checked against tolerances by simply comparing the part to an inspection tool (gage) that is known to be the correct size. If the gage fits, then the part dimension is in tolerance. If the gage does not fit, then the part dimension is out of tolerance. In many situations it is not necessary to know how far the part is out of tolerance, since any amount out of tolerance renders the part unacceptable; however, there may be a possibility of reworking rejected parts to restore them to design specifications. This is done in a secondary rework operation.

In most cases **fixed gages** are used for comparison inspection. These inspection tools do not display a reading; they simply fit or do not fit. Acceptance or rejection of the part feature is decided purely on this basis. Comparison inspection is often referred to as GO and NO-GO or GO/NO-GO gaging. Gages used in this

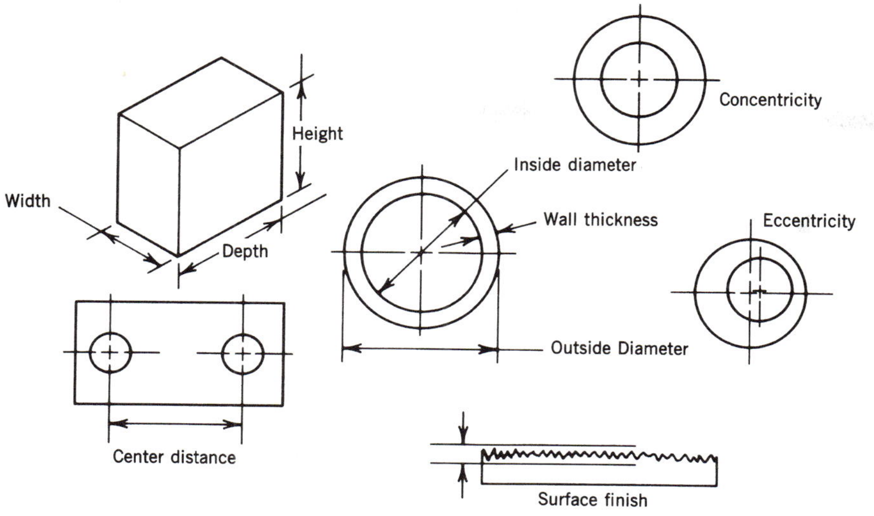

Height
Width
Depth
Concentricity
Inside diameter
Wall thickness
Eccentricity
Outside Diameter
Center distance
Surface finish

FIGURE 20.3
Common size and form specifications.

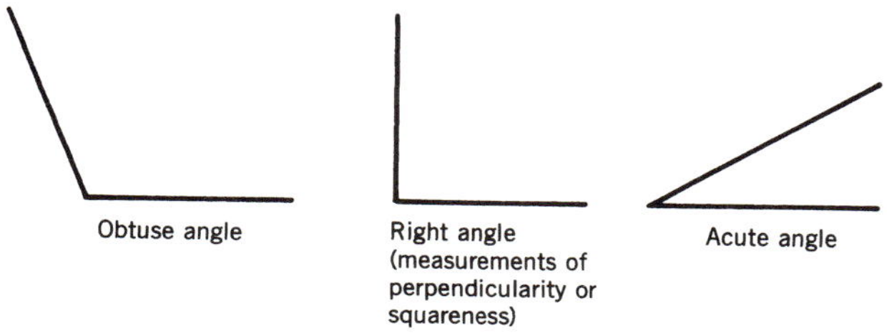

Obtuse angle
Right angle (measurements of perpendicularity or squareness)
Acute angle

FIGURE 20.4
Angular quantities must be measured.

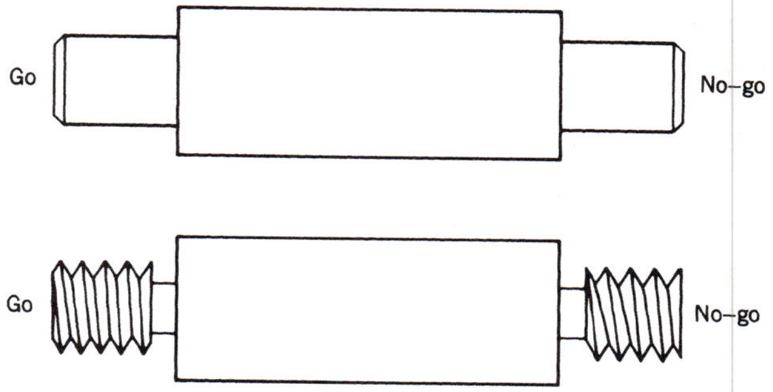

Go
No-go
Go
No-go

FIGURE 20.6
Plug and thread gages.

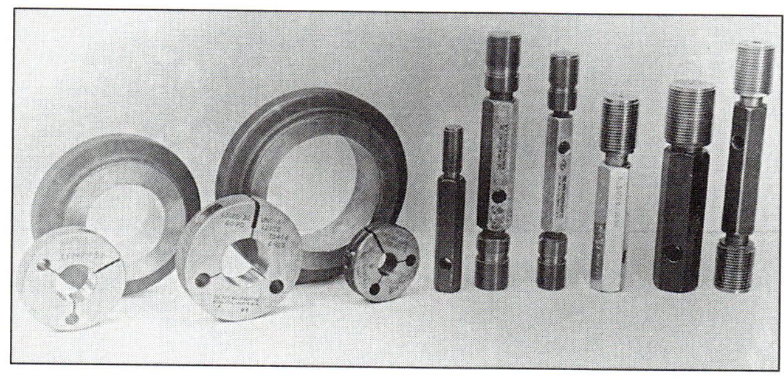

FIGURE 20.5
Fixed GO and NO-GO gages for checking tolerances.

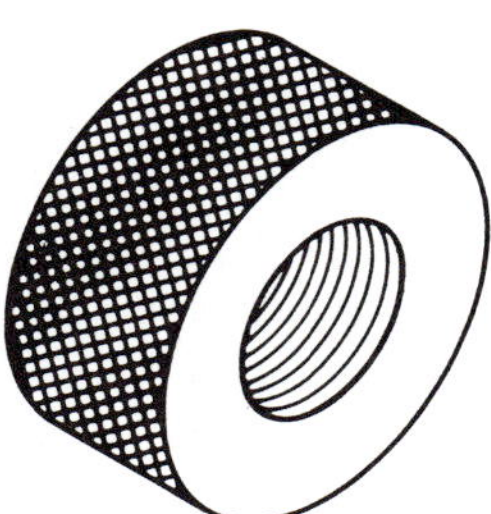

FIGURE 20.7
Ring gage for checking threads.

type of inspection are known as GO and NO-GO gages (Figure 20.5).

Typical GO/NO-GO gages are plug gages (Figure 20.6) used for comparison inspection of internal diameters and ring gages for external diameters (Figure 20.7). An alternative gage for quick GO/NO-GO comparison inspection of outside diameters is the snap gage (Figure 20.8).

Some additional examples of GO/NO-GO gages are thread plug gages and thread ring gages. The thread plug gage is widely used in checking tolerances of internal

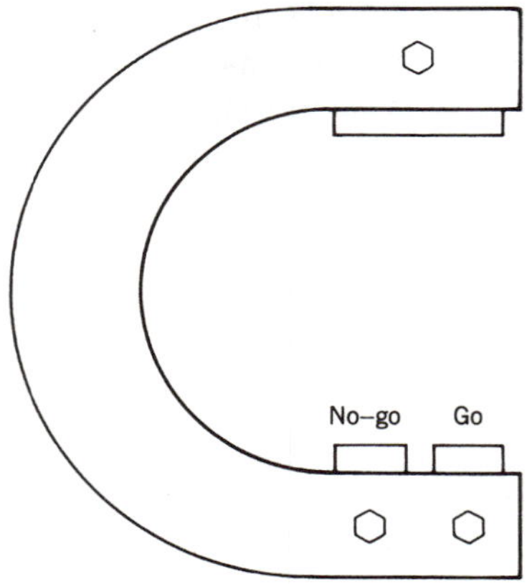

FIGURE 20.8
Snap gage. Such tools are useful for making a quick check of dimensions.

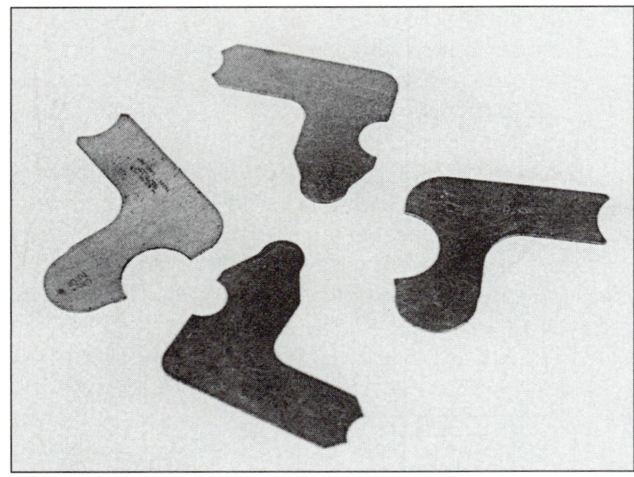

FIGURE 20.10
Radius gages.

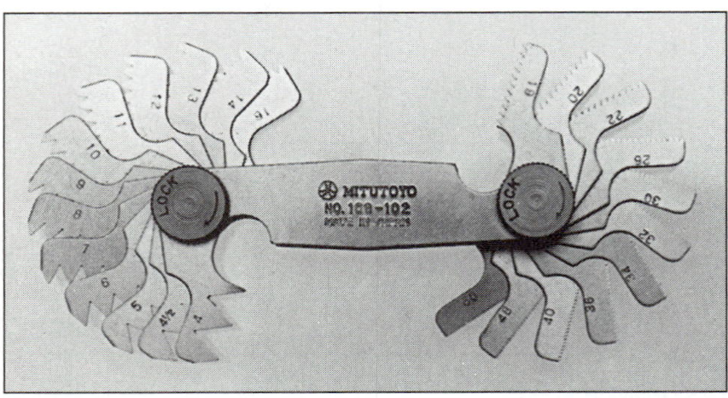

FIGURE 20.9
The screw-pitch gage is used to determine thread pitches (Mitutoyo America Corporation).

threads. One end of the gage, marked the GO end, will screw into the thread if the thread is in tolerance. The other end of the gage, marked NO GO, will not screw into the thread if the thread conforms to the tolerance. The worker has a comparison tool that can be quickly applied to determine whether thread dimensions are within tolerances.

Screw-pitch gages (Figure 20.9) are used to determine thread pitches (distance between threads), and radius gages (Figure 20.10) are used to check the form of arcs of specific radii on manufactured parts.

There are several advantages to comparison inspection. Operation of the gage is extremely simple, so variations introduced by differences in worker skill are minimized. The worker is not required to read or interpret any numeric data, eliminating another source of error. Although precision inspection tools are expensive and often represent a large portion of production tooling cost, the gages used for comparison inspection are simple, durable, and relatively inexpensive when compared with other types of inspection equipment.

A disadvantage of fixed gaging for comparison inspection is that a separate gage is required for each different size that is to be inspected. This may cause a large expense if gages must be purchased for many different dimensions. The primary disadvantage of comparison inspection, however is that very little information about the part is gathered other than whether it is in tolerance (accept) or out of tolerance (reject). A measurement of feature size would be helpful for controlling the manufacturing process that created the feature, but no size reading is obtained in a comparison inspection operation.

MEASUREMENT, PRECISION, AND RESOLUTION

Precision is a complex issue that is critical to any measuring process. Measurement precision involves the resolution of the measuring instrument, the reliability and repeatability of the instrument, the condition of the surfaces of the workpiece, the skill of the worker operating the instrument, the temperature, and any other environmental factor that may affect the reading. *Precision* can be defined as a range of readings within which the process will measure repeatably. Determining the precision of a measuring operation is difficult and requires a statistical study, but precision, or repeatability, can be estimated by measuring a feature on a part several times and figuring the difference between the highest and lowest readings. For example if a part was measured in one location using the same instrument 100 times, and the readings varied from a low of 0.513 in. to a high of 0.519 in., then the precision (repeatability) of this measuring operation could be estimated as 0.519 − 0.513 = 0.006 in. If the instrument or any environmental factors are changed, the precision of the measuring operation may also change.

The *resolution* of a measuring instrument is easy to determine: it is the smallest deviation that the instrument can detect. This is equal to the smallest graduation on the scale or readout of the instrument. The precision of a measuring operation can never be finer than the resolution of the instrument utilized. An instrument with 0.001 in. resolution can never measure to 0.0005 in. precision or any other precision finer than 0.001 in. Do not be fooled, however, into believing that the precision is always equal to the resolution of the instrument. Due to many factors that influence the measuring process, the precision is often more coarse (larger) than the resolution of the instrument.

The precision of a measuring operation must have a very close relationship to the tolerance for the dimension being measured. The commonly used rule requires that the measurement precision must be no larger than one-tenth of the tolerance range. This is a critical rule that must be considered in selecting a measuring instrument and designing a measuring process.

Example:

The length dimension of a part is shown on the blueprint as 0.905 to 0.915 in. (0.910 ± 0.005 in.).

The tolerance range is 0.915 − 0.905 = 0.010 in.
The required measuring precision is 0.010 ÷ 10 = 0.001 in.

The maximum permitted variation in the entire manufacturing process is 0.010 in. precision, and the maximum permitted variation in the measuring process is 0.001 in. precision. This assures that most of the manufacturing variation permitted according to the blueprint will be reserved for the manufacturing process that creates the product feature. The production process may vary 0.010 − 0.001 = 0.009 in., but the measuring process may vary only 0.001 in.

MEASUREMENT OF DIMENSIONS IN MANUFACTURING OPERATIONS

Measurement involves the use of a measuring instrument to obtain a numerical reading of the length, width, thickness, roundness, flatness, and so forth, of a feature of a product. The reading can be compared with the tolerance to judge conformance, and the measurement may be used to adjust the manufacturing process to assure that it continues to produce conforming products. The selection and application of the wide range of tools and equipment used in dimensional measurement is one of the largest and most interesting manufacturing support activities. Machining parts to tolerances of a few millionths of an inch has become commonplace in today's manufacturing environment, and the requirement for precision measurement has grown in lockstep.

Measurement Units

The industrial world, for the most part, has adopted the metric system of measurement known as the Système International d'Unites or the International Metric System (SI). This system is based on the meter (m), which is 39.37 in. long. In manufacturing almost all metric linear dimensions are measured in millimeters (mm). The United States is committed to conversion to the metric system from the English system of feet and inches, and this conversion is being gradually accepted in the United States. Many U.S. manufacturing plants and machine shops continue to use the English system of measurement, so machinists, manufacturing technologists, and engineers often refer to conversion tables or use conversion factors such as the following:

1 in. = 25.4 millimeters (mm)
1 mm = 0.03937 in.

Measuring Tools

Gages and instruments used for measurement of dimensions in manufacturing operations may be separated into three major categories: direct-reading instruments, deviation-type gages, and transfer-type instruments.

Direct-reading instruments have a graduated scale, and measurements are read directly from that scale. Measuring instruments are only as reliable as the user's skill, and accuracy can be affected by the following:

1. Line matching. This requires good eyesight.
2. Parallax in reading. An angled view of matching lines or a measured object can cause an erroneous reading.
3. Deformation of the workpiece or instrument. For example, putting too much pressure on an instrument when measuring a part may stretch or spring the instrument and distort the reading.
4. Temperature effects. A workpiece or measuring tool can expand and contract with temperature changes, thus altering the reading.

Deviation gages have a graduated scale, but the displayed reading does not include the entire part size. The reading represents the difference between the size of a master and the size of the measured feature. For example, if the dimension 0.750 ± 0.010 in. (Figure 20.11) requires

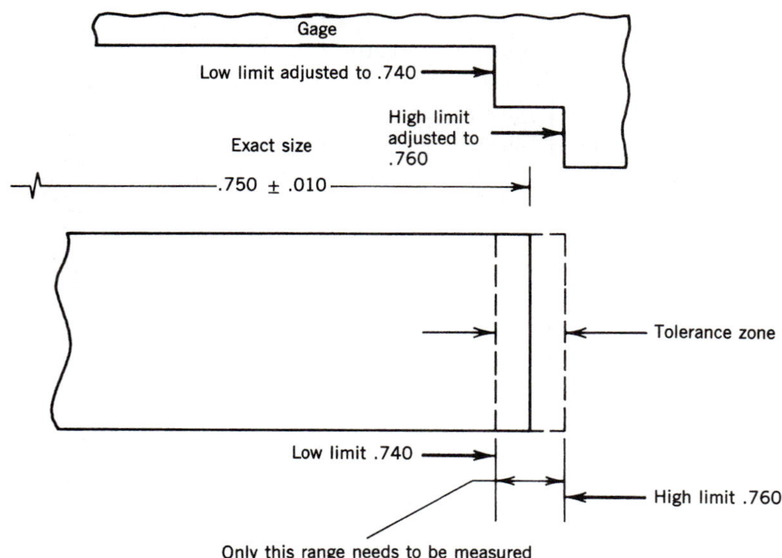

FIGURE 20.11
In production measurement, only the tolerance portion of a dimension must be measured.

Gage
Low limit adjusted to .740
High limit adjusted to .760
Exact size
.750 ± .010
Tolerance zone
Low limit .740
High limit .760
Only this range needs to be measured

FIGURE 20.12
Decimal rule (Kibbe, Neely, Meyer, and White, *Machine Tool Practices,* 7th ed. Reprinted by permission of Pearson Education, Inc. Upper Saddle River, NJ).

On the 50th scale, each inch is divided into 50 equal parts with each part equal to 1/50 or .020 (twenty thousandths of an inch). The scale is also marked at each 1/10 increment for easier reading(1/10= 100 thousandths or .100)
On the 100th scale, each inch is divided into 100 equal parts with each part equal to 1/100 or .010 (ten thousandths). The scale is also marked at each 1/10 increment for easier reading.

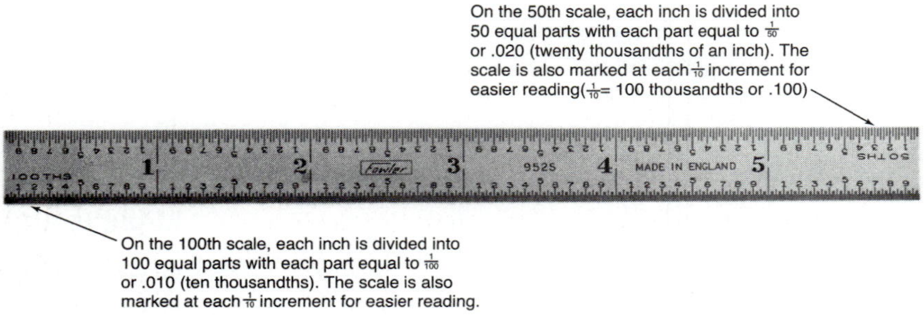

inspection, and the production process yields parts that are all close to this size, it is not necessary to inspect the 0.750-in. portion of the dimension each time. Only the tolerance of ± 0.010 needs inspecting, so it is not necessary to use a measuring tool that has the capability to measure over the full range of the 0.750-in. dimension. Although it would be acceptable to measure this dimension with such a tool, a production gage designed and adjusted to check only the deviation in size from a 0.750-in. master would be much faster, more reliable, and easier to use.

Transfer measuring tools do not have a graduated scale but are used to gage and transfer the dimension to a direct-reading instrument that produces a reading. Telescoping hole gages are good examples of transfer measuring tools.

Semiprecision Measuring Tools

Measuring tools such as machinists' combination sets and machinists' steel rules should be considered to be semiprecision tools. The steel rule (Figure 20.12) can be read with average eyesight to an accuracy of 1/64 in., and a metric rule, 0.5 mm. Inch system rules are made in four fractional divisions, 1/8, 1/16, 1/32, and 1/64 in. in addition to 1/100- and 1/10-in. decimal divisions. Metric rules normally have 1-centimeter (cm), 1-mm, and 1/2-mm divisions. Machinists' rules are often used with a square head or combination set (Figure 20.13). Squareness and angularity are measured with the square head and the protractor of the combination set, respectively. The precision of the protractor is one degree, not minute or second portions of a degree. Square heads vary in precision but often are manufactured within a few minutes of perpendicular.

Thickness gages (Figure 20.14) consist of hardened steel leaves made to specific thicknesses that are marked on each leaf in either inch or metric measure. They are commonly used to determine the dimension of small spaces between product features and can be used in conjunction with a straightedge to determine the flatness of

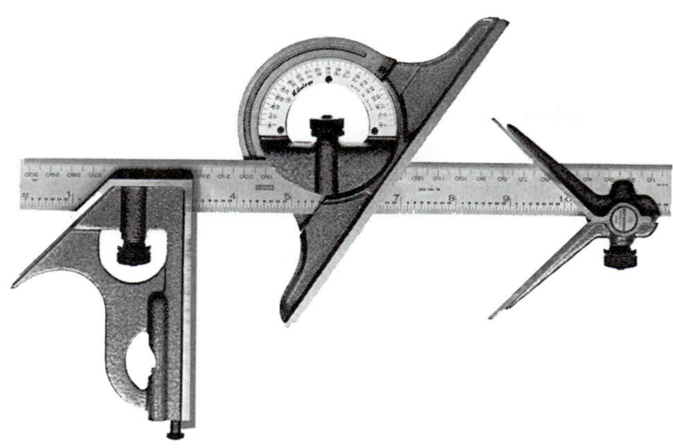

FIGURE 20.13
Combination set. A square head, protractor, or centerhead may be mounted on a graduated blade for various uses (Mitutoyo America Corporation).

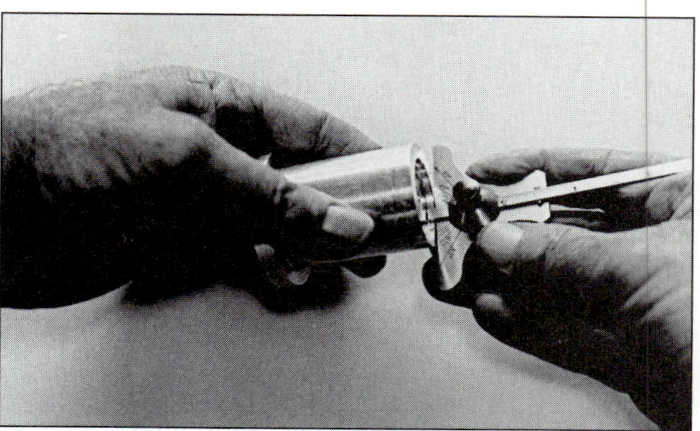

FIGURE 20.15
The rule-depth gage is a semiprecision measuring tool, used here to measure the depth of a counterbore.

FIGURE 20.14
Thickness gages, also called feeler gages, have leaves of various thicknesses (Mitutoyo America Corporation).

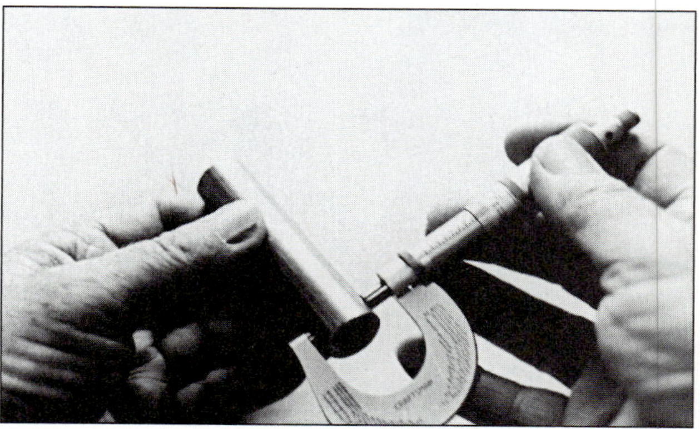

FIGURE 20.16
Small (0 to 1 in.) outside micrometer in use.

a surface. Rule-depth gages (Figure 20.15) are used to determine the depths of grooves and shoulders by the transfer method.

Tools for Precision Measurement

Precision measurement involves direct-reading instruments such as micrometers, calipers, optical comparators, coordinate measuring machines, and surface roughness analyzers. Deviation gages also are used for highly precise measurement. These include indicators, precision height gages, air gages, and electronic gages. Tools such as small hole gages and telescoping gages are required for precision transfer measurements.

Micrometers (commonly called "mikes" by machinists) are perhaps the most widely used precision measuring tools in manufacturing. Outside (caliper-type) micrometers measure lengths and outside diameters, and they typically have a measuring range of 1 in. These can be purchased in sizes ranging from 0 to 1 in. (0 to 25 mm) (Figure 20.16) to very large instruments that can measure dimensions up to several feet (Figure 20.17). Depth micrometers are used to make precision measurements of the depths of slots and shoulders (Figure 20.18). Screw-thread micrometers (Figure 20.19) have interchangeable anvils for different thread pitches and a 60° point on one anvil, making possible the precision measurement of 60°

FIGURE 20.17
Large outside micrometer (The L. S. Starrett Company).

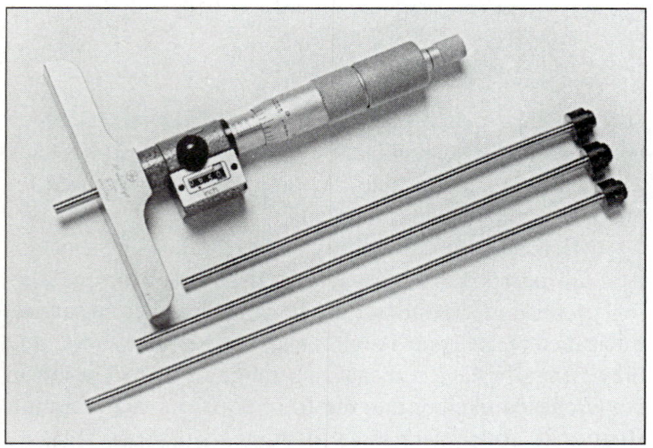

FIGURE 20.18
This depth micrometer reads in thousandths of an inch on the thimble and in a digital readout counter. It has interchangeable rods so the range can be extended (Mitutoyo America Corporation).

FIGURE 20.19
Screw-thread micrometer that uses different interchangeable anvils (Mitutoyo America Corporation).

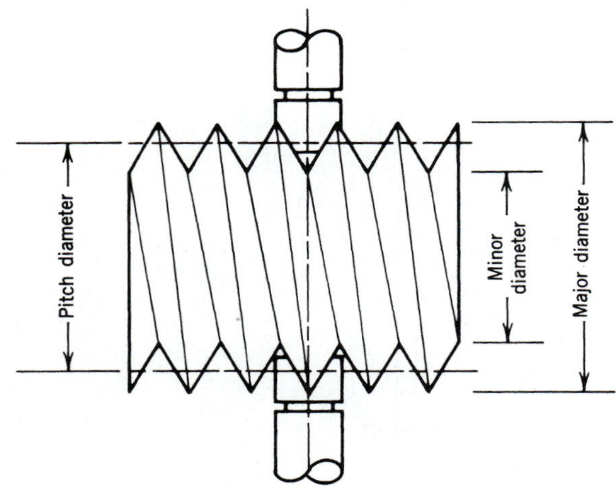

FIGURE 20.20
Thread-pitch diameters are measured on the flank of the threads.

screw-thread **pitch diameters** (Figure 20.20). Inside micrometers (Figure 20.21) have the same barrel and graduated thimble as outside or caliper-type micrometers but are used to measure internal features such as bore diameters.

Micrometers typically measure in 0.001-in. or 0.01-mm increments (resolutions), but digital micrometers and micrometers equipped with a special **vernier scale** can have 0.0001-in. or 1-micron (μm) resolution. The purpose of the vernier is to mechanically magnify tiny variations in size so that they can be read by the human eye. The vernier divides instrument graduations into smaller increments by matching them against a vernier scale.

Dial-type and digital readout **calipers** are quick and easy to read. Calipers are very versatile instruments that

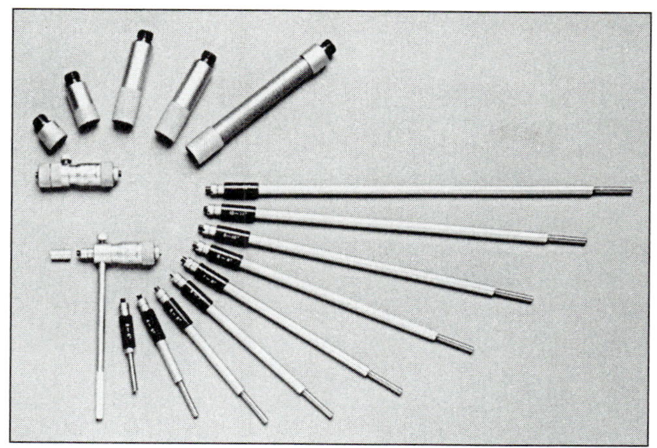

FIGURE 20.21
Inside micrometer set with interchangeable rods that give it a wide range of measurement (Mitutoyo America Corporation).

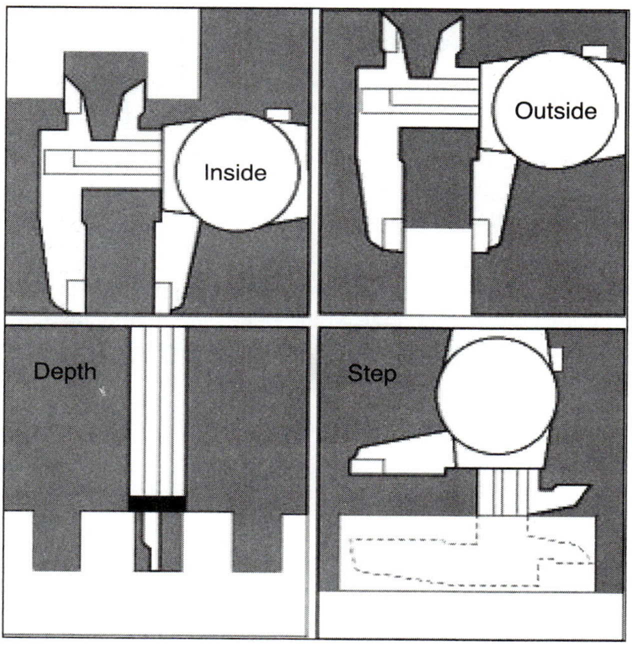

FIGURE 20.22
Dial calipers are used to measure outside diameters (Brown and Sharpe).

can measure outside diameters, inside diameters, and depths, but they cannot measure as precisely as micrometers (Figures 20.22 to 20.24). Dial calipers are capable of discriminating to 0.001 in. or 0.02 mm resolution, and digital calipers read 0.0005-in. or 0.01-mm increments.

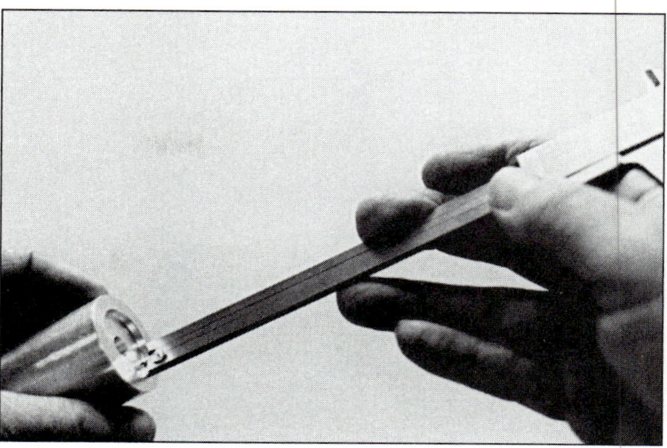

FIGURE 20.23
Caliper measuring depths.

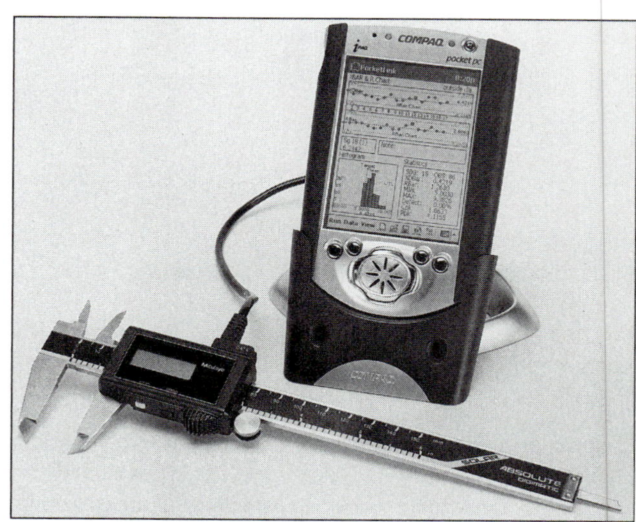

FIGURE 20.24
Digital electronic calipers with data collection device (Mitutoyo America Corporation).

Deviation gages used in precision measurement include mechanical dial indicators (Figures 20.25 and 20.26) and test indicators (Figures 20.27 and 20.28) in addition to electronic digital indicators. Indicators are precision instruments that detect minute movements of a plunger or pivoting needle. Dial and test indicators can detect deviations of 0.001 or 0.0001 in. or equivalent metric measures. Digital indicators typically have finer resolution of 0.00005 in. or one micron (.001 mm). Indicators have many uses, from measuring size dimensions such as height using a surface plate to checking

FIGURE 20.25
Dial indicator (Mitutoyo America Corporation).

FIGURE 20.26
Dial indicators are often mounted on magnetic stands so that they can be fastened where necessary to machine parts for taking measurements (Mitutoyo America Corporation).

form specifications such as **eccentricity** (runout). Indicators are often used with a precision height master (Figure 20.29) or mounted as dial indicator comparator stands (Figures 20.30 and 20.31). Many types of indicating gages are widely used for production measurement (Figure 20.32). The digital comparator, the electronic counterpart of the dial indicator comparator, can measure in millionths of an inch resolution (Figure 20.33).

Air gages (Figures 20.34 and 20.35) and electronic gages are more precise deviation-type gages that are used to measure inside diameters and other precision features. Air gages and electronic gages are used when many parts must be very precisely measured.

Precision transfer-type instruments include telescoping gages (Figures 20.36 and 20.37), small-hole gages for measuring hole diameters (Figure 20.38), and adjustable parallels to gage slot widths that are difficult to measure (Figure 20.39) (p. 386). All these gages adjust to the size of the feature and then are measured using a direct reading instrument to produce a reading.

FIGURE 20.27
Test indicator (Mitutoyo America Corporation).

FIGURE 20.28
A test indicator used in a deviation-type measurement (Griffith, *Geometric Dimensioning and Tolerancing,* 2nd ed., ©2002. Reprinted by permission of Pearson Education, Inc., Upper Saddle River, NJ).

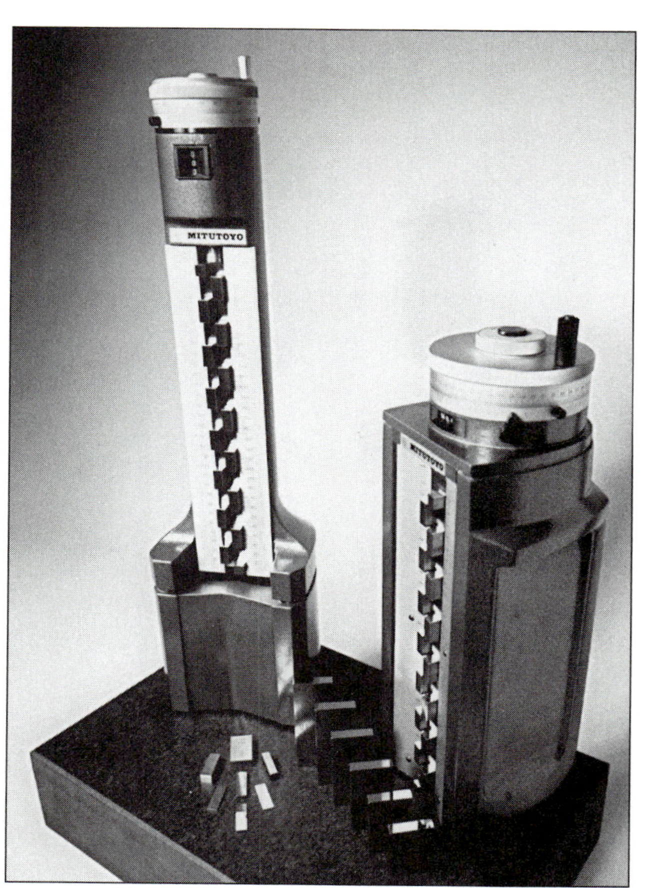

FIGURE 20.29
Precision height master (Mitutoyo America Corporation).

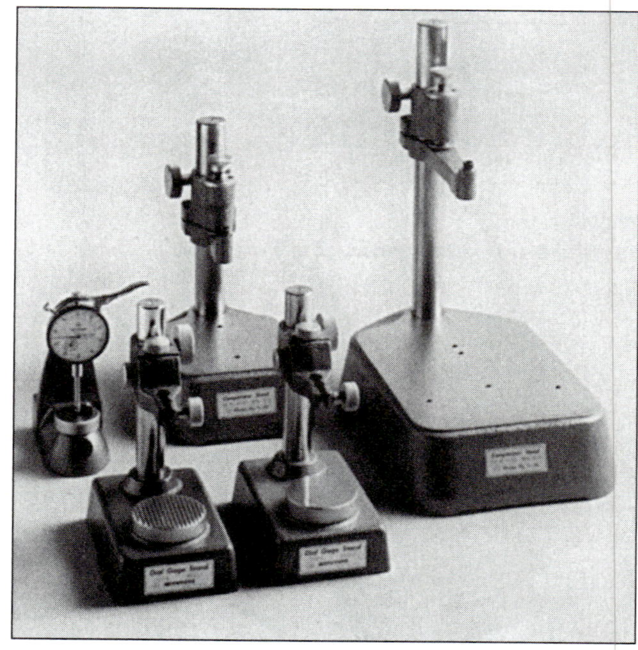

FIGURE 20.30
Mechanical dial indicator comparator stands (Mitutoyo America Corporation).

FIGURE 20.31
Dial indicator comparator stand. Small parts can be quickly measured with this tool (Mitutoyo America Corporation).

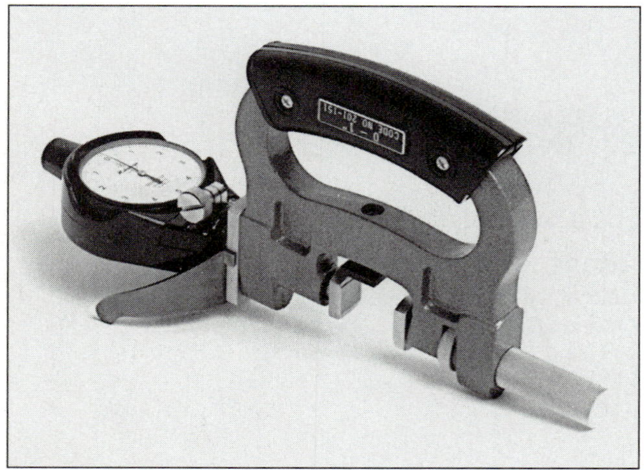

FIGURE 20.32
Dial indicator snap gage for fast and accurate measurement in a production situation (Mitutoyo America Corporation).

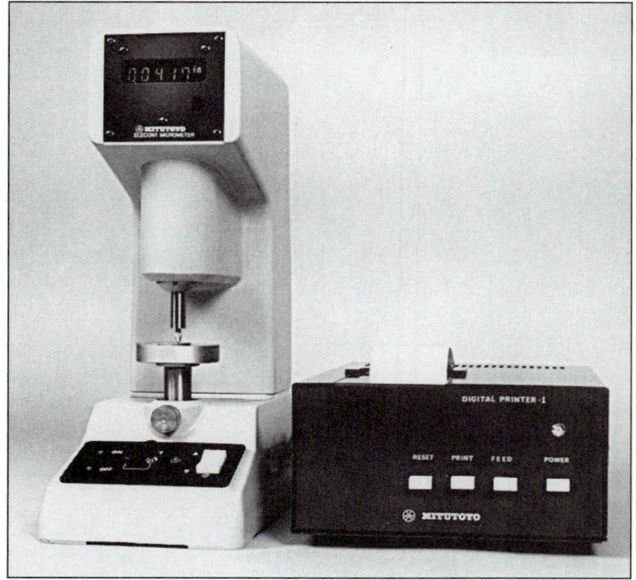

FIGURE 20.33
Digital electronic comparator (Mitutoyo America Corporation).

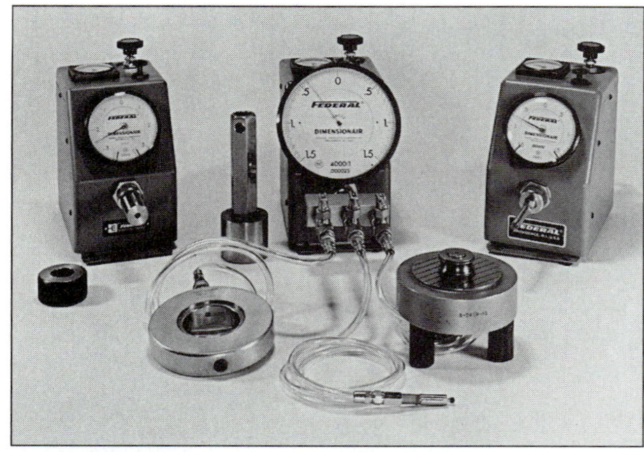

FIGURE 20.34
Air gages measure dimensions as a function of air flow and pressure (Federal Products Corporation).

Form and Location Measurement

The importance of precision measurement of size dimensions is well understood, but measurement of form and location specifications is equally important. An example is two parts mated by a four-hole bolt pattern. Regardless of the size of the holes, if one or more holes are located incorrectly, or if the holes are not

FIGURE 20.35
The air gage measures clearances between bores and nominal shaft sizes. The setting ring is used for adjustment purposes (Federal Products Corporation).

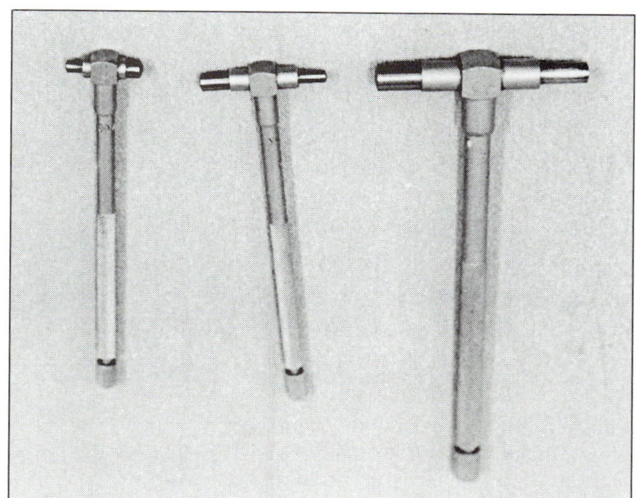

FIGURE 20.36
Telescoping gages.

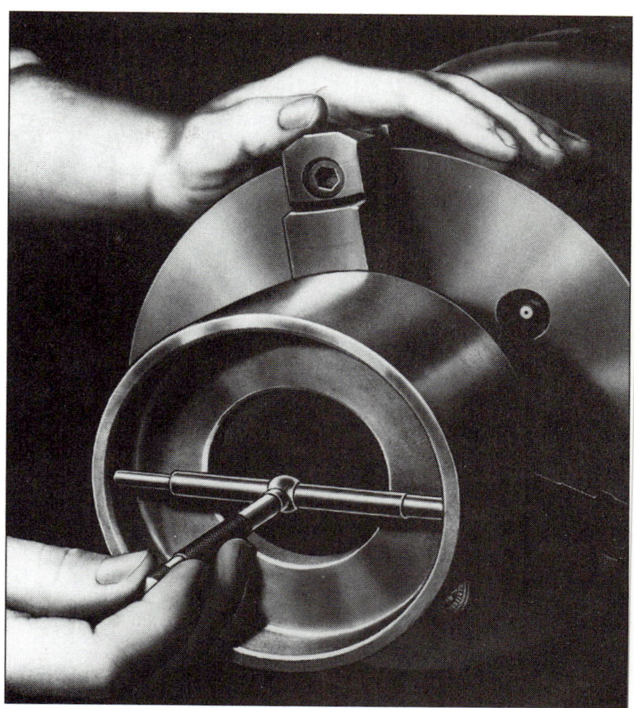

FIGURE 20.37
Telescoping gages used to transfer a dimension from a bore (Kibbe, Neely, Meyer, and White, *Machine Tool Practices,* 7th ed. Reprinted by permission of Pearson Eduction, Inc., Upper Saddle River, NJ).

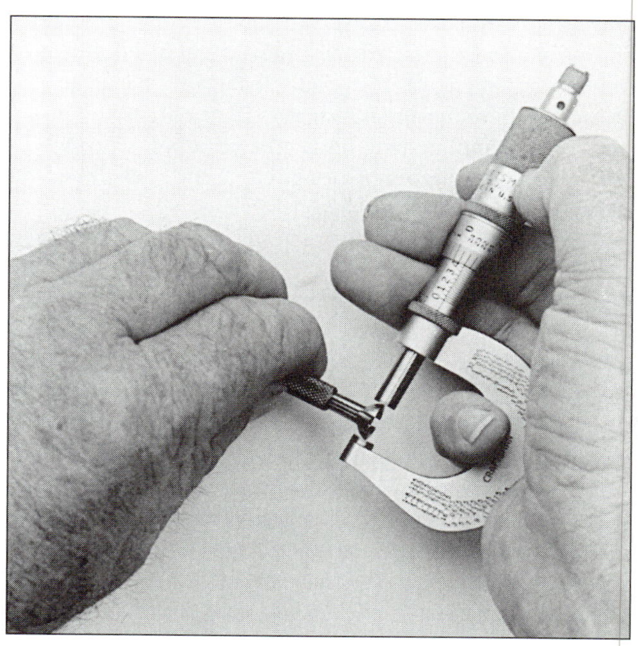

FIGURE 20.38
Small-hole gage being measured with a micrometer.

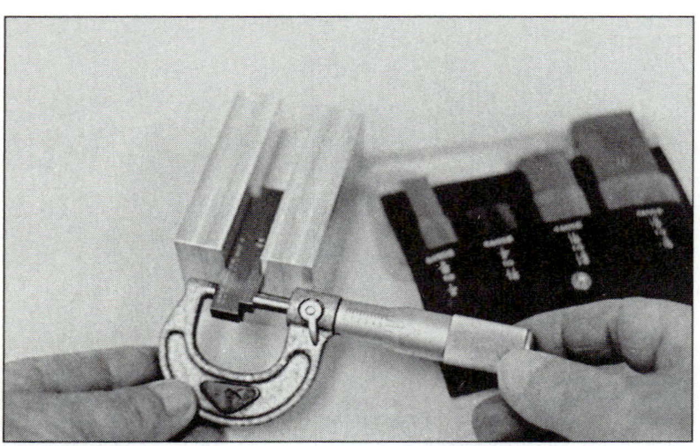

FIGURE 20.39
Adjustable parallel is measured with a micrometer.

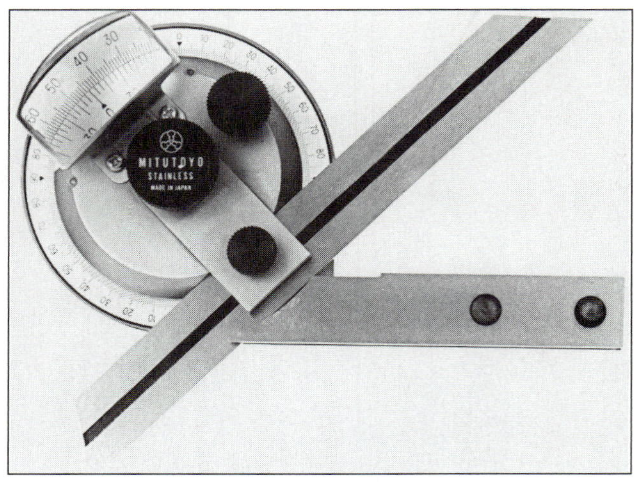

FIGURE 20.40
Universal bevel protractor (Mitutoyo America Corporation).

cylindrical (incorrect form) the parts may not fit together properly. Form measurements often have been shortcut or ignored in the past due to their painstaking, time-consuming, and error-prone nature. The primary tools for form and location specification measurements have been the indicator, surface gage, flat surface plate, and 90° angle plate. The complex setups and procedures required in using these tools to measure perpendicularity, flatness, straightness, roundness, cylindricity, concentricity, feature position, and other specifications were costly and required consummate skill.

The advent of the coordinate measuring machine and other electronic measuring instruments has greatly simplified and increased the precision and reliability of the measurement of form and location specifications. The corresponding improvements in quality and conformance in manufacturing have created a breakthrough in performance, reliability, longevity, and aesthetics in today's products.

The universal bevel protractor is a precision tool used for measuring angles (Figure 20.40). It is more accurate than the combination set protractor, which can measure only in degrees. The bevel protractor has a vernier scale that makes possible a reading within the range of 5 minutes of arc.

The optical comparator (also called a profile projector) (Figure 20.41), in which a magnified image of the part being inspected is projected onto a screen, precisely measures size dimensions that are not accessible to other instruments and is very useful for measuring angles and inspecting form specifications (Figure 20.42).

The toolmaker's microscope has a two-axis stage upon which parts are moved and viewed under high mag-

FIGURE 20.41
The optical comparator projects a greatly magnified image of the part onto a large screen (Mitutoyo America Corporation).

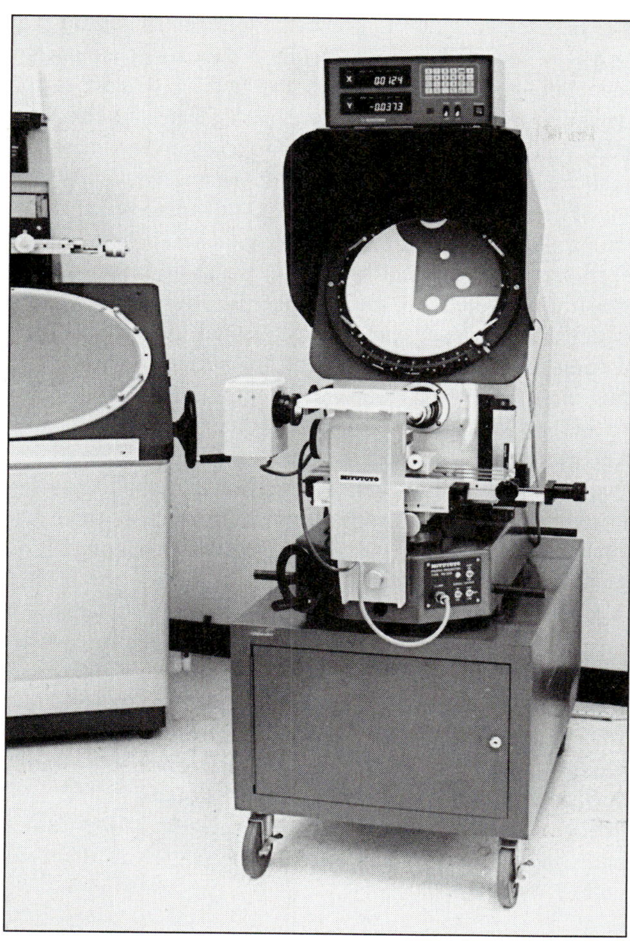

FIGURE 20.42
Optical comparator or profile projector can be used for precision measurement or comparing curves and radii (Mitutoyo America Corporation).

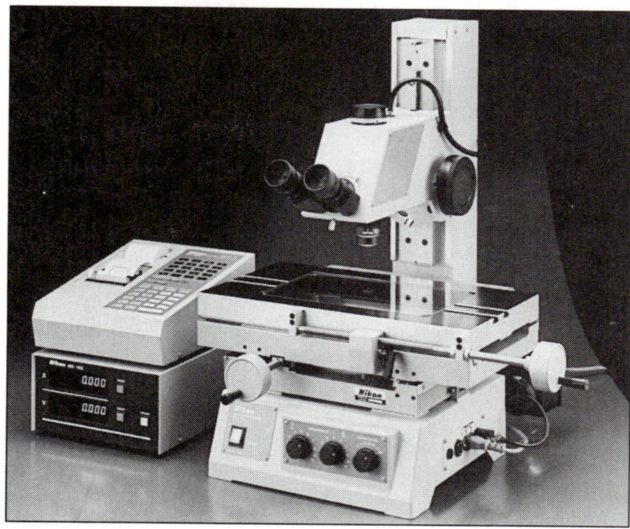

FIGURE 20.43
Toolmaker's microscope, for a magnified look at dimensions (Nikon Inc., Instrument Division).

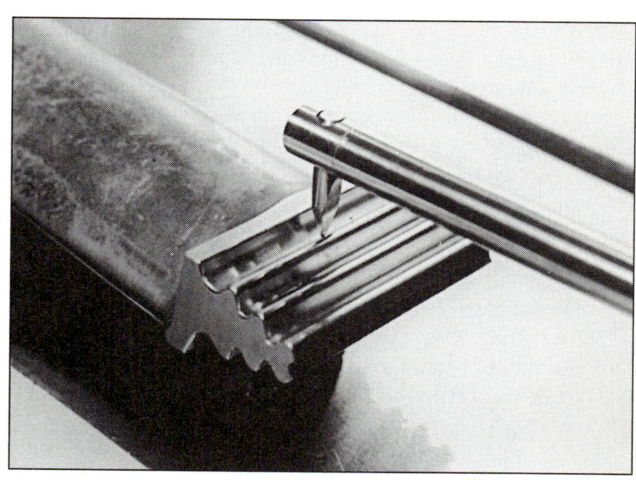

FIGURE 20.44
Measuring profile geometry electronically (Mitutoyo America Corporation).

nification (Figure 20.43). Electronic tracking of the movement of a stylus as it traces the workpiece surface also can be used to measure the two-dimensional (profile) geometry of machined surfaces (Figure 20.43). Other form measurement procedures include inspecting and graphing part geometry such as roundness (Figure 20.44).

The **coordinate measuring machine (CMM)** consists of a probe mounted on sliding ways that permit free motion in three-dimensional space (Figure 20.46). When the probe touches the workpiece the position of each of the three axes is communicated to an attached computer, which calculates the position of the probe. Calculation of the relationship between multiple point locations on the part that has been probed allows features to be measured quickly, easily, and reliably. Feature size may be measured using coordinate measuring machines, but the most important capability of the CMM lies in measuring form and location dimensions such as flatness, roundness, feature position, parallelism, and many other geometric features without the complexity and potential errors of difficult surface plate setups. For example, the measurement probe

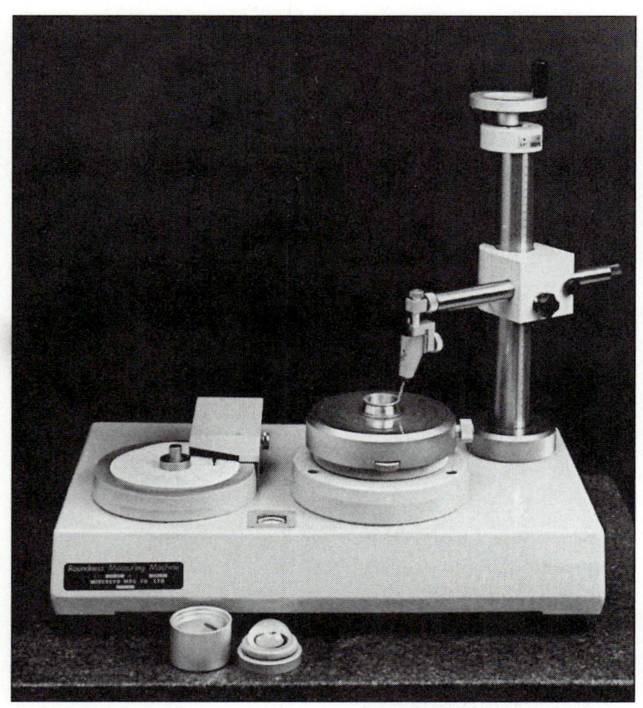

FIGURE 20.45
Measuring and graphing roundness, a form inspection (Mitutoyo America Corporation).

can be used to quickly touch multiple points on the part being inspected. Using the position data for these points, the computer can calculate a location or determine the deviation from a geometrically perfect feature. The process is extremely fast and reliable, and the computer can automatically generate reports of measurement results.

The smoothness of the surface of a machined part is called **surface finish** (surface roughness). When a cross section of a machined surface is magnified, the surface finish looks like a mountain range, a series of peaks with valleys in between, each valley created by the tip of the cutting tool as it removes material. Surface roughness is measured in microns (metric) or microinches (millionths of an inch) from a smooth 32 microinches to a coarse 500 microinches. Ground or polished finishes are smoother than 32 microinches, and a surface with roughness of 4 microinches has a mirrorlike polished finish. Surface finish is vital to many parts such as gears, bearings, and hydraulic sealing surfaces.

Surface roughness can be inspected by comparison or measured directly. In comparison inspection the surface roughness can be estimated by matching a workpiece surface with blocks that display a series of increasingly smooth surfaces (Figure 20.47). The profilometer surface measuring instrument is used for precise measurement of surface finish (surface roughness) (Figures 20.48 and 20.49). The profilometer directly measures surface finish as it traces, records, and displays the height of the peaks and depth of the valleys.

FIGURE 20.46
Three-axis coordinate measurement machine with computer readouts and printer (Sheffield Measurement Division).

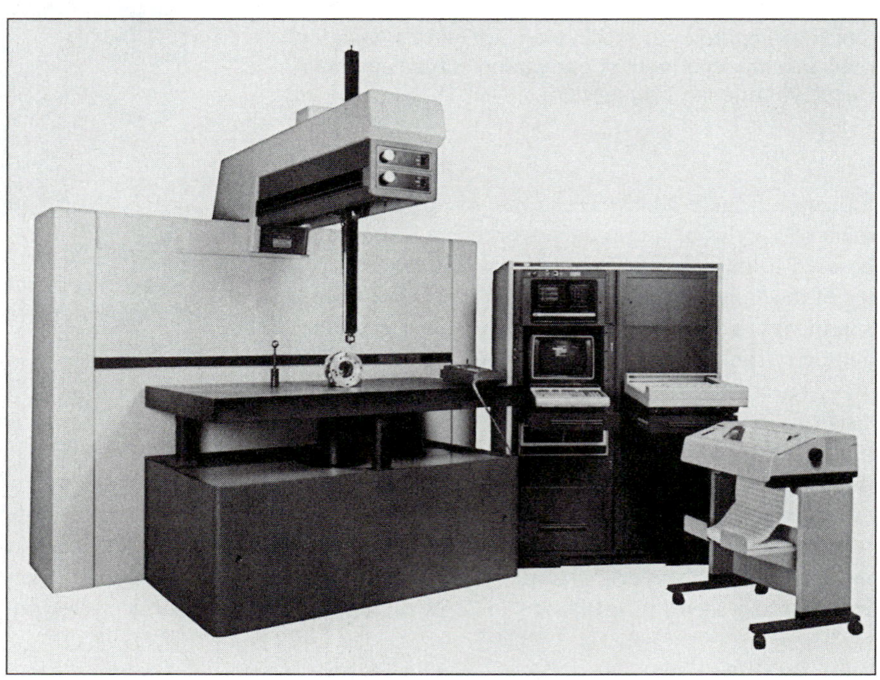

FIGURE 20.47
Roughness comparison standards.

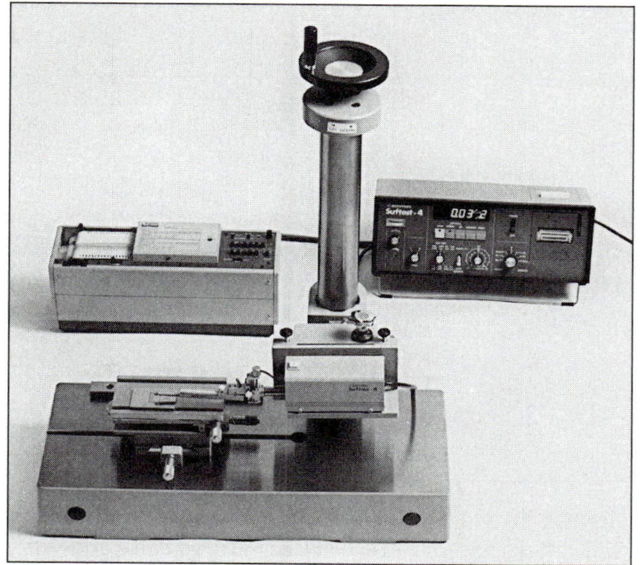

FIGURE 20.48
Surface finish, a measurement of length in millionths of an inch (Mitutoyo America Corporation).

Effects of Temperature Changes

All materials, especially metals, expand or contract with temperature changes, and this size change can affect measurements. Whether the workpiece is made of steel, aluminum, or other materials any precision measurement is meaningless if the measuring temperature is not known, so a standard measuring temperature of 68°F has been established worldwide. Any measurement made at any other temperature has some amount of error owing to temperature effects. Care must be taken when the workpiece and measuring instrument are made of two different materials, because temperature variations will affect the two materials differently.

Extreme precision measurements are performed in temperature-controlled (68°F) gage laboratories with filtered air and controlled humidity. Workpieces and instruments must be allowed to "soak" or stabilize to gage room temperature for a period of time before measurement to assure that the standard measuring temperature has been reached throughout. As a rule of thumb for steel workpieces, if the tolerance is finer than 0.002 in., then normal workplace temperatures can significantly affect common measurements. Because aluminum expands more than steel with temperature changes, measurements of aluminum workpieces with tolerances finer than 0.004 in. are subject to the same temperature concerns. In these circumstances measurement must be performed in an environment where the temperature remains within approximately 10° of the standard 68°F measuring temperature, or better yet in a temperature-controlled environment. For large dimensions and where the temperature differs greatly from 68°F, even more care should be taken to assure that temperature does not affect the validity of measurements.

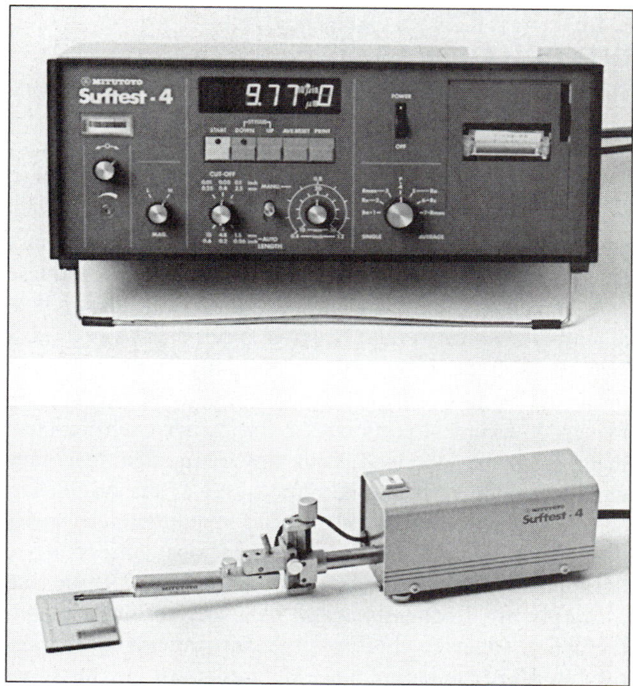

FIGURE 20.49
Surface texture analyzing instruments (Mitutoyo America Corporation).

CALIBRATION, STANDARDS, INTERCHANGEABILITY, AND TRACEABILITY

Manufacturing enterprises depend on the concept of interchangeable parts for modern interchangeable manufacturing. Few complex assemblies are made by single manufacturers. It is more likely that many different makers will participate in the process, each one contributing a part or parts of the final product. It is easy to see how assembly problems could occur if specifications regarding size as well as geometry of assembly parts were not inspected with worldwide consistency to assure conformance. Mechanical parts would not fit, nor could replacement parts be obtained with any assurance that they would fit or have the same dimensions as the original.

The **interchangeability** of manufactured parts depends entirely on workpiece dimensions conforming to design tolerances, and different parts that fit into an assembly are frequently made by many different manufacturers. To make this possible, instruments and gages used by all participating manufacturers must measure consistently, and this is accomplished through frequent calibration. Calibration of measuring tools is extremely important to the control of product quality.

Calibration is the process of comparing measuring tools against standards of known size. The process is a continuous one in almost any manufacturing industry. In a typical manufacturing plant the calibration of inspection tools is the responsibility of the gage laboratory. Calibration equipment is very precise, sensitive, delicate, and expensive and must be used in controlled environments.

Inspection tools used in the factory are sent on a regular basis to an on-site gage laboratory where precision calibration procedures are carried out in a temperature-controlled (68°F) environment. The tools are allowed to stabilize to gage room temperature, they are compared with appropriate standards, adjustments are made to bring the tools to conformity, and then the tools are tagged with a sticker indicating the date of calibration. It is the responsibility of the employees to ensure that the calibration is current for any inspection tools used on the production floor.

Calibration procedures are by no means unique to the measurement of size, location, and form. Calibration is equally important in many other areas such as electronics where frequency standards must be maintained. Calibration of equipment for reading pressures, flow, weight, colors, performance, and other specifications is required for manufactured products.

Gage blocks (Figure 20.50) are among the most common and versatile **measurement standards** used in manufacturing. Gage blocks are used in production measurement, machine setup, and calibration of inspection tools (Figure 20.51). They are used together with sine bars (Figure 20.52) to make angles as precise as 1 to 5 seconds of arc.

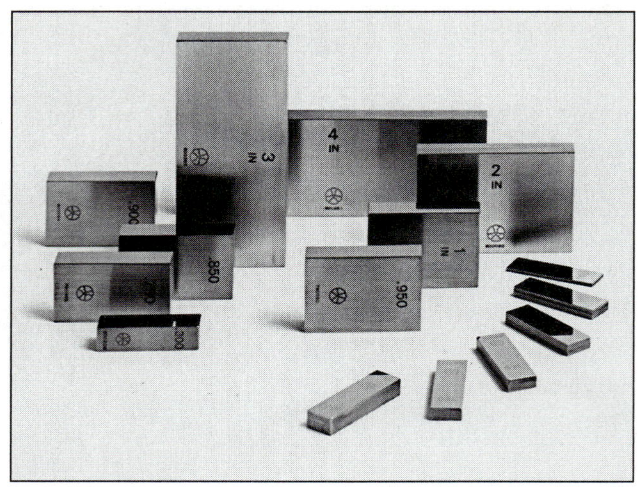

FIGURE 20.50
Gage blocks, one of the most common length measurement standards in manufacturing (Mitutoyo America Corporation).

FIGURE 20.51
Gage blocks assembled into stacks make versatile and highly accurate measuring tools (Mitutoyo America Corporation).

A gage block set (Figure 20.53) consists of a series of blocks of various heights that are made from a hard, wear-resistant material, usually hardened steel. Each block has been manufactured under controlled conditions and is dimensionally very close to its nominal size. A gage block marked 0.500 in. will be, at the standard precision measuring temperature of 68°F, within a few millionths of an inch of its marked size. These sets may be purchased in several grades of precision: the most inexpensive gage blocks are

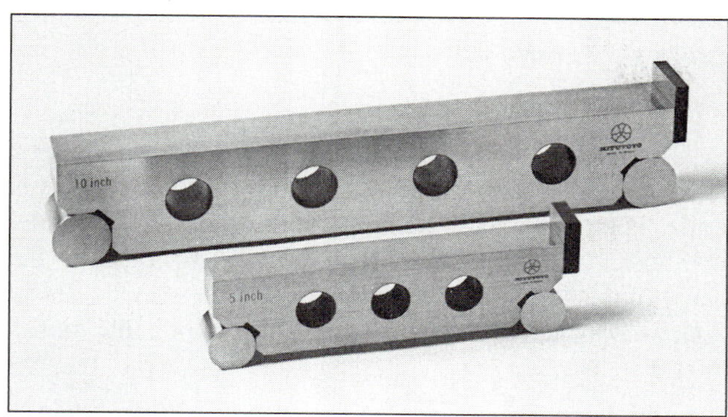

FIGURE 20.52
Precision sine bars (Mitutoyo America Corporation).

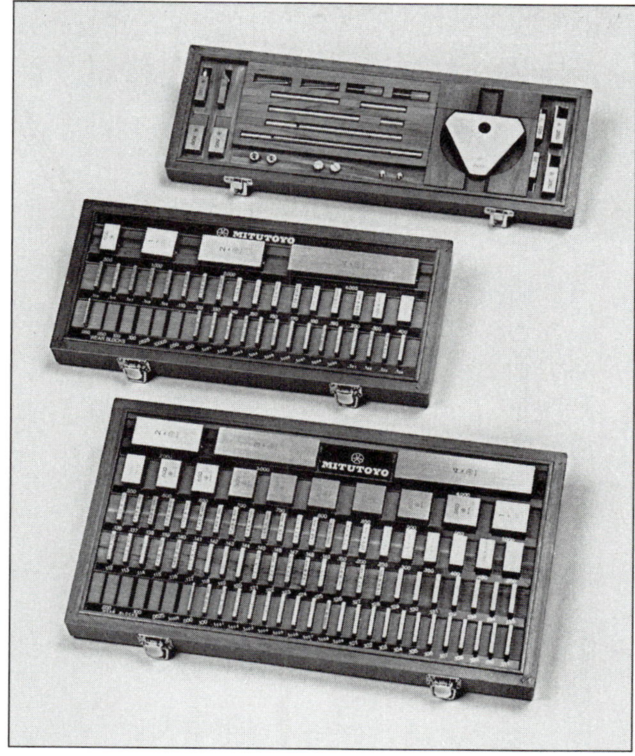

FIGURE 20.53
Gage block sets (Mitutoyo America Corporation).

within 0.000020 in. of nominal size, and laboratory-grade gage blocks will be within 0.000002 in. of nominal. Form tolerances including flatness and parallelism of the working surfaces of the block are also very close.

Nearly perfect dimensions make the gage block a very effective standard, and gage block sets are designed so that several gage blocks can be stacked to make a master of the desired length. If these gage blocks are properly "rung" together with no contaminants or air between blocks, then the stack will be extremely close to the calculated size.

In today's manufacturing environment, parts manufactured on opposite sides of the world may be required to fit together and function effectively in a product assembly. This can be accomplished only if the size standards used for calibration are precisely correlated to one worldwide length standard. This correlation is called "instrument **traceability.**" Master measurement standards are commonly established and maintained by government agencies such as the National Institute of Standards and Technology (NIST), formerly called the National Bureau of Standards (NBS), in the United States. The NIST maintains a length standard that is consistent within atomic precision with the worldwide standard. U.S. manufacturing facilities contract with regional services for calibration of gage blocks and other length standards, and these regional calibration services maintain length standards obtained from NIST. Other regions throughout the world maintain similar systems. All inspection tools used in the world can be consistent because through this procedure they are all traceable to one worldwide standard of length.

Review Questions

1. What is GO and NO-GO gaging?
2. Cite several examples of tools for production gaging.
3. What is the meaning and purpose of calibration?
4. What are used as common shop measurement standards?
5. Who is finally responsible for national measurement standards?
6. What is meant by calibration traceability?
7. How are computer and electronics used in measurement?
8. What is the purpose of final and 100 percent inspection?
9. If you had only an inch-reading micrometer and obtained a measurement on a part of .315, using the conversion factor 1 in. = 25.4 mm, what would be the metric equivalent of your inch measurement?
10. Is a machinists' steel rule considered to be a precision measuring tool, or a semiprecision measuring tool?
11. How can a precision measurement be transferred from an inside diameter of a bore to an outside direct reading instrument such as a caliper-type micrometer?

12. If an inch micrometer can discriminate to 0.0001 in. and a metric micrometer to 0.01 mm, which instrument can measure to the smallest or finest increment? Use the conversion factor 1 in. = 25.4 mm, or 1 mm = 0.03937 in.
13. Since angular measure is in degrees, minutes, and seconds, which instrument, the bevel protractor or the combination set protractor, has better precision?
14. How can surface roughness be measured?
15. When a dimension must be checked quickly by a machine operator at a production machine to determine if it is in tolerance, what type of measuring tool is used?
16. Gage blocks can have an accuracy of (a) ten thousandths, (b) hundred thousandths, or (c) millionths of an inch.

Case Problems

Case: Calibration

A machine operator, using standard production gages, checks parts as they come out of the production process. Parts appear to be in tolerance, but are later rejected in final inspection as being out of tolerance. What could be the cause of this problem and how might it be corrected? Relate your answer to the purpose and procedures of calibration.

PART IV

MANUFACTURING SYSTEMS

Part IV discusses the design, tooling, and production aspects of manufacturing; modern automation methods; how the manufacturing industry functions; job titles and general responsibilities in the manufacturing industry; and how production is accomplished.

Materials and specific processes are, of course, essential to the manufacture of any product; however, actual activities of manufacturing involve a great deal more than simply having material and processes available. In Part IV you will see how the materials of Part I and the specific processes of Part II are brought together to form a complete manufacturing capability. This capability produces the vast variety of products, ranging from individual parts to complex assemblies such as automobiles, aircraft, and consumer products that establish the lifestyle in a modern industrialized society.

Accomplishing the miracles of modern production requires the fully integrated effort of all the manufacturing team members. Products must be designed, prototyped, and tested. Production facilities and special tooling must be created. During the execution of the manufacturing process, complex series of events must be planned and managed. Lastly, the distribution and service of the product is managed.

Manufacturing may be automated so that numbers of varied product units may be made quickly, efficiently, and at competitive prices. As continuing research and development efforts devise new materials and new methods, future manufacturing will be accomplished on an ever more automated level. Individuals staffing manufacturing jobs must understand computers, industrial robotics, and fully automated factories. Training in new methods will play an ever increasing part in supplying the qualified manufacturing technicians and engineers of tomorrow.

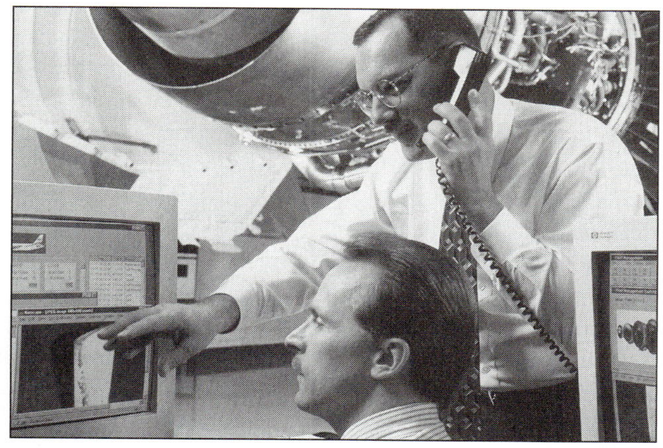

CHAPTER 21

Product and Process Design

Objectives

This chapter will enable you to:

1. Explain the phases of the design process.
2. Describe the use of a CAD system and explain its function as a modern design tool.
3. Describe the functions of production tooling and describe common examples.
4. Describe the flow of a product through the manufacturing process.

Key Words

design engineering	tool design
manufacturing engineering	pattern
industrial engineer	jig
process engineer	fixture
prototype	process routing sheet
rapid prototyping	operation detail sheet

Products seldom just materialize. Most are available only because a great deal of time and planning have been invested behind the scenes in product and process design work. The purpose of this chapter is to survey the major phases of design and to examine some of the modern tools utilized by designers. The design and documentation of the manufacturing process is discussed, along with the development of the tools and equipment necessary to support manufacturing operations.

PRODUCT DESIGN

Design is motivated by the following factors:

1. The need for a specific product
2. The level of available technology
3. Production capability and production cost
4. Safety and reliability
5. Marketability.

Need

Most products are designed and produced to meet a need. With the tremendous capability of modern manufacturing, new products are introduced daily. In fact, needs are often created through the advertising of features and options on products, inducing a prospective customer to buy a new and better model. This is especially true in the consumer goods area. Product designers are engaged in a constant process of designing and redesigning products to fulfill a continuing need.

Capability of Available Technology

The capability of the existing technology is an extremely important consideration in product design and development. The term *state of the art* is frequently heard in product advertising. This means that the product is designed and built to the latest design specifications, and from the latest available materials. As new and better manufacturing processes and materials are developed, the designer naturally makes use of these to improve designs and create advanced products.

For example, during the vacuum-tube age of electronics, handheld calculators and small portable computers with low power requirements were not available. The advent of solid-state electronics completely revolutionized this technology and permitted the electronic equipment designer to create an entirely new selection of products that were previously unavailable.

Production Capability and Costs

Most products made for the general public are successful only if they can be produced and marketed at affordable prices. The best product design is only as good as its ability to be competitively produced. Therefore, a designer must consider the following important questions when a product is being developed:

1. Is the production capability presently available?
2. If not, will sales of the product justify development of new production technology?
3. Can the product be manufactured and marketed at a cost that will return investment and profits?

These questions are complex and often occupy much of engineering and financial managers' time. In some cases the design of a product may involve the parallel development of new production capability necessary to its manufacture. This can be a costly and risky venture, but it also can be enormously profitable if successful. Manufacturing industries often dedicate large portions of their annual budgets to research and development of new products and the manufacturing systems required to produce them.

Safety and Reliability

Safety and reliability are critical factors in the design of products including aircraft, ships, medical equipment, weaponry, automobiles, and consumer products. In recent years product safety considerations have attained greater importance owing to increased public awareness resulting from improved communications and more intense interest in consumer affairs. Personal and public safety are primary concerns of the product design engineer.

Reliability and safety are closely tied. Indeed, product safety is largely determined by reliability. For example a backup electrical generator for emergency power must start and run reliably, as lives may depend on it. Aircraft control surface actuators must work reliably to ensure that the aircraft and passengers are safe, and automotive steering and braking systems must be extremely reliable. Safe products are generally reliable products, and reliable products can be designed for safety as well. These important considerations must govern the work of the designer in every design effort.

Marketability

If the needed manufacturing capability exists or can be developed, then the product may be marketed and distributed. It has often been stated that invention is one thing, but distribution is everything. This adage is very applicable to manufacturing. If a product cannot be properly designed and manufactured in quantities to meet demand and then sold at an acceptable price, it probably will not be successful. The market environments of free enterprise economies are ruthless in selecting only those products that survive the tests of efficient, cost-effective manufacturing.

Costs incurred in marketing encompass advertising, national or international sales organizations, packaging, distributors and agents, distribution costs, and warranty services. The selling price to the end customer must reflect any or all of these expenses while being low enough to support sales and meet competition.

THE PRODUCT DESIGN SEQUENCE

Product design generally begins with an idea that develops from a need determined in many cases by market research. Anyone can design; however, a manufacturer will generally employ individuals with specific engineering backgrounds in the selection and applications of materials and manufacturing specifications. Design engineers must work closely with the marketing and distribution departments in a manufacturing company; thus, product development must be an integrated and coordinated effort.

Design Phases

Product design involves the following phases:

1. Concept
2. Prototype and testing
3. Design evaluation and review
4. Manufacturing review

Concept Phase

The designer often begins by freehand drawing a sketch of the item to be manufactured. Various sketch formats are used, often in pictorial styles or artist's conceptions (Figure 21.1). These enable the designer to see what the product will look like in its finished form. Computer-aided design (CAD) can be used to aid the engineer in the initial design phase. When using a CAD system the

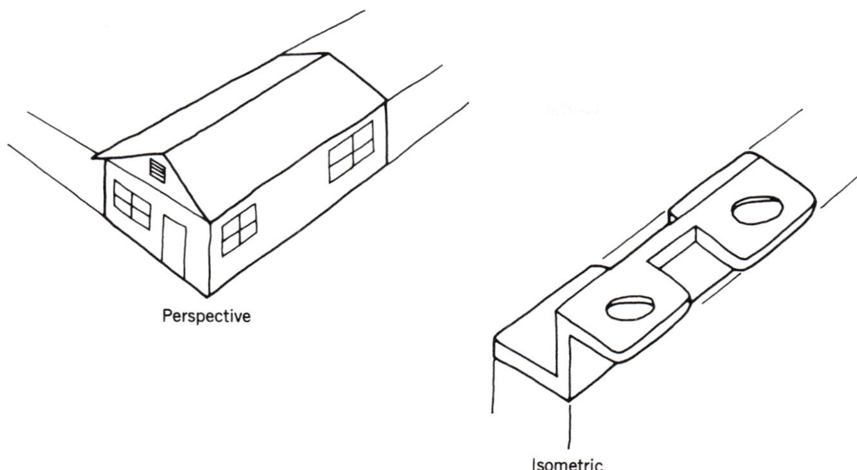

FIGURE 21.1
Freehand pictorial sketch formats.

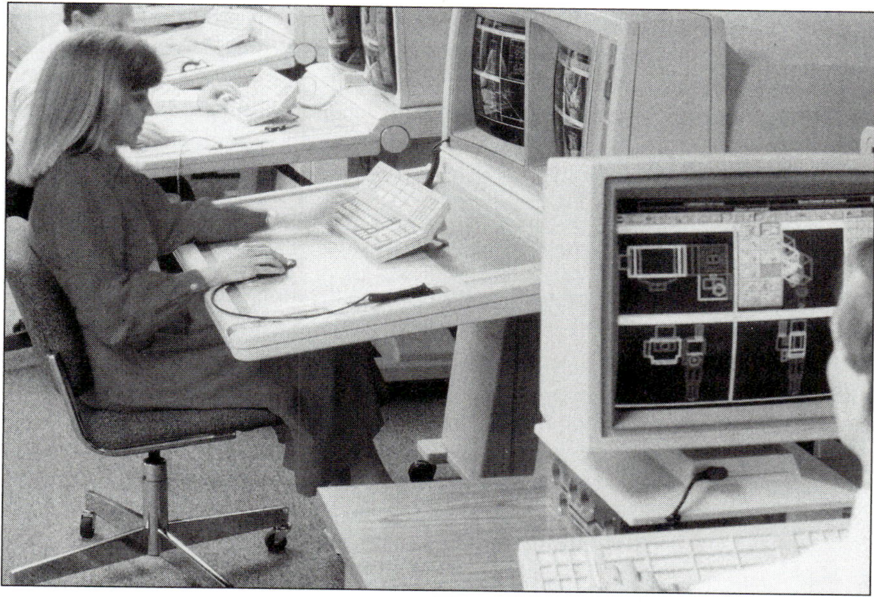

FIGURE 21.2
Dual-screen CAD terminals with digitizers in use (Ross Controls).

designer may sketch ideas directly on the screen of a video display, use a digitizer pad (Figure 21.2) to convert geometry from a sketch into electronic data, or enter data using the mouse and computer keyboard (Figure 21.3). CAD systems have revolutionized traditional design, creating two- and three-dimensional graphics (Figure 21.4) that can be manipulated by the designer in order to make quick adjustments and changes to the design. CAD graphics may be transferred to paper, if required, using a printer or a plotter. Some CAD systems employ dual video screens, and most CAD systems produce screen images in full color.

Designing with a CAD system has many advantages. The computer can electronically generate three-dimensional models that the designer may rotate to view a part or assembly from different perspectives. The

FIGURE 21.3
Computer-aided design (CAD) (Lockheed Martin Co.).

FIGURE 21.4
With the CAD plotter, hard copies of drawings are produced. (Lockheed Martin Co.).

operator may also enlarge the video image of any portion of the model in order to reveal details. Individual layers of the design drawing may be electronically removed and replaced as required to study the interrelationship and position of overlapping systems in the design. This is particularly useful in integrated circuit (IC or chip) electronic design as well as in the layout of piping and electrical systems.

Prototype and Testing Phase

The purpose of prototyping is to build a working model of a newly conceived product for testing to determine if the concept is acceptable. Although the product can be precisely modeled using a CAD system, a working **prototype** is often required. Prototypes may be the same size as the final product or they may be reduced in scale (Figure 21.5). For example, a master model of an automobile may be assembled from prototype parts, each simulating a metal production part, to permit engineers to examine how precisely the parts fit together in the complex auto body assembly (Figure 21.6). The computer-linked probing equipment shown in the illustration can precisely determine how each part fits compared with a computer model of the desired design, permitting an extremely accurate comparison between the prototype and its actual design specifications.

Building a physical prototype can be a very expensive and time-consuming process. Skilled model-makers are often required to create a prototype, and in the case of large and complex products such as buildings, ships, dams, or bridges prototyping would be impractical. Nonetheless, prototyping is an integral part of product design and development.

Rapid Prototyping

Rapid prototyping, the application of modern technologies to the creation of physical prototype models directly from CAD data, is an innovation in product development that has proven to be effective in saving cost and reducing the time required to create prototype models. Typically, these processes involve the manipulation of a three-dimensional solid-model CAD file by dividing the model into thin slices. The rapid prototyping machine then builds the physical part layer by layer, reproducing one slice at a time.

FIGURE 21.5
Prototyping and modeling.

FIGURE 21.6
Plastic master cube used to check auto body parts against design specification.

Many competing technologies are employed in rapid prototyping machines. Some common processes utilize laser technology; these include *stereolithography, laminated object manufacturing,* and *solid ground curing.* The stereolithography apparatus (SLA) begins with a vat of liquid polymer as the raw material. A low-powered laser traces the geometry of one slice of the CAD solid model onto the surface of the liquid polymer, solidifying the polymer as it is exposed and fusing it to the previous layer. An elevator then lowers the solidified prototype part a distance equal to the thickness of one slice, and the process is repeated until the part is completed. The thickness of the slice is a key to the precision of the model. Thinner layers result in smoother, more precise workpiece features. Thinner slices also require the laser hardening of more layers, increasing the time involved in producing a part. Finally, the fragile solidified part must be placed in an oven to complete the curing of the polymer. Parts produced by the SLA process are relatively precise, and they are strong enough to be placed in a functional assembly for physical testing in many applications. The translucent shell of the part also permits viewing of the flow of fluids through the part wall in testing. SLA parts are also used as patterns from which prototype parts can be created by casting.

Other rapid prototyping processes including solid ground curing employ a laser to solidify layers from a powdered raw material instead of a liquid polymer. An advantage of this technology is a wider range of raw material choices including metals, possibly yielding a prototype part that is more practical for use in testing.

Laminated object manufacturing (LOM) uses as raw material a thin sheet of a material similar to paper. As the prototype part is built a sheet is glued to the surface of the growing part, and a laser cuts the profile of the solid model slice into the sheet. When the waste material is pulled away from the completed part it exposes a part with the appearance of wood and similar properties. LOM requires a more powerful laser than those used in other rapid prototyping technologies because it must cut the sheet material. Because the laser cuts only the perimeter of the slice, however, LOM is practical for creating larger parts with thicker sections. LOM parts are often used as patterns for casting prototype parts.

Rapid prototyping machines that do not employ a laser tend to be less expensive. *Fused deposition modeling* (FDM) is a process that extrudes a heated thermoplastic polymer in a thin filament that is deposited, layer by layer, on the surface of the previous slice. The completed thermoplastic part is relatively resilient and can be used in testing for some applications. Wax models also can be created by FDM for use as casting patterns.

Several comparatively inexpensive rapid prototyping processes build each layer by depositing the raw material in particles or small droplets. Parts produced by this technology typically lack strength but can be produced in a variety of colors and materials. These parts function well as prototypes used for testing the fit or form of a product.

The savings in labor and product development time gained though the use of rapid prototyping can be tremendous. Although the machines and materials may be expensive, a complex prototype part can be constructed in a matter of hours compared with the days, weeks, or months required to construct a comparable prototype by conventional model-building means. In addition, complex curves and other profiles are often more faithfully and precisely reproduced by rapid prototyping processes.

Computer-Aided Engineering

CAD-based computer-aided engineering (CAE) systems have the ability to quickly and inexpensively test and evaluate prototype designs by subjecting a computer model

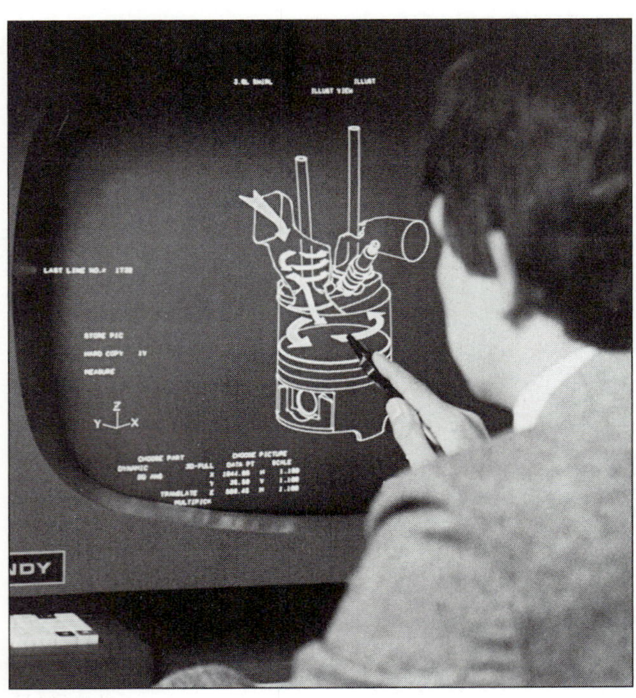

FIGURE 21.7
Computer-aided engineering (CAE) used in evaluating fuel swirling in an internal combustion engine cylinder (Ford Motor Company).

of the design to many conditions that the product could encounter in its service environment. The time and cost of producing and testing prototypes are avoided, and redesign may be accomplished immediately to ensure that the product meets requirements. Examples of design evaluation by computer include such diverse activities as studying the air–fuel mixing patterns in an internal combustion engine (Figure 21.7); evaluation of structural design and identification of weak points (Figure 21.8) using finite element analysis (FEA) software that can evaluate differences in stress in various areas of product components; analysis of the product design or the manufacturing process with heat-transfer and flow-modeling software that displays differences in temperature throughout the product; testing of the product and process for vibration characteristics using modal-analysis, and testing for correct motion of moving components using kinematic analysis. CAE is extremely versatile in the testing of new designs, and the potential savings in time and expense make CAE a valuable tool for modern designers.

Design Evaluation and Refinement Phase

Data gathered from prototype testing or from the evaluation of the computer model using CAE is carefully evaluated (Figure 21.9). In this phase the design is modified according to test results, possibly requiring a new round of prototype testing, to create a functioning, desirable, well-engineered product.

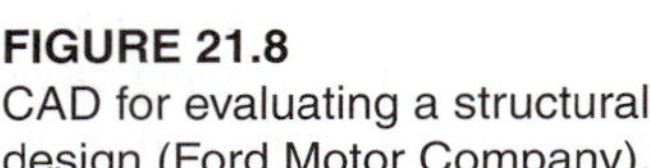

FIGURE 21.8
CAD for evaluating a structural design (Ford Motor Company).

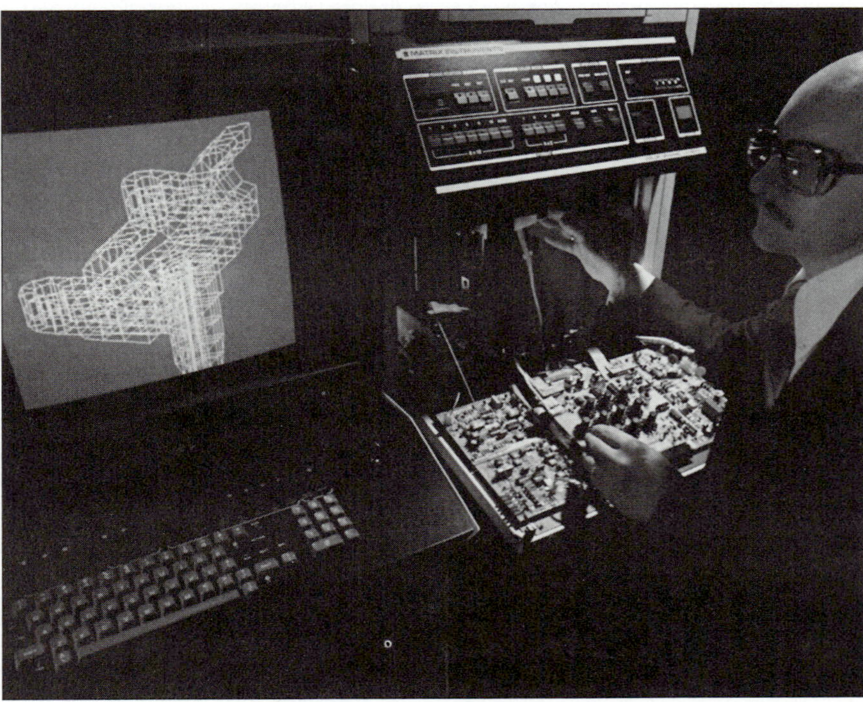

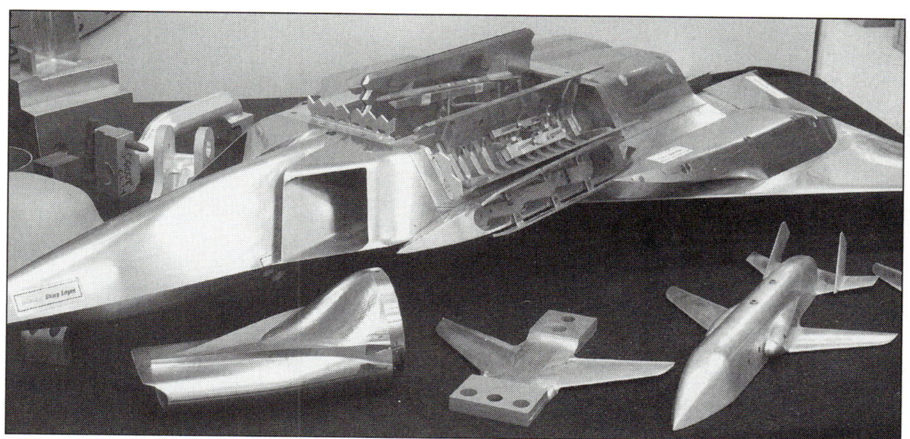

FIGURE 21.9
Prototypes for wind tunnel testing (Lockheed Martin Co.).

Manufacturing Drawing Phase

When the decision is made to release an approved design for production, sketches and drawings are made reflecting the improvements incorporated during the prototype stage. Drawings are refined during the evaluation stage, then converted into finished manufacturing drawings and distributed to the departments responsible for manufacturing the product. In fully integrated CAD systems, electronic drawings are stored in a common database and reproduced on computer video displays or plotted on paper.

THE BLUEPRINT

The manufacturing drawing is called a *blueprint,* a design drawing that is created by product design engineers for the purpose of clearly and concisely communicating the physical specifications of the design to the manufacturing function. Employees throughout the manufacturing organization use the blueprint as an overall reference to the general composition and appearance of the product and as the comprehensive source for the detailed specifications of the product. Manufacturing drawings are commonly produced using CAD. An advantage of the CAD system is that many different part drawings may be permanently stored in the computer database and electronically retrieved for review and revision at any time. This system speeds revised drawing information to the production shop floor.

Blueprints commonly include both drawings of assemblies and detail drawings of individual components. Typical manufacturing drawings are produced in the standard orthographic projection format showing an object in any of several possible views (Figure 21.10). In addition,

FIGURE 21.10
The standard orthographic format of manufacturing drawings.

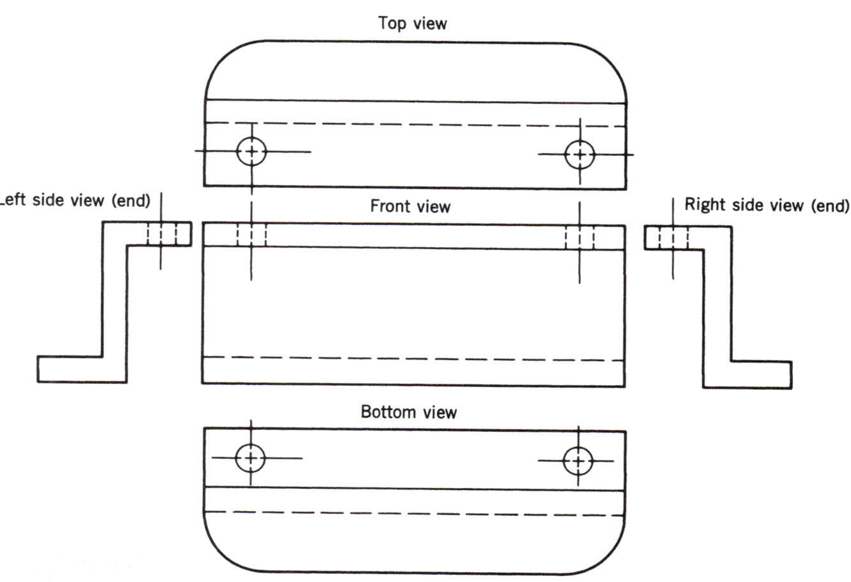

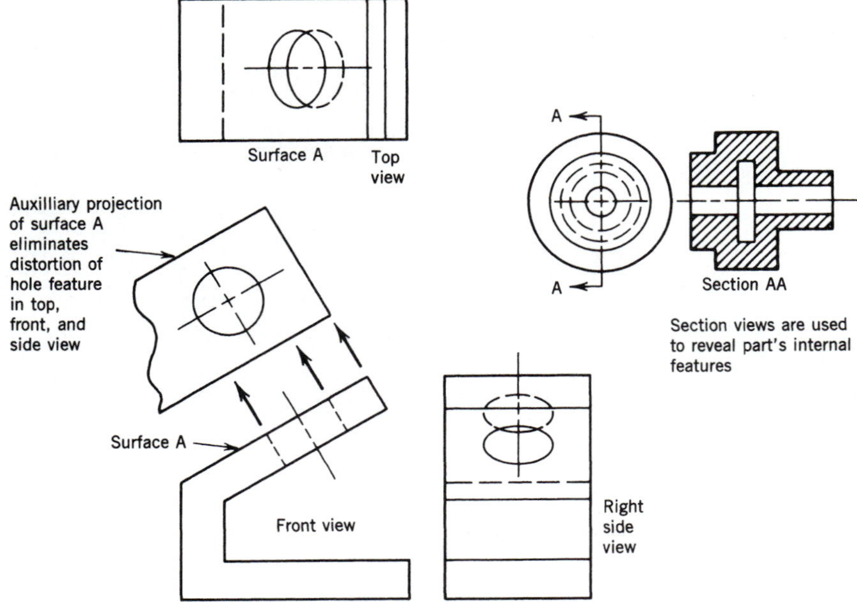

FIGURE 21.11
Auxiliary and section views on orthographic drawings.

several types of auxiliary and section views may be used (Figure 21.11). A pictorial format with no dimensions, such as isometric (Figure 21.12), may be employed in an assembly drawing to show the relationships among the parts in an assembly.

A detail drawing (Figure 21.13) contains all the specifications required for the manufacture of a component. A detail drawing includes the following important information:

1. Part name
2. Drawing number
3. Material specifications
4. Dimensions and locations of part and part features
5. Tolerances for size and geometry
6. Revision codes and dates of most current revisions
7. Any special notes regarding special handling during manufacturing

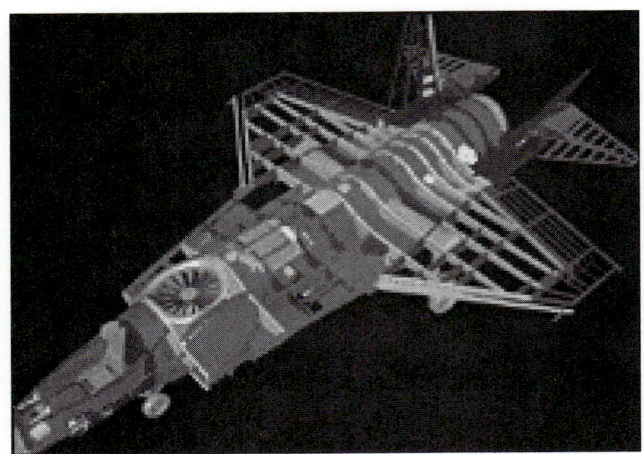

FIGURE 21.12
Isometric pictorial of the F-35 Joint Strike Fighter (Lockheed Martin Co.).

A typical manufacturing drawing contains many dimensions, and each dimension on the blueprint has an associated tolerance. The dimensions on the manufacturing drawing describe the size, form, and location of each feature. Size dimensions such as length or diameter, and form tolerances such as flatness, perpendicularity, and cylindricity are commonplace. The manufacturing engineer must assure that the manufacturing process includes operations that are capable of manufacturing the product within the specified tolerances, whether they involve size, form, or location. These tolerances also concern the production worker, as he or she is charged with executing each operation in a manner that produces parts and assemblies whose dimensions fall within the blueprint tolerances.

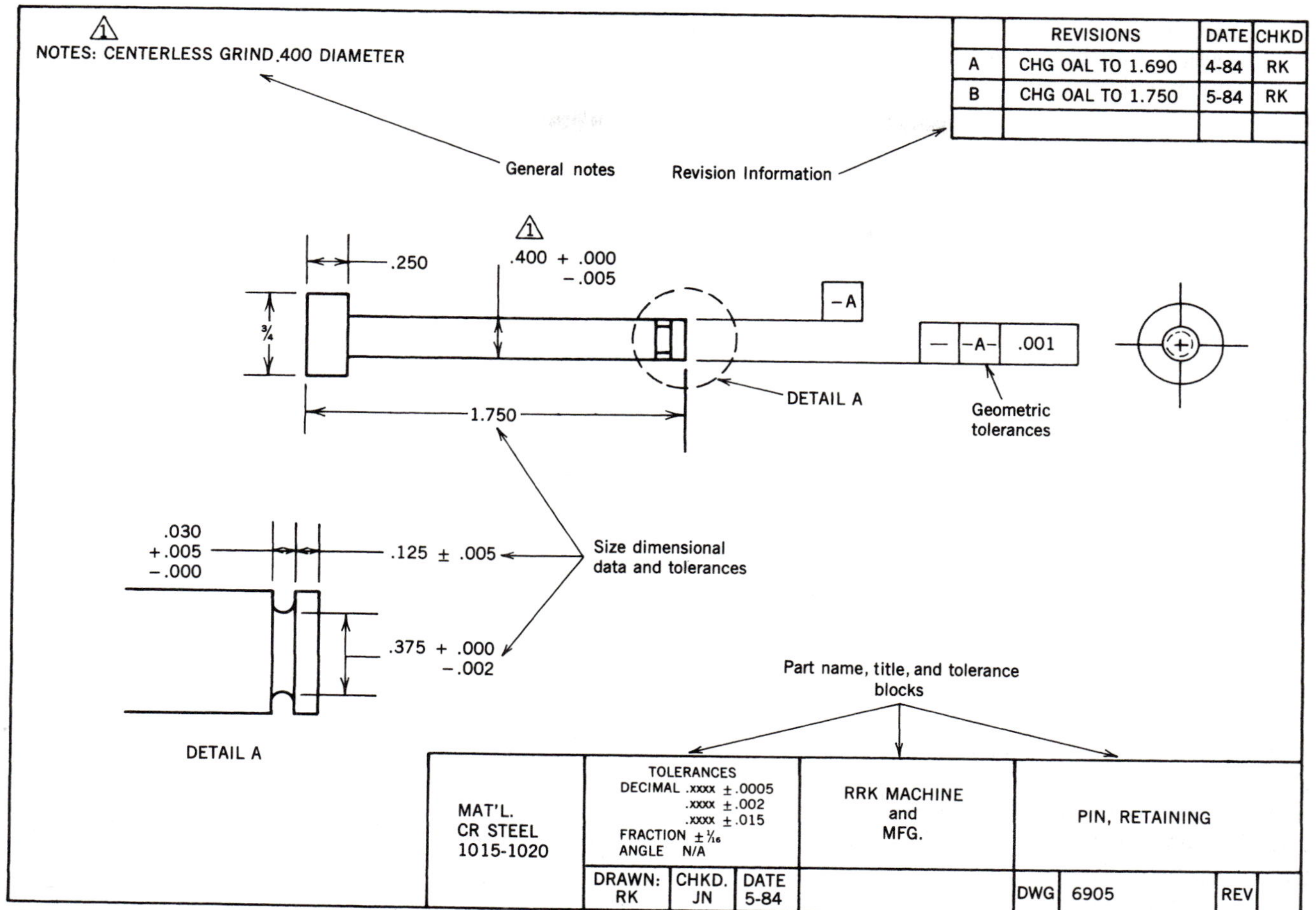

FIGURE 21.13
Detail drawing.

The assembly drawing (Figure 21.14) lists all the components that fit into a given assembly, provides a pictorial view of the assembly, and shows the general location of each component in the assembly. Typically, dimensions found on assembly drawings are shown for general reference only, because the dimensions of the assembly are generally established by the blueprints of the individual components. An assembly drawing includes the following important information:

1. Assembly name
2. Assembly drawing number
3. Bill of materials, a list of individual parts with names and often with cross-referenced numbers to detail drawings
4. Dimensions pertaining to the assembly
5. Revision codes and dates of last revisions
6. Any special notes regarding the assembly

Manufacturing drawings are made to exacting standards. The completed drawings are checked for accuracy and to ensure that no pertinent data have been omitted before blueprints are distributed to the various manufacturing departments or to suppliers. As production progresses, the design may be revised to meet changing design requirements, improve production flow, or solve production difficulties.

A precise comparison of design dimensions and actual dimensions can be made by comparing a mathematical model of the design with the dimensions of an actual production unit that is measured using a *coordinate measuring machine* (CMM) (Figure 21.15). Precise measuring

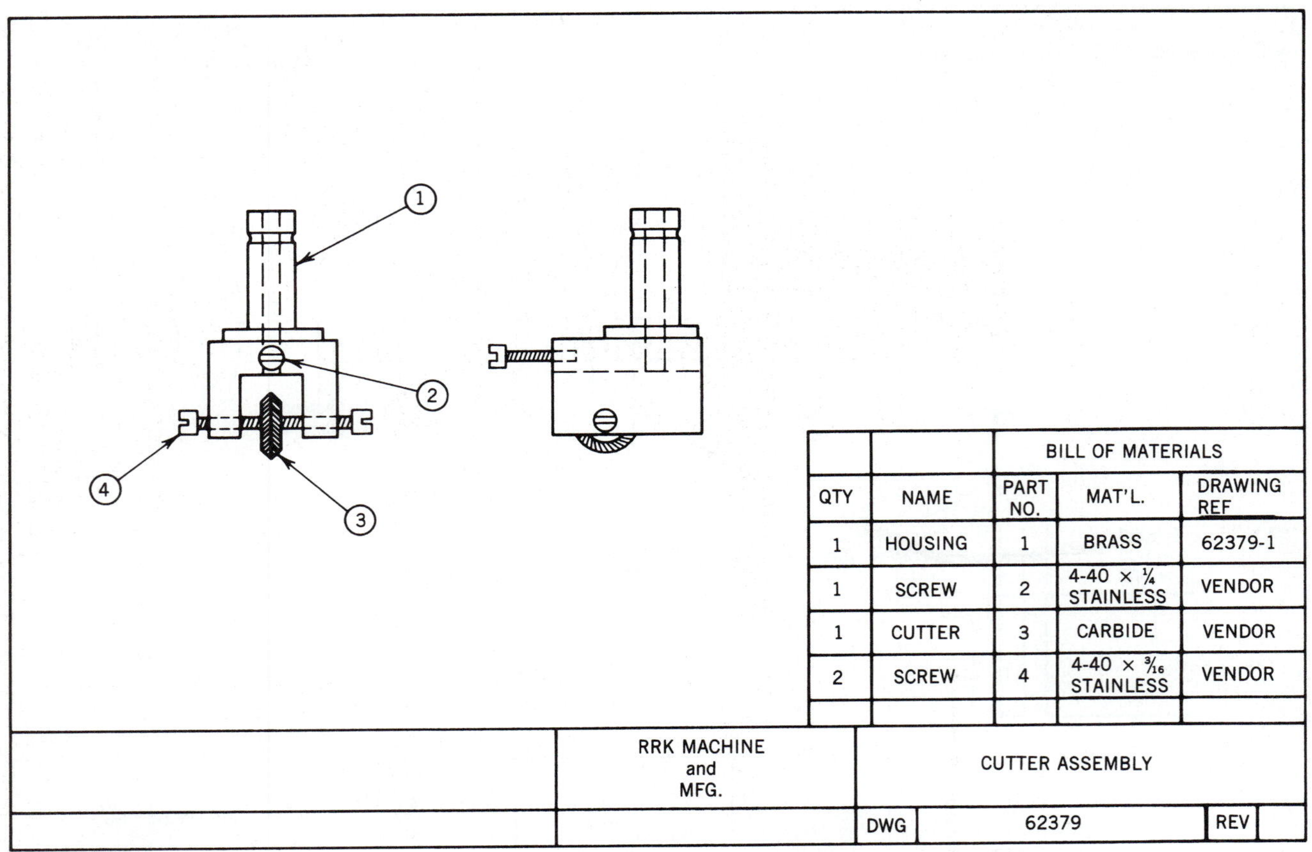

		BILL OF MATERIALS		
QTY	NAME	PART NO.	MAT'L.	DRAWING REF
1	HOUSING	1	BRASS	62379-1
1	SCREW	2	4-40 × 1/4 STAINLESS	VENDOR
1	CUTTER	3	CARBIDE	VENDOR
2	SCREW	4	4-40 × 3/16 STAINLESS	VENDOR

FIGURE 21.14
Assembly drawing with bill of materials.

processes are critical in determining whether the manufactured product conforms to the design specifications on the blueprint.

When CAD is coupled with computer-aided manufacturing (CAM), computerized machines can be used at production stations to control manufacturing operations. In Figure 21.16, CAD and CAM are integrated to enhance aircraft production. The integration of CAD and CAM has, in all its forms, become a major force in modern manufacturing. Integrated CAD and CAM systems and the steps toward complete factory automation will be further discussed in the next chapter.

PREPARING FOR MANUFACTURING

Manufacturing begins when an order is received by the manufacturer. In many cases the order is solicited by the company's sales force as it works in the marketplace to generate business. The sales force is supported by the company's advertising department or an outside advertising agency. The sales force will attend trade shows to demonstrate and promote the company's products and they will make calls on prospective customers.

When an order is received, manufacturing management and financial management make the necessary decisions as to how production will be financed and accomplished. If necessary and justifiable, new production facilities may be built or purchased after careful study to determine the profitability of the proposed production. There is always some risk. It is likely that the manufacturer will have to borrow much of the money needed to finance the production. The sales of the product must return this investment so that the company can repay its loans while at the same time meeting its daily financial requirements. These include meeting payroll and purchasing material for the product being manufactured.

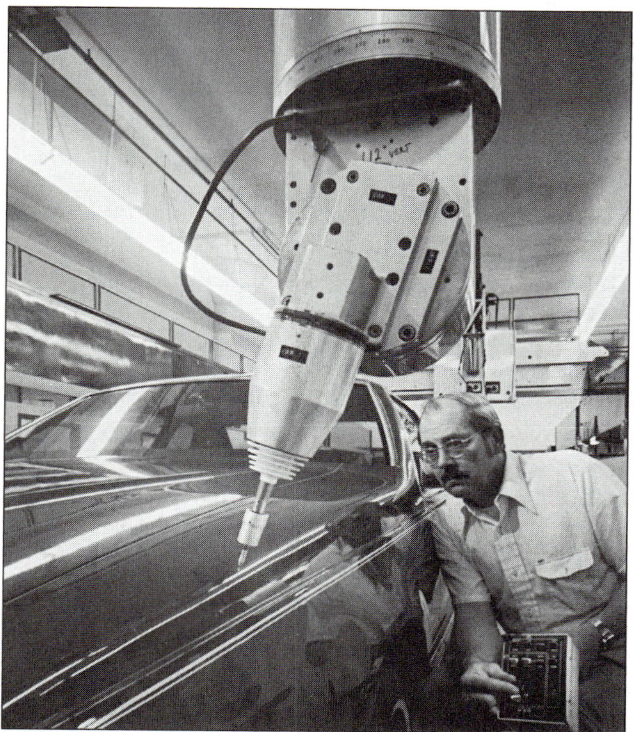

FIGURE 21.15
This coordinate measuring machine takes measurements and compares them against a computer database containing precise design specifications (Ford Motor Company).

Estimating Costs

When the order (Figure 21.17) is received at the manufacturer's facility it is processed, and a price and time of delivery are quoted to the customer. In many cases a manufacturer is asked to bid on a job. Under this system, the customer shops around for the required products and asks several competing companies for their best price for the manufacturing required. Bids are generally provided at no cost to the customer, and the manufacturer must carefully prepare the bid so that it reflects actual production costs plus acceptable profits. Underbidding to get a contract may result in the manufacturer losing money on the job or being unable to complete the work. Both circumstances can result in a business failure.

Cost estimating involves a careful analysis of all the costs involved in production of a product. Major factors influencing costs include:

1. Employee wages, salaries, and benefits
2. Interest on borrowed money
3. Price of material
4. Tooling and production machinery
5. Packaging and shipping
6. Quality control and assurance
7. Plant and equipment maintenance and security
8. Utilities
9. Product liability insurance if required

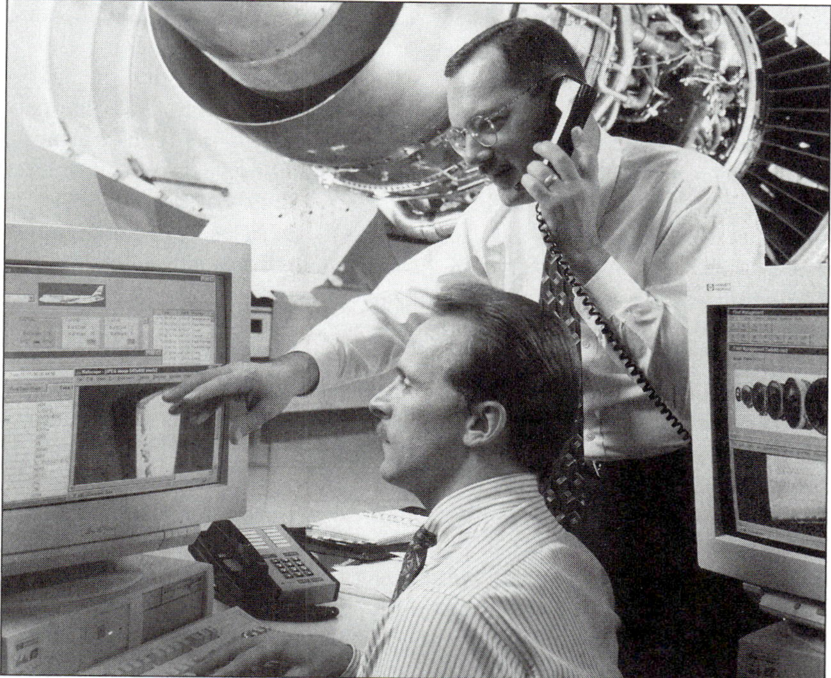

FIGURE 21.16
Integrating CAD and CAM in aircraft production (B. Harris).

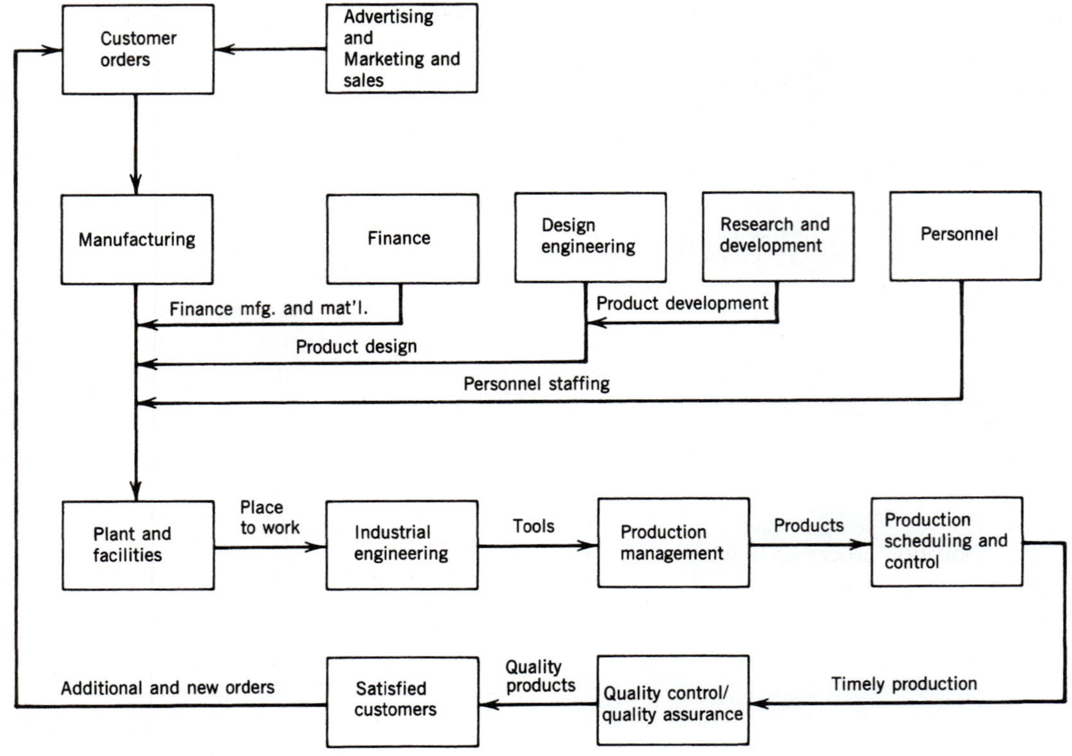

FIGURE 21.17
Manufacturing requires the coordinated efforts of all the team players.

10. Stockholder dividends and other returns to investors
11. Design expense
12. Rent, lease, or construction of manufacturing facilities
13. Guarantees and services included in purchase price

All these factors represent a portion of the end product price. By process of careful cost accounting, each expense can be represented as a percentage of the product price. For example, if employee benefits represent 10 percent of the product price, then $10.00 of a $100.00 product represents that portion of the cost attributed to employee benefits.

Product costs are also influenced by other complex business cycle factors that have direct or indirect influence on the price of a product. Included in these are taxes, depreciation, investment tax credits, and projected sales, often known as forecast sales. If a manufacturer has a strong sales position in a given market and the national economy is in an upturn, the company may be willing to risk venture capital in plant and/or product expansion while at the same time holding its prices in line or maybe even reducing prices. However, in times of economic recession or resulting lowered product demand, high rates of interest on borrowed funds, or foreign as well as domestic competition, a company may take a more conservative approach in new plant and equipment investments. A large debt in a declining economy and/or market can cause a business failure and the unemployment of many workers in the labor force. Accurate cost estimating thus becomes a critical issue in any business, and the manufacturer continually strives to estimate costs efficiently while reducing operating costs at all levels.

One method of cost estimating reduces all expenses to an hourly rate. Production hour rates are often used by service industries or manufacturing concerns that act as suppliers. Job machine shops are an example. When a customer asks for a bid on a job, the price may be quoted as a price per hour of production time plus any materials that are purchased for the job. The production hour rate reflects all the costs of production with the exception of the material.

To establish a production hour rate and then a cost per unit estimate, it is necessary to evaluate carefully all the variables that will affect the final price. These expenses can be expressed in terms of percentages of total operating budgets to determine what a customer will have to be charged per worker hour or per machine hour.

Example

Production hour cost breakouts as percentage of expenses. Cost of the production hour = $200.00 representing:

Labor: 67 percent	(.67)($200.00) = $134.00
Benefits: 12 percent	(.12)($200.00) = $24.00
Utilities: 2 percent	(.02)($200.00) = $4.00
Material: 19 percent	(.19)($200.00) = $38.00
Total: 100 percent	$200.00

When products, particularly those that become parts of more complex assemblies, are manufactured on a continuous basis, it is simpler to estimate and quote prices simply on a cost per unit basis. This method is particularly suited to price quotations on variable quantities. We have all heard the adage "cheaper by the dozen." In production manufacturing, this bit of philosophy usually turns out to be true. Once the production process is established, costs are divided by the number of products produced. The more products produced, the lower the cost for each unit.

Example

$$\begin{aligned} \text{production hour rate} &= \$40.00 \text{ per hr} \\ \text{number of units} &= 10 \text{ units} \\ \text{total time estimate} &= 100 \text{ hr} \\ \text{cost} &= \frac{(\text{time in hours}) \times (\text{production hour rate})}{(\text{number of units})} \\ &= (100 \text{ hr})(\$80.00 \text{ per hr})/(10 \text{ units}) \\ &= \$800.00 \text{ per unit} \end{aligned}$$

The estimation of production hours requires that the estimator is generally informed of the processes involved and then able to estimate accurately the time needed to accomplish the task at average rates of productivity. It is important that all costs are reflected in a price quotation. If they are not, underbidding may occur and a cost overrun above the estimate may result. Contractual arrangements between customers and vendors do not permit restructuring of estimated costs once prices have been agreed upon.

Facilities Design and Purchasing

A production facility must accommodate the required equipment and personnel. Depending on the product, a new factory may have to be built and personnel hired and/or relocated to staff it. Plant facility engineers and *industrial engineers* are responsible for the layout of the factory floor and setup of the production and/or assembly lines in such a way that product flows smoothly through the production process.

The purchasing department, staffed by purchasing agents or buyers, is responsible for procurement of the materials, goods, or services necessary to support the manufacturing operation.

PROCESS DESIGN

Once a product design is complete, a prototype is created and tested, and the decision is made to take the product into manufacturing; then the next phase is the design of the manufacturing process. The process engineering or process planning function is responsible for designing the process and is typically assigned to the **manufacturing engineering** department. **Process engineers** along with tool designers, methods engineers, and numerical control programmers make up the manufacturing engineering department.

Process planning is most effective when it is executed during product design and development, a practice called *concurrent engineering.* The modern concurrent engineering concept assures that much of the development of the manufacturing process occurs in parallel with the design of the product, shortening the time required for the product to reach the market and ensuring manufacturing input during the development of the product design.

Designing a manufacturing process involves analyzing the product blueprints, determining the material, processes, tooling, and machinery required to manufacture the product, and documenting the process. The process engineer begins by extracting the critical features and dimensions from the blueprint in addition to ascertaining the material specifications. Once the critical features and dimensions have been gleaned from the blueprint, the engineer can decide what manufacturing operations will be employed to create the product component. The required manufacturing operations are derived, sequenced, and then grouped into machine setups. As the process is established more detailed raw material requirements evolve, including the size and condition (cast, forged, wrought, etc.) of the materials.

After the sequence of operations is determined, the required workstation, including machine tools and other equipment, is selected for each of the chosen operations. Next, detailed specifications are developed for workholding tooling such as fixtures. Specifications are then established for operating tooling such as cutting tools and forming tools. The tooling specifications are forwarded to the tool design function, and construction of the tooling is done in the tool and die shop. The tooling specifications are sent to purchasing if the tools are to be acquired from an outside vendor. Setup and changeover requirements and the detailed manual and machine tasks are composed, and process parameters such as speeds, feeds, depths of cut, welding amperages, process temperatures, pressures, and heating or cooling times are calculated.

Process Documentation

The manufacturing process for each component and assembly must be documented to provide a permanent engineering record and to facilitate the dissemination of the information to the manufacturing organization. Each manufacturing company develops its own specialized form of process documentation, but vehicles similar to the **process routing sheet** and the **operation detail sheet** described in this text are common.

The process routing sheet that is generated for each manufactured component part typically lists, in sequence as they must occur, the machine setups required to produce the component. Each process step is briefly described. The estimated amount of time required to set up the operation in addition to the *piece time,* the time required to perform the operation on one workpiece, are documented on the sheet. Also included is the workstation where the operation is to be performed. A detailed material specification is listed, including the amount of material required to make one piece. General heading information includes the component name and part number, and when a part enters production a batch identification number and lot quantity (the quantity to be produced) are added to the sheet. The information provided on the process routing sheet does not provide sufficient detail to guide the production worker through the tasks required to complete the operation.

The workstation, operation sequence, and process time information is utilized in the purchase of raw material, to calculate the cost of the product, and to schedule jobs and deliver material to the correct location. Purchasing agents use the detailed material description and lot quantity to order the correct amount of the specific raw material required to supply production. Cost accountants use the operation steps, raw material quantities, and times to estimate the cost of each component and of all assembly operations and then sum them to calculate the expected cost of the product. The production scheduler needs the operation steps and times to determine which machines to schedule for each job, in what sequence, and for how much time.

In medium- to high-volume manufacturing where workers with less training and experience are employed, more detailed design and documentation of the manufacturing process is required. The operation detail sheet documents the detailed information that is useful to production

FIGURE 21.18
Designing in the technology age utilizes CAD (Lockheed Martin Co.).

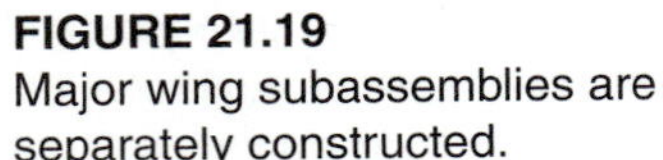

FIGURE 21.19
Major wing subassemblies are separately constructed.

workers and supervisors as a guide in performing the manufacturing operations. This sheet also provides documentation of the process in the manufacturing database. Each step, or operation, in the manufacturing process is described in detail. The detailed process information is often broken down into individual tasks, cuts, and the like. It identifies required work-holding tooling such as fixtures and vises to be employed, specific operating tooling such as cutting tools and forming tools, and process parameters such as speed and feed. As on the process routing sheet, a detailed material specification is listed, including the amount of material required to make one piece.

Methods engineering is generally the responsibility of industrial engineers, who select and copy portions of drawings and generate operations sketches (op. sketches), manufacturing outlines (MOs), or assembly instructions. These documents accompany parts as they flow through a production process. A typical operations sketch shows only a small portion of a production drawing. It emphasizes dimensional data and other pertinent information necessary to a specific manufacturing operation on that portion of the part.

FIGURE 21.20
Major fuselage subassembly is lowered into place (Lockheed Martin Co.).

MANUFACTURING EXECUTION

After design is completed (Figure 21.18), manufacturing facilities are established and staffed, material is procured, and tooling is in place, production manufacturing may begin. As the major assembly components are completed they are fitted together to make the final product (Figures 21.19, 21.20, and 21.21). The sequence of assembly must be established so that all the components will fit correctly into the final assembly.

As the components go together (Figures 21.22 through 21.24), manufacturing engineers are constantly busy solving production problems while designing new tools, jigs, and fixtures. Plant and production maintenance keeps plant and production equipment functioning properly.

FIGURE 21.21
Fuselage sections take shape in a large production facility (Lockheed Martin Co.).

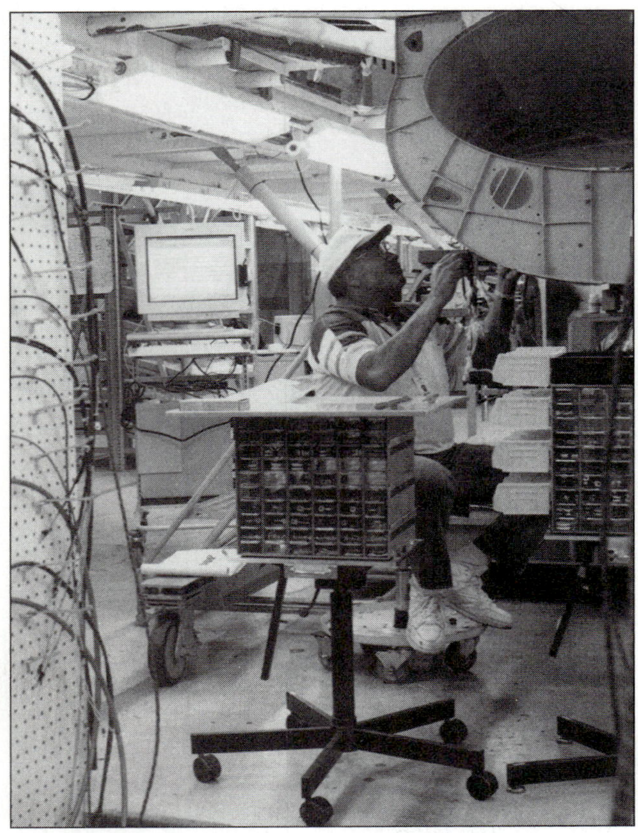

FIGURE 21.22
Assembly details: F-16 wire harness assembly (Lockheed Martin Co.).

FIGURE 21.23
Millions of small parts such as rivets must each be installed with the same care and attention (Lockheed Martin Co.).

FIGURE 21.24
Internal work on major subassemblies takes place simultaneously (Lockheed Martin Co.).

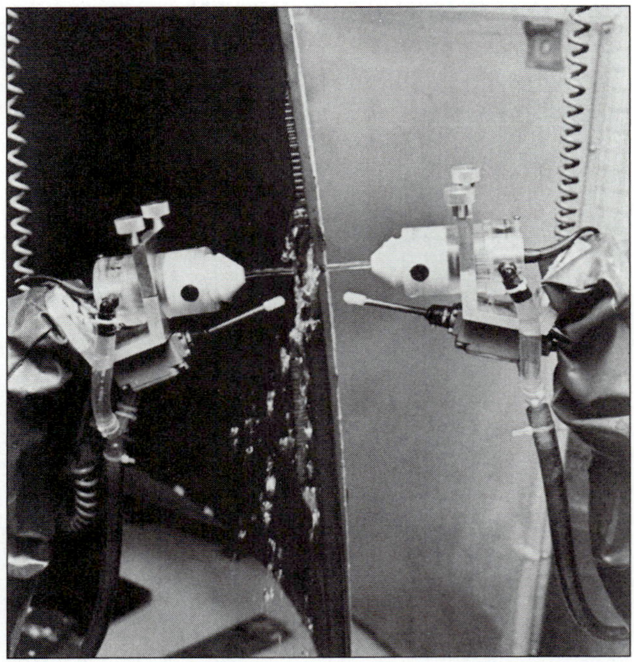

FIGURE 21.25
Robot-controlled water jets inspect epoxy graphite composite bonded joints on a jet engine fan reverser.

FIGURE 21.26
Completed fuselage ready to join wing assemblies (Lockheed Martin Co.).

Production control and production scheduling personnel are responsible for the timely arrival of materials and subassemblies so that the final product may be manufactured with no delays. Quality assurance and control personnel carry out their responsibilities at every stage in production, often assisted by sophisticated instruments (Figure 21.25).

When major subassemblies of a large and complex product are ready for joining (Figures 21.26 and 21.27), the end product begins to take on a familiar shape. Behind the scenes all of the many parts, components, and subassemblies are being manufactured in a similar manner. Other complex subassemblies are shipped from their respective subcontracting manufacturers for installation or assembly in the final product (Figures 21.28 and 21.29).

FIGURE 21.27
The final product, an F-16 fighter jet, takes on a familiar look (Lockheed Martin Co.).

FIGURE 21.28
Major subassemblies are readied for installation on the F-16 airframe (Lockheed Martin Co.).

FIGURE 21.29
Assembly stations for F-16 fighter jets (Lockheed Martin Co.).

FIGURE 21.30
As finished product nears completion, new products begin their trip through the manufacturing process under the same roof (Lockheed Martin Co.).

The product reaches its final assembly stages while at the same time new product units may be beginning their manufacture. The entire manufacturing process may take place under one roof (Figure 21.30). At the end of production the final product leaves the production line ready for final detailing, testing, and delivery to the customer (Figure 21.31).

The production effort is a team effort. Thousands of individuals each contribute a small part in the making of a complex product. Their effort is rewarded by the production of a magnificent example of modern industrial manufacturing capability.

Tool Design

Tool design is among the most important activities supporting the manufacture of almost any product. **Production tooling** includes all the machinery and equipment necessary to support product manufacturing.

Even though a manufactured product may pass through many different processes during its manufacture, these individual processes are each designed to hold and position the part while the specific process is being accomplished. It is frequently necessary to develop special production tooling unique to the particular product being manufactured. One might say that this special tooling adapts the general multiuse material-processing equipment to specific manufacturing operations. Included in special production tooling are the following:

1. Jigs
2. Fixtures
3. Dies

FIGURE 21.31
Final detailing and testing are the culmination of the manufacturing process (Lockheed Martin Co.).

4. Molds
5. Patterns and templates
6. Cutting tools

Jigs Jigs are special tools designed to locate a feature for machining or to align parts for assembly or joining. A simple example is a drill jig used in machining (Figure 21.32). A drill jig consists of a hardened steel sleeve called a drill bushing through which a drill can pass freely but with minimum clearance. The jig aligns the drill, and the resulting hole in the workpiece is drilled in its proper position. Drill jigs may be complex tools designed for drilling many holes in the same workpiece. Jigs may also be used to hold large fabrications in position while welding or assembly occurs.

Alignment of tooling or parts of an assembly may also be determined optically. Optical techniques have many applications in layout, alignment, and assembly of large fabrications such as ships, bridges, and aircraft.

Fixtures Fixtures are special tools designed to hold, locate, support, index, and/or align a workpiece while a machining, joining, or assembly operation is performed. An example of a simple fixture is the common machine tool vise with soft jaws that have been machined to a special shape for the purpose of holding a workpiece in a particular position. In Figure 21.33 special fixtures are used to hold aircraft skin panels. Note the large size of this tooling.

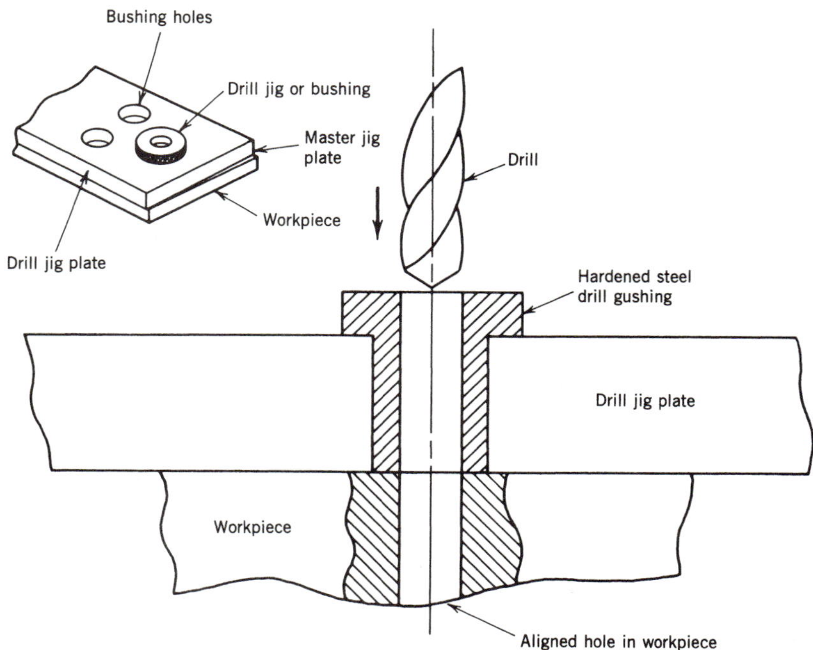

FIGURE 21.32
A drill jig is an example of simple tooling.

FIGURE 21.33
Large tooling fixtures for aircraft skin panels.

Fixtures used for high-volume production tooling can be complex. In Figure 21.34 the entire auto body is held in position while numerous holes are drilled for body panel attachment (Figure 21.35).

Dies Dies include many types of special tools including cavities into which materials are shaped by various processes. Examples discussed in previous chapters include die casting, in which molten metal is poured or pumped into a cavity shaped to the form of the desired product, and forging processes. Dies are also used in machining for thread cutting and in punch press work dies are used in conjunction with punches to cut parts to exact shapes (Figure 21.36). Dies can produce a large variety of useful parts (Figure 21.37).

Dies are also used in metal bending applications. Figure 21.38 shows a press brake bending die, and Figure 21.39 shows the type of products produced by it. Dies are also used extensively in the plastics industry.

Molds Molds are most often used to form products from raw material in a fluid state. Products formed in molds assume the shape of the mold before they change from liquid to solid by cooling or hardening. Examples of molded products are metal castings, fiberglass, and plastic resins. After the material solidifies by cooling or hardening, the mold is opened, removed, or broken, exposing the finished product (Figure 21.40).

FIGURE 21.34
An entire car body is held in a drilling fixture where body panel holes are drilled in one operation.

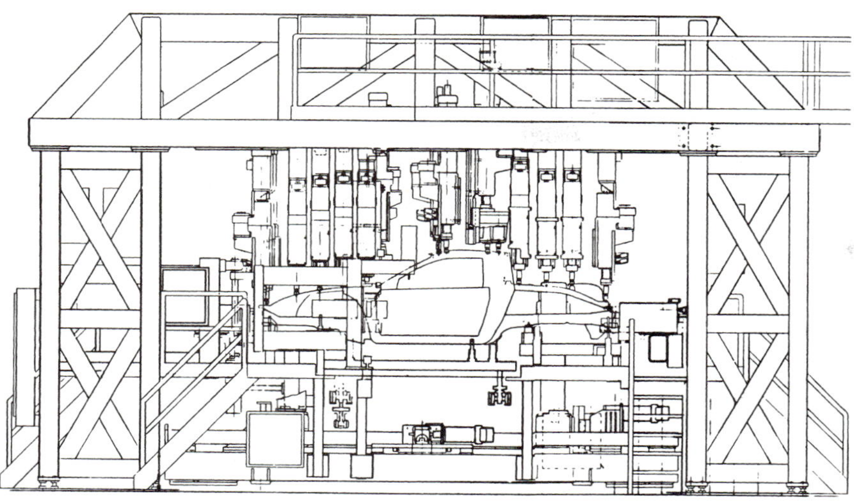

FIGURE 21.35
Body panel drilling operation sequence.

Sequence of Operations

1 Automatic on-line load
2 Clamp body at eight locations
3 Position body per LVDT readings
4 Set back-ups at front, rear, top
5 Verify position for drilling
6 Pierce four slots
7 Drill 39 vertical holes
8 Machine 39 horizontal surfaces
9 Verify operations completed
10 Unclamp and release back-ups
11 Automatic unload to conveyor line

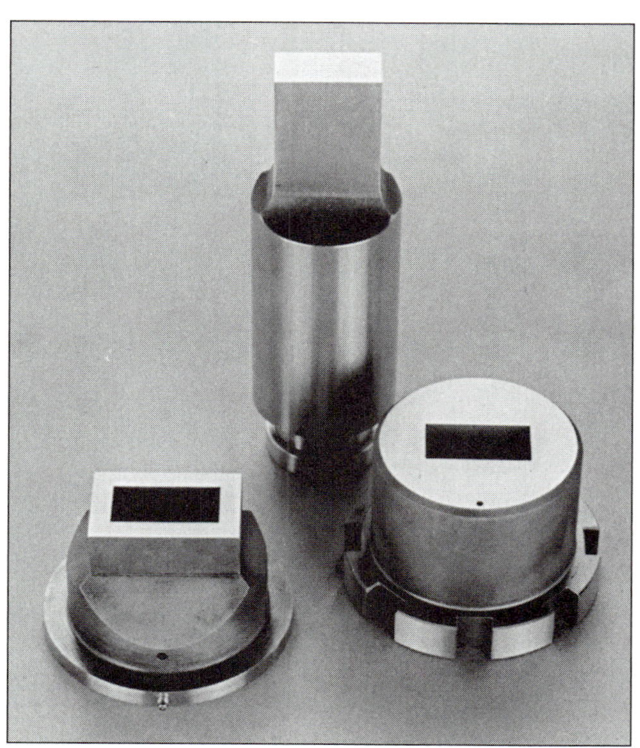

FIGURE 21.36
Punch and die tooling (Baltec Corporation).

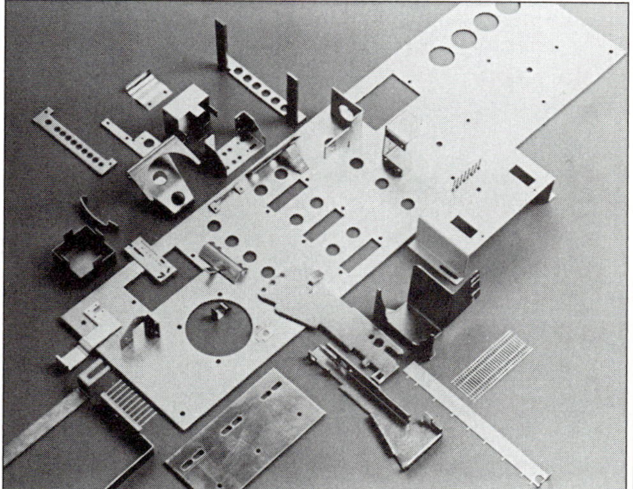

FIGURE 21.37
Punch and die tooling can produce a wide variety of end products (Baltec Corporation).

FIGURE 21.38
Press brake tooling for sheet metal bending (Baltec Corporation).

FIGURE 21.39
Example of press brake products (Baltec Corporation).

Patterns and Templates Patterns and templates cover a wide range of special production tooling. In Figure 21.41 the shape of a bend is checked against a template. In another example, a full-sized template of a hull plate for a ship's hull may be used to guide the fabrication of duplicate hulls of the same size and shape. Patterns are also used in the creation of molds for casting.

Cutting Tools Cutters are numerous and important members of the production tooling family, especially in machining. Cutters generally remove workpiece material by shaving off small pieces or chips. In machining operations such as gear hobbing, a large number and variety of cutters are required (Figure 21.42).

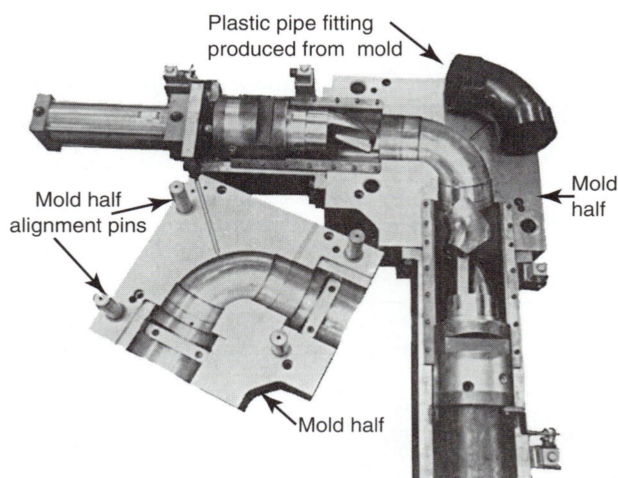

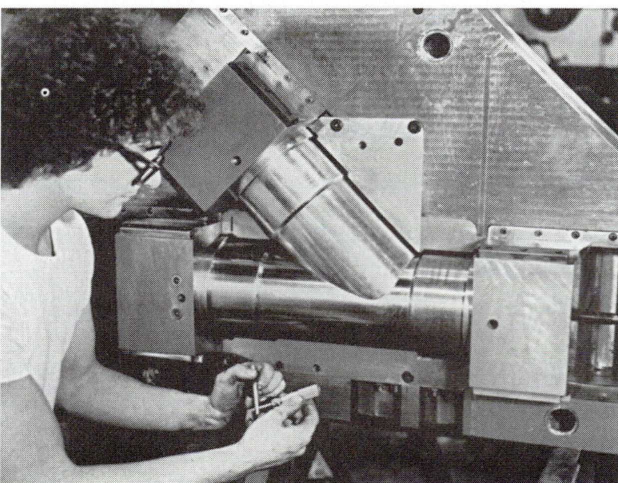

FIGURE 21.40
Mold tooling for manufacturing plastic pipe fittings (Crucible Materials Corp.).

FIGURE 21.41
Inspecting an aircraft panel bend with a template.

Tool Making

Production work-holding tooling must meet exacting dimensional standards but also must be convenient to set up in the manufacturing process. The process operator should be able to easily load and unload production parts during manufacturing. If the tooling is to be used in conventional machining or grinding operations, provisions should be made to clear chips easily and quickly before new parts are loaded in the fixture.

Production tooling often incorporates many heat-treated parts so that accuracy and durability can be preserved despite hard use. Most manufacturers identify and catalog special tooling so that it can be located quickly for use whenever a need arises.

Design and production of tooling often involve as much effort as the design and production of a manufactured product. In many cases the cost of tooling prior to manufacturing represents a major cost of production. In large industries with complex products, the payback for tooling expenses can take several years.

FIGURE 21.42
Many styles and sizes of cutter tooling are required in gear cutting (Foote-Jones/Illinois Gear).

Review Questions

1. What factors motivate design?
2. What are the phases of design?
3. What is CAD and how is it used in design?
4. What are the advantages of CAD?
5. Why are design review and evaluation important?
6. What are the two major types of drawings found in manufacturing?
7. What information is contained on these drawings?
8. Describe the purpose of production tooling.
9. Who is responsible for building production tooling?
10. What major items are included in production tooling?
11. Describe the flow of production in a typical manufacturing industry starting with order placement.

Case Problems

Case 1: Computers or Prototypes

A manufacturer of complex hydraulic equipment decided to reevaluate product design with the intent of branching into a new market requiring lightweight products (automotive and aircraft). One choice would be to build and test a large group of prototypes. This would take about one year and the cost for in-house design work and jobbed-out prototype development would exceed $500,000.

A CAD/CAM system could be purchased for about $750,000 that not only would evaluate designs by computer but also could be interfaced with the manufacturer's existing production capability, making it more efficient and productive. Complete phase-in for the CAD/CAM system to fully support manufacturing would take three years, but the design evaluation capability could be online almost immediately. The market potential for the new lightweight hydraulic equipment is excellent.

If the company must borrow the money for either project, which do you think would pay off in the long run from a standpoint of both efficiency and return on investment?

Case 2: Estimating the Cost

Manufacturer X produces a product at a price of $.0234/production unit. The production rate for this company is 325/hr. Manufacturer Y, a competitor, sells an identical item for $.0205/production unit that can be produced at a rate of 375/hr. Calculate the cost of an order for 15,000 parts from each company, and calculate the time to produce each order. Calculate the total cost of each order and calculate the cost per hour of each company's production.

Case 3: Cost Breakouts

A manufacturer builds and sells a product costing $4500 per production unit. Represented in the product cost are:

Wages	60 percent
Employee benefits	11 percent
Utilities	2 percent
Material	20 percent
Other expenses	7 percent

Calculate each dollar cost as represented by these percentages. If the time involved is 48 worker hours, calculate the worker hour cost in this product. Manufacturing 100 units of the product will result in a 16 percent saving. Calculate the worker hour cost and product price under this condition.

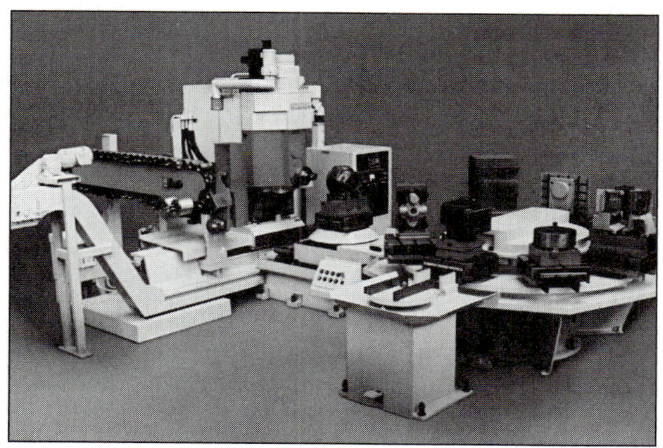

CHAPTER 22

Automation in Manufacturing

Objectives

This chapter will enable you to:

1. Discuss reasons for automating manufacturing.
2. Describe the basic factors required for automation.
3. Discuss computer numerical control automation of manufacturing processes.
4. Discuss CAM, CIM, and flexible manufacturing systems.
5. Discuss the place of industrial robots in manufacturing automation.
6. Describe the advantages and disadvantages of production lines.

Key Words

automation
sensor
mechanized
computer-numerical control (CNC)
computer-aided manufacturing (CAM)
computer-integrated manufacturing (CIM)
distributed or direct numerical control (DNC)
servo system
numerical control (NC)
NC machine axes
NC program
NC programmers
machine control unit (MCU)
machining center
adaptive control
flexible manufacturing systems (FMS)
production line

Most modern manufacturing is automated. This means that the processes are set up in such a way that they can be repeated under computer control. Little or no manual intervention is needed except to start and stop the process or to change workpieces and tooling. Even workpiece and tool changing are fast becoming automated as new computerized manufacturing technology is applied in almost all phases of production. **Automation,** already at a high level in many manufacturing industries, will be refined and applied to an even higher degree in the future. The result of this trend redefines the duties of the typical production worker or machine operator. The purpose of this chapter is to explore the techniques used to automate modern manufacturing and look at the types of automated equipment now being applied to production.

It is necessary to produce in quantity so that the selling price can be low enough to make manufactured products attractive to a wide range of prospective buyers. There will always be demand for one-of-a-kind, handcrafted products, but any product that is built one unit at a time will, by the very nature of the production time involved, be expensive compared with its higher-quantity production counterpart. The difficulty lies in "mass customization," manufacturing a variety of products to suit all needs and desires while keeping costs comparable to high-volume, low-cost "mass production."

FACTORS IN AUTOMATION

Automation would appear to eliminate jobs. The contrary view is that the manufacture, implementation, and maintenance of automated machinery creates many higher-level jobs. Manufacturing, driven by practical economics, will continue to exploit automation in an aggressive manner. Primary reasons for automating manufacturing include the following. Automation:

1. allows production at minimum costs and maximum productivity.
2. produces the required variety.
3. allows standardization and interchangeability of parts.
4. meets product demand and competition.

Costs and Productivity

Cost is a major force behind automation. All manufacturing industries are controlled by the economic forces in play at any given time. Throughout history such factors as inflation, taxation, labor costs, material costs, and costs of marketing and distributing have driven the manufacturing industry to search constantly for better and less expensive ways to accomplish production.

There has been much debate about long-range results of automation including the effect of automation on employment. Advances in manufacturing technology affect employment levels for low-skilled workers, and in the long term there will be fewer low-skilled to semiskilled production jobs as automated manufacturing becomes more and more the standard of industry. As economic conditions place more demands on individuals, employees place more pay demands on their employers who compensate by further automating industries, thus reducing labor intensity. Automation will create more jobs; however, these jobs will be shifted to a higher technical level, and production workers must accept retraining if they expect to remain gainfully employed.

Productivity generally can be defined as the ability to maximize the quantity and quality of production from all phases of manufacturing, including people and machinery. If plant equipment is antiquated, productivity will be low compared with modernized automated manufacturing, so the costs of the product may be higher.

In labor-intensive manufacturing, productivity may be difficult to maintain because of the inconsistencies of the human work force. On the other hand, automatic machines do not become tired or bored and are consistent in their output quality, assuming proper adjustment and maintenance.

Automatic manufacturing equipment can also operate in conditions under which humans could not function. For example, robotic assembly can be done in high temperatures, contaminated environments, or in total darkness, conditions that would not be satisfactory for people.

Standardization and Interchangeability

Few complex manufactured products are made completely by a single manufacturer. Companies often use many subcontractors to manufacture different parts for an assembly and then bring them together at a final assembly point. Examples are autos and aircraft. There are many advantages to manufacturing in this manner. Subcontractors become specialists in the production of individual parts or products. Their expertise develops keenly, and their research and product development can be concentrated in a narrow area, resulting in a product of high quality that may be economically produced in large volumes.

The importance of **standardization** cannot be overemphasized. Because a manufacturer of a general product line such as electrical or hydraulic equipment may act as a supplier to many different prime contractors, standardization of the product is essential. This permits **interchangeability**, the key to successful mass production of complex assemblies. Lack of standardization in manufacturing impedes interchangeability in assembly, making manufacture of complex products time consuming or impossible.

Automation plays an important role in achieving standardization because of the ability to duplicate products in various quantities with little or no variance in part dimensions and part feature locations.

Product Demand and Competition

Demand for a manufactured product will often dictate the degree of automation in production. Low demand dictates low production quantities, making the cost of automation more difficult to justify. For complex products such as ships, offshore oil platforms, and other large and complicated assemblies, demand will be low from a quantity standpoint. These products take time to complete, up to several months or years, and they are fairly labor intensive. Creating highly automated production lines to manufacture these products is complex and costly.

Production Volume

Some products are very well suited to high-volume production. Examples include nails, fasteners, lightbulbs, electronic components, and many common parts for assemblies of all types. These products are manufactured in quantities of billions each year, and the best way to meet volume requirements is to highly automate production. Since products can be made quickly and efficiently using automated manufacturing, their costs can be kept low, making products available at affordable prices.

Competition is a driving force in any free enterprise economy. Rarely do we find a product that is sold by only one maker. Competition tends to weed out producers with low quality, too high prices, or those who are unable to meet the production demand for their product. As the competition automates its manufacturing facilities in order to bring a better, less expensive, and more available product to the customer, then industry must do the same if it expects to compete in the marketplace. There are many examples where a manufacturer's inability or unwillingness to automate production has resulted in business failure.

AUTOMATION SYSTEMS IN MANUFACTURING

Automated control of manufacturing processes may take many forms. It ranges from rudimentary control, perhaps involving a simple computerized device that operates part of a machine, to complex computer-driven feedback systems with considerable decision-making capacity over the process that they are controlling. In a true automatic system, the automation device has an ability to sense conditions of the process and then react by altering its function.

Automation requires the following factors (Figure 22.1):

1. A process to automate
2. A sensing system for making decisions about the process
3. A control action system that operates on the sensed information and then provides a process control action.

Almost any manufacturing process can be automated at any phase of manufacturing, including the processes that inspect and/or package the finished product.

Mechanization and Automation

Consider a simple operation such as the carriage feed on a lathe (Figure 22.2). The motion must be initiated manually by the operator and disengaged manually at the end of the cut. This is not an automated process, but once engaged, it does serve to feed the tool by means of levers, gears, and motors. Thus, the process is **mechanized**.

By setting a feed stop, the operator can further mechanize the process. Initiation is manual, but feed disengagement is mechanically accomplished. Two factors are present here, the process (carriage feed) and the feed stop. This simple system does not have any feedback for further action after feed disengagement. The operator must take over manual control to withdraw and return the tool to the starting point. The operation is mechanized, requiring a manual initiation for each new cycle.

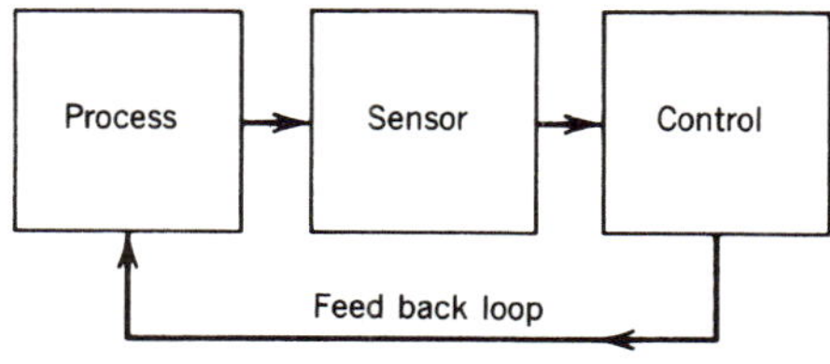

FIGURE 22.1
Factors in process automation.

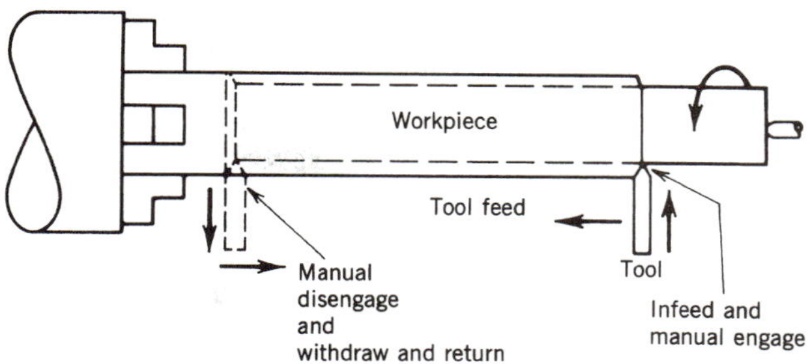

FIGURE 22.2
Control of the carriage feed on a lathe involves manual engage and manual disengage.

On a **computer numerically controlled (CNC)** lathe or turning center, the feed is engaged, disengaged, and the tool automatically withdrawn and returned to the beginning of the cut, or changed to a new tool, all under computer control. All three factors are present here for full automation, the process (tool feed), the **sensor**, and the action based on sensed information (tool withdrawal and return or tool change) (Figure 22.3). Systems such as these employing sensor feedback are called **servo systems.**

Numerical Control Automation

It would be difficult to find a device that has had more influence on manufacturing automation than the computer. Computer control of production processes or **computer-aided manufacturing (CAM)** as well as **computer-integrated manufacturing (CIM),** and their design counterpart (CAD), have profoundly altered manufacturing in dramatic ways. One very important area that has been established for many years and continues to undergo development is that of **numerical control (NC).**

NC is the equipment and methods used to control a process by numeric instructions through computers driving electromechanical or hydraulic actuators. Although NC has been applied to almost every manufacturing process, machining certainly is one area in which its development is unparalleled. Perhaps the reason for this is that machining involves a wide variety of high-precision manufacturing requirements.

All the previously discussed factors are present in an automated numerical control machining system. The processes to be controlled include tool and workpiece positioning, tool changing, speed and feed rates, and special functions unique to the specific process or machine tool. Electromechanical and electronic sensor systems are built into the computer control to make decisions and initiate appropriate actions.

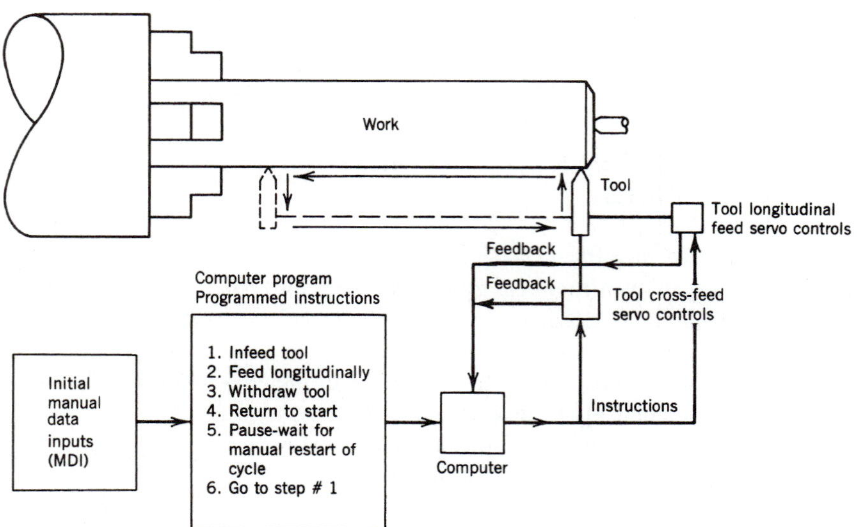

FIGURE 22.3
Full automatic programmable cycling.

NC Systems NC systems include open loop and closed loop. The open-loop NC system does not make use of a feedback system to determine control action. This system lacks precision and the ability to respond to disruptions in the manufacturing process.

Computer control technology permits sophisticated closed-loop NC systems to be used, in which sensed information is fed back to the control, thus initiating required control action. These CNC systems are widely used and very popular in automated manufacturing.

NC Motion and Other Machine Functions

Major components of an NC machine tool move along fundamental NC **machine axes.** The centerline of the machine spindle is the *Z*-axis. If the machine tool is a vertical machining center, the table and saddle components move at right angles to each other in a plane that is perpendicular to the spindle. Table motion is in the *X*-axis, and saddle motion is in the *Y*-axis. If the machine tool is a horizontal spindle design, the spindle centerline is still the *Z*-axis whereas the *X*-*Y* plane becomes vertical (Figure 22.4).

CNC lathe or turning center spindles are the *Z*-axis, whereas cross slide motion is the *X*-axis. Movement of the spindle, table, or saddle along each axis in either direction is defined in the **NC program.** Thus, each point within the entire volume of space in which the spindle may move or the workpiece may be positioned may be specified in the NC program. Modern CNC equipment has the capability to position the machine spindle and/or workpiece to specified locations within a few tenths of thousandths of inches. On large machine tools, this positioning can be repeated accurately over long distances, making possible extremely accurate cuts and location of holes and other features in the workpiece. By simultaneously combining motions by the machine spindle and/or workpiece, the NC machine tool can accomplish angular and contour (continuous path) machining that could not be done by conventional hand operation methods. Thus, the NC machine tool can perform tasks that would be virtually impossible by conventional practices. The mathematics required for continuous path and angular machining is calculated by the control computer, and the appropriate axis commands are directly generated.

The NC Program To accomplish machining tasks, the **machine control unit (MCU)** of the CNC machine must be programmed. This task may be done manually by the NC programmer, or it may be accomplished through the assistance of a computer. In manual programming, the programmer must be able to write the particular NC machine code used to direct each machine axis. In computer-assisted NC programming the programmer uses CAD data and graphics to generate a program, and the computer generates the NC machine code. **NC programmers** must have knowledge of the machining process, the cutting tools required, and the work-holding system, so NC programmers may be experienced CNC machine operators and machinists who have good math skills.

The NC program consists of a set of step-by-step instruction statements describing exactly what the machine is to do regarding positioning the cutting tool

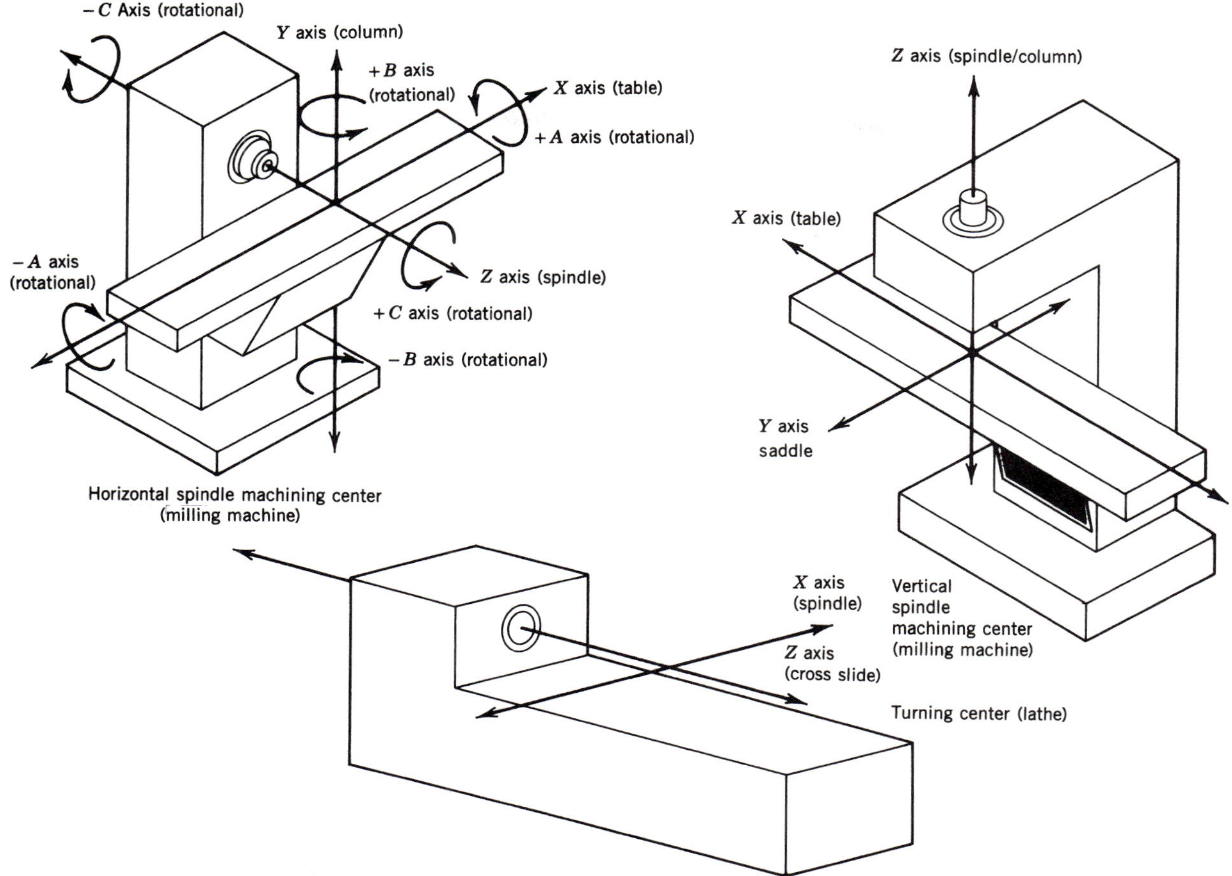

FIGURE 22.4
Reference axes for CNC machine tools.

or workpiece, selecting the required tool, and accomplishing the required cutting operation at the correct speed and feed. Other statements in the NC program initiate machine functions such as coolant on/off, spindle stop/start, drill, tap and bore cycles, tool changes, and workpiece indexing movements.

The NC program resides in the computer memory of the MCU. Programs can be transported to the MCU using diskettes, or NC machine tools may be networked together to receive programs from a single computer. This process, known as **distributed numerical control (DNC),** is common in automated manufacturing. Several programs may reside in the memory of the MCU.

Networking can be extended to information sharing between machine tools and the network, including the Internet. Beyond information sharing, NC machines may be operated remotely over a network; however, care must be taken to ensure that the machine and process are adequately monitored.

Some types of CNC machine tools may be programmed by manually running them through the various operations of a machining task while the MCU memorizes these motions. This process is called *teach programming.* Each manual operation becomes a computer program step, and once the first step is accomplished, the machine tool has been programmed to repeat the job.

Several documents may be required to assist the programming of a CNC machine tool. Generally, the machine operator will need a part drawing (Figure 22.5) indicating dimensions, as well as a setup sheet (Figure 22.6) indicating a list of programmed tools, and a list of gages used to measure the part. Figure 22.7 shows the machine control code generated from the computer-assisted program described in the previous illustration.

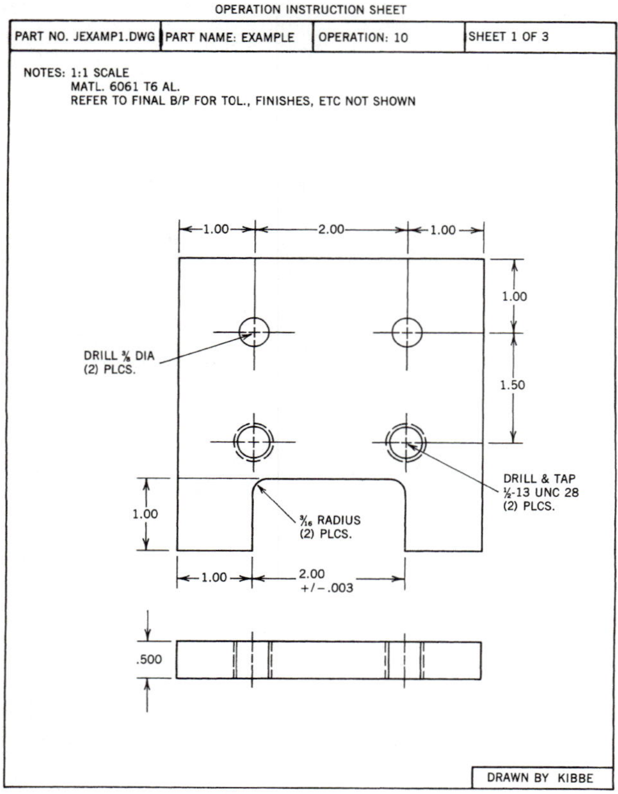

FIGURE 22.5
Part drawing for the CNC machining task.

PART NO. JEXAMP3.DWG	PART NAME: EXAMPLE OPERATION: 10	SHEET 3 OF 3		
SEQ	OPERATION DESCRIPTION	TOOL NO.	RPM	FEED
	SETUP INFO:			
	A) LOAD VISE TO NC MILL TABLE, JAWS PARALLEL TO Y AXIS.			
	B) INDICATE JAWS PARALLEL TO Y AXIS WITHIN .0005 T.I.R.			
	C) SET TOOLS IN APPROPRIATE HOLDERS PER TOOL LIST.			
	D) LOAD TAPE #001 INTO CONTROL MEMORY.			
	E) SET END LOCATING STOP ON VISE. LOAD A PART.			
	F) USING EDGEFINDER, LOCATE UPPER LEFT CORNER OF PART.			
	G) FROM THIS POINT, HANDLE OVER −10.0" IN X AXIS.			
	H) RESET AXIS READOUTS. THIS IS PROG. ORIGIN.			
	I) ZERO RETURN Z AXIS. PROGRAM BEGINS & ENDS HERE.			
	J) SET ALL OFFSET REGISTERS PER G.L.'S & T5 RADIUS.			
	K) USING STANDARD SETUP PROCEDURES (DRY RUN, MACH. LOCK,			
	SING. BLK., ETC,) WORK ALL TOOLS IN TO CUT PART.			
	L) LAYOUT 1ST PART COMPLETE & PRESENT TO INSPECTION.			
	PROG. SEQ.:			
	CENTERDRILL (4) HOLES	1	2000	16IPM
	DRILL 3/8" DIA. HOLES (2) PLCS.	2	2000	10IPM
	DRILL 7/16" MINOR DIA. (2) PLCS.	3	1200	8IPM
	TAP 1/2-13 UNC 2B THREAD (2) PLCS.	4	315	CALC
	MILL 2" POCKET WITH 3/8" E.M.	5	3000	25IPM
	GAGES REQUIRED:			
	.370/.380 DIA. PLUG GAGE			
	.430/.440 DIA. " "			
	1/2-13 UNC 2B THREAD GAGE			
	3/16" RADIUS GAGE			
	S.M.I.			
		DRAWN BY:		

FIGURE 22.6
CNC setup sheet defines job setup requirements, program sequence, tooling, and required inspection gaging tools.

Construction and Capability of NC Machine Tools

Because NC machine tools are designed for repetitive production of precision parts, they are constructed using the latest machine design available. Design effort is directed toward reliability of the mechanical and electronic components as well as long-term maintenance of machine accuracy. This makes the NC machine tool more expensive than its manually operated counterpart; however, its higher initial cost is far outweighed by its precision, productivity, and broad capability. Numerically controlled machine tools, as well as other types of NC material-processing equipment, have established themselves as essential examples of manufacturing automation.

The NC machine tool has capabilities to create precise geometry that cannot be accomplished by manual machines. This capability has reduced or eliminated many constraints previously placed on the product designer. With the capability of computer-controlled production, mass-produced products are appearing that would have been unavailable only a few years ago. Technological advances in computer-aided manufacturing permit new and innovative product designs that profoundly affect us all.

Types of NC Machine Tools Generally speaking, NC machine tools can be classified into two major groups, including machining centers and turning centers. Machining centers are equivalent to milling machines and turning centers perform lathe operations.

Machining Centers CNC machining centers are one of the most common types of automated machine tools. They can have vertical or horizontal spindles, and they can be equipped with a variety of features that increase their versatility. These include automatic tool changers where a number of cutting tools are mounted on a belt or rotary drum. Figure 22.8 shows a CNC machining center with an automatic tool changer. When the machine tool is in operation, the NC computer program selects the correct tool from the belt and mounts it in the machine spindle (Figure 22.9). When the cut is completed the program commands the tools to be removed from the spindle and returned to the tool holder.

```
   TOOL #1 : CENTERDRILL (4) HOLES

N001G91G28Z0T01
N002S63M03
N003T00M03
N004G92Z280000
N005G90G00X110000Y-10000Z160000
N006G81G99Z13900R141000F600
N007X130000
N008X110000Y-25000
N009X130000
N10G80Z161000
N011X0Y0Z280000

   TOOL #2 : DRILL 3/8" DIA. (2) PLCS

N012G91G28Z0T02
N013T01M06
N014G27280000
N015G90G00X110000Y-10000S73M03
N016G45Z160000H02
N017G81G99Z134000R141000F1000
N018X130000
N019G80Z161000
N02G45Z280000H02
N021X0Y0

   TOOL #3 : DRILL 7/16" MINOR DIA. (2) PLCS

N022G91G28Z0T03
N023T02M06
N024G92Z280000
N025G900G00X110000Y-25000S66M03
N026G45Z170000H03
N027G81G99Z144000R151000F800
N028X110000
N029G80Z171000
N030X0Y0

   TOOL #4 : TAP 1/2-13 UNC 2B THREAD (2) PLCS.

N032G91G28Z0T04
N033T03M06N034G92Z280000
N035G90G00X110000Y-25000S53M03
N036G45Z180000H04
N037G84G99Z164000R165000F2423
N038X110000
N039G80Z185000
N040G45Z280000H04
N041X0Y0

   TOOL #5 : MILL 2" POCKET

N042G91G28Z0T05
N043T04M06
N044G927Z2800000
N0455G90G00X111037Y-45000S76M03
N046G45Z260000H05
N047G01Z134000F2500
N048Y-31037
N049X128963
N050Y-46037
N051X110937Y-45000F1000
N052Y-30937F2500
N053X129063
N054Y-45937
N055G00Z154000
N056G45Z28000H05
N057X0Y0
N058G91G28Z0
N059T00
N060M11
N061M16
N062T05
N063M12
N064M30
=&S
```

FIGURE 22.7
The computer output for the CNC program.

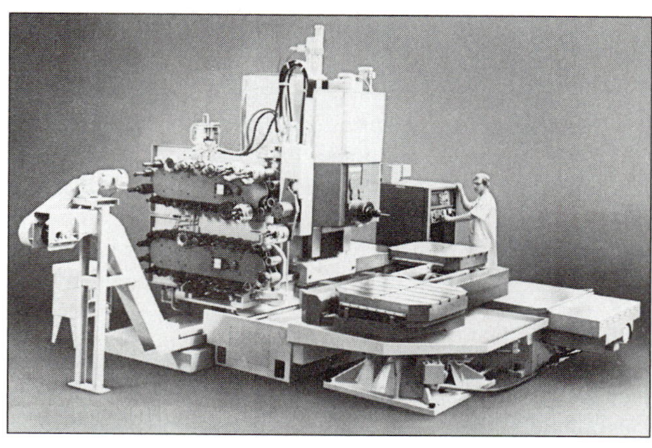

FIGURE 22.8
CNC machining center with twin work-holding pallets and automatic tool changing (Cincinnati Milacron).

FIGURE 22.9
A face milling cutter is delivered from tool changer to spindle for a face milling task (Cincinnati Milacron).

Multiple workpieces may be mounted on several pallets on the same machine tool (Figure 22.10). Pallets may be indexed to present different sides of the workpiece to the cutter for machining (Figure 22.11), and the machining center can be used as a component in a work cell as part of a **flexible manufacturing system (FMS).** In this case, a provision is made to remove a work-holding pallet from the machining center (Figure 22.12) and transport it to another machine or to a new cell or station in the manufacturing system. The machining center pictured in Figure 22.13 is capable of four- and five-axis machining owing to its pivoting head stock.

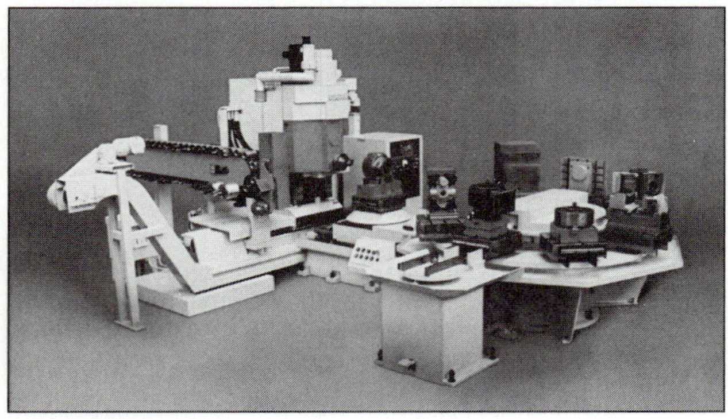

FIGURE 22.10
Multiple workpiece pallets on a CNC machining center with a variety of work-holding methods (Cincinnati Milacron).

FIGURE 22.11
With pallet indexing capability, different sides of a workpiece or different workpieces can be positioned for machining tasks (Cincinnati Milacron).

Turning Centers CNC turning centers (Figure 22.14) are another common type of automated machine tool. As on the machining center, the turning center's capability lies in its ability to bring a variety of tools into use to machine the workpiece.

Turning center tooling is often turret mounted (Figures 22.15 and 22.16). Since the machine spindle holds the workpiece, the tooling for machining on the outside diameter (OD) as well as on the inside diameter (ID) of a workpiece is located on the turret, ready for use during a production run. With four-axis CNC lathes, both ID and OD turning operations may be accomplished simultaneously. An example is inside and outside thread cutting at the same time (Figure 22.17).

The CNC turning center is also suitable for shaft turning operations (Figure 22.18) and can be equipped with a part-handling robot (Figure 22.19) that can automatically load and unload parts from the machine tool.

NC Multispindle Profilers The NC multispindle profiler is another important member of the family of automated machine tools. The profiler is a multiple-spindle version of the vertical or horizontal machining center. These machine tools are often quite large, and work-holding tables can be

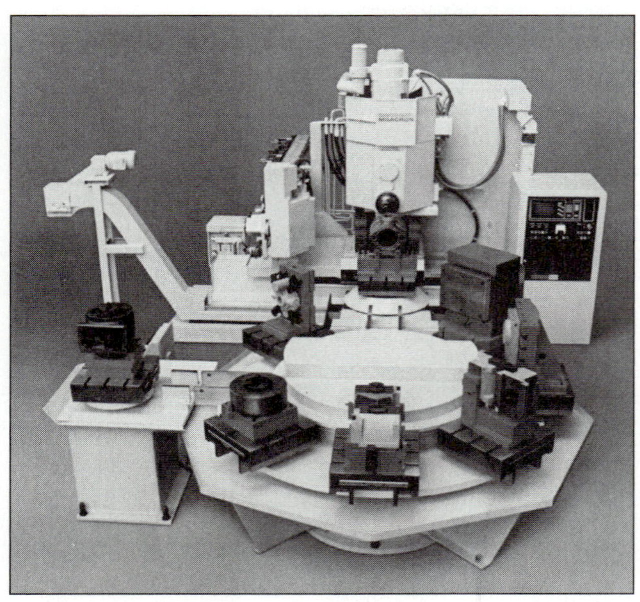

FIGURE 22.12
A workpiece pallet may be off-loaded or loaded when the machining center is a cell in a flexible manufacturing system (Cincinnati Milacron).

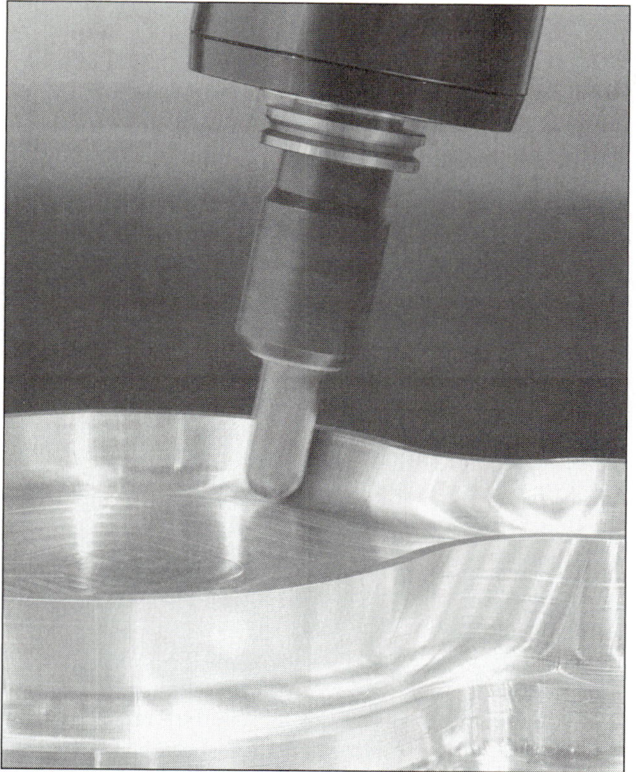

FIGURE 22.13
Four- and five-axis contouring on a CNC machining center (Haas Automation, Inc.).

FIGURE 22.14
CNC turning center (Cincinnati Milacron).

FIGURE 22.15
Turning center turret mounted tooling for ID and OD machining (Cincinnati Milacron).

FIGURE 22.16
Turning center turret tooling for OD shaft turning (Cincinnati Milacron).

FIGURE 22.17
Turning center used for simultaneous OD and ID thread cutting (Cincinnati Milacron).

FIGURE 22.18
OD shaft turning with turret-mounted tooling (Cincinnati Milacron).

FIGURE 22.19
CNC turning center with robot arm for loading and off-loading parts. The automated material-handling system is seen at the left (Cincinnati Milacron).

70 feet long or more. The spindles on the vertical spindle profiler are carried on a gantry supported on each side of the table. On the horizontal spindle model, the work-holding table is vertical, and the spindle gantry is supported on a floor rail.

A common design of profiler is the vertical spindle model with three or more spindles (Figure 22.20). The horizontal spindle model may also have three or more spindles (Figure 22.21). Automatic tool changers are also used on multispindle profilers similar to automatic tool changers commonly found on many NC machining cen-

FIGURE 22.20
This Saxis NC profiler is equipped with three vertical spindles and is engaged in machining complex contour aircraft parts (Lockheed Martin Co.).

FIGURE 22.21
Two views of the horizontal spindle CNC profiler. Workpieces are mounted on the vertical face of the large table (Lockheed Martin Co.).

ters. The NC profiler has numerous advantages including the capability to machine complex parts in multiple batches. The NC profiler is a favorite in the aircraft industry, where complex contour machining of duplicate parts is routinely accomplished.

Trends and Other Applications in Numerical Control
CNC is by no means limited to machining applications. It can be applied in almost any manufacturing process, including fabric cutting, typesetting, flame cutting, punch press operations (Figure 22.22 and 22.23), pipe bending (Figure 22.24), tool and cutter grinding, and nontraditional machining processes such as EDM, as well as measurement, inspection, and industrial robotics.

Numerical control now is and will continue to be a mainstay in manufacturing automation for the foreseeable future. As computer technology advances, NC will advance as well, continuing its important place in manufacturing automation. As computer technology becomes more sophisticated, higher levels of automation can be achieved. One such area now undergoing development is called **adaptive control.** Adaptive control involves sensing one or more conditions existing in the automated process and automatically adjusting the process to improve the performance.

An example of adaptive control is computerized acoustic emission analysis of machining processes. In each machining operation, characteristic acoustic emissions (sounds) are emitted by the machine and tooling during production. These sounds are excellent measures of cutting tool performance. In fact, during conventional manual machining operations, the experienced machinist or machine operator will listen to the

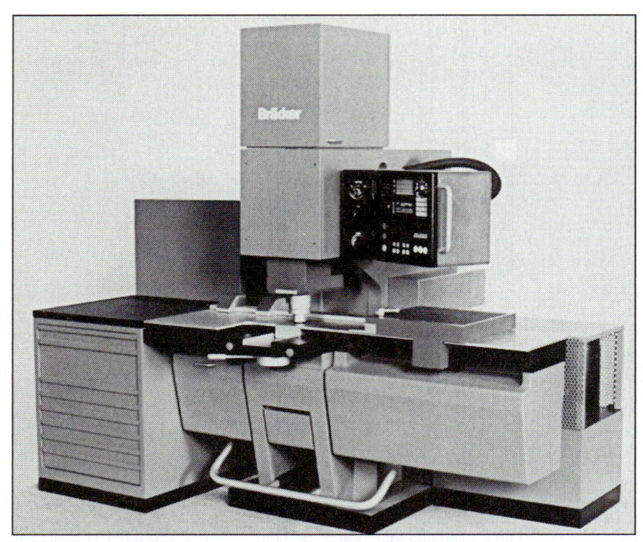

FIGURE 22.22
CNC punch press (Baltec Corporation).

FIGURE 22.23
CNC sheet metal punching is both versatile and productive (Baltec Corporation).

FIGURE 22.24
Numerical control concepts are utilized in the programming of robotic operations such as this drilling process. Note the magazine of drilling tools that can be automatically loaded onto the robotic arm (Lockheed Martin Co.).

sound of the cut as well as observe the visual action. A change in the characteristic sound often will indicate a problem such as a dull or broken tool before the visual aspects of the situation appear. In electronic acoustic analysis, sounds emitted from the production processes are instantly compared with a database in the computer representing the correct characteristic sounds. If the comparison creates a match, then production continues. If the sounds are not comparable, the computer control will stop the machine or take other appropriate corrective action. This is another example of the high degree of sophisticated computer process control that is fast becoming the industry standard.

CIM AND FLEXIBLE MANUFACTURING SYSTEMS FOR THE INTEGRATED FACTORY

CIM

CIM refers to computer-integrated manufacturing. In integrated manufacturing the design and manufacturing engineering functions work together to rapidly create manufacturable products and agile, efficient manufacturing processes. Automated manufacturing equipment is integrated into a system whose individual machines and devices communicate together as a cohesive unit.

In fully integrated CIM systems, once design data have been generated they can be used to generate the programs needed to operate manufacturing equipment such as CNC machining centers (Figure 22.25), flexible manufacturing systems (FMS) or industrial robots.

Flexible Manufacturing Systems

The ultimate in integration is the completely integrated and fully automated manufacturing plant. This concept is called *lights-out manufacturing.* In a lights-out manufacturing facility, workers bring raw materials into the facility, remove finished products, and perform preventive maintenance. At the end of the day, the workers turn off the lights, and the factory operates unmanned until they return the following morning.

The level of automation that is required for lights-out manufacturing is complex and sophisticated. Many human tasks are difficult to automate, plus the software and computer hardware required to manage an entire manufacturing facility is extensive. Even though the cost of this level of automation is difficult to justify economically, lights-out factories do exist, typically as showcases of technology.

Major steps toward full factory automation include the flexible manufacturing system. The flexible manufacturing system consists of manufacturing cells that are equipped with machining centers or other automated manufacturing equipment. Figure 22.26 shows a 10-station FMS with an automated material-handling system. Figure 22.27 shows a manufacturing cell equipped for vertical turning operations. Note how the cell is equipped with pallets for several workpieces. As one part is machined another is ready for loading.

FIGURE 22.25
Integrating CAD and CAM in machining.

FIGURE 22.26
Ten-station flexible manufacturing system or FMS with automated parts transfer between cells (Giddings & Lewis).

FIGURE 22.27
Vertical turning work cell, a component of the FMS (Giddings & Lewis).

FIGURE 22.28
CNC work cell machining center with workpiece pallets (Giddings & Lewis).

FIGURE 22.29
Loading and off-loading parts using an industrial robot (Giddings & Lewis).

FIGURE 22.30
Transporting workpieces mounted on machining pallets between cells in the FMS (Giddings & Lewis).

FIGURE 22.31
Workpiece transfer in a modern fully automated FMS (Giddings & Lewis).

This increases efficiency greatly, as the machining centers in the cell are constantly kept engaged in production (Figure 22.28). Part loading and off-loading can also be accomplished by industrial robots (Figure 22.29).

The entire system of machining centers and cells may be tied together with an automated material handling system designed to move parts between cells and cell workstations (Figure 22.30). Part transfers may be accomplished automatically by automatic guided vehicles (AGVs), which are guided from station to station along a track installed in the factory floor (Figure 22.31). Part inspections are accomplished within the FMS system at

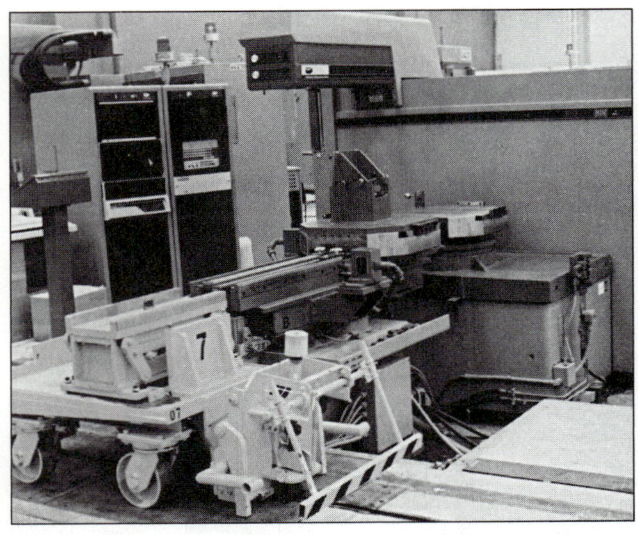

FIGURE 22.32
The high-precision multiaxis workpiece inspection cell in an FMS (Giddings & Lewis).

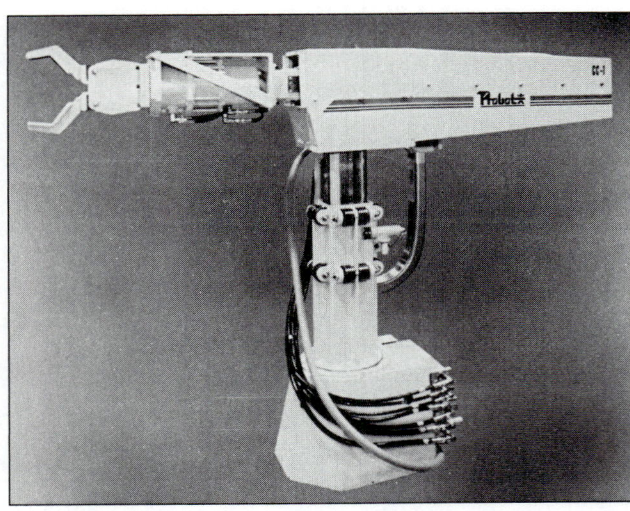

FIGURE 22.34
Programmable teachable industrial robot (Positech Corporation).

FIGURE 22.33
Specialty FMS cell machining differential housings for the auto industry (Giddings & Lewis).

the inspection cell (Figure 22.32). The entire FMS is under the control of a central computer that commands machining center activity as well as material transport and transfer systems.

The FMS is extremely versatile (Figure 22.33) owing to the flexibility of each machining center to do many different jobs. In addition, the production of the system can be reprogrammed easily to handle a different part without the need for expensive retooling or the purchase of new manufacturing equipment.

INDUSTRIAL ROBOTS

The industrial technologist, manufacturing technologist, and manufacturing engineer have always sought better, faster, and less expensive ways to accomplish production. In times past when human labor was easily procured at low cost, industries employed many production workers in all types of repetitive work. In modern times it has become more difficult and certainly more expensive to hire people to do much of the mundane work of manufacturing. Modern production workers not only deserve wages that are in line with the costs of living, they also demand safe and comfortable working conditions and fringe benefits, all of which add to overall labor costs and must be reflected in the final price of manufactured goods.

Although it is unlikely that automation will displace the production worker completely, it is certain that more and more of the less-skilled production work will be taken over by automated equipment. Therefore, the trends toward replacing people with machines will continue. One important contribution toward the trend is the industrial robot (Figure 22.34).

What is a robot? A *robot* is a fully automatic device that can accomplish a specific task such as assembly or loading or unloading parts in a machine tool. Through computer control, the robot may be programmed to do many different functions, thus becoming an extremely versatile, untiring aid to manufacturing automation.

Basic Industrial Robotic Systems

A simple industrial robot may be of the pick-and-place design. These machines, also called *polar robots,* have limited capability and are primarily designed to pick a part from one location, turn, and place the part in another location or on an assembly.

Because industrial robots are built by people to accomplish tasks previously performed by people, it is logical that they have some human characteristics. A robot of the jointed-wrist design is sophisticated and can be programmed with motions in multiple axes. The jointed-wrist capability (Figures 22.35 and 22.36) permits needed flexibility for complex movement necessary in many industrial applications. Such robots have various types of grippers (fingers) that permit them to pick up objects (Figure 22.37). The jointed-wrist robot is very useful for painting, welding, and other tasks requiring the turn, bend, and multiposition capability of wrist motion. Depending on the application, mobility may be an important robot feature, permitting the robot to reach the work that it is intended to accomplish.

FIGURE 22.35
Industrial robot hydraulic jointed-wrist component (Bird-Johnson Fluid Power Division).

The fully automated robotic system has sensors and the ability to feed back information to its computer control, enabling the robot to make decisions regarding its assigned task. This system is to be differentiated from the abilities of a simple positioning device with only one or two functions and no real automated capability. It is also different from a fixed-cycle operation in which a simple task is carried out repeatedly without any variance, and there is no easy way to change or reprogram the function. Sensory capability of industrial robots may include sight, accomplished through use of a video camera. For example, vision-equipped robots can recognize parts in a bin

FIGURE 22.36
Motions of the jointed robotic wrist permit great flexibility in applications (Bird-Johnson Fluid Power Division).

FIGURE 22.37
Styles of robotic finger grippers (Barrington Automation Ltd.).

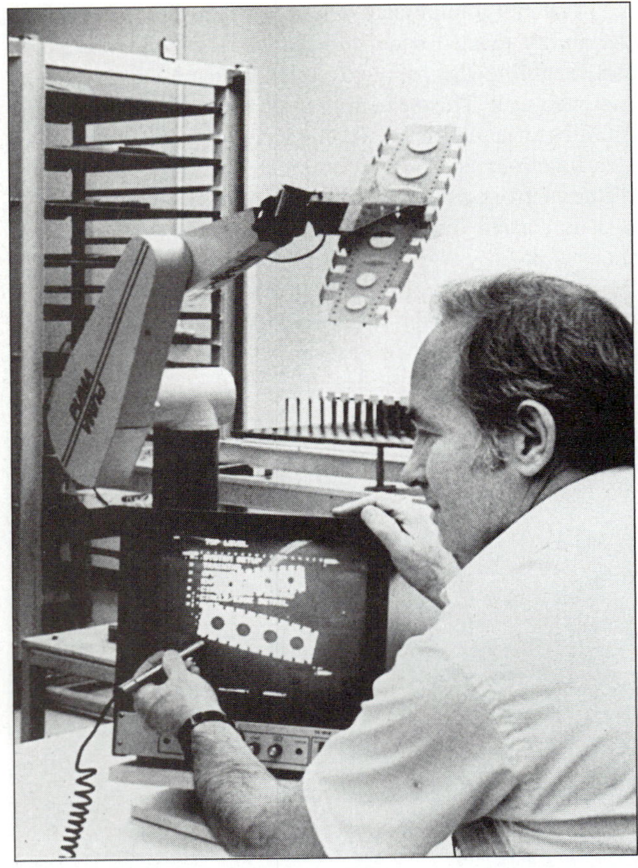

FIGURE 22.38
Sight for industrial robots, a most desirable capability. (Lockheed Martin Co.).

and then select the correct part to be placed in an assembly (Figure 22.38).

The Teachable Robot Programming the teachable industrial robot is accomplished by running the machine through each of its desired positions and functions during an initial manually controlled cycle. The computer remembers each position and function that it is taught. Once this is accomplished, further cycles are repeated automatically. Reprogramming is accomplished by clearing the computer's memory of previous instructions and teaching the robot new instructions. If programming is to be saved for future repeat applications, the programmed cycle may be stored in computer memory.

Industrial Robot Applications

Applications for industrial robots are numerous. Robots can be used to position workpieces in a process or on a machine. They can load and unload parts from a machine (Figure 22.39). Many types of assembly operations are well suited to robotics (Figure 22.40). Robots can move parts to or between workstations in a factory. In Figure 22.41, robotic AGVs follow a wire embedded in the floor for guidance as they carry engine assemblies to the automobile assembly line.

FIGURE 22.39
Using a robot to load and unload parts on a CNC turning center (Cincinnati Milacron).

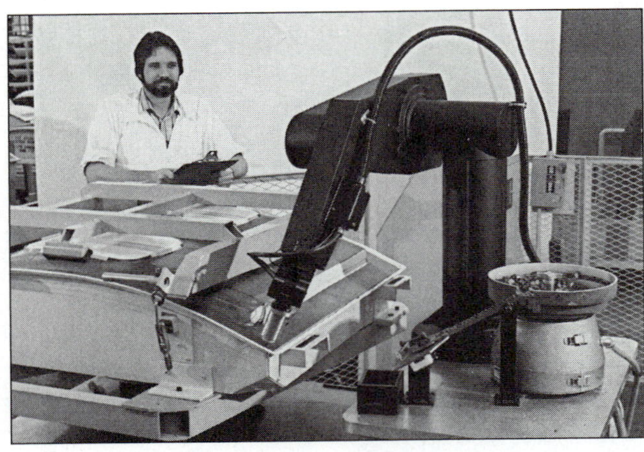

FIGURE 22.40
Using a robot for an assembly operation (Boeing, Inc.).

Welding is a common application for industrial robots (Figure 22.42). Since the automobile industry has shifted to unibody construction in which the car body has become the primary structural unit, robotic spot welding has come into wide use. Robotic welders located on the assembly line permit production spot welding of auto bodies with great accuracy and at a high production rate (Figure 22.43). Other common uses for robots in the automobile industry include painting and the application of caulking-type sealants. Robots are also used for many specialty applications including automatic belt tensioning on automobile engines (Figure 22.44).

THE PRODUCTION LINE AND MASS PRODUCTION

In the earliest examples of mass production it was discovered that work can be accomplished most efficiently by setting up a series of workstations, at which individual processing or assembly operations are performed (Figure 22.45). The product is then moved from one station to the next station, where the next process is accomplished. This sequence is continued until the product arrives at the end of the production line with all manufacturing operations and/or assembly operations completed. Production using these **production lines** is effective for both small and large products (Figure 22.46). Products may be moved from workstation to workstation by hand (Figure 22.47) or on assorted mechanized conveying assemblies (Figure 22.48). A typical system used in the automotive industry uses floor conveyors and specialized pallets to move the product along the production line (Figure 22.49).

There are many advantages to production by this scheme. Each workstation can be highly efficient and equipped with any specialized tools needed for the particular process occurring there. In most production accomplished by this method, only one or two processes are usually done at any one station. Hence, personnel staffing a specific station are specialists at their work. If proper supervision is applied, both efficiency and quality can result.

The specific product to be manufactured dictates the size and capability of the production line required. The estimated demand for the product dictates the required production line capacity.

Any mass-produced product will, of course, require applicable material processing equipment to mill, drill, tap, ream, bore, index, position a workpiece, and perform other functions as well.

Machining Transfer Lines

One specialty type of production line is the machining transfer line, where continuous product flow may be maintained while at the same time the desired multiple machining operations can be accomplished. On the transfer line, a series of machine heads are located and tooled for the desired machining operations. The product is then

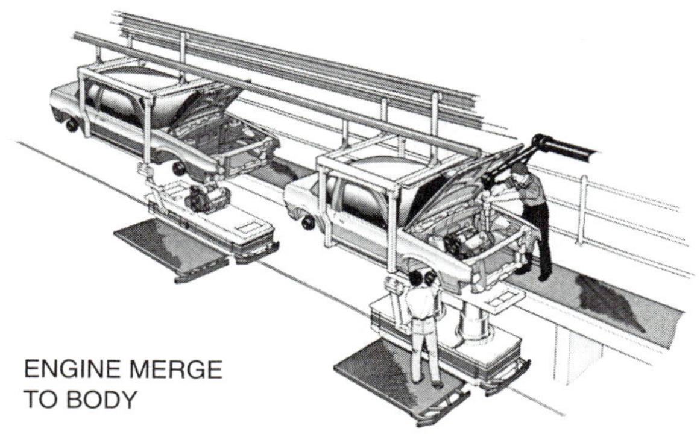

FIGURE 22.41
Robot carriers move engine assemblies to car bodies.

FIGURE 22.42
Robotic welding (PhotoDisc, Getty Images).

FIGURE 22.43
CNC robot welders apply thousands of spot welds to each auto body (PhotoDisc, Getty Images).

FIGURE 22.44
Belt tension adjusting, a specialty application for a specialty type industrial robot.

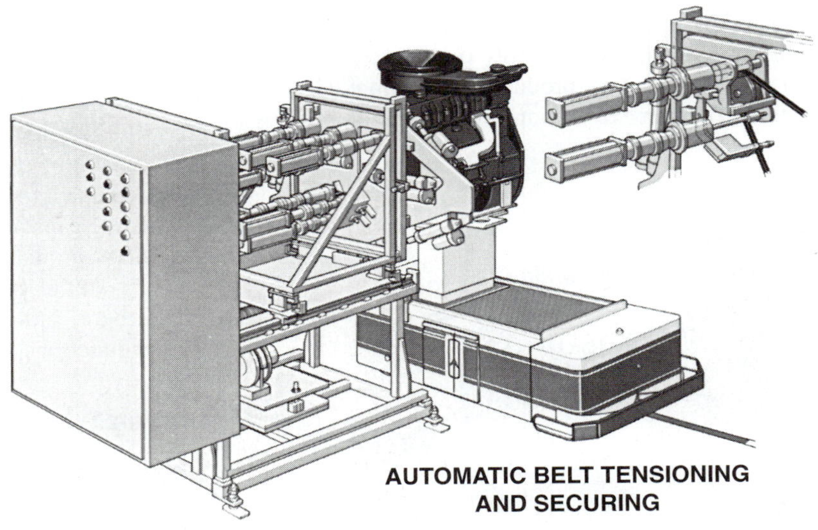

FIGURE 22.45
On the production line, a portion of assembly occus at each station (Corbis Digital Stock).

FIGURE 22.46
Production lines are equally applicable to large products (Lockheed Martin Co.).

FIGURE 22.47
Moving products along the line by hand is also efficient (Getty Images).

FIGURE 22.48
Product transportation using a mechanized conveyor (Allied Uniking).

FIGURE 22.49
Palletized product transportation on the production line.

moved along the line to each station, where required machining is accomplished. The machining transfer line is an example of a machining factory production line, but it is used for specialized machining operations. Transfer lines are very popular in the production of automobile engines and other high-production automotive mechanical systems.

Production Lines, Present and Future

Possible disadvantages of line production are that the work is often repetitive and therefore boring for personnel. Low-quality output can result. Manufacturing industries are experimenting with many techniques to make production line workers more comfortable at their job and keep quality at a high level. There is also a major trend to further automate production lines using industrial robots to replace people in repetitive or hazardous jobs. With the wide capability of the programmable computer-driven robot, it is a certainty that these machines will become ever more present along the production lines of the world's manufacturing industries.

Review Questions

1. What factors prompt manufacturing automation?
2. Are there any negative aspects to automation?
3. What factors are necessary for automation?
4. What is the difference between semiautomatic and full automatic?
5. Explain in general terms what CNC automation is.
6. What are the advantages of CNC?
7. What are the common types of CNC machine tools?
8. What are CAM and CIM and how do they work?
9. What is a flexible manufacturing system?
10. What components does an FMS comprise?
11. What are the advantages of an FMS?
12. What is an industrial robot?
13. What sensory capability might an industrial robot have?
14. What is a teachable robot?
15. List some applications for robotics in manufacturing.
16. What are the major advantages of production lines?
17. What are the disadvantages of production lines?

Case Problems

Case 1: The Level of Automation

A manufacturer of precision mechanical equipment has many varied precision machining requirements including precision hole drilling, milling, and turning. The products produced consist of several precision subassemblies. The production requirements often are varied depending on the particular models being produced and the quantities required. Automation in machining is desirable because of its ability to meet required close tolerances. The manufacturer may equip the factory with several types of automated machine tools. These include machines that have fixed automation cycles, numerical control (NC) machine tools that are easily programmed for variable requirements, and flexible manufacturing systems that not only can be programmed for different tasks but also will provide a greater degree of automation by shifting parts from station to station.

Considering the primary factor of the need to machine many different close-tolerance parts, which automation system would be most applicable—fixed cycle automation, numerical control, or numerical control flexible manufacturing systems? Discuss advantages and disadvantages of each.

Case 2: Robotic Applications

Given the following situations:

A. One automated assembly operation requires merely that certain parts be selected from a supply and then placed in a location for fastening.

B. A second automated assembly operation requires the same parts to be selected from a supply and then placed in a location for fastening; however, in this case, a visual evaluation also must be made to properly reference the part position prior to completing the assembly.

Which operation is suited to robotics: both, only A, only B, neither?

With common readily available robotic technology, which process would be most suited to robotics?

If a common industrial robot was used, what added equipment might make this machine more suited to handling situation B?

Case 3: Automate or Improve Working Conditions

A manufacturer is receiving poor quality work from production lines involving repetitive simple operations on simple parts. The production manager recommends that the company invest in industrial robots to replace all production line workers. The production line foreman suggests that the problem could be solved by rotating production line personnel so that their work is more varied and they take more interest in it. The company gainfully employs several hundred people, but at minimum or very low wages for production line work.

Considering all factors including costs of retooling the factory, the company's public relations image with the community, and the type of product, explain which choice you would make: purchase robots or adjust working conditions for production workers. Also explain why.

Glossary

ABS Acrylonitrile butadiene styrene, an engineering thermoplastic; a terpolymer with very useful properties.

Acicular Needle-shaped particles or structures as found in martensite.

Acoustic emission Machines and cutting tools emit characteristic sounds when operating at optimum efficiency. These sounds are digitized by computer and the data are compared against a database already in the computer. The information is used to determine that the tool is operating as it should. If not, the computer may take action to stop the machine or indicate a problem to the operator.

Acrylonitrile butadiene styrene See *ABS.*

Adaptive control The capability to sense changes in the conditions of an automated process and to react by automatically adjusting the process.

Adhesives Materials or compositions that enable two surfaces to join together. An adhesive is not necessarily a glue, which is considered to be a sticky substance, since many adhesives are not sticky.

ADI Austempered ductile iron, a ductile or nodular cast iron that has been given an austempering heat treatment. As a result it has a matrix of bainite with the free carbon is the form of nodules; it is the toughest of the cast irons.

Aggregate Small particles such as powders that are used for powder metallurgy that are loosely combined to form a whole; also sand and rock as used in concrete.

Aging The process of holding metals at room temperature or at a predetermined temperature for the purpose of increasing their hardness and strength by precipitation; aging is also used to increase dimensional stability in metals such as castings.

AISI American Iron and Steel Institute.

Allotropy The ability of a material to exist in several crystalline forms.

Alloy A substance that has metallic properties and is composed of two or more chemical elements of which at least one is a metal.

Amorphous Noncrystalline, a random orientation of the atomic structure, especially used to describe certain polymers.

Anisotropy A material that has specific physical properties in different directions. Rolled steel is strongest in the direction of rolling.

Anneal In general, the heating of metal to soften it or reduce its strength. There are many types of anneals; see *Full anneal, Partial anneal, Process anneal,* and *Stress relief anneal.* When a metal is heated it goes through three stages; see *Recovery, Recrystallization,* and *Grain growth.*

Anode The metal in a plating process or in a galvanic corrosion cell that is dissolved, removed, or corroded. Electrically it is positive.

Anodic A metal toward the top of the electromotive or galvanic series is *anodic* to those below it; it is more easily corroded by galvanic corrosion.

Anodizing Subjecting a metal to electrolytic action, as takes place at the anode of a cell, in order to coat it with a protective or decorative film; used for nonferrous metals.

Artificial intelligence Computer programs that mimic the rules and logic used by human beings to make decisions.

ASM International A materials society located at Materials Park, Ohio, previously known as The American Society for Metals, publisher of much useful information on materials of all types. See **www.ASM.com**

Assembly drawing One of the two major types of typical manufacturing drawings, showing several parts in an assembly and containing pertinent information regarding the assembly, such as a bill of materials. See *Detail drawing.*

ASTM American Society for Testing and Materials.

Atomic bonding The joining together of atoms with ionic, covalent, or metallic bonds; the molecular or van der Waals bond is a weak bond seen in inert gases, or a secondary bond in polymers.

Atomization The process of converting a liquid into a spray of very small droplets.

Austempered ductile iron See *ADI.*

Austempering A heat-treating process consisting of quenching a ferrous alloy at a temperature above the transformation range in a medium such as molten lead; the temperature of the quenching medium is maintained below that of pearlite and above that of martensite formation to produce a tough, hard, bainitic microstructure.

Austenite A solid solution of iron and carbon and sometimes other elements in which gamma iron, characterized by a face-centered crystal structure, is the solvent.

Austenitizing The process of forming austenite by heating a ferrous alloy above the transformation range.

Autoclave A sealed chamber used to heat and cure parts assembled by adhesive bonding techniques.

Automation Computerized processes and equipment used in manufacturing to accomplish operations with little or no external control or inputs by people.

Bainite An austenitic transformation found in some steels and cast irons. The microstructure consists of ferrite and a fine dispersion of cementite that has the needlelike appearance of martensite.

Basic oxygen furnace See *BOF.*

Beneficiation In processing an ore, improving the return of metal from an ore through some preliminary process.

Bill of materials A list generally appearing on an assembly drawing indicating the names of the parts in the assembly as well as their detail part drawing numbers and possibly the vendor source of the parts.

Binary data Information expressed in terms of zeros and ones called bits. Binary encoded information can be represented as ON (1) or OFF (0) positions of electronic switches within a computer's central processor.

Blanking The operation of cutting a shape with a die from sheet metal stock. The hole material is saved and used for further operations.

Blueprint A manufacturing drawing.

BOF Basic oxygen furnace, an improved method of converting the iron produced by the blast furnace to steel. Older methods used air to reduce the excess carbon; the BOF uses oxygen, sometimes mixed with argon.

Bonding See *Atomic bonding.*

Brazing A process that uses heat to join metal parts below their melting points, joining them mechanically rather than fusing them. Filler metals with melting points above 840°F join the parts.

Breaking point The final rupture of a material being pulled in tension, *after* it has reached its ultimate strength.

Bright anneal The softening process for ductile metals that have been work-hardened. The metal is enclosed in a container with inert gas to avoid scaling and then heated to its recrystallizing temperature. See *Process anneal.*

Brinell hardness The hardness of a metal or alloy measured by pressing a hard ball (usually 10 mm diameter) with a standard load into the specimen. A number is derived by measuring the indentation with a special microscope. The abbreviation *HB* refers to the Brinell hardness number.

Briquette A compacted mass of usually fine material such as metallic powder used in powder metallurgy. Also called a *green compact.*

Brittleness The property of materials to not deform under load but to break suddenly; for example, cast iron and glass are brittle. Brittleness is opposite to plasticity.

BUE, built-up edge A condition in which a metal being machined adheres to the point or cutting face of the tool, becoming in effect a new rough cutting edge of the tool. This situation causes tearing and a rough finish in the workpiece being cut.

CAD Computer-aided design or computer-aided drafting. Drawings are made by computer instead of pencil and paper in this system.

CAE Computer-aided engineering; the use of CAD geometric models to test product designs as a substitute for the building and testing of prototype products. CAE analysis includes determining volume, strength, wind resistance, vibration characteristics, fit, motion, and many other factors.

Calendering A process that involves rolling of the product into sheets to achieve desired surface finishes and thickness.

Calibration The processes of comparing measuring instruments and gages against known measurement standards and then adjusting them to conform with the standards.

CAM Computer-aided manufacturing or computerized control of manufacturing machinery and processes.

Capability The ability of a manufacturing process to create components or products with all dimensions within the tolerances specified on the manufacturing drawing.

Carburizing A process that introduces carbon into a heated solid ferrous alloy by having it in contact with a carbonaceous material. The metal is held at a temperature above the transformation range for a period of time. This is generally followed by quenching to produce a hardened case.

Case hardening A process in which a ferrous alloy is hardened so that the surface layer or case is made considerably harder than the interior or core. Some case-hardening processes are carburizing and quenching, cyaniding, carbonitriding, nitriding, flame hardening, and induction hardening.

Cast iron Iron containing 2 to 4½ percent carbon, silicon, and other trace elements. It is used for casting into molds. Cast iron is somewhat brittle.

Casting A process of producing a metal object by pouring molten metal into a mold.

Catalyst An agent that induces catalysis, which is a phenomenon in which the reaction between two or more substances is influenced by the presence of a third substance (the catalyst) that usually remains unchanged throughout the reaction.

Cathode The metal in a plating process or in a galvanic corrosion cell that is plated onto, or not dissolved, removed, or corroded. Electrically it is negative.

Cathodic A less corrodible metal that is toward the bottom end of the electrochemical or galvanic series.

Cathodic protection The coupling of a very corrodible or more anodic metal, e.g., zinc or magnesium, to a metal to be protected; this metal thus becomes cathodic or protected compared with the more anodic metal.

Cellulose A polysaccharide of glucose units that constitutes the chief part of the cell walls of plants. For example, cotton fiber is over 90 percent cellulose and is the raw material of many manufactured goods such as paper, rayon, and cellophane. In many plant cells, the cellulose wall is strengthened by the addition of lignin, forming lignocellulose.

Cemented carbides Very hard, heat-resistant composite materials composed of metal carbide particles held together by a bonding material.

Cementite Fe_3C. Also known as iron carbide, a compound of iron and carbon.

Centrifuging Casting of molten metals by using centrifugal forces instead of gravity. The mold (or molds) is rotated about a center where molten metal is poured and allowed to follow sprues outward and into the mold cavity.

Ceramics (or ceramic materials) A family of materials that are compounds, traditionally consisting of a metal and an oxide, *but* they may also be carbides, sulfides, nitrides, and intermetallic compounds. Ceramics generally have an ionic bond, are very hard and brittle, and can withstand high temperatures.

CF Cold finished.

Chip A slice of material sheared from a workpiece by a cutting tool in a machining operation.

CIM Computer-integrated manufacturing; software, machines, and devices communicating and working together under computer control in a manufacturing system.

CIP Cold isostatic pressing. The process of compacting a powder by exerting a constant high pressure at room temperature.

Cladding The joining of one metal (usually sheet or plate) to another by using heat and pressure or by an explosive force. With this method, a thin sheet of more expensive metal or one less likely to corrode may be applied to a less expensive metal or one more likely to corrode.

CMC Ceramic matrix composite; a composite material that has a reinforcing material in a ceramic matrix.

CMM Coordinate measuring machine; a computerized measuring instrument that precisely tracks the movement of a probe in three-dimensional space. The CMM records the location of the probe as it touches the workpiece, making the CMM useful for measurement of size, form, and location.

CNC Computer numerical control. The computer is an integral part of the machine control unit in CNC.

Coherent light Light energy that is all the same wavelength and synchronized. A laser is a coherent beam of light.

Coining (embossing) Shaping a piece of metal in a mold or die, often creating raised figures or numbers.

Cold drawing Reducing the cross section of a metal bar or rod by drawing it through a die at a temperature below the recrystallization range, usually room temperature.

Cold finish The surface finish obtained on metal by any of several means of cold working, such as rolling or drawing.

Cold rolling Reducing the cross section of a metal bar in a rolling mill below the recrystallization temperature, usually room temperature.

Cold working Deforming a metal plastically at a temperature below its lowest recrystallization temperature. Strain hardening occurs as a result of this permanent deformation. See *Work hardening.*

Colloid Thick viscous substances, like table jelly or gelatin, that will not pass through a semipermeable membrane.

Compaction The process of making a solid shape in a die by using punches to press together loose powder or a blend of powders; also called molding.

Comparison inspection Inspection accomplished by comparing a known standard with an unknown dimension.

Composite fibers The strands of material used as reinforcement extending through a resin or other matrix in a composite material. An example is carbon fibers in an epoxy matrix. Loads applied to the structure are carried by the fibers.

Composite material A materials category that consists of a matrix or base material reinforced by another material. The materials maintain their original identification but the properties of the combination are improved.

Compound A homogeneous substance combining atoms of different elements in ratios of whole numbers, e.g., H_2O, $CaCO_3$.

Compressive strength (ultimate) The maximum stress that can be applied to a brittle material in compression without fracture.

Compressive strength (yield) The maximum stress that can be applied to a metal in compression without permanent deformation.

Computer graphics Drawings, pictures, graphs, and other images generated by a computer.

Cope In foundry terminology the top half of a split mold used in casting.

Core In foundry terminology, a usually sand and oil form, strengthened by heating, used to create voids (holes, cavities) in castings.

Core print In foundry terminology a portion of the pattern that will create a cavity to position and support a core used to create the casting; the core print is not a portion of the finished casting.

Covalent bond A type of atomic bonding that requires the *sharing* of valence electrons to complete the outer shell; appears principally in gases, liquids, and polymers.

CR Cold rolled.

Crazing Minute surface cracks on the surface of materials, often caused by thermal shock.

Creep Slow plastic deformation in steel and most structural metals caused by prolonged stress under the yield point at elevated temperatures.

Cross-linking Primary, i.e., ionic or covalent, bonds *between* chains of a polymer, so that the polymer takes on a three-dimensional structure; this normally occurs in *thermoset* polymers.

Crystal unit structure or **unit cell** The simplest polyhedron that embodies all the structural characteristics of a crystal and makes up the lattice of a crystal by indefinite repetition.

Crystalline, Crystallinity In solid metals, having a repeating geometric arrangement of atoms. Polymers are said to be crystalline, or exhibit crystallinity, when their chains become aligned into a pattern. The opposite of *amorphous,* having a random structure.

Crystalloid A substance that forms a true solution and is capable of being crystallized.

Curie temperature The temperature above which metals are no longer magnetic. For iron this temperature is 1414°F. (It is named for Pierre Curie, husband of Marie Curie.)

Cutting fluid A term referring to any of several materials used in cutting metals, such as cutting oils, soluble or emulsified oils (water based), and sulfurized oils.

Cutting speed The relative speed between a cutting tool and workpiece in a machining operation. Units are feet per minute (surface feet per minute) or meters per minute.

Decarburization The loss of carbon from the surface of a ferrous alloy as a result of heating it in the presence of a medium such as oxygen that reacts with the carbon.

Deformation Alteration of the form or shape as a result of the plastic behavior of a metal under stress.

Dendrite A crystal characterized by a treelike pattern that is usually formed by the solidification of a metal. Dendrites generally grow inward from the surface of a mold.

Densification The process of increasing the density of a mixture, typically through compaction using pressure and/or heat.

Density Formally, the ratio of a material's mass to unit volume. In SI terms, the units are Kg/m^3, or g/cm^3. In the USA it is more common to use "weight density," with the units of lb/ft^3.

Deoxidizer A substance that is used to remove oxygen from molten metals; for example, ferrosilicon in steelmaking.

Deplating The reverse of the plating process. Ions are removed from a metal object through electrolysis.

Depth of cut The thickness of the layer of material sheared from a workpiece by a cutting tool in a machining operation. The depth of cut determines the width of the chips produced in the machining operation.

Design engineering The function or department that involves the design and documentation of products.

Detail drawing One of the two major types of production drawings showing an individual part. See *Assembly drawing.*

Deviation gages Measuring instruments that do not display a complete dimension but instead display the deviation from the size of a master of known size.

Die casting Casting metal into a mold by using pressure instead of gravity or centrifugal force.

Dielectric oil A liquid that does not conduct electricity. The EDM process takes place with the tool and workpiece submerged under a dielectric oil.

Diffusion The process of slow movement of atoms resulting in the intermingling of atoms or other particles within a metal. In solids, diffusion typically occurs at elevated temperatures but below the melting point. The process may be *(a)* migration of interstitial atoms such as carbon, *(b)* movement of vacancies, or *(c)* direct exchange of atoms to neighboring sites.

Dimension The size of a product feature specified on the manufacturing drawing.

Dimensional measurement Evaluation of the size of a part or form and location of part features.

Direct oxidation The process by which oxygen from the surrounding atmosphere forms a metallic oxide on the surface of a metal. This may be a slow process, as in atmospheric rusting of steel, or a more rapid scaling at high temperature.

Dislocation A region of misaligned atoms (in the solid state) among atoms otherwise in their proper crystal structure.

DNC **1** Direct numerical control. A system in which several machines are controlled by a central computer. **2** Distributed numerical control. The modern form of DNC in which a system of several CNC machines utilizes a central computer for data handling and storage.

Drag In foundry terminology the bottom half of a split mold used in casting.

Draft In foundry terminology, the taper or relief put on the sides of a pattern to enable it to be withdrawn from the molding material, e.g., sand.

Drawing **1** Reducing the diameter of wire or bar stock by pulling through successively smaller dies. **2** Reducing the wall thickness of tubing by drawing it through dies. **3** Forming sheet stock into cuplike parts by pushing it through a die. **4** Tempering of hardened tool steel.

Ductility The property of a material that allows it to deform permanently, or to exhibit plasticity without rupture while under tension.

EBM Electron beam machining. The process by which material is removed by an intense beam of electrons impinging on the workpiece.

Eccentricity A rotating member whose axis of rotation is different or offset from the primary axis of the part or mechanism. Thus, when one turned section of a shaft centers on an axis different from that of the shaft, it is said to be eccentric or to have "runout." For example, the throws or cranks on an engine crankshaft are eccentric to the main bearing axis.

ECDB Electrochemical deburring. The process in which electron flow and chemical action are used to remove burrs from a workpiece.

ECM Electrochemical machining.

EDM Electrodischarge machining.

Elastic deformation The movement or deflection of a material when an external load is applied that is less than the elastic limit.

Elastic limit The extent to which a material can be deformed and still return to the original shape when the load is released. Plastic deformation occurs beyond the elastic limit.

Elastic recovery The tendency for a metal part to return (or relax) somewhat from the form to which it has been bent or deformed when the forces causing the change in form are removed. This tendency is due to the elasticity of the metal and is often referred to as *springback.*

Elasticity The ability of a material to return to its original form after a load within the elastic region has been removed.

Elastomer Any of various elastic substances resembling rubber.

Electrochemical deburring A manufacturing process wherein an electric current flowing through an electrolytic fluid removes (deplates) material from a workpiece, rounding corners and removing sharp edges.

Electrolyte A nonmetallic conductor, usually a fluid, in which electric current is carried by the movement of ions.

Electromechanical Combining electrical or electronics and mechanical principles, such as a solenoid-operated valve in a hydraulic system.

Electron beam welding The fusion of material by energy imparted from an intense beam of electrons.

Electroplating Coating an object with a thin layer of a metal through electrolytic deposition.

ELG Electrolytic grinding. The process in which material is removed by chemical and abrasive processes.

Elongation In a tensile test the strain at rupture or failure, usually expressed in percent, or Elongation = (final length − original length)/original length. Used as an indicator of the ductility of the metal.

Embossing The raising of a pattern in relief on a metal by means of a high pressure on a die plate.

Endurance limit In fatigue testing, the stress below which the test specimen will not fail, regardless of the number of test cycles. Most steels will exhibit an endurance limit. (See *Fatigue strength.*)

Equiaxed grains Grains or crystals formed during freezing with little or no directional cooling such that they freeze "equal on the axes"; contrast with columnar grains.

Equilibrium A condition of balance in which all the forces or processes that are present are counterbalanced by equal and opposite forces. Processes in which the condition appears to be one of rest rather than of change.

Eutectic The alloy composition that freezes at the lowest constant temperature, causing a discrete mixture to form in definite proportions.

Eutectoid The alloy composition that transforms from a high temperature solid into new phases at the lowest constant temperature. In binary (double) alloy systems, it is a mechanical mixture of two phases that forms simultaneously from a solid solution as it cools through the eutectoid ($A_1/A_{3,1}$ in steels) temperature.

Extraction A general term for the process of freeing a metal from its ore. It may include beneficiation processes to improve the percent return from the ore, plus smelting operations that separate the metal from, for example, an oxide.

Extrusion Usually under high pressure and at elevated temperatures, forcing materials through a die containing the shape desired so that a shaped product is produced. Metal and polymer shapes are produced using this method.

Fatigue in metals The tendency of a metal to fail by breaking or cracking under conditions of repeated cyclical stressing that takes place well below the ultimate tensile strength.

Fatigue strength In fatigue testing, the stress that causes failure in a specified number of stress cycles. Many nonferrous metals exhibit fatigue strengths rather than endurance limits! (See *Endurance limit.*)

Feature Any distinct element of a component or product that is shown on the manufacturing drawing. (Examples: holes, grooves, cylinders, tapers surfaces.)

Feed rate The distance that a cutting tool moves in either one revolution (feed per revolution) or in one minute (feed per minute) in a machining operation.

Feedback A signal that is returned to the controller indicating what action has or has not occurred, creating closed-loop control.

Ferrite A magnetic form of iron. A solid solution in which alpha iron is the solvent, characterized by a body-centered cubic crystal structure.

Ferrous From the Latin word *ferrum,* meaning iron. Describes an alloy containing a significant amount of iron.

Fiber **1** The directional property of wrought metals that is revealed by a woodlike appearance when fractured. **2** A preferred orientation of metal crystals after a deformation process such as rolling or drawing. **3** Cellulosic plant cells that are used for manufacturing paper and other products. **4** Strands of materials used as reinforcement in plastic products and other materials.

Fiberglass A resin matrix reinforced with glass fibers for strength. A reinforced plastic manufacturing material with many applications.

Filament winding A composite manufacturing process in which the end product is to have a hollow internal shape. A filament of the fiber is wound around a form, then bonded in place with the resin matrix.

Final inspection The last inspection done on parts at the completion of manufacturing processes.

First-piece inspection Inspection performed at the beginning of a manufacturing process to determine whether the first parts produced are correct.

Fixed gages Instruments used for comparison inspection that do not display a reading.

Fixture Production tooling designed to hold and align parts during manufacturing processes.

Flame hardening A means of surface hardening steel or cast iron by applying flame heat, followed by a quench.

Flash The extrusion of extra material in die casting dies, forging dies, or plastic injection molding dies along the parting line.

Float glass A glass manufacturing process that produces a continuous sheet or ribbon of glass.

Flux A solid, liquid, or gaseous material that is applied to solid or molten metal in order to clean and remove oxides or other impurities.

FMS Flexible manufacturing system. A system of material processing stations or work cells connected by automatic material-handling and transfer equipment, all under computer control.

Forging A method of metalworking in which the metal is hammered into the desired shape, or is forced into a mold by pressure or hammering, usually after being heated to a plastic state. Hot forging requires less force to form a given part than does cold forging, which is usually done at room temperature.

fpm Feet per minute. An English system unit used to measure cutting speed in a machining operation. Also *sfm* and *sfpm.*

Fracture A ruptured surface of metal that shows a typical crystalline pattern. Fatigue fractures, however, often display a smooth, clamshell appearance.

Full annealing Heating a metal to an elevated temperature for a long enough time that any evidence of prior cold working or heat treating is removed, and then cooling at a slow rate; the cooling is often accomplished by letting the metal cool in the furnace with the burners off. Under these conditions the metal will approximate the condition predicted by the phase diagram.

Full automatic Processes in which all phases, once started, are accomplished without the need for further manual input.

Fusion The merging of two materials while in a molten state.

Galvanic corrosion A common type of corrosion process in which a potential difference through an electrolyte causes a deplating (corroding) of one of the metals.

Galvanic series A ranking of metallic elements and their alloys according to the voltage generated when each is electrically connected to another metal in the presence of an electrolyte, usually seawater. Those that generate the most voltage are at the top of the list and are the most easily corroded. Because air and groundwater generally contain salt, the series provides a general ranking of the corrosion resistance of metals.

Galvanized steel Steel, normally anodic in our atmosphere (and therefore subject to corrosion) is coated with zinc, a metal more anodic than steel; this process turns the steel into a cathode and protects it. See *Cathodic protection.*

Gear hobbing Cutting gear teeth using a specially designed milling machine.

Gear shaping Cutting gear teeth using a specially designed gear shaping machine.

Geometric dimension Dimensions that pertain to the form of a part and the position of part features rather than to size.

Geometric tolerance Tolerances pertaining to geometric dimensions of form and position.

Glue laminated beam A structural wood beam made by gluing thinner boards together until a desired dimension for beam thickness is reached. Glue laminated beams will support large loads and can span long distances with only end support.

GMAW Gas metal arc welding. A shield gas welding process also known as *MIG* welding.

Grain growth One of the three stages of heating, it occurs at the highest temperature. Prolonged heating at high temperatures causes the grain size to increase. The driving force seems to be the energy contained in the grain boundaries themselves, so the larger grains tend to consume the smaller ones.

Grain-boundary The outer perimeter of a single grain where it is in contact with an adjoining grain; because atoms are not at their ideal distance apart (and therefore at their lowest energy level), it is a region of higher energy.

Grain In metals, a structure containing atoms of one crystalline orientation. Grains form during the solidification (or crystallization) of the metal; they may be re-formed during recrystallization.

Graphics In this text, a reference to pictures generated by computers.

Graphite fiber Strands of carbon in graphite form used in composite materials as the main load-bearing constituent.

Green sand In foundry terminology a combination of sand, clay, and water; in the proper ratios green sand is ready to be used in molding.

Grinding machines—Cylindrical grinders Machines that use a grinding wheel for the abrasive machining of cylindrical surfaces on rotating workpieces.

Grinding machines—Surface grinders Machines that use a grinding wheel for the abrasive machining of flat surfaces.

GTAW Gas tungsten arc welding. A shield gas welding process also known as *TIG* welding.

Hard-facing When a hard surface is desired on a soft metal, a hard material (usually another metal) is applied to the surface. Some methods employed to do this are arc welding, spray welding, and electroplating.

Hardenability The property that determines the depth and distribution of hardness in a ferrous alloy induced by heating and quenching. Literally, the ability to become hard.

Hardening The process of increasing the hardness of a ferrous alloy by austenitizing and quenching; also, the process of increasing the hardness of some stainless steels and nonferrous alloys by solution heat treatment and precipitation.

Hardness The property of a metal that allows it to resist being permanently deformed. This property is divided into three categories: the resistance to penetration, the resistance to abrasion, and elastic hardness.

Heading Enlarging the end of a rod or bar of metal by hot or cold upset forging.

Heat sink A large mass of metal that has a high thermal conductivity used to stabilize the temperature of a part held in contact with it. In welding, the mass of the base metal often acts as a heat sink to the weld metal.

Hematite The most common of the ores of iron; chemically it is Fe_2O_3.

High-speed steel A specialty steel alloy that retains its hardness at higher temperatures than other steels. Cutting tools made of high-speed steel are commonly used in machining processes.

HIP Hot isostatic pressing. The process of compacting a powder by exerting a constant high pressure at elevated temperatures.

Honing A finishing process that utilizes a rotating and oscillating abrasive tool.

Hot pressing Forming or forging tough metals such as alloy steel at high temperatures.

Hot working A process of forming metals while they are heated above the recrystallization or transformation temperatures.

Hot-short Brittleness in hot metal. The presence of excess amounts of sulfur in steel causes hot-shortness.

HSLA High-strength, low-alloy (steels).

Hydrate The combination of molecular water (H_2O) with another compound.

Hydraulics **1** The dynamics of liquids (hydrodynamics), specifically, the study of the flow of water in pipes. **2** Pertaining to fluid power in which mechanical actions and force are developed by the pumping of fluids (hydraulic oil) at high pressure. Hydraulics may be used to develop large mechanical advantages in many types of mechanisms, such as hydraulic presses and machine tools.

Hypereutectic Used to identify metallic alloys that have a composition *greater than* that of the eutectic composition.

Hypereutectoid Used to identify metallic alloys that have a composition *greater than* that of the eutectoid composition.

Hypoeutectic Used to identify those metallic alloys that have a composition *less than* that of the eutectic composition.

Hypoeutectoid Used to identify those metallic alloys that have a composition *less than* that of the eutectoid composition.

Impact test A test that applies an impact load (by a swinging hammer) to a small notched specimen. The data from the test are in energy units, joules or inch-pounds. The test is performed with the specimens at different temperatures; the results determine the notch toughness of the metals at the temperature tested. There are two types of specimens—Izod and Charpy, with Charpy the one primarily used in the United States.

Impressed current In a corrosion cell (see Figure 18.1) the cathode, the metal that is not corroded, supplies electrons to the electrolyte. If a source of electrons (a DC current) is provided to an otherwise anodic metal, it will become the cathode, and not be corroded. In this way long ferrous-metal pipelines can be protected.

Inclusions Particles of impurities that are usually formed during solidification and are usually in the form of silicates, sulfides, and oxides.

Induction hardening Heating the surface of cast iron or tool steel by means of electromagnetic fields followed by a quench.

Industrial engineer An engineer who places emphasis on manufacturing methods, tooling, and facilities.

Ingot A large block of metal that usually is cast in a metal mold and forms the basic material for further rolling and processing.

Injection molding A process in which the material to be molded is heated sufficiently to become fluid and then *injected* under pressure into a mold cavity. There, it is cooled sufficiently to take the shape of the mold, and then removed from the mold; polymers and ceramics are processed in this manner. Metal die casting is very similar to injection molding.

In-process inspections Part of quality control and assurance in which inspections occur when required during a manufacturing process.

Interchangeability The concept that parts manufactured by many different manufacturers to the same dimensional specifications may be interchanged in assemblies.

Interface **1** A surface that forms a boundary between two phases or systems. **2** That part of manufacturing equipment where different systems or types of equipment connect to one another to make complete functioning systems, such as computer numerical control and flexible manufacturing systems.

Investment casting So named because the pattern is "invested" or "lost" in the process. A wax pattern is cast in a metal mold; the wax is encased in a ceramic coating and heated so that the wax runs out, leaving a cavity of the shape and size of the original pattern. The metal is then poured into the cavity, creating a part of the desired shape and size.

Involute Geometry found in modern gears that permits mating gear teeth to engage each other with rolling rather than sliding friction.

Ion An atom or molecule electrostatically charged by losing or gaining one or more valence electrons.

Ionic bond A type of atomic bonding in which atoms with one or more valence electrons *donate* or *give away* their valence electrons to elements that lack one or more electrons to fill their valence shell; each atom thus becomes an ion; the one donating becomes positive, the one accepting becomes negative. The ionic bond commonly forms compounds between elements widely separated on the periodic table, e.g., Na^+ and Cl^- combine to form NaCl.

IPM Inches per minute feed rate.

IPR Inches per revolution feed rate.

Iron The term *iron* always refers to the element Fe and not to cast iron, steel, or any other alloy of iron.

Isomerism Compounds are said to be isomeric when they have the same elementary composition (i.e., their molecules contain the same numbers and kinds of atoms) but different structures, and hence, properties. It is believed that differences are due to the arrangement of atoms in each molecule.

Isothermal transformation (IT) Transformation in some metals that takes place at a constant temperature.

IT See *Time–temperature transformation.*

Jig Production tooling designed to locate a feature during manufacturing processes.

Kaolin A fine white clay that is used in ceramics and refractories, composed mostly of kaolinite, a hydrous silicate of aluminum. Impurities may cause various colors and tints.

Killed steel Steel that has been deoxidized with agents such as silicon or aluminum to reduce the oxygen content. This prevents gases from evolving during the solidification period.

Lamellar An alternating platelike structure in metals (as in pearlite).

Laminates Composed of multiple layers of the same or different materials.

Lapping A fine finishing process using abrasive particles that are embedded in a piece of softer material called a lap.

Laser "Light Amplification by Stimulated Emission of Radiation." A device in which heat is derived from the intense coherent beam of laser light energy. This intense, narrow beam of light is used in some welding and machining operations.

Lathes—Engine lathes Common mechanical lathes, manually controlled, used for low-volume and tooling work.

Lattice A term that is used to denote a regular array of points in space, for example, the sites of atoms in a crystal. The points of the three-dimensional space lattice are constructed by the repeated application of the basic translations that carry a unit cell into its neighbor.

Lignin A substance that is related to cellulose that with cellulose forms the woody cell walls of plants and the material that cements them together. Methyl alcohol is derived from lignin in the destructive distillation of wood.

Liquidus The temperature at which freezing begins during cooling and ends during heating under equilibrium conditions, represented by a line on a two-phase diagram.

Lost wax A casting process; see *Investment casting.*

Machinability The relative ease of machining.

Machining center A CNC (computer numerically controlled) lathe or milling machine equipped with an automatic tool changer.

Macroscopic Structural details on an object that are large enough to be observed by the naked eye or with low magnification (about 10×).

Macrostructure The structure of metals as revealed by macroscopic examination.

Magnetite A magnetic ore of iron with the chemical composition Fe_3O_4.

Malleability The ability of a metal to deform permanently without rupture when loaded in compression.

Manufacturing drawing A picture of a component or product created by the product design engineering department that contains the specifications followed in manufacturing.

Manufacturing engineering The function or department that involves the design, documentation, and development of manufacturing processes, equipment, and tooling.

Manufacturing process A planned sequence of manufacturing steps that is designed to produce a component or product.

Manufacturing system The machinery, tooling, personnel, and controls that are utilized to manufacture components or products.

Manufacturing technologist A person who is trained in the knowledge, design, and control of manufacturing processes and materials.

Martempering Heat treatment of martensitic material, in which martensite transforms to tempered martensite, composed of stable ferrite and cementite phases.

Martensite Iron phase supersaturated in carbon that is a nonequilibrium product of austenite transformation.

Mass production Continuous production of duplicate product units such as appliances, cars, or any parts of these products.

Matrix In metals, the background phase in which another phase is contained; steel is the matrix in which graphite flakes are present in gray cast iron. In composite materials the matrix is the outer layer of material in which the reinforcement is contained.

MCU Machine control unit; the computerized controller that guides the operation of an automated machine or device.

Measurement standards Known standards to which production gages and other measurement tools are periodically compared to ensure their conformity and accuracy. Gage blocks are an example.

Mechanized Controlled by mechanical cams, stops, and levers, or by hydraulic, pneumatic, or electrical devices. Not computer controlled.

Metal Those materials that occupy the left side of the periodic table are characterized by having one, two, or three valence electrons, and bond with the metallic bond.

Metallic bond In metals, the attractive force between their positive nuclei and inner electron shells (with a net positive charge), and a negatively charged cloud of valence electrons. This type of bonding provides free electrons for electrical and thermal conductivity and permits plastic deformation or cold working.

Metalloid A nonmetal that exhibits some, but not all, of the properties of a metal. Examples are sulfur, silicon, carbon, phosphorus, and arsenic.

Metallurgy The science and study of the behaviors and properties of metals and their extraction from their ores.

Methanol (methyl alcohol, wood alcohol) An alcohol produced by the destructive distillation of wood or made synthetically.

Microalloying The use of very small amounts of certain elements to alloy steels, especially HSLA steels, to prevent grain growth during hot working. The elements used commonly are strong oxide and nitride formers, such as niobium, vanadium, titanium, and aluminum.

Micrometers Instruments that utilize an internal screw thread to perform precision dimensional measurements. Many micrometer configurations exist including outside (caliper-type), inside, and depth micrometers.

Microscopy The use of, or investigation with, the microscope.

Microstructure The structure of polished and etched metal specimens as seen enlarged through a microscope.

MIG Metal inert gas welding. Also known as *GMAW.* A shield gas welding process that uses a continuous wire electrode fed from a reel.

Milling cutters—End milling cutters Cutting tools, used on milling machines and machining centers, that can cut using teeth on the cylindrical surface (peripheral, slab milling) and can cut using teeth on the end (face milling) of the cutter.

Milling cutters—Face milling cutters Cutting tools, used primarily on vertical milling machines, that cut using teeth on the end face (face milling) of the cutter.

Milling cutters—Plain milling cutters Cutting tools, used primarily on horizontal milling machines, that cut using teeth on the cylindrical surface (peripheral, slab milling) of the cutter.

Milling machine—Horizontal milling machine A machine tool with a horizontal spindle to which a milling cutter is attached to perform milling operations.

Milling machine—Vertical milling machine A machine tool with a vertical spindle to which a milling cutter is attached to perform machining operations.

MMC Metal matrix composite; a composite material that has a reinforcing material in a metallic matrix.

Modulus of elasticity The ratio of the unit stress to the unit deformation (strain) of a structural material; a constant as long as the unit stress is below the elastic limit. Shearing modulus of elasticity is often called the *modulus of rigidity.*

Mold A general classification of tooling used to form or shape a product while the material is in a molten or liquid state. Mold tooling is found in many manufacturing processes including metal and plastic casting, injection molding, fiberglassing, and composite part manufacturing.

Monomer A single molecule or a substance consisting of single molecules. The basic unit in a polymer.

Mottled cast iron A cast iron that freezes as a mixture of gray and white cast irons, likely caused by an improper silicon content or cooling rate, or both.

Ms Temperature at which martensite starts to form; in other words, the martensite transformation starting temperature.

Muffle furnace A gas- or oil-fired furnace in which the work is separated from the flame by an inner lining or "muffle."

NC Numerical control; equipment that controls manufacturing processes by the use of computers to execute programmed commands.

NC axes On NC manufacturing equipment, the primary paths along which major components of the machine or workpiece move.

NC program A step-by-step set of instructions written in a format understood by an NC machine and defining all the information necessary to accomplish the manufacturing task required.

NC programmers Persons with knowledge of manufacturing processes plus effective math skills who write NC programs.

NIST National Institute of Standards and Technology. A U.S. government organization that, among other functions, maintains measurement standards for U.S. industry.

Nitriding A process of case hardening in which a special ferrous alloy is heated in an atmosphere of ammonia or is in contact with another nitrogenous material. In this process, surface hardening is achieved by the absorption of nitrogen without quenching.

Nondestructive testing Application of physical principles for detection of flaws or discontinuities in materials without impairing their usefulness.

Nonferrous Metals other than iron and iron alloys.

Normalize To homogenize and produce a uniform structure in alloy steels by heating above the transformation range and cooling in air.

Notch toughness The resistance to fracture of a metal specimen having a notch or groove when subjected to a sudden load, usually tested on an Izod–Charpy testing machine.

Nucleus The positively charged central part of an atom containing the protons and neutrons, and therefore most of the atom's mass.

Open-die drop forging Also called *smith forging.* In open-die forging a drop hammer delivers blows of great force to a heated metal that is shaped by manipulating it under the hammer.

Operation detail sheet A document that provides the detailed information needed to execute each step of a manufacturing process.

Operations sketch A portion of a part drawing used in a specific manufacturing step emphasizing certain dimensional information and indicating the steps to be accomplished.

Orange peel A usually undesirable condition created when metals with a large grain size are stretched, bent, or formed, and large grains roughen the surface, causing it to appear like the skin of an orange.

Ore A mineral or minerals, typically dug from the earth, from which some material, usually a metal, can be profitably mined and extracted.

Oxidation The slow or rapid reaction of oxygen with other elements; burning. In metals, heating under oxidizing conditions often results in permanent damage to metals.

Oxidation–reduction A chemical reaction in which the metal being oxidized loses electrons (to the element doing the oxidizing) and becomes a positive ion. See *Reduction.*

Part drawing A drawing of an individual part; a detail drawing.

Partial anneal A final treatment given to cold-worked products. The temperature and time are closely controlled to reduce the strength and hardness properties of the product to something less than full hard.

Partial (solid) solubility A condition in which two metals are soluble but not in all compositions. It typically occurs when the difference in atom diameters is between 8% and 14%. If metals are only partially soluble, it also means that they are *insoluble* over a range of compositions.

Patterns A general class of production tooling representing the shape and/or size of the parts to be made. Patterns can be used to form molds in casting, and as measurement and gaging tools to check or to guide material cutting and preparation activities.

Pearlite The lamellar mixture of ferrite and cementite in slowly cooled iron–carbon alloys as found in steel and cast iron.

Peel load In metal, plastics, or composites, a force that acts to peel apart joined pieces.

Peening Work hardening the surface of metal by hammering or blasting with shot (small balls). Peening introduces compressive stresses on surfaces that tend to counteract unwanted tensile stresses.

Perforating Piercing many small holes close together.

Performance inspection An inspection designed to evaluate the performance of a part or product. Examples include the burn-in test for electronics and automobile engine tests.

Periphery The perimeter or external boundary of a surface or body.

Permeability In casting of metals, the term is used to define the porosity of foundry sands in molds and the ability of trapped gases to escape through the sand.

Petrochemicals Chemicals derived from petroleum; substances or materials manufactured from a component of crude oil or natural gas.

Phase A portion of an alloy, physically homogeneous throughout, that is separated from the rest of the alloy by distinct boundary surfaces. The following phases occur in the iron–carbon alloy: molten alloy, austenite, ferrite, cementite, and graphite.

Phase change The transformation of a substance from one distinct, separate form to another. When a molten metal solidifies it has changed phase, and depending on the alloy system, it may solidify as one phase (a single phase) or as two phases (a mixture).

Phase diagram For binary (two metals) alloy systems, a diagram of temperature versus percent composition, with 100 percent of one metal on one axis and 100 percent of the second on the other. Lines on the diagram separate single phases from mixtures of phases. Additional time is required to change from one phase to another. In the construction of a phase diagram all the time necessary for phase changes to occur is assumed available; it is thus often termed an *equilibrium* phase diagram. See *Time–temperature transformation.*

Phenolics A group of thermoset polymers based on phenol, an aromatic alcohol. Phenol formaldehyde, the oldest of this family, is still often called Bakelite.

Pickling A process in which metal parts are dipped in acid for the purpose of cleaning and etching prior to plating, painting, or further cold working.

Piercing Cutting (usually small) holes in sheet metal.

Pig iron The product of a blast furnace. It is iron that usually contains about 4.5 percent carbon and impurities such as phosphorus, sulfur, and silicon.

Pitch diameter For threads, the pitch diameter is an imaginary circle, which on a perfect thread occurs at the point where the widths of the thread and groove are equal. On gears, it is the diameter of the pitch circle.

Plasma An ionized gas of extremely high temperature achieved by passing an inert gas through an electric arc. Plasma arcs are used in welding, cutting, and machining processes.

Plastic deformation Deformation that occurs when so much stress is applied to a solid that it does not return to its original condition.

Plasticity The quality of material such that it can be deformed without breaking. Clay is a completely plastic material. Metals exhibit plasticity in varying amounts.

Plastics There is not a universally accepted meaning; in this text an attempt has been made to identify products made from *polymers* as *plastic*. The name suggests that the material should be flexible, but not all plastic products are flexible. See *Polymers*.

Plating The process of depositing a layer of one metal on another, often done electrically, for the purpose of corrosion protection, appearance, improved electrical conductivity, and other engineering requirements.

Plotter A computer-driven drawing machine used to convert computer drawings to paper drawings.

P/M Powder metallurgy. A process of producing a metal object from metal powders by compaction and sintering.

PMC Polymer matrix composite; a composite material that has a reinforcing material in a polymer matrix.

Poisson's ratio The ratio between the strain and the amount of lateral contraction when a rod of elastic material is elongated by stretching (strain).

Polyamide The family of polymers better known as *nylon* that are important because of their widespread applications in industry.

Polyamide–imide A family of polymers that can resist constant temperatures of 500°F for long periods of time, are tough, and have a strength of 32,000 psi.

Polycrystalline The term used to describe the crystalline nature of most metals encountered, i.e., they are made up of more than one metallic crystal, as opposed to being single crystals.

Polyimide A family of polymers resistant to most chemicals and to temperatures of almost 500°F.

Polymer A compound or compounds, usually hydrocarbons, that have been polymerized to form a long chain of repeating unit structures. In this text an attempt has been made to use the term *polymer* as the material from which *plastic* products are made. Polyethylene, used in today's milk jugs, is a polymer. See *Plastics*.

Polymerization A chemical reaction in which two or more small molecules combine to form larger molecules that contain repeating structural units of the original molecules.

Precast concrete Concrete products that have been precast in molds and then removed and transported to construction sites. Examples are precast concrete walls for building, which are often prestressed.

Precipitation hardening A process of hardening an alloy by heat treatment in which a constituent precipitates from a supersaturated solid solution while at room temperature or at some slightly elevated temperature.

Precision Repeatability; the amount of size variation in the component or product features created by a manufacturing process.

Preform To bring to the approximate shape and size before the final forming takes place. Used in forging and powder metallurgy.

Prepreg An intermediate PMC, *preimpregnated* product. Strands of fiber are aligned and coated with a resin, used as a tape or sheet to laminate into the product, and cured by heating.

Primary bond In polymers, the covalent or ionic bond that links the chain and its pendants together. It is a primary bond as compared with a secondary bond; see *van der Waals bond.*

Process anneal An annealing process used within a sequence of cold-working operations that causes the metal to recrystallize, so that additional cold working can be done. See *Bright anneal.*

Process engineer A manufacturing engineer or technologist who designs and documents manufacturing processes.

Process routing sheet The document that lists the steps required, types of machines, and other equipment required in the manufacturing process.

Production line The set of stations where various steps in the assembly or making of parts and products is accomplished in a manufacturing industry.

Production tooling Tools, equipment, and special devices necessary for the manufacture of a product.

Productivity A measure of the amount, quality, and timeliness of products manufactured relative to the amounts of material, labor, and capital that are used.

Programming The preparation of machine control instructions for numerically controlled manufacturing equipment. Programming may be done manually or by computer (CAM).

Proportional limit On the stress–strain diagram, the point where the ratio of stress to strain (i.e., the modulus of elasticity) is no longer a constant, or where that line is no longer straight.

Prototype A working model of a product to be manufactured and used for testing and design evaluation, often full sized.

PSI Pounds per square inch.

Pultrusion A process used in the manufacture of polymer matrix composite parts. In the process fibers are pulled through a liquid resin, which adheres to the fibers, and then is hardened as the composite is pulled through a heated die.

Punching The operation of cutting a hole in sheet metal using a die. The hole material is scrapped.

Quality assurance All activities in manufacturing directed toward ensuring production of a high-quality product.

Quality control One segment of quality assurance often responsible for dimensional inspection during production.

Quenching The process of rapid cooling of metal alloys for the purpose of hardening. *Quenching media* include air, oil, water, molten metals, and fused salts.

Rake angle The angle of the rake face (the surface that the chip rubs across) of a cutting tool relative to the surface of the workpiece. The rake angle greatly affects the shearing action and cutting forces in a machine operation.

Rapid prototyping Any of several automated processes that fabricate a physical prototype directly from a CAD file.

Reciprocate To move back and forth along a given axis.

Recovery The relaxation and reduction of locked-in internal stresses in cold-worked metals.

Recrystallization A process in which the distorted grain structure of metals that are subjected to mechanical deformation is replaced by a new strain-free grain structure during annealing.

Reduction A chemical reaction in which the element being reduced *gains* electrons. When oxygen is removed from a metal oxide, say Al_2O_3, the metal regains the electrons it shared with the oxygen when it was oxidized, so it is reduced. See *Oxidation–reduction.*

Reduction of area In a tensile test, the change in cross-sectional area of the specimen divided by the original area, usually expressed as a percent. When measured after the specimen has failed it is a measure or indicator of ductility.

Refractory Materials that will resist change of shape, weight, or physical properties at high temperatures. These materials are usually silica, fire clay, diaspore, alumina, and kaolin. They are used for furnace linings.

Reinforced concrete Concrete reinforced by internal steel bars or lattice work.

Reinforcements Material added to the matrix of a composite to improve its properties, usually the strength or stiffness.

Residual stress Stresses induced within the structure of a metal by cold working, machining, and heat treatments and remaining in the metal after the treatment is completed.

RIM Reaction injection molding. A chemical reaction between two resins generates the heat required to make the mix a liquid.

Rimmed steel A low-carbon steel (insufficiently deoxidized) that during solidification releases considerable quantities of gases (mainly carbon monoxide). When the mold top is not capped, a side and bottom rim of several inches forms. The solidified ingot has scattered blow holes and porosity in the center but a relatively thick skin free of blow holes.

Riser In metal casting, a vertical opening in the mold that serves as a reservoir to supply molten metal to the casting as it shrinks during solidification.

Robotics Systems and equipment used to operate industrial robots.

Rockwell hardness A hardness measure based on the depth of penetration of a material under standardized conditions. Various penetrators and known weights are used so several scales are able to cover the very soft to the very hard materials. The Rockwell "C" scale is used mostly for hardened steels.

RPM Revolutions per minute.

Rule of ten A general rule that is commonly followed in manufacturing. It specifies that the precision of a measuring process should be no worse than 10 percent of the tolerance range (from the manufacturing drawing) for the feature that is measured.

Runout An eccentricity of rotation, as the off-center rotation of a cylindrical part held in a lathe chuck. The amount of runout of a rotating member is often checked with a dial indicator.

Sacrificial anode A metal slug, usually magnesium or zinc, designed to concentrate galvanic corrosion on itself and thus save a more important structure on which the anode is attached, such as a ship's hull or a buried pipe line.

SAE Society of Automotive Engineers.

Scrap Materials or metals that have lost their usefulness and are collected for reprocessing.

Sealant A sealing agent that has some adhesive qualities: it is used to prevent leakage.

Secondary bond See van der Waal bond.

Semiautomatic A process in manufacturing that requires some degree of manual input but acts without this input for at least part of the cycle.

Sensor A device used to measure some quantity and then send the information back to the control equipment.

Servo A motor or valve that can be controlled by an electrical signal, thus converting electrical inputs to mechanical action outputs.

Servo system An automated control system for machinery involving sensors, feedback that provides information from the sensors to the computerized control unit, and action based on the feedback.

sfm or **sfpm** Surface feet per minute. Also *fpm* (feet per minute).

Shaper A machine tool that utilizes reciprocating motion of the cutting tool rather than the more common rotary motion to make repeated machining cuts.

Shear load A load that tends to force materials apart by application of side-slip action.

Shearing A concentration of forces in which the bending moment is virtually zero and the metal tends to tear or be cut along a transverse axis at the point of applied pressure.

Shell molding A form of gravity casting of metal (usually a high-melting-temperature metal) in which the mold is made of a thin shell of refractory material.

Shield gas Usually an inert gas used to displace air from around a weld zone, thus keeping the weld uncontaminated.

Shot peening A cold-working process in which the surface of a finished part is pelted with finely ground steel shot or glass beads to form a compression layer.

Shrink allowance An amount added to the dimensions of a casting pattern to make *allowance* for the *shrinkage* that will occur when the metal solidifies and cools. This allowance can be added to a scale or ruler so it becomes a "shrink rule."

SI Système Internationale. The metric system of weights and measures.

Sintering A process of fusing compacted material such as metal powders into a solid piece by applying heat sufficient to bond, but not melt, the particles.

Skelp The semifinished steel of which butt-welded pipe is made.

Sketch A simple drawing made without the use of drawing instruments.

Slag (dross) A fused product that occurs in the melting of metals and is composed of oxidized impurities of a metal and a fluxing substance such as limestone. The slag protects the metal from oxidation by the atmosphere, since it floats on the surface of the molten metal.

Slip planes Also called *slip bands.* These are lines that appear on the polished surface of a plastically deformed metal. The slip bands are the result of crystal displacement, defining planes in which shear has taken place.

Slurry A watery mixture of insoluble material, such as mud, lime, or plaster of paris.

Smelting The process of heating ores to a high temperature in the presence of a reducing agent such as carbon (coke).

Smith forging See *Open-die drop forging.*

Soaking A prolonged heating of a metal at a predetermined temperature to create a uniform temperature throughout its mass.

Sodium silicate Na_2SiO_3. Also called water glass.

Soldering A process that uses heat to join metal parts below their melting points, joining them mechanically rather than fusing them. Filler metals with melting points below 840°F join the parts.

Solid solution Found in metals at temperatures below the solidus. In solid solutions the atoms of two or more metals take up atom sites in the lattice of the host metal and are said to be "in solution." Some of the types of solid solutions are continuous, intermediate, interstitial, substitutional, and terminal.

Solid solution strengthening When atoms of one metal take up residency in the lattice of another, they will never have the exactly correct atom diameter; therefore, the atom lattice is strained, increasing the strength of the metal.

Solidification The process in which a liquid metal changes to a solid; in this process heat is removed and the atoms have to fit into an atom lattice.

Solidus Seen as a line on a two-phase diagram, it represents the temperatures at which freezing ends when cooling, or melting begins when heating under equilibrium conditions.

Solubility The degree to which one substance will dissolve in another.

Solute A substance that is dissolved in a solution and is present in minor amounts.

Solution heat treatment A process in which an alloy is heated to a predetermined temperature for a length of time that is suitable to allow a certain constituent to enter into solid solution. The alloy is then cooled quickly to hold the constituent in solution, causing the metal to be in an unstable supersaturated condition. This condition is often followed by age or precipitation hardening.

Solutionizing In heat treating, ensuring that all alloying elements are in solution at the high temperature from which the metal will be quenched.

Solvent A substance that is capable of dissolving another substance and is the major constituent in a solution.

Source inspection Inspection of materials at the supplier's location before they are shipped.

Spalling Breaking small pieces from a surface, often caused by a thermal shock.

Specific gravity A numerical ratio that compares the weight of a given substance with the weight of an equal volume of water. Since water weighs 1.000 grams/cm^3, the weight density of any material expressed in grams/cm^3 is the same as its specific gravity.

Spheroidizing Consists of holding carbon steel for a period of time at just under the transformation temperature. An aggregate of globular carbide is formed from other microstructures such as pearlite.

Spindle speed The rotational speed, measured in rpm (revolutions per minute), of the spindle of the machine tool utilized by a manufacturing process.

Spinning A rotary process using a spinning lathe to form ductile metals into shapes having a symmetry about their rotational axis; for example, cones, pressure-vessel heads, domes.

Springback See *Elastic recovery.*

Sprue In metal casting, a vertical opening in the mold through which the molten metal is poured into the cavity left by the removal of the pattern.

Sputtering To dislodge atoms from the surface of a material by collision with high-energy particles for the purpose of depositing a metallic film on a part.

Stainless steel An alloy of iron containing at least 11 percent chromium and sometimes nickel that resists many forms of rusting and corrosion.

Standardization The practice of designing machinery components with as many like features as practical.

Statistics Mathematically generated data derived from small samples of large batches from which inferences may be made about the specifications of a large batch without the need to inspect every part.

Steel An alloy of iron and less than 2 percent carbon plus some impurities and small amounts of alloying elements is known as plain carbon steel. Alloy steels contain substantial amounts of alloying elements, such as chromium or nickel, besides carbon.

Strain The unit deformation of a material when stress is applied.

Strain hardening An increase in hardness and strength of a metal that has been deformed by cold working or at temperatures lower than the recrystallization range. See *Work hardening.*

Strength The ability of a material to resist external forces; called tensile, compressive, or shear strength, depending on the load. See *Stress.*

Stress The load per unit of area on a stress–strain diagram. *Tensile stress* refers to an object loaded in tension, denoting the longitudinal force that causes the fibers of a material to elongate. *Compressive stress* refers to a member loaded in compression, which gives rise to either a given reduction in volume or a transverse displacement of material. *Shear stress* refers to a force that lies in a parallel plane. The force tends to cause the plane of the area involved to slide on the adjacent planes. *Torsional stress* is a shearing stress that occurs at any point in a body as the result of an applied torque or torsional load.

Stress raiser Can be a notch, nick, weld undercut, sharp change in section, or machining grooves or hairline cracks that provide a concentration of stresses when the metal is under tensile stress. Stress raisers pose a particular problem and can cause early failure in members that are subjected to many cycles of stress reversals.

Stress relief anneal The reduction of residual stress in a metal part by heating it to a given temperature and holding it there for a suitable length of time. This treatment is used to relieve elastic stresses caused by welding, cold working, machining, casting, and quenching.

Surface finish (surface roughness) A form specification that determines the smoothness of the minute peaks and valleys that compose a machined surface.

Surface hardening Usually refers to the hardening of the surface of steel; this can be accomplished two ways: using a steel with sufficient carbon to achieve the hardness desired and heat treating just the surface, or raising the carbon content of the surface and heat treating the whole part. Surface peening will also harden the surface of most metals.

Swaging Compacting or necking down metal bars or tubes by hammering or rotary forming.

Taconite A low-grade iron ore. To reduce the handling cost per the weight of iron, it is crushed, separated, and rolled into balls, in which form it is transported to the blast furnace location.

Temper In ferrous metals, the stress relief of steels that are hardened by quenching for the purpose of toughening them and reducing their brittleness. In nonferrous metals, temper is a condition produced by heat treating or a mechanical treatment such as cold working.

Tensile load A load applied to a part or parts that attempts to pull apart by a stretching action.

Tensile strength See *Ultimate tensile strength.*

Thermal conductivity The quantity of heat that is transmitted per unit time, per unit cross section, per unit temperature gradient through a given substance. All materials are in some measure conductors of heat.

Thermal expansion The increase of the dimension of material that results from the increased movement of atoms caused by increased temperature.

Thermal shock A stress induced on the surface of a material such as carbide tools or fire brick caused by a rapid rate of heating and surface expansion.

Thermal stress Shear stress that is induced in a material due to unequal heating or cooling rates. The difference of expansion and contraction between the interior and exterior surfaces of a metal that is being heated or cooled is an example.

Thermoplastic Capable of softening or fusing when heated and of hardening again when cooled.

Thermosetting Capable of becoming permanently rigid when cured by heating; will not soften by reheating.

TIG Tungsten inert gas welding. See *GTAW.*

Time–temperature transformation, or **TTT,** or **IT** A diagram that shows the transformation of austenite at one temperature when the time for transformation is taken into account. Because the transformation occurs at one temperature it is also termed an IT or isothermal transformation diagram. See *Phase diagram.*

Tolerance The amount of size variation permitted for the size, form, or position of a feature included in a manufacturing drawing.

Tolerance range The total size variation (between the high and low limits) specified by a tolerance on a manufacturing drawing.

Tool and die making The processes of building specialty production tooling to support manufacture of a product.

Tool design The process of designing cutting tools, jigs, and fixtures that will support product manufacturing.

Tool geometry The proper shape of a cutting tool that makes it work effectively for a particular application.

Tool steel A special group of steels that is designed for specific uses, such as heat-resistant steels that can be heat treated to produce certain properties, mainly hardness and wear resistance.

Tooling Generally, any machine tool accessory separate from the machine itself. Tooling includes cutting tools, holders, work-holding accessories, jigs, and fixtures.

Toughness Generally measured in terms of notch toughness, which is the ability of a metal to resist rupture from impact loading when a notch is present. A standard test specimen containing a prepared notch is inserted into the vise of a testing machine. This device, called the Izod–Charpy testing machine, consists of a weight on a swinging arm. The arm or pendulum is released, strikes the specimen, and continues to swing forward. The amount of energy absorbed by the breaking of the specimen is measured by how far the pendulum continues to swing.

Transducer A device by means of which energy can flow from one or more transmission systems to other transmission systems, such as when a mechanical force is converted into electrical energy (or the opposite effect) by means of a piezoelectrical crystal.

Transformation temperature The temperature(s) at which one phase transforms into another phase; for example, where ferrite or alpha iron transforms into austenite or gamma iron.

Transition temperature The temperature at which normally ductile metals become brittle.

Transuranic elements Those elements with an atomic number greater than uranium, or 92; the transuranic elements do not occur in nature.

TTT See *Time–temperature transformation.*

Ultimate tensile strength or **UTS** The stress equal to the maximum load achieved in a tensile test divided by the original area of the specimen.

Ultrasonic Sound frequency above that able to be heard by the human ear.

UM Universal milled (steel plate).

Uncertainty principle A principle in quantum mechanics. It is impossible to assert in terms of the ordinary conventions of geometrical position and of motion that a particle (as an electron) is at the same time at a specified point and moving with a specified velocity.

Upset forging The process of increasing the cross section of stock at the expense of its length.

Vacancy A location in a crystal lattice that should have an atom present but does not.

Valence The capacity of an atom to combine with other atoms to form a molecule. The inert gases have zero valence. Valence considers positive and negative properties of atoms, as determined by the gain or loss of valence electrons.

van der Waals bond A relatively weak bond between molecules, for example, in inert gases. It is a weak secondary bond between adjacent chains of a polymer. The strength of the bond is directly related to the size of the molecules, inversely related to the distance between the molecules, and easily weakened by heat.

Vernier scale An added scale that improves the resolution of micrometers and other precision measuring instruments by mechanically magnifying tiny variations in a dimension.

Video display The television display screen of a computer. Also known as a *CRT* (*cathode ray tube*).

Viscosity The property in fluids, either liquid or gaseous, that may be described as a resistance to flow; also, the capability of continuous yielding under stress.

Visual inspection An inspection made during manufacturing based on simply looking at a product to determine if any deficiencies are visible.

Void A cavity or hole in a substance.

Voltage Electromotive force or a surplus of electrons at a point in an electric circuit. Voltage is equivalent to pressure in a fluid power system.

Vulcanization The process of treating crude or synthetic rubber or similar plastic material chemically to give it useful properties, such as elasticity, strength, and stability.

Warranty A guarantee of product quality provided by the manufacturer promising parts, service, or replacement in the case of a product failure.

Water jet A manufacturing process in which a material is cut by a high-pressure jet of water often containing an abrasive material to enhance cutting action.

Weldment A unit formed by welding together an assembly of pieces.

Work hardening Also called strain hardening and cold working, in which the grains become distorted and elongated in the direction of working (e.g., rolling). This process hardens and strengthens metals but reduces their ductility. Excessive work hardening can cause ultimate brittle failure of the part. A process anneal is often used between periods of working of metals to restore their ductility.

Wrought iron Contains 1 or 2 percent slag, which is distributed through the iron as threads and fibers imparting a tough fibrous structure. Usually contains less than 0.1 percent carbon. It is tough, malleable, and relatively soft.

Yield point The stress at which a marked increase in deformation occurs without an increase in load stress, as seen in mild or medium carbon steel. This phenomenon is not seen in nonferrous metals and other alloy steels.

Yield strength The stress at which a material deviates by a specified amount of strain from the region where stress and strain are proportional.

Index

D

N